Springer Collected Works in Mathematics

For further volumes:
http://www.springer.com/series/11104

Table of contents volume II

Table of contents volume I

si può supporre che $R_k(x)$ sia dello stesso ordine di grandezza del massimo c_ν trascurato, cioè

$$[6] \qquad |R_k(x)| < A_k n^{-\frac{1}{2}\left[\frac{k+3}{3}\right]}$$

con A_k indipendente [1] da n e da x. Questa ipotesi ha avuto un'importanza essenziale nell'applicazione degli sviluppi della forma [3] nella Statistica matematica.

Come ho dimostrato in un lavoro precedente [2], questa ipotesi è soddisfatta quando la funzione di ripartizione $F(x)$ soddisfi ad una condizione assai generale di cui diremo fra poco.

Intanto, come funzione di x limitata e mai decrescente, $F(x)$ ha *quasi dappertutto* una derivata non negativa $F'(x)$ che è integrabile nel senso di Lebesgue da $-\infty$ a $+\infty$ e si ha

$$[7] \qquad \int_{-\infty}^{\infty} F'(x)\,dx \geqq 0,$$

dove il segno di eguaglianza potrà aver luogo soltanto quando *quasi dappertutto* sia $F'(x) = 0$. Ciò si verificherà, per esempio, quando la funzione di ripartizione $F(x)$ risulti crescente solo nei suoi punti di discontinuità.

La condizione a cui abbiamo sopra accennato, alla quale deve soddisfare la $F(x)$, è che *nella* [7] *non deve aversi il segno di eguaglianza. Quando dunque si abbia*

$$[8] \qquad \int_{-\infty}^{\infty} F'(x)\,dx > 0,$$

varrà per ogni $k \geqq 3$, *lo sviluppo* [5] *e per il suo resto si avrà la limitazione* [6]. *Può dimostrarsi con esempi che ciò non si verifica quando non sia soddisfatta la condizione* [8].

---

1) La lettera A e la lettera B che incontreremo in seguito indicheranno sempre *costanti positive* mentre A_k indicherà una grandezza dipendente da k. Le lettere A, ecc. che si incontrerenno in varie formule non dovranno però significare sempre necessariamente la medesima grandezza.

2) H. CRAMÉR, *On the composition of elementary errors*, I, « Skandinavisk Aktuarietidskrift », 1928, pagg. 13–74.

Se ora poniamo $k = 3$ e osserviamo che $\Phi^{(3)}(x)$ è limitata per ogni valore di x, mentre c_3 è dell'ordine di grandezza di $n^{-1/2}$, avremo

$$[9] \qquad |F_n(x) - \Phi(x)| < \frac{A}{\sqrt{n}}$$

con A costante positiva. La condizione [8] permette quindi di migliorare la limitazione [2] di Liapounoff.

2. Supponiamo ora che la funzione di ripartizione $F(x)$ sia continua per ogni valore di x e abbia una derivata $f(x) = F'(x)$ finita. In base a risultati noti della teoria dell'integrale di Lebesgue, per ogni valore di x si avrà

$$F(x) = \int_{-\infty}^{x} f(t)\,dt.$$

La variabile casuale X avrà allora una *densità delle probabilità*, $f(x)$, dovunque finita e da ciò segue, come è noto, che lo stesso dovrà accadere anche per

$$Z_n = \frac{X_1 + X_2 + \cdots + X_n}{\sqrt{n}}.$$

Poniamo

$$f_n(x) = F'_n(x)$$

$$\varphi(x) = \Phi'(x) = \frac{1}{\sqrt{2\pi}} e^{-\frac{x^2}{2}}.$$

Se la derivata $f(x) = F'(x)$, in tutto l'intervallo illimitato $-\infty < x < +\infty$ ha un numero finito di oscillazioni, allora, come ho dimostrato nella Memoria sopra citata [3], si avrà anche per $f_n(x)$ uno sviluppo asintotico che corrisponde allo sviluppo [5]. Avremo cioè, in questo caso, per ogni $k \geq 3$,

$$[10] \qquad f_n(x) = \varphi(x) + \sum_{3}^{k} \frac{c_\nu}{\nu!} \varphi^{(\nu)}(x) + r_k(x),$$

in cui si ha

$$[11] \qquad |r_k(x)| < A_k n^{-\frac{1}{2}\left[\frac{k+3}{3}\right]},$$

3) Cfr. loc. cit. [2]), pag. 63.

con A_k indipendente da n e da x. Da ciò segue in particolare, per n crescente illimitatamente,

$$f_n(x) \to \varphi(x)$$

ed anche

$$[12] \qquad |f_n(x) - \varphi(x)| < \frac{A}{\sqrt{n}}$$

con A costante positiva.

3. Gli sviluppi [5] e [10], con le limitazioni indicate per i rispettivi resti, valgono, come abbiamo visto, sotto condizioni abbastanza generali, per la data funzione di ripartizione $F(x)$ e per la densità di probabilità $f(x)$. Questi sviluppi debbono però essere interpretati solo come *sviluppi asintotici, valevoli quindi per valori di n sufficientemente grandi*. Non possiamo, per esempio, affatto affermare che le serie [5] e [10], sotto le condizioni poste, convergano per $k \to \infty$ rispettivamente verso $F_n(x)$ e $f_n(x)$. Per ottenere serie convergenti, occorre aggiungere nuove condizioni.

Per la convergenza dello sviluppo [3] è sufficiente [4]), per esempio, che l'integrale

$$\int_{-\infty}^{\infty} e^{\frac{x^2}{4}}\, dF(x)$$

converga; la serie [3] convergerà allora per qualsiasi $n \geqq 1$, in ogni punto di continuità, verso $F_n(x)$.

Lo scopo della presente Memoria è quello di dimostrare un teorema [5]), secondo il quale, sotto determinate condizioni, gli sviluppi [5] e [10] *convergono, per* $k \to \infty$, *assolutamente e uniformemente per*

---

[4]) Cfr. loc. cit. [2]), pag. 64. Il criterio di convergenza di M. JACOB («Giornale dell'Istituto Italiano degli Attuari», Anno II, n. 3, luglio 1931-IX), per il quale si richiede soltanto la convergenza dell'integrale $\int_{-\infty}^{\infty} \frac{e^{\frac{x^2}{4}}}{1+|x|}\, dF(x)$, non permette, per quanto mi sembra, una conclusione immediata sulla convergenza dello sviluppo di $F_n(x)$ per $n > 1$.

[5]) Questo teorema fu enunciato, in forma alquanto meno precisa e senza dimostrazione, nel mio lavoro sopra citato, pag. 65. Cfr. anche H. CRAMÉR, *On some classes of series used in mathematical statistics*, «Verhandlungen des 6. Skandinav. Math. -Kongress.», Kopenagen 1925, pag. 399.

tutti i valori reali di x e danno, per $n \to \infty$, rappresentazioni asintotiche nel senso sopra indicato, i resti tendendo verso zero al crescere di $|x|$.

Per rendere possibili tali conclusioni, la data ripartizione di probabilità deve naturalmente soddisfare a condizioni più restrittive di quelle prima considerate, come risulta dal teorema seguente, nel quale le funzioni $f(x)$, $f_n(x)$, $F_n(x)$, ecc., hanno il significato che è stato loro sopra attribuito.

TEOREMA. — *Supponiamo che la densità della probabilità $f(x)$ possegga, per ciascun valore di x, una derivata $f'(x)$ soddisfacente alla disuguaglianza seguente*

$$|f'(x) - \varphi'(x)| < \frac{1}{2\sqrt{\pi}} e^{-\frac{x^2}{4}}. \tag{13}$$

Per ogni intero $n \geqq 3$ si ha allora [6]

$$F_n(x) = \Phi(x) + \sum_3^\infty \frac{c_\nu}{\nu!} \Phi^{(\nu)}(x), \tag{14a}$$

$$f_n(x) = \varphi(x) + \sum_3^\infty \frac{c_\nu}{\nu!} \varphi^{(\nu)}(x), \tag{14b}$$

in cui le serie convergono assolutamente e uniformemente per tutti i valori reali di x e i coefficienti sono dati dalla [4]. *Per il resto*

$$r_k(x) = \sum_{k+1}^\infty \frac{c_\nu}{\nu!} \varphi^{(\nu)}(x),$$

valgono le seguenti disuguaglianze, in cui A e B sono costanti positive,

$$|r_k(x)| < A \left(\frac{2k}{n}\right)^{\frac{1}{2}\left[\frac{k+3}{3}\right]} e^{-\frac{x^2}{4}} \quad \text{per} \quad 2 \leqq k < \frac{n}{e^2}, \tag{15a}$$

$$|r_k(x)| < A e^{-Bn - \frac{x^2}{4}} \quad \text{per} \quad \frac{n}{e^2} < k < 3n, \tag{15b}$$

---

6) Il teorema vale anche per $n = 1$ e $n = 2$. Per evitare ovvie complicazioni, tratteremo soltanto il caso $n \geq 3$.

[15$_c$] $$|r_k(x)| < A\left(\frac{2n}{k}\right)^{\frac{n}{2}-\frac{5}{4}} e^{-\frac{x^2}{4}} \quad \text{per} \quad k \geqq 3n,$$

e per il resto

$$R_k(x) = \sum_{k+1}^{\infty} \frac{c_\nu}{\nu!} \Phi^{(\nu)}(x)$$

valgono le stesse disuguaglianze, nelle quali i membri di destra possono essere anche moltiplicati per il fattore $1/(1+|x|)$.

Se ora in un primo tempo diamo a n un valore fisso $\geqq 3$ e facciamo crescere k indefinitamente, risulta dalla [15$_c$] e dalla corrispondente disuguaglianza per $R_k(x)$ che le serie [14$_a$] e [14$_b$] convergono uniformemente per tutti i valori di x. Dalla dimostrazione che daremo del precedente teorema risulterà inoltre la convergenza assoluta di tali serie.

Se, al contrario, teniamo costante k, mentre n tende all'infinito, risulta dalla [15$_a$] e dalla corrispondente disuguaglianza per $R_k(x)$ che le serie danno sviluppi asintotici delle funzioni $F_n(x)$ e $f_n(x)$ nel senso sopra indicato, l'ordine di grandezza dell'errore essendo determinato dal massimo c_ν trascurato. Inoltre, come è evidente, i resti tenderanno in ogni caso a zero, al crescere di $|x|$.

In particolare, si ha, per $k = 2$,

[16$_a$] $$|F_n(x) - \Phi(x)| < \frac{A}{\sqrt{n}} \cdot \frac{e^{-\frac{x^2}{4}}}{1+|x|},$$

[16$_b$] $$|f_n(x) - \varphi(x)| < \frac{A}{\sqrt{n}} e^{-\frac{x^2}{4}},$$

con A costante positiva; relazioni queste interessanti, perchè più vantaggiose delle [9] e [12].

Per la condizione [13] la derivata $f'(x)$ della densità della probabilità deve essere compresa, per ogni valore di x, tra i limiti

$$\varphi'(x) \pm \frac{1}{2\sqrt{\pi}} e^{-\frac{x^2}{4}} = \pm \frac{1}{\sqrt{2\pi}} e^{-\frac{x^2}{4}} \left(\frac{1}{\sqrt{2}} \mp x e^{-\frac{x^2}{4}}\right),$$

ove vanno presi dappertutto o i segni superiori o quelli inferiori. Le curve che limitano il campo in cui può variare $f'(x)$ sono disegnate nella figura 1.

Le curve $y = \varphi'(x) \pm \frac{1}{2\sqrt{\pi}} e^{-\frac{x^2}{4}}$

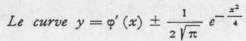

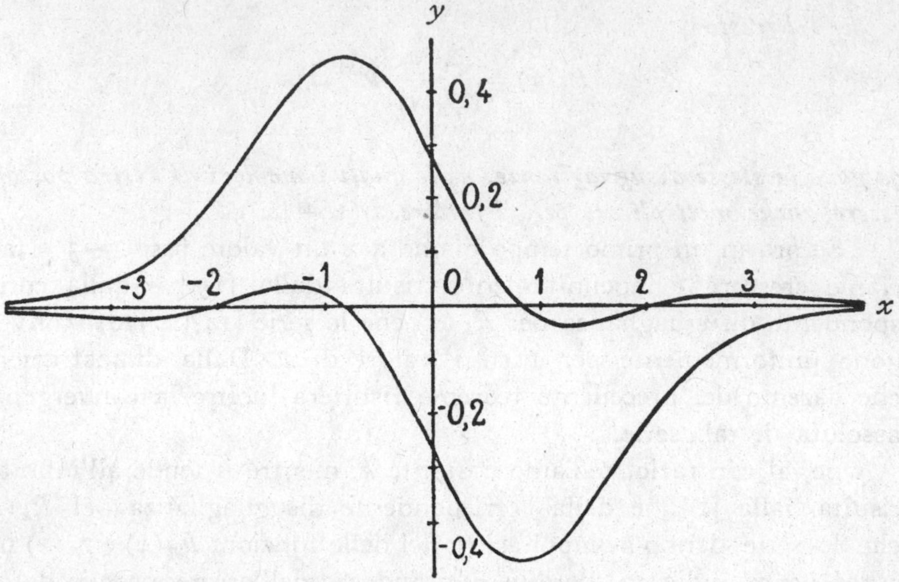

Fig. 1.

Come è facile vedere, la densità della probabilità $f(x)$, deve allora soddisfare, per qualsiasi valore di x, alla disuguaglianza

$$[17] \qquad 0 \leqq f(x) < \frac{A}{1 + |x|} e^{-\frac{x^2}{4}}$$

con A costante positiva [7]).

4. Prima di dimostrare il teorema enunciato al numero precedente, premettiamo il seguente

LEMMA I. - *Ammesso che sia verificata la condizione* [13], *per ogni* $n \geqq 1$ *e* $\nu \geqq 3$, *si ha*

$$\left| \frac{c_\nu}{\nu!} \varphi^{(\nu)}(x) \right| < A \nu^{\frac{1}{4}} e^{-\frac{x^2}{4}} \sum_{j=1}^{\mu} \binom{n}{j} \left(\frac{j}{n}\right)^{\frac{\nu}{2}} \left(\frac{j}{\nu}\right)^{\frac{j}{2}},$$

dove μ *rappresenta il più piccolo dei due numeri* $[\nu/3]$ *e* n; *cioè*

$$[18] \qquad \mu = \mathrm{Min}\left(\left[\frac{\nu}{3}\right], n\right).$$

7) Da ciò, secondo il teorema di convergenza di Jacob [cfr. loc. cit. 4)] si può dedurre la convergenza (condizionata) dello sviluppo in serie di $F_n(x)$ per $n \geq 1$.

Per dimostrare questo lemma introdurremo dapprima le *funzioni caratteristiche* relative alle funzioni di ripartizione $F(x)$ e $F_n(x)$. Indicando tali funzioni caratteristiche con $g(t)$ e $g_n(t)$, avremo

[19] $$g(t)=\int_{-\infty}^{\infty} e^{-itx} f(x)\,dx\,,$$

[20] $$g_n(t)=\int_{-\infty}^{\infty} e^{-itx} f_n(x)\,dx\,.$$

Allora, si ha, come è noto,

[21] $$g_n(t)=\left[g\left(\frac{t}{\sqrt{n}}\right)\right]^n.$$

Nell'intorno del punto $t=0$ si ha inoltre [8], per ogni $k\geqq 3$,

[22] $$e^{\frac{t^2}{2}} g_n(t)=1+\sum_{3}^{k}\frac{c_\nu}{\nu!}(it)^\nu+O(t^{k+1})\,,$$

in cui c_ν è dato dalla [4]. Ciò si deduce senza difficoltà dalla [20] mediante lo sviluppo

$$e^{\frac{t^2}{2}-itx}=\sum_{0}^{\infty}\frac{H_\nu(x)}{\nu!}(-it)^\nu\,,$$

convergente per ogni valore di x e di t.

Consideriamo ora, nelle [19] e [20], t come variabile complessa e poniamo

$$t=\tau e^{i\psi}\,.$$

Segue allora dalle [17] e [19] che $g(t)$ è una funzione trascendente intera di t e, in base alla [21], deve essere tale anche $g_n(t)$.

Lo sviluppo [22] dimostra ora che abbiamo per un t complesso qualunque

[23] $$e^{\frac{t^2}{2}} g_n(t)=1+\sum_{3}^{\infty}\frac{c_\nu}{\nu!}(it)^\nu\,.$$

---

8) Cfr. loc. cit. 2), lemma 5.

Se poniamo

$$e^{\frac{t^2}{2}} g(t) = 1 + h(t),$$

risulta dalla [21]

[24] $$e^{\frac{t^2}{2}} g_n(t) = \left[1 + h\left(\frac{t}{\sqrt{n}}\right)\right]^n.$$

Inoltre, avremo

$$h(t) = e^{\frac{t^2}{2}} \int_{-\infty}^{\infty} e^{-itx} f(x)\, dx - 1 =$$

$$= e^{\frac{t^2}{2}} \int_{-\infty}^{\infty} e^{-itx} [f(x) - \varphi(x)]\, dx =$$

$$= \frac{1}{it} e^{\frac{t^2}{2}} \int_{-\infty}^{\infty} e^{-itx} [f'(x) - \varphi'(x)]\, dx$$

e da ciò risulta, in base alla [13], posto $t = \tau e^{i\psi}$,

[25] $$|h(t)| < \frac{1}{2\tau\sqrt{\pi}} \int_{-\infty}^{\infty} e^{\frac{1}{2}\tau^2 \cos 2\psi + \tau x \operatorname{sen} \psi - \frac{x^2}{4}}\, dx = \frac{1}{\tau} e^{\frac{\tau^2}{2}}.$$

Per le [23] e [24] lo sviluppo in serie di potenze di $h(t)$ non potrà contenere alcun termine di grado inferiore a tre. Se quindi poniamo, per un qualsiasi numero intero j,

$$[h(t)]^j = \sum_{\nu=3j}^{\infty} b_{\nu j} t^\nu,$$

risulterà dalla [25], applicando un noto teorema della teoria delle funzioni,

$$|b_{\nu j}| < \frac{e^{\frac{j\tau^2}{2}}}{\tau^{\nu+j}},$$

per ogni $\tau > 0$. Per $\tau = \sqrt{(\nu + j)/j}$ si ha quindi

[26] $$|b_{\nu j}| < \left(\frac{ej}{\nu + j}\right)^{\frac{\nu+j}{2}}.$$

Ma dalla [24] risulta

$$e^{\frac{t^2}{2}} g_n(t) = \left[1 + h\left(\frac{t}{\sqrt{n}}\right)\right]^n = 1 + \sum_{j=1}^{n} \binom{n}{j} \sum_{\nu=3j}^{\infty} b_{\nu j} \left(\frac{t}{\sqrt{n}}\right)^{\nu} =$$

$$= 1 + \sum_{\nu=3}^{\infty} \left(\frac{t}{\sqrt{n}}\right)^{\nu} \sum_{j=1}^{\mu} \binom{n}{j} b_{\nu j},$$

in cui μ rappresenta il più piccolo dei due numeri $[\nu/3]$ e n e quindi, confrontando con la [23] e introducendo la disuguaglianza [26] per $b_{\nu j}$, si ha

$$\left|\frac{c_\nu}{\nu!}\right| < n^{-\frac{\nu}{2}} \sum_{j=1}^{\mu} \binom{n}{j} \left(\frac{ej}{\nu + j}\right)^{\frac{\nu+j}{2}}.$$

Inoltre abbiamo

$$\varphi^{(\nu)}(x) = \frac{(-1)^\nu}{\sqrt{2\pi}} H_\nu(x)\, e^{-\frac{x^2}{2}}$$

e [9], per ogni valore reale di x,

$$|H_\nu(x)| < 2\sqrt{\nu!}\, e^{\frac{x^2}{4}}.$$

Dalle tre ultime relazioni risulta

$$\left|\frac{c_\nu}{\nu!} \varphi^{(\nu)}(x)\right| < \sqrt{\nu!}\; n^{-\frac{\nu}{2}} e^{-\frac{x^2}{4}} \sum_{j=1}^{\mu} \binom{n}{j} \left(\frac{ej}{\nu + j}\right)^{\frac{\nu+j}{2}}.$$

Per la formula di Stirling si ha

$$(\nu + j)! < A \left(\frac{\nu + j}{e}\right)^{\nu + j} \sqrt{\nu + j}$$

e da ciò segue, essendo j in ogni caso minore di ν,

$$\left|\frac{c_\nu}{\nu!} \varphi^{(\nu)}(x)\right| < A \nu^{\frac{1}{4}} n^{-\frac{\nu}{2}} e^{-\frac{x^2}{4}} \sum_{j=1}^{\mu} \binom{n}{j} j^{\frac{\nu+j}{2}} \sqrt{\frac{\nu!}{(\nu+j)!}} <$$

$$< A \nu^{\frac{1}{4}} e^{-\frac{x^2}{4}} \sum_{j=1}^{\mu} \binom{n}{j} \left(\frac{j}{n}\right)^{\frac{\nu}{2}} \left(\frac{j}{\nu}\right)^{\frac{j}{2}}.$$

Il lemma I è così dimostrato.

---

[9] Una mia dimostrazione di questa disuguaglianza si trova nell'opera di C. V. L. CHARLIER, *Applications de la théorie des probabilités à l'astronomie.* (« Traité du Calcul de probabilités et de ses applications », par Emil Borel, tome II, fasc. 4°, Paris, 1931), pag. 49.

5. Applicando la formula di Stirling, si ha, per $1 \leq j < n$,

$$\binom{n}{j} = \frac{n!}{j!\,(n-j)!} < A \frac{n^n}{j^j (n-j)^{n-j}}$$

e questa relazione vale evidentemente anche per $j = n$, quando si ponga $0^0 = 1$.

Risulta quindi dal lemma I

[27] $$\left| \frac{c_\nu}{\nu!} \varphi^{(\nu)}(x) \right| < A \nu^{\frac{1}{4}} e^{-\frac{x^2}{4}} \sum_{j=1}^{\mu} U_{\nu j}$$

dove

$$U_{\nu j} = \left(\frac{j}{n}\right)^{\frac{\nu}{2}-j} \left(\frac{j}{\nu}\right)^{\frac{j}{2}} \left(1 - \frac{j}{n}\right)^{-(n-j)}.$$

Supponiamo ora n costante e consideriamo $U_{\nu j}$ come funzione di ν e di j, con

$$\nu \geq 3 \quad , \quad 1 \leq j \leq \mathrm{Min}\left(\left[\frac{\nu}{3}\right], n\right).$$

Se introduciamo, mediante le equazioni

$$\nu = \alpha n \ , \ j = \beta n,$$

le nuove variabili α e β, avremo

[28] $$\frac{1}{n} \log U_{\nu j} = \frac{\alpha - \beta}{2} \log \beta - \frac{\beta}{2} \log \alpha - (1-\beta) \log (1-\beta) = \psi(\alpha, \beta).$$

La funzione $\psi(\alpha, \beta)$ ora introdotta dovrà essere considerata nel campo definito dalle disuguaglianze

$$\alpha > 0 \ , \ 0 < \beta \leq \mathrm{Min}\left(\frac{\alpha}{3}, 1\right),$$

per rendere possibile una opportuna valutazione di $U_{\nu j}$ per tutti i valori di ν e di j che vengono considerati nella [27]. Se prendiamo in esame la funzione ψ stessa e le due derivate parziali $\frac{\partial \psi}{\partial \alpha}$ e $\frac{\partial \psi}{\partial \beta}$, si ha, mediante considerazioni elementari che lasciamo al lettore, il seguente

LEMMA II – 1°) *Per* α *costante, soddisfacente alla limitazione* $0<\alpha\leqq 1/e^2$, *la funzione* $\psi(\alpha,\beta)$ *risulta crescente per* β *variabile da zero a* $\alpha/3$; 2°) *per*

$$\frac{1}{e^2}\leqq\alpha\leqq 3\,,\quad 0<\beta\leqq\frac{\alpha}{3}$$

si ha

$$\psi(\alpha,\beta)\leqq\psi\left(\frac{1}{e^2},\frac{1}{3e^2}\right)<-\frac{1}{20};$$

3°) *per*

$$\alpha\geqq 3\,,\ 0<\beta\leqq 1$$

si ha

$$\psi(\alpha,\beta)<\frac{1}{3}-\frac{1}{2}\log\alpha.$$

6. Applicando il lemma II alle relazioni [27] e [28] si hanno immediatamente delle limitazioni per i termini $c_\nu\varphi^{(\nu)}(x)/\nu!$ del nostro sviluppo in serie.

Dal lemma II–1°), risulta che, per ν costante soddisfacente alla limitazione $3\leqq\nu<n/e^2$, il valore $U_{\nu j}$ cresce per j variabile da $j=1$ a $j=\mu=[\nu/3]$. La [27] dà quindi, per $3\leqq\nu<n/e^2$,

[29]
$$\begin{aligned}\left|\frac{c_\nu}{\nu!}\varphi^{(\nu)}(x)\right|&<A\nu^{\frac{5}{4}}\left(\frac{\mu}{n}\right)^{\frac{\nu}{2}-\mu}\left(\frac{\mu}{\nu}\right)^{\frac{\mu}{2}}\left(\frac{n}{n-\mu}\right)^{n-\mu}e^{-\frac{x^2}{4}}<\\&<A\nu^{\frac{5}{4}}\left(\frac{1}{3}\right)^{\frac{\nu}{3}}\left(1+\frac{\mu}{n-\mu}\right)^{n-\mu}\left(\frac{\nu}{n}\right)^{\frac{\nu}{2}-\left[\frac{\nu}{3}\right]}e^{-\frac{x^2}{4}}<\\&<A\nu^{\frac{5}{4}}\left(\frac{e}{3}\right)^{\frac{\nu}{3}}\left(\frac{\nu}{n}\right)^{\frac{\nu}{2}-\left[\frac{\nu}{3}\right]}e^{-\frac{x^2}{4}}<\\&<Ae^{-B\nu}\left(\frac{\nu}{n}\right)^{\frac{\nu}{2}-\left[\frac{\nu}{3}\right]}e^{-\frac{x^2}{4}},\end{aligned}$$

dove B rappresenta, come A, una costante positiva.

Dal lemma II–2°), risulta, sempre mediante le [27] e [28], per $n/e^2<\nu\leqq 3n$,

[30]
$$\left|\frac{c_\nu}{\nu!}\varphi^{(\nu)}(x)\right|<Ae^{-\frac{n}{20}-\frac{x^2}{4}}$$

e infine, dal lemma II–3°), risulta per $\nu\geqq 3n$

[31]
$$\left|\frac{c_\nu}{\nu!}\varphi^{(\nu)}(x)\right|<A\nu^{\frac{1}{4}}\,ne^{\frac{n}{3}}\left(\frac{n}{\nu}\right)^{\frac{n}{2}}e^{-\frac{x^2}{4}}.$$

Se ora diamo a n un valore costante $\geqq 3$ e facciamo crescere ν all'infinito, risulta dalla [31] che la serie $\Sigma c_\nu \varphi^{(\nu)}(x)/\nu!$ converge assolutamente e uniformemente per tutti i valori reali di x.

Valutiamo ora il resto di questa serie,

$$r_k(x) = \sum_{k+1}^{\infty} \frac{c_\nu}{\nu!} \varphi \quad (x).$$

Per $k \geqq 3n$ risulta, in primo luogo, dalla [31]

$$|r_k(x)| < A n^{\frac{n}{2}+1} e^{\frac{n}{3}-\frac{x^2}{4}} \sum_{k+1}^{\infty} \frac{1}{\nu^{\frac{n}{2}-\frac{1}{4}}} <$$

[32] $$< A n^{\frac{5}{4}} e^{\frac{n}{3}-\frac{x^2}{4}} \left(\frac{n}{k}\right)^{\frac{n}{2}-\frac{5}{4}} <$$

$$< A \left(\frac{2n}{k}\right)^{\frac{k}{2}-\frac{5}{4}} e^{-\frac{x^2}{4}}$$

e la [15_c] è dimostrata.

Per $n/e^2 < k < 3n$ risulta poi, per le [30] e [32],

$$|r_k(x)| < A n e^{-\frac{n}{20}-\frac{x^2}{4}} + A \left(\frac{2}{3}\right)^{\frac{n}{2}-\frac{5}{4}} e^{-\frac{x^2}{4}} <$$

[33] $$< A n e^{-\frac{n}{20}-\frac{x^2}{4}} < A e^{-Bn-\frac{x^2}{4}},$$

ossia la disuguaglianza [15_b].

Sia ora $3 \leqq k < n/e^2$. (Non occorre trattare in particolare il caso $k = 2$ considerato nella [15_a], giacchè segue dal caso $k = 3$). Poichè per ogni ν intero positivo vale la disuguaglianza

$$\frac{\nu}{2} - \left[\frac{\nu}{3}\right] \geqq \frac{1}{2}\left[\frac{\nu+2}{3}\right],$$

risulta che per $k < \nu < n/e^2$ si ha

$$\log\left(\frac{\nu}{n}\right)^{\frac{\nu}{2}-\left[\frac{\nu}{3}\right]}-\log\left(\frac{k}{n}\right)^{\frac{1}{2}\left[\frac{k+3}{3}\right]}\leq\frac{1}{2}\left[\frac{\nu+2}{3}\right]\log\frac{\nu}{n}-\frac{1}{2}\left[\frac{k+3}{3}\right]\log\frac{k}{n}=$$

$$=\frac{1}{2}\left[\frac{k+3}{3}\right]\log\frac{\nu}{k}-\frac{1}{2}\left(\left[\frac{\nu+2}{3}\right]-\left[\frac{k+3}{3}\right]\right)\log\frac{n}{\nu}<$$

$$<\frac{\nu-k}{2k}\cdot\frac{k+3}{3}-\frac{\nu+2}{3}+\frac{k+3}{3}+1=-\frac{\nu(k-3)}{6k}+\frac{k+5}{6}<$$

$$<-\frac{k-3}{6}+\frac{k+5}{6}<2$$

e quindi

$$\left(\frac{\nu}{n}\right)^{\frac{\nu}{2}-\left[\frac{\nu}{3}\right]}<A\left(\frac{k}{n}\right)^{\frac{1}{2}\left[\frac{k+3}{3}\right]}.$$

Dalla [29] risulta ora

$$\sum_{\nu=k+1}^{n/e^2}\left|\frac{c_\nu}{\nu!}\varphi^{(\nu)}(x)\right|<Ae^{-Bh}\left(\frac{h}{n}\right)^{\frac{1}{2}\left[\frac{k+3}{3}\right]}e^{-\frac{x^2}{4}},$$

da cui deduciamo, per la [33],

$$[34]\qquad \left|r_k(x)\right|<A\left(\frac{k}{n}\right)^{\frac{1}{2}\left[\frac{k+3}{3}\right]}e^{-\frac{x^2}{4}}+Ane^{-\frac{n}{20}-\frac{x^2}{4}}$$

e poichè, come si può dimostrare senza difficoltà, per n e k interi positivi si ha sempre

$$ne^{-\frac{n}{20}}<A\left(\frac{2k}{n}\right)^{\frac{1}{2}\left[\frac{k+3}{6}\right]},$$

tenuta presente la [34], risulta la [15_a].

7. Abbiamo già dimostrato che la serie

$$[35]\qquad \varphi(x)+\sum_{3}^{\infty}\frac{c_\nu}{\nu!}\varphi^{(\nu)}(x),$$

nella quale i coefficienti sono dati dalla [4], converge assolutamente e uniformemente per tutti i valori reali di x. La serie [35], inoltre, rappresenta effettivamente con la sua somma la funzione $f_n(x)$, come risulta, per esempio, da un teorema generale di Weyl [10]), sui sistemi

[10]) H. WEYL, *Singuläre Integralgleichungen*, «Mathematische Annalen», Band. 66, 1909.

ortogonali di polinomi di Hermite. Ciò può però essere dimostrato facilmente anche in maniera diretta come segue.

Se rappresentiamo la somma della serie [35] con $f_n^*(x)$, e teniamo conto delle limitazioni sopra ottenute per i termini della serie, si ha

$$\left|f_n^*(x)\right| < A e^{-\frac{x^2}{4}} .$$

Se ora formiamo la funzione caratteristica

$$g_n^*(t) = \int_{-\infty}^{\infty} e^{-itx} f_n^*(x)\, dx$$

potremo, sostituita la $f_n^*(x)$ con la serie [35], integrare termine a termine, giacchè tale serie converge assolutamente. Avremo quindi, per t complesso qualunque,

$$e^{\frac{t^2}{2}} g_n^*(t) = 1 + \sum_{3}^{\infty} \frac{c_\nu}{\nu!} (it)^\nu$$

da cui, per la [23], risulta

$$g_n^*(t) = g_n(t) .$$

Le due funzioni $f_n^*(x)$ e $f_n(x)$, integrabili assolutamente da $-\infty$ a $+\infty$, hanno quindi la medesima funzione caratteristica, ossia abbiamo

$$g_n(t) = \int_{-\infty}^{\infty} e^{-itx} f_n(x)\, dx = \int_{-\infty}^{\infty} e^{-itx} f_n^*(x)\, dx .$$

Poichè la data densità di probabilità $f(x)$ possiede, per la [13], una derivata integrabile assolutamente da $-\infty$ a $+\infty$, la funzione caratteristica $g(t)$, per valori reali di t sufficientemente grandi, è dell'ordine di grandezza di $1/|t|$ [11], e, per la [21], $g_n(t)$ è quindi, per n costante, dell'ordine di grandezza di $1/|t|^n$. L'integrale

$$\int_{-\infty}^{\infty} e^{itx} g_n(t)\, dt$$

---

[11]) Cfr. S. BOCHNER, *Vorlesungen über Fourier'sche Integrale*. Leipzig, 1932, pag. 9.

converge allora assolutamente e uniformemente per tutti i valori reali di x e la corrispondente funzione $f_n(x)$ è perciò determinata univocamente da $g_n(t)$ [12], cioè $f_n(x)$ è uguale identicamente a $f_n^*(x)$.

Con ciò è dimostrata la [14_b].

Integrando da $-\infty$ a x avremo poi il corrispondente sviluppo per $F_n(x)$, ossia la [14_a].

Infine, dalle relazioni [5] e [10], risulta

$$R_k(x) = \int_{-\infty}^{x} r_k(t)\,dt = -\int_{x}^{\infty} r_k(t)\,dt$$

e, introducendosi le disuguaglianze sopra indicate per $r_k(x)$, si hanno le corrispondenti disuguaglianze per $R_k(x)$, in cui, evidentemente, può sempre essere introdotto un fattore $1/(1+|x|)$. In particolare risulta in tal modo la convergenza assoluta e uniforme della serie integrata [14_a] e quindi il nostro teorema è ora completamente dimostrato.

8. In questo lavoro abbiamo considerato esclusivamente somme di variabili casuali possedenti tutte la medesima ripartizione di probabilità. Tutti i risultati ottenuti possono essere però generalizzati al caso di somme i cui termini posseggano ripartizioni diverse di probabilità, purchè queste soddisfino a determinate condizioni di regolarità.

---

[12]) Cfr. loc. cit. [11]), pag. 42.

32.

with H. Wold

Mortality variations in Sweden: A study in graduation and forecasting

Skand. Aktuarietidskr. **18**, 161–241 (1935)

Contents.

I. Introduction.

1. The investigation of mortality variations has attracted a great deal of interest in recent years. It is well known that, generally speaking, mortality rates have been continually decreasing during a long period. The detailed study of available experience concerning the structure of this movement, and the analysis of its relations to various factors of social and economic development and of its probable causes, obviously constitute a statistical problem of the highest importance. In the trend of mortality, and the deviations caused by temporary influences, essential features of the history of the population concerned are revealed.

It is a general principle of all practical applications of scientific research that, basing ourselves on knowledge obtained from past experience, we try to form certain ideas as to the way things are going to happen in the future. Applying this principle to the phenomenon of human mortality, we are at once confronted with one of the fundamental problems of Actuarial Science: the problem of »Forecasting Mortality».

Every calculation of a premium or a policy value connected with any kind of life assurance rests upon an hypothesis on the future course of mortality. In the majority of cases, this hypothesis takes the simple form of assuming that past experience will repeat itself without any substantial change. If a further improvement in mortality is to be counted upon, this assumption will be on the safe side as far as ordinary life assurance is concerned. In other cases, such as annuity business, it may easily lead to disastrous results. In any case the problem of making more accurate forecasts will necessarily be of a great practical importance.

Apart from life assurance, there are also other important practical problems where hypotheses of this kind have to be introduced. Modern social development has given rise to a number of questions (*e.g.* questions concerning social insurance, housing conditions, educational problems etc.) where it is highly desirable to obtain some information as to the probable future development of the population of a certain country and its distribution according to age. One of the main factors governing the movement of a population being mortality, it is easily understood that in all such questions some estimate of future mortality will be required.

2. It will appear from a glance at the appended List of References (which does not claim to form a complete list of contributions to our subject) that the problems of secular mortality changes and mortality forecasts have, during the last eight years, been discussed by a considerable number of authors. Several interesting methods have been proposed, but it has been generally felt that owing to the lack of detailed statistical data, covering a sufficiently long period, only preliminary results could be reached.

The object of the present paper is 1) to give a critical analysis of some of the methods hitherto proposed, 2) to give a systematical development of a method which seems to us particularly worthy of attention, and 3) to test this method by means of the unique statistical material furnished by Swedish Population Statistics. We shall be concerned exclusively with the problems of giving a comprehensive account of past mortality changes and (if possible) a reliable forecast of future development. During the course of our work, a number of further interesting questions were raised, some of which will be briefly mentioned in the sequel. The detailed investigation of these questions must, however, be left aside for the moment.

It seems convenient to give already here a highly preliminary account of our method and our statistical material. Our material was taken from the official records concerning the mortality of the Swedish population. For each sex, the material consists of the $12 \cdot 26 = 312$ death rates corresponding to the 12 quinquennial age intervals between ages 30 and 90, and the 26 quinquennial calendar periods between 1800 and 1930. In Diagram 1, these death rates are represented by small black squares. When we consider mortality as a function of age the death rates may, as indicated in the diagram, be arranged by *verticals* or by *diagonals*.[1] If we take the arrangement by verticals, we shall have the mortality of the various *calendar periods* contained in the material; if, on the other hand, we consider the arrangement by diagonals, we shall be concerned with the mortality of successive *generations*. Thus the vertical series of death rates marked out in the diagram corresponds to the mortality during the quinquennial period 1836—1840, while the analogous diagonal series belongs to a generation, whose central birth year is 1790. As will be seen from the diagram, the material contains 26 calendar periods and 15 »complete» generation series. The central birth years of the latter extend from 1770 to 1840.

[1] Of course also other arrangements might be considered, but the above mentioned are those which present themselves most naturally to the mind.

Our method may be briefly described in the following way. We are working throughout with the MAKEHAM formula for the force of mortality

$$\mu_x = \alpha + \beta c^x,$$

which, generally speaking, gives a fairly close agreement to our material for the age interval 30—90, whether we consider the vertical or the diagonal arrangement of our death rates. The constants α, β and $\log c$ are determined for each

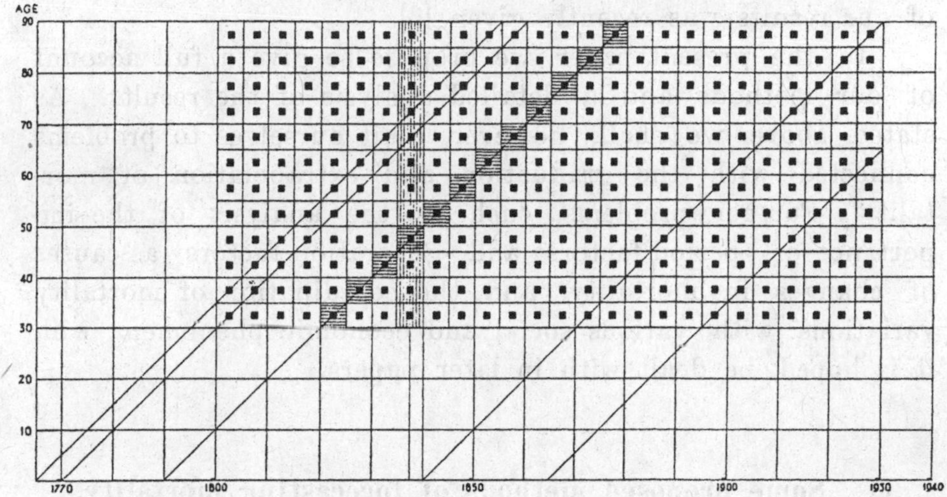

Diagram 1. **Schematic representation of the material. Mortality by periods and by generations.**

calendar period and for each complete generation contained in the material. These constants are then in the first case considered as functions of time (*mortality by periods*), in the second case as functions of the year of birth (*mortality by generations*), and are graduated by means of suitable analytical expressions. In this way, as is readily seen, we obtain two independent *two-dimensional graduations* of the material, which are then compared and used as bases for the extrapolation into the future.

The present investigation has developed out of certain preliminary researches started in 1930 at the Department of Actuarial Mathematics and Mathematical Statistics, University of Stockholm, under the direction of one of the authors (H. C.). The results obtained up to 1931 were communicated to a Committee of experts appointed by the Swedish Government and charged with constructing a new mortality table for annuity assurance.[1] The method applied by us formed the base of one of the main lines followed by the Committee in constructing their table.

Since the publication of the Report of the Committee, we have continued the investigation on an extended scale and with considerably improved methods. A preliminary account of our results was recently given [3].

In the present paper, we propose to give a full account of our methods and a detailed analysis of the results. As stated above we shall, however, limit ourselves to problems connected with the graduation and extrapolation of mortality. Further problems, such as the analysis of the importance of »period factors» and »generation factors» as causes of changes in mortality, and the coordination of mortality variations with various social and economic phenomena will, it is hoped, be dealt with in later papers.

II. Some proposed methods of forecasting mortality.

3. As far as we are aware, the earliest attempt at making an estimate of future mortality by means of an extrapolation of past experience concerning mortality changes is due to the Swedish astronomer H. Gyldén [29, p. 6]. In a lecture delivered before the Swedish Assurance Association in 1875, Gyldén fitted a straight line to the general death rates of the Swedish population (all ages) for the years 1750—1870 and suggested that this line might be considered as giving a probable forecast of future death rates. The latest year contained in his material being 1870 with the death rate 19.8 ‰, he gave in particular the figure 17.7 ‰ as an estimate

[1] Report of the Committee (construction of the mortality table »R 32»): Ref. nr. [28]. Cf. also Åkesson [27] and Meurk [18].

of the death rate for the year 1900. — In reality, the observed death rate for 1900 was 16.8 ‰.

The method used by GYLDÉN is obviously a very primitive one, and apparently no suggestion was made of any immediate use of the forecast obtained for actuarial purposes. — In two reports on certain pension fund problems [19; 20], some practical applications of a similar method were made by THV. RICHARDT. He considered the annuity values a_{60} and a_{65} according to various mortality tables drawn from Norwegian experience and found that, on the average, these values showed an annual decrease amounting to 0.025. This fact was then used for an extrapolation into the future.

4. In these crude methods, the problem is regarded as a one-dimensional extrapolation problem. It is, however, obvious that no adequate treatment of the problem can be obtained in this way. Mortality variations cannot be accurately analyzed without regarding mortality as a function of age, and as soon as this is done, the problem becomes a two-dimensional one. Thus it will be necessary to regard mortality rates as functions of the two variables *age* and *time*, and to study their general behaviour from this point of view.

If we denote by $\mu(x, t)$ the force of mortality for the age x and the calendar year t, the general problem may be stated in the following way. — Statistical data (deaths and exposures) relating to a certain range of values of x and t being available, it is required to find: 1) a two-dimensional graduation of the material, giving the »best-possible» values of $\mu(x, t)$ over the range covered by the data, and 2) an extrapolation of $\mu(x, t)$ to future values of t.

For the practical solution of these problems it is, of course, not necessary *directly* to use two-dimensional graduation and extrapolation methods. As already pointed out in No. 2 with respect to Diagram 1, the observed death rates may in various ways be arranged in one-dimensional sets, each of which may be separately graduated and extrapolated. As we are now concerned with the study of fluctuations in time, the two most natural arrangements of the death rates according to Diagram 1 will be the arrangements by *horizontal rows* and by *diagonals*. Accordingly we shall speak of »horizontal» and »diagonal» methods of graduation and extrapolation. — In the first case, we take the death rates for a given age group and follow their variations with time, thus keeping in $\mu(x, t)$ the age x constant and allowing t to vary. In the second case, on the other hand, we are concerned with the death rates of a given generation. Denoting the birth year by τ, we have

$$\tau = t - x,$$

and thus in the second case we keep $t - x$ constant and regard $\mu(x, t) = \mu(x, \tau + x)$ as a function of x. — A glance at Diagram 1 will help to make these conceptions clear.

In both cases it will, however, be necessary to make certain adjustments of the graduated and extrapolated death rates obtained in the first instance, in order to ensure consistence between the different one-dimensional sets into which the material has been divided. But it is only when a definite principle is adopted for this adjustment (*e. g.* adjustment according to some analytical formula), that we are concerned with a two-dimensional method in the proper sense.

5. In 1912, a mortality table for annuity assurance was constructed by Lindstedt [16], who used data from Swedish population statistics and extrapolated the q_x-values keeping the age constant, thus using a method of the »horizontal» kind. The last year covered by the data available for the construction was 1907, and extrapolation was made graphically over the interval 1907—1915. The result was the table known as the »ÅFK» table, which has been largely used for annuity purposes in Sweden. In his account of the method employed for the construction, Lindstedt explicitly pointed out the necessity of basing a forecast on an extrapolation of past mortality changes. The extrapolated table corresponds fairly well to the actual experience for the Swedish population during the period 1911—1915. (Cf. [32], I, p. 114).

In Great Britain, the problem of mortality forecasting has attracted a great deal of interest. In several investigations, the majority of which have been published since 1923, [31; 8; 30; 21; 12; 8 a], various »horizontal» methods have been applied to annuity experience and to English population statistics. Analytical expressions for the mortality decrease have in some cases been suggested. Thus it has been assumed for the extrapolation of the mortality of annuitants [30], that the rate of mortality $q(x, t)$ for the age x and the calendar year t is of the form

$$q(x, t) = m + nk^t, \tag{1}$$

where m, n and k depend on x but not on t. The same relation has been applied by RICHMOND [21] to English population statistics. GREENWOOD [12] used straight lines for the extrapolation of $q(x, t)$. Reviews of the work made on these lines by British actuaries have been given in this Journal by MALTBY [17] and ELDERTON [7]. In the discussion of the mortality of annuitants, additional difficulties appear which are due to the effect of selection. Questions belonging to this order of ideas fall, however, outside the scope of the present paper.

The relation (1) was applied to German experience by RIEBESELL [22], who found that a linear expression in t would be preferable. The use of (1) was defended by SACHS [24], who suggested the formula[1]

$$q(x, t) = q(x, 1896) \cdot a^{\frac{t-1896}{x+b}}$$

where a and b are constants. Thus in particular he puts $m = 0$ in (1). (Cf also the subsequent discussion in the papers [13; 25; 23].)

Formulae of the type

$$q(x, t) = k + lt + mt^2 + nt^3 \text{ and } \mu(x, t) = k^{1+mt+nt^2}$$

have been applied to Swedish population statistics [9; 28]. Here again, k, l, m and n depend on x but not on t.

[1] Similar expressions were successfully used already by KNIBBS [14, p. 378], who also introduced a quadratic term in t in the exponent.

6. The »diagonal» point of view appears almost simultaneously in two papers by Davidson and Reid [4] and by Derrick [5].

Working with data concerning the population of England and Wales, Derrick considers the curve which, for a given age group, represents the rate of mortality as a function of the time t. The curves for the various age groups appear irregular and unrelated. When, however, the same curves are redrawn with the birth year τ as abscissa, he finds a striking parallelism between the curves in the new diagram. For this remarkable phenomenon, he suggests the following explanation: »Since there is a common association between the records selected diagonally in this way, in that they represent the experience of the same individuals in different stages in their lives, it is not unreasonable to attribute the similarity to a »generation» influence connected in some way with the period of birth». And he claims that the principle »of relating future mortality changes to the changes already observed by the same generation at younger ages, may receive at least equal consideration to that accorded to methods under which the several age curves are unrelated to one another, ...»

The forecasting method proposed by Davidson and Reid is of an essentially »diagonal» character. They criticize existing mortality tables since these do not represent »consecutive human life» (*i. e.* generations), and suggest that »if a law of life could be found actually representing mortality, the constants in the equation to the law varying with, say, the calendar year of birth, it would be possible to forecast mortality in a simple and reliable manner.» Having derived the constants of the law from past data, they propose to »make our estimate of forecasted mortality by projecting the curve onwards from the date of the investigation into the future.» — In their attempts to realize this programme, they work throughout with the Makeham formula and consider the constants α, β and c as functions of the birth year. They try to represent these functions by certain analytical expressions, and in this way it becomes possible to make a forecast of future mortality.

It should be observed that, according to the above, the method of DAVIDSON and REID is a *two-dimensional method* in the proper sense. If MAKEHAM's law holds good for the mortality of the successive generations, and if α, β and c are regarded as known functions of the birth year τ, then the force of mortality becomes a known function of the two variables x and τ. Of course this should not be thought of as an *exact* representation of the force of mortality in a given experience, but rather as a two-dimensional formula for *graduation* and *extrapolation*.

7. The idea of making a systematical use of the MAKEHAM constants for an investigation of mortality changes seems to have been first expressed in 1923 by BLASCHKE [2]. He was working with the ordinary »period» mortality and considered the MAKEHAM constants α, β and c as functions of the calendar year t. This might be characterized as a »vertical» method, since it means that the observed death rates are arranged in vertical sets (cf. Diagram 1), each set corresponding to a given calendar period. Then the MAKEHAM constants are determined for each period and considered as functions of time. — Similar methods have been proposed by KOBI [15], BERNSTEIN [1], and SCHWEER [26].[1]

It is obvious that in this way, just as by the method of DAVIDSON and REID, we obtain a two-dimensional graduation of the observed death rates. The methods are, however, fundamentally different, since the MAKEHAM formula is here assumed to hold for the mortality by *periods*, not for the mortality by *generations*.

8. As we have already pointed out in the Introduction, the present investigation is entirely founded upon a systematical use of the MAKEHAM formula. Throughout the paper, we use the »vertical» and »diagonal» methods simultaneously, thus applying the MAKEHAM formula both to periods and to generations.

---

[1] A two-dimensional method of a quite different character has been used by FUNKE [11].

We are well aware of the fact that this formula has sometimes been severely criticized and we are indeed very far from believing that it represents a universal »Law of Life». It does, however, seem sufficiently remarkable that a mathematical expression of this simple structure has proved capable of serving at least as a first approximation to mortality curves of the most varied provenience. Without being prepared to claim for the MAKEHAM constants any definite biological significance, we think it may be safely maintained that generally these constants, if determined by a sufficiently accurate method, sum up essential features of the corresponding mortality curve in a very convenient way. It will be shown in the sequel that throughout the statistical material used in this investigation, the MAKEHAM formula gives a reasonably good fit to the mortality curves of the various periods and generations for all ages above 30. (It is, of course, the occurrence of tuberculosis that prevents us from getting below 30.) We think it should be concluded that a study of the behaviour of the MAKEHAM constants will be a useful method in order to obtain a general survey of past mortality changes and, with due reservations, a forecast of future mortality.

At the beginning of our investigation in 1930, we did not know that a similar use of the MAKEHAM formula had already been proposed by other authors. Though our method is not entirely new, it seems to us that a full publication of our work may be justified, since the statistical material at our disposal admits of a much more detailed analysis of both the method and the results than has been the case before.

III. Graduation and extrapolation of Swedish mortality.

9. *The statistical material.* — We shall consider the mortality of the Swedish population between ages 30 and 90, during the period $^1/_1$ 1801—$^{31}/_{12}$ 1930. In Tables 1 a—1 b, the exposed to risk n_x are given for quinquennial age groups and calendar periods. The corresponding deaths d_x are set out in Tables 2 a—2 b. These data are taken from official publications, with the exception of the deaths prior to 1850, where the figures given by G. SUNDBÄRG[1] have been used. SUNDBÄRG has discussed the sources of errors to which the Swedish system of population statistics is liable. It does not seem probable that the errors in the death rates are of a magnitude that will to any appreciable degree affect the result of the present investigation.

The death rates $\frac{d_x}{n_x}$, which are found in Tables 3 a—3 b, have been interpreted as approximate values of the force of mortality $\mu(x, t)$ for the central age and central calendar year of the group. In this connection, we refer once more to the schematic representation of the material in Diagram 1.

10. *Determination of the* MAKEHAM *constants.* χ^2*-minimum method.* — A great number of different methods have been proposed for the determination of the constants in the MAKEHAM formula from a given mortality experience. For the purpose of the present investigation, 82 different series of data had to be graduated, and it was necessary to choose a theoretically sound and practically simple method for the determination of the constants and for the testing of the resulting graduations. In both these respects, we have founded our work on the measure of »goodness of fit» introduced by K. PEARSON and usually denoted by χ^2.

Let n_x and d_x be a set of values of exposed to risk and observed deaths, and let μ_x be the corresponding *graduated* values of the force of mortality, so that $n_x \mu_x$ denotes the »calculated deaths». Then the measure of goodness of fit is defined by the expression

$$\chi^2 = \sum_x \frac{(d_x - n_x \mu_x)^2}{n_x \mu_x}, \tag{2}$$

where the sum is extended to all age groups contained in the series under graduation. If the graduation is made according to the MAKEHAM formula, we have

[1] G. SUNDBÄRG: Bevölkerungsstatistik Schwedens 1750—1900. Stockholm, 2nd ed. (1923).

$$\mu_x = \alpha + \beta c^x$$

and in order to obtain the best possible fit, *we have to choose the parameters* α, β *and* c *so as to make* χ^2 *as small as possible.*

A rigorous application of this χ^2-*minimum principle* obviously leads to a very complicated system of equations for the determination of α, β and c, and so we have tried to work out an approximate method of solution.

If we replace in the denominator of (2) the calculated deaths $n_x \mu_x$ by the observed deaths d_x, we obtain the expression

$$(3) \qquad \chi_1^2 = \sum \frac{(d_x - n_x \mu_x)^2}{d_x} = \sum \frac{(d_x - \alpha n_x - \beta c^x n_x)^2}{d_x}.$$

The equations to be solved in order to find the values of α, β and c which minimize the modified expression are obviously much simpler than in the case of (2). Since d_x generally does not differ very much from $n_x \mu_x$ it seems likely that the substitution of χ_1^2 for χ^2 will not cause any substantial change in the resulting values of the parameters. — This will be illustrated by the numerical example given below.

Even in this simplified form the direct solution of α, β and c from the equations obtained by the minimum condition is, however, too complicated for practical work, and thus we have found it necessary to proceed in the following indirect way.

Let us consider the expression (3) *for a fixed value of* c. It is then very easy to determine α and β so as to make χ_1^2 as small as possible. The equations to be solved for this purpose are

$$(4) \qquad \begin{cases} \alpha \sum \dfrac{n_x^2}{d_x} + \beta \sum \dfrac{n_x^2 c^x}{d_x} = \sum n_x, \\ \alpha \sum \dfrac{n_x^2 c^x}{d_x} + \beta \sum \dfrac{n_x^2 c^{2x}}{d_x} = \sum n_x c^x. \end{cases}$$

The resulting values of α and β obviously depend on the given value of c. — Bearing in mind that, in a great number of graduation problems, it has been found that the method of moments gives results which are practically identical with those furnished by the method of least squares, we may, however, also consider the values of α and β which satisfy the simpler system

$$(5)\quad \begin{cases} \alpha\,\Sigma\, n_x + \beta\,\Sigma\, n_x c^x = \Sigma\, d_x, \\ \alpha\,\Sigma\,\Sigma\, n_x + \beta\,\Sigma\,\Sigma\, n_x c^x = \Sigma\,\Sigma\, d_x, \end{cases}$$

that follows from the method of moments. It will be shown in the numerical illustration given below that the differences between the values of α, β and χ^2 obtained from (4) and from (5) are completely insignificant for practical purposes. Thus we have always used the simpler system (5).

If, for a set of values $c_1, c_2, \ldots$ of c, the equations (5) are solved and the corresponding values of χ^2 are calculated from (2) and plotted in a diagram with $\log c$ as abscissa, it is found that the points obtained are extremely well represented by a second degree parabola. If a parabola of the third degree is tried, the fit is so good that the points are represented by the curve with the same exactitude as that given by the calculation. (Cf the numerical illustration below.)

Having in this way calculated α, β and χ^2 for three or four values of $\log c$ near the minimum, it will thus be very simple to find with sufficient accuracy (assuming a parabola of the second or third degree) the value of $\log c$ which corresponds to the minimum, and to calculate the corresponding values of α and β from the equations (5). The values of α, β and c which are thus found are regarded as a sufficiently good approximation to the values that give the absolute minimum of χ^2.

This method of determining the values of the Makeham constants will be referred to in the sequel as the *χ^2-minimum method*. It is sufficiently simple to be applied in practical work. The values of α, β and c found by this method, as well as the corresponding values of χ^2, are given in Tables

4 a and 4 b for all the periods and complete generations contained in our material.

11. *Numerical illustration to χ^2-minimum method.* — In order to give a detailed illustration of the working of our method, we take the experience for 1911—1915, males. All the necessary constants for this period are here given with two decimals more than usually required in practical work. The »best» value of $10^3 \log c$ is very nearly equal to 44, and in order to show the variations of χ^2 and χ_1^2 we shall consider the three values 43.75, 44.00 and 44.25 of $10^3 \log c$. In the first place, we give in Table A the values of α, β, χ^2 and χ_1^2 that are obtained when α and β are determined from the system (4) giving the exact minimum of χ_1^2.

Table A. *1911—15, males. Values of α, β, χ^2 and χ_1^2. α and β determined by the method of least squares.*

$10^3 \log c$	$10^3\,\alpha$	$10^5\,\beta$	χ^2	χ_1^2
43.75	5.0137	3.5721	27.29	27.50
44.00	5.0703	3.4176	25.93	26.02
44.25	5.1262	3.2695	27.52	27.50

If, on the other hand, α and β are determined according to the method of moments by means of the simpler system (5), we obtain the values shown in Table B.

Table B. *1911—15, males. Values of α, β, χ^2 and χ_1^2. α and β determined by the method of moments.*

$10^3 \log c$	$10^3\,\alpha$	$10^5\,\beta$	χ^2	χ_1^2
43.75	5.0014	3.5757	27.41	27.59
44.00	5.0724	3.4181	25.92	26.03
44.25	5.1425	3.2673	27.58	27.62

It appears that the values of corresponding figures in the two tables lie very close together, especially near the minimum. Assuming that, in each table, the three χ^2-values are situated on a second degree parabola, the minimum of this parabola is in both cases found to correspond to the value $10^3 \log c = 43.99$. The corresponding values of α and β are:

	$10^3\,\alpha$	$10^5\,\beta$
Method of least squares	5.0680	3.4236
Method of moments	5.0696	3.4242

Thus it is seen that the results given by the two methods are practically equivalent.

With respect to the substitution of χ_1^2 for χ^2, it appears from Tables A and B that the difference between χ^2 and χ_1^2 is, in every case, small compared to the variations caused by a change in the value of log c.

In the following Table C, some values of χ^2 are given (corresponding to α- and β-values determined by the method of moments) together with graduated values according to parabolas of the second and third degree respectively. The graduation by a second degree parabola is also shown in Diagram 2, where the χ^2-values calculated according to both methods are indicated. It will be seen that the agreement is very good. In the case of the third degree parabola, the deviations fall within the limits of the errors of calculation.

Table C. *1911—15, males. Values of χ^2 by the method of moments. Graduation by parabolas of second and third degree.*

$10^3 \log c$	χ^2	Second degree parabola	Third degree parabola
43.25	40.02	39.968	40.020
43.50	32.09	32.143	32.091
43.75	27.41	27.461	27.409
44.00	25.92	25.921	25.921
44.25	27.58	27.525	27.577
44.50	32.32	32.271	32.323
44.75	40.11	40.161	40.109

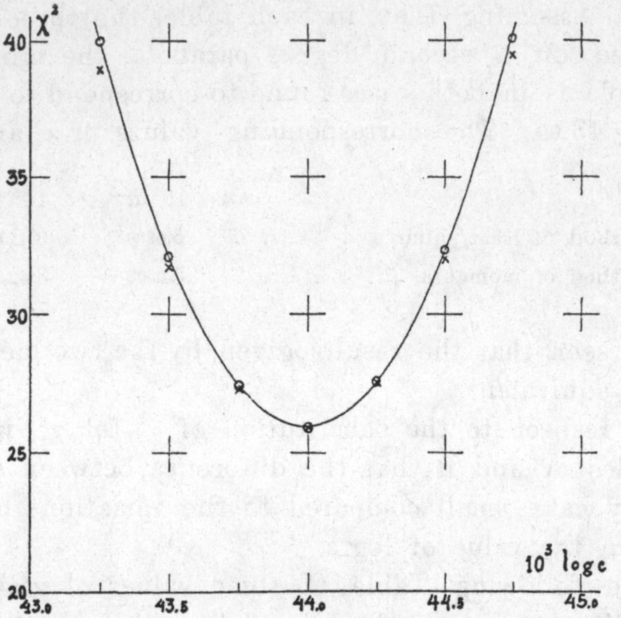

Diagram 2. Period 1911—15. Values of χ^2 for different values of log c.
××××× 1) α and β calculated by least squares.
OOOOO 2) » » » » » moment method.
——— Graduation of 2) by 2nd degree parabola.

From Table C and Diagram 2, it appears that even a comparatively small deviation of log c from the »best» value causes a considerable increase in χ^2. The same thing holds true with respect to other periods and generations. As far as our material goes it thus follows that, contrary to an opinion which has often been expressed in the literature, there exists a well distinguished value of log c giving the best fit of the Makeham curve, and that even moderate deviations from this value may seriously impair the result of the graduation.

12. *The one-dimensional graduations.* — The goodness of fit obtained for the various periods and generations by the method described above is measured by the χ^2 values as given in Tables 4 a—4 b. If the deviations from an *a priori* given Makeham expression μ_x would be wholly fortuitous, and if, in the age group x, the exposed to risk are uniformly distributed over the interval, the mathematical expectation of the quantity $\frac{(d_x - n_x \mu_x)^2}{n_x \mu_x}$ would be equal to unity.[1] As in our case there are twelve age groups, the mathematical expectation of χ^2 would amount to 12. It will be seen from the tables that in practically all cases the χ^2 values are considerably larger, so that the deviations cannot be ascribed to pure chance.[2] For the graduations by periods, the χ^2 values show on the whole a markedly better fit than for generations. This fact is obviously due to those violent fluctuations in mortality which, during certain periods, have affected all ages more or less alike. The existence of fluctuations of this kind must, of course, have a disadvantageous influence on the result of a graduation by generations.

A comparison between observed and calculated deaths shows, however, that according to current standards the fit of our MAKEHAM curves may be regarded as tolerably good, even if certain systematic deviations are discernible. In Table D,

Table D. *Graduations by χ^2-minimum method. Difference between observed and calculated deaths, in percent of calculated deaths.* Males.

Age	Periods			Generations		
	1811—15 $\chi^2=50$	1861—65 $\chi^2=64$	1911—15 $\chi^2=26$	1780 $\chi^2=580$	1810 $\chi^2=588$	1840 $\chi^2=101$
30—40	−1.6	−0.7	−2.1	−3.5	−7.2	−1.4
40—50	+3.7	+2.5	−1.4	+4.9	+9.6	+1.2
50—60	−0.5	+1.6	+1.8	+3.2	+3.9	−0.5
60—70	+0.1	−1.8	−1.8	−6.1	−4.4	−3.0
70—80	+0.6	−1.1	+0.4	−0.6	−4.2	+2.9
80—90	+3.3	+3.6	+0.0	+1.9	+3.7	−2.2

[1] Cf H. CRAMÉR, Sannolikhetskalkylen, Stockholm (1927), p. 156—161.

[2] Unfortunately the χ^2 values given in our preliminary note [3] were wrongly calculated (the values there given should be multiplied by 5), so that the discussion of the corresponding values of the quantity P used in the PEARSON theory must be withdrawn.

a comparison of this kind is given for three periods and three generations chosen as illustrations. The corresponding χ^2 values according to Tables 4 a and 4 b are also given.

A certain regularity may be perceived in the distribution of the signs of the deviations. Thus in the group 30—40 all deviations are negative, while in the groups 40—50 and 80—90 all but one are positive. It would be interesting to make a detailed study of the systematic deviations from the MAKEHAM formula. Considerations of space do not permit us, however, to go further into this matter here.

13. *Graduation of the parameters.* — The values of α, β and $\log c$ found by the χ^2-minimum method are given in Tables 4 a—4 b and shown in graphical form in Diagrams 3 a—5 d. We now proceed to the graduation of these values, considered as functions of the calendar year t or the birth year τ.

We begin with $\log c$, which obviously should be regarded as the most fundamental parameter of the MAKEHAM curve. BERNSTEIN [1] has expressed the opinion that $\log c$ does not show any systematic variation with time. From our tables and diagrams it readily follows, however, that in our material a strongly pronounced variation appears. For periods as well as for generations, a considerable increase has taken place in the values of $\log c$. During the first half of the nineteenth century, the fluctuations in the values for periods are apparently irregular and show no definite tendency. Between 1850 and 1900 there is a steady increase, which after 1900 gradually seems to come to a standstill. The values for men and women run almost parallel to one another. For generations a similar variation is found, the greatest part of the increase being concentrated between the birth years 1800 and 1825.

At an early stage of our investigations, the material was confined to the period 1850—1930, and we endeavoured to graduate the $\log c$ values by various rational and exponential functions. The results were, however, not altogether satisfactory. When, later on, we considered the whole period 1800

—1930, our diagrams suggested at once that *logistic curves* should be used for the graduation. By these curves we have obtained a good fit to the observed log c values for the whole period. Considering in particular the sub-period 1850—1930, the fit is even better than that given by any of the functions previously tried for this shorter period.

The equation of the ordinary logistic curve may be put in the form

$$(6) \qquad y = \frac{A + B\,e^{k\,(t-t_0)}}{1 + e^{k\,(t-t_0)}},$$

where A and B are the ordinates of the asymptotes, t_0 is the abscissa of the inflexion point and k is a constant which, for given A, B and t_0, determines the slope of the curve. We have worked out a method for fitting this curve, according to the method of least squares, to statistical data, which as far as we know seems to be new, and which requires a comparatively small amount of work. An account of this method with a detailed numerical illustration will be found in Appendix 2.

The graduated values of log c are found in Tables 4 a—4 b; the corresponding graduating curves are shown in Diagrams 3 a—3 b. The values of the constants in the final graduations of 10^3 log c are as follows:

	Graduation by periods		Graduation by generations	
	Males	Females	Males	Females
A	35.63	39.92	34.25	39.16
B	44.53	47.58	44.95	47.59
t_0	1871.0	1872.5	1808.0	1813.5
k	0.136	0.124	0.116	0.204

With respect to β, it was found convenient not to work directly with the values of this parameter, but rather with the log β values, which could be well represented by logistic curves. This seems not unreasonable since the formula

$$\log(\mu_x - \alpha) = \log \beta + x \log c$$

suggests that the relation between $\log \beta$ and $\log c$ should be of a simple character.

If the $\log \beta$ and the $\log c$ graduations are quite independently performed, it is obvious that the result of the two-dimensional graduation cannot be expected to be as good as if these graduations are in some convenient way brought into contact with one another. In order to establish such a contact, we have proceeded in the following way. The graduated $\log c$ values having been calculated as shown above, new values of (α and) β were calculated *from the graduated values* of $\log c$ according to the method of moments, and these intermediate values of β were then used for graduation. The original as well as the intermediate β values are shown in Diagrams 4 a—4 b, and it will be found that both series are very well represented by the same graduating curve, especially with respect to the ten last periods and generations, where the intermediate values can hardly be distinguished from the curve.

At the graduation of $\log \beta$ by means of the logistic expression (6) the values obtained for the constant t_0 were almost identical with those previously found in the case of $\log c$. The constants for both graduations were then slightly adjusted in order to have common values of t_0 in both cases. The values of the constants in the final graduations of $\log(10^5 \beta)$ are as follows:

	Graduation by periods		Graduation by generations	
	Males	Females	Males	Females
A	1.39443	1.01023	1.46260	0.99853
B	0.48422	0.18118	0.39951	0.13151
t_0	1871.0	1872.5	1808.0	1813.5
k	0.116	0.108	0.104	0.148

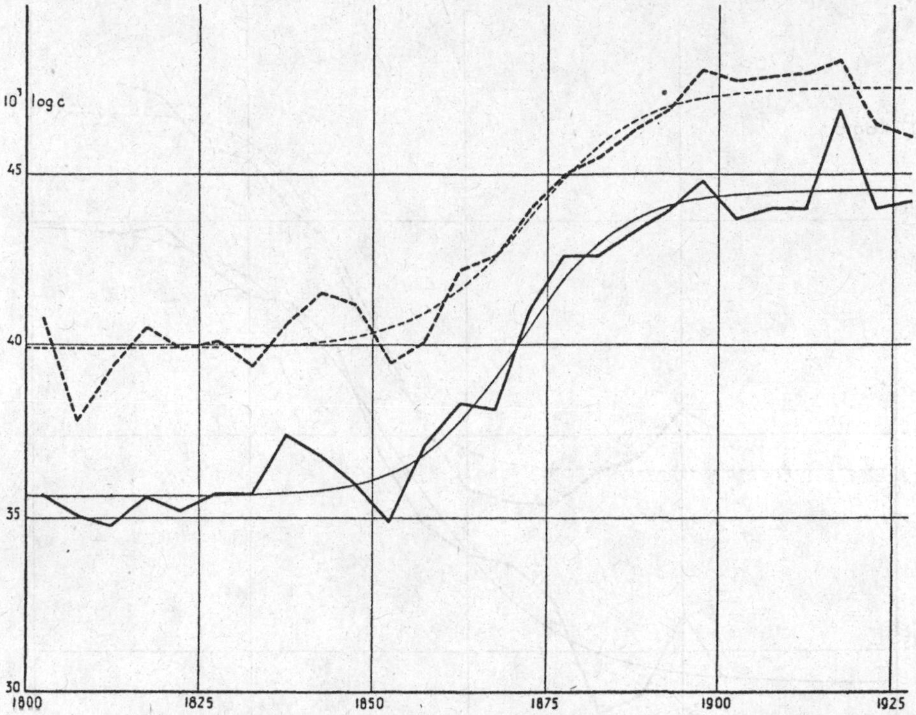

Diagram 3 a. Periods. Values of log c, calculated by χ^2-minimum method and graduated by logistic curve.
Males ———— Females ....

Concerning α we have already mentioned the series of intermediate values obtained by the method of moments from the graduated values of log c. A further series of intermediate values is obtained if, using graduated values of both β and log c, we determine α for each period and generation by the condition that

$$\sum_x \frac{(d_x - n_x \mu_x)^2}{d_x}$$

should be as small as possible. The original α values as well as both series of intermediate values are shown in Diagrams 5 a—5 d. It is obvious that all three series show the same general trend.

The α values have been graduated by straight lines which, however, are not fitted in the usual way to any of the three series of α values, but are determined by the condition that,

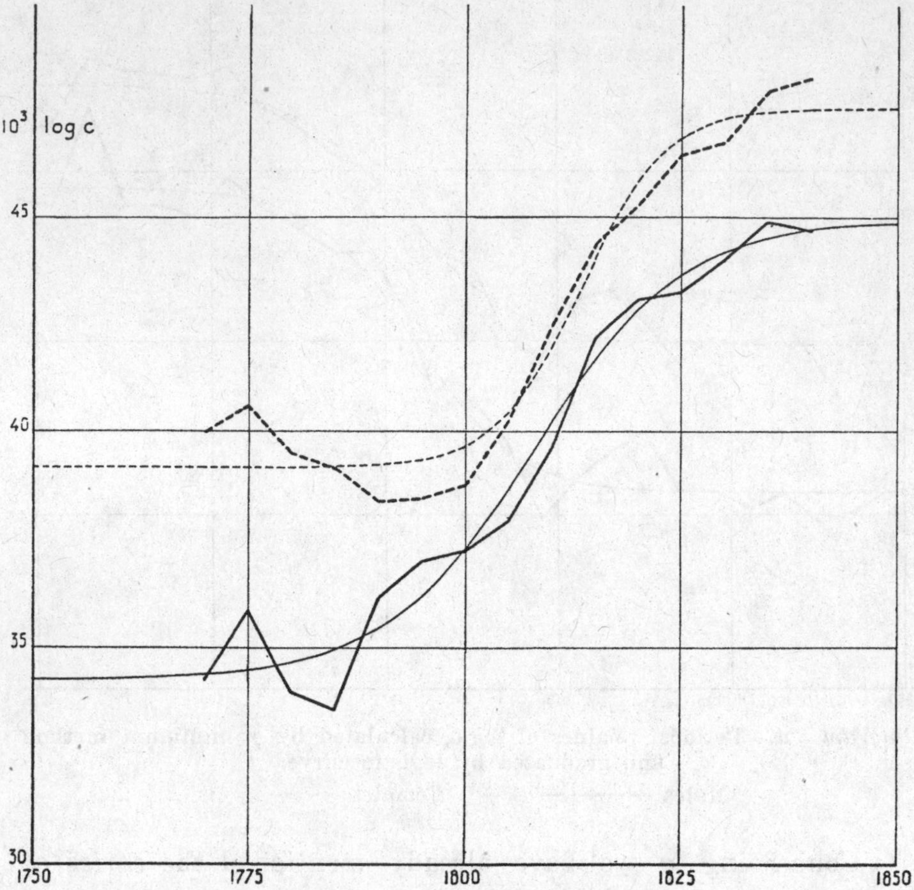

Diagram 3 b. **Generations. Values of log c, calculated by χ^2-minimum method and graduated by logistic curve.**
Males ———— **Females** ------------

for the resulting two-dimensional graduation, the measure of goodness of fit χ_1^2 should be a minimum. Thus *e.g.* at the graduation by periods we consider the expression

$$\chi_1^2 = \sum_t \sum_x \frac{(d_x - n_x \mu_x)^2}{d_x},$$

where the summation is extended to all periods and all age groups contained in the material. Here, μ_x is supposed to follow the MAKEHAM law with β and log c determined according to our graduations. We then put

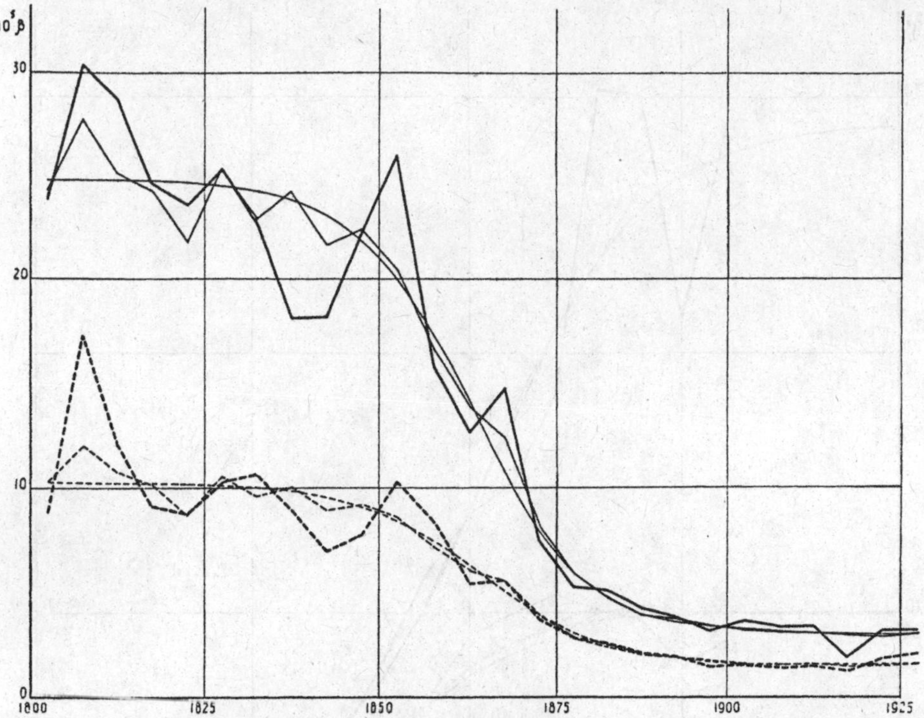

Diagram 4 a. Periods. Values of β, directly calculated by χ^2-minimum method (thick lines), calculated from graduated values of log c (thin lines), and graduated by logistic curve.

Males ———— Females - - - - - - -

$$10^3\,\alpha = r - s\,(t - 1860)$$

and determine r and s from linear equations obtained by the minimum condition. In this way we find the following values:

	Males	Females
r	7.0616	6.5954
s	0.042102	0.031257

For the graduation by generations we proceed in a similar way, putting now

$$10^3\,\alpha = \varrho - \sigma(\tau - 1800).$$

The following values are obtained:

	Males	Females
ϱ	10.6189	8.1009
σ	0.074079	0.037848

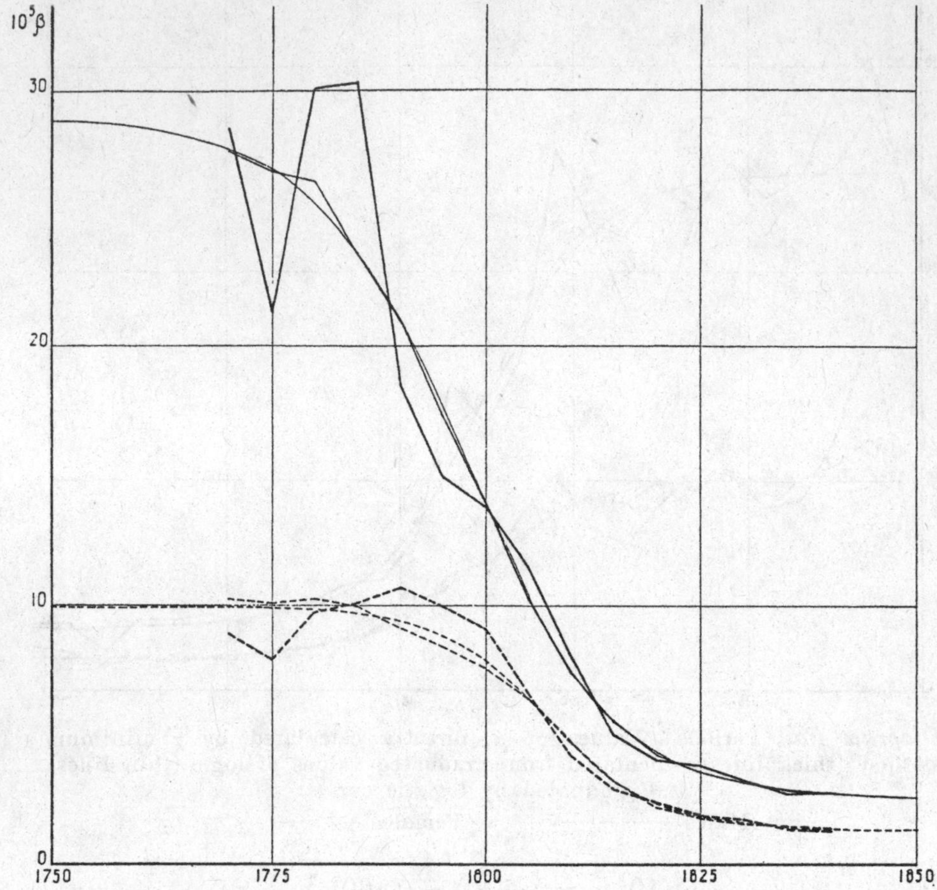

Diagram 4 b. Generations. Values of β, calculated and graduated as in diagram 4 a.

For males, however, the graduation by a straight line could not be extended to the generations born before 1795 (Cf Diagram 5 b). Here, a second degree parabola has been used.

14. *The two-dimensional graduations.* — Before entering upon a discussion of the two-dimensional graduations resulting from our work, we shall briefly mention a problem which naturally presents itself to the mind when we are dealing with mortality by periods and generations from the point of view of the MAKEHAM law. — *Is it possible that the force of mortality* $\mu(x, t)$ *may depend on the age* x *and the calendar*

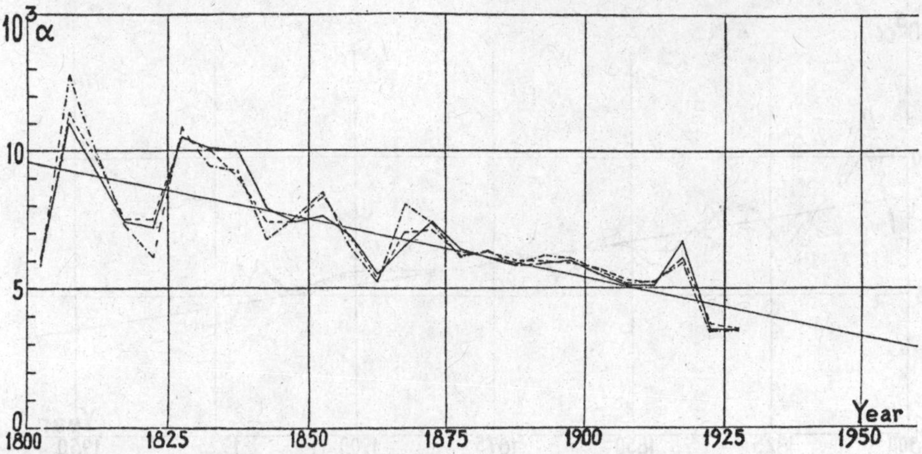

Diagram 5 a. Periods, males. Values of α, directly calculated by χ^2-minimum method (————), calculated from graduated values of log c (··········), from graduated values of log c and β (–·–·–·–) and graduated by straight line.

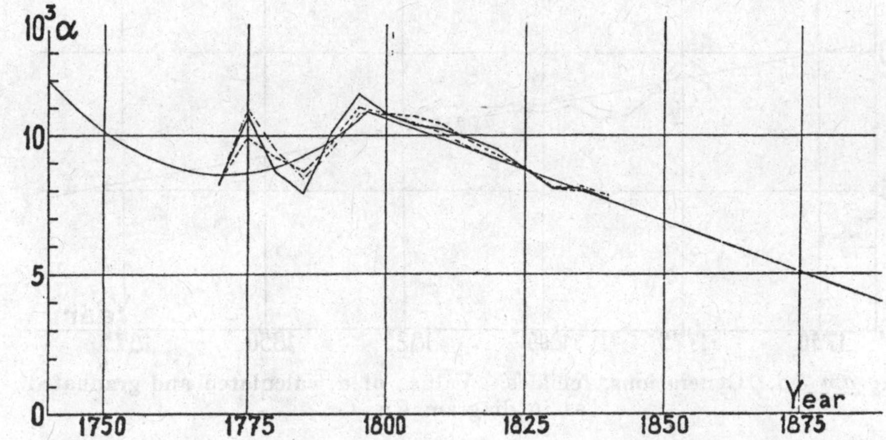

Diagram 5 b. Generations, males. Values of α, calculated and graduated as in diagram 5 a.

year t in such a way that the MAKEHAM *law holds exactly for periods as well as for generations?*

Expressed in mathematical formulae, the problem runs as follows: Is it possible to find a function $\mu(x, t)$ such that we have identically

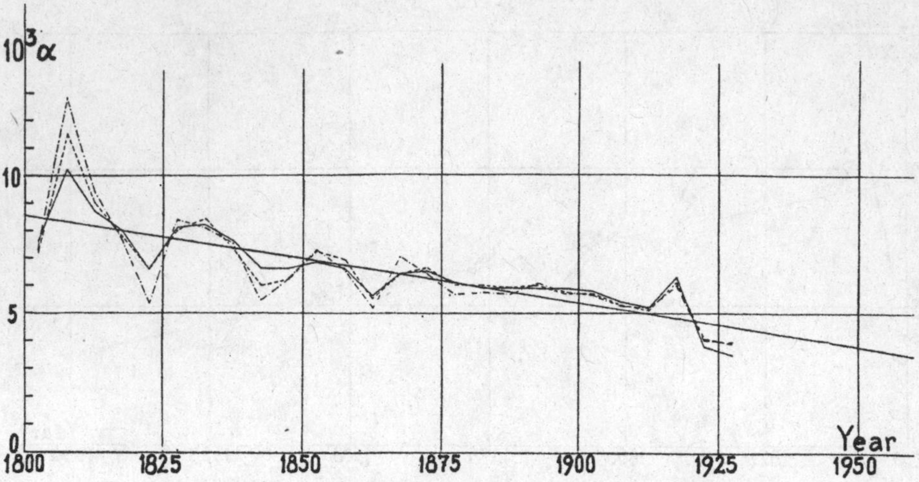

Diagram 5 c. Periods, females, Values of α, calculated and graduated as in diagram 5 a.

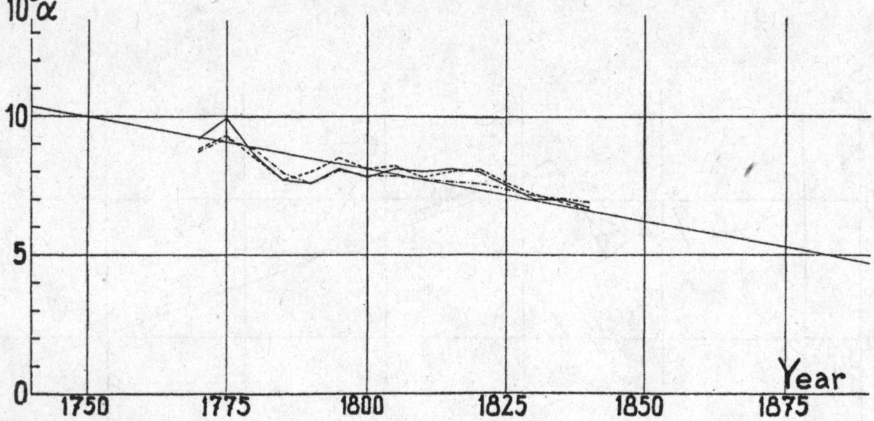

Diagram 5 d. Generations, females. Values of α, calculated and graduated as in diagram 5 a.

$$\mu(x, t) = a + b c^x$$

$$= \alpha + \beta \gamma^x,$$

where a, b and c only depend on t, while α, β and γ only depend on $\tau = t - x$? — If this question can be answered in the affirmative, it would be natural to ask if some function of this character might not be conveniently used for the two-dimensional graduation of a given statistical experience.

In Appendix 1 it will be shown that, if we restrict ourselves to functions a, b and c having three continuous derivatives, the problem has only two different groups of solutions, *viz.*

$$(7)\quad \begin{cases} a = m_1 + m_2 \varkappa^t \\ b = \dfrac{m_3 + m_4 \varkappa^t}{\lambda^t} \\ c = \lambda \end{cases} \qquad \begin{cases} \alpha = m_1 + m_3 \lambda^{-\tau} \\ \beta = \dfrac{\varkappa^\tau (m_4 + m_2 \lambda^\tau)}{\lambda^\tau} \\ \gamma = \varkappa \end{cases}$$

and

$$(8)\quad \begin{cases} a = m_1 \\ b = m_2 \varkappa^t \lambda^{-t^2} \\ c = m_3 \lambda^t \end{cases} \qquad \begin{cases} \alpha = m_1 \\ \beta = m_2 \varkappa^\tau \lambda^{-\tau^2} \\ \gamma = \varkappa m_3 \lambda^{-\tau} \end{cases}$$

where $\varkappa$, λ and $m_1, \ldots m_4$ are constants.

Obviously it would be quite hopeless to attempt to use these functions for a graduation of our material. The logistic type of variation which is found in $\log c$ and β cannot be even approximately represented by (7) or (8). Thus it appears once more that our mathematical expressions must not be regarded as exact representations of the mortality functions, but rather as approximate graduation formulae.

By the method described above we have obtained two fundamentally different two-dimensional graduations of the experience under consideration. The graduated values of the force of mortality are given in Tables 5 a—5 d; the differences $d_x - n_x \mu_x$ between observed and calculated deaths are found in Tables 6 a—6 d. An inspection of the last-mentioned tables shows at once that, as might well be expected, strong systematic deviations are present. Thus for certain periods the differences in all age groups are numerically large and all of the same sign.

The χ^2 values corresponding to the final graduations are given for each period and generation in Tables 4 a—4 b. For the 312 groups contained in the total material χ^2 was calculated according to the formula

$$\chi^2 = \sum_t \sum_x \frac{(d_x - n_x \mu_x)^2}{n_x \mu_x}, \tag{9}$$

and the following values were obtained:

	Graduation by periods	Graduation by generations
Males	15 675	16 588
Females	16 654	24 107

It is interesting to find that, with the exception of the graduation by generations for females, these χ^2 values are all of the same order of magnitude. If, in the expression (9), the summation is restricted to the 180 groups corresponding to the »complete» generations (Cf Diagram 1), the χ^2 values become

	Graduation by periods	Graduation by generations
Males	7 822	7 869
Females	6 938	7 591

Thus in this case all four values are of the same order of magnitude. Possibly this might be interpreted as an indication that the goodness of fit of our graduations is about as good as is compatible with the irregularities of the material. In this connection it should be observed that in the process of graduating $\log c$ and β for generations, only the data belonging to the »complete» generations have been used. For the determination of α, on the other hand, all the 312 groups in the material have been used in making χ_1^2 a minimum.

In Diagrams 6 a—6 d the observed and graduated values of $\mu(x, t)$ are shown for some selected age groups. For both methods of graduation, the agreement seems to be as good as might have been reasonably expected from a two-dimensional graduation of statistical data of this character. Neither graduation seems to deserve to be decidedly preferred to the other.

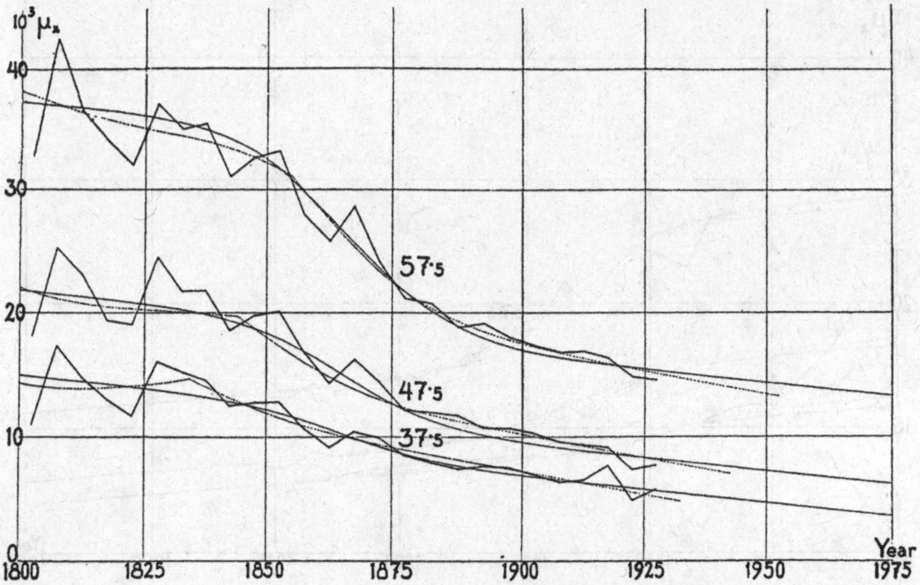

Diagram 6 a. Males. Observed and graduated values of μ_x for ages 37.5, 47.5 and 57.5.

Graduation by periods ————. Graduation by generations ················.

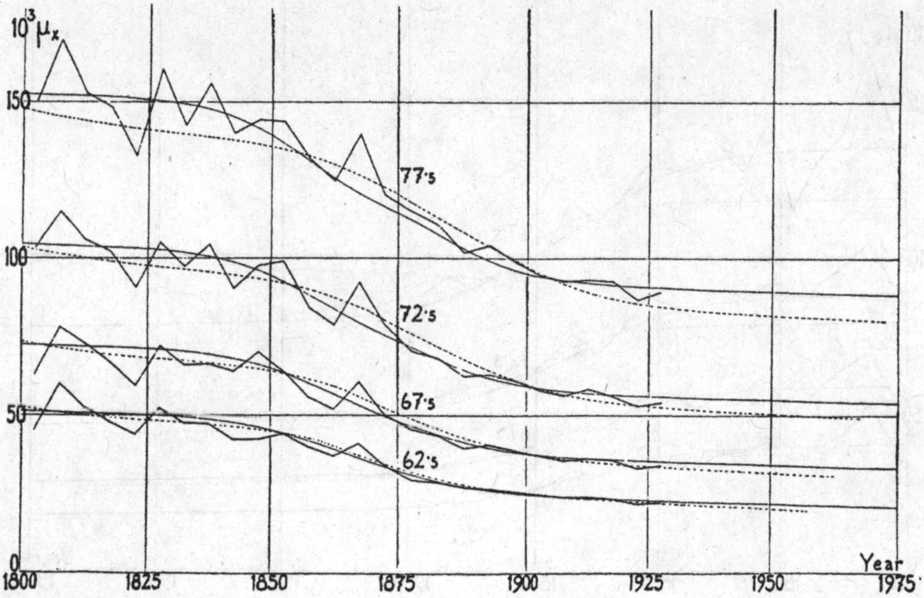

Diagram 6 b. Males. Observed and graduated values of μ_x for ages 62.5, 67.5, 72.5 and 77.5.

Graduation by periods ————. Graduation by generations ················.

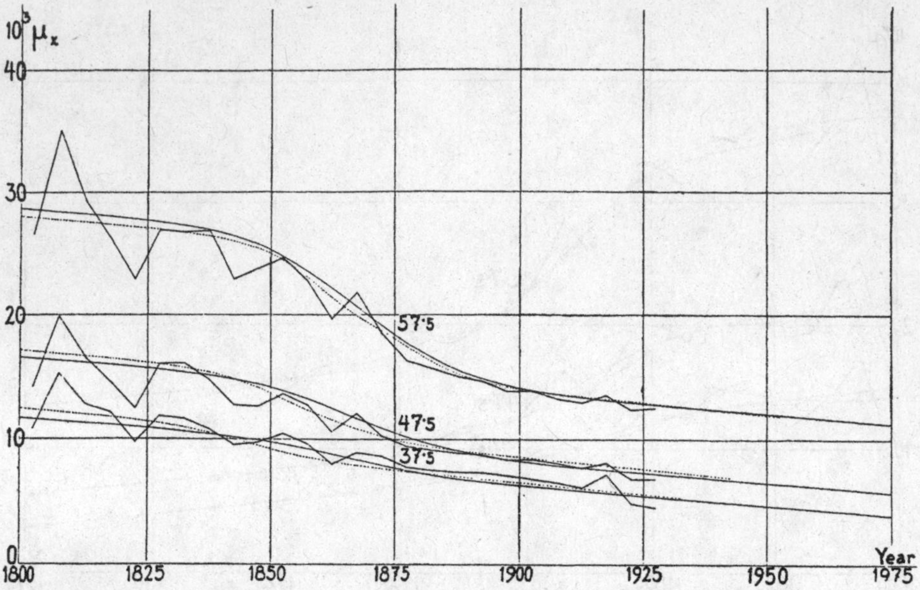

Diagram 6 c. Females. Observed and graduated values of μ_x for ages 37.5, 47.5 and 57.5.

Graduation by periods ————. Graduation by generations ------------.

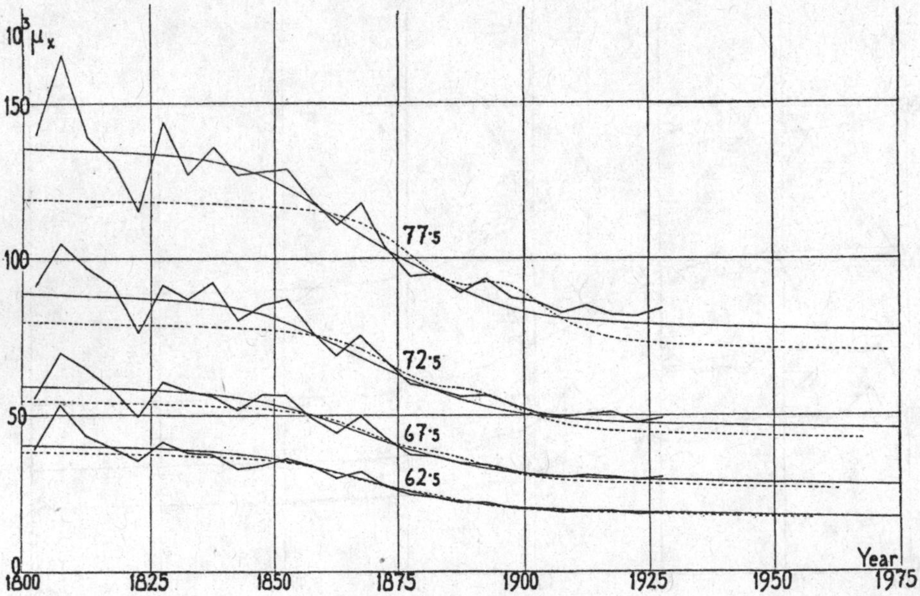

Diagram 6 d. Females. Observed and graduated values of μ_x for ages 62.5, 67.5, 72.5 and 77.5.

Graduation by periods ————. Graduation by generations ------------.

Some further graphical comparisons are given in Diagrams 7 a—7 b, where the μ-curves for males are shown for some selected periods and generations.

15. *Extrapolation.* — By our graduations the force of mortality $\mu(x, t)$ has been approximately represented by two

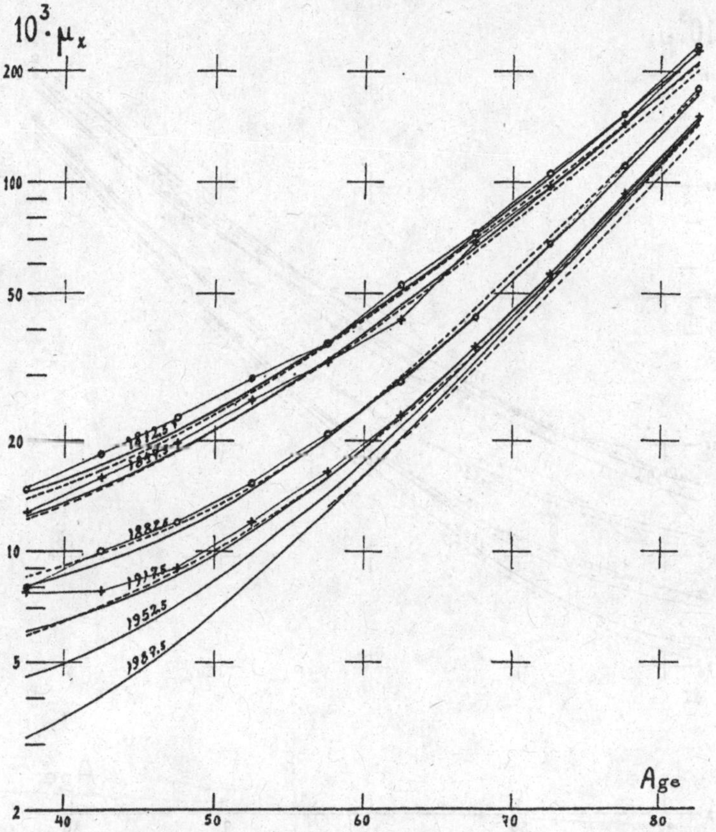

Diagram 7 *a.* $10^3 \mu_x$ for selected periods, males. Logaritmic scale.
Observed values $\substack{\circ\circ\circ\circ \\ ++++}$. Graduation by periods ————.
Graduation by generations -------------.

different analytical expressions, each containing nine constants ($9 = 4 + 3 + 2$, the constant t_0 being the same for $\log c$ and β). By extending the numerical calculation of these expressions to future values of t we obtain an extrapolation of the mortality experience. The extrapolated values of the force of mortality are given in Tables 5 a—5 d for the calendar years $t = 1952.5$ and $t = 1977.5$. In no case, however, the extrapolation has been extended more than 50 years outside the interval covered by the data used for the graduations of $\log c$ and β. It will be seen from Diagram 1 that, for an age group x, the data corresponding to the »complete» generations end at the calendar year $1840 + x$, and so in the case

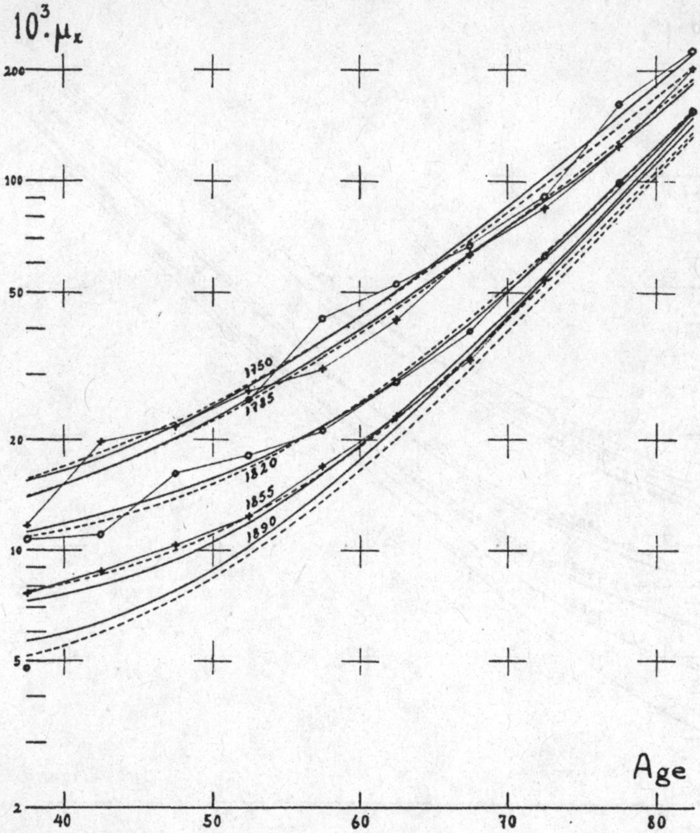

Diagram 7 b. $10^3\mu_x$ for selected generations, males. Logaritmic scale.
Observed values $\circ\circ\circ\circ$ / $++++$. Graduation by periods ————.
Graduation by generations ---------------.

of mortality by generations the extrapolation has not been carried further than the calendar year $1890 + x$.

In graphical form the result of the extrapolation is shown in Diagrams 6 a—6 d and 7 a—7 b. At the inspection of these diagrams, the remark made above with respect to the graduation by generations should be well remembered. For an age group x, the graduation by generations of $\log c$ and β rests only on data for the calendar period between the years $1770 + x$ and $1840 + x$. Thus *e.g.* in Diagrams 6 a and 6 c the dotted curves for the age 37.5 are really to be regarded as the results of a *graduation* only up to the year $t = 1877.5$. From this point on, the curves correspond to values of $\log c$ and β found by an *extrapolation* of our analytical expressions, and it is only the α values that are deduced by a *graduation* of the data. Thus the agreement of these curves to the experience between $t = 1877.5$ and $t = 1927.5$ may be regarded as a preliminary test for the reliability of the extrapolations. For the age group 47.5, the same remark holds with respect to the interval between $t = 1887.5$ and $t = 1927.5$, and so on.

A numerical measure of the fit of the extrapolation has been obtained by summing in (9) only over the 66 values belonging to the incomplete generations after 1840. These sums are given in the following table, showing also for comparison the corresponding χ^2 sums in the period material.

	Graduation by periods	Graduation by generations
Males	3 704	3 742
Females	3 487	4 332

It is interesting to note that the figures for males are practically identical in both cases.

On the other side, that part of the dotted curves for the age group x which fall between $t = 1802.5$ and $t = 1770 + x$ corresponds to a *backward* extrapolation of $\log c$ and β which, however, seems to be less successful at higher ages. The corresponding χ^2-sums may easily be deduced from the tables above.

In the following Table E, some values of the extrapolated force of mortality are compared to the corresponding values according to the ÅFK and R 32 mortality tables.

Table E. Values of $10^3 \mu_x$.

Age x	ÅFK	R 32	Extrapolation for $t = 1952.5$		Extrapolation for $t = 1977.5$
			Periods	Generations	Periods
Males					
32.5	5.6	3.4	4.0	—	3.0
42.5	7.0	4.6	5.4	—	4.5
52.5	11.1	8.0	9.8	—	8.8
62.5	20.7	17.8	21.7	20.1	20.6
72.5	51.5	46.1	54.8	50.3	53.7
82.5	150.0	127.6	147.1	133.9	146.0
Females					
32.5	6.0	3.4	4.2	—	3.5
42.5	6.8	4.4	5.3	—	4.5
52.5	8.8	7.5	8.5	—	7.7
62.5	17.9	16.3	18.0	17.5	17.2
72.5	44.1	41.7	46.4	43.3	45.7
82.5	134.5	115.1	131.5	119.7	130.7

In practically all cases the extrapolated figures lie above the corresponding values taken from the R 32 table. It thus seems that this table, as applied to annuity business, contains a certain safety margin. It is particularly remarkable that, for the higher ages, the extrapolation by periods does not indicate any marked decrease of mortality below the present level.

It will be seen from the tables and diagrams that, especially for the advanced ages, there are considerable differences between the values furnished by our two extrapolation methods. This, of course, lies in the nature of the case and should be regarded as an illustration of the fact that we can never expect to reach a unique solution of a statistical problem such as that of forecasting mortality.

Appendix 1. On MAKEHAM Surfaces.

Let x, t and z denote the rectangular coordinates of a point in three-dimensional space. Then

$$z = \mu(x, t)$$

is the equation of a surface S which represents the force of mortality $\mu(x, t)$ as a function of the age x and the calendar year t. The mortality curve belonging to *a given calendar year* is obtained as the section between S and a plane t=const., while the mortality curve for *a given generation* corresponds to the section with a plane of the class $\tau = t - x =$ const.

We shall consider the following problem: *Is it possible to find a function $\mu(x, t)$ such that both these types of sections will always be* MAKEHAM *curves?* — If such a function is found, the corresponding surface S will be called a MAKEHAM surface.

Thus it is required to find a function $\mu(x, t)$ such that

$$\mu(x, t) = a + b c^x$$

$$= \alpha + \beta \gamma^x,$$

where a, b and c only depend on t, while α, β and γ only depend on $\tau = t - x$. We shall restrict ourselves to the case when all the functions $a, b, \ldots, \gamma$ have three continuous derivatives.

Substituting $t - \tau$ for x and putting

$$f(t, \tau) = \mu(t - \tau, t),$$

$$A = a, \quad B = b c^t, \quad C = -\log c,$$

$$A = \alpha, \quad B = \beta \gamma^{-\tau}, \quad \Gamma = \log \gamma,$$

the problem is transformed in finding a function $f(t, \tau)$ such that

$$f(t, \tau) = A + B e^{C\tau} \tag{10}$$

$$= A + B e^{\Gamma t}$$

where A, B and C only depend on t, while $\mathcal{A}$, $\mathcal{B}$ and Γ only depend on τ.

It will now be shown[1] that any function $f(t, \tau)$ satisfying (10) belongs to one of the types

$$f(t, \tau) = m_1 + m_2 e^{k_1 t} + m_3 e^{k_2 \tau} + m_4 e^{k_1 t + k_2 \tau} \tag{11}$$

or

$$f(t, \tau) = m_1 + m_2\, e^{k_1 t + k_2 \tau + k_3 t \tau}, \tag{12}$$

where the m's and k's are constants. Obviously any function of the type (11) or (12) satisfies (10). Re-introducing the variable x, it is easily seen that we obtain the result already stated in No 14.

Let us first observe that, if B or C is identically equal to zero, it easily follows from (10) that $f(t, \tau)$ is of the form (11) with $m_3 = m_4 = 0$. Thus we may assume during the course of the proof that neither B nor C vanishes identically.

Differentiating (10) three times with respect to τ and putting afterwards $\tau = 0$ we obtain

$$A + B = r_0 + P_0\, e^{kt}, \tag{13}$$

$$B\,C = r_1 + P_1\, e^{kt}, \tag{14 a}$$

$$B\,C^2 = r_2 + P_2\, e^{kt}, \tag{14 b}$$

$$B\,C^3 = r_3 + P_3\, e^{kt}, \tag{14 c}$$

where $r_0, \ldots, r_3$ and k are constants, while P_i is a polynomial in t with constant coefficients, the degree of which does not exceed i.

From the relations (14) we deduce

$$(r_1 + P_1\, e^{kt})(r_3 + P_3\, e^{kt}) = (r_2 + P_2\, e^{kt})^2. \tag{15}$$

For the further discussion we shall distinguish three cases.

---

[1] The method of proof used here was suggested to us by Fil. Mag. G. ARFWEDSON.

First case: $k \neq 0$, $r_1 \neq 0$. — Since (15) must hold identically in t, the coefficients of e^{kt} and e^{2kt} must vanish, so that we have

(16) $$r_1 r_3 = r_2^2,$$

(17) $$r_1 P_3 + r_3 P_1 = 2 r_2 P_2,$$

(18) $$P_1 P_3 = P_2^2.$$

r_1 being different from zero, it follows from (17) that P_3 is at most of the second degree in t. According to (18) this implies, however, that P_2 must be linear, and so by (17) P_1, P_2 and P_3 are all linear polynomials in t. Now, it follows from the formation of the polynomials P at the differentiation of (10) that the coefficient of t in P_1 is obtained by putting $\tau = 0$ in $B\Gamma'$, while the coefficient of t^2 in P_2 is obtained in the same way from $B\Gamma'^2$. Thus if P_2 is linear, P_1 must reduce to a constant, and so by (17) and (18) P_1, P_2 and P_3 are all constants.

r_1 being different from zero, it follows from (16) that there is a constant ϱ such that

(19) $$r_2 = \varrho r_1, \quad r_3 = \varrho^2 r_1.$$

If we have $P_1 = 0$, it follows from (17) and (18) that $P_1 = P_2 = P_3 = 0$. Thus in any case according to (18) we can determine a constant σ such that

(20) $$P_2 = \sigma P_1, \quad P_3 = \sigma^2 P_1.$$

Introducing (19) and (20) into (17), we obtain

$$\varrho = \sigma.$$

(In the particular case when $P_1 = 0$, (20) holds for any value of σ, so that even in this case we may put $\varrho = \sigma$.) From (14 a) and (14 b) we then get

$$C = \varrho,$$

and since r_1 is different from zero, (14 a) shows that $\varrho \neq 0$. Thus by (13) and (14 a) we have

$$A = m_1 + m_2 e^{kt},$$

$$B = m_3 + m_4 e^{kt},$$

so that $f(t, \tau) = A + Be^{C\tau}$ is of the form (11).

Second case: $k \neq 0$, $r_1 = 0$. — Obviously the relations (16), (17) and (18) hold true in this case. From (16) we obtain $r_2 = 0$, and then (17) shows that we have $r_3 P_1 = 0$ for all values of t.

If, in the first place, P_1 is identically equal to zero, it follows from (14 a) that BC vanishes identically. We have already pointed out that in this case $f(t, \tau)$ is of the form (11).

If, however, P_1 is not identically zero, we must have $r_1 = r_2 = r_3 = 0$, and then the relations (14) show that neither P_2 nor P_3 is identically zero. Thus according to (18) there is a linear polynomial Q such that

$$P_2 = QP_1, \quad P_3 = Q^2 P_1,$$

and thus we have $C = Q$. From (14 a) we then obtain

$$B = \frac{P_1}{Q} e^{kt}.$$

Now $\frac{P_1}{Q}$ must be equal to a constant, since otherwise B would become infinite for a finite value of t. Thus it follows from (13) that we have

$$A = m_1 + m_2 e^{kt},$$

$$B = m_3 e^{kt},$$

m_3 being different from zero. Putting $Q = k_1 + k_2 t$, we obtain

$$f(t, \tau) = m_1 + m_2 e^{kt} + m_3 e^{kt + k_1 \tau + k_2 t \tau}.$$

It is readily seen, however, that this expression does not satisfy (10) unless we have either $m_2 = 0$ or $k_2 = 0$. In the former case $f(t, \tau)$ is of the form (12), in the latter case of the form (11).

Third case: $k=0$. — In this case the relation (15) becomes

$$(r_1 + P_1)(r_3 + P_3) = (r_2 + P_2)^2.$$

It follows that any linear factor occurring in $r_2 + P_2$ must also occur in $r_3 + P_3$, and thus according to (14) C must be a linear polynomial in t. If this polynomial does not reduce to a constant, it follows from (13) and (14) that A and B are constants. Introducing these expressions for A, B and C in (10), we find that $f(t, \tau)$ is of the form (12) with $k_1 = 0$. — If, on the other hand, C is equal to a constant, it follows from (13) and (14) that A and B are linear polynomials such that $A + B$ is a constant. Thus we have according to (10)

$$f(t, \tau) = m_1 + m_2 t + (m_3 - m_2 t)\, e^{k_2 \tau}.$$

It is easily seen that this expression does not satisfy (10) unless we have $m_2 = 0$. Then $f(t, \tau)$ is of the form (11) with $m_2 = m_4 = 0$.

Appendix 2. On graduation by logistic curves.

We shall be concerned with the logistic curve as given by the equation

$$y = y(t) = \frac{A + B e^{k(t-\tau)}}{1 + e^{k(t-\tau)}}. \tag{21}$$

Here A and B are the ordinates of the asymptotes, τ is the abscissa of the inflexion point and k is, according to the differential equation

$$y'(t) = k \frac{(y - A)(B - y)}{B - A},$$

a constant connected with the rate of variation of $y(t)$. In the particular case $A = 0$ we obtain the curve originally considered by Verhulst (Cf Yule, [35]). The case $A \neq 0$ was first considered by Pearl and Reed [33].

It is well known that, during the last 15 years, the logistic function has been largely employed for curve fitting, especially in connection with population statistics (For further references, see [34; 35]). The methods hitherto used for the fitting of a logistic curve to given statistical data have, however, been rather unsatisfactory.

During the course of our investigations in Swedish national mortality we have incidentally worked out a method for fitting the logistic curve to statistical data according to the principle of least squares. Such a method has been previously proposed in a recent paper by Wilson and Puffer [34]; that method is, however, characterized by the authors themselves as »extremely laborious». We shall give here a short account of our own method. It will be seen that the amount of work required for its application is fairly moderate.

Let the points $(t_1, y_1), (t_2, y_2), \ldots, (t_n, y_n)$ be given, where $t_\nu = t_0 + \nu h$. It is required to determine the constants in the expression (21) so as to make

$$S = \sum_1^n (y(t_\nu) - y_\nu)^2$$

as small as possible.[1]

Crude approximate values of the asymptotes A and B may be determined by an inspection of a graphical representation of the data. Then the position of the inflexion point τ is approximately determined by the condition

$$y(\tau) = \frac{A+B}{2},$$

and the constant k by the condition

$$y'(\tau) = k\frac{B-A}{4}.$$

[1] The method may be used also when the given ordinates are not equidistant. If, as is frequently the case, we may assume *a priori* $A = 0$, the method will be somewhat simplified.

Having thus found rough approximate values of τ and k, we shall in the first place show how the asymptotes A and B may be so determined that, *τ and k being regarded as fixed*, S becomes as small as possible.

Putting

$$k = \frac{\varkappa}{h}, \quad \tau = t_m + \alpha h,$$

where $-\frac{1}{2} < \alpha \leqq \frac{1}{2}$, we have

$$y(t_\nu) = B - \frac{B-A}{1 + e^{\varkappa(\nu - m - \alpha)}}$$

(22)

$$= B - C f(\nu - m, \alpha, \varkappa),$$

where $C = B - A$ and

$$f(i, \alpha, \varkappa) = \frac{1}{1 + e^{\varkappa(i-\alpha)}}.$$

In order to make S a minimum we have to solve the equations

$$\text{(23)} \qquad \begin{cases} B\Sigma 1 - C\Sigma f = \Sigma y_\nu, \\ B\Sigma f - C\Sigma f^2 = \Sigma y_\nu f, \end{cases}$$

f being written in the place of $f(\nu - m, \alpha, \varkappa)$, and all the sums being extended from $\nu = 1$ to $\nu = n$. B and C being thus determined, we have

$$\text{(24)} \qquad S_{\text{min.}}(k, \tau) = \Sigma y_\nu^2 - B\Sigma y_\nu + C\Sigma y_\nu f.$$

For the numerical calculations, we thus require tables of f, Σf and Σf^2 for various arguments. In Tables 7 a—9 c, these values are given for $\varkappa = 0.52$, 0.56, 0.60, ..., 1.00 and for $\alpha = 0.0$, 0.2 and 0.4. The cases $\alpha = -0.2$ and $\alpha = -0.4$ are also covered by these tables because of the obvious relation

$$f(\nu, \alpha, \varkappa) = 1 - f(-\nu, -\alpha, \varkappa),$$

which enables us to pass from a given value α to $-\alpha$.

Calculating $S_{\text{min.}}(k, \tau)$ according to (24) for the given preliminary value of τ and a number of different values of k

in the vicinity of the preliminary value, we may graphically or by interpolation (assuming a parabola of the 2:d or 3:d degree) determine the value of k which, for this particular value of τ, corresponds to the smallest possible S. Thus we obtain an approximate value of $S_{\text{min.}}(\tau)$, the smallest value of S corresponding to our chosen preliminary τ. The corresponding values of B and C may be determined by interpolation or directly calculated from (23).

Now the whole procedure is repeated for a number of further values of τ (in most cases only two further values are required) in the vicinity of the preliminary value. From the values of $S_{\text{min.}}(\tau)$ thus obtained it is generally possible to determine with sufficient accuracy the value of τ which gives the absolute minimum of S. The corresponding values of k, B and C are then found by interpolation or, in the case of B and C, from (23). — Finally the graduated values $y(t_\nu)$ are calculated from (22).

The method gives good results in cases when the data are not too few in number and show an evident trend of the logistic type. In other cases, the advantage of fitting the logistic curve to statistical data seems to us somewhat doubtful.

Numerical illustration. — Let us consider the values of the MAKEHAM parameter log c obtained at the graduation by periods for males (Table 4 a, col. 2). Here we have $t_0 = 1797.5$, $h = 5$, $n = 26$. By an inspection of the graphical representation of the data, the preliminary values $\tau = 1872.5$ and $k = 0.128$ were chosen. The corresponding values of t_m, α and $\varkappa$ are: $t_m = 1872.5$ ($m = 15$), $\alpha = 0$ and $\varkappa = 0.64$. For $\varkappa = 0.64$, 0.68 and 0.72, B and $A = B - C$ were calculated from (23), and $S_{\text{min.}}(k, \tau)$ from (24). The following values were obtained:

$\varkappa$	A	B	$S_{\text{min.}}(k, \tau)$
0.64	35.686	44.691	17.04
0.68	35.722	44.645	17.01
0.72	35.755	44.604	17.11

For $\tau = 1871.5$ ($m = 15$, $\alpha = -0.2$) the corresponding values are:

$\varkappa$	A	B	$S_{\min.}(k, \tau)$
0.64	35.623	44.615	16.53
0.68	35.660	44.569	16.49
0.72	35.693	44.529	16.54

and for $\tau = 1870.5$ ($m = 15$, $\alpha = -0.4$):

$\varkappa$	A	B	$S_{\min.}(k, \tau)$
0.64	35.561	44.536	16.58
0.68	35.599	44.492	16.47
0.72	35.632	44.453	16.52

For $\tau = 1871.0$, corresponding values have been calculated by means of tables for the case $\alpha = -0.3$, which are not here published. — The resulting values of $S_{\min.}(k, \tau)$ are shown below.

τ	$\varkappa = 0.64$	$\varkappa = 0.68$	$\varkappa = 0.72$
1870.5	16.58	16.47	16.52
1871.0	16.49	16.45	16.50
1871.5	16.53	16.49	16.54
1872.5	17.04	17.01	17.11

Thus we may take $\tau = 1871.0$, $\varkappa = 0.68$ as a sufficient approximation to the values giving the absolute minimum of S. The corresponding values of the other parameters are $k = 0.136$, $A = 35.63$ and $B = 44.53$ in accordance with No 13.

List of References.

[1] F. Bernstein: Säkulare Sterblichkeitsänderung und Prinzip der Gewinnverteilung etc. Bl. Vers.-Math. **2** (1933) p. 390—392.

[2] E. Blaschke: Sulle tavole di mortalità variabili col tempo. Giorn. Mat. Finanz. **5** (1923) p. 1—31.

[3] H. Cramér and H. Wold: On the Development of the Mortality of the Adult Swedish Population since 1800. Nord. Stat. Journ. (1934) p. 3—22.

[4] A. R. DAVIDSON and A. R. REID: On the Calculation of Rates of Mortality. T. F. A. **11** (1927) p. 183—213. Discussion: p. 214—232.

[5] V. P. A. DERRICK: Observations on (1) Errors of Age in the Population Statistics of England and Wales, and (2) the Changes in Mortality indicated by the national records. J. I. A. **58** (1927) p. 117—146. Discussion: p. 146—159.

[6] W. P. ELDERTON: The Analysis of Annuity Experience 1900—1920. J. I. A. Students' Soc. **2** (1926) p. 215—225.

[7] ——: Forecasting Mortality. Skand. Aktuarietidskr. (1932) p. 45—64.

[8] W. P. ELDERTON and H. J. P. OAKLEY: The Mortality of Annuitants, 1900 to 1920. London (1924). The introductory chapters are reprinted from J. I. A. **54** (1923) p. 43—90 and p. 269—284.

[8 a] ——: Mortality Experience of Annuitants 1921—25. London (1929). See also J. I. A. **59** (1928) p. 387—398.

[9] F. ESSCHER: Über die Sterblichkeit in Schweden 1886—1914. Eine Stabilitäts- und Korrelationsuntersuchung. Medd. Lunds Astron. Observ. Ser. II nr 23. Lund (1920).

[10] D. C. FRASER: Notes on recent Reports on the Mortality of Annuitants. J. I. A. **55** (1924) p. 160—177.

[11] G. FUNKE: Die säkularen Sterblichkeitsschwankungen im Deutschen Reiche. Dissert. Dresden (1934). See also report in Bl. Vers.-Math. **3** (1935) p. 175—183.

[12] M. GREENWOOD: The Growth of Population in England and Wales. Metron **5** (1925) p. 66—85. See also P. N. HARVEY: J. I. A. **61** (1928) p. 293—330. Discussion: p. 331—339.

[13] W. IBSCH: Über säkulare Sterblichkeitsänderungen in Deutschland. Bl. Vers.-Math. **2** (1931) p. 31—39.

[14] G. H. KNIBBS: The Mathematical Theory of Population, etc. Melbourne (1917). Cf. also E. CZUBER: Mathematische Bevölkerungstheorie. Berlin (1923).

[15] F. KOBI: Untersuchungen über die Sterblichkeitsänderung, wenn die Überlebensordnungen das MAKEHAMsche Gesetz befolgen, mit besonderer Berücksichtigung schweizerischer Verhältnisse. Festgabe Moser, Bern (1931) p. 49—64.

[16] A. LINDSTEDT: Ålderdomsförsäkringskommitténs betänkande, del II. Stockholm (1912).

[17] C. H. MALTBY: Notes on some present day Questions regarding Mortality. Skand. Aktuarietidskr. (1929) o. 239—253.

[18] B. E. MEURK: New Swedish Mortality Table for Annuity Insurance. Skand. Aktuarietidskr. (1932) p. 251—277.

[19] THV. RICHARDT: Betenkning angaaende valg af dödelighedstabeller for Kristiania kommunale pensionskasse. Dokument no. 48. Oslo (1901).

[20] ——: Dödelighetsundersökelser i Den norske Enkekasse. Oslo (1910).

[21] G. W. RICHMOND: Neue Sterblichkeitserfahrungen in Grossbritannien. Veröff. Deutsch. Ver. Vers.-Wiss. **39** (1926) p. 155—173.

[22] P. Riebesell: Über Sterblichkeitserfahrungen in Deutschland. Z. Ges. Vers.-Wiss. **27** (1927) p. 114—117.

[23] ——: Der Kampf um die Formel für die säkularen Sterblichkeitsschwankungen. Bl. Vers.-Math. **2** (1931) p. 85.

[24] C. W. Sachs: Ein empirisches Gesetz der säkularen Sterblichkeitsschwankungen und Folgerungen daraus für das Rentenversicherungsgeschäft. Bl. Vers.-Math. **1** (1929) p. 219—229.

[25] ——: Nochmals: Säkulare Sterblichkeitsschwankungen und Folgerungen daraus. Bl. Vers.-Math. **2** (1931) p. 39—40.

[26] W. Schweer: Eine graphische Methode zur Berechnung der Konstanten der Makeham-Gompertzschen Formel. Trans. 10th Int. Congr. Actuaries **V** (1934) p. 292—297.

[27] O. A. Åkesson: A research concerning Mortality Assumptions for Life Annuity Insurance. Nord. Stat. Journ. **3** (1931) p. 281—310.

[28] Dödlighetsantaganden för livränteförsäkring. Stockholm (1932).

[29] Försäkringsföreningens Tidskrift. Stockholm (1878).

[30] Mortality Experience of Government Life Annuitants, 1900 to 1920. London (1924). An extract is printed in J. I. A. **55** (1924) p. 144—159.

[31] Welsh Church Act, 1914. London (1914).

[32] 1928 års pensionsförsäkringskommitté. Betänkande, del I (1930), del II (1932), del III (1934). Stockholm.

Appendix 2.

[33] R. Pearl and L. J. Reed: A further Note on the Mathematical Theory of Population Growth. Proc. Nat. Acad. Sci. (1922) p. 365—368.

[34] E. B. Wilson and Ruth Puffer: Least Squares and Laws of Population Growth. Proc. Amer. Acad. Arts Sci. (1933) p. 285—382.

[35] G. U. Yule: The Growth of Population and the Factors which control it. Proc. Roy. Stat. Soc. **88** (1925) p. 1—58.

TABLES

Table 1 a. n_x (Exposed to risk). Males.

Period / Age	1801—1805	1806—1810	1811—1815	1816—1820	1821—1825	1826—1830	1831—1835	1836—1840	1841—1845	1846—1850	1851—1855	1856—1860	1861—1865
30—35	393 070	402 350	412 400	428 205	473 330	510 540	506 565	496 245	520 810	588 180	670 228	704 682	695 808
35—40	367 935	356 130	367 640	393 400	400 500	441 605	469 730	472 015	464 090	491 575	555 828	638 145	680 014
40—45	351 675	342 380	320 400	342 100	365 150	374 355	407 120	436 380	438 440	434 460	459 002	521 395	604 158
45—50	312 100	310 480	300 235	293 785	310 290	328 965	332 415	367 125	394 330	397 640	392 142	418 622	488 385
50—55	277 045	281 070	271 105	270 705	261 940	276 765	288 870	295 410	329 145	355 200	356 288	350 705	381 662
55—60	211 385	229 280	230 475	230 075	228 135	221 900	231 835	245 900	253 230	282 600	305 295	308 105	310 485
60—65	167 170	175 140	183 725	190 115	189 855	188 455	179 800	197 300	215 340	220 375	239 030	254 448	255 413
65—70	121 730	118 990	124 170	136 735	141 810	141 980	138 975	133 185	140 185	151 340	158 322	182 912	206 882
70—75	82 560	79 860	75 790	82 325	93 255	96 925	93 960	92 170	90 000	96 630	103 062	110 150	132 644
75—80	44 160	44 290	39 875	39 710	45 515	51 045	51 525	50 140	49 775	50 040	54 275	59 322	65 038
80—85	18 090	17 965	16 560	16 105	17 110	19 180	20 600	20 420	20 110	20 925	21 185	23 645	27 141
85—90	4 875	4 845	4 670	4 645	4 595	4 865	5 420	5 640	5 705	5 925	6 158	6 282	6 866

Period / Age	1866—1870	1871—1875	1876—1880	1881—1885	1886—1890	1891—1895	1896—1900	1901—1905	1906—1910	1911—1915	1916—1920	1921—1925	1926—1930
30—35	698 606	660 396	677 544	693 847	724 605	756 873	766 184	758 851	872 622	913 248	977 222	1020 915	1095 191
35—40	648 580	638 072	627 501	631 630	647 280	673 908	724 180	717 599	724 646	822 192	874 018	944 200	980 701
40—45	631 658	594 612	603 386	588 496	592 426	609 471	643 482	689 446	686 482	690 797	791 567	847 912	912 059
45—50	555 840	575 737	558 036	562 518	550 033	557 050	577 836	612 874	656 662	657 265	663 060	766 330	816 356
50—55	441 812	500 496	532 488	513 702	521 246	513 448	521 819	545 805	578 570	622 922	624 975	635 515	733 036
55—60	336 390	389 479	452 662	480 417	468 254	478 513	472 109	485 620	505 842	539 260	582 012	578 823	598 421
60—65	262 404	286 128	339 858	396 312	424 098	416 817	427 146	427 272	437 561	459 116	489 448	532 552	538 341
65—70	200 294	209 634	234 820	281 456	333 466	358 301	354 276	367 935	367 702	379 132	398 129	427 810	465 909
70—75	145 472	143 655	155 082	175 473	216 792	259 337	279 156	282 046	292 064	294 346	302 935	322 997	347 247
75—80	78 316	86 787	89 444	98 922	115 567	144 193	175 988	192 221	195 126	204 769	204 408	213 672	229 218
80—85	29 691	34 794	42 174	44 239	50 292	60 170	75 412	95 952	104 605	109 146	112 286	116 056	120 986
85—90	7 407	8 320	10 904	13 901	14 708	17 114	21 536	27 488	36 001	40 454	41 030	44 492	45 912

Table 1 b. n_x (Exposed to risk). Females.

Period / Age	1801—1805	1806—1810	1811—1815	1816—1820	1821—1825	1826—1830	1831—1835	1836—1840	1841—1845	1846—1850	1851—1855	1856—1860	1861—1865
30—35	433 585	444 045	461 030	478 345	511 970	535 360	528 110	519 225	545 100	614 605	699 270	740 355	731 978
35—40	410 505	398 035	407 465	436 785	448 350	483 380	499 880	499 980	491 080	519 640	587 638	671 870	720 266
40—45	388 040	388 235	371 535	391 360	413 365	425 870	455 615	476 345	474 730	468 740	496 208	561 818	641 067
45—50	347 740	349 595	346 095	339 840	358 605	384 605	391 190	422 665	442 620	443 905	439 380	467 950	537 608
50—55	312 995	321 055	319 940	323 405	314 240	332 805	350 870	361 380	392 065	412 600	412 062	405 708	438 480
55—60	248 060	269 235	279 100	285 385	286 775	279 910	290 555	309 835	320 430	351 200	369 245	369 812	371 208
60—65	204 060	214 230	231 690	242 305	245 470	249 005	238 845	260 225	284 965	292 030	310 485	321 092	319 002
65—70	161 015	156 000	161 775	182 090	190 710	193 810	193 600	187 400	197 275	212 815	222 152	248 168	273 978
70—75	115 340	110 720	103 535	114 155	131 140	138 195	136 395	135 500	133 070	144 460	154 765	162 180	187 146
75—80	63 395	63 670	59 080	59 200	68 255	78 035	79 515	79 140	79 830	80 510	87 908	95 552	102 306
80—85	28 470	29 505	26 670	26 185	27 955	32 170	35 520	35 590	35 825	37 535	37 765	42 585	47 839
85—90	8 355	8 885	8 155	8 145	8 785	9 135	10 025	11 375	11 940	12 290	12 658	13 135	14 792

Period / Age	1866—1870	1871—1875	1876—1880	1881—1885	1886—1890	1891—1895	1896—1900	1901—1905	1906—1910	1911—1915	1916—1920	1921—1925	1926—1930
30—35	748 630	733 386	743 680	762 157	797 616	847 262	827 716	809 207	896 541	959 784	1 032 451	1 083 713	1 143 557
35—40	695 205	703 854	700 145	703 163	718 834	756 788	812 810	799 296	778 113	868 913	928 354	1 004 201	1 053 594
40—45	680 775	652 194	670 495	662 222	666 015	683 685	725 678	782 746	768 647	752 508	839 804	899 006	976 041
45—50	603 678	636 844	620 536	633 960	627 762	633 171	654 580	696 297	751 202	741 334	724 888	810 550	870 492
50—55	501 190	559 880	601 018	581 474	597 968	592 570	601 932	622 805	663 669	718 377	710 164	694 583	778 325
55—60	399 848	455 764	518 869	554 738	540 072	555 702	555 694	563 915	585 373	626 132	679 000	670 488	658 543
60—65	325 970	351 734	408 776	464 892	501 366	488 406	508 168	508 020	517 946	539 528	578 620	626 868	622 066
65—70	263 125	269 204	298 850	347 390	400 417	431 600	427 164	445 656	447 989	457 564	478 272	512 851	557 355
70—75	204 624	197 001	208 685	233 337	276 212	318 294	349 008	347 279	365 152	368 374	375 270	394 733	424 267
75—80	118 362	129 122	136 061	141 399	162 368	190 645	225 895	246 566	249 957	264 475	266 166	271 475	287 405
80—85	48 184	60 424	68 906	72 438	78 502	90 317	108 138	129 272	142 472	146 136	153 535	157 166	159 039
85—90	15 755	17 588	22 744	26 633	28 478	30 816	36 392	43 270	53 446	59 792	60 775	65 079	66 071

Table 2 a. d_x (Number of Deaths). Males.

Period / Age	1801—1805	1806—1810	1811—1815	1816—1820	1821—1825	1826—1830	1831—1835	1836—1840	1841—1845	1846—1850	1851—1855	1856—1860	1861—1865
30—35	3 707	6 063	5 314	4 354	4 763	6 774	6 433	5 948	5 213	5 746	7 042	6 571	5 165
35—40	4 063	6 175	5 422	5 127	4 667	7 090	7 161	6 884	5 814	6 268	7 116	6 824	6 207
40—45	5 039	6 928	5 888	5 671	5 766	7 384	7 327	8 006	6 797	6 873	7 585	6 810	6 655
45—50	5 666	7 834	6 930	5 693	5 902	8 047	7 219	7 983	7 328	7 860	7 883	7 036	6 989
50—55	7 114	9 403	7 987	7 110	6 265	8 581	8 434	8 007	8 017	9 130	9 219	7 526	7 236
55—60	6 948	9 754	8 418	7 826	7 304	8 225	8 109	8 705	7 863	9 218	10 123	8 618	8 006
60—65	7 543	10 515	9 654	9 093	8 342	9 831	8 539	9 336	9 092	9 332	10 532	10 141	9 470
65—70	7 674	9 283	9 054	9 136	8 432	10 223	9 161	8 941	8 944	10 438	10 124	10 179	10 740
70—75	8 541	9 174	8 020	8 415	8 459	10 168	9 147	9 615	8 121	9 432	10 211	9 275	10 435
75—80	6 628	7 528	6 091	5 885	6 043	8 201	7 331	7 821	6 970	7 217	7 801	7 819	8 105
80—85	4 097	4 620	3 875	3 563	3 533	4 787	4 659	4 837	4 423	4 775	4 238	4 597	5 290
85—90	1 451	1 889	1 516	1 550	1 343	1 610	1 694	1 972	1 691	1 862	1 823	1 700	2 054

Period / Age	1866—1870	1871—1875	1876—1880	1881—1885	1886—1890	1891—1895	1896—1900	1901—1905	1906—1910	1911—1915	1916—1920	1921—1925	1926—1930
30—35	6 144	5 802	5 069	4 926	4 721	5 149	5 059	4 685	5 153	5 552	8 141	4 656	4 815
35—40	6 747	6 314	5 330	5 007	4 795	5 143	5 451	4 991	4 611	5 348	6 723	4 625	4 679
40—45	8 277	6 936	6 035	5 851	5 341	5 381	5 650	5 677	5 416	5 089	6 165	4 855	5 331
45—50	9 014	8 319	6 772	6 733	6 444	5 976	6 183	6 295	6 301	6 145	6 074	5 638	6 286
50—55	9 418	9 039	8 434	7 862	7 310	7 175	7 021	7 271	7 173	7 730	7 429	6 402	7 418
55—60	9 652	9 388	9 584	9 968	8 823	9 164	8 506	8 418	8 508	9 101	9 499	8 587	8 723
60—65	10 739	9 740	10 027	11 360	11 513	11 076	10 964	10 581	10 462	11 041	11 369	11 590	11 955
65—70	12 124	10 551	10 671	12 204	13 166	14 509	13 651	13 629	13 082	13 706	14 226	14 195	15 841
70—75	13 419	11 339	10 921	11 964	13 549	16 338	17 056	16 343	16 425	17 189	16 990	17 079	18 783
75—80	10 932	10 493	10 280	10 938	11 756	15 011	17 402	18 216	18 057	19 084	18 902	18 525	20 456
80—85	6 012	6 521	7 406	7 923	8 428	10 103	12 705	14 826	15 873	16 559	16 937	16 323	17 844
85—90	2 461	2 428	3 091	3 728	3 915	4 550	5 596	6 778	8 722	9 932	9 805	10 270	10 802

Table 2 b. d_x (Number of Deaths). Females.

Period / Age	1801—1805	1806—1810	1811—1815	1816—1820	1821—1825	1826—1830	1831—1835	1836—1840	1841—1845	1846—1850	1851—1855	1856—1860	1861—1865
30—35	3 973	5 784	5 060	4 464	4 156	5 262	5 184	4 774	4 246	4 972	5 993	6 090	4 947
35—40	4 447	6 073	5 208	5 284	4 382	5 714	5 797	5 393	4 689	5 070	6 143	6 499	5 801
40—45	5 181	6 855	5 425	5 276	4 835	5 882	6 279	6 270	5 376	5 411	6 114	6 109	6 001
45—50	4 926	6 945	5 710	4 944	4 472	6 151	6 282	6 217	5 645	5 604	5 986	5 979	5 700
50—55	6 348	8 686	6 995	6 472	5 165	7 055	7 284	7 128	6 474	7 393	7 457	6 584	6 170
55—60	6 597	9 433	8 147	7 488	6 572	7 538	7 781	8 315	7 339	8 343	9 092	8 426	7 305
60—65	7 997	11 379	10 102	9 598	8 721	10 299	9 089	9 746	9 408	9 942	11 236	10 910	9 566
65—70	8 933	10 873	10 449	10 601	9 427	11 728	11 158	10 498	10 207	12 050	12 515	12 179	12 209
70—75	10 505	11 603	10 055	10 437	10 009	12 635	11 868	12 487	10 691	12 286	13 491	12 457	12 915
75—80	8 876	10 546	8 226	7 739	7 864	11 219	10 106	10 735	10 119	10 286	11 322	11 345	11 344
80—85	6 257	6 887	5 748	5 426	5 005	6 932	7 179	7 385	6 958	7 515	6 954	7 366	8 153
85—90	2 446	3 091	2 553	2 546	2 213	2 807	2 912	3 514	3 176	3 642	3 395	3 192	3 838

Period / Age	1866—1870	1871—1875	1876—1880	1881—1885	1886—1890	1891—1895	1896—1900	1901—1905	1906—1910	1911—1915	1916—1920	1921—1925	1926—1930
30—35	5 598	5 503	5 170	5 140	5 216	5 652	5 262	5 199	5 301	5 530	7 701	4 829	4 917
35—40	6 207	6 074	5 456	5 324	5 321	5 584	5 781	5 475	5 120	5 348	6 696	4 907	4 776
40—45	7 229	6 316	5 641	5 495	5 390	5 506	5 613	5 844	5 412	5 333	6 280	5 127	5 197
45—50	7 288	6 546	5 879	5 820	5 616	5 568	5 495	5 861	6 092	5 739	5 879	5 558	5 978
50—55	7 707	7 663	7 189	6 768	6 484	6 504	6 545	6 461	6 560	7 021	7 293	6 256	7 300
55—60	8 721	8 408	8 462	8 707	8 152	8 168	7 711	7 846	7 762	8 108	9 187	8 302	8 231
60—65	10 448	9 663	10 079	11 026	11 152	10 828	10 379	10 169	9 911	10 451	11 231	11 523	11 684
65—70	13 056	11 527	11 199	12 775	13 918	15 182	13 803	13 709	13 543	14 110	14 511	15 011	16 842
70—75	15 424	13 257	12 517	13 572	15 442	17 907	18 641	17 611	18 032	18 526	19 138	18 810	20 857
75—80	13 965	13 396	12 827	13 410	14 461	17 797	19 728	21 226	20 655	22 418	21 810	22 104	24 148
80—85	9 219	9 660	10 337	10 915	11 409	13 389	15 688	17 995	19 690	20 494	20 726	20 874	21 796
85—90	4 617	4 602	5 382	6 180	6 574	7 215	8 481	9 526	11 612	13 440	12 927	13 877	14 475

Table 3 a. $10^3\ \mu_x$, observed values. Males.

Year / Age	1802.5	1807.5	1812.5	1817.5	1822.5	1827.5	1832.5	1837.5	1842.5	1847.5	1852.5	1857.5	1862.5
32.5	9.43	15.07	12.89	10.17	10.06	13.27	12.70	11.99	10.01	9.77	10.51	9.32	7.42
37.5	11.04	17.34	14.75	13.03	11.65	16.06	15.24	14.58	12.53	12.75	12.80	10.69	9.13
42.5	14.33	20.23	18.38	16.58	15.79	19.72	18.00	18.35	15.50	15.82	16.52	13.06	11.02
47.5	18.15	25.23	23.08	19.38	19.02	24.46	21.72	21.74	18.58	19.77	20.10	16.81	14.31
52.5	25.68	33.45	29.46	26.26	23.92	31.00	29.20	27.10	24.36	25.70	25.88	21.46	18.96
57.5	32.87	42.54	36.52	34 01	32.02	37.07	34.98	35.40	31.05	32.62	33.16	27.97	25.79
62.5	45.12	60.04	52.55	47.83	43.94	52.17	47.49	47.32	42.22	42.35	44.06	39.85	37.08
67.5	63.04	78.01	72.92	66.82	59.46	72.00	65.92	67.13	63.80	68.97	63.95	55.65	51.91
72.5	103.45	114.88	105.82	102.22	90.71	104.91	97.35	104.32	90.23	97.61	99.08	84.20	78.67
77.5	150.09	169.97	152.75	148.20	132.77	160.66	142.28	155.98	140.03	144.22	143.73	131.81	124.62
82.5	226.48	257.17	234 00	221.24	206.49	249.58	226.17	236.88	219.94	228.20	200.05	194.42	194.91
87.5	297.64	339.89	324.63	333.69	292.27	330.94	312.55	349.65	296.41	314.26	296.06	270.59	299.16

Year / Age	1867.5	1872.5	1877.5	1882.5	1887.5	1892.5	1897.5	1902.5	1907.5	1912.5	1917.5	1922.5	1927.5
32.5	8.79	8.79	7.48	7.10	6.52	6.80	6.60	6.17	5.91	6.08	8.33	4.56	4.40
37.5	10.40	9.90	8.49	7.93	7.41	7.63	7.53	6.96	6.86	6.50	7.69	4.90	4.77
42.5	13.10	11.66	10.00	9.94	9.02	8.83	8.78	8.23	7.89	7.37	7.79	5.73	5.85
47.5	16.22	14.45	12.14	11.97	11.72	10.73	10.70	10.27	9.60	9.35	9.16	7.36	7.70
52.5	21.32	18.06	15.84	15.30	14.02	13.97	13.45	13.32	12.40	12.41	11.89	10.07	10.12
57.5	28.69	24.10	21.17	20.75	18.84	19.15	18.02	17.33	16.82	16.88	16.32	14 84	14.58
62.5	40.93	34.04	29.50	28.66	27.15	26.57	25.67	24.76	23.91	24.05	23.23	21.76	22.21
67.5	60.53	50.33	45 44	43.36	39.48	40.49	38.53	37.04	35.58	36.15	35.73	33.18	34.00
72.5	92.24	78.93	70.42	68.18	62.50	63.00	61.10	57.94	56.24	58.40	56.08	52.88	54.09
77.5	139.59	120.91	114.93	110.57	101.72	104.10	98.88	94.77	92.54	93.20	92.47	86.70	89.24
82.5	202.49	187.42	175.61	179.10	167.58	167.91	168.47	154.51	151.74	151.71	150.84	140.65	147.49
87.5	332.25	291.83	283.47	268.18	266.18	265.86	259.84	246.58	242.27	245.51	238.97	230.83	235.28

Table 3 b. $10^3 \mu_x$, observed values. Females.

Year / Age	1802.5	1807.5	1812.5	1817.5	1822.5	1827.5	1832.5	1837.5	1842.5	1847.5	1852.5	1857.5	1862.5
32.5	9.16	13.03	10.98	9.33	8.12	9.83	9.82	9.19	7.79	8.09	8.57	8.23	6.76
37.5	10.83	15.26	12.78	12.10	9.77	11.82	11.60	10.79	9.55	9.76	10.45	9.67	8.05
42.5	13.35	17.66	14.60	13.48	11.70	13.81	13.78	13.16	11.32	11.54	12.32	10.87	9.36
47.5	14.17	19.87	16.50	14.55	12.47	15.99	16.06	14.71	12.75	12.62	13.62	12.78	10.60
52.5	20.28	27.05	21.86	20.01	16.44	21.20	20.76	19.72	16.51	17.92	18.10	16.23	14.07
57.5	26.59	35.04	29.19	26.24	22.92	26.93	26.78	26.84	22.90	23.76	24.62	22.78	19.68
62.5	39.19	53.12	43.60	39.61	35.53	41.36	38.05	37.45	33.01	34.04	36.19	33.98	29.99
67.5	55.48	69.70	64.59	58.22	49.43	60.51	57.63	56.02	51.74	56.62	56.34	49.08	44.56
72.5	91.08	104.80	97.12	91.43	76.32	91.43	87.01	92.15	80.34	85.05	87.17	76.81	69.01
77.5	140.01	165.64	139.23	130.73	115.22	143.77	127.10	135.65	126.76	127.76	128.79	118.73	110.88
82.5	219.78	233.42	215.52	207.22	179.04	215.48	202.11	207.50	194.22	200.21	184.14	172.97	170.43
87.5	292.76	347.89	313.06	312.58	251.91	307.28	290.47	308.92	266.00	296.34	268.22	243.01	259.47

Year / Age	1867.5	1872.5	1877.5	1882.5	1887.5	1892.5	1897.5	1902.5	1907.5	1912.5	1917.5	1922.5	1927.5
32.5	7.48	7.50	6.95	6.74	6.54	6.67	6.36	6.42	5.91	5.76	7.46	4.46	4.30
37.5	8.93	8.63	7.79	7.57	7.40	7.38	7.11	6.85	6.58	6.15	7.21	4.89	4.53
42.5	10.62	9.68	8.41	8.30	8.09	8.05	7.73	7.47	7.04	7.09	7.48	5.70	5.32
47.5	12.07	10.28	9.47	9.18	8.95	8.79	8.39	8.42	8.11	7.74	8.11	6.86	6.87
52.5	15.38	13.69	11.96	11.64	10.84	10.98	10.87	10.37	9.88	9.77	10.27	9.01	9.38
57.5	21.81	18.45	16.31	15.70	15.09	14.70	13.88	13.91	13.26	12.95	13.53	12.38	12.50
62.5	32.05	27.47	24.66	23.72	22.24	22.17	20.42	20.02	19.14	19.37	19.41	18.38	18.78
67.5	49.62	42.82	37.47	36.77	34.76	35.18	32.31	30.76	30.23	30.84	30.34	29.27	30.22
72.5	75.38	67.29	59.98	58.16	55.91	56.26	53.41	50.71	49.38	50.29	51.00	47.65	49.16
77.5	117.99	103.75	94.27	94.84	89.06	93.35	87.33	86.09	82.63	84.76	81.94	81.42	84.02
82.5	191.33	159.87	150.01	150.68	145.33	148.24	145.07	139.20	138.20	140.24	134.99	132.82	137.05
87.5	293.06	261.66	236.64	232.04	230.84	234.14	233.04	220.15	217.27	224.78	212.70	213.23	219.09

Table 4 a. Periods; parameters and χ^2.

Period	$10^3 \log c$				$10^5 \beta$				$10^3 \alpha$				χ^2			
	By χ^2-min.-method		Final graduation		By χ^2-min.-method		Final graduation		By χ^2-min.-method		Final graduation		Minimum value		Final graduation	
	M	F	M	F	M	F	M	F	M	F	M	F	M	F	M	F
(1)	(2)	(3)	(4)	(5)	(6)	(7)	(8)	(9)	(10)	(11)	(12)	(13)	(14)	(15)	(16)	(17)
1801—05	35.7	40.8	35.63	39.92	23.86	8.85	24.780	10.228	6.1	7.6	9.482	8.393	127	158	1 594	846
1806—10	35.1	37.8	35.63	39.92	30.36	17.29	24.766	10.221	11.0	10.2	9.272	8.236	62	125	1 878	4 718
1811—15	34.8	39.3	35.63	39.92	28.73	12.04	24.740	10.209	9.2	8.7	9.061	8.080	50	100	146	492
1816—20	35.6	40.5	35.64	39.92	24.64	9.12	24.694	10.187	7.4	8.0	8.851	7.924	83	159	378	170
1821—25	35.2	39.9	35.64	39.93	23.55	8.71	24.613	10.151	7.2	6.6	8.640	7.767	103	135	1 300	1 950
1826—30	35.7	40.1	35.65	39.94	25.30	10.23	24.469	10 089	10.5	8.1	8.430	7.611	184	99	1 080	292
1831—35	35.7	39.4	35.68	39.97	22.67	10.62	24.215	9.985	10.1	8.2	8.219	7.455	120	63	454	200
1836—40	37.4	40.6	35.72	40.01	18.06	8.97	23.775	9.811	10.0	7.6	8.009	7.299	146	114	436	145
1841—45	36.8	41.5	35.81	40.10	18.16	6.95	23.027	9.528	7.9	6.6	7.798	7.142	126	162	417	906
1846—50	36.0	41.2	35.98	40.25	22.24	7.72	21.802	9.079	7.4	6.6	7.588	6.986	242	176	258	308
1851—55	34.9	39.5	36.29	40.51	25.96	10.23	19.913	8.404	7.6	6.8	7.377	6.830	158	248	482	364
1856—60	37.1	40.1	36.85	40.95	15.83	8.39	17.265	7.469	6.9	6.6	7.167	6.673	60	151	218	207
1861—65	38.3	42.2	37.76	41.64	12.63	5.44	14.032	6.310	5.5	5.6	6.956	6.517	64	80	828	766
1866—70	38.1	42.6	39.04	42.60	14.71	5.60	10.727	5.070	6.5	6.4	6.746	6.361	43	75	1 550	724
1871—75	41.0	44.0	40.53	43.75	7.49	3.74	7.940	3.942	7.4	6.5	6.535	6.205	25	149	257	171
1876—80	42.6	45.0	41.93	44.90	5.26	2.83	5.963	3.065	6.4	6.1	6.325	6.048	36	102	60	345
1881—85	42.6	45.5	42.99	45.86	5.09	2.54	4.721	2.462	6.1	5.9	6.114	5.892	82	128	108	161
1886—90	43.3	46.3	43.68	46.55	4.21	2.12	3.992	2.080	5.8	5.9	5.904	5.736	121	110	136	145
1891—95	43.9	46.9	44.08	46.99	3.83	1.95	3.578	1.849	5.9	5.9	5.693	5.580	17	170	180	393
1896—00	44.8	48.1	44.29	47.25	3.17	1.50	3.345	1.712	6.0	5.9	5.483	5.423	62	109	284	212
1901—05	43.7	47.8	44.41	47.40	3.66	1.51	3.215	1.631	5.5	5.8	5.272	5.267	63	117	151	223
1906—10	44.0	47.9	44.47	47.48	3.36	1.45	3.142	1.584	5.1	5.4	5.062	5.111	57	104	82	135
1911—15	44.0	48.0	44.50	47.53	3.42	1.47	3.101	1.556	5.1	5.2	4.851	4.954	26	118	121	207
1916—20	46.9	48.4	44.51	47.55	1.92	1.30	3.079	1.540	6.7	6.3	4.641	4.798	294	336	1 921	1 600
1921—25	44.0	46.5	44.52	47.57	3.20	1.85	3.066	1.531	3.5	3.8	4.430	4.642	24	64	879	396
1926—30	44.2	46.1	44.53	47.57	3.22	2.09	3.059	1.525	3.5	3.5	4.220	4.486	23	92	475	577

Table 4 b. Generations; parameters and χ^2.

Central Birth Year	$10^3 \log c$				$10^5 \beta$				$10^3 \alpha$				χ^2			
	By χ^2-min.-method		Final graduation		By χ^2-min.-method		Final graduation		By χ^2-min.-method		Final graduation		Minimum value		Final graduation	
	M	F	M	F	M	F	M	F	M	F	M	F	M	F	M	F
(1)	(2)	(3)	(4)	(5)	(6)	(7)	(8)	(9)	(10)	(11)	(12)	(13)	(14)	(15)	(16)	(17)
1770	34.1	40.0	34.37	39.16	30.10	8.88	27.705	9.934	8.3	9.2	8.568	9.236	752	961	755	995
1775	35.8	40.6	34.47	39.16	21.25	7.89	26.873	9.900	10.7	9.9	8.613	9.047	386	688	662	779
1780	34.0	39.5	34.65	39.17	30.06	9.75	25.573	9.828	8.7	8.6	8.833	8.858	580	444	740	541
1785	33.5	39.1	34.94	39.18	30.32	10.08	23.634	9.682	7.9	7.7	9.229	8.669	814	437	841	545
1790	36.1	38.4	35.43	39.23	18.41	10.66	20.935	9.388	10.1	7.6	9.800	8.479	816	575	819	696
1795	37.0	38.4	36.19	39.35	14.91	9.96	17.543	8.828	11.5	8.1	10.547	8.290	467	365	511	452
1800	37.2	38.7	37.28	39.66	13.96	9.07	13.812	7.852	10.8	7.8	10.619	8.101	325	279	336	311
1805	38.0	40.2	38.68	40.42	11.48	6.63	10.311	6.407	10.4	8.1	10.248	7.912	390	176	438	177
1810	39.7	42.5	40.22	41.93	8.06	4.28	7.516	4.730	10.2	8.0	9.878	7.722	588	391	672	402
1815	42.1	44.3	41.66	44.01	5.00	2.96	5.568	3.289	9.9	8.1	9.508	7.533	580	322	621	452
1820	43.1	45.2	42.82	45.82	4.02	2.42	4.331	2.351	9.5	8.0	9.137	7.344	380	194	444	602
1825	43.3	46.4	43.64	46.85	3.84	1.93	3.585	1.842	8.8	7.5	8.767	7.155	368	269	384	345
1830	44.0	46.7	44.18	47.31	3.30	1.79	3.144	1.588	8.1	7.0	8.397	6.965	337	326	348	375
1835	44.9	48.0	44.50	47.48	2.74	1.39	2.884	1.466	8.1	7.0	8.026	6.776	87	243	126	418
1840	44.7	48.3	44.69	47.55	2.82	1.32	2.731	1.407	7.7	6.7	7.656	6.587	101	124	173	501

Table 5 a. $10^3\,\mu_x$, males. Graduation and extrapolation by periods.

Age \ Year	1802.5	1807.5	1812.5	1817.5	1822.5	1827.5	1832.5	1837.5	1842.5	1847.5	1852.5	1857.5	1862.5	1867.5
32.5	13.05	12.84	12.62	12.41	12.18	11.96	11.72	11.45	11.16	10.81	10.39	9.89	9.32	8.74
37.5	14.86	14.64	14.43	14.21	13.98	13.74	13.49	13.20	12.87	12.46	11.95	11.33	10.61	9.87
42.5	17.58	17.37	17.15	16.93	16.69	16.44	16.17	15.85	15.46	14.96	14.32	13.52	12.60	11.64
47.5	21.69	21.47	21.25	21.03	20.78	20.51	20.21	19.83	19.37	18.74	17.92	16.88	15.68	14.42
52.5	27.88	27.66	27.43	27 20	26.93	26.64	26.30	25.85	25.27	24.47	23.39	22.02	20.43	18.77
57.5	37.20	36.98	36.74	36.51	36.21	35.88	35.49	34.93	34.18	33.14	31.69	29.87	27.77	25.59
62.5	51.26	51.03	50.77	50.55	50.20	49.81	49.34	48.62	47.64	46.25	44.30	41.86	39.10	36.29
67.5	72.45	72.21	71.93	71.70	71.28	70.80	70.23	69.27	67.97	66.09	63.44	60.20	56.61	53.06
72.5	104.38	104.12	103.81	103.58	103.06	102.45	101.74	100.44	98.68	96.11	92.52	88.22	83.64	79.34
77.5	152.51	152.22	151.86	151.64	150.96	150.17	149.24	147.46	145.05	141.54	136.68	131.06	125.40	120.53
82.5	225.05	224.71	224.28	224.08	223.16	222.10	220.88	218.40	215,08	210.28	203.74	196.53	189.90	185.10
87.5	334.37	333.97	333.42	333.26	331.98	330.53	328.91	325.42	320.85	314.31	305.58	296.60	289.52	286.31

Age \ Year	1872.5	1877.5	1882.5	1887.5	1892.5	1897.5	1902.5	1907.5	1912.5	1917.5	1922.5	1927.5	1952.5	1977.5
32.5	8.18	7.70	7.29	6.95	6.66	6.40	6.16	5.94	5.72	5.50	5.29	5.08	4.02	2.97
37.5	9.16	8.55	8.05	7.64	7.30	7.01	6.76	6.52	6.30	6.08	5.86	5.65	4.59	3.54
42.5	10.73	9.93	9.28	8.77	8.37	8.03	7.75	7.50	7.27	7.04	6.82	6.61	5.45	4.50
47.5	13.22	12.17	11.31	10.65	10.13	9.73	9.41	9.13	8.88	8.65	8.42	8.21	7.14	6.09
52.5	17.19	15.80	14.65	13.75	13.07	12.56	12.17	11.85	11.58	11.33	11.10	10.88	9.81	8.75
57.5	23.53	21.69	20.11	18.87	17.95	17.26	16.77	16.39	16.08	15.80	15.56	15.34	14.25	13.20
62.5	33.63	31.22	29.07	27.34	26.05	25.09	24.45	23.97	23.59	23.27	23.01	22.79	21.68	20.62
67.5	49.75	46.67	43.77	41.35	39.50	38.14	37.25	36.61	36.13	35.75	35.45	35.22	34.08	33.01
72.5	75.44	71.70	67.89	64.52	61.86	59.86	58.59	57.70	57.06	56.57	56.22	55.98	54.78	53.71
77.5	116.41	112.26	107.45	102.82	98.98	96.03	94.18	92.89	92.00	91.32	90.90	90.65	89.34	88.26
82.5	181.74	177.99	172.34	166.15	160.66	156.25	153.53	151.61	150.31	149.35	148.80	148.53	147.06	145.96
87.5	285.92	284.51	278.80	270.87	263.11	256.54	252.48	249.59	247.65	246.21	245.46	245.19	243.42	242.29

Table 5 b. $10^3\ \mu_x$, males. Graduation and extrapolation by generations.

Age \ Year	1802.5	1807.5	1812.5	1817.5	1822.5	1827.5	1832.5	1837.5	1842.5	1847.5	1852.5	1857.5	1862.5	1867.5
32.5	12.20	12.16	12.25	12.46	12.77	13.18	12.87	12.11	11.40	10.77	10.20	9.71	9.25	8.83
37.5	14.16	13.96	13.88	13.93	14.06	14.26	14.54	14.07	13.16	12.30	11.54	10.88	10.32	9.82
42.5	17.18	16.81	16.57	16.45	16.43	16.45	16.51	16.60	15.92	14.79	13.73	12.79	12.00	11.33
47.5	21.68	21.14	20.74	20.46	20.27	20.15	20.02	19.89	19.73	18.77	17.34	15.99	14.81	13.82
52.5	28.28	27.58	27.02	26.58	26.23	25.95	25.70	25.37	24.97	24.48	23.13	21.31	19.60	18.08
57.5	37.92	37.06	36.33	35.74	35.24	34.80	34.40	33.96	33.36	32.61	31.68	29.84	27.52	25.32
62.5	52.00	50.94	50.07	49.31	48.68	48.11	47.53	46.96	46.28	45.31	44.10	42.61	40.15	37.21
67.5	72.58	71.35	70.26	69.38	68.58	67.89	67.20	66.45	65.63	64.61	63.18	61.37	59.18	55.98
72.5	102.71	101.31	100.05	98.92	98.02	97.16	96.39	95.55	94.54	93.41	91.99	89.89	87.34	84.32
77.5	146.91	145.33	143.91	142.62	141.43	140.52	139.58	138.70	137.64	136.28	134.72	132.76	129.84	126.40
82.5	211.90	210.16	208.56	207.11	205.77	204.49	203.56	202.50	201.47	200.13	198.27	196.15	193.51	189.57
87.5	307.66	305.74	303.98	302.36	300.86	299.44	298.03	297.09	295.87	294.64	292.88	290.36	287.51	284.04

Age \ Year	1872.5	1877.5	1882.5	1887.5	1892.5	1897.5	1902.5	1907.5	1912.5	1917.5	1922.5	1927.5	1952.5	1977.5
32.5	8.43	8.04	7.66	7.28	6.91	6.53	6.16	5.79	5.42	5.05	4.68	4.31	—	—
37.5	9.37	8.95	8.55	8.16	7.78	7.40	7.03	6.65	6.28	5.91	5.54	5.17	—	—
42.5	10.77	10.27	9.82	9.40	9.00	8.61	8.23	7.85	7.48	7.11	6.74	6.36	—	—
47.5	13.01	12.34	11.77	11.28	10.83	10.42	10.01	9.63	9.24	8.87	8.49	8.12	—	—
52.5	16.80	15.77	14.96	14.28	13.72	13.22	12.78	12.36	11.96	11.57	11.20	10.82	—	—
57.5	23.35	21.69	20.35	19.31	18.47	17.79	17.23	16.75	16.30	15.88	15.48	15.10	—	—
62.5	34.41	31.87	29.69	27.91	26.54	25.45	24.62	23.95	23.40	22.90	22.46	22.04	—	—
67.5	52.34	48.86	45.63	42.78	40.40	38.57	37.12	36.03	35.20	34.55	33.97	33.49	20.13	—
72.5	80.29	75.95	71.81	67.86	64.22	61.05	58.58	56.58	55.12	54.03	53.25	52.53	31.48	—
77.5	122.45	117.64	112.81	108.29	103.77	99.32	95.18	91.85	89.07	87.05	85.59	84.58	50.30	—
82.5	185.13	180.28	175.00	170.35	166.24	161.79	156.79	151.59	147.18	143.31	140.47	138.44	81.51	—
87.5	278.87	273.43	268.02	263.12	260.17	258.33	255.52	250.87	244.81	239.19	233.84	229.83	133.90	—
													221.40	219.13

Table 5 c. $10^3 \mu_x$, females. Graduation and extrapolation by periods.

Age \ Year	1802.5	1807.5	1812.5	1817.5	1822.5	1827.5	1832.5	1837.5	1842.5	1847.5	1852.5	1857.5	1862.5	1867.5
32.5	10.42	10.26	10.10	9.94	9.78	9.62	9.44	9.26	9.06	8.83	8.57	8.27	7.94	7.59
37.5	11.60	11.45	11.29	11.12	10.96	10.79	10.60	10.40	10.18	9.92	9.61	9.24	8.82	8.37
42.5	13.48	13.32	13.16	12.99	12.82	12.64	12.44	12.22	11.96	11.65	11.26	10.78	10.23	9.64
47.5	16.45	16.28	16.12	15.95	15.77	15.57	15.36	15.10	14.79	14.40	13.89	13.26	12.51	11.71
52.5	21.14	20.98	20.81	20.63	20.44	20.22	19.98	19.66	19.28	18.77	18.08	17.22	16.20	15.10
57.5	28.59	28.41	28.24	28.04	27.83	27.58	27.30	26.90	26.41	25.71	24.77	23.57	22.16	20.63
62.5	40.37	40.19	39.99	39.77	39.55	39.24	38.89	38.37	37.71	36.75	35.43	33.75	31.78	29.67
67.5	59.02	58.83	58.61	58.35	58.09	57.71	57.27	56.55	55.64	54.29	52.42	50.06	47.32	44.42
72.5	88.56	88.35	88.10	87.77	87.46	86.95	86.37	85.36	84.10	82.18	79.52	76.20	72.42	68.52
77.5	135.33	135 09	134.78	134.35	133.97	133.27	132.49	131.03	129.25	126.50	122.71	118.07	112.96	107.87
82.5	209.39	209.10	208.70	208.12	207.63	206.63	205.55	203.43	200.90	196.95	191.57	185.17	178.44	172.13
87.5	326.66	326.29	325.76	324.91	324.27	322.82	321.31	318.18	314.58	308.92	301.35	292.69	284.19	277.07

Age \ Year	1872.5	1877.5	1882.5	1887.5	1892.5	1897.5	1902.5	1907.5	1912.5	1917.5	1922.5	1927.5	1952.5	1977.5
32.5	7.25	6.93	6.65	6.41	6.20	6.01	5.83	5.66	5.50	5.34	5.18	5.02	4.24	3.46
37.5	7.93	7.53	7.18	6.89	6.65	6.44	6.24	6.07	5.90	5.73	5.57	5.41	4.63	3.85
42.5	9.06	8.53	8.08	7.71	7.42	7.17	6.95	6.76	6.58	6.41	6.25	6.09	5.30	4.52
47.5	10.92	10.21	9.60	9.12	8.73	8.43	8.18	7.96	7.77	7.59	7.42	7.26	6.47	5.69
52.5	14.01	13.03	12.19	11.52	11.00	10.60	10.29	10.04	9.82	9.63	9.45	9.28	8.48	7.70
57.5	19.13	17.75	16.56	15.61	14.89	14.34	13.94	13.62	13.37	13.15	12.96	12.78	11.97	11.19
62.5	27.59	25.67	23.99	22.62	21.57	20.79	20.23	19.81	19.50	19.23	19.03	18.82	17.99	17.21
67.5	41.59	38.95	36.57	34 59	33.04	31.90	31.09	30.50	30.09	29.75	29.53	29.27	28.42	27.64
72.5	64.76	61.22	57.91	55.05	52.75	51.04	49.83	48.97	48.40	47.94	47.67	47.35	46.44	45.66
77.5	103.11	98.56	94.09	90.01	86.61	84.02	82.17	80.87	80.04	79.38	79.06	78.61	77.62	76.84
82.5	166.56	161.17	155.43	149.76	144.77	140.83	137.99	135.99	134.74	133.74	133.32	132.66	131.53	130.75
87.5	271.57	266.17	259.43	251.88	244.67	238.70	234.33	231.19	229.28	227.72	227.15	226.12	224.77	223.99

Table 5 d. $10^3\,\mu_x$, females. Graduation and extrapolation by generations.

Age \ Year	1802.5	1807.5	1812.5	1817.5	1822.5	1827.5	1832.5	1837.5	1842.5	1847.5	1852.5	1857.5	1862.5	1867.5
32.5	11.10	10.90	10.70	10.49	10.25	9.97	9.63	9.23	8.81	8.42	8.07	7.77	7.51	7.29
37.5	12.35	12.16	11.96	11.75	11.52	11.26	10.93	10.51	10.01	9.49	9.00	8.57	8.21	7.91
42.5	14.21	14.02	13.82	13.62	13.40	13.15	12.84	12.44	11.91	11.26	10.59	9.97	9.43	8.96
47.5	17.02	16.83	16.64	16.43	16.22	15.99	15.70	15.33	14.82	14.11	13.24	12.36	11.58	10.87
52.5	21.33	21.14	20.94	20.74	20.53	20.31	20.05	19.71	19.25	18.56	17.59	16.40	15.24	14.26
57.5	27.97	27.78	27.59	27.39	27.19	26.97	26.72	26.43	26.00	25.40	24.45	23.08	21.43	19.91
62.5	38.30	38.11	37.92	37.72	37.52	37.31	37.07	36.79	36.44	35.88	35.05	33.71	31.75	29.43
67.5	54.40	54.21	54.01	53.82	53.63	53.42	53.19	52.93	52.59	52.15	51.39	50.22	48.28	45.43
72.5	79.56	79.37	79.18	78.98	78.79	78.59	78.37	78.13	77.82	77.40	76.82	75.74	74.06	71.20
77.5	118.94	118.75	118.56	118.37	118.17	117.98	117.77	117.54	117.26	116.89	116.33	115.55	113.97	111 49
82.5	180.65	180.46	180.27	180.08	179.89	179.69	179.50	179.27	179.02	178.69	178.21	177.44	176.35	173.98
87.5	277.40	277.21	277.02	276.84	276.65	276.46	276.24	276.05	275.81	275.51	275.11	274.47	273.37	271.79

Age \ Year	1872.5	1877.5	1882.5	1887.5	1892.5	1897.5	1902.5	1907.5	1912.5	1917.5	1922.5	1927.5	1952.5	1977.5
32.5	7.08	6.88	6.69	6.50	6.31	6.12	5.93	5.74	5.55	5.36	5.17	4.98	—	—
37.5	7.66	7.44	7.24	7.04	6.85	6.66	6.47	6.28	6.09	5.90	5.71	5.52	—	—
42.5	8.59	8.30	8.06	7.85	7.65	7.45	7.26	7.07	6.88	6.69	6.50	6.31	—	—
47.5	10.25	9.77	9.42	9.14	8.90	8.69	8.50	8.30	8.11	7.92	7.73	7.54	—	—
52.5	13.33	12.46	11.80	11.33	11.00	10.73	10.51	10.30	10.10	9.91	9.72	9.53	—	—
57.5	18.69	17.48	16.26	15.31	14.65	14.22	13.89	13.64	13.43	13.22	13.03	12.83	—	—
62.5	27.47	26.06	24.52	22.77	21.84	20.38	19.77	19.36	19.07	18.83	18.61	18.41	17.46	—
67.5	42.19	39.72	38.28	36.46	33.93	31.76	30.27	29.39	28.81	28.45	28.18	27.94	26.96	—
72.5	67.03	62.50	59.58	58.56	56.68	53.07	49.70	47.37	46.00	45.16	44.67	44.35	43.26	—
77.5	107.26	101.13	94.85	91.75	92.22	90.96	85.89	80.65	76.90	74.73	73.42	72.72	71.31	—
82.5	170.31	163.97	154.97	146.37	143.89	148.10	149.05	142.18	134.00	127.90	124.39	122.30	119.69	—
87.5	268.21	262.69	253.19	239.96	228.41	228.39	240.85	247.51	238.72	225.98	216.01	210.25	203.37	202.27

Table 6 a. $d_x - n_x\mu_x$, males. Graduation by periods.

Age \ Year	1802.5	1807.5	1812.5	1817.5	1822.5	1827.5	1832.5	1837.5	1842.5	1847.5	1852.5	1857.5	1862.5
32.5	−1 422	899	109	−958	−1 004	670	498	265	−597	−611	80	−397	−1 323
37.5	−1 403	961	118	−463	−933	1 020	823	651	−159	143	475	−404	−1 010
42.5	−1 144	982	394	−120	−329	1 229	743	1 091	20	373	1 013	−242	−960
47.5	−1 103	1 168	551	−484	−545	1 301	500	701	−308	406	856	−32	−669
52.5	−609	1 630	552	−254	−790	1 209	836	371	−300	438	887	−196	−562
57.5	−916	1 276	−49	−575	−957	264	−119	117	−793	−146	449	−584	−616
62.5	−1 027	1 578	325	−517	−1 189	445	−333	−256	−1 167	−860	−56	−511	−517
67.5	−1 145	691	123	−668	−1 677	170	−600	−285	−585	436	79	−832	−971
72.5	−77	859	152	−112	−1 152	238	−412	357	−760	145	675	−443	−660
77.5	−107	786	36	−137	−828	536	−359	427	−250	134	383	44	−51
82.5	26	583	161	−46	−285	527	109	377	98	375	−78	−50	136
87.5	−179	271	−41	2	−182	2	−89	137	−139	0	−59	−163	66

Age \ Year	1867.5	1872.5	1877.5	1882.5	1887.5	1892.5	1897.5	1902.5	1907.5	1912.5	1917.5	1922.5	1927.5
32.5	40	398	−148	−134	−317	107	153	7	−28	330	2 765	−743	−745
37.5	347	467	−37	−76	−149	222	371	140	−116	170	1 411	−910	−862
42.5	924	558	41	387	144	282	480	332	267	70	592	−929	−696
47.5	1 000	708	−22	368	588	331	560	529	305	308	341	−817	−414
52.5	1 125	434	18	339	145	464	469	629	316	518	350	−650	−556
57.5	1 042	224	−233	307	−13	576	357	272	216	431	302	−419	−456
62.5	1 216	116	−583	−161	−83	219	245	134	−25	210	−22	−665	−311
67.5	1 497	122	−287	−116	−623	355	139	−77	−378	8	−5	−972	−568
72.5	1 878	502	−198	51	−438	297	346	−183	−426	393	−146	−1 081	−656
77.5	1 493	390	239	309	−126	738	502	112	−68	246	235	−898	−322
82.5	516	198	−101	299	72	436	921	95	14	153	167	−946	−126
87.5	340	49	−11	−148	−69	47	71	−162	−264	−87	−297	−651	−455

Table 6 b. $d_x - n_x \mu_x$, males. Graduation by generations.

Age \ Year	1802.5	1807.5	1812.5	1817.5	1822.5	1827.5	1832.5	1837.5	1842.5	1847.5	1852.5	1857.5	1862.5
32.5	−1 087	1 171	261	−981	−1 280	46	−85	−63	−726	−586	203	−269	−1 274
37.5	−1 148	1 205	318	−352	−963	792	332	242	−292	221	702	−122	−810
42.5	−1 004	1 172	578	43	−232	1 227	606	761	−185	447	1 284	141	−594
47.5	−1 100	1 269	702	−317	−388	1 419	563	682	−453	398	1 084	341	−244
52.5	−722	1 652	662	−85	−605	1 399	1 011	513	−201	434	976	51	−243
57.5	−1 067	1 258	45	−397	−736	503	135	354	−585	3	450	−577	−539
62.5	−1 151	1 593	455	−282	−901	765	−8	72	−874	−653	−8	−700	−784
67.5	−1 162	793	330	−351	−1 293	584	−179	91	−257	656	121	−1 046	−1 503
72.5	62	1 084	437	272	−682	750	90	808	−388	406	730	−627	−1 151
77.5	141	1 091	352	221	−394	1 028	139	867	119	398	489	−57	−340
82.5	264	844	421	227	12	865	466	702	371	587	38	−41	37
87.5	−49	408	96	146	−39	153	79	296	3	116	20	−125	80

Age \ Year	1867.5	1872.5	1877.5	1882.5	1887.5	1892.5	1897.5	1902.5	1907.5	1912.5	1917.5	1922.5	1927.5
32.5	−26	235	−378	−388	−554	−78	54	10	101	603	3 208	−119	98
37.5	376	334	−286	−393	−487	−99	91	−51	−210	184	1 557	−606	−390
42.5	1 119	533	−163	71	−229	−106	108	2	24	−79	539	−857	−474
47.5	1 332	830	−115	110	240	−58	165	158	−20	70	194	−870	−345
52.5	1 432	628	34	179	−134	132	120	294	21	278	196	−713	−510
57.5	1 135	294	−234	192	−217	327	105	49	35	312	254	−374	−313
62.5	974	−106	−803	−406	−323	13	91	63	−17	296	161	−371	92
67.5	912	−421	−802	−638	−1 101	32	−14	−27	−166	362	469	−337	238
72.5	1 154	−195	−858	−637	−1 162	−317	13	−178	−101	965	621	−119	541
77.5	1 033	−134	−242	−222	−758	48	−78	−80	135	844	1 108	238	1 069
82.5	384	80	−198	181	−140	100	504	−218	16	495	845	21	1 095
87.5	358	108	109	3	45	97	33	−246	−310	28	−9	−134	250

Table 6 c. $d_x - n_x \mu_x$, females. Graduation by periods.

Age \ Year	1802.5	1807.7	1812.5	1817.5	1822.5	1827.5	1832.5	1837.5	1842.5	1847.5	1852.5	1857.5	1862.5
32.5	−546	1 227	401	−293	−852	114	197	−33	−691	−456	−1	−35	−865
37.5	−317	1 517	609	426	−531	501	496	191	−311	−85	498	293	−549
42.5	−49	1 684	537	192	−464	500	609	449	−304	−49	528	52	−557
47.5	−793	1 252	131	−475	−1 183	162	273	−165	−903	−787	−116	−224	−1 028
52.5	−270	1 950	337	−198	−1 258	325	274	22	−1 086	−350	6	−403	−935
57.5	−494	1 783	267	−513	−1 410	−182	−151	−20	−1 122	−687	−54	−292	−921
62.5	−240	2 769	836	−38	−986	528	−201	−238	−1 338	−790	236	72	−573
67.5	−570	1 696	967	−24	−1 652	544	71	−99	−770	496	869	−245	−757
72.5	291	1 821	934	418	−1 461	619	87	921	−500	415	1 184	99	−639
77.5	297	1 945	263	−215	−1 280	819	−429	365	−199	102	534	62	−213
82.5	296	718	182	−24	−799	285	−122	145	−239	123	−281	−520	−384
87.5	−283	192	−104	−100	−636	−142	−309	−105	−580	−155	−420	−652	−365

Age \ Year	1867.5	1872.5	1877.5	1882.5	1887.5	1892.5	1897.5	1902.5	1907.5	1912.5	1917.5	1922.5	1927.5
32.5	−84	189	16	69	101	397	287	479	223	251	2 189	−785	−826
37.5	390	494	185	273	366	552	550	484	400	224	1 374	−689	−927
42.5	668	410	−78	143	252	436	412	401	215	379	894	−493	−746
47.5	217	−411	−456	−269	−108	38	−22	167	111	−22	376	−460	−340
52.5	139	−183	−640	−318	−402	−13	164	52	−101	−35	456	−311	78
57.5	471	−309	−747	−482	−281	−105	−260	−13	−210	−262	260	−390	−182
62.5	777	−41	−413	−125	−188	294	−187	−107	−349	−68	103	−408	−24
67.5	1 367	330	−440	71	68	920	176	−145	−121	342	281	−132	526
72.5	1 403	499	−258	60	237	1 115	827	307	151	697	1 148	−9	768
77.5	1 197	83	−583	106	−154	1 285	749	965	440	1 248	681	643	1 556
82.5	925	−405	−769	−344	−347	314	459	157	316	803	192	−79	698
87.5	252	−175	−672	−730	−599	−325	−207	−613	−744	−269	−912	−906	−465

Table 6 d. $d_x - n_x \mu_x$, females. Graduation by generations.

Age \ Year	1802.5	1807.5	1812.5	1817.5	1822.5	1827.5	1832.5	1837.5	1842.5	1847.5	1852.5	1857.5	1862.5
32.5	−839	943	127	−552	−1 090	−74	99	−19	−558	−202	351	339	−552
37.5	−624	1 234	335	151	−783	273	334	137	−228	139	852	739	−110
42.5	−334	1 412	290	−53	−704	283	428	343	−277	134	861	506	−42
47.5	−994	1 061	−48	−641	−1 345	2	140	−264	−915	−660	168	194	−528
52.5	−327	1 900	295	−237	−1 288	297	249	6	−1 072	−266	210	−69	−513
57.5	−342	1 953	447	−330	−1 226	−11	17	127	−992	−576	63	−108	−649
62.5	182	3 215	1 317	458	−490	1 008	234	173	−975	−535	353	85	−561
67.5	174	2 417	1 711	800	−800	1 375	860	579	−168	951	1 099	−285	−1 020
72.5	1 329	2 816	1 857	1 421	−324	1 774	1 178	1 901	335	1 105	1 601	174	−944
77.5	1 336	2 985	1 221	731	−202	2 012	741	1 433	758	875	1 095	303	−315
82.5	1 114	1 563	940	711	−24	1 152	803	1 005	545	808	224	−190	−284
87.5	128	628	294	291	−217	282	143	374	−117	256	−88	−413	−205

Age \ Year	1867.5	1872.5	1877.5	1882.5	1887.5	1892.5	1897.5	1902.5	1907.5	1912.5	1917.5	1922.5	1927.5
32.5	142	310	52	42	33	308	198	402	156	203	2 166	−775	−780
37.5	708	682	246	236	261	402	371	307	236	59	1 221	−825	−1 039
42.5	1 130	712	73	155	164	279	206	162	−21	157	663	−716	−962
47.5	724	19	−185	−149	−121	−70	−196	−54	−143	−273	139	−707	−586
52.5	562	202	−301	−95	−294	−14	85	−83	−277	−237	255	−494	−117
57.5	761	−111	−608	−312	−114	27	−188	11	−225	−299	210	−432	−219
62.5	853	1	−572	−375	−262	403	23	123	−116	164	335	−145	229
67.5	1 103	170	−672	−522	−680	540	238	217	378	926	906	560	1 270
72.5	854	52	−526	−329	−733	−135	121	350	736	1 580	2 192	1 178	2 043
77.5	769	−453	−933	−2	−437	215	−820	49	496	2 081	1 920	2 171	3 248
82.5	836	−631	−962	−311	−81	393	−328	−1 273	−566	912	1 089	1 324	2 346
87.5	335	−116	−593	−564	−260	177	169	−895	−1 616	−833	−807	−181	584

Table 7 a. $f(i, 0, \varkappa) = \dfrac{1}{1 + e^{\varkappa i}}$.

i \ $\varkappa$	.52	.56	.60	.64	.68	.72	.76	.80	.84	.88	.92	.96	1.00
20	.00003	.00001	.00001	.00000	.00000	.00000	.00000	.00000	.00000	.00000	.00000	.00000	.00000
19	5	2	1	.00001	0	0	0	0	0	0	0	0	0
18	9	4	2	1	0	0	0	0	0	0	0	0	0
17	.00014	7	4	2	.00001	0	0	0	0	0	0	0	0
16	24	.00013	7	4	2	.00001	.00001	0	0	0	0	0	0
15	41	22	.00012	7	4	2	1	.00001	0	0	0	0	0
14	69	39	22	.00013	7	4	2	1	.00001	0	0	0	0
13	.00116	69	41	24	.00014	9	5	3	2	.00001	.00001	0	0
12	195	.00121	75	46	29	.00018	.00011	7	4	3	2	.00001	.00001
11	327	211	.00136	88	56	36	23	.00015	.00010	6	4	3	2
10	549	368	247	.00166	.00111	75	50	34	22	.00015	.00010	7	5
9	919	643	450	314	219	.00153	.00107	75	52	36	25	.00018	.00012
8	.01537	.01121	816	594	432	314	228	.00166	.00121	88	64	46	34
7	2558	1946	.01477	.01121	849	643	487	368	279	.00211	.00159	.00121	91
6	4229	3357	2660	2104	.01663	.01313	.01035	816	643	507	399	314	.00247
5	6914	5732	4743	3917	3230	2660	2188	.01799	.01477	.01213	995	816	669
4	.11106	9622	8317	7176	6180	5315	4565	3917	3357	2875	.02460	.02104	.01799
3	17365	.15710	.14185	.12786	.11507	.10340	9279	8317	7447	6661	5952	5315	4743

2	26115	24601	23148	21755	20424	19155	.17946	.16798	.15710	.14679	.13705	.12786	.11920
1	37285	36355	35434	34525	33626	32739	31865	31003	30153	29318	28496	27688	26894
0	50000	50000	50000	50000	50000	50000	50000	50000	50000	50000	50000	50000	50000
—1	62715	63645	64566	65475	66374	67261	68135	68997	69847	70682	71504	72312	73106
—2	73885	75399	76852	78245	79576	80845	82054	83202	84290	85321	86295	87214	88080
—3	82635	84290	85815	87214	88493	89660	90721	91683	92553	93339	94048	94685	95257
—4	88894	90378	91683	92824	93820	94685	95435	96083	96643	97125	97540	97896	98201
—5	93086	94268	95257	96083	96770	97340	97812	98201	98523	98787	99005	99184	99331
—6	95771	96643	97340	97896	98337	98687	98965	99184	99357	99493	99601	99686	99753
—7	97442	98054	98523	98879	99151	99357	99513	99632	99721	99789	99841	99879	99909
—8	98463	98879	99184	99406	99568	99686	99772	99834	99879	99912	99936	99954	99966
—9	99081	99357	99550	99686	99781	99847	99893	99925	99948	99964	99975	99982	99988
—10	99451	99632	99753	99834	99889	99925	99950	99966	99977	99985	99990	99993	99995
—11	99673	99789	99864	99912	99944	99964	99977	99985	99990	99994	99996	99997	99998
—12	99805	99879	99925	99954	99971	99982	99989	99993	99996	99997	99998	99999	99999
—13	99884	99931	99959	99976	99986	99991	99995	99997	99998	99999	99999	1.0	1.0
—14	99931	99961	99978	99987	99993	99996	99998	99999	99999	1.0	1.0	1.0	1.0
—15	99959	99978	99988	99993	99996	99998	99999	99999	1.0	1.0	1.0	1.0	1.0
—16	99976	99987	99993	99996	99998	99999	99999	1.0	1.0	1.0	1.0	1.0	1.0
—17	99986	99993	99996	99998	99999	1.0	1 0	1.0	1.0	1.0	1.0	1.0	1.0
—18	99991	99996	99998	99999	1.0	1.0	1.0	1.0	1.0	1.0	1.0	1.0	1.0
—19	99995	99998	99999	99999	1.0	1.0	1.0	1.0	1.0	1.0	1.0	1.0	1.0
—20	99997	99999	99999	1.0	1.0	1.0	1.0	1.0	1.0	1.0	1.0	1.0	1.0

Table 7 b. $f(i, 0.2, \varkappa) = \frac{1}{1 + e^{\varkappa(i-0.2)}}$.

i \ $\varkappa$	.52	.56	.60	.64	.68	.72	.76	.80	.84	.88	.92	.96	1.00
20	.00003	.00002	.00001	.00000	.00000	.00000	.00000	.00000	.00000	.00000	.00000	.00000	.00000
19	6	3	1	.00001	0	0	0	0	0	0	0	0	0
18	.00010	5	2	1	.00001	0	0	0	0	0	0	0	0
17	16	8	4	2	1	.00001	0	0	0	0	0	0	0
16	27	.00014	8	4	2	1	.00001	0	0	0	0	0	0
15	45	25	.00014	8	4	2	1	.00001	0	0	0	0	0
14	76	44	25	.00015	8	5	3	2	.00001	.00001	0	0	0
13	.00128	77	46	28	.00017	.00010	6	4	2	1	.00001	0	0
12	216	.00135	84	52	33	20	.00013	8	5	3	2	.00001	.00001
11	363	236	.00153	99	65	42	27	.00018	.00011	7	5	3	2
10	608	412	279	.00188	.00127	86	58	39	27	.00018	.00012	8	6
9	.01019	719	507	357	251	.00177	.00124	88	62	43	30	.00021	.00015
8	1702	.01252	919	675	495	363	266	.00195	.00143	.00104	76	56	41
7	2831	2171	.01663	.01272	972	742	566	432	330	251	.00192	.00146	.00111
6	4671	3740	2989	2385	.01900	.01513	.01203	957	760	604	479	380	302
5	7614	6369	5315	4428	3683	3059	2538	.02104	.01743	.01443	.01194	987	816
4	.12175	.10641	9279	8076	7018	6088	5275	4565	3947	3409	2943	.02538	.02188
3	18908	17250	.15710	.14283	.12966	.11753	.10641	9622	8691	7842	7070	6369	5732

2	28171	26737	25351	24012	22723	21484	20294	.19155	.18064	.17023	.16030	.15084	.14185
1	39747	38984	38225	37472	36726	35985	35252	34525	33805	33093	32388	31691	31003
0	52598	52797	52996	53196	53395	53594	53793	53991	54190	54389	54587	54785	54983
−1	65113	66195	67261	68309	69339	70350	71341	72312	73263	74193	75101	75988	76852
−2	75841	77417	78918	80345	81698	82977	84184	85321	86389	87391	88329	89206	90025
−3	84077	85717	87214	88575	89807	90921	91924	92824	93631	94353	94998	95572	96083
−4	89880	91309	92553	93631	94563	95365	96053	96643	97147	97578	97945	98257	98523
−5	93726	94844	95771	96538	97169	97689	98115	98463	98748	98981	99171	99325	99451
−6	96173	96988	97634	98144	98546	98861	99109	99304	99456	99575	99668	99741	99797
−7	97689	98257	98688	99013	99258	99443	99581	99686	99764	99823	99867	99901	99925
−8	98613	98997	99275	99477	99623	99728	99804	99859	99898	99927	99947	99962	99973
−9	99171	99425	99601	99724	99808	99867	99908	99936	99956	99970	99979	99985	99990
−10	99505	99670	99781	99854	99903	99935	99957	99971	99981	99987	99992	99994	99996
−11	99705	99812	99879	99923	99951	99969	99980	99987	99992	99995	99997	99998	99999
−12	99825	99892	99934	99959	99975	99985	99991	99994	99996	99998	99999	99999	1.0
−13	99896	99938	99964	99979	99987	99993	99996	99997	99998	99999	99999	1.0	1.0
−14	99938	99965	99980	99989	99994	99996	99998	99999	99999	1.0	1.0	1.0	1.0
−15	99963	99980	99989	99994	99997	99998	99999	99999	1.0	1.0	1.0	1.0	1.0
−16	99978	99989	99994	99997	99998	99999	1.0	1.0	1.0	1.0	1.0	1.0	1.0
−17	99987	99993	99997	99998	99999	1.0	1.0	1.0	1.0	1.0	1.0	1.0	1.0
−18	99992	99996	99998	99999	1.0	1.0	1.0	1.0	1.0	1.0	1.0	1.0	1.0
−19	99995	99998	99999	1.0	1.0	1.0	1.0	1.0	1.0	1.0	1.0	1.0	1.0
−20	99997	99999	99999	1.0	1.0	1.0	1.0	1.0	1.0	1.0	1.0	1.0	1.0

Table 7 c. $f(i, 0.4, \varkappa) = \dfrac{1}{1 + e^{\varkappa(i-0.4)}}$.

i \ $\varkappa$	.52	.56	.60	.64	.68	.72	.76	.80	.84	.88	.92	.96	1.00
20	.00004	.00002	.00001	.00000	.00000	.00000	.00000	.00000	.00000	.00000	.00000	.00000	.00000
19	6	3	1	.00001	0	0	0	0	0	0	0	0	0
18	.00011	5	3	1	.00001	0	0	0	0	0	0	0	0
17	18	9	5	2	1	.00001	0	0	0	0	0	0	0
16	30	.00016	9	5	2	1	.00001	0	0	0	0	0	0
15	50	28	.00016	9	5	3	2	.00001	0	0	0	0	0
14	85	49	29	.00017	.00010	6	3	2	.00001	.00001	0	0	0
13	.00143	86	52	31	19	.00011	7	4	3	2	.00001	.00001	0
12	239	.00151	95	60	38	24	.00015	9	6	4	2	1	.00001
11	402	264	.00173	.00113	74	48	32	.00021	.00014	9	6	4	2
10	675	460	314	214	.00146	99	68	46	31	.00021	.00015	.00010	7
9	.01130	803	571	405	288	.00204	.00145	.00103	73	52	37	26	.00018
8	1885	.01398	.01035	766	566	419	309	228	.00169	.00124	92	68	50
7	3131	2422	1871	.01443	.01112	856	659	507	390	299	.00230	.00177	.00136
6	5156	4165	3357	2701	2171	.01743	.01398	.01121	898	719	575	460	368
5	8378	7070	5952	5002	4197	3516	2943	2460	.02055	.01716	.01432	.01194	995
4	.13331	.11753	.10340	9079	7958	6966	6088	5315	4635	4039	3516	3059	.02660
3	20554	18908	17365	.15923	.14579	.13331	.12175	.11106	.10120	9212	8378	7614	6914

2	30322	28987	27688	26425	25200	24012	22864	21755	20685	.19655	.18664	.17712	.16798
1	42263	41678	41096	40516	39939	39365	38794	38225	37660	37098	36540	35985	35434
0	55181	55577	55971	56365	56758	57151	57542	57932	58322	58710	59098	59484	59869
−1	67437	68654	69847	71013	72152	73263	74345	75399	76423	77417	78381	79315	80218
−2	77695	79315	80845	82288	83644	84916	86105	87214	88247	89206	90097	90921	91683
−3	85421	87034	88493	89807	90987	92042	92982	93820	94563	95221	95803	96317	96770
−4	90788	92158	93339	94353	95221	95961	96591	97125	97578	97961	98284	98557	98787
−5	94311	95365	96231	96941	97521	97993	98376	98687	98940	99144	99309	99443	99550
−6	96538	97299	97896	98363	98728	99013	99234	99406	99540	99643	99724	99786	99834
−7	97912	98439	98834	99130	99352	99517	99640	99732	99801	99852	99890	99918	99939
−8	98748	99102	99357	99540	99670	99764	99831	99879	99914	99938	99956	99969	99978
−9	99252	99485	99646	99757	99833	99885	99921	99946	99963	99974	99982	99988	99992
−10	99554	99705	99805	99872	99915	99944	99963	99976	99984	99989	99993	99995	99997
−11	99734	99831	99893	99932	99957	99973	99983	99989	99993	99996	99997	99998	99999
−12	99842	99904	99941	99964	99978	99987	99992	99995	99997	99998	99999	99999	1.0
−13	99906	99945	99968	99981	99989	99994	99996	99998	99999	99999	1.0	1.0	1.0
−14	99944	99969	99982	99990	99994	99997	99998	99999	99999	1.0	1.0	1.0	1.0
−15	99967	99982	99990	99995	99997	99998	99999	1.0	1.0	1.0	1.0	1.0	1.0
−16	99980	99990	99995	99997	99999	99999	1.0	1.0	1.0	1.0	1.0	1.0	1.0
−17	99988	99994	99997	99999	99999	1.0	1.0	1.0	1.0	1.0	1.0	1.0	1.0
−18	99993	99997	99998	99999	1.0	1.0	1.0	1.0	1.0	1.0	1.0	1.0	1.0
−19	99996	99998	99999	1.0	1.0	1.0	1.0	1.0	1.0	1.0	1.0	1.0	1.0
−20	99998	99999	99999	1.0	1.0	1.0	1,0	1.0	1.0	1.0	1.0	1.0	1,0

Table 8 a. $\sum_{\nu=0}^{i} \frac{1}{1+e^{x\cdot\nu}}, (i \geq 0); \quad \sum_{\nu=-1}^{i} \frac{1}{1+e^{x\cdot\nu}}, (i < 0).$

i \ x	.52	.56	.60	.64	.68	.72	.76	.80	.84	.88	.92	.96	1.00
20	1.5938	1.4994	1.4178	1.3464	1.2836	1.2278	1.1779	1.1332	1.0928	1.0561	1.0227	.9922	.9642
19	1.5938	1.4994	1.4178	1.3464	1.2836	1.2278	1.1779	1.1332	1.0928	1.0561	1.0227	.9922	.9642
18	1.5937	1.4994	1.4178	1.3464	1.2836	1.2278	1.1779	1.1332	1.0928	1.0561	1.0227	.9922	.9642
17	1.5936	1.4994	1.4177	1.3464	1.2835	1.2278	1.1779	1.1332	1.0928	1.0561	1.0227	.9922	.9642
16	1.5935	1.4993	1.4177	1.3464	1.2835	1.2278	1.1779	1.1332	1.0928	1.0561	1.0227	.9922	.9642
15	1.5932	1.4992	1.4176	1.3463	1.2835	1.2277	1.1779	1.1332	1.0928	1.0561	1.0227	.9922	.9642
14	1.5928	1.4989	1.4175	1.3463	1.2835	1.2277	1.1779	1.1332	1.0928	1.0561	1.0227	.9922	.9642
13	1.5921	1.4985	1.4173	1.3462	1.2834	1.2277	1.1779	1.1332	1.0928	1.0561	1.0227	.9922	.9642
12	1.5910	1.4979	1.4169	1.3459	1.2833	1.2276	1.1778	1.1331	1.0928	1.0561	1.0227	.9922	.9642
11	1.5890	1.4966	1.4161	1.3454	1.2830	1.2274	1.1777	1.1331	1.0927	1.0561	1.0227	.9922	.9642
10	1.5858	1.4945	1.4148	1.3446	1.2824	1.2271	1.1775	1.1329	1.0926	1.0560	1.0227	.9921	.9641
9	1.5803	1.4909	1.4123	1.3429	1.2813	1.2263	1.1770	1.1326	1.0924	1.0559	1.0226	.9921	.9641
8	1.5711	1.4844	1.4078	1.3398	1.2791	1.2248	1.1759	1.1318	1.0919	1.0555	1.0223	.9919	.9640
7	1.5557	1.4732	1.3996	1.3338	1.2748	1.2216	1.1737	1.1302	1.0907	1.0546	1.0217	.9914	.9636
6	1.5301	1.4538	1.3849	1.3226	1.2663	1.2152	1.1688	1.1265	1.0879	1.0525	1.0201	.9902	.9627
5	1.4878	1.4202	1.3583	1.3016	1.2497	1.2021	1.1584	1.1183	1.0814	1.0475	1.0161	.9871	.9602
4	1.4187	1.3629	1.3108	1.2624	1.2174	1.1755	1.1366	1.1003	1.0667	1.0353	1.0061	.9789	.9536
3	1.3076	1.2667	1.2277	1.1907	1.1556	1.1223	1.0909	1.0612	1.0331	1.0066	.9815	.9579	.9356

2	1.1340	1.1096	1.0858	1.0628	1.0405	1.0189	.9981	.9780	.9586	.9400	.9220	.9047	.8881
1	.8729	.8635	.8543	.8452	.8363	.8274	.8186	.8100	.8015	.7932	.7850	.7769	.7689
0	.5000	.5000	.5000	.5000	.5000	.5000	.5000	.5000	.5000	.5000	.5000	.5000	.5000
−1	.6271	.6365	.6457	.6548	.6637	.6726	.6814	.6900	.6985	.7068	.7150	.7231	.7311
−2	1.3660	1.3904	1.4142	1.4372	1.4595	1.4811	1.5019	1.5220	1.5414	1.5600	1.5780	1.5953	1.6119
−3	2.1924	2.2333	2.2723	2.3093	2.3444	2.3777	2.4091	2.4388	2.4669	2.4934	2.5185	2.5421	2.5644
−4	3.0813	3.1371	3.1892	3.2376	3.2826	3.3245	3.3634	3.3997	3.4333	3.4647	3.4939	3.5211	3.5464
−5	4.0122	4.0798	4.1417	4.1984	4.2503	4.2979	4.3416	4.3817	4.4186	4.4525	4.4839	4.5129	4.5398
−6	4.9699	5.0462	5.1151	5.1774	5.2337	5.2848	5.3312	5.3735	5.4121	5.4475	5.4799	5.5098	5.5373
−7	5.9443	6.0268	6.1004	6.1662	6.2252	6.2784	6.3263	6.3698	6.4093	6.4454	6.4783	6.5086	6.5364
−8	6.9289	7.0156	7.0922	7.1602	7.2209	7.2752	7.3241	7.3682	7.4081	7.4445	7.4777	7.5081	7.5360
−9	7.9197	8.0091	8.0877	8.1571	8.2187	8.2737	8.3230	8.3674	8.4076	8.4441	8.4774	8.5079	8.5359
−10	8.9142	9.0055	9.0852	9.1554	9.2176	9.2729	9.3225	9.3671	9.4074	9.4440	9.4773	9.5079	9.5359
−11	9.9110	10.0034	10.0839	10.1546	10.2170	10.2726	10.3223	10.3669	10.4073	10.4439	10.4773	10.5078	10.5358
−12	10.9090	11.0021	11.0831	11.1541	11.2167	11.2724	11.3221	11.3669	11.4072	11.4439	11.4773	11.5078	11.5358
−13	11.9079	12.0015	12.0827	12.1538	12.2166	12.2723	12.3221	12.3668	12.4072	12.4439	12.4773	12.5078	12.5358
−14	12.9072	13.0011	13.0825	13.1537	13.2165	13.2723	13.3221	13.3668	13.4072	13.4439	13.4773	13.5078	13.5358
−15	13.9068	14.0008	14.0824	14.1537	14.2165	14.2723	14.3221	14.3668	14.4072	14.4439	14.4773	14.5078	14.5358
−16	14.9065	15.0007	15.0823	15.1536	15.2165	15.2722	15.3221	15.3668	15.4072	15.4439	15.4773	15.5078	15.5358
−17	15.9064	16.0006	16.0823	16.1536	16.2165	16.2722	16.3221	16.3668	16.4072	16.4439	16.4773	16.5078	16.5358
−18	16.9063	17.0006	17.0822	17.1536	17.2164	17.2722	17.3221	17.3668	17.4072	17.4439	17.4773	17.5078	17.5358
−19	17.9062	18.0006	18.0822	18.1536	18.2164	18.2722	18.3221	18.3668	18.4072	18.4439	18.4773	18.5078	18.5358
−20	18.9062	19.0006	19.0822	19.1536	19.2164	19.2722	19.3221	19.3668	19.4072	19.4439	19.4773	19.5078	19.5358

Table 8 b. $\sum_{\nu=0}^{i} \frac{1}{1+e^{x(\nu-0.2)}}$, $(i \geq 0)$; $\sum_{\nu=-1}^{i} \frac{1}{1+e^{x(\nu-0.2)}}$, $(i < 0)$.

i \ x	.52	.56	.60	.64	.68	.72	.76	.80	.84	.88	.92	.96	1.00
20	1.7093	1.6162	1.5357	1.4655	1.4039	1.3493	1.3006	1.2570	1.2178	1.1823	1.1501	1.1207	1.0939
19	1.7093	1.6162	1.5357	1.4655	1.4039	1.3493	1.3006	1.2570	1.2178	1.1823	1.1501	1.1207	1.0939
18	1.7093	1.6161	1.5357	1.4655	1.4039	1.3493	1.3006	1.2570	1.2178	1.1823	1.1501	1.1207	1.0939
17	1.7092	1.6161	1.5357	1.4655	1.4038	1.3493	1.3006	1.2570	1.2178	1.1823	1.1501	1.1207	1.0939
16	1.7090	1.6160	1.5356	1.4655	1.4038	1.3492	1.3006	1.2570	1.2178	1.1823	1.1501	1.1207	1.0939
15	1.7087	1.6159	1.5355	1.4655	1.4038	1.3492	1.3006	1.2570	1.2178	1.1823	1.1501	1.1207	1.0939
14	1.7083	1.6156	1.5354	1.4654	1.4038	1.3492	1.3006	1.2570	1.2178	1 1823	1.1501	1.1207	1.0939
13	1.7075	1.6152	1.5352	1.4652	1.4037	1.3492	1.3006	1.2570	1.2178	1.1823	1.1501	1.1207	1.0939
12	1.7062	1.6144	1.5347	1.4650	1.4035	1.3491	1.3005	1.2570	1.2178	1.1823	1.1501	1.1207	1.0939
11	1.7041	1.6131	1.5339	1.4644	1.4032	1.3489	1.3004	1.2569	1.2177	1.1823	1.1501	1.1207	1.0938
10	1.7004	1.6107	1.5323	1.4634	1.4026	1.3484	1.3001	1.2567	1.2176	1.1822	1.1500	1.1207	1.0938
9	1.6944	1.6066	1.5295	1.4616	1.4013	1.3476	1.2995	1.2563	1.2173	1.1820	1.1499	1.1206	1.0938
8	1.6842	1.5994	1.5245	1.4580	1.3988	1.3458	1.2983	1.2554	1.2167	1.1816	1.1496	1.1204	1.0936
7	1.6671	1.5869	1.5153	1.4512	1.3938	1.3422	1.2956	1.2535	1.2153	1.1805	1.1488	1.1198	1.0932
6	1.6388	1.5652	1.4986	1.4385	1.3841	1.3348	1.2900	1.2492	1.2120	1.1780	1.1469	1.1184	1.0921
5	1.5921	1.5278	1.4688	1.4147	1.3651	1.3196	1.2779	1.2396	1.2044	1.1720	1.1421	1.1145	1.0891
4	1.5160	1.4641	1.4156	1.3704	1.3283	1.2890	1.2525	1.2186	1.1870	1.1576	1.1302	1.1047	1.0809
3	1.3942	1.3577	1.3228	1.2896	1.2581	1.2282	1.1998	1.1729	1.1475	1.1235	1.1007	1.0793	1.0590

2	1.2052	1.1852	1.1657	1.1468	1.1284	1.1106	1.0934	1.0767	1.0606	1.0450	1.0300	1.0156	1.0017
1	.9235	.9178	.9122	.9067	.9012	.8958	.8904	.8852	.8800	.8748	.8698	.8648	.8599
0	.5260	.5280	.5300	.5320	.5339	.5359	.5379	.5399	.5419	.5439	.5459	.5479	.5498
—1	.6511	.6620	.6726	.6831	.6934	.7035	.7134	.7231	.7326	.7419	.7510	.7599	.7685
—2	1.4095	1.4361	1.4618	1.4865	1.5104	1.5333	1.5553	1.5763	1.5965	1.6158	1.6343	1.6519	1.6688
—3	2.2503	2.2933	2.3339	2.3723	2.4084	2.4425	2.4745	2.5046	2.5328	2.5594	2.5843	2.6077	2.6296
—4	3.1491	3.2064	3.2595	3.3086	3.3541	3.3961	3.4350	3.4710	3.5043	3.5352	3.5637	3.5902	3.6148
—5	4.0864	4.1548	4.2172	4.2740	4.3258	4.3730	4.4162	4.4556	4.4918	4.5250	4.5554	4.5835	4.6093
—6	5.0481	5.1247	5.1935	5.2554	5.3112	5.3616	5.4073	5.4487	5.4863	5.5207	5.5521	5.5809	5.6078
—7	6.0250	6.1073	6.1804	6.2455	6.3038	6.3561	6.4031	6.4455	6.4840	6.5189	6.5508	6.5799	6.6066
—8	7.0111	7.0972	7.1731	7.2403	7.3000	7.3533	7.4011	7.4441	7.4830	7.5182	7.5503	7.5795	7.6063
—9	8.0028	8.0915	8.1691	8.2375	8.2981	8.3520	8.4002	8.4435	8.4825	8.5179	8.5500	8.5794	8.6062
—10	8.9979	9.0882	9.1670	9.2361	9.2971	9.3514	9.3998	9.4432	9.4823	9.5178	9.5500	9.5793	9.6062
—11	9.9949	10.0863	10.1657	10.2353	10.2966	10.3510	10.3996	10.4431	10.4828	10.5177	10.5499	10.5793	10.6062
—12	10.9932	11.0852	11.1651	11.2349	11.2964	11.3509	11.3995	11.4430	11.4822	11.5177	11.5499	11.5793	11.6061
—13	11.9921	12.0846	12.1647	12.2347	12.2963	12.3508	12.3994	12.4430	12.4822	12.5177	12.5499	12.5793	12.6061
—14	12.9915	13.0843	13.1645	13.2346	13.2962	13.3508	13.3994	13.4430	13.4822	13.5177	13.5499	13.5793	13.6061
—15	13.9911	14.0841	14.1644	14.2345	14.2962	14.3508	14.3994	14.4430	14.4822	14.5177	14.5499	14.5793	14.6061
—16	14.9909	15.0839	15.1644	15.2345	15.2962	15.3508	15.3994	15.4430	15.4822	15.5177	15.5499	15.5793	15.6061
—17	15.9908	16.0839	16.1643	16.2345	16.2962	16.3507	16.3994	16.4430	16.4822	16.5177	16.5499	16.5793	16.6061
—18	16.9907	17.0838	17.1643	17.2345	17.2961	17.3507	17.3994	17.4430	17.4822	17.5177	17.5499	17.5793	17.6061
—19	17.9907	18.0838	18.1643	18.2345	18.2961	18.3507	18.3994	18.4430	18.4822	18.5177	18.5499	18.5793	18.6061
—20	18.9906	19.0838	19.1643	19.2345	19.2961	19.3507	19.3994	19.4430	19.4822	19.5177	19.5499	19.5793	19.6061

Table 8 c. $\sum_{\nu=0}^{i} \frac{1}{1 + e^{\varkappa(i-0.4)}}, \ (i \geq 0); \quad \sum_{\nu=-1}^{i} \frac{1}{1 + e^{\varkappa(i-0.4)}}, \ (i < 0).$

i \ $\varkappa$	.52	.56	.60	.64	.68	.72	.76	.80	.84	.88	.92	.96	1.00
20	1.8299	1.7383	1.6594	1.5908	1.5306	1.4776	1.4304	1.3884	1.3506	1.3166	1.2859	1.2579	1.2325
19	1.8299	1.7383	1.6594	1.5908	1.5306	1.4776	1.4304	1.3884	1.3506	1.3166	1.2859	1.2579	1.2325
18	1.8298	1.7383	1.6594	1.5908	1.5306	1.4776	1.4304	1.3884	1.3506	1.3166	1.2859	1.2579	1.2325
17	1.8297	1.7382	1.6594	1.5908	1.5306	1.4776	1.4304	1.3884	1.3506	1.3166	1.2859	1.2579	1.2325
16	1.8296	1.7382	1.6593	1.5907	1.5306	1.4775	1.4304	1.3884	1.3506	1.3166	1.2859	1.2579	1.2325
15	1.8293	1.7380	1.6592	1.5907	1.5306	1.4775	1.4304	1.3883	1.3506	1.3166	1.2859	1.2579	1.2325
14	1.8288	1.7377	1.6591	1.5906	1.5305	1.4775	1.4304	1.3883	1.3506	1.3166	1.2859	1.2579	1.2325
13	1.8279	1.7372	1.6588	1.5904	1.5304	1.4775	1.4304	1.3883	1.3506	1,3166	1.2859	1.2579	1.2325
12	1.8265	1.7364	1.6583	1.5901	1.5303	1.4773	1.4303	1.3883	1.3506	1.3166	1.2858	1.2579	1.2325
11	1.8241	1.7349	1.6573	1.5895	1.5299	1.4771	1.4302	1.3882	1.3505	1.3165	1.2858	1.2579	1.2325
10	1.8201	1.7322	1.6556	1.5884	1.5291	1.4766	1.4298	1.3880	1.3504	1.3165	1.2858	1.2579	1.2325
9	1.8133	1.7276	1.6525	1.5863	1.5277	1.4756	1.4292	1.3875	1.3501	1.3162	1.2856	1.2578	1.2324
8	1.8020	1.7196	1.6468	1.5822	1.5248	1.4736	1.4277	1.3865	1.3493	1.3157	1.2853	1.2575	1.2322
7	1.7832	1.7056	1.6364	1.5745	1.5191	1.4694	1.4246	1.3842	1.3476	1.3145	1.2843	1.2568	1.2317
6	1.7519	1.6814	1.6177	1.5601	1.5080	1.4608	1.4180	1.3791	1.3438	1.3115	1.2820	1.2551	1.2304
5	1.7003	1.6397	1.5841	1.5331	1.4863	1.4434	1.4040	1.3679	1.3348	1.3043	1.2763	1.2505	1.2267
4	1.6165	1.5690	1.5246	1.4831	1.4443	1.4082	1.3746	1 3433	1.3142	1.2871	1.2620	1.2385	1.2167
3	1.4832	1.4515	1.4212	1.3923	1.3648	1.3386	1.3137	1.2902	1.2679	1.2468	1.2268	1.2079	1.1902

2	1.2777	1.2624	1.2476	1.2331	1.2190	1.2053	1.1920	1.1791	1.1667	1.1546	1.1430	1.1318	1.1210
1	.9744	.9725	.9707	.9688	.9670	.9652	.9634	.9616	.9598	.9581	.9564	.9547	.9530
0	.5518	.5558	.5597	.5637	.5676	.5715	.5754	.5793	.5832	.5871	.5910	.5948	.5987
−1	.6744	.6865	.6985	.7101	.7215	.7326	.7435	.7540	.7642	.7742	.7838	.7931	.8022
−2	1.4513	1.4797	1.5069	1.5330	1.5580	1.5818	1.6045	1.6261	1.6467	1.6662	1.6848	1.7024	1.7190
−3	2.3055	2.3500	2.3919	2 4311	2.4678	2.5022	2.5343	2.5643	2.5923	2.6184	2.6428	2.6655	2.6867
−4	3.2134	3.2716	3.3252	3.3746	3.4200	3.4618	3.5002	3.5356	3.5681	3.5981	3.6256	3.6511	3.6746
−5	4.1565	4.2253	4.2876	4.3440	4.3952	4.4417	4.4840	4.5225	4.5575	4.5895	4.6187	4.6455	4.6701
−6	5.1219	5.1982	5.2665	5.3277	5.3825	5.4319	5.4763	5.5165	5.5529	5.5859	5.6160	5.6434	5.6684
−7	6.1010	6.1826	6.2549	6.3190	6.3760	6.4270	6.4727	6.5138	6.5509	6.5844	6.6149	6.6426	6.6678
−8	7.0885	7.1737	7.2484	7.3144	7.3727	7.4247	7.4711	7.5126	7.5500	7.5838	7.6144	7.6422	7.6676
−9	8.0810	8.1685	8.2449	8.3119	8.3711	8.4235	8.4703	8.5121	8.5497	8.5836	8.6143	8.6421	8.6675
−10	9.0766	9.1656	9.2429	9.3106	9.3702	9.4230	9.4699	9.5118	9.5495	9.5835	9.6142	9.6421	9.6675
−11	10.0739	10.1639	10.2419	10.3100	10.3698	10.4227	10.4697	10.5117	10.5494	10.5834	10.6142	10.6421	10.6675
−12	11.0723	11.1629	11.2413	11.3096	11.3696	11.4226	11.4696	11.5117	11.5494	11.5834	11.6141	11.6421	11.6675
−13	12.0714	12.1624	12.2410	12.3094	12.3695	12.4225	12.4696	12.5117	12.5494	12.5834	12.6141	12.6421	12.6675
−14	13.0708	13.1620	13.2408	13.3093	13.3694	13.4225	13.4696	13.5117	13.5494	13.5834	13.6141	13.6421	13.6675
−15	14.0705	14.1619	14.2407	14.3093	14.3694	14.4225	14.4696	14.5116	14.5494	14.5834	14.6141	14.6421	14.6675
−16	15.0703	15.1618	15.2406	15.3092	15.3694	15.4224	15.4696	15.5116	15.5494	15.5834	15.6141	15.6421	15.6675
−17	16.0702	16.1617	16.2406	16.3092	16.3694	16.4224	16.4696	16.5116	16.5494	16.5834	16.6141	16.6421	16.6675
−18	17.0701	17.1617	17.2406	17.3092	17.3694	17.4224	17.4696	17.5116	17.5494	17.5834	17.6141	17.6421	17.6675
−19	18.0701	18.1617	18.2406	18.3092	18.3694	18.4224	18.4696	18.5116	18.5494	18.5834	18.6141	18.6421	18.6675
−20	19.0700	19.1616	19.2406	19.3092	19.3694	19.4224	19.4696	19.5116	19.5494	19.5834	19.6141	19.6421	19.6675

Table 9 a. $\sum_{\nu=0}^{i}\left(\frac{1}{1+e^{\varkappa\cdot\nu}}\right)^2,\ i\geq 0;\quad \sum_{\nu=-1}^{i}\left(\frac{1}{1+e^{\varkappa\cdot\nu}}\right)^2,\ i<0.$

i \ $\varkappa$	.52	.56	.60	.64	.68	.72	.76	.80	.84	.88	.92	.96	1.00
20	.5073	.4816	.4595	.4402	.4233	.4083	.3951	.3832	.3725	.3629	.3543	.3464	.3392
19	.5073	.4816	.4595	.4402	.4233	.4083	.3951	.3832	.3725	.3629	.3543	.3464	.3392
18	.5073	.4816	.4595	.4402	.4233	.4083	.3951	.3832	.3725	.3629	.3543	.3464	.3392
17	.5073	.4816	.4595	.4402	.4233	.4083	.3951	.3832	.3725	.3629	.3543	.3464	.3392
16	.5073	.4816	.4595	.4402	.4233	.4083	.3951	.3832	.3725	.3629	.3543	.3464	.3392
15	.5073	.4816	.4595	.4402	.4233	.4083	.3951	.3832	.3725	.3629	.3543	.3464	.3392
14	.5073	.4816	.4595	.4402	.4233	.4083	.3951	.3832	.3725	.3629	.3543	.3464	.3392
13	.5073	.4816	.4595	.4402	.4233	.4083	.3951	.3832	.3725	.3629	.3543	.3464	.3392
12	.5073	.4816	.4595	.4402	.4233	.4083	.3951	.3832	.3725	.3629	.3543	.3464	.3392
11	.5073	.4816	.4594	.4402	.4233	.4083	.3951	.3832	.3725	.3629	.3543	.3464	.3392
10	.5073	.4816	.4594	.4402	.4233	.4083	.3951	.3832	.3725	.3629	.3543	.3464	.3392
9	.5072	.4816	.4594	.4402	.4233	.4083	.3951	.3832	.3725	.3629	.3543	.3464	.3392
8	.5072	.4815	.4594	.4402	.4233	.4083	.3951	.3832	.3725	.3629	.3543	.3464	.3392
7	.5069	.4814	.4594	.4401	.4232	.4083	.3950	.3832	.3725	.3629	.3542	.3464	.3392
6	.5063	.4810	.4591	.4400	.4232	.4083	.3950	.3832	.3725	.3629	.3542	.3464	.3392
5	.5045	.4799	.4584	.4396	.4229	.4081	.3949	.3831	.3725	.3629	.3542	.3463	.3392
4	.4997	.4766	.4562	.4380	.4218	.4074	.3944	.3828	.3723	.3628	.3541	.3463	.3391
3	.4874	.4674	.4493	.4329	.4180	.4046	.3924	.3813	.3711	.3619	.3535	.3458	.3388

2	.4572	.4427	.4291	.4165	.4048	.3939	.3837	.3743	.3656	.3575	.3500	.3430	.3365
1	.3890	.3822	.3756	.3692	.3631	.3572	.3515	.3461	.3409	.3360	.3312	.3267	.3223
0	.2500	.2500	.2500	.2500	.2500	.2500	.2500	.2500	.2500	.2500	.2500	.2500	.2500
−1	.3933	.4051	.4169	.4287	.4405	.4524	.4642	.4761	.4879	.4996	.5113	.5229	.5344
−2	.9392	.9736	1.0075	1.0409	1.0738	1.1060	1.1375	1.1683	1.1983	1.2276	1.2560	1.2835	1.3103
−3	1.6221	1.6841	1.7439	1.8016	1.8569	1.9099	1.9606	2.0089	2.0550	2.0988	2.1405	2.1801	2.2176
−4	2.4123	2.5009	2.5845	2.6632	2.7371	2.8064	2.8713	2.9321	2.9889	3.0421	3.0919	3.1384	3.1820
−5	3.2788	3.3895	3.4919	3.5864	3.6736	3.7539	3.8280	3.8964	3.9596	4.0180	4.0721	4.1222	4.1687
−6	4.1960	4.3235	4.4394	4.5447	4.6406	4.7278	4.8074	4.8802	4.9468	5.0079	5.0641	5.1159	5.1637
−7	5.1455	5.2850	5.4101	5.5225	5.6237	5.7150	5.7977	5.8728	5.9412	6.0037	6.0609	6.1135	6.1619
−8	6.1150	6.2627	6.3938	6.5106	6.6150	6.7088	6.7932	6.8695	6.9388	7.0019	7.0596	7.1126	7.1612
−9	7.0967	7.2499	7.3848	7.5043	7.6107	7.7057	7.7910	7.8680	7.9378	8.0012	8.0591	8.1122	8.1610
−10	8.0858	8.2425	8.3799	8.5010	8.6084	8.7042	8.7900	8.8674	8.9373	9.0009	9.0589	9.1121	9.1609
−11	9.0792	9.2383	9.3772	9.4993	9.6073	9.7035	9.7896	9.8671	9.9371	10.0008	10.0588	10.1120	10.1609
−12	10.0753	10.2359	10.3757	10.4984	10.6067	10.7031	10.7894	10.8669	10.9370	11.0007	11.0588	11.1120	11.1608
−13	11.0730	11.2345	11.3749	11.4979	11.6064	11.7030	11.7892	11.8669	11.9370	12.0007	12.0588	12.1120	12.1608
−14	12.0716	12.2337	12.3744	12.4976	12.6063	12.7029	12.7892	12.8668	12.9370	13.0007	13.0588	13.1120	13.1608
−15	13.0708	13.2333	13.3742	13.4975	13.6062	13.7028	13.7892	13.8668	13.9370	14.0007	14.0588	14.1120	14.1608
−16	14.0703	14.2330	14.3740	14.4974	14.6062	14.7028	14.7892	14.8668	14.9370	15.0007	15.0588	15.1120	15.1608
−17	15.0700	15.2329	15.3740	15.4974	15.6062	15.7028	15.7892	15.8668	15.9370	16.0007	16.0588	16.1120	16.1608
−18	16.0699	16.2328	16.3739	16.4973	16.6062	16.7028	16.7892	16.8668	16.9370	17.0007	17.0588	17.1120	17.1608
−19	17.0698	17.2327	17.3739	17.4973	17.6062	17.7028	17.7892	17.8668	17.9370	18.0007	18.0588	18.1120	18.1608
−20	18.0697	18.2327	18.3739	18.4973	18.6062	18.7028	18.7892	18.8668	18.9370	19.0007	19.0588	19.1120	19.1608

Table 9 b. $\sum_{\nu=0}^{i}\left(\frac{1}{1+e^{\varkappa(\nu-0.2)}}\right)^2,\ i \geq 0;\quad \sum_{\nu=-1}^{i}\left(\frac{1}{1+e^{\varkappa(\nu-0.2)}}\right)^2,\ i < 0.$

i \ $\varkappa$	.52	.56	.60	.64	.68	.72	.76	.80	.84	.88	.92	.96	1.00
20	.5738	.5495	.5286	.5107	.4952	.4816	.4698	.4593	.4501	.4419	.4346	.4281	.4224
19	.5738	.5495	.5286	.5107	.4952	.4816	.4698	.4593	.4501	.4419	.4346	.4281	.4224
18	.5738	.5495	.5286	.5107	.4952	.4816	.4698	.4593	.4501	.4419	.4346	.4281	.4224
17	.5738	.5495	.5286	.5107	.4952	.4816	.4698	.4593	.4501	.4419	.4346	.4281	.4224
16	.5738	.5495	.5286	.5107	.4952	.4816	.4698	.4593	.4501	.4419	.4346	.4281	.4224
15	.5738	.5495	.5286	.5107	.4952	.4816	.4698	.4593	.4501	.4419	.4346	.4281	.4224
14	.5738	.5495	.5286	.5107	.4952	.4816	.4698	.4593	.4501	.4419	.4346	.4281	.4224
13	.5738	.5495	.5286	.5107	.4952	.4816	.4698	.4593	.4501	.4419	.4346	.4281	.4224
12	.5738	.5494	.5286	.5107	.4952	.4816	.4698	.4593	.4501	.4419	.4346	.4281	.4224
11	.5738	.5494	.5286	.5107	.4952	.4816	.4698	.4593	.4501	.4419	.4346	.4281	.4224
10	.5738	.5494	.5286	.5107	.4952	.4816	.4698	.4593	.4501	.4419	.4346	.4281	.4224
9	.5737	.5494	.5286	.5107	.4952	.4816	.4698	.4593	.4501	.4419	.4346	.4281	.4224
8	.5736	.5494	.5286	.5107	.4952	.4816	.4698	.4593	.4501	.4419	.4346	.4281	.4224
7	.5734	.5492	.5285	.5107	.4952	.4816	.4697	.4593	.4500	.4419	.4346	.4281	.4224
6	.5725	.5487	.5283	.5105	.4951	.4816	.4697	.4593	.4500	.4419	.4346	.4281	.4224
5	.5704	.5473	.5274	.5099	.4947	.4813	.4696	.4592	.4500	.4418	.4346	.4281	.4224
4	.5646	.5433	.5245	.5080	.4933	.4804	.4689	.4587	.4497	.4416	.4344	.4280	.4223
3	.5497	.5320	.5159	.5015	.4884	.4767	.4661	.4567	.4481	.4405	.4336	.4274	.4218

2	.5140	.5022	.4912	.4811	.4716	.4629	.4548	.4474	.4406	.4343	.4286	.4233	.4186
1	.4346	.4307	.4270	.4234	.4200	.4167	.4136	.4107	.4079	.4053	.4029	.4006	.3984
0	.2767	.2788	.2809	.2830	.2851	.2872	.2894	.2915	.2937	.2958	.2980	.3001	.3023
−1	.4240	.4382	.4524	.4666	.4808	.4949	.5090	.5229	.5367	.5505	.5640	.5774	.5906
−2	.9992	1.0375	1.0752	1.1121	1.1482	1.1834	1.2176	1.2509	1.2831	1.3142	1.3442	1.3732	1.4011
−3	1.7061	1.7723	1.8358	1.8967	1.9548	2.0101	2.0626	2.1125	2.1597	2.2044	2.2467	2.2866	2.3243
−4	2.5139	2.6060	2.6924	2.7734	2.8490	2.9195	2.9853	3.0465	3.1035	3.1566	3.2060	3.2520	3.2950
−5	3.3924	3.5055	3.6096	3.7053	3.7932	3.8738	3.9479	4.0160	4.0786	4.1363	4.1895	4.2386	4.2840
−6	4.3173	4.4462	4.5629	4.6685	4.7643	4.8512	4.9302	5.0021	5.0678	5.1278	5.1829	5.2334	5.2800
−7	5.2716	5.4116	5.5368	5.6489	5.7495	5.8401	5.9218	5.9958	6.0631	6.1243	6.1802	6.2314	6.2785
−8	6.2440	6.3917	6.5224	6.6385	6.7420	6.8346	6.9179	6 9930	7.0610	7.1228	7.1791	7.2307	7.2779
−9	7.2275	7.3802	7.5144	7.6329	7.7382	7.8320	7.9161	7.9917	8.0601	8.1222	8.1787	8.2304	8.2777
−10	8.2177	8.3736	8.5100	8.6300	8.7362	8.8307	8.9152	8.9912	9.0598	9.1219	9.1786	9.2303	9.2776
−11	9.2118	9.3699	9.5076	9.6285	9.7352	9.8301	9.9148	9.9909	10.0596	10.1218	10.1785	10.2302	10.2776
−12	10.2083	10.3677	10.5063	10.6277	10.7347	10.8298	10.9146	10.9908	11.0595	11.1218	11.1785	11.2302	11.2776
−13	11.2062	11.3665	11.5056	11.6272	11.7345	11.8296	11.9145	11.9907	12.0595	12.1218	12.1784	12.2302	12.2776
−14	12.2049	12.3658	12.5052	12.6270	12.7344	12.8295	12.9145	12 9907	13.0595	13.1218	13.1784	13.2302	13.2776
−15	13.2042	13.3654	13.5049	13.6269	13.7343	13.8295	13.9145	13.9907	14.0595	14.1218	14.1784	14.2302	14.2776
−16	14.2038	14.3651	14.5048	14.6268	14.7343	14.8295	14.9145	14.9907	15.0595	15.1218	15.1784	15.2302	15.2776
−17	15.2035	15.3650	15.5048	15.6268	15.7342	15.8295	15.9145	15.9907	16.0595	16.1218	16.1784	16.2302	16.2776
−18	16.2033	16.3649	16.5047	16.6268	16.7342	16.8295	16.9145	16.9907	17.0595	17.1218	17.1784	17.2302	17.2776
−19	17.2033	17.3649	17.5047	17.6268	17.7342	17.8295	17.9145	17.9907	18.0595	18.1218	18.1784	18.2302	18.2776
−20	18.2032	18.3649	18.5047	18.6268	18.7342	18.8295	18.9145	18.9907	19.0595	19.1218	19.1784	19.2302	19.2776

Table 9 c. $\sum_{\nu=0}^{i}\left(\frac{1}{1+e^{x(\nu-0.4)}}\right)^2,\ i \geq 0;\quad \sum_{\nu=-1}^{i}\left(\frac{1}{1+e^{x(\nu-0.4)}}\right)^2,\ i < 0.$

i \ x	.52	.56	.60	.64	.68	.72	.76	.80	.84	.88	.92	.96	1.00
20	.6463	.6238	.6048	.5888	.5752	.5635	.5535	.5450	.5377	.5314	.5261	.5216	.5178
19	.6463	.6238	.6048	.5888	.5752	.5635	.5535	.5450	.5377	.5314	.5261	.5216	.5178
18	.6463	.6238	.6048	.5888	.5752	.5635	.5535	.5450	.5377	.5314	.5261	.5216	.5178
17	.6463	.6238	.6048	.5888	.5752	.5635	.5535	.5450	.5377	.5314	.5261	.5216	.5178
16	.6463	.6238	.6048	.5888	.5752	.5635	.5535	.5450	.5377	.5314	.5261	.5216	.5178
15	.6463	.6238	.6048	.5888	.5752	.5635	.5535	.5450	.5377	.5314	.5261	.5216	.5178
14	.6463	.6238	.6048	.5888	.5752	.5635	.5535	.5450	.5377	.5314	.5261	.5216	.5178
13	.6463	.6238	.6048	.5888	.5752	.5635	.5535	.5450	.5377	.5314	.5261	.5216	.5178
12	.6463	.6238	.6048	.5888	.5752	.5635	.5535	.5450	.5377	.5314	.5261	.5216	.5178
11	.6463	.6238	.6048	.5888	.5752	.5635	.5535	.5450	.5377	.5314	.5261	.5216	.5178
10	.6463	.6238	.6048	.5888	.5752	.5635	.5535	.5450	.5377	.5314	.5261	.5216	.5178
9	.6462	.6238	.6048	.5888	.5752	.5635	.5535	.5450	.5377	.5314	.5261	.5216	.5178
8	.6461	.6237	.6048	.5888	.5751	.5635	.5535	.5450	.5377	.5314	.5261	.5216	.5178
7	.6457	.6235	.6047	.5887	.5751	.5635	.5535	.5450	.5377	.5314	.5261	.5216	.5178
6	.6448	.6229	.6043	.5885	.5750	.5634	.5535	.5450	.5377	.5314	.5261	.5216	.5178
5	.6421	.6212	.6032	.5878	.5745	.5631	.5533	.5448	.5376	.5314	.5261	.5216	.5178
4	.6351	.6162	.5997	.5853	.5728	.5619	.5524	.5442	.5372	.5311	.5259	.5214	.5177
3	.6173	.6024	.5890	.5770	.5664	.5570	.5487	.5414	.5350	.5294	.5246	.5205	.5170

2	.5751	.5666	.5588	.5517	.5452	.5392	.5339	.5291	.5248	.5209	.5176	.5147	.5122
1	.4831	.4826	.4822	.4819	.4817	.4816	.4816	.4817	.4820	.4823	.4828	.4833	.4840
0	.3045	.3089	.3133	.3177	.3222	.3266	.3311	.3356	.3401	.3447	.3493	.3538	.3584
−1	.4548	.4713	.4879	.5043	.5206	.5367	.5527	.5685	.5840	.5993	.6144	.6291	.6435
−2	1.0584	1.1004	1.1415	1.1814	1.2202	1.2578	1.2941	1.3291	1.3628	1.3951	1.4261	1.4557	1.4841
−3	1.7881	1.8579	1.9246	1.9880	2.0481	2.1050	2.1587	2.2093	2.2570	2.3018	2.3439	2.3834	2.4205
−4	2.6123	2.7072	2.7958	2.8782	2.9548	3.0258	3.0917	3.1527	3.2091	3.2615	3.3099	3.3548	3.3964
−5	3.5018	3.6167	3.7218	3.8180	3.9058	3.9861	4.0595	4.1266	4.1881	4.2444	4.2961	4.3437	4.3874
−6	4.4337	4.5634	4.6802	4.7855	4.8805	4.9664	5.0442	5.1147	5.1789	5.2373	5.2906	5.3394	5.3841
−7	5.3924	5.5324	5.6570	5.7682	5.8676	5.9568	6.0370	6.1094	6.1749	6.2343	6.2884	6.3378	6.3829
−8	6.3675	6.5145	6.6442	6.7590	6.8610	6.9521	7.0337	7.1070	7.1732	7.2331	7.2875	7.3371	7.3825
−9	7.3526	7.5042	7.6371	7.7541	7.8577	7.9498	8.0321	8.1059	8.1724	8.2326	8.2872	8.3369	8.3823
−10	8.3437	8.4984	8.6332	8.7516	8.8560	8.9487	9.0313	9.1054	9.1721	9.2324	9.2870	9.3368	9.3822
−11	9.3384	9.4950	9.6311	9.7502	9.8551	9.9481	10.0310	10.1052	10.1720	10.2323	10.2870	10.3368	10.3822
−12	10.3353	10.4931	10.6299	10.7495	10.8547	10.9479	11.0308	11.1051	11.1719	11.2322	11.2870	11.3367	11.3822
−13	11.3334	11.4920	11.6293	11.7491	11.8545	11.9477	12.0308	12.1051	12.1719	12.2322	12.2869	12.3367	12.3822
−14	12.3323	12.4913	12.6289	12.7489	12.8544	12.9477	13.0307	13.1050	13.1719	13.2322	13.2869	13.3367	13.3822
−15	13.3316	13.4910	13.6287	13.7488	13.8543	13.9476	14.0307	14.1050	14.1719	14.2322	14.2869	14.3367	14.3822
−16	14.3312	14.4908	14.6286	14.7488	14.8543	14.9476	15.0307	15.1050	15.1719	15.2322	15.2869	15.3367	15.3822
−17	15.3310	15.4907	15.6286	15.7487	15.8543	15.9476	16.0307	16.1050	16.1719	16.2322	16.2869	16.3367	16.3822
−18	16.3308	16.4906	16.6285	16.7487	16.8543	16.9476	17.0307	17.1050	17.1719	17.2322	17.2869	17.3367	17.3822
−19	17.3308	17.4906	17.6285	17.7487	17.8543	17.9476	18.0307	18.1050	18.1719	18.2322	18.2869	18.3367	18.3822
−20	18.3307	18.4905	18.6285	18.7487	18.8543	18.9476	19.0307	19.1050	19.1719	19.2322	19.2869	19.3367	19.3822

33.

Prime numbers and probability

Proc. 8th Scand. Math. Congr. Stockholm, 1–9 (1934)

1. In investigations concerning the asymptotic properties of various arithmetic functions, it is often possible to make an interesting heuristic use of probability arguments. If, e. g., we are interested in the distribution of a given sequence S of integers, we then consider S as a member of an infinite class C of sequences, which may be concretely interpreted as the possible realizations of some simple game of chance. The methods of the Theory of Probability in many cases allow us to prove that, *with a probability* $=1$, a certain relation R holds in C, i. e. that in a definite mathematical sense »almost all» sequences of C satisfy R. Of course we cannot in general conclude that R holds for the particular sequence S, but often results suggested by considerations of this kind may afterwards be rigorously proved by other methods.

In the first part of this paper, I shall give an account of the application of such arguments to some problems connected with the distribution of prime numbers, and I shall compare the results suggested in this way with the corresponding theorems which are rigorously proved. In the second part of the paper, it will be briefly shown that, if the Riemann hypothesis is assumed, it is possible to prove a certain new result suggested by the heuristic method. — Full proofs are not here given, but are reserved for a forthcoming publication.

I.

2. It was already observed by Gauss that the frequency of prime numbers in the vicinity of a large number x is approximately equal to $\frac{1}{\log x}$. Roughly speaking, we may interpret this fact by saying that *the chance that a given integer n should be prime is approximately* $\frac{1}{\log n}$.

This suggests that by considering the following series of trials we should obtain sequences of integers presenting some kind of analogy with the sequence of ordinary prime numbers.

Let $U_1, U_2, U_3, \ldots$ be an infinite series of urns containing black and white balls, the chance of drawing a white ball from U_n being $\frac{1}{\log n}$ for $n > 2$, while the composition of U_1 and U_2 may be arbitrarily chosen. — We now assume that one ball is drawn from U_1, one from U_2, and so on. The result of this infinite series of drawings will be a sequence of alternately black and white balls, and we are particularly interested in the distribution of white balls in this sequence.

If we denote by P_n the number of the urn from which the n:th white ball in our series was drawn, the numbers $P_1, P_2, \ldots$ will form an increasing sequence of integers. We shall consider the class C of all possible sequences (P_n). Obviously the sequence S of ordinary prime numbers $p_1, p_2, \ldots$ belongs to this class.

3. In the first place, we denote by $\Pi(x)$ the number of those P_n which are $\leq x$, thus forming an analogy to the ordinary notation $\pi(x)$ for the number of primes $p_n \leq x$.

For a given value of x, $\Pi(x)$ constitutes a random variable, and we may ask for the mean value or mathematical expectation of this variable. It is easily proved that, for large values of x, this mean value is asymptotically equal to $Li(x)$, in accordance with the well-known »Prime Number Theorem»:

$$\pi(x) \sim Li(x).$$

It is, however, possible to obtain much more precise information concerning the behaviour of $\Pi(x)$ for large values of x. For this purpose, we apply to our infinite series of trials a theorem connected with the so called *»Uniform Law of Great Numbers»*. It is true that the theorems belonging to this order of ideas which have been proved by Cantelli, Khintchine, Kolmogoroff and Lévy are not sufficiently general to allow of an immediate application to the series of trials now under consideration. It is, however, possible to show that the proofs given by these authors may without difficulty be so modified that they are valid under conditions which are sufficiently general for our purpose. By the application of the theorem thus obtained to our series of trials we obtain the following result which gives a very precise idea of the maximum order of the difference $\Pi(x) - Li(x)$:

With a probability = 1, the relation

$$\limsup_{x\to\infty} \frac{|\Pi(x) - Li(x)|}{\sqrt{x}\,.\sqrt{\frac{\log\log x}{\log x}}} = \sqrt{2} \tag{1}$$

is satisfied.

Let us compare this with the results which are known for the corresponding difference in the theory of prime numbers, $\pi(x) - Li(x)$. It has been proved by v. KOCH that, if the RIEMANN hypothesis is true, we have

$$\limsup_{x\to\infty} \frac{|\pi(x) - Li(x)|}{\sqrt{x}\,.\log x} < \infty,$$

and by LITTLEWOOD that [1]

$$\limsup_{x\to\infty} \frac{|\pi(x) - Li(x)|}{\sqrt{x}\,.\frac{1}{\log x}} = \infty.$$

Thus if we assume the RIEMANN hypothesis the true maximum order of the difference $\pi(x) - Li(x)$ lies »between» the functions $\sqrt{x}\,.\log x$ and $\sqrt{x}\,.\frac{1}{\log x}$, and it is seen that the order of the function occurring in the denominator of (1) falls inside this interval of indetermination. It would, of course, be extremely interesting to know if the latter function represents at least a reasonable approximation to the true maximum order in the prime number problem.

4. The application of probability arguments to certain other famous prime number problems, such as GOLDBACH's problem and the solution of the equation $p_{n+1} - p_n = 2$, has already been made by various authors. We shall here consider the problem of finding the maximum order of the difference between two consecutive prime numbers, $p_{n+1} - p_n$. The corresponding difference $P_{n+1} - P_n$ connected with our infinite series of trials is a random variable, and it is not difficult to prove the following theorem:

[1] For the sake of simplicity, I am quoting this theorem in a slightly less precise form than that given by its author. An analogous remark holds for the theorem of WESTZYNTHIUS quoted below.

With a probability = 1, the relation

$$\limsup_{n\to\infty} \frac{P_{n+1} - P_n}{(\log P_n)^2} = 1 \tag{2}$$

is satisfied.

As in the previous case, we compare this with known results concerning the prime number differences $p_{n+1} - p_n$. I have proved that, if the RIEMANN hypothesis is true, we have

$$\limsup_{n\to\infty} \frac{p_{n+1} - p_n}{\sqrt{p_n}\,.\log p_n} < \infty. \tag{3}$$

On the other hand, it has been proved by WESTZYNTHIUS that

$$\limsup_{n\to\infty} \frac{p_{n+1} - p_n}{\log p_n} = \infty. \tag{4}$$

It is readily seen that in this problem, the interval of indetermination is much larger than in the previous case, there being a considerable gap between the orders of the functions $\sqrt{p_n}\,.\log p_n$ and $\log p_n$ occurring in (3) and (4). Thus even if we assume the truth of the RIEMANN hypothesis, our knowledge of the maximum order of $p_{n+1} - p_n$ is very incomplete.

Of course it is suggested by the relation (2) connected with our probability model that the true maximum order of $p_{n+1} - p_n$ may be considerably lower than the best upper limit hitherto known. In this respect, it is interesting to remark that already in 1884 it was stated by PILTZ that, for every $\varepsilon > 0$, $p_{n+1} - p_n$ is of lower order than p_n^{ε}. But of course this has never been proved. As a matter of fact, it appears to be extremely difficult to prove that the relation (3) holds, if the denominator is replaced by any function of lower order than $\sqrt{p_n}\,.\log p_n$.

In spite of this, it is possible to prove certain results that, to a certain extent, support the conjecture suggested by (2) that the maximum order of $p_{n+1} - p_n$ may be something in the way of a power of $\log p_n$. I shall now pass on to an account of the principal result obtained in this direction.

II.

5. It follows from the relation (3) that, if the RIEMANN hypothesis is assumed, we have for all primes p_n

$$p_{n+1} - p_n < K\sqrt{p_n}\,.\log p_n,$$

where K is a constant. This being the best result of this character hitherto known, we cannot exclude the possibility that there may be an infinity of primes p_n satisfying e. g. the inequality

$$p_{n+1} - p_n > \sqrt{p_n}. \tag{5}$$

On the other hand, it has been suggested by our probability arguments that in reality there should be only a finite number of solutions of (5), and that this should still be the case, even if $\sqrt{p_n}$ were replaced by some function of considerably lower order. Thus it is suggested by (2) that even the inequality

$$p_{n+1} - p_n > (\log p_n)^3 \tag{6}$$

should only have a finite number of solutions.

We are indeed far from being able to prove such a deep result as this. It is, however, possible to prove (always assuming the RIEMANN hypothesis) that, *if there are an infinity of primes satisfying (6), then these primes are, at any rate, relatively scarce.* — This is rather a loose statement and our first object will be to put it in a more precise form.

We shall call any interval of the form

$$p_n \leqq t < p_{n+1}$$

a *prime interval.* Consider all prime intervals such that $p_n \leqq x$. The sum $S(x)$ of the lengths af all these intervals obviously satisfies the relation

$$S(x) \sim x. \tag{7}$$

From the set of prime intervals thus defined, let us now exclude all intervals such that

$$p_{n+1} - p_n \leqq (\log p_n)^3,$$

so that only the *exceptionally large* intervals satisfying (6) remain. For the sum $S_1(x)$ of the lengths of the remaining intervals, we obviously havo

$$S_1(x) < S(x), \tag{8}$$

but it does not seem possible to deduce by elementary methods any upper limit for the order of $S_1(x)$ considerably better than the result immediately obtained by the combination of (7) and (8). Thus in particular it does not seem possible to prove in this way that we have $S_1(x) < cx$ for any $c < 1$ and for all sufficiently large values of x. The theorem with which we are concerned asserts, however, that this relation holds for arbitrarily small positive values of c. As a matter of fact, our theorem may be thus expressed:

If the RIEMANN *hypothesis is true, then* $S_1(x) = o(x)$.

Thus on the RIEMANN hypothesis the exceptionally large prime intervals satisfying (6) are so scarce that, for large values of x, the sum of all the intervals of this kind inside the interval from 0 to x is small compared to the whole interval. Using a terminology introduced by HARDY and RAMANUJAN, we may also say that *almost all integral numbers belong to prime intervals which do not satisfy the relation (6).*

So far, it has not been possible to prove that this theorem holds true if, on the right hand side of (6), the exponent 3 is replaced by any smaller number.

6. The theorem just quoted is only a very particular case of a general theorem concerning the frequency of prime intervals of given magnitude. The proof of this theorem depends, as may be expected, on the theory of the RIEMANN Zeta Function. I shall here only give some general outlines of the proof.

Let us denote by $s = \sigma + i\tau$ a complex variable and by $\varrho = \beta + i\gamma$, $(\gamma > 0)$, a complex zero of $\zeta(s)$, situated in the upper half-plane. By $\Lambda(n)$ we denote the arithmetical function defined by the relations

$$\Lambda(n) = \begin{cases} \log p & \text{for } n = p^m, \ (p \text{ prime}, m \text{ integer}), \\ 0 & \text{otherwise.} \end{cases}$$

We shall consider a certain DIRICHLET series with complex exponents connected with the zeros ϱ, viz. the series

$$F(s) = \sum_{\gamma > 0} e^{\varrho i s} = \sum_{\gamma > 0} e^{-(\gamma - i\beta)s},$$

the sum being extended to all zeros ϱ in the upper half-plane. It is obvious that the series is absolutely convergent for $\sigma > 0$ and represents a function which is regular in every point of this half-plane.

Let us consider the function $F(s)$ in the domain D defined by the inequalities

$$0 < \sigma \leqq 1, \ \tau \geqq 1.$$

It then follows from a theorem previously given by me [1] that we have

$$(9)\qquad 2\pi F(s) = \pi - \sum_{2}^{\infty} \frac{\Lambda(n)}{n}\left(\frac{1}{s - i\log n} + \frac{1}{i\log n}\right) + g(s),$$

where $g(s)$ denotes a function regular in D and such that

$$\Re g(s) \to 0$$

uniformly in D, as $\tau \to \infty$.

Solving (9) with respect to the sum occurring in the right member and taking the real parts of both members we obtain

$$(10)\qquad \sum_{2}^{\infty} \frac{\Lambda(n)}{n} \cdot \frac{\sigma}{\sigma^2 + (\tau - \log n)^2} = \pi - 2\pi \Re F(s) + o(1),$$

the last term tending to zero as $\tau \to \infty$, uniformly for $0 < \sigma \leq 1$.

In (10), we take σ very small and τ very large, and we consider the positive quantity

$$\frac{\sigma}{\sigma^2 + (\tau - \log n)^2}$$

where n runs through all integral values. It is then readily seen that this quantity is large for values of n lying near e^τ, but becomes small as soon as n differs considerably from e^τ. Thus in the sum in the first member of (10) the values of n lying near e^τ occur with a *greater weight* than the other values. As $\Lambda(n)$ only differs from zero when n is a power of a prime number, and the contribution of the squares and higher powers can be shown to be relatively small, this makes it possible to obtain from (10) some information as to the prime numbers in a certain neighbourhood of e^τ.

The relations (9) and (10) are independent of the Riemann hypothesis, but of course the results obtained will be much more precise if we assume the truth of this hypothesis. *Thus on the* Riemann *hypothesis* it is possible to deduce from (10) the following result:

To every given $\delta > 0$ it is possible to determine $\tau_0 = \tau_0(\delta)$ such that for

$$(11)\qquad \begin{cases} \tau > \tau_0, \\ \tau e^{-\tau} < \sigma < \dfrac{1}{\tau^2}, \end{cases}$$

[1] Mathematische Zeitschrift, Vol. 4, 1919.

we have

$$\pi(e^{\tau}+\delta\sigma\tau e^{\tau})-\pi(e^{\tau}-\delta\sigma\tau e^{\tau})>1-\pi e^{-\frac{1}{2}\tau}\,\Re\sum_{\gamma>0}e^{-\gamma s}.$$

Thus as soon as we have, for values of σ and τ satisfying (11),

$$\varkappa(\sigma,\tau)=\pi e^{-\frac{1}{2}\tau}\,\Re\sum_{\gamma>0}e^{-\gamma s}<1, \tag{12}$$

there must be at least one prime number between the limits

$$e^{\tau}(1\pm\delta\sigma\tau).$$

The smaller we can here take σ, considered as a function of τ, the smaller becomes the interval on both sides of the point e^{τ}, within which we may assert the existence of at least one prime number. Unfortunately we cannot prove that, for such forms of σ as would enable us to improve the best previously known upper limit for the order of $p_{n+1}-p_n$, the relation (12) holds for *all* sufficiently large values of τ. We can, however, get a less precise result by arguing in the following way.

Putting

$$\sigma=\frac{1}{2\delta}\tau^2e^{-\tau}, \tag{13}$$

(11) is obviously satisfied for all sufficiently large τ, and it is possible to show that for all sufficiently large τ_1 we have

$$\int_{\tau_1}^{\tau_1+1}\big(\varkappa(\sigma,\tau)\big)^2\,d\tau<K\delta,$$

K being an absolute constant. It follows that, σ being defined by (13), the measure of the set of points[1] belonging to the interval $\tau_1<\tau<\tau_1+1$ and such that (12) does *not* hold, must be less than $K\delta$. If τ_1 is sufficiently large, (11) and (12) are both satisfied in all the remaining points τ of the interval, and thus for all these values of τ there is at least one prime number between the limits

$$e^{\tau}\pm\frac{1}{2}\tau^3.$$

[1] It is readily seen that this set consists of a finite number of intervals and (perhaps) isolated points, so that we may talk of its measure without introducing any delicate questions of principle.

Putting $e^{r} = x$, these limits become

$$x \pm \frac{1}{2}(\log x)^{3}. \tag{14}$$

It now follows that, for all sufficiently large x_1, the measure of the set of points belonging to the interval $x_1 < x < ex_1$ and such that there is no prime between the limits (14), is *small compared to* x_1. From this result, it is easy to deduce the theorem stated above.

34.

Über die Vorausberechnung der Bevölkerungsentwicklung in Schweden

Skand. Aktuarietidskr. 18, 35–54 (1935)

Einleitung.

1. Die andauernde Abnahme der Geburtenzahlen hat in den letzten Jahren ein lebhaftes Interesse für die Frage nach der künftigen Entwicklung der Bevölkerungszahl und der Altersgliederung der Bevölkerung eines Landes hervorgerufen. Es wurden hierüber einerseits eine grosse Anzahl von theoretischen Arbeiten verfasst, andererseits wurden auch wiederholt numerische Berechnungen gemacht, bei denen man von bestimmten Annahmen über die zukünftige Entwicklung der Sterblichkeit und der Geburtenhäufigkeit ausging und daraus die Altersgliederung der Bevölkerung zu verschiedenen Zeitpunkten ableitete.[1]

Bei solchen numerischen Berechnungen handelt es sich nun in erster Linie darum, die Einwirkung des Geburtenrückganges auf die Entwicklung der Bevölkerungszahl und der Altersgliederung zu erläutern. Der Mechanismus dieses Zusammenhanges ist zwar durch die genannten theoretischen Arbeiten formelmässig geklärt; die numerischen Berechnungen geben uns aber eine konkretere Vorstellung von dem Verlauf der Entwicklung und sind auch für viele praktische Zwecke wichtig, z. B. für Abschätzungen über die zukünftigen Kosten

[1] Für derartige Berechnungen über die schwedische Bevölkerung vgl. H. Cramér: Besparingskommittén och pensionsförsäkringen, Stockholm 1925, S. D. Wicksell: Sveriges framtida befolkning under olika förutsättningar, Ekonomisk Tidskrift 1926; 1928 års pensionsförsäkringskommitté: Betänkande, del I (1930), del II (1932), del III (1934).

für irgend eine Sozialversicherung oder über den zukünftigen Bedarf an Wohnungen, Unterrichtsmitteln u. dgl.

Die theoretischen Arbeiten haben in mehreren Fällen interessante Resultate von *asymptotischer Bedeutung* gegeben. Es wurde so z. B. gezeigt, dass bei geeigneten Annahmen über Sterblichkeit und Geburtenhäufigkeit, die relative Altersverteilung nach langer Zeit gegen eine bestimmte Endlage strebt. Demgegenüber sollen die numerischen Rechnungen die Verhältnisse *während der nächsten Dezennien* behandeln. Sie sollen uns vor allem zeigen, in welcher Weise der Geburtenrückgang die Entwicklung während dieser Zeit voraussichtlich beeinflussen wird.

Um eine konkrete Vorstellung von dem Einfluss des Geburtenrückganges zu gewinnen, wird man sich nicht damit begnügen, numerische Angaben über die Entwicklung nur bei einer einzigen Annahme über den Verlauf der Geburtenzahlen zu erhalten. Man wird vielmehr entsprechende Angaben für eine Reihe von verschiedenen Annahmen verlangen, um so in die Lage gesetzt zu werden, die Bedeutung einer beliebigen Annahme über den Geburtenrückgang einigermassen genau abzuschätzen. Bei der üblichen Methode, die genannten Berechnungen auszuführen, wird aber der hierzu nötige Arbeitsaufwand sehr erheblich, und man kann daher genötigt werden, die Anzahl der Alternativen zu beschränken und sich so mit wenig durchsichtigen Ergebnissen zu begnügen.

Es dürfte demnach nicht ohne Interesse sein eine Methode anzugeben, die, von gewissen vereinfachenden Annahmen ausgehend, eine sehr bequeme Ausführung der Vorausberechnung gestattet. Es handelt sich hier nur um eine Näherungsmethode, die aber sehr geeignet scheint eine Vorstellung davon zu geben, wie sich eine Bevölkerung entwickelt, wenn die Geburtenhäufigkeit eine zeitlang stark abnimmt und dann in eine stationäre Endlage übergeht.

Im Folgenden werden wir uns ausschliesslich mit der schwedischen Bevölkerung beschäftigen. Wir benutzen dabei die Bemerkung, dass die Altersverteilung dieser Bevölkerung um 1910, unmittelbar vor dem Beginn der starken Abnahme der Geburtenzahlen, mit guter Annäherung durch eine einfache mathematische Formel dargestellt werden kann. Es wird in der Tat gezeigt werden, dass sich diese Verteilung an die gewöhnlichen D_x-Zahlen einer gewissen Sterbetafel gut anschliesst. Durch geeignete Voraussetzungen über den Verlauf der Geburtenzahlen u. dgl. wird dann erreicht, dass die Altersverteilung zu einer beliebigen Zeit ebenfalls durch einfache Formeln erfasst werden kann. Die ganze Rechenarbeit kann dann mit Hilfe einer Reihe von Grundtabellen für dieselbe Sterbetafel und eine Reihe verschiedener (z. T. negativer) Werte des Zinsfusses mühelos ausgeführt werden.

Wir werden im Folgenden zuerst die Methode beschreiben und dann einige Anwendungen auf die Bevölkerungsentwicklung sowie auf gewisse Fragen der Sozialversicherung geben.[1]

Beschreibung der Methode.

2. Die Altersverteilung der Bevölkerung zur Zeit t sei durch die Funktion $L(x, t)$ gegeben, so dass

$$L(x, t)\,dx$$

die Anzahl der zur Zeit t im Altersintervall $(x, x + dx)$ vorhandenen Personen bezeichnet. $L(x, t)$ soll als durchweg stetige Funktion von x und t angenommen werden.

In Bezug auf die Sterblichkeit nehmen wir an, dass sie für beide Geschlechter mit der *Frauensterblichkeit* nach der Tabelle R 32[2] zusammenfällt. Unsere Methode könnte zwar auch unter Annahme verschiedener Sterblichkeiten für Männer und Frauen durchgeführt werden; der Einfachheit wegen wollen wir aber hier davon absehen. Die Bezeichnungen μ_x, l_x, D_x u. s. w. beziehen sich im Folgenden immer auf die genannte Sterbetafel. Die Wanderungen werden nicht berücksichtigt.

Es gilt nun

---

[1] Eine vorläufige Darstellung der Methode und deren Anwendungen auf die Sozialversicherung wurde im Gutachten der Pensionsversicherungskommission gegeben: Betänkande, Stockholm 1934, S. 186.

[2] Vgl. diese Zeitschrift, 1932, S. 251.

$$(1)\qquad L(x,t)=\begin{cases}\dfrac{l_x}{l_{x-t}}\,L(x-t,0) & \text{für}\quad x\geqq t,\\[2ex] \dfrac{l_x}{l_0}\,L(0,t-x) & \text{»}\quad x\leqq t.\end{cases}$$

Um die Funktion $L(x, t)$ für alle positiven Werte von x und t angeben zu können, genügt es also nach unseren Voraussetzungen die beiden Funktionen $L(x, 0)$ und $L(0, t)$ zu kennen. Die erstere gibt die Altersverteilung der Initialbevölkerung an, während die letztere den Verlauf der Geburtenzahlen bestimmt.

3. *Bestimmung von* $L(x, 0)$. — Als Nullpunkt der Zeitachse wählen wir das Ende des Kalenderjahres 1910, unmittelbar vor dem Beginn des starken Geburtenrückganges der letzten Dezennien. Die Altersverteilung der schwedischen Bevölkerung in diesem Zeitpunkt lässt sich mit ziemlich guter Annäherung durch die einfache Formel

$$(2)\qquad L(x,0)=a\,l_x\,k_0^{-x}=a\,D_x^{(0)}$$

darstellen. Hier sind a und k_0 Konstanten, und $D_x^{(0)}$ bezeichnet die gewöhnliche diskontierte Zahl der Lebenden, die dem Zinsfaktor

$$k_0 = 1 + i_0$$

entspricht. Dies ergibt sich aus Fig. 1, wo die Werte des Quotienten $L(x, 0) : l_x$ (aus Durchschnittswerten von $L(x, 0)$ und l_x für fünfjährige Altersintervalle bestimmt) in logaritmischem Masstab eingezeichnet sind. Die Punkte lassen sich, wie man sieht, recht gut durch eine Gerade ausgleichen. Nach der Methode der kleinsten Quadrate findet man

$$k_0 = 1.015,$$

und die Altersverteilung schliesst sich also den D_x-Zahlen für den Zinsfuss 1.5 % an, wie sich weiter aus der folgenden Tabelle ergibt.

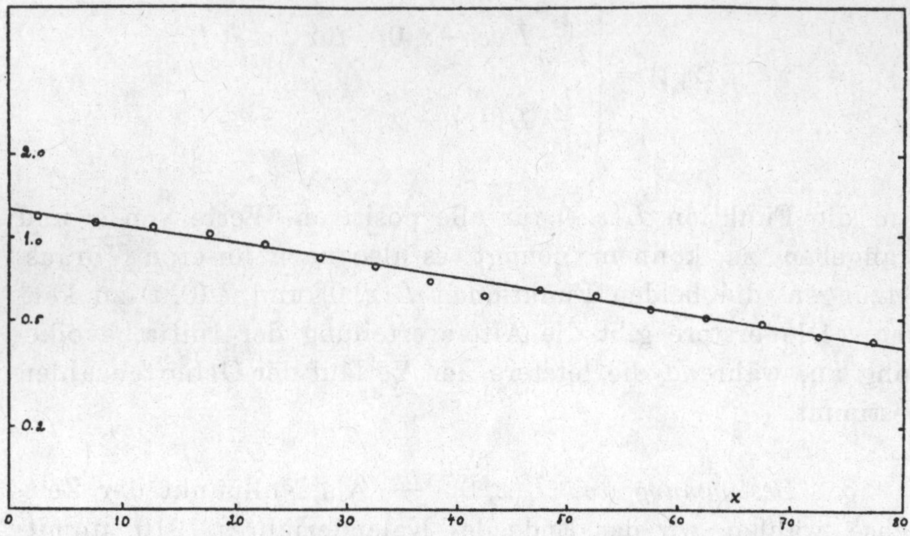

Fig. 1. Schwedische Bevölkerung Ende 1910. Beobachtete Werte des Quotienten $L(x, 0) : l_x$. Logarithmischer Masstab auf der Ordinatenachse.

Tabelle 1. *Relative Altersverteilung der schwedischen Bevölkerung Ende 1910.*

Alter	Beobachtet	Berechnet nach der Formel $L(x, 0) = a D_x^{(0)}$
0—10	218	218
10—20	192	185
20—30	156	155
30—40	125	129
40—50	102	107
50—60	88	86
60—70	66	65
70—80	40	40
>80	13	15
Summe	1 000	1 000

Wir wollen demnach im Folgenden annehmen, dass die Altersverteilung $L(x, 0)$ der Initialbevölkerung durch die Formel (2) mit $k_0 = 1.015$ gegeben ist. Die Konstante a wird später festgelegt werden.[1]

4. *Einführung des Reproduktionsmasses* r_t. *Bestimmung von* $L(0, t)$. — Die Anzahl der in einem Zeitelement $(t, t+dt)$ geborenen Kinder, deren Mütter dem Altersintervall $(x, x+dx)$ angehören, bezeichnen wir durch

$$\lambda(x, t)\, L(x, t)\, dx\, dt .$$

$\lambda(x, t)$ bedeutet also die *Geburtenintensität* der x-jährigen Bevölkerung zur Zeit t. Da nun

$$L(0, t)\, dt$$

die Gesamtzahl der Geburten im Zeitelement bezeichnet, so ist offenbar

$$L(0, t) = \int_0^\infty \lambda(x, t)\, L(x, t)\, dx$$

und nach dem ersten Mittelwertsatz der Integralrechnung

(3) $$L(0, t) = L(\xi, t) \int_0^\infty \lambda(x, t)\, dx .$$

Durch die letzte Gleichung ist ξ als ein Mittelwert der Alter der gebärenden Mütter zur Zeit t definiert, der im Allgemeinen von t abhängt. Die Funktionen $L(x, t)$ und $\lambda(x, t)$ können nun für jede gegebene Zeit t durch die Angaben der Bevölkerungsstatistik angenähert bestimmt werden, und es zeigt sich dabei, dass der Mittelwert ξ nur in geringem Masse von t abhängig ist. Es wurden z. B. die folgenden Werte von ξ gefunden:

[1] Nach der Formel (2) verhält sich die Bevölkerung von 1910 genau so, als ob die Sterblichkeit immer der Tabelle R 32 gefolgt wäre und das unten eingeführte »Reproduktionsmass« r_t immer den Wert $r_t = 1.015$ gehabt hätte. — Tatsächlich trifft natürlich keine von diesen Voraussetzungen zu.

Kalenderjahr	ξ
1900	29.1
1910	31.6
1920	29.4
1930	30.4

Trotz der starken Abnahme der Geburtenzahlen während dieser Zeit hat sich also der Mittelwert ξ nur wenig geändert. Wir werden daher im Folgenden einfach

$$\xi = 30$$

für alle Werte von t annehmen. Wenn wir dann weiter die positive Grösse r_t durch die Gleichung

$$r_t^{30} = \frac{l_{30}}{l_0} \int\limits_0^\infty \lambda(x, t)\, dx$$

definieren, so folgt aus (3)

$$\frac{l_{30}}{l_0} L(0, t) = r_t^{30} L(30, t), \tag{4}$$

und hieraus mit Hilfe von (1)

$$L(30, t+30) = r_t^{30} L(30, t). \tag{5}$$

Die letzte Beziehung zeigt, dass die Grösse r_t die Rolle eines einjährigen Reproduktionsmasses spielt. Wenn die zur Zeit t Geborenen nach Verlauf von 30 Jahren das »mittlere Alter der gebärenden Mütter» erreichen, hat sich die Grösse der Altersklasse, die für die Geburtenzahl massgebend ist, nach (5) in der Tat mit r_t^{30} multipliziert. Ist während einer längeren Zeit immer $r_t = 1$, so bekommen wir demnach eine stationäre Bevölkerung; ist aber r_t konstant und von 1 verschieden, so wird die Bevölkerung in geometrischer Progression zu- oder abnehmen, je nachdem $r_t > 1$ oder $r_t < 1$ ist. $100\,(r_t - 1)$

ist dabei der in Prozent ausgedrückte »Zinsfuss», der für das Anwachsen oder Abnehmen der Bevölkerung massgebend ist.

Wir wählen r_t — oder $100(r_t - 1)$ — als Mass der Reproduktivität der Bevölkerung zur Zeit t. Um die Einwirkung des Geburtenrückganges zu verfolgen, wollen wir r_t von $t = 0$ an zuerst abnehmen und dann in eine stationäre Endlage übergehen lassen. Zu diesem Zweck setzen wir

$$(6) \qquad r_t = \begin{cases} k_0^{1-\frac{t}{30}} k_1^{\frac{t}{30}} & \text{für} \quad 0 < t \leqq 30, \\ k_1 & \text{»} \qquad t \geqq 30. \end{cases}$$

Hier bezeichnet $k_0 = 1.015$ dieselbe Konstante, die wir oben in (2) bei der Bestimmung von $L(x, 0)$ eingeführt haben, und k_1 ist eine neue positive Konstante $< k_0$. Es folgt aus (6), dass r_t von k_0 bis k_1 abnimmt, wenn t von 0 bis 30 wächst (zwischen 1910 und 1940), sowie dass r_t für grössere Werte von t konstant $= k_1$ wird. Wir setzen

$$k_1 = 1 + i_1,$$

wo also i_1 den für die Endlage charakteristischen »Zinsfuss» darstellt, der natürlich sowohl positiv als negativ sein kann. Jedem Wert von i_1 entspricht also eine Annahme über den zukünftigen Verlauf der Geburtenhäufigkeit.

In Fig. 2 zeigen wir den Verlauf unseres hypothetischen r_t für die folgenden Werte von i_1 die bei den späteren numerischen Anwendungen berücksichtigt werden:

$$\begin{aligned} 100\, i_1 = {} & 0\,, \\ & -0.5, \\ & -1.0, \\ & -1.5, \\ & -2.0. \end{aligned}$$

Die Exponentialkurven lassen sich bei dem gewählten Massstab nicht von geraden Linien unterscheiden. Zum Vergleich geben wir in der selben Figur auch die nach (4) bestimmten

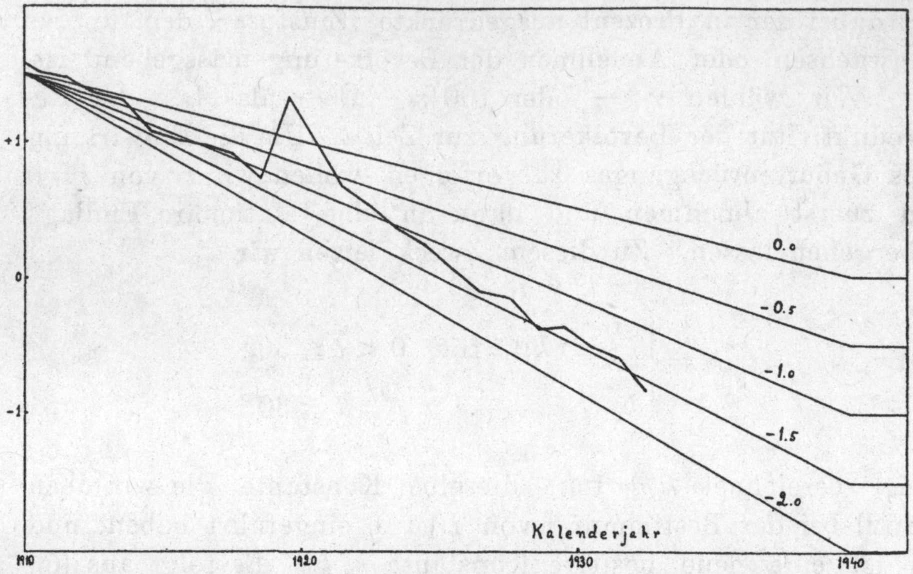

Fig. 2. Beobachteter und hypothetischer Verlauf des Reproduktionsmasses $100(r_t-1)$. $100\,i_1 = 0.0, -0.5, -1.0, -1.5, -2.0$

Werte von r_t an, die den tatsächlichen Beobachtungen an der schwedischen Bevölkerung entsprechen.

Die Bedeutung der hypothetischen Beziehungen (6) für r_t liegt nun darin, dass wir mit ihrer Hilfe eine besonders einfache Darstellung der Funktion $L(0, t)$ erhalten. Wie wir jetzt zeigen werden, folgt nämlich aus unseren Annahmen

$$L(0, t) = a\,k_1^t\,l_0 \tag{7}$$

für alle $t > 0$.

Für $0 < t \leqq 30$ gilt zuerst nach (4) und (6)

$$L(0, t) = \frac{l_0}{l_{30}}\,k_0^{30-t}\,k_1^t\,L(30, t)$$

und weiter nach (1) und (2)

$$L(0, t) = \frac{l_0}{l_{30-t}}\,k_0^{30-t}\,k_1^t\,L(30-t, 0)$$
$$= a\,k_1^t\,l_0.$$

Wir nehmen jetzt an, dass (7) für $0 < t \leqq 30\,n$ bewiesen worden ist (n ganz $\geqq 1$). Für $30\,n < t \leqq 30(n+1)$ haben wir dann

$$\begin{aligned} L(0,t) &= \frac{l_0}{l_{30}}\, k_1^{30}\, L(30,t) \\ &= k_1^{30}\, L(0, t-30) \\ &= k_1^{30} \cdot a\, k_1^{t-30}\, l_0 \\ &= a\, k_1^t\, l_0, \end{aligned}$$

womit (7) allgemein bewiesen ist.

5. *Zusammenfassung.* — Aus (1), (2) und (7) erhalten wir die allgemeine Darstellung der Funktion $L(x,t)$ in der folgenden Form

$$L(x,t) = \begin{cases} a\, k_0^{t-x}\, l_x = a\, k_0^t\, D_x^{(0)} & \text{für} \quad x \geqq t \\ a\, k_1^{t-x}\, l_x = a\, k_1^t\, D_x^{(1)} & \text{»} \quad x \leqq t \end{cases} \tag{8}$$

wo

$$k_0 = 1 + i_0 = 1.015,$$

$$k_1 = 1 + i_1,$$

und $D_x^{(0)}$ bezw. $D_x^{(1)}$ die dem (eventuell negativen) Zinsfuss i_0 bezw. i_1 entsprechende diskontierte Zahl bedeutet.

Die zur Zeit t innerhalb eines beliebigen Altersintervalles (x_1, x_2) vorhandene Bevölkerung ist gleich

$$\int_{x_1}^{x_2} L(x,t)\, dx$$

und lässt sich also unmittelbar durch die gewöhnlichen $\overline{N}_x$-Zahlen ausdrücken.

Sobald wir über D_x- und $\overline{N}_x$-Tabellen für die Werte $i_0 (= 0.015)$ und i_1 des Zinsfusses verfügen, können wir also die Altersverteilung der Bevölkerung mühelos berechnen.

Um unsere hypothetischen Formeln zuerst mit der beobachteten Verteilung der Bevölkerung von 1930 zu vergleichen und dann für Vorausberechnungen zu verwenden, bestimmen wir schliesslich die multiplikative Konstante a in solcher Weise, dass für 1930 die berechnete Bevölkerungszahl oberhalb 20 Jahre (die ja von unserer Annahme über r_t unabhängig ist) mit der wirklichen übereinstimmt. Wir erhalten dann

$$a\, k_0^{20}\, \overline{N}_{20}^{(0)} = 4\,058\,129$$

und hieraus

$$a = 1.1508.$$

Da die wirkliche Sterblichkeit zwischen 1910 und 1930 nicht unbedeutend grösser als die hier vorausgesetzte war, wird bei dieser Bestimmung von a die berechnete Bevölkerungszahl für 1910 etwas zu klein, nämlich 5.1 Millionen anstatt 5.5 Millionen. Dass trotzdem die *relative Altersverteilung*, die ja von a unabhängig ist, auch für 1910 mit den Formeln sehr gut übereinstimmt, folgt aus der oben mitgeteilten Tabelle 1.

Anwendungen.

6. *Bevölkerungszahl.* — Die Bevölkerungszahl zur Zeit t ergibt sich aus (8) gleich

$$\int_0^\infty L(x, t)\, dx = a\, k_0^t\, \overline{N}_t^{(0)} + a\, k_1^t \left(\overline{N}_0^{(1)} - \overline{N}_t^{(1)}\right).$$

Für grosse Werte von t (etwa $t > 90$) wird dieser Ausdruck mit guter Annäherung durch das Glied $a\, k_1^t\, \overline{N}_0^{(1)}$ dargestellt; die Bevölkerungszahl wird also asymptotisch wie eine Exponentialfunktion wachsen oder abnehmen, je nachdem $i_1 > 0$ oder $i_1 < 0$ ist. In Tabelle 2 und Figur 3 geben wir eine Übersicht über die Entwicklung der Bevölkerungszahl für die oben angegebenen speziellen Werte von i_1. Die Wirkliche Bevölkerungszahl Ende 1930 betrug 6.1 Millionen, was mit der für $100\, i_1 = -1.5$ berechneten Zahl übereinstimmt.

Tabelle 2. *Berechnete Bevölkerungszahl in Millionen.*

Jahr	100 i_1				
	0.0	−0.5	−1.0	−1.5	−2.0
1930	6.4	6.3	6.2	6.1	6.0
1940	7.0	6.7	6.5	6.3	6.1
1950	7.4	7.0	6.6	6.3	6.0
1960	7.8	7.2	6.6	6.1	5.6
1970	8.1	7.2	6.4	5.7	5.1
1980	8.3	7.0	6.0	5.2	4.5
1990	8.3	6.8	5.6	4.5	3.8
2000	8.4	6.5	5.1	3.9	3.1
2010	8.4	6.2	4.6	3.4	2.5

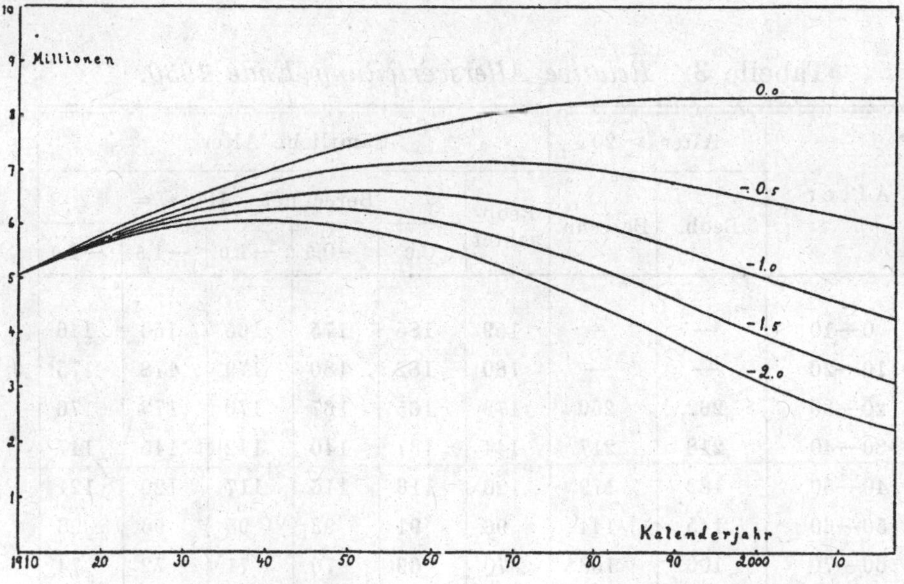

Fig. 3. Berechnete Entwicklung der Bevölkerungszahl.
$100\,i_1 = 0.0, -0.5, -1.0, -1.5, -2.0$

Für jeden negativen Wert von i_1 hat die Bevölkerungszahl ein Maximum, und dieses liegt für $100\,i_1 = -0.5$ etwa bei 1965 und für $100\,i_1 = -2.0$ schon bei 1940.

7. *Relative Altersverteilung. Geburtenrate und Sterblichkeitsrate.* — Die Übereinstimmung der berechneten und der beobachteten Altersverteilung von 1910 wurde oben durch Tabelle 1 erwiesen. Wenn man für 1930 ($t = 20$) die Altersverteilung der Bevölkerung *oberhalb 20 Jahre* betrachtet, die ja von der Annahme über r_t unabhängig ist, so bleibt, wie sich aus dem ersten Teil von Tabelle 3 ergibt, die Übereinstimmung fast ebenso gut.

In dem zweiten Teil von Tab. 3 geben wir die beobachtete und die berechnete Altersverteilung 1930 für sämtliche Alter. Die berechnete Verteilung hängt hierbei selbstverständlich etwas von dem angenommenen Verlauf von r_t zwischen 1910 und 1930 ab. Es folgt aus der Tabelle, dass sich die beobachtete Verteilung am besten an die für $100\, i_1 = -1.5$ berechnete Verteilung anschliesst, was ja auch mit Rücksicht auf Fig. 2 zu erwarten war.

Tabelle 3. *Relative Altersverteilung Ende 1930.*

Alter	Alter > 20		Sämtliche Alter					
	Beob.	Berechn.	Beob-achtet	Berechnet. $100\, i_1 =$ 0.0	−0.5	−1.0	−1.5	−2.0
0—10	—	—	159	185	175	165	155	146
10—20	—	—	180	182	180	179	178	175
20—30	262	260	173	165	167	170	173	176
30—40	218	217	144	137	140	142	145	147
40—50	182	179	120	113	115	117	120	121
50—60	145	144	96	91	93	95	96	98
60—70	106	108	70	69	70	71	72	74
70—80	66	66	44	42	43	44	44	45
>80	21	26	14	16	17	17	17	18
Summe	1 000	1 000	1 000	1 000	1 000	1 000	1 000	1 000

Die zukünftige Entwicklung der berechneten relativen Altersverteilung geht aus den Figuren 4—7 hervor, welche die beobachtete Verteilung für 1930 und die berechneten für

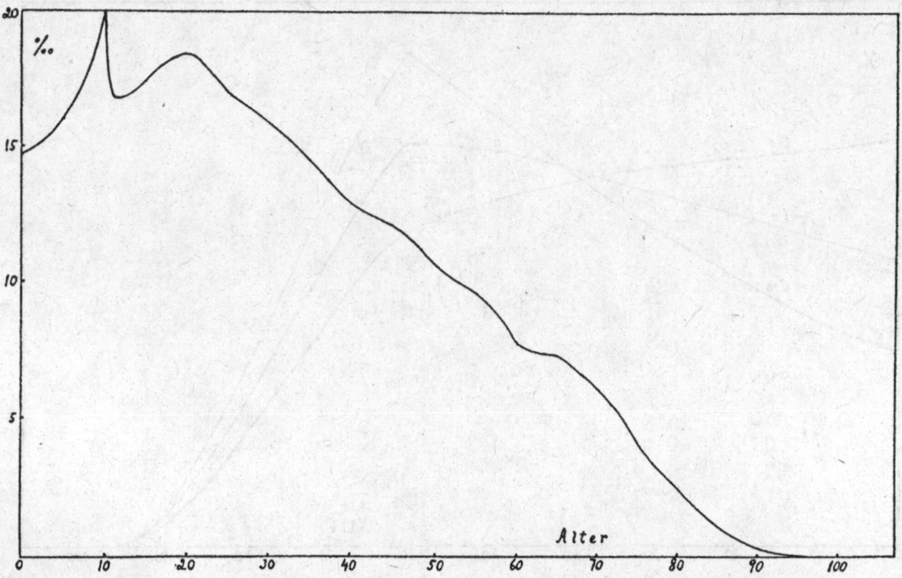

Fig. 4. Beobachtete relative Altersverteilung Ende 1930.

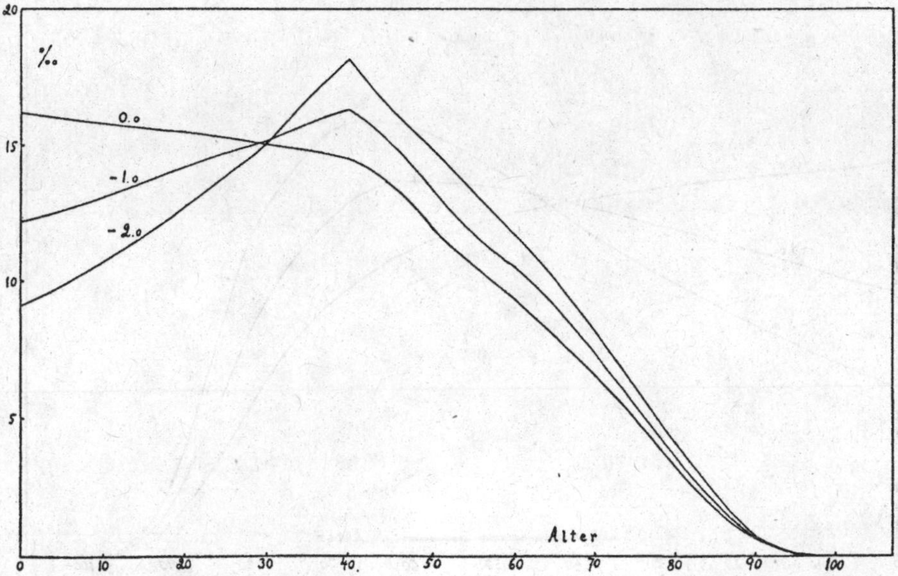

Fig. 5. Berechnete relative Altersverteilung Ende 1950.
$100\, i_1 = 0.0, \ -1.0, \ -2.0.$

1950, 1970 und 1990 veranschaulichen. In diesen Figuren ist die Fläche jeder Verteilungskurve gleich 1 000, so dass die Ordinaten die Grösse von einjährigen Altersklassen in ‰ der

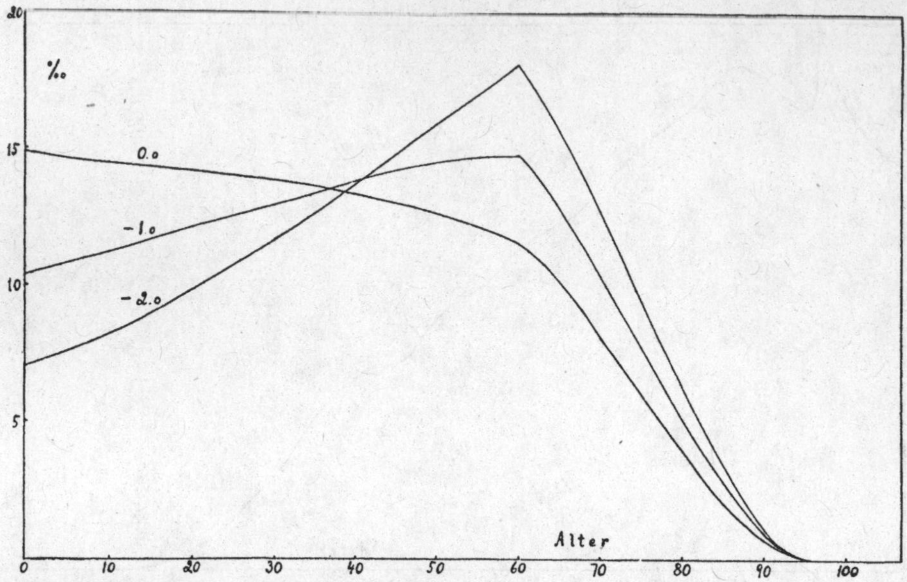

Fig. 6. Berechnete relative Altersverteilung Ende 1970.
$100\, i_1 = 0.0, \ -1.0, \ -2.0.$

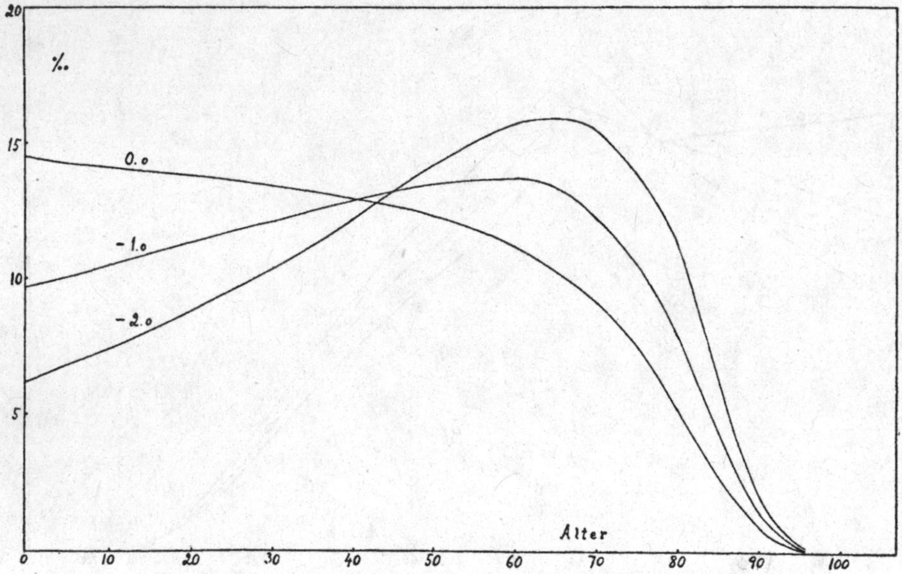

Fig. 7. Berechnete relative Altersverteilung Ende 1990.
$100\, i_1 = 0.0, \ -1.0, \ -2.0.$

Gesamtbevölkerung angeben. Von 1990 an verändern sich die nach unseren Voraussetzungen berechneten Verteilungen nur wenig. Der grösseren Übersichtlichkeit wegen haben wir in diesen Figuren nur die Werte $100\,i_1 = 0$, -1 und -2 berücksichtigt.

Die grosse Veränderung der Altersverteilung, die durch den Geburtenrückgang verursacht wird, tritt aus den Figuren deutlich hervor. Schon bei der Annahme $100\,i_1 = 0$, die ja einer wesentlich grösseren Geburtenhäufigkeit als der jetzigen entspricht, wächst der Anteil der höheren Alter stark, während derjenige der jüngeren Alter abnimmt. Dies wird auch durch die folgende Tabelle 4 veranschaulicht.

Tabelle 4. *Relative Verteilung der Bevölkerung auf die drei Altersklassen* 0—15, 15—65 *und* > 65 *Jahre.*

Jahr		0—15	15—65	>65	Summe
Beobachtet:					
	1850	329	622	49	1 000
	1870	340	605	55	1 000
	1890	333	590	77	1 000
	1910	317	599	84	1 000
	1930	248	660	92	1 000
Berechnet:					
$100\,i_1 =$ 0.0.	1950	238	658	104	1 000
	1970	219	652	129	1 000
	1990	213	635	152	1 000
$100\,i_1 = -0.5$.	1950	215	675	110	1 000
	1970	190	664	146	1 000
	1990	182	637	181	1 000
$100\,i_1 = -1.0$.	1950	193	690	117	1 000
	1970	164	673	163	1 000
	1990	154	634	212	1 000
$100\,i_1 = -1.5$.	1950	173	703	124	1 000
	1970	140	678	182	1 000
	1990	128	626	246	1 000
$100\,i_1 = -2.0$.	1950	154	716	130	1 000
	1970	119	680	201	1 000
	1990	106	612	282	1 000

Es folgt aus dieser Tabelle, dass sich die Veränderungen der ältesten und der jüngsten Altersklassen gewissermassen gegenseitig kompensieren, so dass der relative Anteil der eigentlich produktiven Altersklassen nicht so stark variiert, wie man vielleicht erwarten könnte. Diese Tatsache stimmt mit den bekannten von G. SUNDBÄRG gemachten Beobachtungen wohl überein.

Die allgemeine *Geburtenrate* der Bevölkerung zur Zeit t ist gleich

$$\frac{L(0,t)}{\int\limits_0^\infty L(x,t)\,dx} = \frac{k_1^t\, l_0}{k_0^t\, \overline{N}_t^{(0)} + k_1^t\,(\overline{N}_0^{(1)} - \overline{N}_t^{(1)})}.$$

Für $t=0$ hat dieser Ausdruck den Wert $\frac{1}{\bar{a}_0^{(0)}}$ und nimmt für wachsendes t monoton gegen den Grenzwert $\frac{1}{\bar{a}_0^{(1)}}$ ab.

Die allgemeine *Sterblichkeitsrate* ist, wie man durch eine leichte Rechnung findet, gleich

$$\frac{\int\limits_0^\infty \mu_x\, L(x,t)\,dx}{\int\limits_0^\infty L(x,t)\,dx} = \frac{k_0^t\, \overline{M}_t^{(0)} + k_1^t\,(\overline{M}_0^{(1)} - \overline{M}_t^{(1)})}{k_0^t\, \overline{N}_t^{(0)} + k_1^t\,(\overline{N}_0^{(1)} - \overline{N}_t^{(1)})},$$

und ist eine beständig wachsende Funktion von t, die für $t=0$ den Wert $\frac{1}{\bar{a}_0^{(0)}} - \delta_0$ annimmt und für $t \to \infty$ gegen den Grenzwert $\frac{1}{\bar{a}_0^{(1)}} - \delta_1$ strebt.

Die berechneten Werte der Geburtenrate und der Sterblichkeitsrate für $t=0$ und $t \to \infty$ sind in Taballe 5 gegeben.

Tabelle 5. *Anfangs- und Endwerte der Geburtenrate und der Sterblichkeitsrate.*

$100\,i_1$	Geburtenrate ‰		Sterblichkeitsrate ‰	
	$t=0$	$t\to\infty$	$t=0$	$t\to\infty$
0.0	24.0	14.5	9.1	14.5
−0.5	24.0	11.9	9.1	16.9
−1.0	24.0	9.7	9.1	19.7
−1.5	24.0	7.8	9.1	22.8
−2.0	24.0	6.2	9.1	26.2

Die beobachteten Werte der Geburtenrate für 1910 und 1911 waren 24.7 bezw. 24.0 ‰, was mit dem für $t=0$ berechneten Werte gut übereinstimmt. Die beobachteten Sterblichkeitsraten waren dagegen bedeutend höher als die berechneten, nämlich 14.0 bezw. 13.8 ‰, was ja bei der von uns für die Vorausberechnung angenommenen niedrigen Sterblichkeit ganz natürlich ist.

8. *Prämien einer Altersrentenversicherung.* — Wir betrachten eine Bevölkerung, die sich nach den vorstehenden Formeln entwickelt. Von dem Zeitpunkt $t=0$ an sei in dieser Bevölkerung eine Altersrentenversicherung eingeführt, wobei jede Person, die ein bestimmtes Alter z noch nicht erreicht hat, das Recht auf eine bei dem Alter z beginnende Altersrente von dem konstanten Betrage 1 erhält. Diejenigen, die für $t=0$ das Alter z schon überschritten haben, werden nicht in die Versicherung aufgenommen. Beitragsplichtig sind alle Personen zwischen y und z Jahren.

Der Zinsfuss, mit welchem gerechnet wird, sei $100\,i\,\%$; die entsprechenden Funktionen D_x, $\overline{N}_x$ u. s. w. schreiben wir ohne obere Indizes, während die oberen Indizes 0 bezw. 1 dieselbe Bedeutung haben wie oben. Die Verzinsungsintensitäten, welche i, i_0 und i_1 entsprechen, bezeichnen wir mit

$$\delta = \log(1+i),$$
$$\delta_0 = \log(1+i_0) = \log 1.015,$$
$$\delta_1 = \log(1+i_1).$$

Die Prämie nach dem *Umlagesystem* wird dann im Zeitpunkt t

$$P_1(t) = \frac{\int_z^{z+t} L(x,t)\,dx}{\int_y^z L(x,t)\,dx},$$

während die Prämie nach dem *Kapitaldeckungssystem*.

$$P_2(t) = \frac{L(z,t)\,\bar{a}_z}{\int_y^z L(x,t)\,dx}$$

beträgt. Unter Benutzung von (8) kann man unmittelbar die Prämien $P_1(t)$ und $P_2(t)$ durch D_x- und $\overline{N}_x$-Funktionen ausdrücken.

Wir können auch eine *konstante Durchschnittsprämie* für diese Versicherung berechnen, die von $t=0$ an von jeder Person zwischen den Altern y und z mit konstantem jährlichem Betrage zahlbar ist. Wir erhalten dann

$$P_3 = \frac{\bar{a}_z \int_0^\infty L(z,t)\,v^t\,dt}{\int_y^z L(x,0)\,\bar{a}_{x,\,\overline{z-x|}}\,dx + \bar{a}_{y,\,\overline{z-y|}} \int_0^\infty L(y,t)\,v^t\,dt},$$

und ferner nach einiger Rechnung

$$P_3 = \frac{(\delta - \delta_1) D_z^{(0)} + (\delta_1 - \delta_0) D_z}{(\delta - \delta_1)(\overline{N}_y^{(0)} - \overline{N}_z^{(0)}) + (\delta_1 - \delta_0)(\overline{N}_y - \overline{N}_z)} \bar{a}_z.$$

Auch für die Grösse der Reserven nach den verschiedenen Systemen können ohne Schwierigkeit ähnliche Formeln entwickelt werden, was jedoch hier nicht näher ausgeführt werden soll.[1] In der folgenden Tabelle 6 geben wir einige numerische Beispiele für die Prämien $P_1(t)$, $P_2(t)$ und P_3. Der Zinsfuss i ist hierbei gleich 0.04 gewählt.

Tabelle 6. *Jährliche Prämien für eine Altersrente vom Betrage* 100.

Jahr	100 i_1				
	0.0	−0.5	−1.0	−1.5	−2.0
Umlagesystem:					
1910	0	0	0	0	0
1930	13.2	13.2	13.2	13.3	13.3
1950	15.8	16.4	16.9	17.5	18.0
1970	19.7	21.9	24.2	26.5	29.0
1990	24.0	28.4	33.4	39.2	45.7
Kapitaldeckungssystem:					
1910	11.5	11.5	11.5	11.5	11.5
1930	11.5	11.6	11.6	11.6	11.6
1950	12.9	13.4	13.8	14.3	14.7
1970	16.1	17.9	19.7	21.7	23.7
1990	17.3	19.7	22.2	24.9	27.9
Konstante Durchschnittsprämie:	12.7	13.0	13.2	13.4	13.6

[1] Unter sehr allgemeine Voraussetzungen über die Gruppe der versicherten Personen wurden derartige Formeln für Prämien und Reserven von F. Tricomi aufgestellt: Giornale dell'Istituto Italiano degli Attuari, Vol. 1, p. 36 und 170. Die oben angeführten Prämienformeln ergeben sich aus seinen durch Spezialisierung.

35.

Über eine Eigenschaft der normalen Verteilungsfunktion

Math. Z. **41**, 405–414 (1936)

Zweck dieser Arbeit ist, einen lange vermuteten[1]) Satz über die normale Verteilungsfunktion[2]) zu beweisen und auf den n-dimensionalen Raum zu verallgemeinern. In Abschnitt I werden wir den Satz zuerst in rein analytischer Fassung aussprechen und dann wahrscheinlichkeitstheoretisch deuten. Der Beweis des Satzes erfolgt in Abschnitt II; Abschnitt III bringt die n-dimensionale Verallgemeinerung, die hier zum ersten Male ausgesprochen und bewiesen wird.

I.

Jede nirgends abnehmende Funktion der reellen Veränderlichen x, die für $x \to -\infty$ dem Grenzwert 0, für $x \to +\infty$ dem Grenzwert 1 zustrebt, wird bekanntlich als eine *Verteilungsfunktion* bezeichnet[3]). Die „Faltung" $F(x)$ von zwei Verteilungsfunktionen $F_1(x)$ und $F_2(x)$:

$$F(x) = \int_{-\infty}^{\infty} F_1(x-t)\, dF_2(t)$$

ist bekanntlich wieder eine Verteilungsfunktion.

Normale Verteilungsfunktion nennen wir jede Funktion von der Form $\Phi\left(\frac{x-m}{\sigma}\right)$, wo

$$\Phi(x) = \frac{1}{\sqrt{2\pi}} \int_{-\infty}^{x} e^{-\frac{t^2}{2}}\, dt,$$

---

[1]) Vgl. P. Lévy [7], S. 366; [8], S. 381—388. (Zahlen in eckigen Klammern verweisen auf das angefügte Literaturverzeichnis.) Nach Angabe von Herrn Lévy (a. a. O.) hat er den Satz schon seit 1931 vermutet.

[2]) Es scheint noch kaum hinreichend bekannt zu sein, daß die Normalverteilung schon von De Moivre (Miscellanea Analytica, 2. Suppl., 1733, sowie Doctrine of Chances, 3. Aufl., 1756, S. 243—254) bei seiner Behandlung des klassischen Bernoullischen Problems gefunden wurde. Die übliche Benennung nach Gauss oder Laplace ist also nicht historisch begründet.

[3]) Eine besondere Verabredung über die Werte der Funktion in etwaigen Unstetigkeitspunkten ist für unsern Zweck nicht nötig.

und $\sigma \geqq 0$ und m zwei reelle Konstanten sind. Der Fall $\sigma = 0$ entspricht der *ausgearteten* Normalfunktion $E(x-m)$, wo

$$E(x) = \begin{cases} 0 & \text{für} \quad x < 0, \\ 1 & \text{,,} \quad x > 0, \end{cases}$$

und $E(0)$ irgendein Wert im Intervalle $0 \leqq E(0) \leqq 1$ zugeteilt werden kann.

Es handelt sich hier zunächst um den folgenden Satz:

Satz 1. *Wenn die Faltung von zwei Verteilungsfunktionen normal ist, so sind auch beide Komponenten normal. — Genauer: Die Integralgleichung*

$$(1) \qquad \int_{-\infty}^{\infty} F_1(x-t)\, dF_2(t) = \Phi\left(\frac{x-m}{\sigma}\right),$$

wo $F_1(x)$ und $F_2(x)$ unbekannte Verteilungsfunktionen sind, hat nur die Lösungen

$$F_1(x) = \Phi\left(\frac{x-m_1}{\sigma_1}\right), \qquad F_2(x) = \Phi\left(\frac{x-m_2}{\sigma_2}\right),$$

wo die Konstanten den Bedingungen

$$m = m_1 + m_2, \qquad \sigma^2 = \sigma_1^2 + \sigma_2^2$$

genügen.

Wenn σ, σ_1 oder σ_2 gleich Null ist, so hat die entsprechende Verteilungsfunktion nur eine einzige Wachstumsstelle (nämlich $x = m$, m_1 bzw. m_2), und der Satz ist dann trivial. Wir können also diesen Fall von der Betrachtung ausschließen und werden demnach im folgenden erstens $\sigma > 0$ voraussetzen, zweitens nur solche Lösungen der Integralgleichung (1) berücksichtigen, welche *Verteilungsfunktionen mit wenigstens zwei Wachstumsstellen* sind. In der Sprache der Wahrscheinlichkeitsrechnung besagt unser Satz nach dieser Einschränkung:

Wenn die Summe von zwei unabhängigen nicht-konstanten stochastischen Variablen normal verteilt ist, so ist auch jede Komponente für sich normal verteilt[4]).

Dieser Satz, der sich natürlich unmittelbar auf eine beliebige Gliederzahl erweitern läßt, wurde von Herrn P. Lévy[1]) vermutet. Wie Herr Lévy zeigt, ergeben sich aus dem Satze interessante Folgerungen in bezug auf die Gültigkeitsbedingungen des sogenannten zentralen Grenzwertsatzes der Wahrscheinlichkeitsrechnung. Man erhält z. B. leicht den

---

[4]) In einer vorläufigen Mitteilung (Cramér [2]) wurde der Beweis dieses Satzes kurz angedeutet.

folgenden Satz[5]): *Es seien $x_1, x_2, \ldots$ unabhängige stochastische Variablen und $a_1, a_2, \ldots$ eine beschränkte Folge positiver Zahlen. Damit die Verteilungsfunktion von*

$$\frac{x_1 + x_2 + \ldots + x_n}{a_n}$$

für $n \to \infty$ gegen $\Phi(x)$ konvergiert, ist notwendig und hinreichend, daß 1) $\lim a_n = a > 0$ *existiert, und* 2) *jedes x_i normal verteilt ist, wobei die Summe der ersten Momente gegen Null, die Summe der Streuungen gegen a^2 konvergiert.*

II.

Man sieht unmittelbar ein, daß wir uns auf den Speziallfall $m = 0$, $\sigma = 1$ von Satz 1 beschränken können, wo die Integralgleichung (1) die Form

$$\text{(1a)} \qquad \int_{-\infty}^{\infty} F_1(x-t)\, dF_2(t) = \Phi(x)$$

annimmt. Wenn nämlich (1) eine Lösung hat, die nicht von der Normalform ist, so zeigt eine einfache Variabeltransformation, daß dies auch von (1a) gilt. Wir werden also versuchen, alle Verteilungsfunktionen $F_1(x)$ und $F_2(x)$ mit wenigstens je zwei Wachstumsstellen anzugeben, welche (1a) genügen.

Wir zeigen zuerst, daß die Momente zweiter Ordnung

$$\int_{-\infty}^{\infty} x^2\, dF_1(x) \quad \text{und} \quad \int_{-\infty}^{\infty} x^2\, dF_2(x)$$

endlich sein müssen. Dies ergibt sich aus der für beliebige positive a und b geltenden Ungleichung

$$(F_1(a) - F_1(-a)) \int_{-b}^{b} y^2\, dF_2(y) = \int_{-a}^{a} \int_{-b}^{b} y^2\, dF_1(x)\, dF_2(y)$$

$$\leqq \int_{-a}^{a} \int_{-b}^{b} 2((x+y)^2 + x^2)\, dF_1(x)\, dF_2(y) \leqq 2 \int_{-\infty}^{\infty} \int_{-\infty}^{\infty} (x+y)^2\, dF_1(x)\, dF_2(y) + 2a^2.$$

Aus (1a) folgt aber

$$\int_{-\infty}^{\infty} \int_{-\infty}^{\infty} (x+y)^2\, dF_1(x)\, dF_2(y) = \int_{-\infty}^{\infty} x^2\, d\Phi(x) = 1,$$

[5]) Vgl. hierzu W. Feller [3], S. 531.

und wir erhalten somit, wenn wir für a irgendeinen festen Wert wählen, für den $F_1(a) - F_1(-a) > 0$ wird,

$$\int_{-b}^{b} y^2 dF_2(y) \leqq \frac{2(1+a^2)}{F_1(a) - F_1(-a)}$$

für beliebige $b > 0$. Das zweite Moment von F_2 ist also endlich, und das gleiche Resultat für F_1 ergibt sich natürlich in entsprechender Weise. A fortiori sind auch die beiden Momente erster Ordnung endlich, und es folgt unmittelbar aus (1a), daß ihre Summe gleich dem ersten Moment von $\Phi(x)$, d. h. gleich Null ist. Da nun mit $F_1(x)$, $F_2(x)$ auch zugleich das Funktionenpaar $F_1(x-k)$, $F_2(x+k)$ für beliebige k der Gleichung (1a) genügt, können wir ohne Beschränkung der Allgemeinheit voraussetzen, daß beide Momente erster Ordnung gleich Null sind. Wenn dann die Momente zweiter Ordnung mit σ_1^2 bzw. σ_2^2 bezeichnet werden, so folgt aus (1a) $\sigma_1^2 + \sigma_2^2 = 1$.

Wir behaupten nun weiter, daß die beiden Integrale

$$\int_{-\infty}^{\infty} e^{\frac{x^2}{2}} dF_1(x) \quad \text{und} \quad \int_{-\infty}^{\infty} e^{\frac{x^2}{2}} dF_2(x) \tag{2}$$

konvergent sind. Hieraus folgt unter anderem die Endlichkeit der Momente beliebig hoher Ordnung.

Aus dem (vorausgesetzten) Verschwinden des ersten Momentes von $F_2(x)$ folgt, daß es ein $\lambda > 0$ gibt, so daß

$$F_2(-\lambda) > 0$$

ist. Für jedes $x < 0$ gilt dann nach (1a)

$$\begin{aligned}\Phi(x-\lambda) &\geqq \int_{-\infty}^{-\lambda} F_1(x-\lambda-t)\, dF_2(t) \\ &\geqq F_1(x) \cdot F_2(-\lambda),\end{aligned}$$

und somit

$$F_1(x) \leqq \frac{\Phi(x-\lambda)}{F_2(-\lambda)} < A e^{-\frac{x^2}{2} + \lambda x}, \tag{3}$$

wo A von x unabhängig ist. In entsprechender Weise erhalten wir für $x > 0$

$$1 - F_1(x) < B e^{-\frac{x^2}{2} - \mu x}, \tag{4}$$

wo B und μ positive von x unabhängige Zahlen sind. Aus (3) und (4), sowie aus den entsprechenden Ungleichungen für $F_2(x)$, folgt die Konvergenz der Integrale (2).

Wir führen jetzt die *charakteristischen Funktionen*

$$(5) \qquad f_1(t) = \int_{-\infty}^{\infty} e^{itx}\, dF_1(x), \qquad f_2(t) = \int_{-\infty}^{\infty} e^{itx}\, dF_2(x)$$

ein. Aus der Konvergenz der Integrale (2) folgt offenbar, daß $f_1(t)$ und $f_2(t)$ ganze Funktionen der komplexen Veränderlichen t sind. Aus (1a) folgt aber bekanntlich

$$(6) \qquad f_1(t) \cdot f_2(t) = \int_{-\infty}^{\infty} e^{itx}\, d\Phi(x) = e^{-\frac{t^2}{2}}.$$

Wir schließen hieraus, daß $f_1(t)$ und $f_2(t)$ *ganze Funktionen ohne Nullstellen* sind, also von der Form $e^{g_1(t)}$ bzw. $e^{g_2(t)}$ mit ganzen $g_1(t)$ und $g_2(t)$.

Aus der Ungleichung $2|tx| \leqq |t^2| + |x^2|$ folgt weiter

$$|f_1(t)| < e^{\frac{|t|^2}{2}} \int_{-\infty}^{\infty} e^{\frac{x^2}{2}}\, dF_1(x).$$

Das Integral auf der rechten Seite ist konvergent und von t unabhängig; die Ordnung der ganzen Funktion $f_1(t) = e^{g_1(t)}$ ist also höchstens gleich 2. Nach der klassischen Hadamardschen Theorie muß dann $g_1(t)$ ein Polynom (höchstens) zweiten Grades sein. Entsprechendes gilt für $g_2(t)$.

Nach unseren Voraussetzungen über die Momente folgt aber aus (5) durch Entwicklung nach Potenzen von t

$$f_1(t) = e^{g_1(t)} = 1 - \frac{\sigma_1^2}{2} t^2 + \dots,$$

$$f_2(t) = e^{g_2(t)} = 1 - \frac{\sigma_2^2}{2} t^2 + \dots.$$

Hieraus folgt schließlich

$$f_1(t) = e^{-\frac{\sigma_1^2 t^2}{2}}, \qquad f_2(t) = e^{-\frac{\sigma_2^2 t^2}{2}},$$

was bekanntlich[6]) mit

$$F_1(x) = \Phi\left(\frac{x}{\sigma_1}\right), \qquad F_2(x) = \Phi\left(\frac{x}{\sigma_2}\right)$$

äquivalent ist. Die gesuchte allgemeinste Lösung von (1a) ist also von der Form

$$F_1(x) = \Phi\left(\frac{x-k}{\sigma_1}\right), \qquad F_2(x) = \Phi\left(\frac{x+k}{\sigma_2}\right),$$

wo k beliebig und $\sigma_1^2 + \sigma_2^2 = 1$ ist. Durch eine einfache Variabeltransformation folgt daraus die Behauptung des Satzes 1 über die allgemeinste Lösung von (1).

---

[6]) Vgl. z. B. Lévy [6], Kap. 2.

III.

Es sei R_n ein n-dimensionaler Euklidischer Raum mit dem variablen Punkt $x = (x_1, x_2, \ldots, x_n)$. Für die Verallgemeinerung der obigen Betrachtungen auf diesen Raum ist es zweckmäßig, die Verteilungsfunktionen als *Mengenfunktionen* zu behandeln[7]). Als *Verteilungsfunktion in* R_n bezeichnen wir dann jede nicht-negative, absolut additive Mengenfunktion $F(M)$, die für alle Borelschen Mengen M von R_n definiert ist und für $M = R_n$ den Wert 1 annimmt. Wählt man speziell für M die Menge $M(x_r \leqq z_r,\ r = 1, 2, \ldots, n)$, so erhält man bei variablen $z_1, \ldots, z_n$ eine Punktfunktion in R_n, welche die n-dimensionale Verallgemeinerung der oben betrachteten eindimensionalen Verteilungsfunktionen darstellt. Mengenfunktionen und Punktfunktionen der angegebenen Art entsprechen einander eineindeutig.

Die *charakteristische Funktion* einer Verteilung $F(M)$ ist eine Funktion $f(t_1, \ldots, t_n)$ der n Variablen $t_1, \ldots, t_n$, definiert durch das über den ganzen Raum R_n erstreckte Lebesgue-Radonsche Integral

$$f(t_1, \ldots, t_n) = \int_{R_n} e^{i(t_1 x_1 + \cdots + t_n x_n)}\, dF. \tag{7}$$

Die Variablen t_r können wir hier immer reell voraussetzen. Die Verteilung $F(M)$ wird durch ihre charakteristische Funktion eindeutig bestimmt.

Die *Faltung* zweier Verteilungen F_1 und F_2 ist eine neue Verteilung F, definiert durch das Integral

$$F(M) = \int_{R_n} F_1(M - x)\, dF_2,$$

wo $M - x$ die aus M durch die Translation $-x$ entstandene Menge bezeichnet. Zwischen den entsprechenden charakteristischen Funktionen besteht wie im eindimensionalen Fall die Beziehung

$$f = f_1 \cdot f_2. \tag{8}$$

Eine normale Verteilungsfunktion in R_n wird gewöhnlich in direkter Verallgemeinerung des eindimensionalen Falles als Integral über eine Frequenzfunktion definiert, wobei im Exponenten eine positiv-definite quadratische Form der Koordinaten auftritt. Für die Verallgemeinerung von Satz 1 brauchen wir aber hier einen etwas umfassenderen Begriff, der gewisse ausgeartete Fälle mit einschließt, welche Grenzfälle von eigentlichen Normalverteilungen sind. Wir definieren daher:

[7]) Für die im folgenden benutzten Sätze über Mengenfunktionen vergleiche man z. B. Bochner [1], Haviland [4], Jessen-Wintner [5].

$F(M)$ ist eine *normale Verteilungsfunktion in* R_n, wenn die entsprechende charakteristische Funktion von der Form

$$f(t_1, \ldots, t_n) = e^{iL - \frac{1}{2}Q} \tag{9}$$

ist, wo L eine reelle Linearform und

$$Q = \sum_{r,s} a_{rs} t_r t_s \qquad (r, s = 1, 2, \ldots, n;\ a_{rs} = a_{sr})$$

eine nicht-negative quadratische Form in den Variablen t_r bezeichnet. Wenn Q eine positiv-definite Form ist, heißt $F(M)$ *eigentlich normale Verteilungsfunktion in* R_n; wenn aber Q nur semi-definit ist, nennen wir $F(M)$ eine *ausgeartete normale Verteilungsfunktion in* R_n.

Durch eine geeignete Translation des Koordinatensystems (wobei der Anfangspunkt in den „Schwerpunkt" der Verteilung kommt) kann man immer die Linearform L zum Verschwinden bringen, ohne die quadratische Form Q zu verändern. Die charakteristische Funktion erscheint also nach Ausführung der Translation in der Form

$$f(t_1, \ldots, t_n) = e^{-\frac{1}{2}Q}. \tag{9a}$$

Jede eigentlich normale Verteilung $F(M)$ läßt sich nach Ausführung der genannten Translation durch das über die Borelsche Menge M erstreckte Lebesguesche Integral

$$F(M) = \frac{1}{(2\pi)^{\frac{n}{2}}\sqrt{A}} \int_M \cdots \int e^{-\frac{1}{2}P}\, dx_1 \ldots dx_n$$

darstellen, wo $A = |a_{rs}| > 0$ und

$$P = \sum_{r,s} \frac{A_{rs}}{A} x_r x_s$$

die zu Q reziproke Form ist. Die Funktion

$$\frac{1}{(2\pi)^{\frac{n}{2}}\sqrt{A}}\, e^{-\frac{1}{2}P}$$

heißt dann die *n-dimensionale Frequenzfunktion* von $F(M)$.

Für eine ausgeartete Normalverteilung ist aber $A = 0$, und der obige Ausdruck der Frequenzfunktion wird daher sinnlos. Die entsprechende Verteilungsfunktion läßt sich überhaupt nicht als das Integral einer n-dimensionalen Frequenzfunktion darstellen. Sie verschwindet für jede Menge M, die außerhalb der durch die Gleichung $\sum_{r,s} A_{rs} x_r x_s = 0$ (also durch eine oder mehrere lineare Gleichungen zwischen den x_r) definierten Menge M^* liegt, und könnte in M^* als das Integral einer Frequenzfunktion mit weniger als n Variablen dargestellt werden[8]). Jede ausgeartete Normalverteilung läßt sich, wie man leicht sieht, als Grenzfunktion eigentlicher Normalverteilungen darstellen.

Aus der Form der charakteristischen Funktion einer Normalverteilung folgt, daß die Faltung zweier Normalverteilungen F_1 und F_2 wieder eine Normalverteilung F ergibt. Da die entsprechenden quadratischen Formen Q sich hierbei nach (8) additiv zusammensetzen, muß F notwendig eigentlich normal werden, so bald wenigstens eine der Komponenten eigentlich ist. Die Faltung zweier ausgearteten Normalverteilungen kann aber auch eigentlich normal sein, da ja die Summe von zwei semi-definiten Formen sehr wohl definit sein kann.

Nach diesen Vorbereitungen können wir jetzt Satz 1 in der folgenden Weise verallgemeinern:

Satz 2. *Wenn die Faltung $F(M)$ von zwei Verteilungsfunktionen eine normale Verteilungsfunktion in R_n ist, so haben auch die Komponenten $F_1(M)$ und $F_2(M)$ dieselbe Eigenschaft.*

Dieser Satz kann natürlich genau wie Satz 1 wahrscheinlichkeitstheoretisch gedeutet werden, etwa in der folgenden Weise: *Wenn die Summe von zwei unabhängigen stochastischen Vektoren in R_n normal verteilt ist, so ist auch jede Komponente für sich normal verteilt.*

Wir beweisen diesen Satz durch Zurückführung auf den eindimensionalen Fall[9]). Ohne Beschränkung der Allgemeinheit können wir annehmen, daß für die Verteilung F die obengenannte Translation schon ausgeführt ist, so daß die charakteristische Funktion $f(t_1, \ldots, t_n)$ durch (9a) gegeben ist.

Es seien nun den Veränderlichen $t_1, \ldots, t_n$ feste reelle Werte erteilt, die nicht sämtlich verschwinden. M_z bezeichne für jedes reelle z die durch

$$t_1 x_1 + \ldots + t_n x_n \leqq z \tag{10}$$

definierte Menge in R_n. Dann sind $F(M_z)$, $F_1(M_z)$ und $F_2(M_z)$ nirgends abnehmende Funktionen der reellen Veränderlichen z, die für $z \to -\infty$

---

[8]) Ein einfaches Beispiel einer ausgearteten Normalverteilung in R_2 erhalten wir in der folgenden Weise: Es sei z eine normal verteilte stochastische Variable. Der Vektor $(x = z,\ y = kz)$ hat dann für jedes konstante k eine Verteilung von jener Art. Der Punkt (x, y) liegt notwendig auf der Geraden $y = kx$, und die Verteilung kann durch eine auf dieser Geraden definierte eindimensionale normale Frequenzfunktion bestimmt werden.

[9]) Die hier benutzte Methode der Zurückführung wurde von H. Wold und dem Verfasser mit Hinblick auf verschiedene andere Fragen in der Theorie der n-dimensionalen Verteilungsfunktionen ausgearbeitet. Einige Beispiele derartiger Anwendungen sollen demnächst veröffentlicht werden.

dem Grenzwert 0, für $z \to +\infty$ dem Grenzwert 1 zustreben. Alle drei Funktionen sind also Verteilungsfunktionen im Sinne von Abschnitt I. Die entsprechenden charakteristischen Funktionen sind nach (7) und (10)

$$\int_{-\infty}^{\infty} e^{itz} d_z F(M_z) = \int_{R_n} e^{it\Sigma t_r x_r} dF = f(t t_1, \ldots, t t_n),$$

und entsprechend für F_1 und F_2. Aus (8) und (9a) folgt aber

$$f(t t_1, \ldots, t t_n) = e^{-\frac{t^2}{2} Q(t_1, \ldots, t_n)},$$

$$f(t t_1, \ldots, t t_n) = f_1(t t_1, \ldots, t t_n) \cdot f_2(t t_1, \ldots, t t_n).$$

Die eindimensionale Verteilungsfunktion $F(M_z)$ ist nach der ersten Gleichung normal, nach der zweiten Gleichung die Faltung von $F_1(M_z)$ und $F_2(M_z)$. Nach Satz 1 sind also auch $F_1(M_z)$ und $F_2(M_z)$ als Funktionen von z normale Verteilungsfunktionen. Man hat demnach

$$f_1(t t_1, \ldots, t t_n) = e^{Rit - S\frac{t^2}{2}}, \tag{11}$$

wo R und S Funktionen der t_r sind. Die Momente erster und zweiter Ordnung der Funktion $F_1(M_z)$ sind somit

$$\int_{-\infty}^{\infty} z\, d_z F_1(M_z) = R,$$

$$\int_{-\infty}^{\infty} z^2 d_z F_1(M_z) = R^2 + S.$$

Aus der Definition (10) der Menge M_z folgt aber

$$\int_{-\infty}^{\infty} z\, d_z F_1(M_z) = \int_{R_n} (t_1 x_1 + \ldots + t_n x_n)\, dF_1,$$

$$\int_{-\infty}^{\infty} z^2 d_z F_1(M_z) = \int_{R_n} (t_1 x_1 + \ldots + t_n x_n)^2 dF_1,$$

wo die Konvergenz der rechten Seiten aus der (bereits feststehenden) Konvergenz der linken Seiten folgt. Durch Vergleich ergibt sich nun, daß R eine lineare und S eine quadratische Form in den t_r ist. Da (11) eine charakteristische Funktion in t ist, muß die quadratische Form S nicht-negativ sein. Setzt man dann in (11) $t = 1$, so folgt, daß $f_1(t_1, \ldots, t_n)$ von der Form (9) ist. Nach unserer Definition ist daher die entsprechende Verteilungsfunktion $F_1(M)$ eine normale Verteilungsfunktion in R_n. Entsprechendes gilt natürlich für $F_2(M)$.

Literaturverzeichnis.

[1] S. Bochner:
Monotone Funktionen, Stieltjessche Integrale und harmonische Analyse. Math. Annalen 108 (1933), S. 378—410.

[2] H. Cramér:
Sur une propriété de la loi de Gauss. C. R. Acad. Sci., Paris, 202 (1936), S. 615—616.

[3] W. Feller:
Über den zentralen Grenzwertsatz der Wahrscheinlichkeitsrechnung. Math. Zeitschr. 40 (1935), S. 521—559.

[4] E. K. Haviland:
On the theory of absolutely additive distribution functions. Amer. Journ. Math. 56 (1934), S. 625—658.

[5] B. Jessen and A. Wintner:
Distribution functions and the Riemann zeta function. Trans. Amer. Math. Soc. 38 (1935), S. 48—88.

[6] P. Lévy:
Calcul des probabilités. Paris 1925.

[7] Sur les intégrales dont les éléments sont des variables aléatoires indépendantes. Annali Scuola norm. sup. Pisa (II) 3 (1934), S. 337—366.

[8] Propriétés asymptotiques des sommes de variables aléatoires indépendantes ou enchainées. Journ. Math. pures appl. (VII) 14 (1935), S. 347—402.

(Eingegangen am 28. Februar 1936.)

36.

with H. Wold

Some theorems on distribution functions

J. London Math. Soc. **11**, 290–294 (1936)

Let R_n be a Euclidean space of n dimensions with the variable point $\mathbf{x} = (x_1, \ldots, x_n)$. A completely additive, non-negative set function $F(E)$, defined for all Borel sets E of R_n and such that $F(R_n) = 1$, will be called a *distribution function in* R_n. The *characteristic function* (or "Fourier-Stieltjes transform") of $F(E)$ is the function $f(\mathbf{t})$ defined for all points $\mathbf{t} = (t_1, \ldots, t_n)$ in R_n by the Lebesgue-Radon integral

$$(1) \qquad f(\mathbf{t}) = \int_{R_n} e^{i(t_1 x_1 + \ldots + t_n x_n)}\, dF.$$

It is known that a distribution function is uniquely determined by its characteristic function‡.

In this note we consider a relation (4) between the characteristic functions of an arbitrary distribution function in R_n and a certain associated distribution function in R_1. By means of this relation, it is often possible to find simple proofs of theorems on n-dimensional distributions, when the corresponding theorems in one dimension are known. We give here four applications of the method, while further applications will be given elsewhere§. Our applications III and IV are simple proofs of previously known theorems, while I and II seem to be new¶. The so called *central limit theorem* for probability distributions in R_n may be proved with the aid of IV.

Let $F(E)$ be a given distribution function in R_n and let $\mathbf{t} = (t_1, \ldots, t_n)$ be a point in R_n such that $\mathbf{t} \neq (0, \ldots, 0)$. Further, let s be a given Borel set in a one-dimensional space R_1, and let $S_\mathbf{t}$ denote the set which consists of all points $\mathbf{x}$ in R_n such that $\mathbf{tx} = t_1 x_1 + \ldots + t_n x_n$ belongs to the set s. Then a distribution function $G_\mathbf{t}(s)$ in R_1 may be defined by putting

$$G_\mathbf{t}(s) = F(S_\mathbf{t}).$$

Now it is well known that the one-dimensional distribution function $G_\mathbf{t}$ is completely defined if, for every real z, we know its value for the set

---

† Received 21 March, 1936; read 23 April, 1936.

‡ *Cf.* Lévy [7], Bochner [1], Haviland [5].

§ *Cf.* Cramér [3].

¶ Prof. Carleman kindly informs us that Theorem I is known to him. His method of proof is, however, quite different from ours.

s which consists of all real numbers not greater than z. The corresponding set $S_{\mathbf{t}}$ in R_n is the half-space $S_{\mathbf{t},z}$ defined by the inequality

$$(2) \qquad t_1 x_1 + \ldots + t_n x_n \leqslant z.$$

Thus the distribution function $G_{\mathbf{t}}$ in R_1 may be replaced by the function

$$(3) \qquad F(S_{\mathbf{t},z}),$$

which for fixed $\mathbf{t}$ is a never decreasing function of the real variable z, tending to 0 as $z \to -\infty$ and to 1 as $z \to +\infty$. The corresponding characteristic function is, by (1),

$$(4) \qquad \int_{-\infty}^{\infty} e^{iuz} d_z F(S_{\mathbf{t},z}) = \int_{R_n} e^{iu(t_1 x_1 + \ldots + t_n x_n)} d_{\mathbf{x}} F = f(u\mathbf{t}),$$

where u is a real variable. By means of (4) we obtain immediately the following uniqueness theorem.

I. *If* $F_1(E)$ *and* $F_2(E)$ *are distribution functions in* R_n, *such that* $F_1 = F_2$ *for every half-space* (2), *then* $F_1 \equiv F_2$.

Proof. By hypothesis $F_1(S_{\mathbf{t},z}) = F_2(S_{\mathbf{t},z})$ for all real z, so long as

$$\mathbf{t} \neq (0, \ldots, 0).$$

Then, for $u = 1$, the relation (4) gives $f_1(\mathbf{t}) = f_2(\mathbf{t})$, and obviously this holds also for $\mathbf{t} = (0, \ldots, 0)$. Thus, by the uniqueness theorem for characteristic functions, we have $F_1 \equiv F_2$.

Remark. It follows from the proof that it would be sufficient to assume that, for any given $\mathbf{t} \neq (0, \ldots, 0)$, the two distributions coincide for *almost all* real z. This remark will be used below.

We next consider the problem of moments for distribution functions in R_n.

II.† *Denote by* $\mu_{p_1, \ldots, p_n}$ $(p_i = 0, 1, \ldots)$ *the moments*

$$\mu_{p_1, \ldots, p_n} = \int_{R_n} x_1^{p_1} \ldots x_n^{p_n} dF$$

---

† Less precise results in this direction have been given by Romanovsky [8] and Haviland [5].

of a distribution function $F(E)$ in R_n, and put

$$\lambda_k = \mu_{k,0,\ldots,0} + \mu_{0,k,0,\ldots,0} + \cdots + \mu_{0,0,\ldots,k}.$$

Then, if the series $$\sum_{k=1}^{\infty} \lambda_{2k}^{-1/2k}$$

is divergent, the distribution $F(E)$ is uniquely determined by its moments.

Proof. For $n=1$ the theorem reduces to the classical Carleman criterion (*cf.* [2]), so that we need consider only the case $n>1$. The $2k$-th moment of the one-dimensional distribution (3) is

$$\begin{aligned}\nu_{2k} &= \int_{-\infty}^{\infty} z^{2k}\, d_z F(S_{\mathbf{t},z}) \\ &= \int_{R_n} (t_1 x_1 + \ldots + t_n x_n)^{2k}\, dF \\ &\leqslant n^{k-1} |\mathbf{t}|^{2k} \int_R (x_1^{2k} + \ldots + x_n^{2k})\, dF \\ &= n^{k-1} |\mathbf{t}|^{2k} \lambda_{2k}.\end{aligned}$$

Hence the series $\sum_1^{\infty} \nu_{2k}^{-1/2k}$ is divergent, and by the Carleman criterion the one-dimensional distribution (3) is uniquely determined by the moments μ for every fixed $\mathbf{t} \neq (0, \ldots, 0)$. Any two different distribution functions in R_n, having the same moments μ, thus coincide for every half-space (2). Consequently, by Theorem I, they are identical.

In the following (previously known) Theorems III and IV it will be assumed that the theorems are already known for $n=1$, and then the truth of the theorems for $n>1$ will be proved by our method.

III. *Convolution theorem for characteristic functions*†. *$F_1(E)$ and $F_2(E)$ being given distribution functions in R_n, the convolution ("Faltung") of F_1 and F_2 is the distribution function $F(E)$ defined by*

$$F(E) = \int_{R_n} F_1(E - \mathbf{x})\, dF_2, \tag{5}$$

† *Cf.* Bochner [1], Haviland [5].

where $E-\mathbf{x}$ denotes the set obtained from E by the translation $-\mathbf{x}$. Then the corresponding characteristic functions satisfy the relation

$$f(\mathbf{t}) = f_1(\mathbf{t}) f_2(\mathbf{t}). \tag{6}$$

Proof. If we consider the function (3) and the corresponding functions formed with F_1 and F_2, it follows from (5) that

$$F(S_{\mathbf{t},z}) = \int_{-\infty}^{x} F_1(S_{\mathbf{t},z-w})\, d_w F_2(S_{\mathbf{t},w}).$$

Thus, if in (4) we put $u = 1$, we obtain (6) from the convolution theorem for one-dimensional distributions.

IV. *Continuity theorem for characteristic functions*†. *Let $\{F_k(E)\}$ be a sequence of distribution functions in R_n and let $\{f_k(\mathbf{t})\}$ be the corresponding sequence of characteristic functions. If the sequence $\{f_k(\mathbf{t})\}$ converges uniformly in every finite sphere $|\mathbf{t}| < A$, then the limit function $f(\mathbf{t})$ is the characteristic function of a distribution $F(E)$, and the sequence $\{F_k(E)\}$ converges to the limit $F(E)$ for every Borel set E which is a continuity set of $F(E)$*‡.

Proof. For every fixed $\mathbf{t}$, the sequence $\{f_k(u\mathbf{t})\}$ converges, by hypothesis, uniformly in every finite interval of the real variable u. Thus, by (4) and the continuity theorem in one dimension, the sequence $\{F_k(S_{\mathbf{t},z})\}$ converges for every fixed $\mathbf{t} \neq (0, \dots 0)$ and for almost all real z to a definite limit $L_{\mathbf{t},z}$, which is a never increasing function of z, tending to 0 as $z \to -\infty$ and to 1 as $z \to +\infty$. It also follows from the one-dimensional theorem that the limit function $f(u\mathbf{t})$ satisfies the relation

$$f(u\mathbf{t}) = \int_{-\infty}^{\infty} e^{iuz}\, d_z L_{\mathbf{t},z}. \tag{7}$$

It is always possible§ to choose from the sequence $\{F_k(E)\}$ a sub-sequence which converges to a completely additive, non-negative set

† One dimensional case: Lévy [7]; $n > 1$: Romanovsky [8], Bochner [1], Haviland [6].

‡ E is a continuity set of $F(E)$ if $F(E') = F(E'')$, where E' is the set of all interior points of E and E'' is the closure of E. For any given $F(E)$ and $\mathbf{t} \neq (0, \dots, 0)$, the half-space $S_{\mathbf{t},z}$ is a continuity set of $F(E)$ for all real z with the exception of at most an enumerable set.

§ This is proved in practically the same way as in the one-dimensional case. *Cf. e.g.* Frostman [4].

function $F^*(E)$ in all continuity sets of the latter. Obviously, we have $0 \leqslant F^*(E) \leqslant 1$. We shall first show that $F^*(E)$ is always a distribution function, *i.e.* that $F^*(R_n) = 1$. This follows immediately from the fact that, for every fixed $\mathbf{t} \neq (0, \ldots, 0)$ and for almost all real z, we must have $F^*(S_{\mathbf{t},z}) = L_{\mathbf{t},z} \to 1$ as $z \to +\infty$.

Further, by the remark to Theorem I, the limit functions $F^*(E)$ corresponding to any two convergent sub-sequences must be identical, since, for every fixed $\mathbf{t} \neq (0, \ldots, 0)$, they coincide for almost all half-spaces (2). This obviously implies the convergence of the sequence $\{F_k(E)\}$. Finally, it follows from (7) that the limit function $f(\mathbf{t})$ is the characteristic function of the distribution $\lim_{k \to \infty} F_k(E)$.

Remark. It has been pointed out above that Theorem IV may be applied in proving the so-called *central limit theorem in probability.* As a matter of fact, let $F(E)$ be a given distribution function in R_n such that the moments of the first order disappear, while the moments of the second order are finite. Then it is easily shown that the expression $\{f(\;/\sqrt{k})\}^k$ tends, as $k \to \infty$, to a definite limit which is the characteristic function of a certain Gaussian distribution in R_n. Thus, according to Theorem IV, the k-fold convolution of $F(E)$ with itself gives a distribution which, after an appropriate linear substitution, tends to a Gaussian distribution. This theorem for n-dimensional distributions is thus by our method deduced directly from the corresponding limit theorem in one dimension.

References.

[1] Bochner, S., " Monotone Funktionen, Stieltjessche Integrale und harmonische Analyse ", *Math. Annalen*, 108 (1933), 378–410.

[2] Carleman, T., *Les fonctions quasi-analytiques*, (Paris, 1926), Collection Borel.

[3] Cramér, H., " Über eine Eigenschaft der normalen Verteilungsfunktion ", *Math. Zeitschrift*, 41 (1936), 405–414.

[4] Frostman, O., " Potentiel d'équilibre et capacité des ensembles ", *Meddel. Lunds Univ. Mat. Seminar.*, 3 (1935), 1–118.

[5] Haviland, E. K., " On the theory of absolutely additive distribution functions ", *American J. of Math.*, 56 (1934), 625–658.

[6] ————, " On the inversion formula for Fourier-Stieltjes transforms in more than one dimension ", *American J. of Math.*, 57 (1935), 94–100 and 382–388.

[7] Lévy, P., *Calcul des probabilités*, (Paris, 1925).

[8] Romanovsky, V., " Sur un théorème limite du calcul des probabilités ", *Rec. Soc. Math. Moscou*, 36 (1929), 36–64.

Department of Mathematical Statistics,
University of Stockholm.

Printed by C. F. Hodgson & Son, Ltd., Newton St., London, W.C.2.

37.

On the order of magnitude of the difference between consecutive prime numbers

Acta Arith. **2** (1), 23–46 (1936)

Introduction.

Let p_n denote the n: th prime number. It has been proved by Hoheisel [8] [1]) that we have

(1) $$p_{n+1} - p_n = O(p_n^{1-\delta})$$

for some $\delta > 0$. On the other hand, it is known (Westzynthius [11]) that the relation

(2) $$p_{n+1} - p_n = O(\log p_n)$$

is certainly *not* true. Thus with respect to the maximum order of the difference $p_{n+1} - p_n$ there remains a large domain of uncertainty.

If the Riemann hypothesis is assumed, it is possible (Cramér [4]) to improve (1) to

(3) $$p_{n+1} - p_n = O(\sqrt{p_n} \log p_n),$$

but obviously even in this case a comparatively wide gap is still left open between (2) and (3). It has been conjectured by Piltz [9] that we have for every $\varepsilon > 0$

$$p_{n+1} - p_n = O(p_n^{\varepsilon}),$$

but this has never been proved.

---

1) Numbers in brackets refer to the appended list of references.

1. Odbitka.

In the first section of the present paper, an heuristic methot founded on probability arguments is briefly exposed. It is suggested that the true maximum order of $p_{n+1}-p_n$ should be equal to $(\log p_n)^2$, so that we should be able to replace (1) and (3) by [2])

$$p_{n+1}-p_n=O((\log p_n)^2). \tag{4}$$

In the second section it will be shown that, *if the Riemann hypothesis is assumed*, a number of results may be proved which, roughly speaking, may be interpreted in the following way. Let us consider the primes p_n, such that the difference $p_{n+1}-p_n$ is *exceptionally large, i. e.* larger than some function $f(p_n)$ increasing more rapidly than $(\log p_n)^2$. *Then the frequency of such primes* p_n *is small.*

We shall here only mention two particular theorems belonging to this order of ideas. (For preliminary results cf. Cramér [4], [5], [6].)

1) Consider the sums

$$S(x)=\sum_{p_n<x}(p_{n+1}-p_n)$$

and

$$S_1(x)=\sum_{p_n<x}{}'(p_{n+1}-p_n),$$

the first of which is extended to all primes $p_n<x$, while in the second the summation is restricted to those $p_n<x$ which satisfy

$$p_{n+1}-p_n>(\log p_n)^3.$$

We then obviously have, as x tends to infinity,

$$S(x)\sim x,$$

while it will be shown that on the Riemann hypothesis we have

$$S_1(x)=O\left(\frac{x}{\log\log x}\right)=o(x).$$

This is only a very particular case of our theorem II, which gives an upper limit for the frequency of „prime intervals" (p_n, p_{n+1}) satisfying an inequality of the form $p_{n+1}-p_n>p_n^{\alpha}(\log p_n)^{\beta}$.

2) If the relation (4) could be proved, it is immediately seen that the series

---

2) Cf. the numerical data given by Western [10].

$$\sum_{n=2}^{\infty} \frac{(p_{n+1}-p_n)^2}{p_n(\log p_n)^{\lambda}}$$

would be convergent for $\lambda > 4$. It will be shown that this is actually the case, if the Riemann hypothesis is true. (For $\lambda \leqq 2$ the series is certainly divergent.)

The proofs of the theorems of Section II are founded on a number of Lemmas, some of which are independent of the Riemann hypothesis. In particular, we would draw the attention to Lemma 3, from which *i. a.* a proof of Hoheisel's theorem (1) may be obtained.[3])

I. Results suggested by probability arguments.

In investigations concerning the asymptotic properties of arithmetic functions, it is often possible to make an interesting heuristic use of probability arguments. If, *e. g.*, we are interested in the distribution of a given sequence S of integers, we then consider S as a member of an infinite class C of sequences, which may be concretely interpreted as the possible realizations of some game of chance.[4]) It is then in many cases possible to prove that, *with a probability* $= 1$, a certain relation R holds in C, *i. e.* that in a definite mathematical sense „almost all" sequences of C satisfy R. Of course we cannot in general conclude that R holds for the particular sequence S, but results suggested in this way may sometimes afterwards be rigorously proved by other methods.

With respect to the ordinary prime numbers, it is well known that, roughly speaking, we may say that the chance that a given inte-

[3]) While the present paper was being printed, N. Tchudakoff has published a theorem (C. R. Acad. Sci. U. R. S. S., vol. I, 1936, p. 201) on the zeros of the function $\zeta(s)$, from which he states (without proof) that it is possible to deduce the relation $p_{n+1} - p_n = O\left(p_n^{\frac{3}{4}+\varepsilon}\right)$ for every $\varepsilon > 0$. This deduction can be performed by means of our Lemma 3.

[4]) Arguments of this character being frequently misunderstood, it will be convenient to make the following remarks. By the methods of the modern theory of probability, the class C may be defined in a purely analytic way as an abstract space without any reference to concrete interpretation. The term „almost all" is then interpreted in the sense of the Lebesgue measure theory. Up to this point, the developments indicated in the text are thus mathematically exact. The heuristic part of the argument does not come in until it is suggested that the relation R may hold for the particular sequence S. The present Section I being of an introductory character, we shall not enter upon all details of the proofs. The theorems on probability required in the sequel will be found in a convenient form *e. g.* in Cantelli [2], p. 334 and 336.

ger n should be a prime is approximately $\frac{1}{\log n}$. This suggests that by considering the following series of independent trials we should obtain sequences of integers presenting a certain analogy with the sequence of ordinary prime numbers p_n.

Let $U_1, U_2, U_3, \ldots$ be an infinite series of urns containing black and white balls, the chance of drawing a white ball from U_n being $\frac{1}{\log n}$ for $n > 2$, while the composition of U_1 and U_2 may be arbitrarily chosen. We now assume that one ball is drawn from each urn, so that an infinite series of alternately black and white balls is obtained. If P_n denotes the number of the urn from which the n: th white ball in the series was drawn, the numbers $P_1, P_2, \ldots$ will form an increasing sequence of integers, and we shall consider the class C of all possible sequences (P_n). Obviously the sequence S of ordinary prime numbers (p_n) belongs to this class.

We shall denote by $\Pi(x)$ the number of those P_n which are $\leqq x$, thus forming an analogy to the ordinary notation $\pi(x)$ for the number of primes $p_n \leqq x$. Then $\Pi(x)$ is a random variable, and if we denote by z_n a variable taking the value 1 if the n: th urn gives a white ball and the value 0 in the opposite case, we have

$$\Pi(x) = \sum_{n \leqq x} z_n,$$

and it is easily seen that the mean value of $\Pi(x)$ is, for large values of x, asymptotically equal to $\mathrm{Li}(x)$. It is, however, possible to obtain much more precise information concerning the behaviour of $\Pi(x)$ for large values of x. As a matter of fact, it may be shown (cf. Cramér [6]) that, *with a probability* = 1, the relation

$$\limsup_{x \to \infty} \frac{|\Pi(x) - \mathrm{Li}(x)|}{\sqrt{2x} \cdot \sqrt{\frac{\log \log x}{\log x}}} = 1 \tag{5}$$

is satisfied. With respect to the corresponding difference $\pi(x) - \mathrm{Li}(x)$ in the prime number problem, it is known that, if the Riemann hypothesis is assumed, the true maximum order of this difference lies between the functions $\frac{\sqrt{x}}{\log x}$ and $\sqrt{x} \cdot \log x$. It is interesting to find that the order of the function occurring in the denominator of (5) falls inside this interval of indetermination.

We shall now consider the order of magnitude of the difference $P_{n+1}-P_n$. Let $c>0$ be a given constant and let E_m denote the event that black balls are obtained from all urns $U_{m+\nu}$ with $1\leqq\nu\leqq c(\log m)^2$. Then it is seen that the following two events have the same probability: a) The inequality

$$P_{n+1}-P_n>c(\log P_n)^2 \tag{6}$$

is satisfied for an infinity of values of n, and b) An infinite number of the events E_m are realized.

If ε_m denotes the probability of the event E_m, we have

$$\varepsilon_m=\prod_{\nu=1}^{c(\log m)^2}\left(1-\frac{1}{\log(m+\nu)}\right)$$

and it is easily shown that we can find two positive constants A and B such that for all sufficiently large values of m

$$\frac{A}{m^c}<\varepsilon_m<\frac{B}{m^c}. \tag{7}$$

Thus if $c>1$ the series $\Sigma\,\varepsilon_m$ is convergent, and consequently the probability of the realization of an infinite number of events E_m is equal to zero. (Cf. Cantelli [2], p. 334.)

On the other hand, suppose $c<1$ and let us consider the events $E_{m_1}, E_{m_2}, \ldots\ldots$, where $m_1=2$ and

$$m_{r+1}=m_r+[c.(\log m_r)^2]+1.$$

It is then shown without difficulty that we have for some constant K and for all sufficiently large r

$$m_r<Kr(\log r)^2,$$

and thus according to (7) the series $\Sigma\,\varepsilon_{m_r}$ is divergent if $c<1$. The events E_{m_r} being mutually independent, we conclude that with a probability $=1$ an infinite number of these events will be realized. (Cf. Cantelli [2], p. 336.)

Thus the probability of an infinite number of solutions of the inequality (6) is equal to zero if $c>1$ and to one if $c<1$. Combining these two results, we obtain the following theorem: *With a probability* $=1$, *the relation*

$$\limsup_{n\to\infty}\frac{P_{n+1}-P_n}{(\log P_n)^2}=1$$

is satisfied. — Obviously we may take this as a suggestion that, for the particular sequence of ordinary prime numbers p_n, some similar relation may hold.

II. Some theorems concerning the difference $p_{n+1} - p_n$.

We shall begin by proving a series of Lemmas, the three first of which are independent of the Riemann hypothesis. — Let us denote by $s = \sigma + i\tau$ a complex variable and by $\rho = \beta + i\gamma$, $(\gamma > 0)$, a complex zero of $\zeta(s)$, situated in the upper half-plane. By $\Lambda(n)$ we denote the arithmetical function defined by the relations

$$\Lambda(n) = \begin{cases} \log p & \text{for} \quad n = p^m \quad (p \text{ prime, } m \text{ integer}), \\ 0 & \text{otherwise.} \end{cases}$$

We shall consider the following two analytic functions:

$$F(s) = \sum_{\gamma > 0} e^{\rho i s} = \sum_{\gamma > 0} e^{-(\gamma - i\beta)s}, \tag{8}$$

the sum being extended to all zeros ρ in the upper half-plane, and

$$G(s) = \sum_{n=2}^{\infty} \frac{\Lambda(n)}{n} \left(\frac{1}{s - i\log n} + \frac{1}{i\log n} \right). \tag{9}$$

Obviously the Dirichlet series with complex exponents representing $F(s)$ is absolutely convergent for $\sigma > 0$, and $F(s)$ is regular in every point of this half-plane. $G(s)$ is a meromorphic function with simple poles in the points $s = i\log p^m$. Putting

$$H(s) = 2\pi F(s) + G(s), \tag{10}$$

it can be shown (cf. Cramér [3]) that, if a cut is made in the s-plane along the negative imaginary axis from $s = 0$ to $s = -i\infty$, $H(s)$ is regular and uniform in every finite domain which has no point common with the cut. In this paper we shall, however, only consider the function $H(s)$ in the domain D defined by the inequalities

$$0 < \sigma \leqq 1, \qquad \tau > 1.$$

In the first place, the following Lemma will be proved.

Lemma 1. *We have*

$$\Re H(s) = \pi + O\left(\frac{1}{\tau}\right)$$

uniformly in D.

According to a theorem which I have previously given (Cramér [3], formula (13), p. 114), we have for $\sigma > 0$

$$2\pi F(s) = e^{is} G(-s) - G(s) + \left(\frac{\pi}{2} + bi\right)\left(1 + \frac{1}{si}\right) - i\frac{\Gamma'}{\Gamma}\left(\frac{s}{\pi}\right) +$$

$$+ \pi e^{is} - s \int_0^1 e^{isv} \log |\zeta(v)|\, dv + \frac{1}{s} \int_0^{\infty e^{i\alpha}} \frac{z}{e^z - 1} \cdot \frac{dz}{z + is},$$

where b denotes a real constant and the last integral is taken along the vector $\arg z = \alpha$ with $0 < \alpha < \frac{\pi}{2}$. If, now, we suppose that s belongs to the domain D, we get by some easy calculation

$$2\pi F(s) = -G(s) + \frac{\pi}{2} + bi - i\frac{\Gamma'}{\Gamma}\left(\frac{s}{\pi}\right) -$$

$$- s \int_0^1 e^{isv} \log |\zeta(v)|\, dv + O\left(\frac{1}{\tau}\right).$$

Throughout the proof of this Lemma, all O's hold uniformly in D. By well-known properties of the Gamma function we have in D

$$\frac{\Gamma'}{\Gamma}\left(\frac{s}{\pi}\right) = \log \frac{\tau}{\pi} + \frac{\pi i}{2} + O\left(\frac{1}{\tau}\right).$$

Thus we obtain by (10)

$$\Re H(s) = \pi - \Re s \int_0^1 e^{isv} \log |\zeta(v)|\, dv + O\left(\frac{1}{\tau}\right).$$

We find, however, easily

$$\Re s \int_0^1 e^{isv} \log |\zeta(v)|\, dv = O\left(\int_0^1 (1 + v\tau) e^{-v\tau} \log \frac{2}{1 - v}\, dv\right) = O\left(\frac{1}{\tau}\right),$$

and thus Lemma 1 is proved.

Introducing the definition of $G(s)$ according to (9), we obtain from Lemma 1

$$\sum_{n=2}^{\infty} \frac{\Lambda(n)}{n} \cdot \frac{\sigma}{\sigma^2 + (\tau - \log n)^2} = \pi - 2\pi \Re F(s) + O\left(\frac{1}{\tau}\right). \tag{11}$$

If in this relation we take σ very small and τ very large, and consider the quantity

$$\frac{\sigma}{\sigma^2 + (\tau - \log n)^2}, \quad (n = 2, 3, \ldots),$$

it is readily seen that this quantity is large for values of n lying near e^τ, but becomes small as soon as n differs considerably from e^τ. This makes it possible to show that the value of the sum in the first member of (11) is dominated by the terms corresponding to values of n in a certain vicinity of e^τ. As $\Lambda(n)$ differs from zero only when n is a power of a prime number, and the influence of the squares and higher powers can be estimated without difficulty, we can in this way obtain some information as to the occurrence of primes in a given interval. This will be shown by Lemma 3 below. For the proof of this Lemma, the following elementary Lemma 2 will be required.

Lemma 2. *Two positive constants a and b being given, we can always determine $C = C(a, b)$ such that*

$$\sum_{x < n \leqq x+h} \frac{\Lambda(n)}{n} < C \frac{h \log x}{x \log h}$$

holds for $x > 2$, $h > 2$, $a \log x < h < b x$.

Denoting by $f(x) = \pi(x) + \frac{1}{2}\pi\left(x^{\frac{1}{2}}\right) + \ldots$ the well-known prime number function introduced by Riemann, we have

$$\sum_{x < n \leqq x+h} \frac{\Lambda(n)}{n} < \frac{\log(b+1)x}{x} (f(x+h) - f(x))$$

$$= \frac{\log(b+1)x}{x} \sum_{r=1}^{\frac{\log(x+h)}{\log 2}} \frac{1}{r}\left(\pi\left((x+h)^{\frac{1}{r}}\right) - \pi\left(x^{\frac{1}{r}}\right)\right)$$

$$< C \frac{\log x}{x} \sum_{r=1}^{C \log x} \frac{1}{r}\left(\pi\left(x^{\frac{1}{r}} + h x^{\frac{1}{r}-1}\right) - \pi\left(x^{\frac{1}{r}}\right)\right).$$

Throughout the proof of this Lemma, the letter C will be used to denote an unspecified constant depending only on a and b. — We have further (cf. Brun [1] p. 32—35, Hardy-Littlewood [7] p. 69)

$$\pi(x+h)-\pi(x)<C\frac{h}{\log h},$$

$$\pi\left(x^{\frac{1}{r}}+hx^{\frac{1}{r}-1}\right)-\pi\left(x^{\frac{1}{r}}\right)<hx^{\frac{1}{r}-1}+1\leqq\frac{h}{\sqrt{x}}+1,\ (r\geqq 2),$$

and thus we obtain

$$\sum_{x<n\leqq x+h}\frac{\Lambda(n)}{n}<C\frac{\log x}{x}\left(\frac{h}{\log h}+\left(\frac{h}{\sqrt{x}}+1\right)\sum_{r=2}^{C\log x}\frac{1}{r}\right)$$

$$<C\frac{\log x}{x}\left(\frac{h}{\log h}+\left(\frac{h}{\sqrt{x}}+1\right)\log\log x\right)$$

$$<C\frac{h\log x}{x\log h},$$

so that Lemma 2 is proved. — We now proceed to the proof of the fundamental Lemma 3.

Lemma 3. *It is possible to find two positive absolute constants* λ *and* τ_0 *such that the inequality*

$$\pi(e^{\tau}+\Delta)-\pi(e^{\tau}-\Delta)>\frac{\sigma e^{\tau}}{\tau}(1-3\,\Re F(s)),$$

where

$$\Delta=\frac{\lambda\sigma\tau}{\tau+\log\sigma},$$

holds for

(12) $$\begin{cases}\tau>\tau_0,\\ \tau\log^2\tau\, e^{-\tau}<\sigma<\dfrac{1}{\tau^2}.\end{cases}$$

In order to prove this Lemma, we shall consider (11). Putting

$$Z_n=\frac{\Lambda(n)}{n}\cdot\frac{\sigma}{\sigma^2+(\tau-\log n)^2},$$

$$\varphi=\frac{\tau}{\tau+\log\sigma}=\frac{\Delta}{\lambda\ \sigma},$$

we shall first show that λ and τ_0 may be so determined that we have, subject to the conditions (12),

(13) $$S=\sum{}' Z_n<\frac{1}{3},$$

the sum being extended to all $n\geqq 2$ such that

$$|\tau - \log n| > \lambda\varphi\sigma.$$

We put

(14)
$$S = S_1 + S_2 + S_3 + S_4,$$

the sums $S_1, \ldots, S_4$ containing the groups of terms Z_n defined by the inequalities:

$$\begin{aligned} S_1: &\quad \log n < \tau - 1, \\ S_2: &\quad \tau - 1 \leqq \log n < \tau - \lambda\varphi\sigma, \\ S_3: &\quad \tau + \lambda\varphi\sigma < \log n \leqq \tau + 1, \\ S_4: &\quad \tau + 1 < \log n. \end{aligned}$$

One or more of these sums may be empty, if the corresponding inequalities are not satisfied by any integral value of n. We shall assume from the beginning $\tau > \tau_0 > 10$. and the letter K will be used to denote an unspecified absolute constant. From (12) we obtain easily

(15)
$$1 < \varphi < \frac{\tau}{\log \tau}.$$

We now proceed to the evaluation of the sums S_ν. In the first place, we have by (12)

(16)
$$S_1 < \sigma \sum_{n=2}^{e^\tau} \frac{\Lambda(n)}{n} < K\sigma\tau < \frac{K}{\tau},$$

Further, if we put

$$S_4 = \sum_{\nu=1}^{\infty} U_\nu$$

with

$$U_\nu = \sum_{\tau+\nu < \log n \leqq \tau+\nu+1} Z_n$$

$$< \frac{\sigma}{\nu^2} \sum_{\tau+\nu < \log n \leqq \tau+\nu+1} \frac{\Lambda(n)}{n} < \frac{K\sigma}{\nu^2},$$

we have

(17)
$$S_4 < K\sigma < \frac{K}{\tau^2}.$$

We shall now consider S_3. Putting

$$V_\nu = \sum_{\tau+(\lambda\varphi+\nu)\sigma < \log n \leqq \tau+(\lambda\varphi+\nu+1)\sigma} Z_n,$$

we have

(18) $$S_3 \leqq \sum_{\nu=0}^{r} V_\nu,$$

r being determined by the condition

(19) $$(\lambda\varphi + r)\sigma < 1 \leqq (\lambda\varphi + r + 1)\sigma.$$

We have further

(20) $$V_\nu < \frac{1}{\sigma(\lambda\varphi+\nu)^2} \cdot \sum_{\tau+(\lambda\varphi+\nu)\sigma < \log n \leqq \tau+(\lambda\varphi+\nu+1)\sigma}' \frac{\Lambda(n)}{n}.$$

In order to estimate the sum in the second member of (20) by means of Lemma 2, we put in the inequality stated in this Lemma

$$x = e^{\tau+(\lambda\varphi+\nu)\sigma},$$

$$h = (e^\sigma - 1)\,x.$$

Then we have by (19) for $\nu = 0, 1, \ldots, r$

$$e^\tau < x < e^{\tau+1}.$$

Further we obtain by (12), observing the assumption $\tau_0 > 10$,

$$h > \sigma x > \sigma e^\tau > \tau \log^2 \tau$$

$$> \tau > \frac{1}{2} \log x,$$

and

$$h < 2\sigma x < 2x.$$

For $\tau_0 > 10$ we thus have $x > 2$, $h > 2$ and $\frac{1}{2}\log x < h < 2x$, so that according to Lemma 2 we obtain from (20)

$$V_\nu < K \frac{2\sigma x \cdot \log x}{x \cdot \log(2\sigma x)} \cdot \frac{1}{\sigma(\lambda\varphi+\nu)^2} < K \frac{\log x}{\log(\sigma x)} \cdot \frac{1}{(\lambda\varphi+\nu)^2}$$

$$< K \frac{\tau}{\log(\sigma e^\tau)} \cdot \frac{1}{(\lambda\varphi+\nu)^2} = \frac{K\varphi}{(\lambda\varphi+\nu)^2}.$$

Then (18) gives us

(21) $$S_3 < K\varphi \sum_{\nu=0}^{\infty} \frac{1}{(\lambda\varphi+\nu)^2}.$$

Now we have for $c > 1$

$$\sum_{\nu=0}^{\infty} \frac{1}{(c+\nu)^2} < \frac{2}{c}.$$

If $\lambda > 1$ we have by (15) $\lambda\varphi > 1$ and so we obtain from (21)

$$S_3 < \frac{K}{\lambda}. \tag{22}$$

In exactly the same way it can be shown that S_2 satisfies an inequality of the same form, and thus we conclude from (14), (16), (17) and (22)

$$S < K\left(\frac{1}{\tau} + \frac{1}{\tau^2} + \frac{1}{\lambda}\right).$$

Here K is an absolute constant, and thus it is possible to choose λ and τ_0 such that for $\tau > \tau_0$ we have $S < \frac{1}{3}$, *i. e.* (13) is proved.

The value of λ determined in this way will be regarded as definitely fixed, while obviously the value of τ_0 may without inconvenience be further increased. From (11) and (13) it follows that if τ_0 is sufficiently large we have, always subject to the conditions (12),

$$\sum_{|\tau - \log n| \leq \lambda\varphi\sigma} Z_n > \pi - \frac{2}{3} - 2\pi\,\Re\,F(s). \tag{23}$$

The terms Z_n occurring in (23) are different from zero only when n is a power of a prime number, $n = p^m$, and in this case we have

$$Z_n = \frac{\Lambda(n)}{n} \cdot \frac{\sigma}{\sigma^2 + (\tau - \log n)^2} < \frac{\tau + \lambda\varphi\sigma}{\sigma e^{\tau - \lambda\varphi\sigma}} \cdot \frac{1}{m}. \tag{24}$$

It follows from (12) and (15) that $\varphi\sigma < \frac{1}{\tau \log \tau}$, and thus if τ_0 is sufficiently large the right hand side of (24) is less than

$$\frac{2\pi}{3} \cdot \frac{\tau}{\sigma e^{\tau}} \cdot \frac{1}{m}.$$

This being so, we obtain from (23), $f(x)$ denoting the Riemann function (cf. p.)

$$\frac{2\pi}{3} \cdot \frac{\tau}{\sigma e^{\tau}}\left(f(e^{\tau + \lambda\varphi\sigma}) - f(e^{\tau - \lambda\varphi\sigma})\right) > \frac{3\pi}{4} - 2\pi\,\Re\,F(s),$$

$$f(e^{\tau + \lambda\varphi\sigma}) - f(e^{\tau - \lambda\varphi\sigma}) > \frac{\sigma e^{\tau}}{\tau}\left(\frac{9}{8} - 3\,\Re\,F(s)\right). \tag{25}$$

We shall now estimate the contribution to the left hand side of

(25) which is due to the squares and higher powers of prime numbers. If τ_0 is sufficiently large, we have by (12) and (15)

$$\sum_{m=2}^{\frac{\tau+\lambda\varphi\sigma}{\log 2}} \frac{1}{m}\left(\pi\left(e^{\frac{\tau+\lambda\varphi\sigma}{m}}\right)-\pi\left(e^{\frac{\tau-\lambda\varphi\sigma}{m}}\right)\right)$$

$$<\sum_{2}^{K\tau}\frac{1}{m}\left(e^{\frac{\tau+\lambda\varphi\sigma}{m}}-e^{\frac{\tau-\lambda\varphi\sigma}{m}}+1\right)<K\left(\lambda\varphi\sigma e^{\frac{\tau}{2}}\sum_{2}^{K\tau}\frac{1}{m^2}+\log\tau\right)$$

$$<K\frac{\sigma e^{\tau}}{\tau}\left(\lambda\varphi\tau e^{-\frac{\tau}{2}}+\frac{\tau\log\tau}{\sigma e^{\tau}}\right)<K\frac{\sigma e^{\tau}}{\tau}\left(\lambda e^{-\frac{\tau}{4}}+\frac{1}{\log\tau}\right)$$

$$<\frac{\sigma e^{\tau}}{8\tau},$$

and thus we obtain from (25), observing that $\lambda\varphi\sigma=\Delta$,

$$\pi(e^{\tau+\lambda\varphi\sigma})-\pi(e^{\tau-\lambda\varphi\sigma})>\frac{\sigma e^{\tau}}{\tau}(1-3\,\Re F(s)).$$

Thus Lemma 3 is proved.

Lemma 3 gives a lower limit for the number of primes in a certain interval. For a fixed value of τ, it is easily seen that the length of this interval is a steadily increasing function of σ between the limits imposed by (12). Let us now consider σ as a function of τ which for all sufficiently large τ satisfies the second relation (12). If, for a certain form of this function, it can be proved that

$$\Re F(s)<\frac{1}{3} \tag{26}$$

for all sufficiently large values of τ, it follows from Lemma 3 that there is at least one prime p in the interval $e^{\tau}(1-2\Delta)<p\leqq e^{\tau}(1+2\Delta)$. The smaller we can take the order of the function $\sigma=\sigma(\tau)$, the smaller becomes the order of magnitude of this interval. The principal difficulty of the problem consists in proving (26) for functions $\sigma(\tau)$ of sufficiently small order.

Putting in particular $\sigma=e^{-\delta\tau}$, it is possible to show that, if $\delta>0$ is sufficiently small, (26) holds for all sufficiently large τ. According to Lemma 3 it follows that, from a certain value of τ on, there is at least one prime p in the interval $e^{\tau}-\frac{2\lambda}{1-\delta}e^{(1-\delta)\tau}<p\leqq e^{\tau}+\frac{2\lambda}{1-\delta}e^{(1-\delta)\tau}$. Sub-

stituting here x for $e^{\tau} - \frac{2\lambda}{1-\delta} e^{(1-\delta)\tau}$ and c for $\frac{5\lambda}{1-\delta}$, we conclude that for all sufficiently large x, there is at least one prime p in the interval $x < p \leqq x + c x^{1-\delta}$. Taking $x = p_n$, we thus obtain a new proof of Hoheisel's relation (1). The detailed proof of (26) in this case will, however, not be given here[5]).

Up to this point, everything has been independent of the Riemann hypothesis. We shall now develop some consequences of this hypothesis, which in the sequel will be referred to as „the R. h." In the first place, it will be shown that by the aid of Lemma 3 we obtain a simple proof of the following theorem, first proved in 1919 (Cramér [4]).

Theorem I. *If the R. h. is true, then*

$$p_{n+1} - p_n = O\left(\sqrt{p_n} \log p_n\right).$$

If the Riemann hypothesis is true, every complex zero ρ of $\zeta(s)$ has the real part $1/2$, and thus we have

$$\Re F(s) \leqq |F(s)| < \sum_{\gamma>0} \left| e^{\left(\frac{1}{2}+i\gamma\right)is} \right| = e^{-\frac{1}{2}\tau} \sum_{\gamma>0} e^{-\gamma\sigma}. \tag{27}$$

Now, it is known that the number $N(T)$ of zeros satisfying the inequality $0 < \gamma < T$ is of the form

$$N(T) = \frac{T}{2\pi}\left(\log \frac{T}{2\pi} - 1\right) + O(\log T), \tag{28}$$

and hence we deduce for $\sigma \to 0$

$$\sum_{\gamma>0} e^{-\gamma\sigma} = \sigma \int_0^\infty N(v)\, e^{-v\sigma}\, dv \backsim \frac{1}{2\pi\sigma} \log \frac{1}{\sigma}. \tag{29}$$

Putting

$$\sigma = \tau e^{-\frac{1}{2}\tau}, \tag{30}$$

we conclude from (29)

$$e^{-\frac{1}{2}\tau} \sum_{\gamma>0} e^{-\gamma\sigma} \to \frac{1}{4\pi}$$

[5]) From Tchudakoff's theorem (cf. footnote [2])) it follows that we can here choose for δ any positive number $< \frac{1}{4}$.

as $\tau \to \infty$, and thus by (27) the relation (26) is certainly satisfied for all sufficiently large τ. Putting in Lemma 3 $\sigma = \tau e^{-\frac{1}{2}\tau}$ it thus follows that, from a certain value of τ on, there is at least one prime p in the interval $e^\tau - 2\lambda\tau e^{\frac{1}{2}\tau} < p \leqq e^\tau + 2\lambda\tau e^{\frac{1}{2}\tau}$. Substituting x for $e^\tau - 2\lambda\tau e^{\frac{1}{2}\tau}$ we conclude that for all sufficiently large x there is at least one prime p in the interval $x < p \leqq x + 5\lambda\sqrt{x}\log x$. Taking $x = p_n$, we obtain Theorem I.

As soon as we choose for $\sigma = \sigma(\tau)$ any function of lower order than (30), it seems very difficult to prove that (26) holds for *all* sufficiently large values of τ. If the R. h. is assumed we can, however, in certain cases prove that (26) holds *on the average*, as will be shown by the following Lemma 4.

Lemma 4. *Let $\sigma = \sigma(\tau)$ denote a function tending to zero as τ tends to infinity, such that for all $\tau > m > 0$, $\sigma(\tau)$ is steadily decreasing and satisfies the inequality $0 < \sigma(\tau) < 1$. Then if the R. h. is true we can find an absolute constant K such that for all $t > m$*

$$\int_t^{t+1} |F(\sigma + i\tau)|^2 d\tau < K e^{-t} \frac{\sigma(t)}{\sigma^2(t+1)} \log^2 \frac{1}{\sigma(t+1)}.$$

Putting *e. g.* $\sigma(\tau) = e^{-c\tau}$ with $\frac{1}{2} < c < 1$, we have

$$\int_t^{t+1} |F(\sigma + i\tau)|^2 d\tau < K t^2 e^{-(1-c)t},$$

and it is seen that, although in this case $\sigma(\tau)$ is of lower order than (30), $|F(s)|$ and thus *a fortiori* also $|\Re F(s)|$ is small *on the average* for large values of τ.

Throughout the proof we shall suppose $t > m$, and as before we shall use the letter K to denote an unspecified absolute constant. — Putting

$$f(s) = \sum_{\gamma > 0} e^{-\gamma s},$$

we have on the R. h. for $t < \tau < t+1$

$$F(s) = e^{\frac{1}{2}is} f(s) = e^{\frac{1}{2}i\sigma - \frac{1}{2}\tau} f(s).$$

$$|F(s)|^2 = e^{-\tau}|f(s)|^2,$$

(31) $$\int_t^{t+1} |F(\sigma+i\tau)|^2 d\tau < e^{-t}\int_t^{t+1} |f(\sigma+i\tau)|^2 d\tau.$$

Putting

$$\Phi(v,\tau) = \sum_{\gamma<v} e^{-\gamma i\tau},$$

we have further

$$f(\sigma+i\tau) = \sum_\gamma e^{-\gamma(\sigma+i\tau)} = \sigma\int_0^\infty \Phi(v,\tau)\, e^{-v\sigma}\, dv,$$

$$|f(\sigma+i\tau)|^2 \leqq \sigma^2\left(\int_0^\infty |\Phi(v,\tau)|\, e^{-v\sigma}\, dv\right)^2$$

$$\leqq \sigma^2 \int_0^\infty e^{-v\sigma}\, dv \cdot \int_0^\infty |\Phi(v,\tau)|^2 e^{-v\sigma}\, dv$$

$$= \sigma \int_0^\infty |\Phi(v,\tau)|^2 e^{-v\sigma}\, dv,$$

(32) $$\int_t^{t+1} |f(\sigma+i\tau)|^2 d\tau < \sigma(t)\int_0^\infty e^{-v\sigma(t+1)}\, dv \int_t^{t+1} |\Phi(v,\tau)|^2 d\tau.$$

Denoting by $g(x)$ the function

$$g(x) = 2 - |x|,$$

we have $g(x) > 0$ for $-2 < x < 2$ and $g(x) > 1$ for $0 < x < 1$. Thus we have

(33) $$\int_t^{t+1} |\Phi(v,\tau)|^2 d\tau \leqq \int_{-2}^{2} g(x)\, |\Phi(v, t+x)|^2 dx$$

$$\leqq \sum_{\substack{\gamma<v\\ \gamma'<v}} \left| \int_{-2}^{2} g(x)\, e^{i(\gamma-\gamma')(t+x)}\, dx \right|$$

$$\leqq 2 \sum_{\gamma \leqq \gamma' < v} \left| \int_{-2}^{2} g(x)\, e^{(\gamma-\gamma') i x}\, dx \right|$$

$$= 4 \sum_{\gamma \leqq \gamma' < v} \frac{1 - \cos 2(\gamma - \gamma')}{(\gamma - \gamma')^2}.$$

We have, however,

$$\frac{1-\cos 2(\gamma-\gamma')}{(\gamma-\gamma')^2} \leqq 2 \operatorname{Min}\left(1, \frac{1}{(\gamma-\gamma')^2}\right),$$

and hence

$$\sum_{\gamma \leqq \gamma' < v} \frac{1-\cos 2(\gamma-\gamma')}{(\gamma-\gamma')^2} <$$

$$< 2 \sum_{\gamma < v} \left(N(\gamma+1) - N(\gamma) + \frac{N(\gamma+2) - N(\gamma+1)}{1^2} + \ldots + \frac{N(v) - N(\gamma + [v-\gamma])}{[v-\gamma]^2} \right)$$

$$< K \sum_{\gamma < v} \left(\log \gamma + \frac{\log(\gamma+1)}{1^2} + \ldots + \frac{\log(\gamma + [v-\gamma])}{[v-\gamma]^2} \right)$$

$$< K \log v\, N(v) < K v \log^2 v.$$

Thus we obtain from (32) and (33)

$$\int_{t}^{t+1} |f(\sigma + i\tau)|^2\, d\tau < K \sigma(t) \int_{0}^{\infty} v \log^2 v\, e^{-v\sigma(t+1)}\, dv$$

$$< K \frac{\sigma(t)}{\sigma^2(t+1)} \log^2 \frac{1}{\sigma(t+1)}.$$

Finally, the truth of Lemma 4 follows from (31).

We are now in a position to prove a theorem which gives an upper limit for the frequency of certain exceptionally large „prime intervals" (p_n, p_{n+1}). We shall first introduce some new notations. Let α and β be constants such that

(34) $$0 \leqq \alpha \leqq \frac{1}{2}, \qquad \beta \geqq 0.$$

Putting

(35) $$h = h(x) = x^{\alpha} \log^{\beta} x,$$

we denote by $S_{\alpha,\beta}(x)$ the sum

(36)
$$S_{\alpha,\beta}(x) = \sum_{\substack{p_n \leqq x \\ p_{n+1}-p_n > h(p_n)}} (p_{n+1} - p_n)$$

which is equal to the *total length* of all prime intervals (p_n, p_{n+1}) such that $p_n \leqq x$ and

(37)
$$p_{n+1} - p_n > h(p_n).$$

Further, we denote by $N_{\alpha,\beta}(x)$ the *number* of primes $p_n \leqq x$ satisfying (37), so that

(38)
$$N_{\alpha,\beta}(x) = \sum_{\substack{p_n \leqq x \\ p_{n+1}-p_n > h(p_n)}} 1.$$

It is then trivial that we have

$$S_{\alpha,\beta}(x) = O(x),$$

and hence it can be simply deduced that we have

$$N_{\alpha,\beta}(x) = O\left(\frac{x}{h}\right).$$

If the R. h. is true, these results can be considerably improved, as shown by the following Theorem II.

Theorem II. *If the R. h. is true, the functions* $S_{\alpha,\beta}(x)$ *and* $N_{\alpha,\beta}(x)$ *defined by (36) and (38) satisfy the relations*

(39a)
$$S_{\alpha,\beta}(x) = O\left(\frac{x \log^3 x}{h \log h}\right) \quad \text{for} \quad 0 \leqq \alpha < \frac{1}{2}, \quad \beta \geqq 0,$$

$$\text{and for} \quad \alpha = \frac{1}{2}, \ 0 \leqq \beta \leqq 1,$$

(39b)
$$S_{\alpha,\beta}(x) = O(1) \quad \text{for} \quad \alpha = \frac{1}{2}, \quad \beta > 1,$$

and

(40a)
$$N_{\alpha,\beta}(x) = O\left(\frac{x \log^3 x}{h^2 \log h}\right) \quad \text{for} \quad 0 \leqq \alpha < \frac{1}{2}, \quad \beta \geqq 0,$$

(40b)
$$N_{\alpha,\beta}(x) = O(\log^{3-2\beta} x) \quad \text{for} \quad \alpha = \frac{1}{2}, \ 0 \leqq \beta \leqq 1,$$

(40c) $N_{\alpha,\beta}(x) = O(1)$ for $\alpha = \frac{1}{2}$, $\beta > 1$.

As soon as $h(x)$ increases for large values of x at least as rapidly as $\log^3 x$, (39) and (40) give better results than the trivial relations given above. Putting *e. g.* in (39a) $h(x) = \log^3 x$, we get the result

$$S_{0,3}(x) = \sum_{\substack{p_n \leq x \\ p_{n+1} - p_n > \log^3 p_n}} (p_{n+1} - p_n) = O\left(\frac{x}{\log \log x}\right)$$

stated in the Introduction.

Putting on the other hand $h(x) = \sqrt{x} \log x$, it follows from (40b) that the number of primes $p_n \leq x$, satisfying the inequality

$$p_{n+1} - p_n > \sqrt{p_n} \log p_n,$$

is at most of the form $O(\log x)$. If the second member of the last inequality is replaced by $C\sqrt{p_n} \log p_n$, it follows from Theorem I that the constant C may be so determined that the modified inequality is not satisfied by any prime number p_n.

In the case $\alpha = 0$, $0 \leq \beta \leq 2$, (39a) and (40a) are trivial. (39b) and (40c) follow immediately from Theorem I. Thus it remains to prove (39a), (40a) and (40b) in the following cases:

(41a) $\alpha = 0$, $\beta > 2$;

(41b) $0 < \alpha < \frac{1}{2}$, $\beta \geq 0$;

(41c) $\alpha = \frac{1}{2}$, $0 \leq \beta \leq 1$.

We now proceed to the proof of (39) and (40) in these cases. For a later purpose we shall, however, in the case $\alpha = \frac{1}{2}$ until further notice consider also values of $\beta > 1$.

We put in Lemmas 3 and 4

(42) $$\sigma = \sigma(\tau) = \frac{1}{24\lambda} \tau^{\beta-1} e^{(\alpha-1)\tau} (\alpha\tau + (\beta - 1) \log \tau).$$

Bearing in mind that in the case $\alpha = 0$ we have $\beta > 2$, it is then seen that for all sufficiently large values of τ, say for $\tau > M$, the conditions of Lemmas 3 and 4 are both satisfied. (It is even seen that if β has a fixed value >2, the value of M can be chosen independently of α for $0 \leqq \alpha \leqq \frac{1}{2}$. This remark will be used in the proof of the following Theorem III.).

Let us now consider the interval $t < \tau < t+1$, where $t > M$. Putting

$$x = e^t, \qquad \xi = e^\tau,$$

we establish a one-to-one correspondence between the intervals $(t, t+1)$ and (x, ex). Let (p_n, p_{n+1}) be a prime interval on the ξ-axis such that

$$x < p_n < p_{n+1} < e x, \qquad p_{n+1} - p_n > h(p_n), \tag{43}$$

$h(x)$ being defined by (35). The number of intervals (p_n, p_{n+1}) satisfying (43) is obviously greater than

$$N_{\alpha,\beta}(2x) - N_{\alpha,\beta}(x) \tag{44}$$

as soon as M is sufficiently large.

For the length of the corresponding interval $(\log p_n, \log p_{n+1})$ on the τ-axis, we have the inequality

$$\log p_{n+1} - \log p_n > \frac{h(p_n)}{2 p_n} = \frac{1}{2} p_n^{\alpha-1} \log^\beta p_n \tag{45}$$

$$> \frac{1}{2e} t^\beta e^{(\alpha-1)t}$$

if M is sufficiently large.

Further, we have in the notation of Lemma 3

$$\Delta = \frac{\lambda \sigma \tau}{\tau + \log \sigma} = \frac{\frac{1}{24} \tau^\beta e^{(\alpha-1)\tau} (\alpha \tau + (\beta - 1)\log \tau)}{\alpha \tau + (\beta - 1)\log \tau + \log(\alpha \tau + (\beta - 1)\log \tau) - \log(24\lambda)}.$$

Thus as soon as M is sufficiently large we have by (45)

$$\Delta < \frac{1}{24} \tau^\beta e^{(\alpha-1)\tau} < \frac{1}{4} (\log p_{n+1} - \log p_n). \tag{46}$$

From (46) it follows that for every value of τ between the limits

$$\frac{\log p_{n+1} + \log p_n}{2} \pm \frac{\log p_{n+1} - \log p_n}{4}, \tag{47}$$

the interval $(\tau - \Delta, \tau + \Delta)$ falls entirely in the interior of the interval $(\log p_n, \log p_{n+1})$. Thus for every τ between the limits (47) we have

$$\pi(e^{\tau+\Delta}) - \pi(e^{\tau-\Delta}) = 0,$$

and so obtain from Lemma 3

(48) $$\Re F(s) \geqq \frac{1}{3}.$$

The distance between the limits (47) is according to (45) greater than

$$\frac{1}{12} t^{\beta} e^{(\alpha-1)t}.$$

The number of different intervals $(\log p_n, \log p_{n+1})$ satisfying (43) being greater than the quantity (44), we have by (48) for all $t > M$

(49) $$\int_t^{t+1} (\Re F(\sigma + i\tau))^2 d\tau > \frac{1}{9} \cdot \frac{1}{12} t^{\beta} e^{(\alpha-1)t} (N_{\alpha,\beta}(2x) - N_{\alpha,\beta}(x)).$$

Introducing the expression (42) for σ into Lemma 4 we obtain, however,

(50) $$\int_t^{t+1} |F(\sigma + i\tau)|^2 d\tau < K \frac{t^3}{t^{\beta} e^{\alpha t} (\alpha t + (\beta - 1)\log t)}$$

if M is sufficiently large, K being always an absolute constant. From (49) and (50) we obtain, since in the case $\alpha = 0$ we have $\beta > 2$,

$$N_{\alpha,\beta}(2x) - N_{\alpha,\beta}(x) < K \frac{t^3 e^t}{(t^{\beta} e^{\alpha t})^2 (\alpha t + \beta \log t)}.$$

Substituting x for e^t, we obtain

(51) $$N_{\alpha,\beta}(2x) - N_{\alpha,\beta}(x) < K \frac{x \log^3 x}{h^2 \log h}$$

for all $x > e^M$, if M is sufficiently large.

(It will be seen without difficulty that, during all the calculations leading up to (51), the remark made above with respect to the value of M for a fixed $\beta > 2$ holds true.)

From (51) we deduce

(52) $$S_{\alpha,\beta}(2x) - S_{\alpha,\beta}(x) < K\frac{x\log^3 x}{h\log h}.$$

So far, we have disregarded the condition $\beta \leqq 1$ in the case (41c). Henceforth we shall suppose α and β so chosen that one of the cases (41a) — (41c) is present.

Substituting in (51) and (52) successively $\frac{x}{2}, \frac{x}{2^2}, \ldots, \frac{x}{2^{\left[\frac{\log \frac{x}{M}}{\log 2}\right]}}$ for x and adding the results, we obtain (39a), (40a) and (40b). Thus Theorem II is proved.

We shall now consider the convergence problem for the series

(53) $$\sum_{n=1}^{\infty}\frac{p_{n+1}-p_n}{p_n\log^{\lambda} p_n}$$

and

(54) $$\sum_{n=1}^{\infty}\frac{(p_{n+1}-p_n)^2}{p_n\log^{\lambda} p_n}.$$

Theorem III. a) *The series (53) is divergent for* $\lambda \leqq 1$ *and convergent for* $\lambda > 1$. — b) *The series (54) is divergent for* $\lambda \leqq 2$. *If the R. h is true, (54) is convergent for* $\lambda > 4$.

a) and the first part of b) are almost obvious. We need only observe that $p_{n+1} - p_n$ is on the average of the order $\log p_n$, and that the series

$$\sum\frac{1}{p_n\log^{\lambda} p_n}$$

is divergent for $\lambda \leqq 0$, convergent for $\lambda > 0$.

Thus it only remains to prove the convergence of (54) for $\lambda > 4$. For a fixed β such that $2 < \beta < 3$, it follows from the remark made above in connection with the relation (51) that we have for $0 \leqq \alpha \leqq \frac{1}{2}$ and for all $x > e^M$, where M may depend on β but not on α,

$$Q(\alpha, x) = N_{\alpha,\beta}(2x) - N_{\alpha,\beta}(x) < K x^{1-2\alpha}\frac{\log^{3-2\beta} x}{\alpha\log x + \beta\log\log x},$$

K being an absolute constant.

We have further for $x > e^M$

$$\sum_{\substack{x < p_n \leqq 2x \\ p_{n+1}-p_n \leqq \log^\beta p_n}} (p_{n+1}-p_n)^2 < K \frac{x}{\log x} \cdot \log^{2\beta} x = K x \log^{2\beta-1} x \tag{55}$$

and

$$\sum_{\substack{x < p_n \leqq 2x \\ p_{n+1}-p_n > \log^\beta p_n}} (p_{n+1}-p_n)^2 < K \int_0^{\frac{1}{2}} (x^\alpha \log^\beta x)^2 \, d_\alpha (-Q(\alpha, x)) \tag{56}$$

$$< K Q(0, x) \log^{2\beta} x + K \log^{2\beta+1} x \int_0^{\frac{1}{2}} x^{2\alpha} Q(\alpha, x) \, d\alpha$$

$$< K \left(\frac{x \log^3 x}{\log \log x} + x \log^3 x \int_0^{\frac{1}{2}} \frac{d\alpha}{\alpha + \beta \frac{\log \log x}{\log x}} \right) < K x \log^3 x \log \log x.$$

From (55) and (56) we obtain, since $\beta > 2$,

$$\sum_{x < p_n \leqq 2x} (p_{n+1}-p_n)^2 < K x \log^{2\beta-1} x \tag{57}$$

for all sufficiently large x. Hence we obtain

$$\sum_{x < p_n \leqq 2x} \frac{(p_{n+1}-p_n)^2}{p_n \log^\lambda p_n} < \frac{K}{(\log x)^{1+\lambda-2\beta}}.$$

Substituting here $2x$, $2^2 x, \ldots$ for x, it follows that (54) is convergent for $\lambda > 2\beta$. Since β may be taken as near to 2 as we please, (54) converges for all $\lambda > 4$, and Theorem III is proved.

From (57) we can also obtain other similar relations, as *e. g.*

$$\sum_{p_n \leqq x} (p_{n+1}-p_n)^2 = O(x \log^{3+\varepsilon} x)$$

and

$$\sum_{p_n \leqq x} \left(\frac{p_{n+1}-p_n}{\log p_n} \right)^2 = O(x \log^{1+\varepsilon} x)$$

for every $\varepsilon > 0$, which hold if the Riemann hypothesis is true.

REFERENCES.

[1] V. Brun, Le crible d'Eratosthène et le théorème de Goldbach. Norske Vid. Selsk. Skr., Mat.-Naturv. Kl. (1920), no. 3.

[2] F. P. Cantelli, Considerazioni sulla legge uniforme dei grandi numeri etc. Giorn. Inst. Ital. Attuari, vol. 4 (1933), p. 327.

[3] H. Cramér, Studien über die Nullstellen der Riemannschen Zetafunktion, Math. Z., vol. 4 (1919), p. 104.

[4] H. Cramér, Some theorems concerning prime numbers. Ark. Mat. Astron. Fys., vol. 15 (1920), no. 5.

[5] H. Cramér, On the distribution of primes. Proc. Cambridge Philos. Soc., vol 20 (1921), p. 272.

[6] H. Cramér, Prime numbers and probability. 8:de Skandinav. Mat.-Kongr. 1934, Förhandl., p. 107.

[7] G. H. Hardy and J. E. Littlewood, Some problems of partitio numerorum, III: On the expression of a number as a sum of primes. Acta math., vol. 44 (1922), p. 1.

[8] G. Hoheisel, Primzahlprobleme in der Analysis. S.-B. preuss. Akad. Wiss. (1930), n:o 33.

[9] A. Piltz, Über die Häufigkeit von Primzahlen in arithmetischen Progressionen etc., Habilitationsschrift, Jena (1884).

[10] A. E. Western, Note on the magnitude of the difference between successive primes. J. London Math. Soc., vol. 9 (1934), p. 276.

[11] E. Westzynthius, Über die Verteilung der Zahlen, die zu den n ersten Primzahlen teilerfremd sind. Soc. Sci. Fennicae Comment. Phys.-Math., vol. 5 (1931). n:o 25.

(Received december 9, 1935.)

38.

Sur un nouveau théorème-limite de la théorie des probabilités

Actual. Sci. Indust. (736), 5–23 (1938)

CHAPITRE PREMIER

Considérons une suite $Z_1, Z_2, \ldots$ de variables aléatoires indépendantes ayant toutes la même fonction de répartition $V(x)$, et telles que

$$(1) \qquad E(Z_n) = 0, \qquad E(Z_n^2) = \sigma^2 > 0.$$

Désignons par $W_n(x)$ la fonction de répartition de la somme

$$Z_1 + \cdots + Z_n,$$

et par $F_n(x)$ la fonction de répartition de la variable

$$\frac{Z_1 + \cdots + Z_n}{\sigma\sqrt{n}}.$$

On a donc

$$F_n(x) = \text{Prob}\,(Z_1 + \cdots + Z_n \leqslant \sigma\, x\sqrt{n})$$

et

$$(2) \qquad F_n(x) = W_n(\sigma x\sqrt{n}).$$

D'après le théorème limite classique de Laplace-Liapounoff (dans sa forme moderne précisée par Lindeberg et par M. Paul Lévy) on a alors pour chaque valeur réelle fixe de x

$$(3) \qquad \lim_{n\to\infty} F_n(x) = \Phi(x) = \frac{1}{\sqrt{2\pi}}\int_{-8}^{x} e^{-\frac{t^2}{2}}\,dt.$$

Par ce théorème, on a donc une expression asymptotique (pour $n \to \infty$) de la probabilité $F_n(x)$ de l'inégalité

$$Z_1 + \cdots + Z_n \leqslant \sigma x \sqrt{n}$$

ou, ce qui revient au même, de la probabilité $1 - F_n(x)$ de l'inégalité

$$Z_1 + \cdots + Z_n > \sigma x \sqrt{n}$$

x étant toujours un nombre réel *indépendant de* n.

Il est alors naturel de se demander ce que deviennent ces probabilités *lorsque* x *peut varier avec* n, *en tendant vers* $+\infty$ *ou vers* $-\infty$ *quand* n *croît indéfiniment.*

Dans ces conditions, la relation (3) ne donne que le résultat évident

$$\lim_{n\to\infty} F_n(x) = \begin{cases} 1 & \text{quand } x \to +\infty, \\ 0 & \quad « \quad x \to -\infty. \end{cases}$$

qui exprime seulement que $F_n(x)$ tend vers les mêmes limites que $\Phi(x)$ lorsque $\to \pm\infty$.

Pour savoir si l'équivalence asymptotique de $F_n(x)$ et $\Phi(x)$ subsiste dans les conditions indiquées, on pourrait se proposer d'étudier les rapports

$$(4a) \qquad \frac{1 - F_n(x)}{1 - \Phi(x)} \quad \text{pour } x \to +\infty,$$

et

$$(4b) \qquad \frac{F_n(x)}{\Phi(x)} \quad \text{pour } x \to -\infty.$$

Si x est indépendant de n, il suit de (3) que ces rapports tendent tous les deux vers l'unité lorsque n tend vers l'infini ; il s'agit maintenant de savoir ce qui arrive quand x devient infini avec n.

On sait que le théorème de Liapounoff fournit, sous certaines conditions, une borne supérieure du module $|F_n(x) - \Phi(x)|$ qui est du même ordre de grandeur que $\frac{\log n}{\sqrt{n}}$ (voir le chap. III). Quand ce théorème est applicable, on trouve sans difficulté que les rapports (4) tendent encore vers l'unité lorsque $|x|$ reste inférieur à $\left(\frac{1}{2} - \varepsilon\right)\sqrt{\log n}$, où $\varepsilon > 0$. Cependant ce résultat semble bien insuffisant, notre problème étant d'étudier le comportement des rapports (4) dans un domaine beaucoup plus étendu, par exemple pour des valeurs de $|x|$ qui sont du même ordre de grandeur qu'une puissance de n.

Avant d'aborder ce problème, observons qu'on ne peut guère espérer *a priori* d'obtenir des expressions asymptotiques à la fois simples et générales qu'en se bornant aux valeurs de x qui sont de la forme $o(\sqrt{n})$. En effet, la fonction de répartition $V(x)$, qui représente les données du problème, peut être choisie de manière que toute sa variation reste comprise dans un intervalle fini $(-\mu\sigma, \mu\sigma)$. Pour la fonction $F_n(x)$, toute la variation sera alors comprise dans l'intervalle $(-\mu\sqrt{n}, \mu\sqrt{n})$, ce qui montre que les rapports (4) s'annuleront identiquement pour $x > \mu\sqrt{n}$ et pour $x < -\mu\sqrt{n}$ respectivement.

Dans ce qui va suivre, nous allons étudier le comportement asymptotique des rapports (4) en imposant toujours à la fonction $V(x)$ la *condition A* qui va être formulée à l'instant, et en supposant x de la forme $o\left(\frac{\sqrt{n}}{\log n}\right)$. On verra cependant plus tard (Chap. V) que, si la fonction $V(x)$ satisfait à une certaine condition additionnelle B, on peut même considérer des valeurs de x qui sont du même ordre de grandeur que $\sqrt{n}$.

CONDITION A. — *Il existe un nombre* $A > 0$ *tel que l'intégrale*

$$R = \int_{-\infty}^{\infty} e^{hy} dV(y) \tag{5}$$

converge pour $|h| < A$.

En supposant que cette condition soit satisfaite, nous allons établir entre autres le résultat fondamental suivant. *Pour* $x > 1$, $x = o\left(\frac{\sqrt{n}}{\log n}\right)$, *on a*

$$\frac{1 - F_n(x)}{1 - \Phi(x)} = e^{\frac{x^3}{\sqrt{n}} \lambda\left(\frac{x}{\sqrt{n}}\right)} \left[1 + O\left(\frac{x \log n}{\sqrt{n}}\right)\right],$$

$$\frac{F_n(-x)}{\Phi(-x)} = e^{-\frac{x^3}{\sqrt{n}} \lambda\left(-\frac{x}{\sqrt{n}}\right)} \left[1 + O\left(\frac{x \log n}{\sqrt{n}}\right)\right],$$

où

$$\lambda(z) = \sum_0^\infty c_\nu z^\nu$$

est une série de puissances convergente pour toute valeur suffisamment petite de $|z|$.

Ce théorème, dont nous déduirons plusieurs corollaires importants, sera démontré dans le chapitre III. Par l'introduction de la condition additionnelle B, nous parviendrons dans le chapitre V à des résultats encore plus précis. Enfin, dans le dernier chapitre, nous donnerons des théorèmes analogues aux précédents pour le cas d'un processus stocastique homogène.

CHAPITRE II

Supposons que la condition A soit satisfaite et choisissons un nombre réel h situé dans le domaine de convergence de l'intégrale (5). On s'assure facilement que la fonction (¹)

$$\overline{\mathrm{V}}(x) = \frac{1}{\mathrm{R}} \int_{-\infty}^{x} e^{hy} d\mathrm{V}(y) \tag{6}$$

possède toutes les propriétés essentielles d'une fonction de répartition.

Considérons donc une suite $\overline{\mathrm{Z}}_1, \overline{\mathrm{Z}}_2, \ldots$ de variables aléatoires indépendantes ayant toutes la même fonction de répartition $\overline{\mathrm{V}}(x)$. Posons

$$\mathrm{E}(\overline{\mathrm{Z}}_n) = \overline{m}, \qquad \mathrm{E}(\overline{\mathrm{Z}}_n^2) = \overline{\sigma}^2. \tag{7}$$

Désignons encore par $\overline{\mathrm{W}}_n(x)$ la fonction de répartition de la somme

$$\overline{\mathrm{Z}}_1 + \cdots + \overline{\mathrm{Z}}_n,$$

et par $\overline{\mathrm{F}}_n(x)$ la fonction de répartition de la variable

$$\frac{\overline{\mathrm{Z}}_1 + \cdots + \overline{\mathrm{Z}}_n - \overline{m}n}{\overline{\sigma}\sqrt{n}}.$$

(¹) Une transformation analogue a été employée par F. Esscher : On the probability function in the collective theory of risk, *Skandinavisk Aktuarietidskrift*, 15 (1932), p. 175.

On a alors

$$\overline{F}_n(x) = \overline{W}_n(\overline{m}n + \overline{\sigma}\, x\sqrt{n}). \tag{8}$$

Le but de ce chapitre est d'établir une relation entre les fonctions $F_n(x)$ et $\overline{F}_n(x)$ définies par (2) et (8) respectivement.

Introduisons les fonctions caractéristiques $v(z)$, $\overline{v}(z)$, $w_n(z)$ et $\overline{w}_n(z)$, où

$$v(z) = \int_{-\infty}^{\infty} e^{izy} dV(y),$$

tandis que $\overline{v}(z)$, $w_n(z)$ et $\overline{w}_n(z)$ sont définies par des expressions analogues où figurent les fonctions $\overline{V}$, W_n et $\overline{W}_n$ respectivement. Considérons ici z comme une variable complexe et posons $z = \xi + i\eta$. D'après la condition A et la relation (6), la fonction $v(z)$ est holomorphe à l'intérieur de la bande $|\eta| < A$, et l'on a

$$R = v(-ih),$$

$$\overline{v}(z) = \frac{1}{R} v(z - ih).$$

D'autre part, les variables Z_n ainsi que les variables $\overline{Z}_n$ étant mutuellement indépendantes, il est bien connu qu'on a

$$w_n(z) = [v(z)]^n,$$

$$\overline{w}_n(z) = [\overline{v}(z)]^n,$$

d'où

$$\overline{w}_n(z) = \frac{1}{R^n} w_n(z - ih), \tag{9}$$

les deux membres de cette dernière relation étant des fonctions holomorphes de z à l'intérieur de la bande $|\eta - h| < A$. En posant ici $z = ih$, il vient

$$\frac{1}{R^n} = \overline{w}_n(ih) = \int_{-\infty}^{\infty} e^{-hy} d\overline{W}_n(y).$$

Remplaçons dans (9) z par $z + ih$; nous avons alors

$$w_n(z) = R^n \overline{w}_n(z + ih) = \frac{\overline{w}_n(z + ih)}{\overline{w}_n(ih)}. \tag{10}$$

Or cette relation équivaut à

$$W_n(x) = R^n \int_{-\infty}^{x} e^{-hy} d\overline{W}_n(y). \tag{11}$$

En effet, les deux membres de (11) sont des fonctions de répartition en x, dont les fonctions caractéristiques, comme on le voit sans difficulté, sont égales à $w_n(z)$ et $R^n\overline{w}_n(z+ih)$ respectivement. Donc, d'après (10), ces fonctions de répartition sont identiques.

En introduisant dans (11) les fonctions $F_n(x)$ et $\overline{F}_n(x)$ définies par (2) et (8), on obtient au moyen d'une substitution simple

$$F_n(x) = R^n e^{-\bar{h}\bar{m}n} \int_{-\infty}^{\frac{\sigma x - \bar{m}\sqrt{n}}{\bar{\sigma}}} e^{-\bar{h}\bar{\sigma}\sqrt{n}y} d\overline{F}_n(y). \tag{12a}$$

Si l'on fait ici tendre x vers l'infini positif on obtient de plus, en faisant la différence,

$$1 - F_n(x) = R^n e^{-\bar{h}\bar{m}n} \int_{\frac{\sigma x - \bar{m}\sqrt{n}}{\bar{\sigma}}}^{\infty} e^{-\bar{h}\bar{\sigma}\sqrt{n}y} d\overline{F}_n(y). \tag{12b}$$

Les équations (12a) et (12b) expriment la relation entre $F_n(x)$ et $\overline{F}_n(x)$ que nous nous sommes proposés d'établir.

Or, d'après la définition (8) de $\overline{F}_n(x)$, le théorème de Laplace-Liapounoff peut encore être appliqué à cette fonction. Ce théorème nous apprend que $\overline{F}_n(x)$ tend, pour $n \to \infty$, vers la fonction de répartition normale $\Phi(x)$ définie par (3). En remplaçant dans (12a) et (12b) $\overline{F}_n(y)$ par $\Phi(y)$, on doit donc obtenir des expressions approchées de la fonction $F_n(x)$. C'est en précisant ce raisonnement que nous parviendrons, dans les chapitres suivants, à des résultats nouveaux concernant l'allure asymptotique de $F_n(x)$.

CHAPITRE III

Considérons maintenant h comme une variable réelle, et posons

$$\log R = \log \int_{-\infty}^{\infty} e^{hy} dV(y) = \sum_{2}^{\infty} \frac{\gamma_\nu}{\nu!} h^\nu, \tag{13}$$

où, d'après la condition A, la série converge certainement pour toute valeur suffisamment petite de $|h|$. Les coefficients γ_ν sont les semi-invariants de la fonction de répartition $V(x)$, et il suit de (1) qu'on a $\gamma_1 = 0$, $\gamma_2 = \sigma^2$. Par (6) et (7), on a de plus

$$\overline{m} = \frac{d}{dh} \log R = \sum_{2}^{\infty} \frac{\gamma_\nu}{(\nu - 1)!} h^{\nu-1}, \tag{14}$$

$$\overline{\sigma}^2 = \frac{d\overline{m}}{dh} = \sum_{2}^{\infty} \frac{\gamma_\nu}{(\nu - 2)!} h^{\nu-2}. \tag{15}$$

Dans le voisinage du point $h = 0$, $\overline{m}$ est donc une fonction continue et toujours croissante de la variable réelle h.

Soit maintenant z un nombre réel donné. L'équation

$$\sigma z = \overline{m} = \sum_{2}^{\infty} \frac{\gamma_\nu}{(\nu - 1)!} h^{\nu-1} \tag{16}$$

admet alors, pour tout z de module suffisamment petit, une seule racine réelle h, qui a le même signe que z et tend vers zéro avec z. Réciproquement, cette racine h peut être développée en série de puissances de z, convergente pour tout z de module suffisamment petit. Les premiers termes de ce développement sont d'ailleurs

$$h = \frac{z}{\sigma} - \frac{\gamma_3}{2\sigma^4} z^2 - \frac{\sigma^2\gamma_4 - 3\gamma_3^2}{6\sigma^7} z^3 + \cdots. \tag{17}$$

De (13) et (14) on tire

$$h\overline{m} - \log R = \sum_{2}^{\infty} \frac{(\nu - 1)\gamma_\nu}{\nu!} h^\nu.$$

En remplaçant ici h par son développement (17), il vient

$$h\overline{m} - \log R = \frac{1}{2} z^2 - \frac{\gamma_3}{6\sigma^3} z^3 - \frac{\sigma^2\gamma_4 - 3\gamma_3^2}{24\sigma^6} z^4 + \cdots.$$

Donc, en observant que $z^2 = \left(\frac{\overline{m}}{\sigma}\right)^2$ et en posant

$$\frac{\overline{m}^2}{2\sigma^2} - h\overline{m} + \log \mathrm{R} = z^3\lambda(z), \tag{18}$$

$\lambda(z)$ admet un développement en série de puissances

$$\lambda(z) = c_0 + c_1 z + c_2 z^2 + \cdots \tag{19}$$

convergent pour tout z de module suffisamment petit, et l'on a

$$c_0 = \frac{\gamma_3}{6\sigma^3}, \qquad c_1 = \frac{\sigma^2\gamma_4 - 3\gamma_3^2}{24\sigma^6}, \ldots. \tag{20}$$

Ceci posé, nous pouvons énoncer notre théorème fondamental de la manière suivante.

Théorème 1. — *Supposons que la condition A soit satisfaite. Soit x un nombre réel qui peut dépendre de n, tel que $x > 1$ et $x = o\left(\frac{\sqrt{n}}{\log n}\right)$ lorsque n tend vers l'infini. Pour la fonction de répartition $\mathrm{F}_n(x)$ introduite dans le chapitre I, on a alors*

$$\frac{1 - \mathrm{F}_n(x)}{1 - \Phi(x)} = e^{\frac{x^3}{\sqrt{n}}\lambda\left(\frac{x}{\sqrt{n}}\right)}\left[1 + \mathrm{O}\left(\frac{x \log n}{\sqrt{n}}\right)\right]$$

et

$$\frac{\mathrm{F}_n(-x)}{\Phi(-x)} = e^{-\frac{x^3}{\sqrt{n}}\lambda\left(-\frac{x}{\sqrt{n}}\right)}\left[1 + \mathrm{O}\left(\frac{x \log n}{\sqrt{n}}\right)\right]$$

$\lambda(z)$ étant la fonction définie par (18) *et qui admet le développement* (19), *dont les premiers coefficients sont donnés par* (20).

Les démonstrations des deux relations énoncées étant tout à fait analogues, nous nous bornerons à la démonstration de la première relation.

Si dans (12b) nous prenons $x = \frac{\overline{m}\sqrt{n}}{\sigma}$, nous aurons

$$1 - \mathrm{F}_n\left(\frac{\overline{m}\sqrt{n}}{\sigma}\right) = \mathrm{R}^n e^{-h\overline{m}n}\int_0^\infty e^{-h\overline{\sigma}\sqrt{n}y}d\overline{\mathrm{F}}_n(y) \tag{21}$$

pour toute valeur réelle de h telle que $|h| < \mathrm{A}$. Posons maintenant

$$\overline{\mathrm{F}}_n(y) = \Phi(y) + \mathrm{Q}_n(y). \tag{22}$$

D'après le théorème de LIAPOUNOFF, on a [1] alors pour tout $n > 1$ et pour tout y réel

$$|\,Q_n(y)\,| < k\frac{\log n}{\sqrt{n}},$$

où

$$k = \frac{3}{\sigma^3}\int_{-\infty}^{\infty} |\,y - \overline{m}\,|^3 dV(y).$$

Le nombre k ainsi défini dépend évidemment de h. Or il suit de (14), (15) et de la condition A qu'on peut déterminer un nombre positif $A_1 < A$ tel que pour $|\,h\,| < A_1$, on ait $k < K$ et par conséquent

$$|\,Q_n(y)\,| < K\frac{\log n}{\sqrt{n}}, \tag{23}$$

où K est indépendant de h, n et y.

Dès maintenant, nous considérons h comme une variable essentiellement *positive.* Faisons tendre n vers l'infini et h vers zéro, de manière que le produit $h\sqrt{n}$ ait une borne inférieure positive.

Nous avons alors par (14) et (15)

$$h\overline{\sigma}\sqrt{n} = \frac{\overline{m}\sqrt{n}}{\sigma} + O(h^2\sqrt{n}).$$

De (22) et (23) on déduit au moyen de calculs faciles

$$\int_0^{\infty} e^{-h\overline{\sigma}\sqrt{n}y} d\overline{F}_n(y)$$

$$= \frac{1}{\sqrt{2\pi}}\int_0^{\infty} e^{-h\overline{\sigma}\sqrt{n}y - \frac{1}{2}y^2} dy - Q_n(0) + h\overline{\sigma}\sqrt{n}\int_0^{\infty} e^{-h\overline{\sigma}\sqrt{n}y} Q_n(y)dy$$

$$= \frac{1}{\sqrt{2\pi}}\int_0^{\infty} e^{-h\overline{\sigma}\sqrt{n}y - \frac{1}{2}y^2} dy + O\left(\frac{\log n}{\sqrt{n}}\right)$$

$$= \frac{1}{\sqrt{2\pi}}\int_0^{\infty} e^{-h\overline{\sigma}\sqrt{n}y - \frac{1}{2}y^2} dy\cdot[1 + O(h\log n)]$$

$$= \frac{1}{\sqrt{2\pi}}\int_0^{\infty} e^{-\frac{\overline{m}\sqrt{n}}{\sigma}y - \frac{1}{2}y^2} dy\cdot[1 + O(h\log n)]$$

$$= e^{\frac{n\overline{m}^2}{2\sigma^2}}\left[1 - \Phi\left(\frac{\overline{m}\sqrt{n}}{\sigma}\right)\right]\cdot[1 + O(h\log n)].$$

---

[1] Voir H. CRAMÉR, Random variables and probability distributions, *Cambridge Tracts in Mathematics*, N° 36, Cambridge 1937, p. 77.

En introduisant la dernière expression dans (21), on aura

$$(24)\qquad \frac{1 - F_n\left(\frac{\bar{m}\sqrt{n}}{\sigma}\right)}{1 - \Phi\left(\frac{\bar{m}\sqrt{n}}{\sigma}\right)} = e^{n\left(\frac{\bar{m}^2}{2\sigma^2} - h\bar{m} + \log R\right)}[1 + O(h \log n)].$$

Soit maintenant x un nombre réel qui peut dépendre de n, tel que $x > 1$ et $x = o\left(\frac{\sqrt{n}}{\log n}\right)$. Formons l'équation

$$(25)\qquad x = \frac{\bar{m}\sqrt{n}}{\sigma},$$

qui peut aussi s'écrire

$$\frac{\sigma x}{\sqrt{n}} = \bar{m} = \sum_2^\infty \frac{\gamma_\nu}{(\nu-1)!} h^{\nu-1},$$

et qui, par la substitution $z = \frac{x}{\sqrt{n}}$, devient identique à l'équation (16). Pour tout n suffisamment grand, l'équation (25) admet donc une seule racine positive h qui tend vers zéro lorsque n tend vers l'infini. D'après (18) on a

$$\frac{\bar{m}^2}{2\sigma^2} - h\bar{m} + \log R = \left(\frac{x}{\sqrt{n}}\right)^3 \lambda\left(\frac{x}{\sqrt{n}}\right),$$

où $\lambda(z)$ est défini par (19)-(20). Le produit $h\sqrt{n}$ est bien borné inférieurement, car on déduit de (17)

$$h \backsim \frac{z}{\sigma} = \frac{x}{\sigma\sqrt{n}},$$

et nous avons supposé $x > 1$.

Dans (24), on peut donc prendre h égal à la racine de (25). On obtient ainsi

$$\frac{1 - F_n(x)}{1 - \Phi(x)} = e^{\frac{x^3}{\sqrt{n}}\lambda\left(\frac{x}{\sqrt{n}}\right)}\left[1 + O\left(\frac{x \log n}{\sqrt{n}}\right)\right],$$

et le théorème 1 est démontré.

Du théorème 1, on peut déduire plusieurs corollaires intéressants. Démontrons d'abord le théorème suivant.

THÉORÈME 2. — *Si la condition A est satisfaite, on a pour* $x > 1$, $x = O\left(n^{\frac{1}{6}}\right)$,

$$\frac{1 - F_n(x)}{1 - \Phi(x)} = e^{\frac{c_0 x^3}{\sqrt{n}}} + O\left(\frac{x \log n}{\sqrt{n}}\right),$$

$$\frac{F_n(-x)}{\Phi(-x)} = e^{-\frac{c_0 x^3}{\sqrt{n}}} + O\left(\frac{x \log n}{\sqrt{n}}\right).$$

Ceci est une conséquence immédiate du théorème 1, si l'on remarque que, pour $x = O(n^{1/6})$, l'exposant $\frac{x^3}{\sqrt{n}} \lambda\left(\frac{x}{\sqrt{n}}\right)$ reste borné lorsque n tend vers l'infini. On voit en particulier que, si $x = o(n^{1/6})$, les deux rapports considérés tendent vers l'unité quand n tend vers l'infini.

En observant que l'on a pour $x > 1$,

$$1 - \Phi(x) < \frac{1}{x\sqrt{2\pi}} e^{-\frac{x^2}{2}},$$

$$\Phi(-x) < \frac{1}{x\sqrt{2\pi}} e^{-\frac{x^2}{2}},$$

on obtient aussi sans difficulté le théorème suivant qui se rattache immédiatement au théorème de Liapounoff.

THÉORÈME 3. — *Si la condition A est satisfaite, on a pour* $x > 0$, $x = O(n^{1/6})$,

$$1 - F_n(x) = [1 - \Phi(x)] e^{\frac{c_0 x^3}{\sqrt{n}}} + O\left(\frac{\log n}{\sqrt{n}} e^{-\frac{x^2}{2}}\right),$$

$$F_n(-x) = \Phi(-x) e^{-\frac{c_0 x^3}{\sqrt{n}}} + O\left(\frac{\log n}{\sqrt{n}} e^{-\frac{x^2}{2}}\right).$$

Si, tout en restant dans les conditions du théorème 1, x est d'un ordre de grandeur plus élevé que celui de $n^{1/6}$, le théorème 1 fournit encore des expressions asymptotiques des probabilités $1 - F_n(x)$ et $F_n(-x)$. Si, par exemple, le coefficient $c_0 = \frac{\gamma^3}{6\sigma^3}$ est différent de zéro, on voit que l'exposant $\frac{x^3}{\sqrt{n}} \lambda\left(\frac{x}{\sqrt{n}}\right)$ est asymptotiquement équivalent à $\frac{c_0 x^3}{\sqrt{n}}$. Lorsque $\frac{x}{n^{1/6}}$ tend vers l'infini positif, cet exposant tend donc vers $+\infty$ ou vers $-\infty$ selon le signe de c_0. D'après le théorème 1, le rapport $\frac{1-F_n(x)}{1-\Phi(x)}$ tend vers $+\infty$ ou vers 0 suivant le cas. Par le même raisonnement, le rapport $\frac{F_n(-x)}{\Phi(-x)}$ tend alors vers 0 ou vers $+\infty$ respectivement. — Si $c_0 = 0$, c'est évidemment le premier coefficient $c_\nu \neq 0$ qui va dominer la question, sans qu'il soit nécessaire d'en préciser ici tous les détails.

En dernier lieu, un calcul simple permet de déduire du théorème 1 la généralisation suivante d'un théorème dû à M. Khintchine (cf. le chapitre suivant).

THÉORÈME 4. — *Soit c une constante positive. Si la condition A est satisfaite, les deux expressions*

$$\frac{F_n\left(x+\frac{c}{x}\right)-F_n(x)}{1-F_n(x)} \quad \text{et} \quad \frac{\Phi\left(x+\frac{c}{x}\right)-\Phi(x)}{1-\Phi(x)},$$

tendent, pour $n\to\infty$, $x\to\infty$, $x = 0\left(\frac{\sqrt{n}}{\log n}\right)$, *vers une même limite, à savoir vers la quantité* $1-e^{-c}$.

Il y a évidemment un théorème correspondant pour les valeurs négatives de la variable.

CHAPITRE IV

Si, en particulier, on choisit les variables aléatoires Z_n introduites au début de ce travail telles que, pour chaque Z_n, il n'y ait que deux valeurs possibles :

$$Z_n = \begin{cases} 1-p & \text{avec la probabilité } p, \\ -p & \text{» » » } q = 1-p, \end{cases}$$

on voit qu'on arrive au *cas des épreuves répétées.* On sait que, dans ce cas particulier, la fonction de répartition $F_n(x)$ peut être interprétée de la manière suivante.

Supposons qu'on fasse une série de n tirages indépendants

d'une urne, la probabilité d'amener une boule blanche étant toujours égale à p. Désignons par ν le nombre des boules blanches obtenues au cours des n tirages. Alors nous avons

$$F_n(x) = \text{Prob}\left(\nu \leqq np + x\sqrt{npq}\right),$$

et il est bien connu que

$$\lim_{n\to\infty} F_n(x) = \Phi(x),$$

pour tout x réel indépendant de n.

La quantité $\frac{\nu - np}{\sqrt{npq}}$ est (avec un léger changement formel) ce que, d'après M. Borel, on appelle l'*écart relatif*. Pour tout x réel indépendant de n, la probabilité d'avoir un écart relatif $\leqq x$ tend donc vers la limite $\Phi(x)$ lorsque n tend vers l'infini.

Cependant, il peut souvent être important de connaître le comportement asymptotique des *probabilités des grands écarts relatifs*, c'est-à-dire le comportement de $F_n(x)$ quand x peut varier avec n, en tendant vers $+\infty$, ou vers $-\infty$ quand n croît indéfiniment. Ce cas particulier du problème qui nous occupe dans ce travail a été considéré par plusieurs auteurs [1]. La plupart des résultats trouvés dans cette direction rentrent dans les théorèmes démontrés dans le chapitre précédent.

Ainsi M. Smirnoff a démontré un théorème qui peut s'exprimer par la relation

$$\frac{1 - F_n(x)}{1 - \Phi(x)} = 1 + o\left(\frac{1}{x^{2s}}\right),$$

pour $x = o\left(n^{\frac{1}{4s+6}}\right)$, $s = 0, 1, 2, \ldots$, et par une relation analogue pour les écarts négatifs. Comme on a pour les valeurs indiquées de x

$$e^{\frac{c_0 x^3}{\sqrt{n}}} + O\left(\frac{x \log n}{\sqrt{n}}\right) = 1 + O\left(\frac{x^3 + x \log n}{\sqrt{n}}\right) = 1 + o\left(\frac{1}{x^{2s}}\right),$$

ce résultat est contenu dans notre théorème 2.

---

[1] Voir p. ex. N. Smirnoff, Uber Wahrscheinlichkeiten grosser Abweichungen, *Rec. Soc. Math.* Moscou, 40 (1933), p. 441 ; A. Khintchine, Uber einen neuen Grenzwertsatz der Wahrscheinlichkeitsrechnung, *Math. Annalen* 101 (1929), p. 745 ; M. Fréchet, Recherches théoriques modernes sur le calcul des probabilités, Paris, 1937, p. 222 ; P. Lévy, Théorie de l'addition des variables aléatoires, Paris 1937, p. 284.

D'autre part, M. LÉVY a donné, pour le cas des épreuves répétées, la relation (sur laquelle nous reviendrons dans le chapitre suivant)

$$\log(1 - F_n(x)) \sim \log(1 - \Phi(x)) \sim -\frac{x^2}{2},$$

qui est une conséquence de notre théorème 1, et enfin notre théorème 4 a été démontré pour le même cas particulier par M. KHINTCHINE. Dans ces théorèmes de MM. LÉVY et KHINTCHINE, notre condition $x = o\left(\frac{\sqrt{n}}{\log n}\right)$ se trouve remplacée par la condition un peu moins restreinte (¹) $x = o(\sqrt{n})$.

CHAPITRE V

Retournons au problème général posé dans le chapitre I. Jusqu'ici, nous avons assujetti la fonction de répartition donnée $V(x)$ à la seule condition A ; nous allons maintenant introduire une condition additionnelle B, qui nous permettra d'aller plus loin dans l'étude du comportement de la fonction $F_n(x)$ pour de grandes valeurs de n et x.

La fonction $V(x)$ peut, d'une manière unique, être mise sous la forme (²)

$$V(x) = \beta U_1(x) + (1 - \beta) U_2(x) \tag{26}$$

avec $0 \leqq \beta \leqq 1$, où $U_1(x)$ et $U_2(x)$ sont deux fonctions de répartition telles qu'on ait *presque partout*

$$U_1(x) = \int_{-\infty}^{x} U_1'(y)dy,$$

$$U_2'(x) = 0.$$

(¹) Dans le cas des épreuves répétées, la fonction de répartition $V(x)$ ne satisfait pas à la condition B, qui va être introduite dans le chapitre suivant et qui nous permettra de remplacer dans nos théorèmes la condition $x = o\left(\frac{\sqrt{n}}{\log n}\right)$ par la condition $x = o(\sqrt{n})$. Les résultats cités de MM. KHINTCHINE et LÉVY ne sont donc pas entièrement contenus dans nos résultats.

(²) Voir H. CRAMÉR, *l. c.*, p. 17.

CONDITION B. — *Dans la décomposition* (26) *de* V (x), *on a* $\beta > 0$.

Si V(x) satisfait à la condition B, on voit immédiatement qu'il en est de même pour la fonction $\overline{V}(x)$ définie par (6). Pour la fonction $Q_n(y)$ définie par (22) on a alors [1]

$$| Q_n(y) | < \frac{K}{\sqrt{n}}$$

pour tout $n > 1$, pour tout y réel et pour tout h de module suffisamment petit, la constante K étant indépendante de n, y et h.

En introduisant ce résultat dans les calculs du chapitre III, on voit tout de suite que le facteur log n, qui intervient dans les évaluations, peut être partout omis. De même la condition $x = o\left(\frac{\sqrt{n}}{\log n}\right)$, dont le seul but est d'assurer la relation $\frac{x \log n}{\sqrt{n}} \to 0$ peut être remplacée par $x = o\left(\sqrt{n}\right)$. On a donc le théorème suivant.

THÉORÈME 5. — *Si les deux conditions A et B sont satisfaites, on peut remplacer dans les théorèmes* 1 *et* 4 *la condition* $x = o\left(\frac{\sqrt{n}}{\log n}\right)$ *par* $x = o(\sqrt{n})$. *On peut aussi omettre le facteur* log n *qui apparaît dans l'évaluation du reste dans les théorèmes* 1, 2 *et* 3.

On peut cependant aller plus loin et considérer aussi des valeurs de x qui sont du même ordre de grandeur que $\sqrt{n}$. Considérons en effet la condition A, et désignons par A_1 et $-A_2$ les bornes supérieures et inférieures des valeurs de h telles que l'intégrale (5) converge. A_1 et A_2 sont certainement des quantités positives, qui peuvent être finies ou non. En tenant compte des relations (14) et (15) on voit que, pour $-A_2 < h < A_1$, la quantité $\overline{m}$ définie par (7) est une fonction continue et toujours croissante de h, qui s'annule pour $h = 0$. Les deux limites

$$\lim_{h \to A_1 - 0} \overline{m} = \sigma C_1, \qquad \lim_{h \to -A_2 + 0} \overline{m} = -\sigma C_2,$$

---

[1] Voir H. CRAMÉR, *l. c.*, p. 81. Il ne résulte pas immédiatement du théorème cité que la constante K peut être prise indépendante de h. En parcourant la démonstration du théorème on s'assure cependant sans difficulté qu'il en est bien ainsi.

existent donc, C_1 et C_2 ayant des valeurs positives, finies ou non. Pour tout c donné dans l'intervalle $-C_2 < c < C_1$, l'équation

$$\bar{m} = \sigma c \tag{27}$$

a une seule racine h dans l'intervalle $-A_2 < h < A_1$, dont le signe est le même que celui de c.

Soit maintenant h un nombre quelconque donné dans l'intervalle $-A_2 < h < A_1$, et considérons l'identité (21), où h entre comme paramètre. Les conditions A et B étant satisfaites, nous avons pour la fonction $Q_n(y)$ définie par (22) le développement suivant [1] :

$$Q_n(y) = \left(\frac{p_2(y)}{n^{\frac{1}{2}}} + \frac{p_5(y)}{n} + \cdots + \frac{p_{3k-1}}{n^{\frac{k}{2}}}\right) e^{-\frac{y^2}{2}} + O\left(\frac{1}{n^{\frac{k+1}{2}}}\right),$$

où k est un entier arbitraire, tandis que les p_ν sont des polynomes dont le degré coïncide avec l'indice, et tels que les $p_{2\nu}$ sont des polynomes pairs, les $p_{2\nu-1}$ des polynomes impairs.

On en déduit, en refaisant les calculs du chapitre III,

$$\int_0^\infty e^{-h\bar{\sigma}\sqrt{n}y}\, d\bar{F}_n(y) = \frac{1}{\sqrt{n}}\left[b_0 + \frac{b_1}{n} + \cdots + \frac{b_{k-1}}{n^{k-1}} + O\left(\frac{1}{n^k}\right)\right]$$

pour tout entier positif k, les coefficients b_ν dépendant de h. On a d'ailleurs $b_0 = \dfrac{1}{h\bar{\sigma}\sqrt{2\pi}}$. En introduisant dans (21), on obtient donc

$$1 - F_n\left(\frac{\bar{m}\sqrt{n}}{\sigma}\right) = \frac{1}{\sqrt{n}} e^{-(h\bar{m} - \log R)n}\left[b_0 + \frac{b_1}{n} + \cdots + \frac{b_{k-1}}{n^{k-1}} + O\left(\frac{1}{n^k}\right)\right]. \tag{28}$$

Soit maintenant c un nombre donné tel que $0 < c < C_1$, et prenons h égal à la racine (unique) positive de l'équation (27). En introduisant cette valeur dans (28) et en posant

$$\alpha = h\bar{m} - \log R \tag{29}$$

(où l'on voit facilement que α est toujours positif), on a le théorème suivant.

THÉORÈME 6. — *Si les deux conditions A et B sont satisfaites,*

---

[1] Voir H. CRAMÉR, *l. c.*, p. 81.

on peut trouver un nombre positif C_1 *(fini ou non) tel que, pour tout c dans l'intervalle* $0 < c < C_1$, *on ait*

$$1 - F_n(c\sqrt{n}) = \frac{1}{\sqrt{n}} e^{-\alpha n}\left[b_0 + \frac{b_1}{n} + \cdots + \frac{b_{k-1}}{n^{k-1}} + O\left(\frac{1}{n^k}\right)\right],$$

où α *est donné par* (27) *et* (29). *Ici k est un entier positif arbitraire, et les* b_ν *sont indépendants de n, mais dépendent de c. En particulier, on a toujours* $b_0 > 0$.

Il y a évidemment une expression analogue pour $F_n(-c\sqrt{n})$ où $-C_2 < c < 0$.

Ce théorème donne lieu à une remarque intéressante. Si les conditions A et B sont satisfaites, il suit du théorème 1 (avec les compléments apportés par le théorème 5) que l'on a, pour $x \to \infty$, $x = o(\sqrt{n})$,

$$\log[1 - F_n(x)] - \log[1 - \Phi(x)] = x^2 o(1) = o(\log[1 - \Phi(x)]),$$

d'où

$$\log[1 - F_n(x)] \sim \log[1 - \Phi(x)].$$

D'autre part, pour $x = c\sqrt{n}$, on déduit du théorème 6

$$\log[1 - F_n(x)] \sim -\frac{\alpha}{c^2}x^2 \sim \frac{2\alpha}{c^2}\log[1 - \Phi(x)],$$

où, en général, la constante $\frac{2\alpha}{c^2}$ diffère de l'unité.

CHAPITRE VI

Considérons maintenant une variable aléatoire Z_t, fonction d'un paramètre continu t, qu'on peut interpréter comme signifiant par exemple le temps. Supposons que l'accroissement $\Delta Z_t = Z_{t+\Delta t} - Z_t$ soit toujours une variable aléatoire indépendante de Z_t, et que la loi de répartition de ΔZ_t ne dépende ni de t ni de Z_t, mais seulement de Δt. On dit alors que la variable Z_t définit un *processus stocastique homogène.* Supposons encore que la valeur moyenne $E(Z_t)$ s'annule pour tout t, et que $E(Z_t^2)$ soit toujours fini.

Il résulte alors d'un théorème de M. KOLMOGOROFF (1) qu'on peut assigner une constante $\sigma_0^2 \geqq 0$ et une fonction $S(x)$ bornée, jamais décroissante et continue au point $x = 0$, avec les propriétés suivantes. Posons

$$\sigma_1^2 = S(+\infty) - S(-\infty),$$
$$\sigma^2 = \sigma_0^2 + \sigma_1^2,$$
$$F(x, t) = \text{Prob}\,(Z_t \leqq \sigma x\sqrt{t}),$$
$$f(z, t) = \int_{-\infty}^{\infty} e^{izy} dF(y, t).$$

(Les différentielles devront toujours être prises par rapport à la variable y). Alors on a

$$E(Z_t^2) = \sigma^2 t$$

et

$$\log f(z, t) = -\frac{\sigma_0^2}{2\sigma^2} z^2 + \frac{1}{\sigma^2}\int_{-\infty}^{\infty} \frac{e^{izy} - 1 - izy}{y^2} dS(\sigma y\sqrt{t}).$$

Il s'ensuit sans difficulté

$$\lim_{t \to \infty} f(z, t) = e^{-\frac{z^2}{2}},$$

ce qui implique

$$\lim_{t \to \infty} F(x, t) = \Phi(x)$$

pour tout x réel indépendant de t.

Ici encore, on peut donc poser le problème d'étudier le comportement de $F(x, t)$ lorsque x peut varier avec t, en tendant vers $+\infty$ ou vers $-\infty$ quand t tend vers l'infini. Ce problème n'est en réalité qu'un cas particulier du problème dont nous nous sommes occupés dans les chapitres précédents.

Supposons que l'intégrale

$$\int_{-\infty}^{\infty} e^{hy} dS(y) \tag{30}$$

converge pour tout h de module suffisamment petit, et posons

$$\overline{S}(x) = \int_{-\infty}^{x} e^{hy} dS(y)$$

---

(1) Sulla forma generale di un processo stocastico omogeneo. *Rend. R. Accad. Lincei*, (6), 15 (1932), p. 805 et p. 866.

Alors il existe une variable $\overline{Z}_t$ liée à un processus stocastique homogène, dont la répartition est définie au moyen de σ_0^2 et $\overline{S}(x)$ de la même manière que la répartition de Z_t a été définie par σ_0^2 et $S(x)$. Désignons par $\overline{\sigma}_1^2$, $\overline{\sigma}^2$, $\overline{F}(x, t)$ et $\overline{f}(z, t)$ les quantités analogues aux précédentes formées en partant de σ_0^2 et $\overline{S}(x)$. Définissons ici les quantités $\overline{m}$ et R en posant

$$\overline{m} = \sigma_0^2 h + \int_{-\infty}^{\infty} \frac{e^{hy} - 1}{y} dS(y),$$

$$\log R = \frac{1}{2}\sigma_0^2 h^2 + \int_{-\infty}^{\infty} \frac{e^{hy} - 1 - hy}{y^2} dS(y).$$

On démontre alors par un calcul analogue à celui du chapitre II l'identité suivante qui a lieu pour tout h réel appartenant au domaine de convergence de l'intégrale (30) :

$$1 - F\left(\frac{\overline{m}\sqrt{t}}{\sigma}, t\right) = R^t e^{-h\overline{m}t} \int_0^{\infty} e^{-h\overline{\sigma}\sqrt{t}y} d\overline{F}(y, t).$$

Cette identité est, comme on le voit, tout à fait analogue à l'identité (21). On peut aussi s'en servir d'une manière absolument analogue, en démontrant des théorèmes sur la fonction $F(x, t)$ qui sont parfaitement analogues aux théorèmes 1-6 sur la fonction $F_n(x)$. La seule différence est que le paramètre discontinu n a été remplacé par le paramètre continu t. Dans les conditions A et B, on doit remplacer la fonction $V(x)$ par la fonction $S(x)$ considérée dans ce chapitre (1).

---

(1) La condition B peut être remplacée par une autre condition plus générale que celle obtenue de la manière indiquée. Voir H. Cramér, *l. c.*, p. 99. — Un théorème contenu dans notre théorème 6 a été énoncé, pour un cas particulier important du processus homogène, par F. Lundberg, Försäkringsteknisk riskutjämning, Stockholm 1926-1928, et démontré par F. Esscher, *l. c.*

39.

On the representation of a function by certain Fourier integrals

Trans. Amer. Math. Soc. **46**, 191–201 (1939)

1. **Introduction.** Let us consider a complex-valued function $f(t)$ of the real variable t, which is *bounded for all real t and integrable in the Lebesgue sense over every finite interval.* It is proposed to investigate the conditions under which $f(t)$ admits a representation of one of the following types:

$$\text{(F)} \qquad f(t) = \int_{-\infty}^{\infty} e^{itx} dF(x),$$

where $F(x)$ is real, bounded and never decreasing;

$$\text{(G)} \qquad f(t) = \int_{-\infty}^{\infty} e^{itx} dG(x),$$

where $G(x)$ is of bounded variation in $(-\infty, \infty)$; and

$$\text{(g)} \qquad f(t) = \int_{-\infty}^{\infty} e^{itx} g(x) dx,$$

where $g(x)$ is absolutely integrable over $(-\infty, \infty)$. The functions $G(x)$ and $g(x)$ are not necessarily real.

We shall say that a representation of one of these types *exists*, whenever $f(t)$ is represented by the corresponding expression for *almost all* real t. If, in addition, we know a priori that $f(t)$ is continuous, it readily follows from elementary properties of the above integrals that our representation holds for *all* real t.

Now let us denote by $\mu(t)$ a function which satisfies the following conditions (1) and (2):

$$\text{(1)} \qquad \int_{-\infty}^{\infty} | \mu(t) | \, dt \text{ is finite},$$

$$\text{(2a)} \qquad \mu(t) = \int_{-\infty}^{\infty} e^{itx} m(x) dx,$$

where $m(x)$ is real and never negative, and

$$\text{(2b)} \qquad \mu(0) = \int_{-\infty}^{\infty} m(x) dx = 1.$$

* Presented to the Society, February 25, 1939; received by the editors January 31, 1939.

The functions $\mu(t)=e^{-t^2/2}$, $\mu(t)=e^{-|t|}$, and

$$\mu(t) = \begin{cases} 1 - |t|, & |t| \leqq 1, \\ 0, & |t| \geqq 1, \end{cases} \tag{3}$$

are examples of functions satisfying these conditions. The corresponding $m(x)$-functions are, respectively,

$$\frac{1}{(2\pi)^{1/2}} e^{-x^2/2}, \quad \frac{1}{\pi(1+x^2)}, \quad \frac{1-\cos x}{\pi x^2}.$$

For any positive ϵ we denote by $g_\epsilon(x)$ the function defined for all real x by the absolutely convergent integral

$$g_\epsilon(x) = \frac{1}{2\pi}\int_{-\infty}^{\infty} e^{-itx}\mu(\epsilon t)f(t)dt. \tag{4}$$

Obviously $g_\epsilon(x)$ is bounded and everywhere continuous.

We then have for any particular $\mu(t)$ *satisfying* (1) *and* (2) *the following necessary and sufficient conditions for the existence of a representation of* $f(t)$ *according to* (F), (G), *or* (g):

Type (F). $g_\epsilon(x)$ *should be real and never negative for* $0<\epsilon<1$ *and for all real* x.

Type (G). $\int_{-\infty}^{\infty}|g_\epsilon(x)|dx<$const. *for* $0<\epsilon<1$.

Type (g). $g_\epsilon(x)$ *should satisfy the condition for type* (G), *and further*

$$\lim_{\substack{\epsilon\to 0\\ \epsilon'\to 0}} \int_{-\infty}^{\infty} |g_\epsilon(x) - g_{\epsilon'}(x)|dx = 0.$$

If a given function $f(t)$ satisfies one of these conditions for *one* particular function $\mu(t)$, it follows that the same condition is automatically satisfied for *all* $\mu(t)$ satisfying (1) and (2).

Proofs of the conditions will be given in §§3–5. In §7, it will be shown that similar conditions hold for functions $f(t_1, \cdots, t_k)$ of any number of variables.

2. **A particular case.** Choosing for $\mu(t)$ the particular function given by (3), we obtain, writing $A=1/\epsilon$,

$$\begin{aligned} g_\epsilon(x) &= \frac{1}{2\pi}\int_{-A}^{A}\left(1-\frac{|t|}{A}\right)f(t)e^{-itx}dt \\ &= \frac{1}{2\pi A}\int_0^A\int_0^A f(t-u)e^{-ix(t-u)}dt\,du. \end{aligned} \tag{5}$$

In this particular case, our conditions are analogous to those given by

Hausdorff [4] with respect to the problem of representing a sequence of numbers c_k, $(k=0, \pm 1, \pm 2, \cdots)$, in the form

$$c_k = \int_0^{2\pi} e^{ikx} dF(x)$$

or in one of the similar forms corresponding to (G) or (g).

Our condition for type (F) constitutes, in the particular case when $\mu(t)$ is given by (3), a simplified form of a well known theorem due to Bochner (cf. §7). For type (G), Bochner [2] and Schoenberg [6] have given a necessary and sufficient condition which is, however, fundamentally different from ours.

Some applications of our conditions to the theory of random processes will be given in a forthcoming paper.

3. **Representation of type** (F). In the case of a representation

$$\text{(F)} \qquad f(t) = \int_{-\infty}^{\infty} e^{itx} dF(x)$$

with a real, bounded, and never decreasing $F(x)$, it is almost obvious that our condition is *necessary*. We obtain, in fact, from (4)

$$\begin{aligned} g_\epsilon(x) &= \frac{1}{2\pi}\int_{-\infty}^{\infty} e^{-itx}\mu(\epsilon t)dt \int_{-\infty}^{\infty} e^{ity} dF(y) \\ &= \frac{1}{2\pi}\int_{-\infty}^{\infty} dF(y) \int_{-\infty}^{\infty} e^{-it(x-y)}\mu(\epsilon t)dt, \end{aligned}$$

the inversion of the order of integration being justified by the absolute convergence of the integrals. According to (1) and (2) we have, however, almost everywhere

$$m(x) = \frac{1}{2\pi}\int_{-\infty}^{\infty} e^{-itx}\mu(t)dt,$$

so that $g_\epsilon(x)$ is given by the "Faltung"

$$g_\epsilon(x) = \frac{1}{\epsilon}\int_{-\infty}^{\infty} m\left(\frac{x-y}{\epsilon}\right) dF(y),$$

which is obviously real and never negative.

In order to show that the condition is also *sufficient*, we consider the identity

$$\int_{-2A}^{2A} \left(1 - \frac{|x|}{2A}\right) e^{-itx} dx = 2A\left(\frac{\sin At}{At}\right)^2,$$

which holds for every $A>0$. Multiplying by $(2\pi)^{-1}\mu(\epsilon t)f(t)dt$ and integrating with respect to t over $(-\infty, \infty)$, we obtain, according to (4),

$$\int_{-2A}^{2A}\left(1-\frac{|x|}{2A}\right)g_\epsilon(x)dx=\frac{1}{\pi}\int_{-\infty}^{\infty}\left(\frac{\sin t}{t}\right)^2\mu(\epsilon t/A)f(t/A)dt.$$

Now $|\mu(t)|\leqq 1$ by (2), and $f(t)$ is bounded by hypothesis; say $|f(t)|\leqq c$. Thus if $g_\epsilon(x)$ is real and never negative, we conclude

$$\int_{-2A}^{2A}\left(1-\frac{|x|}{2A}\right)g_\epsilon(x)dx\leqq c$$

for $0<\epsilon<1$ and for all positive A. This obviously implies

$$\int_{-\infty}^{\infty}g_\epsilon(x)dx\leqq c \tag{6}$$

for $0<\epsilon<1$.

From (4) and (6) we then obtain for almost all values of x

$$\mu(\epsilon t)f(t)=\int_{-\infty}^{\infty}e^{itx}g_\epsilon(x)dx. \tag{7}$$

Now, since both $\mu(t)$ and the integral are continuous functions of t, it follows that it is possible to find a continuous function $f^*(t)$ which coincides with $f(t)$ for almost all real t. We then have

$$\mu(\epsilon t)f^*(t)=\int_{-\infty}^{\infty}e^{itx}g_\epsilon(x)dx$$

for *all* real t and for $0<\epsilon<1$.

Consider now the last relation for a sequence of values of ϵ tending to zero. As $\mu(0)=1$, the left-hand side tends to $f^*(t)$ uniformly in every finite interval. According to a fundamental theorem on characteristic functions due to Lévy [5] (cf. also Bochner [1]), we then have for all real t

$$f^*(t)=\int_{-\infty}^{\infty}e^{itx}dF(x)$$

where $F(x)$ is real and never decreasing. As $f^*(t)=f(t)$ for almost all t, this proves our assertion.

4. **Representation of type** (G).† If we have for almost all t

$$f(t)=\int_{-\infty}^{\infty}e^{itx}dG(x), \tag{G}$$

† The author is indebted to Mr. E. Frithiofson of Lund for a remark leading to a simplification of the condition for this type.

where $G(x)$ is of bounded variation in $(-\infty, \infty)$, we obtain as in the preceding paragraph

$$g_\epsilon(x) = \frac{1}{\epsilon}\int_{-\infty}^{\infty} m\left(\frac{x-y}{\epsilon}\right) dG(y)$$

and thus, $m(x)$ being never negative,

$$\int_{x_1}^{x_2} |g_\epsilon(x)|\,dx \leqq \int_{-\infty}^{\infty} |dG(y)| \int_{(x_1-y)/\epsilon}^{(x_2-y)/\epsilon} m(x)dx. \tag{8}$$

Hence we obtain, using (2b),

$$\int_{-\infty}^{\infty} |g_\epsilon(x)|\,dx \leqq \int_{-\infty}^{\infty} |dG(y)|$$

for $0<\epsilon<1$. Thus our condition is *necessary*.

In order to show that the condition is also *sufficient* we observe that, owing to the convergence of $\int_{-\infty}^{\infty}|g_\epsilon(x)|\,dx$, the relation (4) may be converted into (7) for almost all real t. As in the preceding section, it follows that there is a continuous function $f^*(t)$ which coincides with $f(t)$ for almost all real t. We then have as before

$$\mu(\epsilon t)f^*(t) = \int_{-\infty}^{\infty} e^{itx} g_\epsilon(x)dx$$

for all real t and for $0<\epsilon<1$. Putting

$$G_\epsilon(x) = \int_{-\infty}^{x} g_\epsilon(y)dy,$$

we may write this as

$$\mu(\epsilon t)f^*(t) = \int_{-\infty}^{\infty} e^{itx} dG_\epsilon(x).$$

When ϵ tends to zero, the left-hand side of this relation tends to $f^*(t)$ uniformly in every finite interval. On the other hand, $\int_{-\infty}^{\infty}|g_\epsilon(x)|\,dx$ is uniformly bounded for $0<\epsilon<1$, so that $G_\epsilon(x)$ is of uniformly bounded variation in $(-\infty, \infty)$. It is well known that we can always find a sequence $\epsilon_1, \epsilon_2, \cdots$ tending to zero and a function $G(x)$ of bounded variation in $(-\infty, \infty)$ such that

$$G(x) = \lim_{n\to\infty} G_{\epsilon_n}(x) = \lim_{n\to\infty} \int_{-\infty}^{x} g_{\epsilon_n}(y)dy \tag{9}$$

in every point of continuity x of $G(x)$. It then follows from a lemma given by

Bochner [2, p. 274], that we have for all real t

$$f^*(t) = \int_{-\infty}^{\infty} e^{itx} dG(x).$$

As $f^*(t) = f(t)$ for almost all t, this proves our assertion.

For a later purpose it will now be shown that, if our condition for type (G) is satisfied, then the integral

$$\int_{-\infty}^{\infty} | g_\epsilon(x) | dx \tag{10}$$

is *uniformly convergent* for $0 < \epsilon < 1$. If the condition is satisfied, we already know that $f(t)$ admits a representation of type (G). Now let $\delta > 0$ be given. We can then choose $y_0 > 0$ and $x_0 > y_0$ such that

$$\int_{y_0}^{\infty} | dG(y) | < \delta, \qquad \int_{x_0 - y_0}^{\infty} m(x) dx < \delta.$$

Obviously x_0 and y_0 can be chosen independently of ϵ. For $x_2 > x_1 > x_0$ and for $0 < \epsilon < 1$, we then conclude from (8) and (2b) that

$$\int_{x_1}^{x_2} | g_\epsilon(x) | dx < \delta \int_{-\infty}^{y_0} | dG(y) | + \int_{y_0}^{\infty} | dG(y) |$$

$$< \delta \left[1 + \int_{-\infty}^{\infty} | dG(y) | \right].$$

A similar inequality evidently holds for negative values of x_1 and x_2, and thus the uniform convergence of (10) is established.

5. **Representation of type** (g). As in the preceding cases, we begin by proving that our condition is *necessary*. Any representation of type (g) being a particular case of type (G), it is obvious that the first part of the condition is necessary. It thus remains to show that, if

$$f(t) = \int_{-\infty}^{\infty} e^{itx} g(x) dx \tag{g}$$

holds for almost all t, where $g(t)$ is absolutely integrable over $(-\infty, \infty)$, then

$$\lim_{\substack{\epsilon \to 0 \\ \epsilon' \to 0}} \int_{-\infty}^{\infty} | g_\epsilon(x) - g_{\epsilon'}(x) | dx = 0.$$

As $| g_\epsilon - g_{\epsilon'} | \leqq | g - g_\epsilon | + | g - g_{\epsilon'} |$, it is only necessary to prove that

$$\lim_{\epsilon \to 0} \int_{-\infty}^{\infty} | g(x) - g_\epsilon(x) | dx = 0. \tag{11}$$

According to the preceding section, it follows from the first part of the condition that the integral (10) converges uniformly for $0<\epsilon<1$. Given $\delta>0$, we can thus choose $x_0=x_0(\delta)$ such that

$$\int_{|x|>x_0} |g(x)-g_\epsilon(x)|\,dx<\delta \tag{12}$$

for $0<\epsilon<1$.

We now choose a function $g^*(x)$, *bounded and continuous* for all real x, such that

$$\int_{-\infty}^{\infty} |g(x)-g^*(x)|\,dx<\delta \tag{13}$$

with

$$|g^*(x)|<K=K(\delta),$$

and we put

$$g_\epsilon^*(x)=\frac{1}{\epsilon}\int_{-\infty}^{\infty} m\left(\frac{x-y}{\epsilon}\right)g^*(y)dy.$$

We then have

$$\begin{aligned}\int_{-x_0}^{x_0} |g(x)-g_\epsilon(x)|\,dx &\leqq \int_{-x_0}^{x_0}|g(x)-g^*(x)|\,dx+\int_{-x_0}^{x_0}|g^*(x)-g_\epsilon^*(x)|\,dx\\ &\quad+\int_{-x_0}^{x_0}|g_\epsilon^*(x)-g_\epsilon(x)|\,dx.\end{aligned} \tag{14}$$

According to (13), the first term on the right-hand side is less than δ. We further have, using (2b),

$$g^*(x)-g_\epsilon^*(x)=\frac{1}{\epsilon}\int_{-\infty}^{\infty} m\left(\frac{x-y}{\epsilon}\right)(g^*(x)-g^*(y))dy. \tag{15}$$

Now, $g^*(x)$ is uniformly continuous in every finite interval. The numbers δ and x_0 being given, we can thus choose $h=h(\delta,x_0)$ such that for $|x|<x_0$, $|x-y|<h$ we have

$$|g^*(x)-g^*(y)|<\delta/x_0.$$

We can further choose $y_0=y_0(\delta,x_0,K)$ such that

$$\int_{|y|>y_0} m(y)dy<\frac{\delta}{2Kx_0}\cdot$$

For any ϵ such that $0<\epsilon<h/y_0$, we then obtain from (15)

$$|g^*(x) - g_\epsilon^*(x)| < \frac{\delta}{x_0} + 2K \int_{|y|>h/\epsilon} m(y)dy < \frac{2\delta}{x_0}$$

and

(16) $$\int_{-x_0}^{x_0} |g^*(x) - g_\epsilon^*(x)|\, dx < 4\delta.$$

Finally, we have

$$g_\epsilon^*(x) - g_\epsilon(x) = \frac{1}{\epsilon} \int_{-\infty}^{\infty} m\left(\frac{x-y}{\epsilon}\right)(g^*(y) - g(y))dy,$$

and hence by (13)

(17) $$\begin{aligned} \int_{-x_0}^{x_0} |g_\epsilon^*(x) - g_\epsilon(x)|\, dx &\leqq \int_{-\infty}^{\infty} |g^*(y) - g(y)|\, dy \int_{(-x_0-y)/\epsilon}^{(x_0-y)/\epsilon} m(x)dx \\ &\leqq \int_{-\infty}^{\infty} |g^*(y) - g(y)|\, dy < \delta. \end{aligned}$$

From (12), (14), (16), and (17) we then obtain

$$\int_{-\infty}^{\infty} |g(x) - g_\epsilon(x)|\, dx < 7\delta$$

for all sufficiently small $\epsilon > 0$, so that (11) is proved.

We now have to show that our condition is *sufficient*. From the first part of the condition, it follows by the preceding paragraph that we have for almost all real t

(G) $$f(t) = \int_{-\infty}^{\infty} e^{itx} dG(x),$$

where $G(x)$ is of bounded variation in $(-\infty, \infty)$, and according to (9)

(18) $$G(x) = \lim_{n\to\infty} G_{\epsilon_n}(x) = \lim_{n\to\infty} \int_{-\infty}^{x} g_{\epsilon_n}(y)dy$$

in every point of continuity x of $G(x)$.

From the second part of the condition it follows, however, that there is a function $g(x)$, absolutely integrable over $(-\infty, \infty)$, such that

$$\lim_{n\to\infty} \int_{-\infty}^{\infty} |g(y) - g_{\epsilon_n}(y)|\, dy = 0.$$

Hence we obtain for all real x

$$\lim_{n\to\infty} G_{\epsilon_n}(x) = \lim_{n\to\infty} \int_{-\infty}^{x} g_{\epsilon_n}(y)dy = \int_{-\infty}^{x} g(y)dy.$$

It then finally follows from (18) that

$$G(x) = \int_{-\infty}^{x} g(y)dy$$

for almost all x, and

$$f(t) = \int_{-\infty}^{\infty} e^{itx} g(x)dx$$

for almost all t, so that the proof is completed.

If the first part of our condition for type (g) is replaced by the condition given above for type (F), it is readily seen that we obtain a necessary and sufficient condition for representation of type (g) with a real and non-negative $g(x)$.

It may be worth while to point out that the first part of our condition for type (g) is *not* contained in the second part. This is shown by the example

$$f(t) = \begin{cases} i(-1-t), & -1 < t < 0, \\ i(1-t), & 0 < t < 1, \\ 0, & t = 0, \ |t| \geqq 1. \end{cases}$$

In the particular case when $\mu(t)$ is given by (3), this function yields for $0<\epsilon<1$

$$g_\epsilon(x) = \frac{x - \sin x}{\pi x^2} + \epsilon \frac{x \sin x - 2(1-\cos x)}{\pi x^3},$$

so that the second part of the condition is satisfied but not the first part. Accordingly, no representation of any of our three types exists, which is also directly seen from the behaviour of $f(t)$ near $t=0$.

6. **The case of an unbounded** $f(t)$. In all the preceding paragraphs it has been a priori assumed that $f(t)$ is *bounded*. It will, however, be seen that this assumption has only been used on two occasions; namely (a) to ensure the absolute convergence of the integral (4) which defines $g_\epsilon(x)$, and (b) for the proof that our condition for type (F) is sufficient.

Let us now omit this assumption and consider the class of all functions $f(t)$ which are integrable over any finite interval. Let us further choose for $\mu(t)$ the particular function given by (3). As this function is equal to zero for $|t| \geqq 1$, it is obvious that the integral (4) will still be absolutely convergent for any positive ϵ.

Thus the conditions for types (G) *and* (g) *remain true under the present con-*

ditions, while in the condition for type (F) *it will have to be explicitly stated that* $|f(t)|$ *should be less than a constant K for almost all values of t.*

The necessity of this addition to the condition for type (F) is shown by the example $f(t)=|t|^{-\alpha}$, $(0<\alpha<1)$, where obviously no K can be found such that $|f(t)|<K$ for almost all t. The corresponding function $g_\epsilon(x)$ can be shown to be positive for $0<\epsilon<1$ and for all real x, although evidently no representation of type (F) exists.

7. **Functions of several variables.** So far we have only considered functions $f(t)$ of a single variable t. All our considerations can, however, be extended to functions $f(t_1, \cdots, t_k)$ of any finite number of real variables. This requires only a straightforward generalization of our above arguments, based on the elementary properties of Fourier integrals in several variables. The only delicate point arising in this connection is the generalization to several variables of Bochner's lemma used in the proof of our condition for type (G). This generalization is, however, easily performed by means of a general induction method due to Cramér and Wold (Cramér [3, p. 104]).

We obtain in this way direct generalizations of our above conditions, the auxiliary functions $\mu(t)$ and $g_\epsilon(x)$ being replaced by the functions of k variables obtained when, in (1), (2), and (4), we regard x, t, and ϵt as abbreviations for $(x_1, \cdots, x_k)$, $(t_1, \cdots, t_k)$, and $(\epsilon t_1, \cdots, \epsilon t_k)$, respectively, and put $tx=t_1x_1+\cdots+t_kx_k$, the integrals being taken over the k-dimensional euclidean space R_k. Moreover, in the definition (4) of $g_\epsilon(x)$ the factor $1/2\pi$ has to be replaced by $1/(2\pi)^k$.

For $\mu(t_1, \cdots, t_k)$ we may, for example, choose any function of the form $\mu(t_1)\mu(t_2)\cdots\mu(t_k)$, where $\mu(t)$ satisfies the conditions (1) and (2). The definition (4) of $g_\epsilon(x)$ will then be replaced by

$$g_\epsilon(x_1, \cdots, x_k) = \frac{1}{(2\pi)^k}\int_{R_k} e^{-i(t_1x_1+\cdots+t_kx_k)}\mu(\epsilon t_1)\cdots\mu(\epsilon t_k)f(t_1, \cdots, t_k)dt_1\cdots dt_k. \tag{19}$$

In particular, we obtain in this way the following new characterization of the class of *positive definite functions* of k variables as defined by Bochner [1, p. 406]. Bochner has established the identity of this class with the class of functions represented for all real t_r by the expression

$$f(t_1, \cdots, t_k) = \int_{R_k} e^{i(t_1x_1+\cdots+t_kx_k)}dF(x_1, \cdots, x_k),$$

where F is real, bounded, and, for each particular x_r, never decreasing. The class of positive definite functions such that $f(0, \cdots, 0)=1$ is thus identical with the class of characteristic functions of k-dimensional random variables in the sense of the theory of probability (cf. Cramér [3]). Using our generalized condition for type (F) we then conclude:

For any particular $\mu(t)$ *satisfying* (1) *and* (2) *a necessary and sufficient condition that a given bounded and continuous function* $f(t_1, \cdots, t_k)$ *should be positive definite is that* $g_\epsilon(x_1, \cdots, x_k)$ *as defined by* (19) *should be real and never negative for* $0<\epsilon<1$ *and for all real* $x_1, \cdots, x_k$.

Choosing, in particular, in (19) the special function $\mu(t)$ given by (3), we obtain in analogy with (5), writing $A=1/\epsilon$,

$$g_\epsilon(x_1, \cdots, x_k) = \frac{1}{(2\pi)^k}\int_{-A}^{A}\cdots\int_{-A}^{A} f(t_1, \cdots, t_k) \cdot \prod_{r=1}^{k}\left[\left(1-\frac{|t_r|}{A}\right)e^{-it_rx_r}\right]dt_1\cdots dt_r$$

$$= \frac{1}{(2\pi A)^k}\int_0^A\cdots\int_0^A f(t_1-u_1, \cdots, t_k-u_k) \cdot\exp\left(-i\sum_1^k x_rt_r\right)\cdot\exp\left(i\sum_1^k x_ru_r\right)dt_1\cdots dt_k du_1\cdots du_k.$$

Now Bochner's original condition for a positive definite function requires that

$$\int_a^A\cdots\int_a^A f(t_1-u_1, \cdots, t_k-u_k)\rho(t_1, \cdots, t_k) \cdot\overline{\rho(u_1, \cdots, u_k)}dt_1\cdots dt_k du_1\cdots du_k \geqq 0$$

for all real a, A and for *all continuous functions* $\rho(t_1, \cdots, t_k)$. Thus our condition, with the particular choice of $\mu(t)$ according to (3), involves a considerable simplification.

References

1. S. Bochner, *Monotone Funktionen, Stieltjessche Integrale und harmonische Analyse*, Mathematische Annalen, vol. 108 (1933), p. 378.

2. ———, *A theorem on Fourier-Stieltjes integrals*, Bulletin of the American Mathematical Society, vol. 40 (1934), p. 271.

3. H. Cramér, *Random Variables and Probability Distributions*, Cambridge, 1937.

4. F. Hausdorff, *Momentprobleme für ein endliches Intervall*, Mathematische Zeitschrift, vol. 16 (1923), p. 220.

5. P. Lévy, *Calcul des Probabilités*, Paris, 1925.

6. I. J. Schoenberg, *Remark on the preceding note by Bochner*, Bulletin of the American Mathematical Society, vol. 40 (1934), p. 277.

University of Stockholm,
Stockholm, Sweden

40.

On the theory of stationary random processes

Ann. Math. **41** (1), 215–230 (1940)

I. Introduction

1. Let t be a real parameter which may be regarded as representing time and let, for every t, $X(t)$ denote a real-valued random variable.

Following Khintchine [6], we shall say that $X(t)$ is attached to a *one-dimensional random process* if, for any finite set of real values $t_1, \cdots, t_k$, the simultaneous probability distribution[1] of the variables $X(t_1), \cdots, X(t_k)$ is known. This distribution may, *e.g.*, be characterized by means of the corresponding k-dimensional distribution function $F(x_1, \cdots, x_k; t_1, \cdots, t_k)$, which represents the probability of the set of inequalities

$$X(t_\nu) \leqq x_\nu, \qquad (\nu = 1, 2, \cdots, k).$$

F is a distribution function in the variables $x_1, \cdots, x_k$, with $t_1, \cdots, t_k$ as parameters.

Any system of distribution functions $F(x_1, \cdots, x_k; t_1, \cdots, t_k)$, where $k = 1, 2, \cdots$, such that every F is symmetric in all pairs (x_ν, t_ν) and satisfies the *consistency relation*

$$F(x_1, \cdots, x_j, \infty, \cdots, \infty; t_1, \cdots, t_k) = F(x_1, \cdots, x_j; t_1, \cdots, t_j)$$

for any $j < k$, defines a random process.

For certain purposes this definition is not complete enough, and Doob [3] has given more elaborate definitions which meet the requirements of these cases. In the present paper we shall, however, only deal with problems where the simple definition given above is sufficient.

The definition is immediately extended to an *n-dimensional process*, if an n-dimensional random variable $Z(t) = \{X_1(t), \cdots, X_n(t)\}$ is chosen as the basic variable of the process. The simultaneous distribution of the variables $Z(t_1), \cdots, Z(t_k)$ is then characterized by a distribution function in a space of nk dimensions, which represents the probability of the set of inequalities

$$X_\mu(t_\nu) \leqq x_{\mu\nu}, \qquad (\mu = 1, 2, \cdots, n; \nu = 1, 2, \cdots, k).$$

The corresponding extension of the conditions of symmetry and consistency is obvious.

---

[1] Cf the writer's Cambridge Tract [1] on "Random Variables and Probability Distributions."

Finally, we may evidently assume the n components $X_\mu(t)$ of the basic variable $Z(t)$ to be *complex random variables*

$$X_\mu(t) = U_\mu(t) + i\, V_\mu(t), \tag{1}$$

and so obtain a *complex multi-dimensional process*. The simultaneous distribution of $Z(t_1), \cdots, Z(t_k)$ is then characterized by a distribution function F in a space of $2nk$ dimensions, which represents the probability of the set of inequalities

$$U_\mu(t_\nu) \leqq u_{\mu\nu},$$

$$V_\mu(t_\nu) \leqq v_{\mu\nu},$$

where $\mu = 1, 2, \cdots, n$; $\nu = 1, 2, \cdots, k$. F is a distribution function in the arguments $u_{\mu\nu}$ and $v_{\mu\nu}$, with $t_1, \cdots, t_k$ as parameters.

2. Consider a random process defined by

$$Z(t) = \{X_1(t), \cdots, X_n(t)\}$$

where, in the general case, the $X_\mu(t)$ are complex random variables defined according to (1). It evidently amounts to the same thing if we consider *simultaneously* the n processes defined by the components $X_1(t), \cdots, X_n(t)$, and we shall apply these two points of view without distinction in the sequel.

We shall always assume that the mean values $E\{|X_\mu(t)|^2\}$ are finite for all μ and t. It then follows that the mean values $E\{X_\mu(t)\}$ and $E\{X_\mu(t)\overline{X_\nu(u)}\}$ are finite for all μ, ν, t and u.

Generalizing a definition introduced by Khintchine [6][2], we shall say that $Z(t)$ defines a *stationary random process*, if the following two conditions are satisfied for all μ and ν:

(A) $E\{X_\mu(t)\} = m_\mu$ *is independent of* t;

(B) $E\{X_\mu(t)\overline{X_\nu(u)}\} = R_{\mu\nu}(t-u)$ *is a function of the difference* $t - u$.

If these conditions are satisfied, we shall also say that the n processes defined by the variables $X_1(t), \cdots, X_n(t)$ are *stationary* and *stationarily correlated*.

Obviously we may without restriction of generality assume

$$\text{(C)} \qquad \begin{cases} E\{X_\mu(t)\} = m_\mu = 0, \\ E\{|X_\mu(t)|^2\} = R_{\mu\mu}(0) = \sigma_\mu^2 > 0 \end{cases}$$

for all μ. This being assumed, we shall denote the $R_{\mu\nu}(t)$ as the *correlation functions* of the stationary processes attached to the variables $X_1(t), \cdots, X_n(t)$. More particularly, we shall call $R_{\mu\mu}(t)$ the *autocorrelation function* of the process defined by $X_\mu(t)$, while $R_{\mu\nu}(t)$ and $R_{\nu\mu}(t)$ will, for $\mu \neq \nu$, be called the *mutual*

[2] Khintchine introduces also another, narrower, concept of stationarity which, however, we shall have no occasion to use in the present paper.

correlation functions of the processes connected with $X_\mu(t)$ and $X_\nu(t)$. It follows from (B) that we always have

$$R_{\nu\mu}(-t) = \overline{R_{\mu\nu}(t)}. \tag{2}$$

In the particular case when the variables $X_\mu(t)$ are real, $R_{\mu\nu}(t)/(\sigma_\mu\sigma_\nu)$ is evidently identical with the ordinary correlation coefficient of $X_\mu(t_0 + t)$ and $X_\nu(t_0)$.

We shall finally assume that

$$\lim_{t\to 0} R_{\mu\mu}(t) = R_{\mu\mu}(0) = \sigma_\mu^2 \tag{D}$$

for $\mu = 1, 2, \cdots, n$. It will be shown later (Lemma 1) that, if this condition is satisfied, all the correlation functions $R_{\mu\nu}(t)$ are continuous for all real values of t. The process defined by each $X_\mu(t)$, as well as the one defined by the composite variable $Z(t)$, will then be called a *continuous stationary random process*. This concept has been introduced by Khintchine [6], who has justly emphasized its great importance for various branches of applications.

3. The main purpose of the present paper is to study the mutual relations of the correlation functions $R_{\mu\nu}(t)$. In Chapter III, we shall deduce a necessary and sufficient condition that a set of n^2 *a priori* given functions $R_{\mu\nu}(t)$ may be correlation functions of n complex random processes satisfying the conditions (A)–(D). It will be shown to be necessary and sufficient that the $R_{\mu\nu}(t)$ should, for all real t, be given by Fourier-Stieltjes integrals of the form

$$R_{\mu\nu}(t) = \int_{-\infty}^{\infty} e^{itx}\, dF_{\mu\nu}(x),$$

where the $F_{\mu\nu}(x)$ are functions of bounded variation in $(-\infty, \infty)$ which satisfy certain conditions as specified by Theorem 1. In particular, it follows from these conditions that the functions $F_{\mu\mu}(x)$ corresponding to the autocorrelation functions $R_{\mu\mu}(t)$ are real and never decreasing, with a total variation equal to σ_μ^2, as already proved by Khintchine [6].

In Chapter IV, some further properties of the correlation functions will be deduced, and the particular case of random processes defined by *real* variables $X_1(t)$ and $X_2(t)$ will be considered. Finally, in Chapter V we shall show that an analogous theory may be developed for the case when our random processes are considered for *integral values* only of the parameter t.

II. Auxiliary Theorems

4. In this Chapter we shall, for the sake of convenience, give as Lemmas a number of results (partly previously known) wanted in the sequel. In the first place, we shall prove the following result indicated in the Introduction.

Lemma 1. *Let $R_{\mu\nu}(t)$ be the n^2 correlation functions of n random processes satisfying the conditions* (A)–(D) *of paragraph* 2. *Then $R_{\mu\nu}(t)$ is continuous for all real t, and we have*

$$| R_{\mu\nu}(t) | \leqq \sigma_\mu\sigma_\nu . \tag{3}$$

The inequality (3) follows immediately from (B) and (C) by means of the Schwarz inequality. Further, we have

$$R_{\mu\nu}(t + \Delta t) - R_{\mu\nu}(t) = E\{\overline{X_\nu(0)}[X_\mu(t + \Delta t) - X_\mu(t)]\}$$

and thus by the Schwarz inequality

$$\begin{aligned} | R_{\mu\nu}(t + \Delta t) - R_{\mu\nu}(t) | &\leqq \sigma_\nu \sqrt{E\{| X_\mu(t + \Delta t) - X_\mu(t) |^2\}} \\ &= \sigma_\nu \sqrt{2\sigma_\mu^2 - R_{\mu\mu}(\Delta t) - R_{\mu\mu}(-\Delta t)}. \end{aligned}$$

According to (D), this tends to zero with Δt.

5. We next consider some convergence properties of sequences of random variables. If $X_1, X_2, \cdots$ and X are real or complex random variables, we shall say that X_n *converges in the mean* to X, and write

$$\underset{n\to\infty}{\text{l.i.m.}}\, X_n = X,$$

whenever the mean value

$$E\{| X_n - X |^2\}$$

is finite for all sufficiently large n and tends to zero as $n \to \infty$.

If the mean value

$$E\{| X_m - X_n |^2\}$$

tends to zero as m and n tend independently to infinity, then it is known[3] that there is a variable X such that X_n converges in the mean to X. If, in addition, $E\{| X_n |^2\}$ is finite for all sufficiently large n, it immediately follows that $E\{| X |^2\}$ is finite. Further, if X' and X'' are variables such that X_n converges in the mean both to X' and to X'', then X' and X'' are equivalent, i.e.

$$Pr\{X' = X''\} = 1.$$

LEMMA 2. *Let $X_1, X_2, \cdots$ and $Y_1, Y_2, \cdots$ be two sequences of random variables such that $E\{| X_n |^2\}$ and $E\{| Y_n |^2\}$ are finite for all n. Suppose that the mean values $E\{| X_m - X_n |^2\}$ and $E\{| Y_m - Y_n |^2\}$ both tend to zero as m and n tend independently to infinity, so that there are two variables X and Y such that $E\{| X |^2\}$ and $E\{| Y |^2\}$ are finite, and* l.i.m. $X_n = X$, l.i.m. $Y_n = Y$. *Then*

$$\begin{aligned} &E(X_n) \to E(X), \qquad E(Y_n) \to E(Y), \\ &E\{| X_n |^2\} \to E\{| X |^2\}, \qquad E\{| Y_n |^2\} \to E\{| Y |^2\}, \\ &E(X_n Y_n) \to E(XY). \end{aligned}$$

[3] Cf. e.g. Fréchet [4], Lévy [7].—It is also known that convergence in the mean implies convergence "in probability," which is expressed by the relation $Pr\{|X_n - X| > \epsilon\} \to 0$ for any $\epsilon > 0$.

In fact, we obtain by means of the Schwarz inequality

$$| E(X_n) - E(X) |^2 \leqq E^2\{| X_n - X |\} \leqq E\{| X_n - X |^2\},$$

and

$$\begin{aligned}[E\{| X_n |^2\} - E\{| X |^2\}]^2 &= E^2\{(| X_n | - | X |)(| X_n | + | X |)\} \\ &\leqq E\{(| X_n | - | X |)^2\} \cdot E\{(| X_n | + | X |)^2\} \\ &\leqq E\{| X_n - X |^2\} \cdot E\{(| X_n - X | + 2 | X |)^2\} \\ &\leqq 2E\{| X_n - X |^2\} \cdot E\{| X_n - X |^2 + 4 | X |^2\}\end{aligned}$$

Hence we immediately conclude that $E(X_n) \to E(X)$ and $E\{| X_n |^2\} \to E\{| X |^2\}$. Applying the last result to the sequences $X_n + \bar{Y}_n$ and $X_n + i\bar{Y}_n$, we find that $E(X_nY_n) \to E(XY)$.

6. In the sequel, we shall repeatedly have to deal with integrals containing a random variable attached to a continuous stationary random process. Integrals of this kind may be very simply defined in complete analogy to the elementary definition of an ordinary Riemann integral, as shown by the following Lemma.

LEMMA 3. *Let $X(t)$ be a (complex) random variable, attached to a continuous stationary random process, and let $\varphi(t)$ be a continuous function of t for $a \leqq t \leqq b$. Consider a sequence of random variables $Z_1, Z_2, \cdots$, defined by*

$$Z_n = \sum_{\nu=1}^{n} \varphi(t_\nu^{(n)})X(t_\nu^{(n)})(t_\nu^{(n)} - t_{\nu-1}^{(n)}),$$

where $a = t_0^{(n)} < t_1^{(n)} < \cdots < t_{n-1}^{(n)} < t_n^{(n)} = b$. Suppose that

$$\operatorname*{Max}_{\nu=1,2,\cdots,n} (t_\nu^{(n)} - t_{\nu-1}^{(n)}) \to 0$$

as $n \to \infty$. Then Z_n converges in the mean to a random variable Z. Further, if $\{Z_n^*\}$ is another sequence formed with the same $X(t)$ and $\varphi(t)$, but with another system of points $t_\nu^{(n)}$ satisfying the above conditions, and converging in the mean to the limit Z^*, then Z^* and Z are equivalent. For the limit Z we write, by definition,

$$Z = \operatorname*{l.i.m.}_{n\to\infty} Z_n = \int_a^b \varphi(t)X(t)\,dt,$$

so that the last integral is a random variable, which is hereby completely defined but for equivalence.

It is evident that we may assume $E\{X(t)\} = 0$.

Putting

$$Z' = \sum_{\nu=1}^{m} \varphi(t'_\nu)X(t'_\nu)(t'_\nu - t'_{\nu-1}),$$

$$Z'' = \sum_{\nu=1}^{n} \varphi(t''_\nu)X(t''_\nu)(t''_\nu - t''_{\nu-1}),$$

it will obviously only be necessary to show that the mean value $E\{|Z' - Z''|^2\}$ can be made as small as we please by taking the length δ of the maximum interval in the two divisions corresponding to the points t'_ν and t''_ν sufficiently small. In fact, this point being established, it directly follows that any sequence $\{Z_n\}$ or $\{Z^*_n\}$ converges in the mean, and that the mean value $E\{|Z_n - Z^*_n|^2\}$ tends to zero. Hence the limits Z and Z^* must be equivalent.

From the current text-book theory of the ordinary Riemann integral, it is well known that we may restrict ourselves to the case when the division points t'_ν $(\nu = 0, 1, \cdots, m)$ *are all contained among the points* t''_ν $(\nu = 0, 1, \cdots, n)$. This being assumed, we may write

$$Z' - Z'' = \sum_{\nu=1}^{n} \{\varphi(\tau_\nu)X(\tau_\nu) - \varphi(t''_\nu)X(t''_\nu)\}(t''_\nu - t''_{\nu-1})$$

where

$$|\tau_\nu - t''_\nu| < \delta.$$

Denoting by $R(t)$ the autocorrelation function of our continuous stationary process, we then have

$$\begin{aligned} E\{|Z' - Z''|^2\} \\ &= \sum_{\mu,\nu} \varphi(\tau_\mu)\{\overline{\varphi(\tau_\nu)}R(\tau_\mu - \tau_\nu) - \overline{\varphi(t''_\nu)}R(\tau_\mu - t''_\nu)\}(t''_\mu - t''_{\mu-1})(t''_\nu - t''_{\nu-1}) \\ &\quad - \sum_{\mu,\nu} \varphi(t''_\mu)\{\overline{\varphi(\tau_\nu)}R(t''_\mu - \tau_\nu) - \overline{\varphi(t''_\nu)}R(t''_\mu - t''_\nu)\}(t''_\mu - t''_{\mu-1})(t''_\nu - t''_{\nu-1}). \end{aligned}$$

$\varphi(t)$ and $R(t)$ being both continuous functions, it easily follows that the last expression can be made less than any given ϵ by taking δ sufficiently small. Thus the Lemma is proved.

It is easily shown that the elementary properties of the ordinary Riemann integral hold for the integral of a random variable as defined by Lemma 3. Integrals of this type are themselves random variables, and we shall have occasion to consider in the sequel certain mean values connected with such variables. In this respect, we shall require the following Lemma.

LEMMA 4. *Let $X_1(t)$ and $X_2(t)$ be random variables attached to continuous stationary processes satisfying the conditions* (A)–(D) *of paragraph* 2, *and let $\varphi_1(t)$ and $\varphi_2(t)$ be continuous functions of t for $a \leqq t \leqq b$. Then*

$$(4) \qquad E\left\{\int_a^b \varphi_1(t)X_1(t)\,dt \cdot \int_a^b \varphi_2(u)\overline{X_2(u)}\,du\right\} = \int_a^b\int_a^b \varphi_1(t)\varphi_2(u)R_{12}(t-u)\,dt\,du.$$

In particular, for $X_1(t) = X_2(t)$, $\varphi_1(t) = \overline{\varphi_2(t)}$, we have

$$E\left\{\left|\int_a^b \varphi_1(t)X_1(t)\,dt\right|^2\right\} = \int_a^b\int_a^b \varphi_1(t)\overline{\varphi_1(u)}R_{11}(t-u)\,dt\,du.$$

The last expression, being the mean value of a real and non-negative variable, is evidently real and non-negative.

In order to prove this Lemma, we have only to express the integrals in the first member of (4) as limits of finite sums of random variables according to Lemma 3. The mean value of the expression thus obtained is equal to a Riemann sum corresponding to the double integral in the second member of (4). An application of the last relation of Lemma 2 then immediately yields the desired result.

III. Main Theorem on the Correlation Functions

7. We now proceed to the proof of the following theorem, which constitutes the main result of the present paper.[4]

THEOREM 1. *a) Let $X_1(t), \cdots, X_n(t)$ be complex random variables attached to n continuous stationary processes satisfying the conditions* (A)–(D) *of paragraph 2. Then the correlation functions defined by* (B) *are for all real t given by Fourier-Stieltjes integrals of the form*

$$R_{\mu\nu}(t) = \int_{-\infty}^{\infty} e^{itx}\, dF_{\mu\nu}(x), \tag{5}$$

where the $F_{\mu\nu}(x)$ are functions of bounded variation in $(-\infty, \infty)$, which we may always assume to be everywhere continuous to the right. Further, denoting by $\Delta F_{\mu\nu}$ the increase $F_{\mu\nu}(\beta) - F_{\mu\nu}(\alpha)$ of $F_{\mu\nu}(x)$ over the real interval $\alpha \leqq x \leqq \beta$, the form

$$H(z_1, \cdots, z_n) = \sum_{\mu,\nu=1}^{n} z_\mu \bar{z}_\nu \Delta F_{\mu\nu} \tag{6}$$

is for any closed interval (α, β) a non-negative Hermite form. (In particular, putting all the z-variables except z_μ equal to zero, it follows that $F_{\mu\mu}(x)$ is real and never decreasing.)

b) Conversely, let $R_{\mu\nu}(t)$ be a set of n^2 functions given by the Fourier-Stieltjes integrals (5), *the functions $F_{\mu\nu}(x)$ being of bounded variation in $(-\infty, \infty)$ and everywhere continuous to the right, and suppose that the form* (6) *is, for any closed real interval (α, β), a non-negative Hermite form. Then it is possible to find n continuous stationary random processes having the correlation functions $R_{\mu\nu}(t)$.*

8. In order to prove the first part of the theorem, we shall begin by deducing the expression (5) for the correlation functions. We put, for any real x and any $A > 1$,

$$g_{\mu\nu}(x, A) = \frac{1}{2\pi A} \int_0^A \int_0^A R_{\mu\nu}(t-u) e^{-ix(t-u)}\, dt\, du. \tag{7}$$

[4] Professor A. Kolmogoroff has kindly informed the author that the same theorem has been found by him.

According to Lemma 1, $R_{\mu\nu}(t)$ is a bounded and continuous function, and it is then known[5] that a necessary and sufficient condition for a representation of $R_{\mu\nu}(t)$ of the form (5), where $F_{\mu\nu}(x)$ is of bounded variation in $(-\infty, \infty)$, is

$$\int_{-\infty}^{\infty} | g_{\mu\nu}(x, A) | \, dx < \text{const.} \tag{8}$$

for all $A > 1$. Moreover, a necessary and sufficient condition for a representation of the form (5) with a real and never decreasing $F_{\mu\nu}(x)$ is that $g_{\mu\nu}(x, A)$ should be real and non-negative for all real x and all $A > 1$. If the latter condition is satisfied, then (8) holds *a fortiori.*

Now according to Lemma 4 we have

$$\begin{aligned} &\frac{1}{2\pi A} E\left\{\int_0^A X_\mu(t)e^{-ixt}\,dt \cdot \int_0^A \overline{X_\nu(u)}e^{ixu}\,du\right\} \\ &\qquad = \frac{1}{2\pi A}\int_0^A \int_0^A R_{\mu\nu}(t-u)e^{-ix(t-u)}\,dt\,du = g_{\mu\nu}(x, A). \end{aligned} \tag{9}$$

Thus in particular we have for $\mu = \nu$

$$g_{\mu\mu}(x, A) = \frac{1}{2\pi A} E\left\{\left|\int_0^A X_\mu(t)e^{-ixt}\,dt\right|^2\right\},$$

which shows that $g_{\mu\mu}(x, A)$ is real and non-negative, so that we have

$$R_{\mu\mu}(t) = \int_{-\infty}^{\infty} e^{ixt}\,dF_{\mu\mu}(x),$$

where $F_{\mu\mu}(x)$ is real and never decreasing. From $| R_{\mu\mu}(t) | \leqq \sigma_\mu^2$ it then follows (Cramér [2], formula (6)) that

$$\int_{-\infty}^{\infty} g_{\mu\mu}(x, A)\,dx \leqq \sigma_\mu^2. \tag{10}$$

On the other hand, we obtain from (9), using the Schwarz inequality,

$$\begin{aligned} | g_{\mu\nu}(x, A) |^2 &= \frac{1}{(2\pi A)^2}\left| E\left\{\int_0^A X_\mu(t)e^{-ixt}\,dt \cdot \int_0^A \overline{X_\nu(u)}e^{ixu}\,du\right\}\right|^2 \\ &\leqq \frac{1}{2\pi A} E\left\{\left|\int_0^A X_\mu(t)e^{-ixt}\,dt\right|^2\right\} \cdot \frac{1}{2\pi A} E\left\{\left|\int_0^A X_\nu(t)e^{-ixt}\,dt\right|^2\right\} \\ &= g_{\mu\mu}(x, A) \cdot g_{\nu\nu}(x, A). \end{aligned}$$

Hence we deduce by (10), using once more the Schwarz inequality,

$$\int_{-\infty}^{\infty} | g_{\mu\nu}(x, A) | \, dx \leqq \sigma_\mu \sigma_\nu .$$

[5] Cramér [2].

Thus the condition (8) is satisfied, and the representation (5) is established. Obviously, we may always assume that the functions $F_{\mu\nu}(x)$ are everywhere continuous to the right, since a modification of $F_{\mu\nu}(x)$ in an enumerable set of points does not affect the value of the integral in the second member of (5).

It remains to prove the assertion concerning the form $H(z_1, \cdots, z_n)$ defined by (6). Obviously it is only necessary to show that, for arbitrary complex $z_1, \cdots, z_n$, this form is always real and non-negative.

Let (α, β) be a given real interval and put

$$\varphi(t) = \frac{e^{-it\alpha} - e^{-it\beta}}{2\pi it}.$$

According to Lemma 4, the mean value

$$H_M(z_1, \cdots, z_n) = E\left\{\left|\int_{-M}^{M} \varphi(t) \sum_{\mu=1}^{n} z_\mu X_\mu(t)\, dt\right|^2\right\}$$

exists for any $M > 0$ and for arbitrary complex z_μ, and is always real and non-negative. We have, however, by Lemma 4

$$H_M(z_1, \cdots, z_n) = \sum_{\mu,\nu} z_\mu \bar{z}_\nu E\left\{\int_{-M}^{M} \varphi(t) X_\mu(t)\, dt \cdot \int_{-M}^{M} \overline{\varphi(u) X_\nu(u)}\, du\right\}$$

$$(11) \qquad = \sum_{\mu,\nu} z_\mu \bar{z}_\nu \int_{-M}^{M}\int_{-M}^{M} \varphi(t)\overline{\varphi(u)} R_{\mu\nu}(t-u)\, dt\, du$$

$$= \sum_{\mu,\nu} z_\mu \bar{z}_\nu \int_{-\infty}^{\infty} dF_{\mu\nu}(x) \left|\int_{-M}^{M} \varphi(t) e^{itx}\, dt\right|^2.$$

Let us first suppose that α and β have been chosen such that the functions $F_{\mu\nu}(x)$ are all continuous for $x = \alpha$ and for $x = \beta$. Given any $\epsilon > 0$, the expression

$$\int_{-M}^{M} \varphi(t) e^{itx}\, dt = \frac{1}{\pi}\int_0^M \frac{\sin t(x-\alpha)}{t}\, dt - \frac{1}{\pi}\int_0^M \frac{\sin t(x-\beta)}{t}\, dt$$

tends for indefinitely increasing M to the limit zero, uniformly for $x < \alpha - \epsilon$ and for $x > \beta + \epsilon$, and to the limit unity, uniformly for $\alpha + \epsilon < x < \beta - \epsilon$. In the remaining intervals $|x - \alpha| \leqq \epsilon$ and $|x - \beta| \leqq \epsilon$, we have

$$\left|\int_{-M}^{M} \varphi(t) e^{itx}\, dt\right| < 2.$$

It then follows from (6) and (11) that we have

$$\lim_{M\to\infty} H_M(z_1, \cdots, z_n) = H(z_1, \cdots, z_n).$$

$H_M(z_1, \cdots, z_n)$ being, according to the above, real and non-negative, the same holds true for $H(z_1, \cdots, z_n)$. As the functions $F_{\mu\nu}(x)$ are everywhere continuous to the right, this result is immediately extended to arbitrary real values of α and β, and the first part of the theorem is proved.

9. Let $R_{\mu\nu}(t)$ be a set of n^2 given functions satisfying the conditions of the second part of the theorem. We then have to show that we can find n continuous stationary random processes having the functions $R_{\mu\nu}(t)$ for their correlation functions. It will, in fact, be shown that n processes with the given correlation functions can be defined by random variables of the form

$$X_\mu(t) = U_\mu(t) + iV_\mu(t), \qquad (\mu = 1, 2, \cdots, n) \tag{12}$$

in such a way that the simultaneous probability distribution of the $2nk$ real random variables

$$U_\mu(t_r), \qquad V_\mu(t_r), \qquad (\mu = 1, 2, \cdots, n; r = 1, 2, \cdots, k), \tag{13}$$

is always a *normal* distribution.

A normal distribution in any number of variables[6] is completely defined by its first and second order moments. The functions $R_{\mu\nu}(t)$ being given, we now define the normal distribution of the $2nk$ variables (13) by putting

$$\begin{cases} E\{U_\mu(t_r)\} = E\{V_\mu(t_r)\} = 0, \\ E\{U_\mu(t_r)U_\nu(t_s)\} = E\{V_\mu(t_r)V_\nu(t_s)\} = \dfrac{R_{\mu\nu}(t_r - t_s) + R_{\nu\mu}(t_s - t_r)}{4}, \\ E\{U_\mu(t_r)V_\nu(t_s)\} = \dfrac{-R_{\mu\nu}(t_r - t_s) + R_{\nu\mu}(t_s - t_r)}{4i}. \end{cases} \tag{14}$$

From our hypotheses concerning the given functions $R_{\mu\nu}(t)$, it easily follows that the relation (2) is satisfied, so that the second order moments defined by (14) are real. In order to show that these moments really define a normal distribution in the variables (13), it is sufficient to show[7] that the quadratic form $Q(u, v)$ in the $2nk$ real variables $u_{\mu r}$, $v_{\mu r}$, defined by

$$\begin{aligned} Q(u, v) &= E\{[\sum_{\mu,r} u_{\mu r} U_\mu(t_r) + v_{\mu r} V_\mu(t_r)]^2\} \\ &= \sum_{\mu,r,\nu,s} \Big\{(u_{\mu r} u_{\nu s} + v_{\mu r} v_{\nu s}) \frac{R_{\mu\nu}(t_r - t_s) + R_{\nu\mu}(t_s - t_r)}{4} \\ &\qquad + (u_{\mu r} v_{\nu s} - u_{\nu s} v_{\mu r}) \frac{-R_{\mu\nu}(t_r - t_s) + R_{\nu\mu}(t_s - t_r)}{4i}\Big\} \end{aligned}$$

is non-negative. If this is so, the moments (14) define a normal distribution with the characteristic function

$$e^{-\frac{1}{2}Q(u,v)}.$$

Using (2) we obtain, however, by some easy transformations

$$Q(u, v) = \tfrac{1}{2} \sum_{\mu,r,\nu,s} (u_{\mu r} - iv_{\mu r})(u_{\nu s} + iv_{\nu s}) R_{\mu\nu}(t_r - t_s).$$

[6] Cf. Cramér [1], p. 109.

[7] Cf. Cramér, l.c.

Substituting here the expression (5) for $R_{\mu\nu}(t)$, we obtain further

$$Q(u, v) = \tfrac{1}{2} \int_{-\infty}^{\infty} \sum_{\mu,\nu} z_\mu \bar{z}_\nu \, dF_{\mu\nu}(x),$$

where

$$z_\mu = z_\mu(x) = \sum_r (u_{\mu r} - iv_{\mu r})e^{it_r x}.$$

It now follows from our hypothesis concerning the form (6) that $Q(u, v)$ is never negative.

Thus the moments (14) define a normal distribution in the $2nk$ variables (13). Evidently this distribution satisfies the necessary symmetry and consistence conditions, so that the corresponding complex variables (12) define a multi-dimensional random process. From (14) we obtain, however,

$$\begin{aligned} E\{X_\mu(t_r)\overline{X_\nu(t_s)}\} &= E\{U_\mu(t_r)U_\nu(t_s) + V_\mu(t_r)V_\nu(t_s) - iU_\mu(t_r)V_\nu(t_s) + iU_\nu(t_s)V_\mu(t_r)\} \\ &= R_{\mu\nu}(t_r - t_s). \end{aligned}$$

Since t_r and t_s are arbitrary, this shows that the n random processes defined by the variables (12) are stationary and stationarily correlated processes, the correlation functions of which are identical with the given functions $R_{\mu\nu}(t)$. Thus our theorem is completely proved.

IV. Further Properties of the Correlation Functions. The Case of Real Processes

10. Let us first observe that the form H defined by (6) is non-negative and Hermitian for any interval (α, β) if, and only if, the same holds true when $\Delta F_{\mu\nu}$ denotes the total increase

$$\sum_r [F_{\mu\nu}(\beta_r) - F_{\mu\nu}(\alpha_r)]$$

of $F_{\mu\nu}(x)$ over any finite or enumerable set of non-overlapping closed intervals (α_r, β_r). Thus Theorem 1 remains valid if we introduce this more general definition of $\Delta F_{\mu\nu}$.

Without restriction of generality, we may assume $F_{\mu\nu}(-\infty) = 0$. Any $F_{\mu\nu}(x)$ being of bounded variation in $(-\infty, \infty)$ and everywhere continuous to the right, it is known that its points of discontinuity x_r form at most an enumerable set, that a finite derivative $F'_{\mu\nu}(x)$ exists almost everywhere, and that we may write for all x

$$F_{\mu\nu}(x) = F^{(a)}_{\mu\nu}(x) + F^{(d)}_{\mu\nu}(x) + F^{(s)}_{\mu\nu}(x). \tag{15}$$

Here

$$F^{(a)}_{\mu\nu}(x) = \int_{-\infty}^{x} F'_{\mu\nu}(t) \, dt$$

(the integral being taken in the Lebesgue sense) is the *absolutely continuous component* of $F_{\mu\nu}$;

$$F_{\mu\nu}^{(d)}(x) = \sum_{x_r \leq x} p_r$$

is the *discontinuous component* (p_r being the saltus of $F_{\mu\nu}$ at $x = x_r$); finally $F_{\mu\nu}^{(s)}(x)$ is the *singular component*. $F_{\mu\nu}^{(s)}(x)$ is everywhere continuous and has almost everywhere a derivative equal to zero. There is a set Z of measure zero such that the total variation of $F_{\mu\nu}^{(s)}(x)$ over Z is equal to the total variation of $F_{\mu\nu}^{(s)}(x)$ over $(-\infty, \infty)$. The three components in the representation of $F_{\mu\nu}(x)$ according to (15) are uniquely determined by $F_{\mu\nu}(x)$.

Substituting the expression (15) for $F_{\mu\nu}(x)$ in the definition (6) of the form $H(z_1, \cdots, z_n)$, we obtain $H = H_a + H_d + H_s$, where

$$H_a(z_1, \cdots, z_n) = \sum_{\mu,\nu=1}^{n} z_\mu \bar{z}_\nu \Delta F_{\mu\nu}^{(a)}$$

and similarly for H_d and H_s. We then have the following theorem.

THEOREM 2. *If the form H is, for any interval (α, β), a non-negative Hermite form, then the same holds true for each of the forms H_a, H_d and H_s. The converse is obvious.*

Denoting by x any point such that all derivatives $F'_{\mu\nu}(x)$ exist, and observing that, by hypothesis, the form H corresponding to the interval $(x, x + dx)$ is non-negative, we find that

$$\sum_{\mu,\nu=1}^{n} z_\mu \bar{z}_\nu F'_{\mu\nu}(x)$$

is, for almost all x, a non-negative Hermite form. Integrating between the limits α and β, it follows that the same holds true for $H_a(z_1, \cdots, z_n)$.

Similarly we find, denoting by $p_{\mu\nu}$ the saltus of $F_{\mu\nu}$ in a certain point x and considering the form H corresponding to the interval $(x - \epsilon, x + \epsilon)$, that the form $\sum_{\mu,\nu=1}^{n} z_\mu \bar{z}_\nu p_{\mu\nu}$ is a non-negative Hermite form. Adding together the analogous forms for all discontinuities x belonging to (α, β), we obtain the desired result for H_d.

Finally, it is seen without difficulty that, to any closed interval (α, β) and any given $\epsilon > 0$, we can find a sub-set E of (α, β) consisting of a finite or enumerable number of non-overlapping closed intervals of total length $< \epsilon$, such that the total variation of any $F_{\mu\nu}^{(s)}$ over E differs by less than ϵ both from the total variation of $F_{\mu\nu}^{(s)}$ over (α, β) and from the total variation of $F_{\mu\nu}$ over E. Denoting by $\Delta F_{\mu\nu}$ the increase of $F_{\mu\nu}$ over E, the corresponding form H is, according to the above, a non-negative Hermite form. Allowing now ϵ to tend to zero, we obtain the form H_s corresponding to the interval (α, β), and the proof of the theorem is completed.

11. Consider now the particular case $n = 2$ of our previous theorems. We then have two continuous stationary processes, defined by the random variables

$X_1(t)$ and $X_2(t)$. For $n = 2$, the form H defined by (6) is a non-negative Hermite form if, and only if, F_{11} and F_{22} are real and never decreasing, $\Delta F_{12} = \overline{\Delta F_{21}}$, and

$$|\Delta F_{12}|^2 \leqq \Delta F_{11} \cdot \Delta F_{22}. \tag{16}$$

From Theorems 1 and 2, we then immediately obtain the following result.

THEOREM 3. *Four given functions $R_{\mu\nu}(t)$, $(\mu, \nu = 1, 2)$, are the correlation functions of two continuous stationary processes if, and only if, the $R_{\mu\nu}(t)$ are given by the Fourier-Stieltjes integrals* (5), *where F_{11} and F_{22} are real and never decreasing, $F_{21} = \bar{F}_{12}$, and*

$$|\Delta F_{12}^{(a)}|^2 \leqq \Delta F_{11}^{(a)} \cdot \Delta F_{22}^{(a)}, \tag{17a}$$

$$|\Delta F_{12}^{(d)}|^2 \leqq \Delta F_{11}^{(d)} \cdot \Delta F_{22}^{(d)}, \tag{17d}$$

$$|\Delta F_{12}^{(s)}|^2 \leqq \Delta F_{11}^{(s)} \cdot \Delta F_{22}^{(s)}, \tag{17s}$$

for any interval (α, β), the notation being the same as that used in connection with (15).

The condition (17a) is obviously equivalent to the condition that almost everywhere

$$|F'_{12}|^2 \leqq F'_{11} F'_{22}.$$

Further, (17d) is equivalent to the condition that the discontinuity points of F_{12} form a sub-set of the set of points which are discontinuity points both of F_{11} and of F_{22}, and that the saltuses $p_{\mu\nu}$ corresponding to any discontinuity point satisfy the inequality

$$|p_{12}|^2 \leqq p_{11} p_{22}.$$

Finally, it should be observed that the three conditions (17) are equivalent to the single condition (16).

12. Let us now assume that the two random processes considered in the preceding paragraph are *real*, i.e. that they are defined by two *real* random variables $X_1(t)$ and $X_2(t)$.

Any correlation function $R_{\mu\nu}(t)$ being in this case real, its expression as a Fourier-Stieltjes integral may be written

$$R_{\mu\nu}(t) = \int_{-\infty}^{\infty} e^{itx} dF_{\mu\nu}(x)$$

$$= \int_0^{\infty} [\cos tx \, df_{\mu\nu}(x) + \sin tx \, dg_{\mu\nu}(x)],$$

where $f_{\mu\nu}(x)$ and $g_{\mu\nu}(x)$ are *real* functions of bounded variation in $(0, \infty)$. We may, in fact, take

$$f_{\mu\nu}(x) = \mathfrak{R}[F_{\mu\nu}(x) + F_{\mu\nu}(-x)],$$

$$g_{\mu\nu}(x) = \mathfrak{J}[-F_{\mu\nu}(x) + F_{\mu\nu}(-x)].$$

Since F_{11} and F_{22} are real and never decreasing, we then have $g_{11} = g_{22} = 0$, while f_{11} and f_{22} are never decreasing. $f_{\mu\nu}$ and $g_{\mu\nu}$ may be resolved into components of the types (a), (d) and (s) in the same way as $F_{\mu\nu}$ according to (15). Theorem 3 now immediately yields the following result.

THEOREM 4. *Four given functions $R_{\mu\nu}(t)$, $(\mu, \nu = 1, 2)$, are the correlation functions of two real continuous stationary processes if, and only if, the $R_{\mu\nu}(t)$ are given by the expressions*

$$R_{11}(t) = \int_0^\infty \cos tx \, df_{11}(x),$$

$$R_{22}(t) = \int_0^\infty \cos tx \, df_{22}(x),$$

$$R_{12}(t) = R_{21}(-t) = \int_0^\infty [\cos tx \, df_{12}(x) + \sin tx \, dg_{12}(x)],$$

where the $f_{\mu\nu}$ and $g_{\mu\nu}$ are real functions of bounded variation in $(0, \infty)$ such that f_{11} and f_{22} are never decreasing, while

(18a) $$(\Delta f_{12}^{(a)})^2 + (\Delta g_{12}^{(a)})^2 \leqq \Delta f_{11}^{(a)} \cdot \Delta f_{22}^{(a)},$$

(18d) $$(\Delta f_{12}^{(d)})^2 + (\Delta g_{12}^{(d)})^2 \leqq \Delta f_{11}^{(d)} \cdot \Delta f_{22}^{(d)},$$

(18s) $$(\Delta f_{12}^{(s)})^2 + (\Delta g_{12}^{(s)})^2 \leqq \Delta f_{11}^{(s)} \cdot \Delta f_{22}^{(s)},$$

for any interval (α, β) with $0 < \alpha < \beta$.

As in the preceding paragraph, the conditions (18a) and (18d) are equivalent to the conditions

$$(f'_{12})^2 + (g'_{12})^2 \leqq f'_{11} f'_{22}$$

and

$$p_{12}^2 + q_{12}^2 \leqq p_{11} p_{22}$$

respectively, $p_{\mu\nu}$ and $q_{\mu\nu}$ denoting here the saltuses of $f_{\mu\nu}$ and $g_{\mu\nu}$ at any given point x.

Finally, the three conditions (18) are evidently equivalent to the single condition

$$(\Delta f_{12})^2 + (\Delta g_{12})^2 \leqq \Delta f_{11} \cdot \Delta f_{22} .$$

V. ON DISCRETE STATIONARY PROCESSES

13. Let us consider a sequence of (real or complex) random variables $X(t)$, where $t = 0, \pm 1, \pm 2, \cdots$, such that the simultaneous probability distributions of the variables $X(t_1), \cdots, X(t_k)$ are determined for any finite set of integral $t_1, \cdots, t_k$. If these probability distributions satisfy the conditions of symmetry and consistency stated in paragraph 1, we shall say that the sequence $X(t)$ defines a *discrete random process*.

Let, now, the variables $X_1(t), \cdots, X_n(t)$ define n discrete random processes such that for all integral values of t and u

$$(19) \qquad \begin{aligned} E\{X_\mu(t)\} &= 0, \\ E\{X_\mu(t)\overline{X_\nu(u)}\} &= R_{\mu\nu}(t-u), \end{aligned}$$

so that the latter mean value is a function of the difference $t - u$. We shall then say that the n processes defined by the $X_\mu(t)$ are *stationary* and *stationarily correlated*. For the general theory of the discrete stationary process, reference may be made to Wold [8].

In this case, our correlation functions $R_{\mu\nu}(t)$ become sequences of numbers, depending on the (positive or negative) integral-valued variable t. Without restriction of generality we may assume

$$(20) \qquad R_{\mu\mu}(0) = \sigma_\mu^2 = 1.$$

The quantity $R_{\mu\nu}(t)$ is then, in the particular case when $X_\mu(t)$ and $X_\nu(t)$ are real, identical with the ordinary correlation coefficient between $X_\mu(t_0 + t)$ and $X_\nu(t_0)$. Even in the case of complex variables $X_\mu(t)$, we shall call the $R_{\mu\nu}(t)$ the correlation coefficients of our n processes. For $\mu = \nu$, $R_{\mu\mu}(t)$ yields the *autocorrelation coefficients* of the process defined by $X_\mu(t)$, while $R_{\mu\nu}(t)$ and $R_{\nu\mu}(t)$ constitute, for $\mu \neq \nu$, the *mutual correlation coefficients* of the processes connected with $X_\mu(t)$ and $X_\nu(t)$. We obviously still have $R_{\mu\nu}(t) = \overline{R_{\nu\mu}(-t)}$.

For the correlation coefficients of a group of n discrete stationary processes, we can prove theorems which are completely analogous to our above theorems for the continuous case. In fact, the only modification required is to replace in the Fourier-Stieltjes integral (5) the infinite interval of integration by an interval of length 2π and, at the same time, to restrict the variable t to integral values. Theorem 1 will then be replaced by the following theorem.

THEOREM 5. *a) Let* $X_1(t), \cdots, X_n(t)$, *where* $t = 0, \pm 1, \pm 2, \cdots$, *be* n *sequences of complex random variables attached to n discrete stationary processes satisfying the conditions* (19) *and* (20). *Then the correlation coefficients* $R_{\mu\nu}(t)$ *defined by* (19) *are for all positive and negative integral t given by the Fourier-Stieltjes integrals*

$$(21) \qquad R_{\mu\nu}(t) = \int_{-\pi}^{\pi} e^{itx}\, dF_{\mu\nu}(x),$$

where the $F_{\mu\nu}(x)$ *are functions of bounded variation in* $(-\pi, \pi)$, *which we may always assume to be everywhere continuous to the right. Further, the form* $H(z_1, \cdots, z_n)$ *defined by* (6) *is, for any closed interval* (α, β) *with* $-\pi \leqq \alpha < \beta \leqq \pi$, *a non-negative Hermite form. (In particular, it follows that* $F_{\mu\mu}(x)$ *is real and never decreasing, with a total variation equal to unity.)*

b) Conversely, let $R_{\mu\nu}(t)$, *for* $t = 0, \pm 1, \pm 2, \cdots$, *be a set of* n^2 *sequences of quantities given by the Fourier-Stieltjes integrals* (21), *the functions* $F_{\mu\nu}(x)$ *being of bounded variation in* $(-\pi, \pi)$ *and everywhere continuous to the right, and suppose that* $R_{\mu\mu}(0) = 1$ *and that the form* $H(z_1, \cdots, z_n)$ *is, for any closed interval*

(α, β) with $-\pi \leqq \alpha < \beta \leqq \pi$, a non-negative Hermite form. Then it is possible to find n discrete stationary processes having the correlation coefficients $R_{\mu\nu}(t)$.

The proof of this theorem is quite similar to the proof of Theorem 1. We only have to replace the definition (7) of the auxiliary function $g(x, A)$ by

$$g_{\mu\nu}(x, A) = \frac{1}{2\pi A} \sum_{t=0}^{A} \sum_{u=0}^{A} R_{\mu\nu}(t-u) e^{-ix(t-u)}.$$

According to Hausdorff [5], a necessary and sufficient condition for a representation of $R_{\mu\nu}(t)$ of the form (21), where $F_{\mu\nu}(x)$ is of bounded variation in $(-\pi, \pi)$, is

$$\int_{-\pi}^{\pi} | g_{\mu\nu}(x, A) | \, dx < \text{const.}$$

for all $A > 1$. Moreover, a necessary and sufficient condition for a representation of the form (21) with a real and never decreasing $F_{\mu\nu}(x)$ is that $g_{\mu\nu}(x, A)$ should be real and non-negative for all real x and all $A > 1$. These conditions are, of course, completely analogous to those given above for the case of a continuous variable t and a representation of the form (5), and all the rest of the proof runs on quite similar lines.

In order to obtain, for the case of discrete processes, analogues of our theorems 2, 3 and 4, we only have to replace everywhere the infinite Fourier-Stieltjes integral (5) by the corresponding finite integral (21). In theorems 3 and 4, the interval $(0, \infty)$ is then replaced by $(0, \pi)$. Otherwise everything remains unchanged.

UNIVERSITY OF STOCKHOLM,
STOCKHOLM, SWEDEN

REFERENCES

[1] H. Cramér, *Random variables and probability distributions*, Cambridge Tracts in Mathematics, No. 36, Cambridge 1937.

[2] ———, *On the representation of functions by certain Fourier integrals*, Trans. Amer. Math. Soc., vol. 46 (1939), p. 191.

[3] J. L. Doob, *Stochastic processes depending on a continuous parameter*, Trans. Amer. Math. Soc., vol. 42 (1937), p. 107.

[4] M. Fréchet, *Recherches théoriques modernes sur la théorie des probabilités*, premier livre, Paris 1937.

[5] F. Hausdorff, *Momentprobleme für ein endliches Intervall*, Math. Zeitschrift, vol. 16 (1923), p. 220.

[6] A. Khintchine, *Korrelationstheorie der stationären stochastischen Prozesse*, Math. Annalen, vol. 109 (1934), p. 604.

[7] P. Lévy, *Théorie de l'addition des variables aléatoires*, Paris, 1937.

[8] H. Wold, *A study in the analysis of stationary time series*, Inaugural Dissertation (Stockholm), Uppsala 1938.

41.

On harmonic analysis in certain functional spaces

Ark. Mat. Astr. Fys. **28 B** (12), 1–7 (1942)

Read January 29th 1942.

1. In generalized harmonic analysis as developed by WIENER [7, 8] and BOCHNER [1] we are concerned with a measurable complex-valued function $f(t)$ of the real variable t (which may be thought of as representing time), and it is assumed that the limit

$$\varphi(t) = \lim_{T\to\infty} \frac{1}{2T}\int_{-T}^{T} f(\tau + t)\overline{f(\tau)}\, d\tau \tag{1}$$

exists for all real t. If, in addition, $\varphi(t)$ is continuous at the particular point $t = 0$, it is continuous for all real t and may be represented by a FOURIER-STIELTJES integral

$$\varphi(t) = \int_{-\infty}^{\infty} e^{itx}\, d\Phi(x), \tag{2}$$

where $\Phi(x)$ is real, bounded and never decreasing.

Denoting by x_0 a fixed continuity point of $\Phi(x)$, the infinite integral

$$F(x) = -\frac{1}{2\pi i}\int_{-\infty}^{\infty} \frac{e^{-itx} - e^{-itx_0}}{t} f(t)\, dt \tag{3}$$

exists, by the PLANCHEREL theorem, as a »limit in the mean» over $-\infty < x < \infty$. Conversely, we have for almost all real t

$$f(t) = \lim_{\varepsilon \to 0} \int_{-\infty}^{\infty} e^{itx} \frac{F(x+\varepsilon) - F(x-\varepsilon)}{2\varepsilon} dx, \tag{4}$$

where the integral should be similarly interpreted as in (3). In the particular case when $F(x)$ as defined by (3) is of bounded variation in $(-\infty, \infty)$, the expression (4) reduces to an ordinary FOURIER-STIELTJES integral:

$$f(t) = \int_{-\infty}^{\infty} e^{itx} dF(x). \tag{4 a}$$

In the general case, (4) may be regarded as a *generalized form* of the FOURIER-STIELTJES representation (4 a).

Further, we have for any real numbers a and b, which are continuity points of $\Phi(x)$,

$$\Phi(b) - \Phi(a) = \lim_{\varepsilon \to 0} \int_{a}^{b} \frac{|F(x+\varepsilon) - F(x-\varepsilon)|^2}{2\varepsilon} dx. \tag{5}$$

2. All the results so far quoted are concerned with the harmonic analysis of an individual function $f(t)$. On the other hand, in many important applications it is required to deal simultaneously with an infinite class of functions $f(t)$ and to deduce results, analogous to those quoted above, which are true *almost always*, with reference to some appropriate measure defined in the class of functions under consideration.

In this respect, a particular interest is attached to the case when the measure considered is *invariant for translations in time,* in the sense that whenever S is a measurable set of functions $f(t)$, then the set S_h obtained by substituting $f(t+h)$ for $f(t)$ is also measurable and has the same measure for any real h. WIENER [9] has already obtained certain results concerning this case, and it is the purpose of the present note to give a further contribution to the study of the problem.

The subject-matter of this note is intimately related to the theory of *stationary random processes* developed by KHINTCHINE [3] and other authors, and indeed the theorem given below could be stated in the terminology of this theory. In this preliminary note we shall, however, avoid every reference to the theory of probability, reserving a further discussion of these questions, as well as of various generalizations and applications, for a later publication,

3. In the sequel, we shall consider the space $\mathfrak{S}$ of all finite complex-valued functions $f(t) = g(t) + i\,h(t)$ of the real variable t. A set of functions $f(t)$ defined by a finite number of pairs of inequalities of the form

$$\alpha_\nu < g(t_\nu) \leqq \beta_\nu,$$

$$\gamma_\nu < h(t_\nu) \leqq \delta_\nu,$$

will be called an *interval* in $\mathfrak{S}$. Let $\mathfrak{C}$ denote the smallest additive class of sets in $\mathfrak{S}$ containing all intervals, and let $P(S)$ denote a non-negative measure defined for all sets S belonging to $\mathfrak{C}$, and such that $P(\mathfrak{S})$ is finite. According to a theorem of KOLMOGOROFF [4, p. 27], $P(S)$ is uniquely defined by its values on all intervals.

For any functional $G(f)$ which is measurable ($\mathfrak{C}$), the LEBESGUE-STIELTJES integral with respect to the measure $P(S)$ over a measurable set S_0 will be denoted by[1]

$$\int_{S_0} G(f)\,dP.$$

If t_0 is any given real number, $G(f) = |f(t_0)|^2$ is a measurable functional, and we shall always assume that the integral

$$\int_{\mathfrak{S}} |f(t_0)|^2\,dP$$

is finite.

If the measure $P(S)$ is invariant for translations in time in the sense defined above, we shall then obviously have

$$(6)\qquad \begin{cases} \displaystyle\int_{\mathfrak{S}} f(t_0)\,dP = K, \\ \displaystyle\int_{\mathfrak{S}} f(t_0+t)\overline{f(t_0)}\,dP = \varphi(t), \end{cases}$$

where K is a constant and $\varphi(t)$ is a function of t only[2].

[1] For the definition of this integral, cf. SAKS [5], Ch. I.

[2] In order to emphasize the analogy with ordinary harmonic analysis as reviewed in section 1, we use systematically the same letters to denote corresponding functions. It is, however, obvious that the integrals occurring

Conversely, we may say that the relations (6) imply *invariance of the measure P for time translations, as far as moments of the first and second orders are concerned.*

If $\varphi(t)$ is continuous at the particular point $t = 0$, it can be shown[1] that $\varphi(t)$ is uniformly continuous for all real t and representable in the form (2), with a real, bounded and never decreasing $\Phi(x)$.

Whenever a measure $P(S)$ satisfies the relations (6) with a continuous $\varphi(t)$, we shall say, in accordance with a definition introduced by KHINTCHINE[1], that *$P(S)$ is attached to a stationary and continuous distribution in function space $\mathfrak{S}$.*

4. *A net N on the real axis* will be defined as a sequence of divisions $D_1, D_2, \ldots$ of that axis, such that D_n consists of a finite number k_n of points

$$\alpha_1^{(n)} < \alpha_2^{(n)} < \cdots < \alpha_{k_n}^{(n)}.$$

Denoting by δ_n the maximum length of the intervals $\alpha_{\nu+1}^{(n)} - \alpha_\nu^{(n)}$ occurring in D_n, we shall require that

$$\alpha_1^{(n)} \to -\infty, \quad \alpha_{k_n}^{(n)} \to +\infty, \quad \delta_n \to 0,$$

as $n \to \infty$.

If, for a given function $F(x)$ and a given real t, the limit

$$\Lambda = \lim_{n \to \infty} \sum_{\nu=1}^{k_n - 1} e^{i t \alpha_\nu^{(n)}} \left[F(\alpha_{\nu+1}^{(n)}) - F(\alpha_\nu^{(n)})\right]$$

exists, we shall call this limit a *generalized* FOURIER-STIELTJES *integral associated with the net N*, and we shall write

$$\Lambda = \int_{-\infty}^{\infty} e^{i t x} \, d^{(N)} F(x).$$

We shall finally say that a net N is *efficient with respect to the measure $P(S)$*, if I) every $\alpha_\nu^{(n)}$ is a point of continuity of $\Phi(x)$, and II) the three series with positive terms

e.g. in (1) and (6) are fundamentally different. (1) defines an average over the time axis for an individual function $f(t)$, while (6) deals with averages over function space $\mathfrak{S}$.

[1] KHINTCHINE [3]. Cf. also CRAMÉR [2].

$$\sum_n [\Phi(\alpha_1^{(n)}) - \Phi(-\infty)], \quad \sum_n \left[\Phi(\infty) - \Phi\left(\alpha_{k_n}^{(n)}\right)\right], \quad \sum_n \delta_n^2$$

are all convergent.

5. After these preliminaries, we can now state the following theorem, where the expression »almost all functions $f(t)$» will have to be interpreted with respect to P-measure, while the expression »almost all real t» refers to ordinary LEBESGUE measure.

Theorem. *Let $P(S)$ be a measure attached to a stationary and continuous distribution in $\mathfrak{S}$. It is then possible to find a transformation*

$$T[f] = F,$$

by which to every element $f(t)$ of $\mathfrak{S}$ corresponds another uniquely defined element $F(t)$ of the same space, so that the following properties hold:

I) *If N is any net efficient with respect to $P(S)$, then almost all functions $f(t)$ satisfy the relation*

$$f(t) = \int_{-\infty}^{\infty} e^{itx} d^{(N)} F(x) \tag{7}$$

for almost all real t.

II) *For any real numbers $a < b$ which are continuity points of $\Phi(x)$, we have*

$$\Phi(b) - \Phi(a) = \int_{\mathfrak{S}} |F(b) - F(a)|^2 \, dP. \tag{8}$$

The analogy between the relations (7) and (8) of the theorem and the corresponding relations (4), (4 a) and (5) of ordinary harmonic analysis is obvious. A proposition recently given by SLUTSKY [6] is contained in the particular case of the theorem when $\varphi(t)$ is an almost periodic function. — Owing to restrictions of space, only the main lines of the proof can be indicated here. Denoting by x_0 a fixed continuity point of $\Phi(x)$, the expression

$$F_{m,n}[x, f(t)] = -\frac{1}{2\pi i} \sum_{\nu=-mn}^{mn} \frac{e^{-\frac{\nu i x}{n}} - e^{-\frac{\nu i x_0}{n}}}{\nu} f\left(\frac{\nu}{n}\right), \quad (\nu \neq 0), \tag{9}$$

is, for any fixed integers m, n and any fixed real x, a $\mathfrak{S}$-measurable functional in $\mathfrak{S}$. Obviously $|F_{m,n}|^2$ has a finite integral with respect to P over $\mathfrak{S}$. The structural analogy between (9) and (3) is evident.

Now, let E denote the enumerable set of discontinuity points of $\Phi(x)$. For any x which does not belong to E, it can be shown that the repeated limit in the mean over $\mathfrak{S}$

$$F[x, f(t)] = \underset{m \to \infty}{\text{l.i.m.}} \left\{ \underset{n \to \infty}{\text{l.i.m.}}\ F_{m,n}[x, f(t)] \right\}$$

exists. In other words, there are functionals $F_m[x, f(t)]$ and $F[x, f(t)]$ such that

$$\lim_{n \to \infty} \int_{\mathfrak{S}} |F_{m,n} - F_m|^2 \, dP = 0,$$

$$\lim_{m \to \infty} \int_{\mathfrak{S}} |F_m - F|^2 \, dP = 0.$$

For any x which does not belong to E, and for almost all $f(t)$, the functional $F[x, f(t)]$ is determined as the limit of a convergent sub-sequence of $F_{m,n}$. In all other cases, we put $F = 0$, so that F is completely defined for all real x and all functions $f(t)$ in $\mathfrak{S}$. The integral of $|F|^2$ with respect to P over $\mathfrak{S}$ is finite for every fixed x.

We now define the transformation $T[f]$ by putting for any $f(t)$ in $\mathfrak{S}$

$$T[f] = F(x) = F[x, f(t)].$$

It can then be shown without difficulty that this function $F(x)$ satisfies part II of the theorem.

On the other hand, if we consider the expression

$$h_n[t, f] = \sum_{\nu} e^{it\alpha_\nu^{(n)}} [F(\alpha_{\nu+1}^{(n)}) - F(\alpha_\nu^{(n)})],$$

where the $\alpha_\nu^{(n)}$ belong to an efficient net N, we obtain for any fixed n (integer) and t (real)

$$\int_{\mathfrak{S}} |f - h_n|^2 \, dP < t^2 \delta_n^2 + \Phi(\infty) - \Phi\left(\alpha_{k_n}^{(n)}\right) + \Phi(\alpha_1^{(n)}) - \Phi(-\infty).$$

Hence, considering the product space $\mathfrak{T}\mathfrak{S}$, where $\mathfrak{T}$ denotes the real t-axis, it follows from the properties of an efficient net that almost all functions $f(t)$ satisfy the relation

$$f(t) = \underset{n \to \infty}{\text{l\"{i}m}}\, h_n[t, f]$$

for almost all real t. According to our definition of a generalized FOURIER-STIELTJES integral, this proves the remaining part of the theorem.

References.

1) **Bochner, S.,** Monotone Funktionen, Stieltjessche Integrale und harmonische Analyse. Math. Ann. 108 (1933), p. 378. — 2) **Cramér, H.,** On the theory of stationary random processes. Ann. of Math., 41 (1940), p. 215. — 3) **Khintchine, A.,** Korrelationstheorie der stationären stochastischen Prozesse. Math. Ann. 109 (1934), p. 604. — 4) **Kolmogoroff, A.,** Grundbegriffe der Wahrscheinlichkeitsrechnung. Berlin 1933. — 5) **Saks, S.,** Theory of the integral, 2:d ed., Warszawa-Lwów 1937. — 6) **Slutsky, E.,** Sur les fonctions aléatoires presque périodiques etc. Actualités scientifiques et industr., 738 (1938), p. 33. — 7) **Wiener, N.,** Generalized harmonic analysis. Acta Math. 55 (1930), p. 117. — 8) ——, The Fourier integral. Cambridge 1933. — 9) ——, The homogeneous chaos. American J. of Math., 60 (1938), p. 897.

Tryckt den 2 mars 1942.

Uppsala 1942. Almqvist & Wiksells Boktryckeri-A.-B.

42.

A contribution to the theory of statistical estimation

Skand. Aktuarietidskr. **29**, 85–94 (1946)

1. The Estimation Problem.

In problems of statistical estimation, we are concerned with certain variables assumed to be random variables having more or less unknown probability distributions. A number of observed values of these variables are given, and it is required to use these values to learn something about the unknown distributions.

The classical theory of errors of observation deals with a special group of problems of this type. In its modern, more general form, the theory of statistical estimation was founded by R. A. Fisher, to whom the main ideas and results of the theory are due.[1] Fisher considers in the first place the problems which arise when the mathematical form of the distributions is assumed to be known, the only unknown element being a certain number of parameters, the numerical values of which it is required to estimate by means of the data. In the present paper, we shall be concerned with a general problem of this type, restricting ourselves to the case of distributions of the continuous type (Mathematical Methods, 22.1). An obvious

[1] With respect to the literature, I refer to my recent book »Mathematical Methods of Statistics», Uppsala 1945 (an American edition will appear in the Princeton Mathematical Series), where the theory of estimation is treated in Chs 32—34. In the sequel, we shall use the terminology and the notations of this book, which will be referred to as »Mathematical Methods». I take this opportunity of quoting the following important papers on estimation theory, which were unfortunately omitted from the List of References of the book: J. L. Doob, »Probability and Statistics», and »Statistical Estimation», Trans. American Math. Soc., 36, p. 759 (1934), and 39, p. 410 (1936). M. Fréchet, »Sur l'extension de certaines évaluations statistiques au cas de petits échantillons», Rev. Inst. Intern. de Statistique, 1943, p. 182.

modification of the method used below will suffice to give the analogous results for distributions of the discrete type.

We shall consider the following problem. — The random variables $x_1, \ldots, x_n$ *have a joint distribution of the continuous type, defined by the probability density* $f(x_1, \ldots, x_n; \alpha_1, \ldots, \alpha_k)$ *of known mathematical form, but involving k unknown parameters* $\alpha_1, \ldots \alpha_k$, *where* $k < n$. *We observe one point* $\boldsymbol{x} = (x_1 \ldots, x_n)$ *in the space* $\boldsymbol{R}_n$ *of the variables. It is required to use the observed values* $x_1, \ldots, x_n$ *to form estimates of the unknown parameters* $\alpha_1, \ldots, \alpha_k$.

It should be observed that, throughout this paper, the number n of the variables will be regarded as a fixed, finite number. The problems that arise when n tends to infinity will thus not be considered here.

If, in particular, $x_1, \ldots, x_n$ are independent random variables, all having the same probability distribution, the above problem reduces to an ordinary »sampling problem» with k unknown parameters. The general problem covers, however, also other interesting cases, such as cases when the x_i are correlated, or consist of several independent samples from different distributions (cf Mathematical Methods, 32.8). Even in the general case, it is often convenient to talk of the point $\boldsymbol{x} = (x_1, \ldots, x_n)$ as a *sample point*, which is represented in the *sample space* $\boldsymbol{R}_n$.

2. Statement of a General Theorem.

The parameters $\alpha_1, \ldots, \alpha_k$ are not to be regarded as random variables, but as unknown constants which have to be estimated from the data. We may regard the α_j as the unknown coordinates of a point $\boldsymbol{\alpha} = (\alpha_1, \ldots, \alpha_k)$ in a *parametric space* $\boldsymbol{P}_k$ of k dimensions. We shall suppose that the probability density $f(x_1, \ldots, x_n; \alpha_1 \ldots, \alpha_k)$ is given for all points $\boldsymbol{\alpha}$ belonging to a certain non-degenerate interval A in the space $\boldsymbol{P}_k$, and that the unknown parametric point $\boldsymbol{\alpha}$ belongs to this interval.

Consider now a set of k functions of the x_i

$$\alpha_j^* = \alpha_j^* (x_1, \ldots, x_n), \qquad (j = 1, \ldots, k),$$

which do not involve the unknown parameters α_j, and let us propose to use α_j^* as an estimate of the unknown constant α_j. The α_j^* are functions of the random variables x_i, and are thus themselves random variables, having a certain k-dimensional joint distribution. In order to avoid unnecessary complications, we shall suppose throughout that the α_j^* have been chosen such that each α_j^* is an *unbiased estimate* of the corresponding α_j, i. e. such that

$$(1) \qquad E(\alpha_j^*) = \alpha_j, \qquad (j = 1, \ldots, k),$$

whatever be the values of the α_j within the interval A.

This being assumed, we now ask how the α_j^* should be chosen in order that their joint probability distribution should have *the greatest possible concentration about the point* $\alpha = (\alpha_1, \ldots, \alpha_k)$, this being a natural way of interpreting the general desideratum of obtaining the »best possible» estimates of the unknown α_j (cf Mathematical Methods, 32,1).

According to (1), the »probability mass» in the joint distribution of the α_j^* has its centre of gravity in the point α. We further assume that the joint distribution of the α_j^* is non-singular, i. e. that none of the functions α_j^* reduces to a linear combination of the others.

It follows that the *moment matrix* of the distribution, $\Lambda = \{\lambda_{ij}\}$, the elements of which are the central moments

$$(2) \qquad \lambda_{ij} = E(\alpha_i^* - \alpha_i)(\alpha_j^* - \alpha_j),$$

is definite positive. Thus the reciprocal matrix $\Lambda^{-1} = \{\lambda^{ij}\}$ exists. The equation

$$(3) \qquad \sum_{i,j} \lambda^{ij}(u_i - \alpha_i)(u_j - \alpha_j) = k + 2,$$

where i and j both run from 1 to k, while $u_1, \ldots, u_k$ are the coordinates of a point in P_k, then represents a k-dimensional ellipsoid U with its centre in the point $\alpha = (\alpha_1, \ldots, \alpha_k)$. If a unit of mass is uniformly distributed over the interior of the ellipsoid U, it is known that this distribution will have the same centre of gravity α and the same second order central moments λ_{ij} as the distribution of the α_j^* (Mathematical Methods, 22.7). Accordingly U will be called the *concentration ellipsoid* of the α^*-distribution, and may serve as a suitable geometrical representation of the concentration of this distribution about the point α.

We are now interested in choosing the estimates α_j^* so as to render the concentration ellipsoid U as small as possible, thus realizing the greatest possible concentration of the joint distribution about the point α. However, it is not possible to reach in this way an arbitrarily high degree of concentration.

In fact, it will be proved that, subject to certain general conditions of regularity, there is a fixed ellipsoid U_0 depending only on the given probability density f, such that U_0 lies entirely within the concentration ellipsoid U of any set of estimates $\alpha_1^*, \ldots, \alpha_k^*$. The equation of U_0 is

$$\sum_{i,j} \varkappa_{ij} (u_i - \alpha_i)(u_j - \alpha_j) = k + 2, \tag{4}$$

where

$$\varkappa_{ij} = \mathbf{E}\left(\frac{\partial \log f}{\partial \alpha_i} \frac{\partial \log f}{\partial \alpha_j}\right) = \int_{-\infty}^{\infty} \cdots \int_{-\infty}^{\infty} \frac{1}{f} \frac{\partial f}{\partial \alpha_i} \frac{\partial f}{\partial \alpha_j} dx_1 \ldots dx_n. \tag{5}$$

Thus if, in a particular case, we have succeeded in finding a set of estimates $\alpha_1^*, \ldots, \alpha_k^*$, the concentration ellipsoid of which coincides with U_0, we have attained the greatest possible concentration in our estimation problem, and we may denote this set as a set of *joint efficient estimates* of the parameters $\alpha_1, \ldots, \alpha_k$.

The general lines of a proof of the above theorem have been indicated in Mathematical Methods, 32.6—8, and it is the purpose of the present paper to complete this proof.

In the particular case referred to at the end of paragraph 1, when the x_i are independent random variables all having the same distribution, we have to replace the probability density $f(x_1, \ldots, x_n; \alpha_1, \ldots, \alpha_k)$ by an expression of the form

$$f(x_1; \alpha_1, \ldots, \alpha_k) \ldots f(x_n; \alpha_1, \ldots, \alpha_k),$$

where $f(x; \alpha_1, \ldots, \alpha_k)$ is a one-dimensional probability density.

The expression (5) of the coefficients $\varkappa_{ij}$ in the equation of the optimum ellipsoid U_0 then reduces to

$$\varkappa_{ij} = n\,\mathbf{E}\left(\frac{\partial \log f}{\partial \alpha_i}\frac{\partial \log f}{\partial \alpha_j}\right) = n\int_{-\infty}^{\infty}\frac{1}{f}\frac{\partial f}{\partial \alpha_i}\frac{\partial f}{\partial \alpha_j}\,dx.$$

3. Proof of a Lemma.

We shall require the following simple Lemma, which may be regarded as a generalized form of the well-known SCHWARZ inequality.

Let $\varphi_1, \ldots, \varphi_k$ and $\psi_1, \ldots, \psi_k$ be two sequences of real functions of a variable point z in any number of dimensions, all of which belong to L_2 over a certain set S in the space of z. Write

$$\lambda_{ij} = \int \varphi_i \varphi_j\, dz,$$

$$\mu_{ij} = \int \psi_i \psi_j\, dz,$$

$$\nu_{ij} = \int \varphi_i \psi_j\, dz,$$

all integrals being extended over S, and let Λ, $\mathbf{M}$ and $\mathbf{N}$ denote the matrices formed by the elements λ_{ij}, μ_{ij} and ν_{ij} respectively. Suppose that Λ is non-singular, so that the reciprocal matrix $\Lambda^{-1} = \{\lambda^{ij}\}$ exists, and consider the matrix

$$\boldsymbol{P} = \{\varrho_{ij}\} = \boldsymbol{N}' \Lambda^{-1} \boldsymbol{N},$$

where $\boldsymbol{N}'$ denotes the transpose of $\boldsymbol{N}$. For any real values of the variables $u_1, \ldots, u_k$ we then have

(6) $$\sum_{i,j} \mu_{ij} u_i u_j \geqq \sum_{i,j} \varrho_{ij} u_i u_j.$$

In the particular case $k = 1$, the inequality (6) reduces to the SCHWARZ inequality

$$\int \varphi_1^2\, dz \cdot \int \psi_1^2\, dz \geqq \left(\int \varphi_1 \psi_1\, dz\right)^2.$$

In the further particular case when k is arbitrary, while $\boldsymbol{N}$ is the unit matrix $\boldsymbol{I}$, we have $\boldsymbol{P} = \Lambda^{-1}$, and (6) reduces to

$$\sum_{i,j} \mu_{ij} u_i u_j \geqq \sum_{i,j} \lambda^{ij} u_i u_j. \tag{7}$$

It is this last inequality that we shall have occasion to apply in the sequel.

In order to prove the Lemma, we observe that we have for any values of the real variables u_i and v_j

$$\int (v_1 \varphi_1 + \cdots + v_k \varphi_k)(u_1 \psi_1 + \cdots + u_k \psi_k)\, dz = \sum_{ij} \nu_{ij} v_i u_j = \boldsymbol{v}' N \boldsymbol{u},$$

where $\boldsymbol{u}$ and $\boldsymbol{v}$ are conceived as column vectors. By the SCHWARZ inequality we thus obtain

$$(\boldsymbol{v}' N \boldsymbol{u})^2 \leqq \int (v_1 \varphi_1 + \cdots + v_k \varphi_k)^2 dz \cdot \int (u_1 \psi_1 + \cdots + u_k \psi_k)^2 dz$$
$$= \boldsymbol{v}' \Lambda \boldsymbol{v} \cdot \boldsymbol{u}' M \boldsymbol{u}.$$

Taking here $\boldsymbol{v} = \Lambda^{-1} N \boldsymbol{u}$, we have

$$(\boldsymbol{u}' P \boldsymbol{u})^2 \leqq \boldsymbol{u}' P \boldsymbol{u} \cdot \boldsymbol{u}' M \boldsymbol{u}.$$

Both the quadratic forms occurring in the last relation being non-negative, this is equivalent with (6), so that the Lemma is proved.

4. Conditions of Regularity.

Consider a transformation of variables in the sample space, replacing the »old» variables $x_1, \ldots, x_n$ by »new» variables $\alpha_1^*, \ldots, \alpha_k^*, \xi_1, \ldots, \xi_{n-k}$ such that the equations of transformation do not involve the parameters $\alpha_1, \ldots, \alpha_k$. The new variables are functions of the x_i, and may thus be regarded as random variables, having a joint probability distribution determined by the distribution of the x_i. Denote by $g(\alpha_1^*, \ldots, \alpha_k^*; \alpha_1, \ldots, \alpha_k)$ the joint probability density of the α_j^*, and by $h(\xi_1, \ldots, \xi_{n-k} \,|\, \alpha_1^*, \ldots, \alpha_k^*; \alpha_1, \ldots, \alpha_k)$ the conditional probability density of $\xi_1, \ldots, \xi_{n-k}$, relative to given values of $\alpha_1^*, \ldots, \alpha_k^*$. We then have (cf Mathematical Methods, 32.6)

$$Jf = gh, \tag{8}$$

where J is a Jacobian not involving the parameters α_j, and further

(9) $$f\,dx_1 \ldots dx_n = g\,h\,d\alpha_1^* \ldots d\alpha_k^*\,d\xi_1 \ldots d\xi_{n-k}.$$

We now introduce the following *conditions of regularity*. — Suppose that, for every α in A, and for every $i = 1, \ldots, k$, the partial derivatives $\frac{\partial f}{\partial \alpha_i}$, $\frac{\partial g}{\partial \alpha_i}$ and $\frac{\partial h}{\partial \alpha_i}$ exist[1] and satisfy the conditions

$$\left|\frac{\partial g}{\partial \alpha_i}\right| < G_i(\alpha_1^*, \ldots, \alpha_k^*),$$

$$\left|\frac{\partial h}{\partial \alpha_i}\right| < H_i(\xi_1, \ldots, \xi_{n-k};\ \alpha_1^*, \ldots, \alpha_k^*),$$

where G_i and $\alpha_j^*\,G_i$ are integrable over $\boldsymbol{P}_k$, while H_i is integrable over the space of the variables $\xi_1, \ldots, \xi_{n-k}$.

Suppose further that the integrals

$$\int \left(\frac{\partial \log f}{\partial \alpha_i}\right)^2 f\,d\boldsymbol{x}, \quad \int \left(\frac{\partial \log g}{\partial \alpha_i}\right)^2 g\,d\boldsymbol{\alpha}^*, \quad \int \left(\frac{\partial \log h}{\partial \alpha_i}\right)^2 g\,h\,d\boldsymbol{\alpha}^*\,d\boldsymbol{\xi},$$

are all finite. Here $d\boldsymbol{x}$ stands as an abbreviation for $dx_1 \ldots dx_n$, and similarly for $d\boldsymbol{\alpha}^*$ and $d\boldsymbol{\xi}$, while each integral is extended over the whole space of the variables concerned. The same abbreviated notation will be used in the sequel.

Assuming these conditions to be satisfied, it follows that the relations

(10) $$\int g\,d\boldsymbol{\alpha}^* = \int h\,d\boldsymbol{\xi} = 1,$$

which hold for every point $\alpha = (\alpha_1, \ldots, \alpha_k)$ in the non-degenerate interval A, may be differentiated under the integral signs with respect to every α_i, so that we obtain

(11) $$\int \frac{\partial g}{\partial \alpha_i}\,d\boldsymbol{\alpha}^* = \int \frac{\partial h}{\partial \alpha_i}\,d\boldsymbol{\xi} = 0.$$

[1] It should be observed that this condition is generally *not* satisfied in cases when the probability density f is identically zero in a domain, the extension of which is variable with the parameters $\alpha_1, \ldots, \alpha_k$.

According to the condition (1), we further have, differentiation under the integral being still legitimate,

$$\int \alpha_i^* g\, d\alpha = \alpha_i, \text{ and } \int \alpha_i^* \frac{\partial g}{\partial \alpha_j}\, d\alpha^* = \varepsilon_{ij}, \tag{12}$$

where as usual $\varepsilon_{ij} = 1$ for $i = j$, while $\varepsilon_{ij} = 0$ for $i \neq j$. From (11) and (12) we obtain

$$\begin{aligned} &\int \frac{\partial \log h}{\partial \alpha_i} h\, d\xi = 0, \\ &\int (\alpha_i^* - \alpha_i)\sqrt{g}\cdot \frac{\partial \log g}{\partial \alpha_j}\sqrt{g}\, d\alpha^* = \varepsilon_{ij}. \end{aligned} \tag{13}$$

5. Proof of the Theorem.

The relation (8) holds for all points α in the interval A. Taking the logarithms of both sides, and forming the total differentials with respect to $\alpha_1, \ldots, \alpha_k$, we obtain, the Jacobian J being independent of the α_i,

$$\sum_i \frac{\partial \log f}{\partial \alpha_i} d\alpha_i = \sum_i \frac{\partial \log g}{\partial \alpha_i} d\alpha_i + \sum_i \frac{\partial \log h}{\partial \alpha_i} d\alpha_i.$$

We now square both members of the last relation, multiply by (9) and integrate over the whole space $\boldsymbol{R}_n$. Using (10) and (13) we obtain

$$\begin{aligned} \sum_{i,j} d\alpha_i\, d\alpha_j \int \frac{\partial \log f}{\partial \alpha_i}\frac{\partial \log f}{\partial \alpha_j} f\, d\boldsymbol{x} = \\ = \sum_{i,j} d\alpha_i\, d\alpha_j \int \frac{\partial \log g}{\partial \alpha_i}\frac{\partial \log g}{\partial \alpha_j} g\, d\alpha^* + \\ + \int \left(\sum_i \frac{\partial \log h}{\partial \alpha_i} d\alpha_i\right)^2 g\, h\, d\alpha^*\, d\xi \\ \geqq \sum_{i,j} d\alpha_i\, d\alpha_j \int \frac{\partial \log g}{\partial \alpha_i}\frac{\partial \log g}{\partial \alpha_j} g\, d\alpha^*. \end{aligned}$$

The differentials $d\alpha_i$ are arbitrary, and thus according to the homogeneity we have for any real $u_1, \ldots, u_n$

(14) $$\sum_{i,j} \varkappa_{ij} u_i u_j \geqq \sum_{i,j} \mu_{ij} u_i u_j,$$

where

$$\varkappa_{ij} = \int \frac{\partial \log f}{\partial \alpha_i} \frac{\partial \log f}{\partial \alpha_j} f\, d\boldsymbol{x},$$

(15)

$$\mu_{ij} = \int \frac{\partial \log g}{\partial \alpha_i} \sqrt{g} \cdot \frac{\partial \log g}{\partial \alpha_j} \sqrt{g}\, d\alpha^*.$$

We now apply the Lemma of paragraph 3, taking

$$\varphi_i = (\alpha_i^* - \alpha_i)\sqrt{g}, \quad \psi_i = \frac{\partial \log g}{\partial \alpha_i} \sqrt{g},$$

while the set S is the whole space of the variables α_j^*. Then

$$\int \varphi_i \varphi_j\, d\alpha^* = \boldsymbol{E}(\alpha_i^* - \alpha_i)(\alpha_j^* - \alpha_j) = \lambda_{ij},$$

where the λ_{ij} are the elements of the moment matrix $\Lambda = \{\lambda_{ij}\}$ defined by (2). Further by (13) and (15)

$$\int \psi_i \psi_j\, d\alpha^* = \mu_{ij}, \quad \int \varphi_i \psi_j\, d\alpha^* = \varepsilon_{ij}.$$

According to the corollary (7) of the Lemma, we then have

(16) $$\sum_{i,j} \mu_{ij} u_i u_j \geqq \sum_{i,j} \lambda^{ij} u_i u_j,$$

where the λ^{ij} are the elements of the reciprocal matrix Λ^{-1}. From (14) and (16) it follows that

(17) $$\sum_{i,j} \varkappa_{ij} u_i u_j \geqq \sum_{i,j} \lambda^{ij} u_i u_j.$$

Now the concentration ellipsoid of the α^*-distribution has the equation (3), and thus (17) shows that the ellipsoid (4) lies wholly within the concentration ellipsoid, so that the theorem is proved.

6. Complementary Remarks.

An inspection of the above proof will show that, in order that $\alpha_1^*, \ldots, \alpha_k^*$ should be a set of joint efficient estimates, it is necessary and sufficient that the following two conditions should be satisfied:

1. The conditional probability density h should be independent of the parameters $\alpha_1, \ldots, \alpha_k$.

2. Any derivative $\frac{\partial \log g}{\partial \alpha_i}$ should be a linear combination of the differences $\alpha_1^* - \alpha_1, \ldots, \alpha_k^* - \alpha_k$, the coefficients of which are independent of the α_j^* but may depend on the α_j.

These conditions form a straightforward generalization of the corresponding conditions in the case of a single unknown parameter (cf Mathematical Methods, 32.3).

Finally, it may occur that we are only interested in the estimation of one particular parameter, say α_1, while $\alpha_2, \ldots, \alpha_k$ play the rôle of uninteresting »nuisance parameters». It can then be shown by the same argument as in the case of two parameters (cf Mathematical Methods, 32.6) that no set of estimates satisfying our regularity conditions yields a smaller standard deviation of the estimate α_1^* than does a set of joint efficient estimates.

(Received April 1946.)

43.

Problems in probability theory

Ann. Math. Statist. **18** (2), 28–39 (1947)

1. Introduction. The following survey of problems in probability theory has been written for the occasion of the Princeton Bicentennial Conference on "The Problems of Mathematics," Dec. 17–19, 1946. It is strictly confined to the purely mathematical aspects of the subject. Thus all questions concerned with the philosophical foundations of mathematical probability, or with its ever increasing fields of application, will be entirely left out.

No attempt to completeness has been made, and the choice of the problems considered is, of course, highly subjective. It is also necessary to point out explicitly that the literature of the war years has only recently—and still far from completely—been available in Sweden. Owing to this fact, it is almost unavoidable that this paper will be found incomplete in many respects.

I. FUNDAMENTAL NOTIONS

2. Probability distributions. From a purely mathematical point of view, probability theory may be regarded as the theory of certain classes of *additive set functions*, defined on spaces of more or less general types. The basic structure of the theory has been set out in a clear and concise way in the well-known treatise by Kolmogoroff [53]. We shall begin by recalling some of the main definitions. Note that the word *additive*, when used in connection with sets or set functions, will always refer to a *finite or enumerable* sequence of sets.

Let ω denote a variable point in an entirely arbitrary space Ω, and consider an additive class C of sets in Ω, such that the whole space Ω itself is a member of C. Further, let $P(S)$ be an additive set function, defined for all sets S belonging to the class C, and suppose that

$$P(S) \geqq 0 \text{ for all } S \text{ in } C,$$
$$P(\Omega) = 1.$$

We shall then say that $P(S)$ is a *probability measure*, which defines a *probability distribution in* Ω. For any set S in C, the quantity $P(S)$ is called the *probability* of the *event* expressed by the relation $\omega \subset S$, i.e. the event that the variable point ω takes a value belonging to S. Accordingly we write

$$P(S) = P(\omega \subset S).$$

Suppose now that $\omega' = g(\omega)$ is a function of the variable point ω, defined throughout the space Ω, the values ω' being points of another arbitrary space Ω'. Let S' be a set in Ω' and denote by S the set of all points ω such that $\omega' = g(\omega)$ belongs to S'. Whenever S belongs to C, we define a set function $P'(S')$ by writing

$$P'(S') = P(S).$$

It is then easy to see that $P'(S')$ is defined for all S' belonging to a certain additive class C' in the new space Ω', and that $P'(S')$ is a probability measure in Ω', such that $P'(S')$ signifies the probability of the event $\omega' \subset S'$ (which is equivalent to $\omega \subset S$). We shall say that $P'(S')$ is attached to the probability distribution in Ω' which is *induced* by the given distribution in Ω and the function $\omega' = g(\omega)$.

3. Random variables. Consider in particular the case when ω' is a real number ξ, such that $\xi = g(\omega)$ is a real-valued C-measurable function of the argument ω. Then C' includes the class B_1 of all Borel sets S' of the space $\Omega' = R_1$ of all real numbers, and we shall call ξ a *one-dimensional real random variable.* The probability of the event $\xi \subset S'$ is uniquely defined for any Borel set S' of R_1, as soon as the function

$$F(x) = P(\xi \leqq x)$$

is known for all real x. $F(x)$ is called the *distribution function* (*d.f.*) of the random variable ξ. If the function $\xi = g(\omega)$ is integrable over Ω with respect to the measure $P(S)$, we write

$$E\xi = \int_\Omega g(\omega)\, dP = \int_{-\infty}^{\infty} x\, dF(x),$$

and denote this expression as the *expectation* or *mean value* of the random variable ξ. Any real-valued B-measurable function $\eta = h(\xi)$ is also a random variable with the probability distribution induced by the original ω-distribution and the function $\eta = h(g(\omega))$. If η is integrable over Ω with respect to P, its mean value may be written in the form

$$E\eta = Eh(\xi) = \int_\Omega h(g(\omega))\, dP = \int_{-\infty}^{\infty} h(x)\, dF(x).$$

More generally, if $\omega' = (\xi_1, \cdots, \xi_n)$ is a point in an n-dimensional Euclidean space R_n, while C' includes the class B_n of all Borel sets of R_n, we are concerned with an *n-dimensional real random variable.* The distribution of this variable, which is also called the joint distribution of the n one-dimensional variables $\xi_1, \cdots, \xi_n$, is uniquely defined, as soon as the joint d.f.

$$F(x_1, \cdots, x_n) = P(\xi_1 \leqq x_1, \cdots, \xi_n \leqq x_n)$$

is known for all real $x_1, \cdots, x_n$.

The variables $\xi_1, \cdots, \xi_n$ are said to be *independent*, if $F(x_1, \cdots, x_n) = F_1(x_1) \cdots F_n(x_n)$, where $F_\nu(x_\nu)$ is the d.f. of the variable ξ_ν.

The extension to *complex random variables* is obvious. Suppose e.g. that $\xi = g(\omega)$ and $\eta = h(\omega)$ are two one-dimensional real variables, and consider the complex variable $\xi + i\eta = g(\omega) + ih(\omega)$. By definition, we identify the distribution of this variable with that of the two-dimensional real variable (ξ, η), and we put

$$E(\xi + i\eta) = E\xi + iE\eta.$$

Joint distributions of several complex variables are introduced in a corresponding way.

4. Characteristic functions. If ξ is a one-dimensional real random variable, the mean value

$$\varphi(z) = Ee^{iz\xi} = \int_{-\infty}^{\infty} e^{izx}\, dF(x)$$

exists for all real z, and we have

$$|\varphi(z)| \leqq 1, \qquad \varphi(0) = 1.$$

$\varphi(z)$ is called the *characteristic function* (*c.f.*) of the distribution corresponding to the variable ξ. The reciprocal formula (Lévy)

$$F(x) - F(y) = -\frac{1}{2\pi i} \lim_{Z\to\infty} \int_{-Z}^{Z} \frac{e^{-izx} - e^{-izy}}{z}\, \varphi(z)\, dz,$$

which holds for any continuity points x and y of F, shows that there is a one-one correspondence between the d.f. $F(x)$ and the c.f. $\varphi(z)$. As we shall see below, the c.f. provides a powerful analytical tool for operations with probability distributions.

When a complex-valued function $\varphi(z)$ of the real variable z is given, it is often important to be able to decide whether $\varphi(z)$ is or is not the c.f. of some distribution. If we assume a priori that $\varphi(0) = 1$, each of the following conditions is necessary and sufficient for $\varphi(z)$ to be a c.f.

A. $\varphi(z)$ should be bounded and continuous for all z, and such that the integral

$$\int_0^A \int_0^A \varphi(z - u) e^{ix(z-u)}\, dz\, du$$

is real and non-negative for all real x and all $A > 0$ (Cramér [11], in simplification of an earlier result due to Bochner, [4]).

B. There should exist a sequence of functions $\psi_1(z), \psi_2(z), \cdots$ such that

$$\varphi(z) = \lim_{n\to\infty} \int_{-\infty}^{\infty} \psi_n(x + z)\overline{\psi_n(x)}\, dx$$

holds uniformly in every finite z-interval (Khintchine, [45]).

These general theorems are not always easy to apply in practice. Among less general results which are more easily applicable, we mention the almost trivial fact that a function $\varphi(z)$ which near $z = 0$ is of the form $\varphi(z) = 1 + o(z^2)$ cannot be a c.f. unless $\varphi(z) = 1$ for all z, and the two following theorems:

1) An integral function $\varphi(z)$ of order $\gamma < 1$ can never be a c.f. (Lévy, [64]), and

2) an integral function $\varphi(z)$ of finite order $\gamma > 2$ cannot be a c.f. unless the convergence exponent of its zeros is equal to γ (Marcinkiewicz, [72]). The latter result shows e.g. that no function of the form $e^{g(z)}$, where $g(z)$ is a polynomial of degree > 2, can be a c.f.

It would be highly desirable to obtain further results in this direction.

The c.f. of the joint distribution of n real random variables $\xi_1, \cdots, \xi_n$ is the function $\varphi(z_1, \cdots, z_n)$ defined by the relation

$$\varphi(z_1, \cdots, z_n) = Ee^{i(z_1\xi_1+\cdots+z_n\xi_n)}.$$

Most of the above results for c.f. in one variable can be directly generalized to the multi-variable case.

5. Random sequences and random functions. Let t be a variable point in an arbitrary space T, and consider the space Ω, where each point ω is a real-valued function $\omega = x(t)$ of the variable argument t. Let $t_1, \cdots, t_n$ be any finite set of distinct points t. The set of all functions $\omega = x(t)$ satisfying the inequalities

$$a_j < x(t_j) \leqq b_j, (j = 1, \cdots, n),$$

will be called an *interval* in the space Ω. The Borel sets in Ω will be defined as the smallest additive class B of sets in Ω containing all intervals.

Suppose now that, for any choice of n and the t_j, the variables $x(t_1), \cdots, x(t_n)$ are random variables having a known n-dimensional joint distribution. If the family of all distributions corresponding in this way to finite sequences $t_1, \cdots, t_n$ satisfies certain obvious consistency conditions, a fundamental theorem due to Kolmogoroff asserts that this family determines a unique probability distribution in the space Ω of all functions $x(t)$. The corresponding probability

$$P(S) = P(x(t) \subset S)$$

is uniquely defined for all Borel sets S of Ω.

Consider in particular the case where T is the set of non-negative integers $t = 0, 1, 2, \cdots$. The space Ω then is the space of all sequences $(x_0, x_1, \cdots)$ of real numbers. As soon as the joint distribution of any finite number of variables $x_{\nu_1}, \cdots, x_{\nu_n}$ is defined, and these distributions are mutually consistent, it then follows that there is a unique probability distribution of the *random sequence* $(x_0, x_1, \cdots)$, the corresponding probability being defined for every Borel set of the space Ω of sequences. Similarly we may consider the doubly infinite sequence $(\cdots, x_{-1}, x_0, x_1, \cdots)$.

Consider further the more general case when T is any set of real numbers. Then Ω is the space of all real-valued functions $\omega = x(t)$ defined on the set T, and as before the knowledge of the distributions for all finite sets of variables $x(t_1), \cdots, x(t_n)$ permits us to determine a probability distribution in the space Ω of *random functions* $x(t)$, the probability $P(S) = P(x(t) \subset S)$ being always defined for all Borel sets S in Ω.

The generalization of the above considerations to *complex-valued* random sequences and functions is immediate.

6. Various modes of convergence. Consider a sequence $F_1(x)$, $F_2(x)$, ... of d.f:s, and let the corresponding c.f:s be $\varphi_1(t)$, $\varphi_2(t)$, $\cdots$ In order that $F_n(x)$

converge to a d.f. $F(x)$, in every continuity point of the latter, it is necessary and sufficient[1] that $\varphi_n(t)$ converge for every real t to a limit $\varphi(t)$ which is continuous at $t = 0$. Then $\varphi(t)$ is the c.f. corresponding to the d.f. $F(x)$.

Further, let x and $x_1, x_2, \cdots$ be complex-valued random variables, such that the random sequence $(x, x_1, x_2, \cdots)$ has a well defined distribution. We shall be concerned with various modes of convergence of x_n to x.

A. When $P(|x_n - x| > \epsilon) \to 0$ as $n \to \infty$, for any $\epsilon > 0$, we shall say that x_n *converges to x in probability.*

B. When $E|x_n - x|^\gamma \to 0$, as $n \to \infty$, where $\gamma > 0$ is fixed, we shall say that x_n converges to x in the mean of order γ. Unless otherwise stated we shall in the sequel always consider the case $\gamma = 2$, and in this case we shall use the notation

$$\underset{n\to\infty}{\text{l.i.m.}}\, x_n = x.$$

C. When $P(\lim_{n\to\infty} x_n = x) = 1$, we shall say that x_n converges with probability one, or *converges almost certainly to x.*

With respect to the last definition, we may remark that the set defined by the relation $\lim x_n = x$ is always a Borel set in the space of our random sequence, so that the probability of this relation is well defined. In fact, this probability is given by the expression

$$\lim_{m\to\infty} \lim_{n\to\infty} \lim_{p\to\infty} P\left(|x_\nu - x| < \frac{1}{m} \quad \text{for} \quad \nu = n, n+1, \cdots, n+p\right)$$

where the limit process applies to a probability attached to a Borel set in a finite number of dimensions. The case of almost certain convergence is precisely the case when this expression takes the value 1.

Convergence in the mean of any positive order, as well as almost certain convergence, both imply convergence in probability, which may be written symbolically $B \to A$ and $C \to A$. Between B and C, there is no simple relation of this kind. Further, A and B both imply almost certain convergence for any partial sequence $x_{n_1}, x_{n_2}, \cdots$ such that the subscripts n_k increase sufficiently rapidly with k.

II. PROBLEMS CONNECTED WITH THE ADDITION OF INDEPENDENT VARIABLES

7. During the early development of the theory of probability, the majority of problems considered were connected with gambling. The gain of a player in a certain game may be regarded as a random variable, and his total gain in a

[1] As I have already stated in a paper published in 1938, there is an error in the statement of this theorem given in my Cambridge Tract [9] *Random Variables and Probability Distributions*. For the truth of the theorem, it is essential that $\varphi_n(t)$ should be supposed to converge to $\varphi(t)$ *for every real t*. However, in the particular case when the limit $\varphi(t)$ is analytic and regular in the vicinity of $t = 0$, it can be proved that it is sufficient to assume convergence in some interval $|t| < a$.

sequence of repetitions of the game is the sum of a number of independent variables, each of which represents the gain in a single performance of the game. Accordingly a great amount of work was devoted to the study of the probability distributions of such sums. A little later, problems of a similar type appeared in connection with the theory of errors of observation, when the total error was considered as the sum of a certain number of partial errors due to mutually independent causes. At first only particular cases were considered, but gradually general types of problems began to arise, and in the classical work of Laplace several results are given concerning the general problem to study the distribution of a sum

$$z_n = x_1 + \cdots + x_n$$

of independent variables, when the distributions of the x_j are given. This problem may be regarded as the very starting point of a large number of those investigations by which the modern Theory of Probability was created. The efforts to prove certain statements of Laplace, and to extend his results further in various directions, have largely contributed to the introduction of rigorous foundations of the subject, and to the development of the analytical methods. At the same time, more general types of problems have developed from the original problem, and the number and importance of practical applications have been steadily increasing.

8. Composition of distributions. Let x_1 and x_2 be two independent variables, with the d.f.'s F_1 and F_2, and the c.f.'s φ_1 and φ_2, and let the sum $x_1 + x_2$ have the d.f. F and the c.f. φ. Then

$$F(x) = \int_{-\infty}^{\infty} F_1(x - y)\, dF_2(y) = \int_{-\infty}^{\infty} F_2(x - y)\, dF_1(y).$$

We shall say that F is the *composition* of F_1 and F_2, and write this as a symbolical multiplication:

$$F = F_1 * F_2 = F_2 * F_1 .$$

To this symbolical multiplication of the d.f:s corresponds a real multiplication of the c.f.'s:

$$\varphi(z) = \varphi_1(z)\varphi_2(z).$$

The operation of composition is both commutative and associative, so that any symbolical product $F = F_1 * F_2 \cdots * F_n$ is uniquely defined and independent of the order of the components. When at least one of the components is continuous (absolutely continuous), the same holds for the composite, and in many cases it is true that the composite is at least as regular as the most regular of the components (Lévy, [58], [63], etc.). However, this general statement does not hold generally, as is shown by an interesting example due to Raikov, [77], where F_1 and F_2 are integral analytic functions, while the composite $F = F_1 * F_2$ is not regular at the origin.

It seems to be an important unsolved problem to find convenient restrictions ensuring the validity of the above statements of the "smoothing effect" of the operation of composition.

When $F = F_1 * F_2$, we may say that F is "divisible" by each component F_1 and F_2, and it seems natural to try to develop a theory of symbolical factorization for d.f.'s. In this connection, it is important to note that symbolical division is not unique. In fact, Khintchine has shown by an example that it is possible to find the d.f.'s F, F_1, F_2, and F_3 such that

$$F = F_1 * F_2 = F_1 * F_3,$$

while $F_2 \neq F_3$. Another fundamental problem belonging to this order of ideas is to decide whether a given d.f. F is decomposable or not. F is called decomposable, if there is at least one representation of the form $F = F_1 * F_2$, where each component F_ν has more than one point of increase. So far, this problem has only been solved in very special cases, and the general problem still remains open for research. A particular case of some interest would be to know if there exists an absolutely continuous and indecomposable d.f., such that $F(a) = 0$ and $F(b) = 1$ for some finite a and b.

As soon as we restrict ourselves to certain special classes of distributions, it is possible to reach results of a more definite character concerning the factorization problems. Some results of this type will be considered below.

9. Closed families of distributions. The fact that certain families of distributions are closed with respect to the operation of composition has played an important part in many applications. If F_1 and F_2 belong to a family of this character, so does the symbolical product $F = F_1 * F_2$. We first give some simple examples of such families.

The normal distribution. The d.f. F has the form $F = \phi\left(\frac{x-m}{\sigma}\right)$, where $\sigma > 0$, and

$$\phi(x) = \frac{1}{\sqrt{2\pi}} \int_{-\infty}^{x} e^{-(t^2/2)}\, dt.$$

The c.f. corresponding to F is $e^{miz - \frac{1}{2}\sigma^2 z^2}$, and it follows that for any real m_1, m_2 and any positive σ_1, σ_2 we have

$$\phi\left(\frac{x-m_1}{\sigma_1}\right) * \phi\left(\frac{x-m_2}{\sigma_2}\right) = \phi\left(\frac{x-m}{\sigma}\right),$$

where

$$m = m_1 + m_2, \qquad \sigma^2 = \sigma_1^2 + \sigma_2^2.$$

The Poisson distribution. Here the d.f. is $F = F(x; \lambda, m, a)$ where $\lambda > 0$, $a \neq 0$, and F is a step-function with a jump equal to $\frac{\lambda^\nu}{\nu!} e^{-\lambda}$ in the point $x = m + \nu a$, where $\nu = 0, 1, \cdots$. The corresponding c.f. is $e^{miz + \lambda(e^{aiz} - 1)}$, and it follows that for any fixed a we have

$$F(x; \lambda_1, m_1, a) * F(x; \lambda_2, m_2, a) = F(x; \lambda_1 + \lambda_2, m_1 + m_2, a).$$

The Pearson Type III distribution. $F = F(x;\, \alpha,\, \lambda) = \dfrac{\alpha^\lambda}{\Gamma(\lambda)} \displaystyle\int_0^x t^{\lambda-1} e^{-\alpha t}\, dt$, $(x > 0)$. The corresponding c.f. is $\left(1 - \dfrac{iz}{\alpha}\right)^{-\lambda}$, and for any fixed $\alpha > 0$ and any positive λ_1 and λ_2 we have

$$F(x;\, \alpha,\, \lambda_1) * F(x;\, \alpha,\, \lambda_2) = F(x;\, \alpha,\, \lambda_1 + \lambda_2).$$

Stable distributions. We shall say that a closed family is stable, when all its members are of the form $F(ax + b)$, where F is a d.f., while $a > 0$ and b are constants. Obviously the normal family is an example of a stable family. It has been shown by Lévy and Khintchine [49], that a d.f. $F(x)$ generates a stable family when and only when the logarithm of its c.f. is of the form

$$\log \varphi(z) = \beta iz - \gamma \,|\, z \,|^\alpha \left(1 + i\delta \frac{z}{|z|}\, \omega\right), \tag{9.1}$$

where $\alpha, \beta, \gamma, \delta$ are real constants such that

$$0 < \alpha \leqq 2, \qquad \gamma > 0, \qquad |\, \delta \,| \leqq 1,$$

while

$$\omega = \begin{cases} tg\, \dfrac{\alpha\pi}{2} & \text{for } \alpha \neq 1 \\[2mm] \dfrac{2}{\pi} \log |\, z \,| & \text{for } \alpha = 1. \end{cases}$$

For $\alpha = 2$ we obtain the normal family.

A more general and very important closed family is the family I of *infinitely divisible distributions.* A d.f. F belongs to I if to every $n = 1, 2, \cdots$ there exists a d.f. G such that $F = G^{[n]}$, where $G^{[n]}$ denotes the symbolical nth power of G. Obviously the family I is a closed family which contains all the families mentioned above. Lévy [60], [63], has shown that F is infinitely divisible when and only when the logarithm of its c.f. is of the form

$$\begin{aligned} \log \varphi(z) = \beta iz - \gamma z^2 &+ \int_{-\infty}^0 \left(e^{izu} - 1 - \frac{izu}{1+u^2}\right) dM(u) \\ &+ \int_0^\infty \left(e^{izu} - 1 - \frac{izu}{1+u^2}\right) dN(u), \end{aligned} \tag{9.2}$$

where β and $\gamma > 0$ are real constants, while $M(u)$ and $N(u)$ are non-decreasing functions such that

$$M(-\infty) = N(+\infty) = 0,$$

$$\int_{-a}^0 u^2\, dM(u) < \infty \quad \text{and} \quad \int_0^a u^2\, dN(u) < \infty$$

for any finite $a > 0$. When M and N reduce to zero, we obtain the normal family. When $\gamma = 0$ and one of the functions M and N reduces to zero, while the other is a step-function with a single jump equal to λ at the point $x = a$, we obtain a Poisson family. Generally, it follows from (9.2) that any infinitely divisible distribution may be regarded as a product of a normal distribution and a finite, enumerable or continuous set of Poisson distributions.

The representation of $\log \varphi(z)$ in the form (9.2) is unique. It follows that the problem of finding all possible factorizations of an infinitely divisible d.f. F can be completely solved, as long as we restrict ourselves to factors which are themselves infinitely divisible. In fact, in order that

$$F = F_1 * F_2,$$

where all three d.f.'s belong to I, it is necessary and sufficient that the logarithms of the corresponding c.f.'s should be of the form (9.2), with

$$\beta = \beta_1 + \beta_2, \qquad \gamma = \gamma_1 + \gamma_2,$$
$$M = M_1 + M_2, \qquad N = N_1 + N_2.$$

In the two simple cases of the normal and the Poisson distributions, the decompositions obtained in this way remain the only possible, even if we remove the restriction that the factors should belong to I. Thus in any factorization of a normal distribution, all factors are normal (Cramér, [8]), while in any factorization of a Poisson distribution, all factors belong to the Poisson family (Raikov, [75]). For the type III distribution, and the non-normal stable distributions, however, the corresponding property does not hold.

In some cases, an infinitely divisible distribution may be represented as a product of indecomposable distributions, or as a product of an indecomposable distribution and another infinitely divisible distribution. The results so far obtained in this direction (Lévy, [63], [64], Khintchine, [46], [47]; Raikov, [76]) are all concerned with more or less particular cases, and the general factorization problem for infinitely divisible distributions still remains unsolved. A particular case of some interest would be the case when the functions M and N are both absolutely continuous. There does not seem to have been given any example of this type, where a factor not belonging to I may occur.[2]

Finally we mention a general theorem due to Khintchine, [46], which asserts that an arbitrary d.f. F may be represented in one of the forms

$$F = G, \qquad F = H \text{ or } F = G * H,$$

where G is infinitely divisible, while H is a finite or infinite product of indecomposable factors. This seems to be practically the only result so far known concerning the factorization of a general distribution.

A certain number of the results mentioned above have been generalized to multi-dimensional distributions.

---

[2] While the present paper was being printed, I have proved that such factors do occur, as soon as at least one of the derivatives M' and N' is bounded away from zero in some interval $(-a, 0)$ or $(0, a)$.

10. The Laws of large numbers. In modern terminology, the classical Bernoulli theorem may be expressed in the following way. Let $x_1, x_2, \cdots$ be a sequence of independent variables, such that each x_ν may only assume the values 1 and 0, the corresponding probabilities being p and $q = 1 - p$. Then the arithmetic mean

$$\frac{z_n}{n} = \frac{x_1 + \cdots + x_n}{n} \tag{10.1}$$

converges in probability to p, as $n \to \infty$.

Both classical and modern authors have laid down much work on the generalization of this simple result in various directions. Generally, we shall say that a sequence of random variables $x_1, x_2, \cdots$ satisfies the *Weak Law of Large Numbers* if there exist two sequences of constants $a_1, a_2, \cdots$ and $b_1, b_2, \cdots$, such that $a_n > 0$, and

$$\frac{z_n - b_n}{a_n} = \frac{x_1 + \cdots + x_n - b_n}{a_n}$$

converges in probability to zero.

Let $x_1, x_2 \cdots$ be independent variables, such that x_ν has the d.f. $F_\nu(x)$. It has been shown by Feller [27] that *for any given sequence $a_1, a_2, \cdots$, the conditions*

$$\begin{aligned} &\sum_{\nu=1}^{n} \int_{|x|>a_n} dF_\nu(x) = o(1), \\ &\sum_{\nu=1}^{n} \int_{|x|<a_n} x^2\, dF_\nu(x) = o(a_n^2), \end{aligned} \tag{10.2}$$

are sufficient for the validity of the weak law of large numbers, and that the corresponding sequence $b_1, b_2, \cdots$ can be defined by

$$b_n = \sum_{\nu=1}^{n} \int_{|x|<a_n} x\, dF_\nu(x).$$

When there is a constant $c > 0$ such that for all ν

$$F_\nu(+0) > c, \qquad F_\nu(-0) < 1 - c, \tag{10.3}$$

the conditions are also necessary. This theorem contains as particular cases all previously known results in this direction. A simple *NS* condition for the existence of at least one sequence $a_1, a_2, \cdots$ such that 10.2 holds does not seem to be known.

When the weak law is satisfied, this means that, for any given $\epsilon > 0$ and *for any fixed large n*, there is a probability very near to 1 that the sum $z_n = x_1 + \cdots + x_n$ will fall between the limits $b_n \pm \epsilon a_n$. The more stringent condition that, with a probability tending to 1 as $n \to \infty$, z_ν will fall between the limits $b_\nu \pm \epsilon a_\nu$ *for all values of $\nu \geqq n$* is equivalent to the condition that $\frac{z_n - b_n}{a_n}$ converges *almost certainly* to zero. When this holds, we shall say that the variables x_ν satisfy the *Strong Law of Large Numbers.* The most important result so far known in this connection is concerned with the case $a_n = n$, and is expressed by the following theorem (Kolmogoroff, [52], [55]):

When the x_ν are independent and (10.3) holds, a sufficient condition for the validity of the strong law with $a_n = n$ consists in the simultaneous convergence of the two series

$$\Sigma \int_{|x|>n} dF_n(x) \qquad \text{and} \qquad \Sigma \frac{1}{n^2} \int_{|x|<n} x^2 \, dF_n(x).$$

Some improved conditions of this type have been given by Marcinkiewicz and Zygmund, [73], but the problem of finding a *NS* condition for the strong law is still unsolved, even in the case $a_n = n$.

Important generalizations of the laws of large numbers to cases when the x_ν are not assumed to be independent have been given i.a. by Khintchine [44], Lévy [62], [63] and Loève [67].

11. The central limit theorem and allied theorems. It was already known to De Moivre that, in the case 10.1 of the Bernoulli distribution, the d.f. of the normalized sum

$$\frac{x_1 + \cdots + x_n - np}{\sqrt{npq}}$$

tends, as $n \to \infty$, to the normal d.f. $\phi(x)$. Considerably more general results in this direction were stated by Laplace. After a long series of more or less successful attempts, a rigorous proof of the main statements of Laplace was given in 1901 by Liapounoff, [65]. More general cases were later considered i.a. by Lindeberg [66], Lévy [61], [63], Khintchine [43] and Feller, [25]. The following final form of the *Central Limit Theorem* is due to Feller.

Consider the expression

$$u_n = \frac{z_n - b_n}{a_n} = \frac{x_1 + \cdots + x_n - b_n}{a_n}, \tag{11.1}$$

where the x_ν are independent variables. We shall say that the x_ν obey the central limit law, if the sequences $\{a_\nu\}$ and $\{b_\nu\}$ can be found such that the d.f. of u_n tends to $\phi(x)$ as $n \to \infty$. In order to avoid unnecessary complications, we shall restrict ourselves to sequences $\{a_\nu\}$ such that

$$a_\nu \to +\infty, \qquad \frac{a_{\nu+1}}{a_\nu} \to 1,$$

and we shall assume that the conditions (10.3) are satisfied. Then Feller's theorem runs as follows:

The independent variables $x_1, x_2, \cdots$ obey the central limit law if, and only if, there exists a sequence $q_n \to \infty$ such that simultaneously

$$(11.2)\qquad \begin{aligned} &\sum_{\nu=1}^{n} \int_{|x|>q_n} dF_\nu(x) \to 0, \\ &\frac{1}{q_n^2} \sum_{\nu=1}^{n} \int_{|x|<q_n} x^2\, dF_\nu(x) \to \infty. \end{aligned}$$

When these conditions are satisfied, explicit expressions for the a_n and b_n can be obtained.

Feller's theorem gives a complete solution of the problem. However, we might still try to express in a more direct way the condition that the q_n should exist. We may also ask what happens when the conditions (11.2) are not satisfied. Some particular cases of the latter question will be considered below. However, very few general results are known in this direction.

The central limit theorem has been extended in various directions. Bernstein [3], Lévy [62], [63], Loève [67] and others have considered cases where the x_ν are not assumed to be independent. Important results have been reached but still much remains to be done.

On the other hand, several authors have considered symmetrical functions, other than sums, of n independent random variables. The problem of investigating the asymptotic behaviour of the distributions of such functions, as n tends to infinity, is of great importance in the theory of statistical sampling distributions. It is known (c.f. e.g. Cramér, [15]) that under certain general regularity conditions there exists a normal limiting distribution. However, it is also known that it is possible to give examples of particular functions (such as e.g. the function which is equal to the largest of the n variables), where there exist limiting distributions which are non-normal. The conditions under which this phenomenon may occur seem to deserve further study.

A further problem belonging to the same order of ideas is to find a closer asymptotic representation of the d.f. of the standardized sum z_n than that provided by the normal function $\phi(x)$. Consider e.g. the simple case when the x_ν are independent variables all having the same d.f. $F(x)$ with a finite mean m, a finite variance σ^2, and finite moments up to a certain order $k \geqq 3$. Let $G_n(x)$ be the d.f. of the variable

$$\frac{x_1 + \cdots + x_n - nm}{\sigma\sqrt{n}}.$$

It then follows from a theorem of Cramér [5], [9] that, as soon as the d.f. $F(x)$ contains an absolutely continuous component, there is an asymptotic expansion

$$(11.3)\qquad G_n(x) = \phi(x) + \sum_{\nu=1}^{k-3} \frac{p_\nu(x)}{n^{\nu/2}} e^{-x^2/2} + O(n^{-(k-2)/2}),$$

where the constant implied by the O is independent of n and x. Cramér has also given similar expansions in more general cases, and his results have been further extended by P. L. Hsu [39], who deduces analogous expansions also for other functions than sums. The most general conditions under which expansions of this type exist are still unknown.

It follows from (11.3) that the difference $G_n(x) - \phi(x)$ is, for any fixed x, of the order $n^{-\frac{1}{2}}$ as $n \rightarrow \infty$. It is often important to know the asymptotic behaviour of $G_n(x)$ when n and x increase simultaneously, and in that case (11.3) yields only a trivial result. This case has been investigated by Cramér [10], and Feller [29], and the results so far obtained permit important applications to the so called law of the iterated logarithm (cf. below). However, it seems likely that similar results may be obtained in considerably more general cases than those hitherto investigated.

A further interesting type of problems belonging to this order of ideas may be approached in the following way. Consider the variables (11.1) in the particular case when $x_1, x_2, \cdots$ are independent variables all having the same d.f. $F(x)$. When the a_n and b_n can be found such that the d.f. of the normalized sum u_n tends to $\phi(x)$, we shall say that F belongs to the *domain of attraction* of the normal law. Feller's theorem gives a NS condition that this should be so. Now when this condition is not satisfied, it may still occur that the a_n and b_n can be so chosen that the d.f. of u_n tends to a limiting d.f. $\Psi(x)$, which is necessarily different from $\phi(x)$. Then it is easily seen that $\Psi(x)$ must be a stable distribution, with its c.f. defined by (9.1), and it is natural to say that F belongs to the domain of attraction of Ψ. NS and sufficient conditions that this should hold have been given by Doeblin [16], and Gnedenko [34]. When the a_n and b_n cannot be found such that the d.f. of the normal sum u_n converges to a limit, it may still be possible to obtain a limiting d.f. by considering only a partial sequence $u_{n_1}, u_{n_2}, \cdots$. Khintchine [47] has proved the interesting theorem that the totality of limiting d.f.'s that may be obtained in this way coincides with the class of infinitely divisible d.f.'s defined by (9.2). There are also further results in the same direction given by Bawly [2], Khintchine [44], Lévy, [61]–[63], and Gnedenko, [35].

12. The law of the iterated logarithm. Consider a sequence of independent variables $x_1, x_2, \cdots$, such that the mean $Ex_n = 0$ for all n, while the variances $Ex_n^2 = \sigma_n^2$ are finite. Put $s_n^2 = \sigma_1^2 + \cdots + \sigma_n^2$, and suppose that the variables obey the central limit law with $a_n = s_n$, $b_n = 0$. (In particular this will be the case when all x_n have the same distribution.) For any function $\psi(n)$ tending to infinity with n we then have

$$\lim_{n \rightarrow \infty} P(|z_n| > s_n \psi(n)) = 0. \tag{12.1}$$

On the other hand, if $\psi(n)$ tends to a finite limit > 0, the same probability has a positive limit.

It seems natural to consider the relation within the brackets in (12.1) not only for a single large value of n, but to require the probability that this relation holds simultaneously for *an infinite number of values of n*. The development of this problem has led to the so called law of the iterated logarithm.

We shall in this respect use the following terminology due to Lévy. A non-decreasing positive function $\psi(n)$ will be said to belong to the *lower class* with respect to the variables x_n if, with a probability equal to one, there are infinitely many n such that

$$| z_n | > s_n\psi(n).$$

On the other hand, $\psi(n)$ will be said to belong to the *upper class* if the probability of the same property is equal to zero.

Every $\psi(n)$ belongs to one of these two classes. This is a special case of the so called *null-or-one law*: if S is a Borel set in the space of the independent random variables $x_1, x_2, \cdots$, such that any two points differing at most in a finite number of coordinates either both belong to S or both belong to the complementary set, then $P(S)$ can only assume the values 0 or 1.

It was proved by Kolmogoroff [51] that, subject to certain restrictions, the function

$$\psi(n) = \sqrt{c \log \log s_n}$$

belongs to the lower class for any $c < 2$, and to the upper class for any $c > 2$, which may be expressed by the relation

$$P\left(\limsup \frac{z_n}{s_n \sqrt{2 \log \log s_n}} = 1\right) = 1. \tag{12.1}$$

More general results were proved by Feller [30], who proved i.a. that, subject to certain restrictions, $\psi(n)$ belongs to the lower or upper class according as

$$\Sigma \frac{\sigma_n^2}{s_n^2} \psi(n)e^{-(\psi^2(n)/2)} \tag{12.2}$$

is divergent or convergent (in certain special cases, this had been previously found by Kolmogoroff and Erdös [24]. Feller also proved a more complicated result, which contains the above as a particular case, and from which it follows that the simple criterion (12.2) no longer holds when the restrictions imposed in its proof are removed.

13. Convergence of series. For any sequence of random variables x_n, the probability

$$P\left(\sum_1^\infty x_n \text{ converges}\right)$$

has a uniquely determined value. When the x_n are independent, it follows from the null-or-one law that this probability is either 0 or 1. By a theorem of Khintchine and Kolmogoroff [48], the value 1 is assumed when and only when the three series

$$\sum_n \int_{|x_n|>1} dF_n, \quad \sum_n Ey_n, \quad \sum_n \sigma^2 y_n$$

are convergent, where

$$y_n = \begin{cases} x_n & \text{when} \quad |x_n| \leqq 1. \\ 0 & \text{when} \quad |x_n| > 1. \end{cases}$$

For the case when the x_n are not assumed to be independent, various results have been given by Lévy [63] and others, but our knowledge of the properties of these series is still not very advanced.

14. Generalizations. In several instances it has been pointed out above that the results concerning sums of independent variables may, to a certain extent, be extended to cases when the variables are not independent. Generally the independence condition has then to be replaced by some condition restricting the degree of dependence. Results of this type were first give by Bernstein [3], and then in more general cases by Lévy [62], [63], and Loève [67]. However, this field has so far only been very incompletely explored.

Similar remarks apply to the generalization of the various theorems quoted above to cases of variables and distributions in more than one dimension.

III. STOCHASTIC PROCESSES

15. The theory of random variables in a finite number of dimensions is able to deal adequately with practically all problems considered in classical probability theory. However, during the early years of the present century, there appeared in the applications various problems, where it proved necessary to consider probability relations bearing on infinite sequences of numbers, or even on functions of a continuous variable.

The mathematical set-up required for the study of such problems involves the introduction of probability distributions in spaces of random sequences or random functions (cf. 5 above). Generally, any process in nature which can be analyzed in terms of probability distributions in spaces of these types will be called a *stochastic process*. It is convenient to apply this name also to the probability distribution used for the study of the process. We shall thus say, e.g., that a certain random function $x(t)$ is attached to the stochastic process which is defined by the probability distribution of $x(t)$. In the majority of applications, the variable t will represent the time, and we shall often use a terminology directly referring to this case. However, there are also other types of problems in the applications (t may e.g. be a spatial variable in an arbitrary number of dimensions), and it is obvious that the purely mathematical problems connected with these classes of probability distributions will have to be considered quite independently of any concrete interpretation of the variable t or the funcion $x(t)$.

A well-known example of this type of problems is afforded by the Brownian movement. Let $x(t)$ be the abscissa at the time t of a small particle immersed in a liquid, and subject to molecular impacts. In every instant, the quantity $x(t)$ receives a random impulse, and the problem arises to study the behaviour of $x(t)$. According as we are content to consider $x(t)$ for a discrete sequence of t-points, say for $t = 0, 1, 2, \cdots$, or we wish to consider all positive values of t, we shall then have to introduce a probability distribution in the space of the random sequence $x(0)$, $x(1)$, $\cdots$, or in the space of the random function $x(t)$, where $t > 0$. We may then discuss such questions as the distribution of $x(t)$ for a given value of t, the joint and conditional distributions of $x(t)$ for two or more values of t, and, in the case of a continuous variable t, continuity, differentiability and other similar properties of the random function $x(t)$.

Wiener [82], [83] (cf. also Paley and Wiener [74]) was the first to give a rigorous treatment of this process. He proved in 1923 that it is possible to define a probability distribution in a suitably restricted functional space, such that the increment $\Delta x(t) = x(t + \Delta t) - x(t)$ is independent of $x(t)$ for any $\Delta t > 0$. With a probability equal to 1, the function $x(t)$ is continuous for all $t > 0$, and for any fixed $t > 0$, the random variable $x(t)$ is normally distributed.

Another example of stochastic processes studied at this stage occurs in the theory of risk of an insurance company. Let $x(t)$ denote the total amount of claims up to the time t in a certain insurance company. As in the case of the Brownian movement, it may seem natural to assume that the increment $\Delta x(t)$ is independent of $x(t)$. On the other hand, $x(t)$ is in this case an essentially discontinuous function, which is never decreasing, and increases only by jumps of varying magnitudes occurring for certain discrete values of t, which are not a priori known. Processes of this type were studied by F. Lundberg [69], [70], H. Cramér [6] and others.

Further examples of particular processes were discussed in connection with various applications, but no general theory of the subject existed until 1931, when Kolmogoroff published a basic paper [53] dealing with the class of stochastic processes which will here be denoted as Markoff processes (Kolmogoroff uses the term "stochastically definite processes"), of which the two examples mentioned above form particular cases. The theory of this class of processes was further developed by Feller [26], [28]. In 1934, Khintchine [42] introduced another important class of processes known as stationary processes. From 1937, the general theory of the subject was subjected to a penetrating analysis in a series of important works by Doob [18]–[22].[3]

16. Probability distributions in functional spaces. We have seen in 5 above how a probability distribution in the space of all functions $x(t)$ may be defined, when t varies in an arbitrary space T. Generally, we shall here content ourselves to consider the cases when T is the set of all real numbers, or the set of all non-negative real numbers. Most results obtained for these cases will be readily generalized to cases when t varies in a Euclidean space of a finite number of dimensions. On the other hand, when T is enumerable, say consisting of the points $t = 0, \pm 1, \pm 2, \cdots$, so that we are concerned with a random sequence $x(0)$, $x(\pm 1)$, $\cdots$, the results for the continuous case will generally hold and assume a simpler form which will not be particularly stated here.

[3] A further interesting paper by Doob has appeared while the present paper was being printed: "Probability in function space", *Bull. Amer. Math. Soc.*, Vol. 53 (1947), pp. 15–30.

The case when T is a space of an infinite number of dimensions does not seem to have been considered so far.

In the present paragraph, it will be convenient to assume the function $x(t)$ to be real-valued, but the generalization to a complex-valued $x(t)$ requires only obvious modifications. In the sequel we shall sometimes consider the real-valued and sometimes the complex-valued case, according as the occasion requires.

Let now X be the space of all real-valued functions $x(t)$ of the real variable t, where $-\infty < t < \infty$. According to 5, a probability measure $P(S)$ is uniquely defined for all Borel sets S in X by means of the family of joint distributions of all finite sequences $x(t_1), \cdots, x(t_n)$. In fact, $P(S)$ can be defined for a more general class of sets than the Borel sets. For any set S in X, we may define an outer P-measure $\bar{P}(S)$ as the lower bound of $P(Z)$ for all sums Z of finite or enumerable sequences of intervals, such that $S \subset Z$. Further, the inner P-measure $\underline{P}(S)$is defined by the relation $\underline{P}(S) = 1 - \bar{P}(X - S)$. When the outer and inner measures are equal, S is called P-measurable, and $P(S)$ is defined as their common value. Any P-measurable set differs from a Borelset by a set of P-measure zero.

In many cases, this definition will be sufficient for an adequate treatment of the problems that we wish to consider. However, in other cases we encounter certain characteristic difficulties, which make it desirable to consider the possibility of amending the basic definition. Thus it often occurs that we are interested in the probability that the random function $x(t)$ satisfies certain regularity conditions in a non-enumerable set of points t. We may, e.g., wish to consider the probability that $x(t)$ is continuous for all t, that $x(t)$ should be Lebesque-measurable for all t, that $x(t) \leqq k$ for all t, etc. Let S denote the set of all functions satisfying a condition of this type. It can then be shown that the inner measure $\underline{P}(S)$ is always equal to zero so that S is never measurable, except in the (usually trivial) case when $P(S) = 0$.

Consequently many interesting probabilities are left undetermined by the general definition of a probability distribution in X given above. The possibility of modifying the definition so as to enable us to study probabilities of this type has been thoroughly investigated by Doob [18]. He considers a subspace X_0 of the general functional space X, where X_0 is chosen so as to contain only, or almost only, "desirable" functions, i.e. functions satisfying such regularity conditions as seem natural with respect to the problem under investigation. We start from a given probability measure $P(S)$ in X, and ask if it is possible to define a probability measure in the restricted space X_0, which corresponds in some natural way to the given distribution in X. Let S_0 be a set in X_0, and suppose that it is possible to find a P-measurable set S in X such that $SX_0 = S_0$. According to Doob, a probability measure P_0 in X_0 is then uniquely defined by the relation

$$P_0(S_0) = P(S)$$

if and only if the condition

$$\overline{P}(X_0) = 1$$

is satisfied.

The problem is thus reduced to finding a subspace X_0 of outer P-measure 1, such that X_0 contains only functions of sufficiently regular behaviour. When this can be done, we can restrict ourselves to consider only functions $x(t)$ belonging to X_0, the probability distribution in this space being defined by the measure P_0. We shall then say that $x(t)$ is a random function, attached to a stochastic process with the restricted space X_0. Doob has obtained a great number of interesting results in this connection, e.g. with respect to the problem of choosing X_0 such that it contains almost only Lebesque-measurable functions, or such that the probability of the relation $x(t) \leqq k$ has a well-defined value for all k. In particular he has shown that the last problem can be solved for any given P-measure. However, our knowledge of the various possibilities which exist with respect to the choice of X_0 is still very incomplete, and it seems likely that further important results may be reached along this line of research.

An alternative method of introducing probability distributions in functional spaces has been used by Wiener [82], [83], (cf. also Paley and Wiener, [74]). Consider a given probability measure Π in an arbitrary space Ω, defined for all sets Σ of an additive class C. Let $x(t, \omega)$ denote a function (real- or complex-valued, as the case may be) of the arguments t (real) and ω (point in Ω), such that $x(t, \omega)$ for every fixed t becomes a C-measurable function of ω. On the other hand, when ω is fixed, $x(t, \omega) = x(t)$ reduces to a function of the real variable t. Let X_0 denote the set of all functions $x(t)$ corresponding in this way to points of Ω. Further, let $S_0 = SX_0$, where S is a Borel set in X, and let Σ denote the set of all points ω such that $x(t, \omega) \subset S_0$. Then Σ belongs to C, and a probability measure P_0 in the functional space X_0 is uniquely defined by the relation

$$P_0(S_0) = \Pi(\Sigma). \tag{16.1}$$

The relations between the two modes of definition have been discussed by Doob and Ambrose [23] who have shown that they are largely equivalent. However, it seems likely that in particular problems the one or the other procedure may sometimes be the more advantageous, and further investigations on this subject seem desirable.

17. Processes with a finite mean square. Consider a stochastic process defined by a probability measure $P(S)$ in the space X of all complex-valued functions $x(t)$ of the real variable t. For any fixed t_0, the random variable $x(t_0)$ is then a complex-valued function of the variable point $x(t)$ in the space X, i.e. a point Q_{t_0} in the space Ω of all complex-valued functions defined on X. When t_0 varies, the point Q_{t_0} describes a "curve" in Ω, which then corresponds to our stochastic process.

Suppose, in particular, that the mean square

$$E \mid x(t) \mid^2 = \int_X \mid x(t) \mid^2 dP$$

is finite for any fixed value of t. This implies that for fixed t the function $x(t)$ belongs to L_2 over X, relative to the probability measure P. The random variable $x(t)$ may then be regarded as an element of the Hilbert space H of all complex-valued functions f belonging to L_2 over X, the inner product (f, g) of two elements f and g being defined by the relation

$$(f, g) = \int_X f\bar{g}\, dP = E(f\bar{g}).$$

The stochastic process to which $x(t)$ is attached then corresponds to a "curve" in H (Kolmogoroff, [56], [57]), so that the well-known theory of Hilbert space is available for the study of the process. In particular, convergence in the usual metric of Hilbert space is equivalent to convergence in the mean of order 2 for random variables.

Let H_x be the smallest closed linear subspace of H which contains all elements of the form $a_1x(t_1) + \cdots + a_nx(t_n)$. If the *covariance function*

$$r(t, u) = (x(t), x(u)) = E(x(t)\overline{x(u)})$$

is continuous for all real values of t and u, then $x(t) \to x(t_0)$ in the mean, as $t \to t_0$, and we shall say that the process $x(t)$ is *continuous*. For any continuous process, H_x is separable. When $g(t)$ is a continuous non-random function of t, and $x(t)$ is attached to a continuous stochastic process, the Riemann-Darboux sums formally associated with the integral

$$\int_a^b g(t)x(t)\, dt$$

are easily shown to tend to a limit y, which is an element of H_x, i.e. a random variable. By definition, we may identify the integral with this variable y, and this integral will possess the essential properties of the ordinary Riemann integral (Cramér, [12]).

The application of the theory of Hilbert space to stochastic processes seems to open very interesting possibilities. Some applications to particular classes of stochastic processes will be mentioned below. Futher important results belonging to this order of ideas will be given in a work by K. Karhunen [40], which is in course of publication.

18. Relations to ergodic theory. There is a close connection between the theory of stochastic processes and ergodic theory. In ergodic theory, as summarized e.g. in the treatise of E. Hopf [38], we consider an arbitrary space Ω, and a probability measure Π, defined for all sets Σ belonging to the additive class C. We further consider a one-parameter group of one-one transformations of Ω into itself (a "flow" in Ω) such that the transformation corresponding to the parameter value t takes the point $\omega = \omega_0$ into ω_t, while $(\omega_t)_u = \omega_{t+u}$. Let $f(\omega)$ be a given function, defined throughout Ω, and such that $f(\omega_t)$ is C-measurable for every fixed t. The well-known ergodic theorems due to von Neumann, Birkhoff, Khintchine and others are then concerned with the asymptotic behaviours of mean values, which in the classical cases are of the types

$$\frac{f(\omega_0) + f(\omega_1) + \cdots + f(\omega_{n-1})}{n}$$

or

$$\frac{1}{T}\int_0^T f(\omega_t)\,dt,$$

as n or T tends to infinity. (In the case of the latter expression, it is necessary to introduce some additional condition implying measurability in t.)

Writing $x(t, \omega) = f(\omega_t)$, it is seen that to a given transformation group $\omega \to \omega_t$ and a given function $f(\omega)$, there corresponds a stochastic process in the sense of Wiener's definition (cf. 16). The space X_0 of this process consists of all functions $x(t)$ representable in the form $x(t) = f(\omega_t)$, when $\omega = \omega_0$ varies over Ω. The corresponding probability measure P_0 is defined by (16.1).

Thus any of the above-mentioned ergodic theorems may be expressed as a theorem concerning "temporal" mean values of the types

$$\frac{x(0) + x(1) + \cdots + x(n-1)}{n}$$

or

$$\frac{1}{T}\int_0^T x(t)\,dt.$$

If, according to some reasonable convergence definition, we may assign a limit to either of these expressions, as n or T tends to infinity, this limit will be a random variable, and it is important to find conditions which imply that this variable has a constant value for "almost all" functions $x(t)$, i.e. for all $x(t)$ except at most a set of P_0-measure zero.

In the particular case when $x(0)$, $x(1)$, $\cdots$ are independent variables all having the same distribution, the classical ergodic theorems yield simple cases of the laws of large numbers (cf. 10). The mean ergodic theorem of von Neumann gives the weak law, while the Birkhoff-Khintchine theorem gives the strong law. Some more general results belonging to this order of ideas will be mentioned in the sequel.

It will be seen that the two theories are largely equivalent, and it seems likely that further comparative studies of the methods will be of great value to both sides.

19. Markoff processes. Consider now a stochastic process, defined by a probability measure $P(S)$ in the space X of all real-valued functions $x(t)$ of the real variable t. For any $t_1 < t_2$, there is a certain conditional probability $P(x(t_2) \subset S \mid x(t_1) = a_1)$ of the relation $x(t_2) \subset S$, relative to the hypothesis that $x(t_1)$ assumes the given value a_1. Suppose now that this conditional probability is independent of any additional hypothesis concerning the behaviour of $x(t)$ for $t < t_1$, so that we have e.g. for any $t_0 < t_1 < t_2$ and for any a_0

$$P(x(t_2) \subset S \mid x(t_1) = a_1) = P(x(t_2) \subset S \mid x(t_1) = a_1, x(t_0) = a_0).$$

In this case the process is called a *Markoff process.*

The general theory of this type of processes, which forms a natural generalization of the classical concept of Markoff chains, has been studied in basic works by Kolmogoroff [53] and Feller [26], [28]. Writing

$$P(x(t) \leqq \xi \mid x(t_0) = a_0) = F(\xi;\ t,\ a_0,\ t_0),$$

where $t_0 < t$, F will be the distribution function of the random variable $x(t)$, relative to the hypothesis $x(t_0) = a_0$. Then F satisfies the Chapman-Kolmogoroff equation

$$F(\xi; t, a_0, t_0) = \int_{-\infty}^{\infty} F(\xi; t, \eta, t_1)\, d_\eta F(\eta; t_1, a_0, t_0), \tag{19.1}$$

which expresses that, starting from the state $x(t_0) = a_0$, the state $x(t) \leqq \xi$ must be reached by passing through some intermediate state $x(t_1) = \eta$, where $t_0 < t_1 < t$. Subject to certain general conditions, it is possible to show that any solution of this equation satisfies certain integro-differential equations, which in some important cases reduce to partial differential equations of parabolic type, and that the d.f. F is uniquely determined by these equations. However, the general conditions mentioned above are in many cases difficult to apply to particular classes of processes, and it would be important to have further investigations concerning these questions.

Markoff processes (not belonging to the subclass of differential processes, which will be considered in the following paragraph) appear in several important applications, e.g. in the theory of cosmic radiation, in certain genetical problems, in the theory of insurance risk etc. In these cases, we are often concerned with the class of *purely discontinuous* Markoff processes, where the function $x(t)$ only changes its value by jumps. If, in addition, there are only a finite or enumerable set of possible values for $x(t)$, the Chapman-Kolmogoroff equation (19.1) reduces to

$$\pi_{ik}(t_0, t) = \sum_j \pi_{ij}(t_0, t_1)\pi_{jk}(t_1, t), \tag{19.2}$$

where $\pi_{ik}(t_0, t)$ denotes the "transition probability", i.e. the probability that $x(t)$ will be in the kth state at the time t, when it is known to have been in the ith state at the time t_0. In matrix form, this equation may be written

$$\Pi(t_0, t) = \Pi(t_0, t_1)\Pi(t_1, t), \tag{19.3}$$

where Π denotes the matrix of the π_{ik}.

When only a sequence of discrete values of t are considered, we have here the classical case of Markoff chains, which has received a detailed treatment in the well-known book by Fréchet [32] (cf. also Doob, [19]). The case when t is a continuous variable has been treated by Feller [28], O. Lundberg [71], Arley [1], and other authors. Some of the most important problems of this branch of the subject are concerned with the existence of a unique system of solutions of (19.2) or (19.3), and with the asymptotic behaviour of the solutions for large values of $t - t_0$. Though important results have been reached, there still remains much to be done here, and the same thing holds a fortiori with respect to the analogous problems for general Markoff processes.

20. Differential processes. A particularly interesting case of a Markoff process arises when, for any $\Delta t > 0$, the increment $\Delta x(t) = x(t + \Delta t) - x(t)$ is independent of $x(\tau)$ for $\tau \leqq t$. The process is then called a *differential process.* Some of the earliest studied stochastic processes belong to this class, which contains in particular the two examples discussed above in 15. Further cases of such processes arise e.g. in the theory of radioactive disintegration and in telephone technique.

Let us suppose that $x(0)$ is identically equal to zero, and that the process is uniformly continuous in probability in every finite interval $0 \leqq t \leqq T$, i.e. that for any fixed positive ϵ

$$P(\mid x(t + \Delta t) - x(t) \mid > \epsilon) \to 0$$

as $\Delta t \to 0$, uniformly for $0 \leqq t \leqq T$. Then it follows from the works of Lévy, [60], [63], Khintchine [47] and Kolmogoroff [54] that, for any $t > 0$, the random variable $x(t)$ has an infinitely divisible distribution, with a characteristic function $\varphi(z; t)$ given by (9.2), where β, γ, $M(u)$ and $N(u)$ may depend on t.

In the particularly important case when the distribution of the increment $x(t + \Delta t) = x(t)$ does not involve t, but only depends on the length Δt of the interval, we say that the process is *temporally homogeneous*, and in this case we have

$$\log \varphi(z; t) = t \log \varphi(z; 1),$$

so that we obtain the general formula for $\varphi(z; t)$ simply by replacing in (9.2) β, γ, $M(u)$ and $N(u)$ by $t\beta$, $t\gamma$, $tM(u)$ and $tN(u)$ respectively.

When $t \to \infty$, or $t \to 0$, the appropriately normalized distribution of $x(t)$ tends, under certain conditions, to a stable distribution (Cramér [7], Gnedenko [36]). When this limiting distribution is normal, there are sometimes even asymptotic expansions analogous to (11.3). Still, the problem of the asymptotic behaviour of the distribution for large t does not seem to be definitely cleared up.

Khintchine [41] and Gnedenko [37] have given interesting generalizations of the law of the iterated logarithm (cf. 12) to processes of the type considered here.

The continuous process discussed in 15 in connection with the Brownian movement corresponds to the temporally homogeneous case when β, $M(u)$ and $N(u)$ all reduce to zero, so that

$$\varphi(z) = e^{-\gamma t z^2},$$

which shows that the distribution of $x(t)$ is normal, with mean zero and variance $2\gamma t$.

On the other hand, in the applications to the theory of insurance risk, γ is zero, while $M(u)$ and $N(u)$ are connected with the distribution of the various magnitudes of claims. In this type of applications, it is often very important to find the probability that $x(t)$ satisfies an inequality of the form

$$x(t) < a + bt$$

for all values of t. It follows from the discussion in 16 that the definition of a probability of this type is somewhat delicate. The problem, which can be regarded as an extended form of the classical problem of "the gambler's ruin," has been solved in certain particular cases. It leads to integral equations, which in the simplest case are of the Volterra, in other cases of the Wiener-Hopf type (Cramér [6], [13], Segerdahl [79], Täcklind [81]).

21. Orthogonal processes. Consider now the case of a complex-valued $x(t)$, and suppose that $E\,|\,x(t)\,|^2$ is finite for all t. Without restricting the generality, we may assume that $Ex(t) = 0$ for all t.

Suppose now that instead of requiring, as in the case of a differential process, that the variables $x(\tau)$ and $\Delta x(t)$ should be *independent* when $\tau \leqq t$, we only lay down the less stringent condition that these variables should be *non-correlated*, i.e. that

$$E(x(\tau)\overline{\Delta x(t)}) = 0.$$

We then obtain a process which is no longer necessarily of the Markoff type. The condition implies that, for any two disjoint intervals (t_1, t_2) and (t_3, t_4), we have

$$E[(x(t_2) - x(t_1))(\overline{x(t_4)} - \overline{x(t_3)})] = 0,$$

so that the "chords" corresponding to two disjoint "arcs" of the curve in Hilbert space representing the process are always orthogonal (Kolmogoroff [56], [57]). A process of this type may accordingly be called an *orthogonal process*.

For a process of this type we have, writing $E\,|\,x(t)\,|^2 = F(t)$, $F(t + \Delta t) - F(t) = E\,|\,x(t + \Delta t) - x(t)\,|^2$, so that $F(t)$ is a never decreasing function of t. If $F(t)$ is bounded for all t, we shall say that the orthogonal process is bounded. For a bounded orthogonal process, the Stieltjes integral

$$\int_{-\infty}^{\infty} g(t)\, dx(t),$$

where $g(t)$ is bounded and continuous, may be defined as the limit in the mean of sums of the form

$$\sum_{\nu} g(t_\nu)(x(t_\nu) - x(t_{\nu-1})).$$

22. Stationary processes. When we are concerned with a process representing the temporal development of a system governed by laws which are invariant under a translation in time, it seems natural to assume that the joint distribution of any group of variables of the form

$$x(t_1 + \tau), \cdots, x(t_n + \tau) \tag{22.1}$$

is independent of τ. A process satisfying this condition will be called a *stationary* process. If a stochastic process is defined by means of a "flow" $\omega \to \omega_t$ in a space Ω (cf. 18), the process will be stationary when and only when the corresponding flow is *measure-preserving*, i.e. if the transformation $\omega \to \omega_t$ changes any C-measurable set S into a set S_t of the same measure.

Under appropriate conditions with respect to the measurability of $x(t)$, the Birkhoff-Khintchine ergodic theorem holds for a stationary process, i.e. there exists a random variable y such that we have

$$P_0\left(\lim_{T\to\infty} \frac{1}{T}\int_0^T x(t)\,dt = y\right) = 1, \tag{22.2}$$

where P_0 is the probability measure in a suitably restricted space in the sense of Doob. Further work seems to be required here, in order to make the situation quite clear, also with regard to metric transitivity.

For a stationary process, any finite moment of the joint distribution of the variables (22.1) is obviously independent of τ. Suppose now that we only require that this invariance under translations in time should hold for moments of the first and second order of the joint distributions, which are assumed to be finite. The wider class of processes obtained in this way may be called *stationary of the second order*. Processes of this type have been studied for the first time by Khintchine [42]. We shall assume that $x(t)$ is complex-valued. Without restricting the generality, we may further assume that $Ex(t) = 0$ for all t. The product moment $E(x(t)\overline{x(u)})$ will then be a function of the difference $t - u$:

$$E(x(t)\overline{x(u)}) = R(t - u). \tag{22.3}$$

Assuming, in addition, that $R(t)$ is continuous at $t = 0$, it follows that $R(t)$ is continuous for all t, and the process is continuous in the sense of 17. It was shown by Khintchine that a NS condition that a given function $R(t)$ should be associated with a second order stationary and continuous process by means of the relation (22.3) is that we should have

$$R(t) = \int_{-\infty}^{\infty} e^{itx}\,dF(x) \tag{22.4}$$

for all t, where the spectral function $F(x)$ is real, never decreasing and bounded. In particular, we have

$$F(+\infty) - F(-\infty) = R(0) = E \mid x(t) \mid^2 = \sigma^2.$$

Khintchine's condition for $R(t)$ was generalized by Cramér to the case of an arbitrary number of processes $x_1(t), \cdots, x_n(t)$, such that the product moments $E(x_i(t)\overline{x_j(u)})$ are functions of the difference $t - u$. The corresponding spectral functions $F_{ij}(x)$ are in general complex-valued and of bounded variation. Further, the expression (Cramér, [12])

$$\sum_{i,j=1}^{n} z_i \bar{z}_j \Delta F_{ij},$$

where $\Delta F_{ij} = F_{ij}(b) - F_{ij}(a)$ is, for any $a < b$, a non-negative Hermite form in the variables z_i. This result is closely connected with a theorem on Hilbert space considered by Kolmogoroff and Julia. It is further shown that, to any given functions $F_{ij}(x)$, $(i, j = 1, \cdots, n)$, satisfying these conditions, we can always find n processes $x_1(t), \cdots, x_n(t)$ such that the joint distribution of any set of variables $x_i(t_j)$ is always *normal*, while the covariance functions $R_{ij}(t - u) = E(x_i(t)\overline{x_j(u)})$ are given by the expression

$$R_{ij}(t) = \int_{-\infty}^{\infty} e^{itx}\, dF_{ij}(x).$$

For a process $x(t)$ which is continuous and stationary of the second order, with $Ex(t) = 0$ for all t, we have the mean ergodic theorem

$$\underset{T\to\infty}{\text{l.i.m.}} \frac{1}{2T} \int_{-T}^{T} e^{-\lambda i t} x(t)\, dt = y \tag{22.5}$$

for any real λ. The random variable y has the mean 0 and the variance $F(\lambda + 0) - F(\lambda - 0)$, where F is the spectral function appearing in (22.4). If λ is a point of continuity for F, it thus follows that $y = 0$ with a probability equal to 1. On the other hand, if λ is a discontinuity, y has a positive variance. Let $\lambda_1, \lambda_2, \cdots$ be all the discontinuities of $F(x)$, and let $\sigma_1^2, \sigma_2^2, \cdots$ be the corresponding saltuses, while $y_1, y_2, \cdots$ are the limits in the mean obtained from (22.5) for $\lambda = \lambda_1, \lambda_2, \cdots$. Then two different y_j are always orthogonal: $E(y_j\bar{y}_k) = 0$ for $j \neq k$, and we have

$$x(t) = \sum_{\nu} y_\nu e^{\lambda_\nu i t} + \xi(t), \tag{22.6}$$

where $E\xi(t) = 0$ and

$$E \mid \xi(t) \mid^2 = \sigma^2 - \sum_{\nu} \sigma_\nu^2.$$

If $F(x)$ is a step-function, we have $\sigma^2 = \sum_\nu \sigma_\nu^2$, and it follows that $\xi(t) = 0$ with a probability equal to 1, so that (22.6) gives a "stochastic Fourier expansion" of $x(t)$ (Slutsky, [80]).

Even when $F(x)$ is arbitrary, we can obtain a spectral representation of $x(t)$ generalizing (22.6). In fact, it can be shown (Cramér, [14]) that $x(t)$ can always be represented by a Fourier-Stieltjes integral

$$x(t) = \int_{-\infty}^{\infty} e^{itu}\, dz(u), \tag{22.7}$$

where $z(u)$ is a random function attached to a bounded orthogonal process (cf. 21), such that

$$E\,|\,z(u + \Delta u) - z(u)\,|^2 = F(u + \Delta u) - F(u).$$

Conversely, we have

$$z(u + \Delta u) - z(u) = -\int_{-\infty}^{\infty} \frac{e^{-it(u+\Delta u)} - e^{-itu}}{2\pi i t}\, x(t)\, dt, \tag{22.8}$$

so that there is a one-one correspondence between $x(t)$ and $\Delta z(u)$. The integrals (22.7) and (22.8) are defined as limits in the mean, as shown above in 17 and 21. These results are in close correspondence with generalized harmonic analysis for an arbitrary function, as developed by Wiener [83] and Bochner [4]. The spectral representation of a stochastic process has important applications, some of which will be considered in a forthcoming paper by Karhunen [40]. An extension of the spectral representation to a more general class of processes has been given by Loève [68].

When, in particular, the $x(t)$ process is such that the joint distribution of any group of variables $x(t_1), \cdots, x(t_n)$ is normal, it follows that any increment $\Delta z(u)$ is normally distributed. Since two uncorrelated normally distributed variables are always independent, it follows that in this case the $z(u)$ process is a differential process with normally distributed increments. Important results for this case have recently been given by Doob [22].

The properties of continuity, differentiability etc. for processes of the type here considered are still incompletely known, and further work is required. A further group of important unsolved problems are connected with an interesting decomposition theorem by Wold [84], which holds for processes with a discrete time variable. The generalization of this theorem to the continuous case does not seem to have so far been given in a final form.

REFERENCES

[1] N. ARLEY, "On the theory of stochastic processes and their applications to the theory of cosmic radiation," Thesis, Copenhagen, 1943.

[2] G. M. BAWLY, "Ueber eine Verallgemeinerung der Grenzwertsätze der Wahrscheinlichkeitsrechnung," *Rec. Math.* (*Mat. Sbornik*), *N. S.*, Vol. 1 (1936), pp. 917–929.

[3] S. N. BERNSTEIN, "Sur l'extension du théorème limite du calcul des probabilités aux sommes de quantités dépendantes," *Math. Ann.*, Vol. 97 (1927), pp. 1–59.

[4] S. BOCHNER, "Monotone Funktionen, Stieltjessche Integrale und harmonische Analyse," *Math. Ann.*, Vol. 108 (1933), pp. 378–410.

[5] H. CRAMÉR, "On the composition of elementary errors," *Skand. Aktuarietidskr.*, Vol. 11 (1928), pp. 13–74, 141–180.

[6] ———, "On the mathematical theory of risk." Published by the Insurance Company Skandia, Stockholm, 1930.

[7] ———, "Sur les propriétés asymptotiques d'une classe de variables aléatoires," *C. R. Acad. Sci. Paris*, Vol. 201 (1935), pp. 441–443.

[8] ———, "Ueber eine Eigenschaft der normalen Verteilungsfunktion," *Math. Zeit.*, Vol. 41 (1936), pp. 405–414.

[9] ———, *Random variables and probability distributions*, Cambridge Tracts in Math., Cambridge, 1937.

[10] ———, "Sur un nouveau théorème—limite de la théorie des probabilités," *Actualités Scientifiques*, Paris, No. 736 (1938), pp. 5–23.

[11] ———, "On the representation of a function by certain Fourier integrals," *Trans. Amer. Math. Soc.*, Vol. 46 (1939), pp. 191–201.

[12] ———, "On the theory of stationary random processes," *Ann. of Math.*, Vol. 41 (1940), pp. 215–230.

[13] ———, "Deux conférences sur la theorie des probabilités," *Skand. Aktuarietidskr.*, 1941, pp. 34–69.

[14] ———, "On harmonic analysis in certain functional spaces," *Ark. Mat. Astr. Fys.*, Vol. 28B (1942), pp. 1–7.

[15] ———, *Mathematical Methods of Statistics.* Princeton Univ. Press, Princeton, 1946.

[16] W. Doeblin, "Premiers éléments d'une étude systématique de l'ensemble de puissances d'une loi de probabilités," *C. R. Acad. Sci. Paris*, Vol. 206 (1938), pp. 306–308.

[17] ———, "Sur un théorème du calcul des probabilités," *C. R. Acad. Sci. Paris*, Vol. 209 (1939), pp. 742–743.

[18] J. L. Doob, "Stochastic processes depending on a continuous parameter," *Trans. Amer. Math. Soc.*, Vol. 42 (1937), pp. 107–140.

[19] ———, "Stochastic processes with an integral-valued parameter," *Trans. Amer. Math. Soc.*, Vol. 44 (1938), pp. 87–150.

[20] ———, "Regularity properties of certain families of chance variables," *Trans. Amer. Math. Soc.*, Vol. 47 (1940), pp. 455–486.

[21] ———, "The law of large numbers for continuous stochastic processes," *Duke Math. Jour.*, Vol. 6 (1940), pp. 290–306.

[22] ———, "The elementary Gaussian processes," *Annals of Math. Stat.*, Vol. 15 (1944), pp. 229–282.

[23] J. L. Doob and M. Ambrose, "On the two formulations of the theory of stochastic processes depending upon a continuous parameter," *Ann. of Math.*, Vol. 41 (1940), pp. 737–745.

[24] P. Erdös, "On the law of the iterated logarithm," *Ann. of Math.*, Vol. 43 (1942), pp. 419–436.

[25] W. Feller, "Ueber den zentralen Grenzwertsatz der Wahrscheinlichkeitsrechnung," *Math. Zeit.*, Vol. 40 (1935), pp. 521–559.

[26] ———, "Zur Theorie der stochastischen Processe (Existenz- und Eindeutigkeitssatze)," *Math. Ann.*, Vol. 113 (1936), pp. 113–160.

[27] ———, "Ueber das Gesetz der Grossen Zahlen," *Acta. Univ. Szeged.*, Vol. 8 (1937), pp. 191–201.

[28] ———, "On the integro-differential equations of purely discontinuous Markoff processes," *Trans. Amer. Math. Soc.*, Vol. 48 (1940), pp. 488–575.

[29] ———, "Generalization of a probability limit theorem of Cramér," *Trans. Amer. Math. Soc.*, Vol. 54 (1943), pp. 361–372.

[30] ———, "The general form of the so-called law of the iterated logarithm," *Trans. Amer. Math. Soc.*, Vol. 54 (1943), pp. 373–402.

[31] ———, "The fundamental limit theorems in probability," *Bull. Amer. Math. Soc.*, Vol. 51 (1945), pp. 800–832.

[32] M. Fréchet, *Recherches théoriques modernes sur la théorie des probabilités*, Vol. 2, Paris, 1937.

[33] B. V. Gnedenko, "Sur les fonctions caractéristiques," *Bull. Math. Univ. Moscou*, Vol. 1, (1937), pp. 16–17.

[34] ———, "On the theory of the domains of attraction of stable laws," *Učenye Zpiski, Moskovskoga gosudaistvenogo Univaziteta*, Vol. 30 (1939), pp. 61–81.

[35] ———, "On the theory of limit theorems for sums of independent random variables," *Bull. Acad. Sci. URSS*, Vol. 30 (1939), pp. 181–232, 643–647.

[36] ———, "On locally stable probability distributions," *C. R. Acad. Sci. URSS*, Vol. 35 (1942), pp. 263–266.

[37] ———, "Investigation on the growth of homogeneous random processes," *C. R. Acad. Sci. URSS*, Vol. 36 (1942), pp. 3–41.

[38] E. Hopf, *Ergodentheorie*, Ergebnisse der Mathematik, Vol. 5, No. 2, Berlin, 1937.

[39] P. L. Hsu, "The approximate distribution of the mean and of the variance of independent variates," *Annals of Math. Stat.*, Vol. 16 (1945), pp. 1–29.

[40] K. Karhunen, Paper on stochastic processes, to appear in the *Acta Soc. Sci. Fennicae.*

[41] A. Khintchine. *Asymptotische Gesetze der Wahrscheinlichkeitsrechnung*, Ergebnisse der Mathematik, Vol. 2, No. 4, Berlin, 1933.

[42] ———, "Korrelationstheorie der stationären stochastischen Prozesse," *Math. Ann.*, Vol. 109 (1934), pp. 604–615.

[43] ———, "Sul dominio di attrazione della legge di Gauss," *Giorn. Ist. Ital. Attuari*, Vol. 6 (1935), pp. 378–393.

[44] ———, "Su una legge dei grandi numeri generalizzata," *Giorn. Ist. Ital. Attuari*, Vol. 7 (1936), pp. 365–377.

[45] ———, "Zur Kennzeichnung der characteristischen Funktionen," *Bull. Math. Univ. Moscou*, Vol. 1 (1937), pp. 1–31

[46] ———, "Contribution à l'arithmetique des lois de distribution," *Bull. Math. Univ. Moscou*, Vol. 1 (1937), pp. 6–17.

[47] ———, "Zur Theorie der unbeschränkt teilbaren Verteilungsgesetze," *Rec. Math. N. S.*, Vol. 2 (1937), pp. 79–117.

[48] A. Khintchine and A. Kolmogoroff, "Ueber Konvergenz von Reihen, deren Glieder durch den Zufall bestimmt werden," *Rec. Math.*, Vol. 32 (1925), pp. 668–677.

[49] A. Khintchine and P. Lévy, "Sur les lois stables," *C. R. Acad. Sci. Paris*, Vol. 202 (1936), pp. 374–376.

[50] A. Kolmogoroff, "Ueber die Summen durch den Zufall bestimmter unabhängiger Grössen," *Math. Ann.*, Vol. 99 (1928) and Vol. 102 (1929), pp. 484–489.

[51] ———, "Ueber das Gesetz des iterierten Logarithmus," *Math. Ann.*, Vol. 101 (1929), pp. 126–135.

[52] ———, "Sur la loi forte des grands nombres," *C. R. Acad. Sci. Paris*, Vol. 191 (1930), pp. 910–911.

[53] ———, "Ueber die analytischen Methoden der Wahrscheinlichkeitsrechnung," *Math. Ann.*, Vol. 104 (1931), pp. 415–458.

[54] ———, "Sulla forma generale di un processo stocastico omogeneo," *Atti. Acad. naz. Lincei, Rend.*, Vol. 15 (1932), pp. 805–828, 866–869.

[55] ———, *Grundbegriffe der Wahrscheinlichkeitsrechnung*, Ergebnisse der Mathematik, Vol. 2, No. 3, Berlin, 1933.

[56] ———, "Wiener Spiralen und einige andere interessante Kurven im Hilbertschen Raum," *C. R. Acad. Sci. URSS*, Vol. 26 (1940), pp. 115–118.

[57] ———, "Kurven im Hilbertschen Raum, die gegenüber einer ein parametrigen Gruppe von Bewegungen invariant sind," *C. R. Acad. Sci. URSS*, Vol. 26 (1940), pp. 6–9.

[58] P. Lévy, *Calcul des probabilités*, Paris, 1925.

[59] ———, "Sur les séries dont les termes sont des variables éventuelles indépendantes," *Studia Math.*, Vol. 3 (1931), pp. 117–155.

[60] ———, "Sur les intégrales dont les éléments sont des variables aléatoires indépendantes," *Ann. Scuola norm. super. Pisa*, (*2*), Vol. 3 (1934), pp. 337–366.
[61] ———, "Propriétés asymptotiques des sommes de variables aléatoires indépendantes," *Ann. Scuola norm. super. Pisa* (*2*), Vol. 3 (1934), pp. 347–402.
[62] ———, "La loi forte des grands nombres pour les variables aléatoires enchainées," *Jour. Math. Pures Appl.*, Vol. 15 (1936), pp. 11–24.
[63] ———, *Théorie de l'addition des variables aléatoires*, Paris, 1937.
[64] ———, "L'arithmetique des lois de probabilités," *Jour. Math. Pures Appl.*, Vol. 17 (1938), pp. 17–39.
[65] A. M. Liapounoff, "Nouvelle forme du théorème sur la limite de la probabilité," *Mémoires Acad. Saint-Petersbourg s. 8*, Vol. 12 (1901).
[66] J. W. Lindeberg, "Eine neue Herleitung des Exponentialgesetzes in der Wahrscheinlichkeitsrechnung," *Math. Zeft.*, Vol. 15 (1922), pp. 211–225.
[67] W. Loéve, "Étude asymptotique des sommes de variables aléatoires liées," *Jour. Math. Pures Appl.*, Vol. (9) 24 (1945), pp. 249–318.
[68] ———, "Analyse harmonique générale d'une fonction aléatoire," *C. R. Acad. Sci Paris*, Vol. 220 (1945), pp. 380–382.
[69] F. Lundberg, "Zur Theorie der Rückversicherung," *Verhandlungen Kongr. f. Versicherungsmathematik*, Wien, 1909.
[70] ———, *Försäkringsteknisk riskutjämning*, Stockholm, 1927.
[71] O. Lundberg, "On random processes and their application to sickness and accident statistics," Thesis, Stockholm, 1940.
[72] J. Marcinkiewicz, "Sur une propriété de la loi de Gauss," *Math. Zeit.*, Vol. 4 (1938), pp. 612–618.
[73] J. Marcinkiewicz and A. Zygmund, "Sur les fonctions indépendantes," *Fund. Math.*, Vol. 29 (1937), pp. 60–90.
[74] R. E. A. C. Paley and N. Wiener, *Fourier Transforms in the Complex Domain*, Amer. Math. Soc. Colloquium Publ., Vol. 19, New York, 1934.
[75] D. Raikov, "On the composition of Poisson laws," *C. R. Acad. Sci. URSS*, Vol. 14 (1937), pp. 9–11.
[76] ———, "On the decomposition of Gauss and Poisson laws," *Bull. Acad. Sci. URSS, Ser. Math.*, 1938, pp. 91–120.
[77] ———, "On the composition of analytic distribution functions," *C. R. Acad. Sci. URSS*, Vol. 23 (1939), pp. 511–514.
[78] I. J. Schoenberg, "Metric spaces and positive definite functions," *Trans. Amer. Math. Soc.*, Vol. 44 (1938), pp. 522–536.
[79] C. O. Segerdahl, "On homogeneous random processes and collective risk theory," Thesis, Stockholm, 1939.
[80] E. Slutsky, "Sur les fonctions aléatoires presque périodiques et sur la décomposition des fonctions aléatoires stationaires en composantes," *Actualités Scientifiques*, No. 738, Paris, 1938, pp. 33–55.
[81] S. Täcklind, "Sur le risque de ruine dans des jeux inéquitables," *Sknad. Aktuarietidskr.*, 1942, pp. 1–42.
[82] N. Wiener, "Differential space," *Jour. Math. Phys. M. I. T.*, Vol. 2 (1923), pp. 131–172.
[83] ———, "Generalized harmonic analysis," *Acta Math.*, Vol. 55 (1930), pp. 117–258.
[84] H. Wold, "A study in the analysis of stationary time series," Thesis, Stockholm, 1938.

44.

On the factorization of certain probability distributions

Ark. Mat. 1 (7), 61–65 (1949)

1. The object of this Note is to show that a large class of probability distributions of the type known as infinitely divisible have divisors which are not infinitely divisible. Among these distributions are, in particular, the Pearson Type III distribution and all non-normal stable distributions.

In the first two paragraphs, we briefly recall some known definitions and results. The main theorem will then be stated and proved in paragraph 3.

If ξ is a random variable, the probability of the relation $\xi \leqq x$ is a function $F(x)$ of the real variable x. The function $F(x)$, which determines the *probability distribution* of the variable ξ, is known as the *distribution function* (d. f.) of ξ. In order that a given function $F(x)$ should be the d. f. of some random variable, it is necessary and sufficient that $F(x)$ should be never decreasing and everywhere continuous to the right, and such that $F(-\infty) = 0$ and $F(+\infty) = 1$. The function

$$\varphi(t) = \int_{-\infty}^{\infty} e^{itx}\, d\, F(x)$$

is called the *characteristic function* (c. f.) of the distribution. To every d. f. $F(x)$, there is one and only one c. f. $\varphi(t)$, uniquely defined for all real t.

Suppose now that the variables ξ_1 and ξ_2 are independent in the ordinary probability sense, and consider the sum $\xi = \xi_1 + \xi_2$. Let the d. f:s of the variables ξ, ξ_1 and ξ_2 be F, F_1 and F_2, while the corresponding c. f:s are φ, φ_1 and φ_2. We then have

$$F(x) = \int_{-\infty}^{\infty} F_1(x-t)\, d\, F_2(t) = \int_{-\infty}^{\infty} F_2(x-t)\, d\, F_1(t).$$

We shall say that F is *composed* of F_1 and F_2, and use the abbreviated notation

$$F = F_1 * F_2 = F_2 * F_1.$$

To this symbolic multiplication of the d. f:s corresponds a real multiplication of the c. f:s. We have, in fact, $\varphi(t) = \varphi_1(t)\,\varphi_2(t)$. The operation of composition is commutative and associative, so that any symbolic product $F = F_1 * \cdots * F_n$ is uniquely defined and independent of the order of the factors. The symbolic product of n identical factors will be written as a symbolic nth power: $F^{[n]}$.

Thus the class of all d. f:s forms a *semigroup* with respect to the operation of composition. The *unit element* of this semigroup is the d. f.

$$E(x) = \begin{cases} 0 & \text{for} \quad x < 0, \\ 1 & \text{for} \quad x \geqq 0, \end{cases}$$

which is the d. f. of a »variable» ξ which always takes the value 0. The only divisors of the unit are the d. f:s of the form $E(x-m)$, where m is a constant. We have, in fact $E(x) = E(x-m) * E(x+m)$. The corresponding relation between the c. f:s is $1 = e^{mit} \cdot e^{-mit}$.

A *factorization* of the probability distribution defined by $F(x)$ is any representation $F = F_1 * \cdots * F_n$, where $n > 1$, and no F_r is a divisor of the unit. Each F_r is then said to be a divisor of F. When F has no factorization, it is called *indecomposable.* A simple case of an indecomposable d. f. is

$$F(x) = p\,E(x) + q\,E(x-1),$$

where p and q are positive constants such that $p + q = 1$.

2. The probability distribution defined by $F(x)$ is called *infinitely divisible,* if to every $n = 1, 2, \ldots$ we can find a d. f. G such that $F = G^{[n]}$. The properties of infinitely divisible distributions have been studied by several authors, notably by PAUL LÉVY, who has shown (cf e. g. 3) that the logarithm of the c. f. $\varphi(t)$ of any infinitely divisible distribution can be written in the form

$$\text{(1)} \qquad \log \varphi(t) = m i t - \tfrac{1}{2}\sigma^2 t^2 + \int_{-\infty}^{\infty} \left(e^{itx} - 1 - \frac{itx}{1+x^2}\right) dM(x),$$

where m and $\sigma^2 \geqq 0$ are real constants, while $M(x)$ is real and never decreasing in each of the intervals $(-\infty, 0)$ and $(0, +\infty)$, and such that $M(-\infty) = M(+\infty) = 0$, $\int_{-1}^{1} x^2\, dM(x) < \infty$. Further, any choice of m, σ and $M(x)$ consistent with these conditions yields an infinitely divisible distribution, and the representation of $\log \varphi(t)$ in the form (1) is unique.

The class of infinitely divisible distributions includes many important types of distributions occurring in various applications. Thus for $M(x) = 0$ we have the normal distribution, and for $\sigma = 0$, $M(x) = \lambda E(x-1) - \lambda E(x)$ the Poisson distribution. For $\sigma = 0$, $M(x) = 0$ for $x < 0$ and $M'(x) = \lambda x^{-1} e^{-\alpha x}$ for $x > 0$, we obtain the Pearson Type III distribution, and finally for $\sigma = 0$, $M'(x) = A\,|x|^{-\alpha-1}$ for $x < 0$ and $M'(x) = B x^{-\alpha-1}$ for $x > 0$, where $0 < \alpha < 2$, we obtain the non-normal stable distributions.

It follows from the above that the problem of finding all possible factorizations of an infinitely divisible d. f. F can be completely solved, as long as we restrict ourselves to factors which are themselves infinitely divisible. In fact, if $F = F_1 * \cdots * F_n$, where every F_r is infinitely divisible, the logarithm of the c. f. φ_r corresponding to F_r can be written in the form (1), and we must have $m = \sum m_r$, $\sigma^2 = \sum \sigma_r^2$, $M(x) = \sum M_r(x)$. Conversely, any choice of the m_r,

σ_r and $M_r(x)$ consistent with these conditions yields a factorization of F, such that all the factors are infinitely divisible.

In the two particular cases of the normal and the Poisson distributions, it is known (CRAMÉR, 1, p. 52, and RAIKOV, 5) that the factorizations obtained in this way remain the only possible ones, even if we remove the restriction that the factors F_r should be infinitely divisible. Thus any divisor of a normal distribution is itself normal, and similarly for the Poisson distribution.

On the other hand, it has been shown by examples (cf e. g. LÉVY, 3 and 4) that *there exist infinitely divisible d. f:s F such that F can be factorized into $F = F_1 * F_2$, where at least one of the factors is not infinitely divisible.* This case occurs e. g. when F is the symbolic product of a certain number of appropriately chosen Poisson distributions.

The object of this Note is to show that, in fact, this property holds for a large class of infinitely divisible distributions.

3. Consider an infinitely divisible d. f. F, and let the logarithm of the corresponding c. f. φ be represented in the form (1). The function $M(x)$, being never decreasing, has almost everywhere a finite derivative $M'(x) \geqq 0$. We now proceed to prove the following theorem:

Suppose that we can find two positive constants k and c such that $M'(x) > k$ almost everywhere in at least one of the two intervals $(-c, 0)$ and $(0, c)$. Then F has a divisor which is not infinitely divisible.

Thus in particular the Type III distribution, as well as any non-normal stable distribution, has a divisor which is not infinitely divisible.

In order to prove the theorem, we may restrict ourselves to the case when the condition is satisfied in the interval $(0, c)$. It then follows from the preceding paragraph that F has an infinitely divisible divisor G defined by taking in the expression (1) of the c. f. $m = \sigma = 0$, $M'(x) = k$ for $0 < x < c$, and $M'(x) = 0$ elsewhere. Thus we have only to show that G has a non-infinitely divisible divisor.

Denoting by $\gamma(t)$ the c. f. corresponding to G, we have according to (1)

$$\log \gamma(t) = k \int_0^c \left(e^{itx} - 1 - \frac{itx}{1+x^2}\right) dx = \log \gamma_1(t) + \log \gamma_2(t,), \tag{2}$$

where we take

$$\log \gamma_r(t) = k \int_0^c \left(e^{itx} - 1 - \frac{itx}{1+x^2}\right) a_r(x)\, dx, \qquad (r = 1, 2), \tag{3}$$

$$a_1(x) = \begin{cases} -\varepsilon \text{ for } (\tfrac{1}{2} - \varepsilon)c < x < (\tfrac{1}{2} + \varepsilon)c, \\ 1 \quad \text{elsewhere in } (0, c), \\ 0 \quad \text{outside } (0, c), \end{cases}$$

$$a_2(x) = \begin{cases} 1 + \varepsilon \text{ for } (\tfrac{1}{2} - \varepsilon)c < x < (\tfrac{1}{2} + \varepsilon)c, \\ 0 \quad \text{elsewhere.} \end{cases} \tag{4}$$

Here ε denotes a constant such that $0 < \varepsilon < \tfrac{1}{2}$.

It follows from (1) that $\gamma_2(t)$ is the c. f. of an infinitely divisible d. f. G_2. We shall now show that, if ε is sufficiently small, $\gamma_1(t)$ will be the c. f. of a non-infinitely divisible d. f. G_1. Then by (2) we have $G = G_1 * G_2$, and our theorem will be proved.

For every real x, we define the functions $\beta_1(x)$, $\beta_2(x)$, ... by writing

$$\beta_1(x) = k\,\alpha_1(x),$$

$$\beta_2(x) = \int_0^x \beta_1(x-t)\,\beta_1(t)\,dt,$$

$$\cdots\cdots\cdots\cdots$$

$$\beta_n(x) = \int_0^x \beta_{n-1}(x-t)\,\beta_1(t)\,dt.$$

$$\cdots\cdots\cdots\cdots$$

We then have $\beta_n(x) = 0$ everywhere outside $(0, nc)$, and it is well known that we have

$$\int_0^{nc} e^{itx}\beta_n(x)\,dx = k^n\left(\int_0^c e^{itx}\alpha_1(x)\,dx\right)^n.$$

Further, $|\beta_n(x)| \leqq k^n c^{n-1}$ for all x, and it follows that

$$\int_0^\infty e^{itx}\sum_1^\infty \frac{\beta_n(x)}{n!}\,dx = \sum_1^\infty \frac{1}{n!}\int_0^{nc} e^{itx}\beta_n(x)\,dx = \exp\left(k\int_0^c e^{itx}\alpha_1(x)\,dx\right) - 1. \tag{5}$$

Writing

$$\varkappa = k\int_0^c \alpha_1(x)\,dx, \qquad \lambda = k\int_0^c \frac{x}{1+x^2}\alpha_1(x)\,dx,$$

we now obtain from (3) and (5)

$$\gamma_1(t) = e^{-\varkappa-\lambda i t}\left(1 + \int_0^\infty e^{itx}\sum_1^\infty \frac{\beta_n(x)}{n!}\,dx\right) = \int_{-\infty}^\infty e^{itx}\,dG_1(x),$$

where

$$G_1(x) = e^{-\varkappa}\left[E(x+\lambda) + \int_{-\infty}^x \sum_1^\infty \frac{\beta_n(u+\lambda)}{n!}\,du\right].$$

Obviously $G_1(-\infty) = 0$, while from (3) we obtain $G_1(+\infty) = \gamma_1(0) = 1$. In order to prove that $G_1(x)$ is a d. f., we now only have to show that it is never decreasing, i. e. that $\sum_1^\infty \beta_n(x)/n! \geqq 0$ for all x.

Consider first $\beta_1(x)$, which is everywhere non-negative, except between the limits $(\frac{1}{2} \pm \varepsilon)\, c$, where it takes the value $-k\varepsilon$. Further, as $\varepsilon \to 0$, it is obvious that $\beta_2(x)$ tends to $k^2(c - |c - x|)$ uniformly in $0 < x < 2c$. Also it is easily seen that for all sufficiently small ε we have $\beta_2(x) \geqq 0$ and $\beta_3(x) \geqq 0$ for all x. It follows that we can find ε_0 such that for $\varepsilon = \varepsilon_0$ and for all x

$$\beta_1(x) + \frac{1}{2!}\beta_2(x) + \frac{1}{3!}\beta_3(x) \geqq 0,$$

$$\beta_4(x) = \int_0^x \beta_2(x-t)\,\beta_2(t)\,dt \geqq 0,$$

$$\beta_5(x) = \int_0^x \beta_3(x-t)\,\beta_2(t)\,dt \geqq 0,$$

$$\cdots\cdots\cdots\cdots\cdots$$

and hence $\sum_1^\infty \frac{\beta_n(x)}{n!} \geqq 0$. Thus G_1 is a d. f.

It now only remains to show that G_1 is not infinitely divisible. In fact, if G_1 were infinitely divisible, $\log \gamma_1(t)$ could be expressed in the form (1) with a never decreasing $M(x)$. Now since the derivative $M'(x)$ exists almost everywhere and is $\geqq 0$, and since the representation (1) is unique, it follows from (2) and (3) that we should have almost everywhere in $(0, c)$

$$k = M'(x) + k\,\alpha_2(x) \geqq k\,\alpha_2(x),$$

but this is obviously inconsistent with (4). Thus G_1 cannot be infinitely divisible, and our theorem is proved.

4. By a theorem due to KHINTCHINE (2), any d. f. which is not infinitely divisible has an indecomposable divisor. Writing the function G of the preceding paragraph in the form $G = G^{[\frac{1}{2}]} * G^{[\frac{1}{4}]} * \cdots$, and applying first the argument of the preceding paragraph and then KHINTCHINE's theorem to each factor, we obtain the following result:

Any infinitely divisible d. f. F satisfying the conditions of the preceding theorem is divisible by the product of an infinite sequence of indecomposable d. f:s.

REFERENCES: 1. **Cramér, H.**, Random variables and probability distributions. Cambridge Tracts in Mathematics, Cambridge 1937. — 2. **Khintchine, A.**, Contribution à l'arithmétique des lois de distribution. Bull. Math. Univ. Moscou. 1 (1937), p. 6. — 3. **Lévy, P.**, Théorie de l'addition des variables aléatoires. Paris 1937. — 4. ——, L'arithmétique des lois de probabilités. Journ. Math. Pures et Appl., 17 (1938), p. 17. — 5. **Raikov, D.**, On the composition of Poisson laws. C. R. Acad. Scient. URSS, 14 (1937), p. 9.

Tryckt den 19 april 1949

Uppsala 1949. Almqvist & Wiksells Boktryckeri AB

45.

A contribution to the theory of stochastic processes

Proc. Second Berkeley Symp. Math. Statist. Prob., University of California Press, Berkeley and Los Angeles, 329–339 (1951)

1. Introduction

Let ω denote a point or element of an arbitrary space Ω, where a probability measure $\Pi(\Sigma)$ is defined for every set Σ belonging to a certain additive class of sets in Ω, the Π-measurable sets. The probability distribution in Ω defined by $\Pi(\Sigma)$ will be referred to as the *probability field* (Π, Ω). The points ω will be denoted as the *elementary events* of the field, while any set Σ corresponds to an *event*, the probability of which is equal to $\Pi(\Sigma)$.

A complex valued Π-measurable function

$$x = g(\omega)$$

constitutes a *random variable*, defined on the field (Π, Ω). The *mean value* of x is defined by the relation

$$Ex = \int_{\Omega} g(\omega)\, d\Pi. \tag{1}$$

Throughout the paper, we shall always assume that, for every random variable considered, we have

$$Ex = \int_{\Omega} g(\omega)\, d\Pi = 0, \qquad E|x|^2 = \int_{\Omega} |g(\omega)|^2 d\Pi < \infty.$$

The first condition introduces some formal simplification, but does not imply any restriction of the generality of our considerations, while the second condition is essential. Two variables x and y are considered as identical, if $E|x - y|^2 = 0$.

Consider a complex valued function $x(t, \omega)$ such that, for every fixed t belonging to some specified set T, the function $x(t, \omega)$ is a Π-measurable function of ω, and thus defines a random variable $x(t)$ on the field (Π, Ω). When t ranges over T, we thus obtain a family of random variables, depending on the parameter t. On the other hand, to any fixed elementary event ω there corresponds a function

$$x(t) = x(t, \omega),$$

defined for all t belonging to T, and to any event Σ there corresponds a set of functions $x(t)$ having the probability $\Pi(\Sigma)$. The function $x(t)$ will be denoted as a *random function*, defined on the field (Π, Ω).

Throughout this paper, the set T will be assumed to be the real axis, $-\infty < t < +\infty$. However, most of our considerations may easily be extended to more general spaces.

In the majority of applications, t will represent the time, and $x(t)$ will then denote some variable quantity attached to a system, the temporal development of which is subject to random influences. With reference to this type of applications, the random function $x(t)$, or the family of random variables $x(t)$, will be said to constitute a *random* or *stochastic process*.

According to (1), the mean value

$$r(t, u) = E x(t) \overline{x(u)}$$

exists for all t and u. The function $r(t, u)$ is known as the *covariance function* of the process. We always have

$$r(u, t) = \overline{r(t, u)}, \qquad r(t, t) \geqq 0 .$$

Consider the family $L(x)$ of all random variables of the form

$$c_1 x(t_1) + c_2 x(t_2) + \ldots + c_n x(t_n), \tag{2}$$

where the c_i are complex constants. Closing the set $L(x)$ with respect to convergence in the mean, we obtain an extended set $L_2(x)$. The elements of $L_2(x)$ are random variables expressible in the form (2), or as limits in the mean of random variables of the form (2). If the inner product of two arbitrary elements y and z of $L_2(x)$ is defined by the relation

$$(y, z) = E y \bar{z},$$

it is known, Karhunen [5], Cramér [3], that $L_2(x)$ is a Hilbert space. We shall call $L_2(x)$ the *linear space* of the process. If, for any t, we have

$$\underset{t_n \to t}{\text{l.i.m.}}\ x(t_n) = x(t),$$

we shall say that the process is continuous in the mean. A necessary and sufficient condition for continuity in the mean is that the covariance function $r(t, u)$ should be continuous in every point of the line $t = u$. If this condition is satisfied, $r(t, u)$ is continuous for all t and u, and the space $L_2(x)$ is separable.

Integrals of the types

$$J_1 = \int_a^b g(t) x(t) dt, \qquad J_2 = \int_a^b g(t) dx(t),$$

where $g(t)$ denotes a nonrandom function, can be defined for example as limits in the mean of certain sequences of approximating sums formally associated with the integrals. See, for example, Doob [4], Cramér [1], Karhunen [5], Loève [6]. Both integrals are random variables with zero mean values, and we have

$$E|J_1|^2 = \int_a^b \int_a^b g(t) \overline{g(u)} r(t, u) dt du, \tag{3}$$

$$E|J_2|^2 = \int_a^b \int_a^b g(t) \overline{g(u)} d^2 r(t, u). \tag{4}$$

Suppose first that (a, b) is a finite interval. As soon as the double integral occurring in the expression for $E|J_i|^2$ exists $(i = 1, 2)$, the corresponding integral J_i

exists and possesses the ordinary formal properties of an integral. If the double integral converges when extended over the whole plane, the corresponding J_i will converge in the mean as $a \to -\infty$, $b \to +\infty$, and the limit will be defined as the integral over $(-\infty, \infty)$.

In the present paper, we shall first investigate the properties of certain *additive random set functions* attached to a class of stochastic processes. We shall then deduce a theorem concerning the representation of a stochastic process by means of integrals of the type J_2, and show that this theorem includes several previously known representations as particular cases.

2. Additive random set functions

In many types of applications, it is natural to consider the value $x(t)$ assumed by a random function at the instant t as built up additively by the successive increments of the function during the development of the process up to the instant t. In such a case we may conceive the process as an *impulse process*, the value of the impulse received during the time interval $(t, t + \Delta t)$ being $\Delta x(t) = x(t + \Delta t) - x(t)$. Thus the impulse corresponding to any time interval is a random variable associated with the interval, or a *random interval function.*

For any finite half open interval $I = (t_1, t_2)$, this interval function $X(I)$ is defined by the relation

$$X(I) = x(t_2) - x(t_1)\,. \tag{5}$$

If $I_1, I_2, \ldots, I_n$ are finite disjoint half open intervals, and $I = I_1 + \ldots + I_n$, this definition may be consistently extended by writing

$$X(I) = X(I_1) + X(I_2) + \ldots + X(I_n)\,. \tag{6}$$

It is natural to ask if the definition could be further extended, so that we could define the impulse $X(S)$ received, for example, during an arbitrary Borel set S of time points. Such an impulse function $X(S)$ should obviously possess the additive property which is a direct generalization of (6), and we should even like to extend this property in some reasonable way to the sum of an infinite sequence of disjoint sets.

It will be convenient in the first place to restrict our considerations to *bounded* sets. In this connection, we shall lay down the following definition. An *additive random set function* is a family of random variables such that:

1) For every bounded Borel set S of real numbers, $X(S)$ is a uniquely defined random variable.

2) If $S_1, S_2, \ldots$ are disjoint Borel sets, such that $S = S_1 + S_2 + \ldots$ is bounded, then $X(S) = X(S_1) + X(S_2) + \ldots$, where the series converges in the mean.

As usual we assume $EX(S) = 0$ and $E|X(S)|^2 < \infty$. Our problem is now if, given a stochastic process $x(t)$, we can find an additive random set function $X(S)$ such that, whenever S is a half open interval $I = (t_1, t_2)$, we shall have $X(S) = X(I)$ as defined by (5).

Consider first the case of a bounded set $I = I_1 + I_2 + \ldots$, where the I_n are disjoint half open intervals, $I_n = (t_n, t_n + h_n)$. Writing

$$X(I) = X(I_1) + X(I_2) + \dots, \tag{7}$$

we must then require that the series should converge in the mean. We have

$$E\left|\sum_{\mu=m}^{n} X(I_\mu)\right|^2 = \sum_{\mu,\nu=m}^{n} EX(I_\mu)\overline{X(I_\nu)} = \sum_{\mu,\nu=m}^{n} \Delta^2_{I_\mu * I_\nu} r(t, u) \tag{8}$$

where

$$\Delta^2_{I_\mu * I_\nu} r(t, u) = r(t_\mu + h_\mu, t_\nu + h_\nu) - r(t_\mu, t_\nu + h_\nu) - r(t_\mu + h_\mu, t_\nu) + r(t_\mu, t_\nu).$$

Suppose now that the covariance function $r(t, u)$ is of bounded variation in every finite domain D, in the sense that

$$\sum_{k=1}^{N} |\Delta^2_{i_k * j_k} r(t, u)| < C \tag{9}$$

for any finite sequences of half open one dimensional intervals i_k and j_k such that the $i_k * j_k$ are disjoint two dimensional intervals belonging to D.

It then follows from (8) that the series $\Sigma X(I_\mu)$ converges in the mean, so that (7) defines a random variable $X(I)$. We shall see below that this definition is unique.

Under the condition (9), the covariance function $r(t, u)$ determines a complex valued additive set function $R(W)$ defined for all bounded Borel sets W in the (t, u)-plane, and such that

$$R(i * j) = \Delta^2_{i * j} r(t, u) \tag{10}$$

for any pair of finite half open one dimensional intervals i, j. We have

$$R = R_1^{(+)} - R_1^{(-)} + i(R_2^{(+)} - R_2^{(-)}),$$

where $R_1^{(+)}, \dots$ are nonnegative additive set functions, which are finite for any bounded Borel set W. For an *unbounded* W, the functions $R_1^{(+)}, \dots$ may be infinite, and R may become indeterminate.

Let now S be a bounded one dimensional Borel set, and let $\epsilon_1, \epsilon_2, \dots$ be a decreasing sequence of positive numbers tending to zero. We can then always find a sequence of bounded one dimensional sets $I_1, I_2, \dots$, such that

$$I_n = i_{n1} + i_{n2} + \dots, \tag{11}$$

where the i_{nk} are disjoint half open intervals, while

$$I_1 \supset I_2 \supset \dots \supset S, \tag{12}$$

$$R_1^{(+)}(S * S) \leq R_1^{(+)}(I_n * I_n) < R_1^{(+)}(S * S) + \epsilon_n, \tag{13}$$

and similar relations for $R_1^{(-)}$, $R_2^{(+)}$ and $R_2^{(-)}$. Writing

$$z_n = X(I_n) = \sum_{k=1}^{\infty} x(i_{nk}),$$

we obtain by some calculation, using (10) and (11),

$$E|z_m - z_n|^2 = E\left|\sum_{k=1}^{\infty} [x(i_{mk}) - x(i_{nk})]\right|^2$$
$$= R(I_m * I_m) - R(I_m * I_n) - R(I_n * I_m) + R(I_n * I_n).$$

We have, however, according to (12) and (13),

$$|R(I_m * I_n) - R(S * S)| < \epsilon_q \sqrt{2},$$

where $q = \inf(m, n)$, and hence it follows that the sequence $\{z_n\}$ converges in the mean. A similar argument shows that the random variable l.i.m. z_n is independent of the choice of the approximating sequence $\{I_n\}$, the representation (11), and the sequence $\{\epsilon_n\}$. Thus if we write

$$X(S) = \underset{n\to\infty}{\text{l.i.m.}}\ z_n = \underset{n\to\infty}{\text{l.i.m.}} \sum_{k=1}^{\infty} x(i_{nk}),$$

the random variable $X(S)$ will be uniquely defined for all bounded Borel sets S. Obviously $X(S) = X(I)$ whenever S is a bounded half open interval I.

It follows easily from the definition that we have $EX(S) = 0$, and

$$EX(S_1)\overline{X(S_2)} = R(S_1 * S_2) = \int_{S_1}\int_{S_2} d^2 r(t, u).$$

In particular,

$$E|X(S)|^2 = R(S * S) = \int_S \int_S d^2 r(t, u).$$

Hence we immediately obtain for any bounded and disjoint S_1 and S_2

$$E|X(S_1 + S_2) - X(S_1) - X(S_2)|^2 = 0,$$

that is,

$$X(S_1 + S_2) = X(S_1) + X(S_2).$$

Suppose now $S = S_1 + S_2 + \ldots$, where S is bounded, and the S_k are disjoint. Then,

$$X(S) = X(S_1) + \ldots + X(S_{n-1}) + X(S_n + \ldots),$$

$$E|X(S) - X(S_1) - \ldots - X(S_{n-1})|^2 = E|X(S_n + \ldots)|^2$$
$$= R[(S_n + \ldots) * (S_n + \ldots)] \to 0,$$

as the limiting set of $(S_n + \ldots) * (S_n + \ldots)$ is empty. Thus

$$X(S_1 + S_2 + \ldots) = X(S_1) + X(S_2) + \ldots$$

in the sense of convergence in the mean. We have thus proved the following theorem.

If the covariance function $r(t, u)$ of the stochastic process $x(t)$ is of bounded variation in every finite domain, in the sense expressed by (9), *there exists an additive random set function $X(S)$, uniquely defined for all bounded Borel sets S, and such that $X(S) = x(t_2) - x(t_1)$ when S is a half open interval (t_1, t_2).*

Conversely, suppose that an additive random set function $X(S)$ is defined for all bounded Borel sets, and is such that $EX(S) = 0$ and

$$EX(S_1)\overline{X(S_2)} = \int_{S_1}\int_{S_2} d^2 r(t, u)$$

where $r(t, u)$ is of bounded variation in every finite domain. Writing

$$x(t) = \begin{cases} X[(t_0, t)], & t > t_0 \\ -X[(t, t_0)], & t < t_0, \end{cases} \tag{14}$$

we have $Ex(t) = 0$, and the covariance function becomes

$$Ex(t)\overline{x(u)} = r(t, u) - r(t_0, u) - r(t, t_0) + r(t_0, t_0).$$

The definition of the integral

$$J_2(S) = \int_S g(t)\,dx(t) = \int_S g(t)\,dX$$

can now be extended to any bounded Borel set S, and we find that this integral is an additive random set function such that $EJ_2(S) = 0$ and

$$EJ_2(S_1)\overline{J_2(S_2)} = \int_{S_1}\int_{S_2} g(t)\overline{g(u)}\,d^2 r(t, u).$$

We shall now apply our results to some simple particular cases. Consider first a process which is *orthogonal to its increments*, so that

$$Ex(t)\overline{[x(t+h) - x(t)]} = 0$$

for all t and all $h > 0$. Writing $E|x(t)|^2 = F(t)$, where $F(t)$ is a nonnegative and never decreasing function, the covariance function becomes

$$r(t, u) = Ex(t)\overline{x(u)} = F(\inf t, u), \tag{15}$$

and the R "distribution" is a distribution of real and positive mass on the line $t = u$, such that the segment of the line consisting of the points (τ, τ) with $\tau \leqq t$ contains the mass $F(t)$. The corresponding set function

$$X(S) = \int_S dx(t)$$

is defined for all Borel sets S having a finite upper bound, and is such that

$$EX(S_1)\overline{X(S_2)} = \int_{S_1S_2} dF(t).$$

Thus in particular for two disjoint sets S_1 and S_2, the random variables $X(S_1)$ and $X(S_2)$ are orthogonal. Random set functions of this type have been used by Karhunen [5].

Consider finally a *stationary process*, that is, a process such that the covariance function $r(t, u)$ is a function of the difference $t - u$:

$$E x(t) \overline{x(u)} = r(t, u) = r(t-u).$$

It is then well known that we have

$$r(t) = \int_{-\infty}^{\infty} e^{it\lambda} dF(\lambda),$$

where $F(\lambda)$ is real, never decreasing and bounded. Let us suppose

$$\int_{-\infty}^{\infty} \lambda^2 dF(\lambda) < \infty.$$

Then,

$$R(W) = \iint_W d^2 r(t-u) = -\iint_W r''(t-u)\, dt du$$

$$= \iint_W dt du \int_{-\infty}^{\infty} \lambda^2 e^{i(t-u)\lambda} dF(\lambda).$$

Thus the "density" of the R "distribution" at the point (t, u) is

$$-r''(t-u) = \int_{-\infty}^{\infty} \lambda^2 e^{i(t-u)\lambda} dF(\lambda),$$

which is constant on every line $t - u = \text{const}$. Let, for example, S denote the rectangle bounded by the lines $t + u = a$, $t + u = b$ and $t - u = \pm h$, then it will be seen that the "mass" in this rectangle is equal to

$$(b-a) \int_{-\infty}^{\infty} \lambda \sin h\lambda\, dF(\lambda).$$

On the other hand, the infinite rectangle $t \leqq t_0$, $u \leqq t_0$ carries the constant "mass"

$$r(t, t) = r(0) = \int_{-\infty}^{\infty} dF(\lambda).$$

The corresponding random set function $X(S)$ is such that

$$EX(S_1) \overline{X(S_2)} = -\int_{S_1}\int_{S_2} r''(t-u)\, dt du.$$

3. The linear space of an additive random set function

Suppose that we are given an additive random set function $Z(S)$, defined for all bounded Borel sets S, and such that

$$EZ(S_1) \overline{Z(S_2)} = \int_{S_1}\int_{S_2} d^2 \rho(\lambda, \mu)$$

where $\rho(\lambda, \mu)$ is of bounded variation in every finite domain. Denote by $L(Z)$ the set of all random variables of the form $c_1 Z(S_1) + \ldots + c_n Z(S_n)$, where the S_k are bounded Borel sets, and the c_k are complex constants, and let $L_2(Z)$ denote the closure of $L(Z)$ with respect to convergence in the mean. We define the inner product of two arbitrary elements z and u of $L_2(Z)$ by the relation $(z, u) = Ez\bar{u}$, and denote $L_2(Z)$ as the *linear space* of the set function $Z(S)$. Like $L_2(x)$ in section 1, $L_2(Z)$ is a Hilbert space.

For any $g(\lambda)$ such that

$$\int_{-\infty}^{\infty}\int_{-\infty}^{\infty} g(\lambda)\overline{g(\mu)}\,d^2\rho(\lambda,\mu) = \lim_{\substack{a_1,a_2\to-\infty\\ b_1,b_2\to\infty}} \int_{a_1}^{b_1}\int_{a_2}^{b_2} g(\lambda)\overline{g(\mu)}\,d^2\rho(\lambda,\mu) \tag{16}$$

is finite, the integral

$$z = \int_{-\infty}^{\infty} g(\lambda)\,dZ \tag{17}$$

will be a well defined element in $L_2(Z)$, and we shall have

$$E z_1 \overline{z_2} = \int_{-\infty}^{\infty}\int_{-\infty}^{\infty} g_1(\lambda)\overline{g_2(\mu)}\,d^2\rho(\lambda,\mu). \tag{18}$$

Consider now the set $\Lambda_2(\rho)$ of all measurable complex valued functions $g(\lambda)$, defined for $-\infty < \lambda < \infty$, and such that the integral (16) exists. Two functions g_1 and g_2 will be considered as identical, if

$$\int_{-\infty}^{\infty}\int_{-\infty}^{\infty} [g_1(\lambda) - g_2(\lambda)][\overline{g_1(\mu)} - \overline{g_2(\mu)}]\,d^2\rho(\lambda,\mu) = 0.$$

If we define the inner product of two elements in $\Lambda_2(\rho)$ by the integral in the second member of (18), the set $\Lambda_2(\rho)$ will have all the properties of Hilbert space, except possibly the completeness property. In fact, it is obvious that (g_1, g_2) has the ordinary bilinear and Hermite symmetric properties, and further

$$(g, g) = \int_{-\infty}^{\infty}\int_{-\infty}^{\infty} g(\lambda)\overline{g(\mu)}\,d^2\rho(\lambda,\mu) = E\left|\int_{-\infty}^{\infty} g(\lambda)\,dZ\right|^2 \geqq 0,$$

and $(g, g) = 0$ only when

$$\int_{-\infty}^{\infty}\int_{-\infty}^{\infty} g(\lambda)\overline{g(\mu)}\,d^2\rho(\lambda,\mu) = 0,$$

that is, when $g(\lambda)$ is identical with 0. It follows that we have the Schwarz inequality,

$$|(g_1, g_2)|^2 \leqq (g_1, g_1)(g_2, g_2).$$

If the space $\Lambda_2(\rho)$ is not complete, it can be made complete by adjunction, von Sz. Nagy [7, p. 4], and we shall suppose that this has been done, so that $\Lambda_2(\rho)$ is a Hilbert space.

To any function $g(\lambda)$ in $\Lambda_2(\rho)$ there corresponds by (17) a uniquely defined element z in $L_2(Z)$, and it follows from (18) that we have $(z_1, z_2) = (g_1, g_2)$ and hence, defining the norms in the usual way,

$$\|z_1 - z_2\| = \|g_1 - g_2\|. \tag{19}$$

An "ideal" element in $\Lambda_2(\rho)$ is a sequence of functions $g_1, g_2, \ldots$ convergent in norm, and according to (19) the sequence $z_1, z_2, \ldots$ of the corresponding elements in $L_2(Z)$ converges in the mean, and thus defines a unique limiting element z in $L_2(Z)$. Consequently to every element in $\Lambda_2(\rho)$ there corresponds one element in $L_2(Z)$, uniquely defined by (17). The transformation is obviously linear and, according to (19), isometric.

We shall show that, conversely, to every element in $L_2(Z)$, there corresponds one and only one element in $\Lambda_2(\rho)$, so that (17) defines a one to one linear and isometric correspondence between the two spaces. Suppose first that $z = c_1 Z(S_1) + \ldots + c_n Z(S_n)$ is an element of $L(Z)$. Then obviously the unique corresponding element in $\Lambda_2(\rho)$ is the function $c_1 e_{S_1}(\lambda) + \ldots + c_n e_{S_n}(\lambda)$, where $e_S(\lambda)$ denotes the characteristic function of the set S. Now any other element of $L_2(Z)$ is the limit in the mean of a sequence $z_1, z_2, \ldots$ of elements in $L(Z)$, and according to (19) the corresponding elements $g_1, g_2, \ldots$ in $\Lambda_2(\rho)$ will converge in norm, and will thus define an element of $\Lambda_2(\rho)$ such that (19) holds. It finally follows from (19) that this element is unique.

Obviously the elements $Z(S)$ form a base of the space $L_2(Z)$, and similarly the elements $e_S(\lambda)$ form a base of the space $\Lambda_2(\rho)$. It follows that whenever we are concerned with a one to one linear correspondence between the two spaces, such that $Z(S)$ and $e_S(\lambda)$ are always corresponding elements, this will coincide with the transformation defined by (17). We shall use this remark in the following paragraph.

4. Integral representation of a stochastic process

Consider now a function $g(t, \lambda)$ such that, for every fixed real t, $g(t, \lambda)$ belongs to $\Lambda_2(\rho)$. The integral

$$x(t) = \int_{-\infty}^{\infty} g(t, \lambda)\, dZ \tag{20}$$

is then defined for every real t, and we have

$$r(t, u) = Ex(t)\overline{x(u)} = \int_{-\infty}^{\infty}\int_{-\infty}^{\infty} g(t, \lambda)\overline{g(u, \mu)}\, d^2\rho(\lambda, \mu). \tag{21}$$

Conversely, if we know that the covariance function of a given stochastic process is of the form (21), it can be shown that the random function associated with the process can be expressed in the form (20). We have, in fact, the following theorem.

Let $x(t)$ be the random function associated with a stochastic process such that $Ex(t) = 0$ and the covariance function $r(t, u)$ is given by the expression (21), *where $\rho(\lambda, \mu)$ is known to be a covariance function which is of bounded variation in every finite domain, in the sense expressed by* (9). *Then there exists an additive random set function $Z(S)$ such that $EZ(S) = 0$ and*

$$EZ(S_1)\overline{Z(S_2)} = \int_{S_1}\int_{S_2} d^2\rho(\lambda, \mu),$$

and such that (20) *holds for every real t. Further, we have $L_2(Z) = L_2(x)$ when and only when there does not exist any element in $\Lambda_2(\rho)$ different from zero, which is orthogonal to $g(t, \lambda)$ for all real t.*

According to (21), the correspondence $x(t) \rightleftarrows g(t, \lambda)$ defines a one to one linear and isometric correspondence between the space $L_2(x)$ and the subspace of $\Lambda_2(\rho)$

spanned by the functions $g(t, \lambda)$ when the parameter t ranges through the whole real axis.

Suppose first that the set of functions of λ obtained from $g(t, \lambda)$ when t ranges through the real axis forms a base of the space $\Lambda_2(\rho)$. Then the correspondence defined by $x(t) \rightleftarrows g(t, \lambda)$ will extend to the whole spaces $L_2(x)$ and $\Lambda_2(\rho)$. Let S be any bounded Borel set, and let $Z(S)$ denote the element of $L_2(x)$ that corresponds to the function $e_S(\lambda)$ in $\Lambda_2(\rho)$. Then $EZ(S) = 0$, and by (21) we have

$$EZ(S_1)\overline{Z(S_2)} = \int_{-\infty}^{\infty}\int_{-\infty}^{\infty} e_{S_1}(\lambda)\, e_{S_2}(\mu)\, d^2\rho(\lambda, \mu) = \int_{S_1}\int_{S_2} d^2\rho(\lambda, \mu).$$

It is also easily seen that $Z(S)$ is a completely additive function of S, in the sense of convergence in the mean. Thus $Z(S)$ defines an additive random set function. Since $Z(S)$ is an element of $L_2(x)$, we have $L_2(Z) \subset L_2(x)$. On the other hand, the elements $e_S(\lambda)$ form a base of $\Lambda_2(\rho)$, and thus the corresponding elements $Z(S)$ in $L_2(x)$ form a base of the latter space, so that we have $L_2(Z) = L_2(x)$. Thus the correspondence $x(t) \rightleftarrows g(t, \lambda)$ defines a one to one linear correspondence between $L_2(Z)$ and $\Lambda_2(\rho)$, such that $Z(S)$ and $e_S(\lambda)$ are corresponding elements. According to the concluding remark of the preceding paragraph, the element $x(t)$ in $L_2(Z) = L_2(x)$ corresponding to the element $g(t, \lambda)$ in $\Lambda_2(\rho)$ is then for every t given by the relation (20).

In the case when the $g(t, \lambda)$ do not form a base of $\Lambda_2(\rho)$, we have to adjoin a conveniently chosen set of functions $h(u, \lambda)$ in order to obtain a base, and the proof can then be completed in the same manner as in the case treated by Karhunen [5, p. 47]. It should be observed that, in this case, we have to perform an extension of the basic probability field in order to prove the existence of $Z(S)$, and accordingly we have $L_2(Z) \supset L_2(x)$.

The truth of the last assertion of the theorem follows now from the fact that a nonzero element of $\Lambda_2(\rho)$ which is orthogonal to all the $g(t, \lambda)$ exists when and only when the $g(t, \lambda)$ do not form a base of the space $\Lambda_2(\rho)$.

5. Applications to some particular cases

Taking first $g(t, \lambda) = e^{it\lambda}$, and supposing that $\rho(\lambda, \mu)$ is of bounded variation over the whole plane, we obtain the class of stochastic processes denoted by Loève [6] as *harmonizable processes*, and (20) becomes the spectral representation

$$x(t) = \int_{-\infty}^{\infty} e^{it\lambda} dZ$$

of a process belonging to this class. In this case, the functions $e^{it\lambda}$ obviously form a base of the space $\Lambda_2(\rho)$, and we have $L_2(x) = L_2(Z)$.

Suppose, on the other hand, that $\rho(\lambda, \mu)$ in (21) is the covariance function of an orthogonal process. (Compare section 2.) The relation (21) then reduces to

$$r(t, u) = \int_{-\infty}^{\infty} g(t, \lambda)\overline{g(u, \lambda)}\, dF(\lambda), \tag{22}$$

and (20) gives the representation due to Karhunen [5] of a stochastic process having a covariance function of the form (22). In the particular case when $g(t, \lambda) = e^{it\lambda}$

and $F(\lambda)$ is bounded, this reduces to the well known spectral representation of a stationary process.

Consider in particular the case of a stationary process with an absolutely continuous *spectral function* $F(\lambda)$. Then (22) can be written in the form

$$r(t, u) = r(t-u) = \int_{-\infty}^{\infty} e^{i(t-u)\lambda} F'(\lambda)\, d\lambda$$
$$= \int_{-\infty}^{\infty} e^{it\lambda} \sqrt{F'(\lambda)} \cdot \overline{e^{iu\lambda} \sqrt{F'(\lambda)}}\, d\lambda .$$

where $F'(\lambda)$ and $\sqrt{F'(\lambda)}$ are nonnegative. Since $\sqrt{F'(\lambda)}$ is quadratically integrable over the real axis, it has a Fourier transform which we denote by $g(\tau)$. Then $g(t+\tau)$ is the Fourier transform of $e^{it\lambda}\sqrt{F'(\lambda)}$, and the Parseval formula gives

$$r(t, u) = \int_{-\infty}^{\infty} e^{it\lambda} \sqrt{F'(\lambda)}\, \overline{e^{iu\lambda} \sqrt{F'(\lambda)}}\, d\lambda = \int_{-\infty}^{\infty} g(t+\tau)\, \overline{g(u+\tau)}\, d\tau .$$

Hence we obtain by the theorem of the preceding paragraph

$$x(t) = \int_{-\infty}^{\infty} g(t+\tau)\, dZ$$

where [5, p. 72] $Z(S)$ is an additive random set function such that

$$EZ(S_1)\overline{Z(S_2)} = m(S_1 S_2),$$

where m denotes the ordinary Lebesgue measure of the product set S_1S_2.

REFERENCES

[1] H. Cramér, "On the theory of stationary random processes," *Annals of Math.*, Vol. 41 (1940), pp. 215–230.

[2] ———, "On harmonic analysis in certain functional spaces," *Ark. Mat. Astr. Fys.*, Vol. 28 B (1942), pp. 1–7.

[3] ———, "Problems in probability theory," *Annals of Math. Stat.*, Vol. 18 (1947), pp. 165–193.

[4] J. L. Doob, "Stochastic processes depending on a continuous parameter," *Trans. Amer. Math. Soc.*, Vol. 42 (1937), pp. 107–140.

[5] K. Karhunen, "Über lineare Methoden in der Wahrscheinlichkeitsrechnung," *Ann. Acad. Scient. Fennicae*, Ser. AI, No. 37 (1947), pp. 1–79.

[6] M. Loève, "Fonctions aléatoires du second ordre," supplement to P. Lévy, *Processus Stochastiques et Mouvement Brownien*, Gauthier-Villars, Paris, 1948. This work contains references to the earlier papers on the subject by M. Loève.

[7] B. von Sz. Nagy, *Spektraldarstellung linearer Transformationen des Hilbertschen Raumes*, Springer, Berlin, 1942.

46.

On some questions connected with mathematical risk

Univ. California Publ. Stat. 2 (5), 99–123 (1954)

1. Introduction. 1.1. The questions of mathematical risk that arise in connection with various economic activities, such as insurance, gambling, and so forth, are particular cases of the general problem of choice from among a set of possible probability distributions on a space of sure prospects. Consider, for example, the case of an individual who contemplates taking out insurance. From his point of view, each possible insurance corresponds to a certain probability distribution on the scale of incomes, and he has to make up his mind which of these he prefers, taking account also, of course, of the possible action of taking no insurance at all. On the other hand, looking at the question from the point of view of an insurance company, this company is engaged in a continuous game of chance with the totality of its policyholders. The gains or losses incurred by the company during successive periods in the course of this game may be considered as random variables, and the probability distributions of these variables will depend on a number of different factors, such as the number and types of the insurances involved, the reinsurance policy of the company, and so forth. It should be in the interest of the company to study the effect of variations of these factors on the relevant probability distributions, and then so organize its business as to obtain the most favorable probability distributions possible.

The general question of choice from among a set of probability distributions has recently been discussed by von Neumann and Morgenstern, in their book [**20**] on the theory of games, and by a number of other authors. Generally speaking, the main results of this discussion can be expressed in the following way.

Let S be a set of sure prospects x. The x's may be, for example, sums of money, or bundles of commodities. If P and Q are probability distributions on the space S, and α is a number such that $0 \leqq \alpha \leqq 1$, we denote by $\alpha P + (1 - \alpha)Q$ the probability distribution formed in the obvious way as a linear combination of P and Q. Let C denote any convex class of probability distributions on S, that is, a class containing $\alpha P + (1 - \alpha)Q$ as soon as it contains P and Q. Suppose that there exists a *preference pattern* on C, so that, for any two probability distributions P and Q belonging to C, we always have *at least one* of the relations $P \geqq Q$ or $Q \geqq P$, the first of which should be read: *P is preferred or indifferent to Q.*

Then, subject to conditions which only impose very mild restrictions on the properties of the preference pattern defined by the relation $\geqq$, it is possible to prove the existence of a real-valued function $f(x)$, known as the *measurable utility* of x, which is defined for any sure prospect x in S, and is such that $P \geqq Q$ if and only if

$$E_P f(x) \geqq E_Q f(x) ,$$

This paper was prepared with the support of the Office of Ordnance Research, U. S. Army under Contract DA–04–200–ORD–171. A summary of it formed the substance of the 1953 Rietz lecture, which was delivered before the Institute of Mathematical Statistics at its meeting in Washington, D.C., on December 28, 1953.

where E_P and E_Q denote the mathematical expectations corresponding to the distributions P and Q respectively.[1]

According to this result, the question of choice among probability distributions is reduced to finding the distribution that maximizes expected utility. It would thus seem that, beyond the information required for calculating the expectation $Ef(x)$, no further knowledge of the behavior of any proposed probability distribution would be called for.

However, the utility function $f(x)$ is usually unknown, and has to be determined from a study of the actual behavior of individuals and organizations in situations of choice. In order to analyze the preference pattern actually operating in such a situation, it will be necessary to possess very detailed and accurate knowledge of the behavior of the distributions involved. In particular, when we are concerned with probability distributions on the scale of monetary incomes, it will be important to study the extreme tail values of the distributions, which correspond to exceptionally large gains or losses. It is, in fact, precisely these large gains or losses that give rise to the well-known paradoxes connected with gambling, insurance, and so forth, and a careful study of the behavior of the various possible distributions in these domains seems to be essential before a given situation of choice can be analyzed.

1.2. As far as probability distributions connected with insurance problems are concerned, such a study of the behavior of the distributions is one of the main objects of the mathematical theory of insurance risk. In particular, this is true of the form of that theory known as *collective risk theory*, which was founded by the Swedish actuary Dr. F. Lundberg. In a series of works ([**15**] to [**19**]), the first of which was published in 1909, Lundberg studied the risk business of an insurance company as a continuous game of chance between the company on one side and the totality of its policyholders on the other. In modern terminology, the mathematical setup used by Lundberg for this purpose would be denoted as a *stochastic process with independent increments.* The particular process involved is a generalized Poisson process of a kind that will be strictly defined in section 2.2 below. For this process, Lundberg has given important results concerning the asymptotic behavior of the probability distributions of the gains, and also concerning the *ruin problem,* which may be regarded as a generalized form of the classical problem of the gambler's ruin, adapted to the continuous game of an insurance company.

The work of Lundberg was pioneering, and his methods did not always conform to modern standards of mathematical rigor. It is interesting to compare this work with the similar pioneering work on the Brownian-movement process due to Bachelier in France. Both types of processes are concerned with the random walk of a certain particle; in the Brownian-movement process the path of the particle is almost surely continuous, while in the generalized Poisson process studied by Lundberg the movement is essentially discontinuous. We know today that the Brownian-movement process and the generalized Poisson process form two essential building stones of the general stochastic process with independent increments.

[1] Under appropriate conditions, this result will hold true even for certain nonconvex classes of probability distributions. Suppose, for example, that C is the class of all *two-point distributions* on S, namely, the class of all distributions assigning the probability 1 to some set consisting of two points x. If a preference pattern is given on this class C, the existence of a measurable utility $f(x)$ can be proved, subject to conditions which differ only slightly from the conditions for a convex class as given, for example, by Herstein and Milnor [**14**].

A considerable amount of work has been done, mainly by Swedish authors, in order to reconstruct the Lundberg theory on solid mathematical foundations, and extend its results. Some of the main works belonging to this group are quoted in the appended list of references. By means of the modern theory of stochastic processes, it is now possible to present the whole theory from a unifying point of view, and to give rigorous proofs of all its main results. To a certain degree, this has already been achieved by the work of the Swedish school mentioned above, but most of these works are not easily available outside Scandinavia,[2] and some parts of the theory still do not seem to be definitely cleared up, while others can be considerably simplified. A survey of the theory with proofs of some of its most important results will be given in the present paper.

The basic definitions will be given in part 2, which also contains a study of the asymptotic behavior of the probability distributions associated with the stochastic process under investigation. In this study we shall, however, only give a brief sketch of the mathematical proofs involved, since the main purpose of this paper is to give a simplified account of the results so far obtained in the case of the *ruin problem*. This problem will be discussed in part 3, where it will be shown that its solution depends on a certain integral equation of the Wiener-Hopf type. Explicit formulas for the solution of this equation will be given, together with some important results concerning the asymptotic behavior of the solution.

2. The risk process. 2.1. *The amount of claims.*—Consider the totality of all insurances in force with a given insurance company. In this totality, certain random events known as *claims* will arise from time to time. We shall assume that the probability that exactly one claim occurs within the time interval $(\tau, \tau + \Delta\tau)$ is, for small values of $\Delta\tau$,

$$\lambda_\tau \Delta\tau + o(\Delta\tau),$$

with $k < \lambda_\tau < K$, where k and K are positive constants, while the probability of more than one claim during the same time interval is $o(\Delta\tau)$. We shall also assume that these probabilities are independent of any information with respect to the claims that may have occurred before the time τ. This assumption will no doubt involve a certain idealization of observed facts, but may nevertheless be regarded as sufficiently realistic for a first approximation. Introducing a transformed time t by means of the equation

$$t = \int_0^\tau \lambda_u \, du\,,$$

we find that the probability of exactly one claim in the interval $(t, t + \Delta t)$ will be

$$\Delta t + o(\Delta t)\,,$$

while the probability of more than one claim will be $o(\Delta t)$. It is well known that, under these assumptions, the number of claims will generate a *Poisson process*, the probability that exactly n claims will occur in any time interval of length t being given by the Poisson expression $t^n/n!\, e^{-t}$.

We further assume that each claim gives rise to a certain payment from the

[2] A survey of the theory has recently been given by Dubourdieu [10].

company to some policyholder (or vice versa; see the following paragraph). The amount y of this payment will be regarded as a random variable having a given distribution function $P(y)$. It will be assumed that the distribution function (d.f.) $P(y)$ is independent of time, and is also independent of the number and amounts of claims that have previously occurred. This, of course, represents a further idealization of observed facts. Certain parts of the theory can be developed without introducing this assumption, but this possibility will not be considered here.

In most forms of insurance, the amount y paid by the company under each claim will be essentially positive, so that the given d.f. $P(y)$ will have its total range of variation on the positive y axis. However, there are certain forms of insurance, such as annuity insurance, where the circumstances are reversed. In the case of an ordinary whole-life annuity, the company makes continuous payments as long as the insured person is alive, but when the random event of the death of that person occurs, the reserve value of the annuity becomes available to the company, which means that the company *receives* a certain payment, so that in this case the amount y will be negative. Accordingly, we shall in the sequel generally regard y as a random variable capable of assuming both positive and negative values, so that the d.f. $P(y)$ will have the whole real axis as its possible range of variation. The cases when y is always positive, or always negative, will constitute interesting particular cases that will be separately considered.

Let $Y(t)$ denote the total amount of claims (each y being counted with its proper sign) occurring in the time interval $(0, t)$. In particular, we shall then have

$$Y(0) = 0 .$$

For any $t > 0$, $Y(t)$ will be a random variable, and we denote by $G(y, t)$ the d.f. of this variable, so that

$$G(y, t) = \Pr(Y(t) \leqq y) .$$

In order to find an expression of $G(y, t)$, we first consider the conditional probability of the relation $Y(t) \leqq y$, relative to the hypothesis that exactly n claims have arisen. Clearly this will coincide with the probability of the relation

$$y_1 + \cdots + y_n \leqq y ,$$

where the y_i are independent random variables having the common d.f. $P(y)$. This probability is the n-fold convolution of $P(y)$ with itself:

$$\Pr (y_1 + \cdots + y_n \leqq y) = P^{n^*}(y) ,$$

where

$$P^{0^*}(y) = \epsilon(y) = \begin{cases} 0 , & y < 0 , \\ 1 , & y \geqq 0 , \end{cases}$$

$$P^{1^*}(y) = P(y) ,$$

and for $n > 1$

$$P^{n^*}(y) = \int_{-\infty}^{\infty} P^{(n-1)^*}(y - z) dP(z) .$$

Since the events that exactly 0, 1, 2, · · · claims have occurred are mutually exclusive, it now follows that

$$(1) \qquad G(y, t) = \sum_0^\infty \frac{t^n}{n!} e^{-t} P^{n*}(y) .$$

The corresponding characteristic function (c.f.) will be written in the form

$$(2) \qquad E\, e^{sY(t)} = \int_{-\infty}^{\infty} e^{sy} d_y G(y, t) = \sum_0^\infty \frac{t^n}{n!} e^{-t} \int_{-\infty}^{\infty} e^{sy} dP^{n*}(y)$$

$$= \sum_0^\infty \frac{t^n}{n!} e^{-t} \left(\int_{-\infty}^{\infty} e^{sy} dP(y) \right)^n = e^{t(p(s)-1)} ,$$

where

$$(3) \qquad p(s) = \int_{-\infty}^{\infty} e^{sy} dP(y) ,$$

while $s = \sigma + i\tau$ is a complex variable. The above integrals will always be absolutely convergent for $\sigma = 0$, and will then yield the c.f. of the random variable $Y(t)$ in its usual form. However, we shall later have to consider this function also for certain values of $\sigma \neq 0$, and shall then have to introduce certain restrictive assumptions with respect to the given d.f. $P(y)$, which will render the above integrals absolutely convergent also for these values of σ.

It now follows from the assumptions introduced above that the increments of $Y(t)$ over disjoint time intervals will always be independent, and that the increment $Y(t_0 + t) - Y(t_0)$ over any time interval of length t will have the same distribution as the variable $Y(t) = Y(t) - Y(0)$ considered above, which corresponds to the case $t_0 = 0$.

Thus $Y(t)$ will be a random function defined for $t \geqq 0$, and associated with a *stochastic process with stationary and independent increments* (see Doob [9], p. 419, Ex. 3).

Assuming that the first order moment

$$p_1 = \int_{-\infty}^{\infty} y dP(y)$$

exists, it easily follows from (1) or (2) that we have

$$(4) \qquad EY(t) = p_1 t .$$

2.2. *The sample functions.*—The obvious natural form of a sample function $Y(t)$ corresponding to the stochastic process defined above is as follows: We start from $t = 0$ with the value $Y(0) = 0$. As t increases, the function $Y(t)$ then remains constantly equal to zero until the occurrence of the first claim, say at time $t = \tau_1$, when there is a jump of a certain (positive or negative) length y_1. The function then remains constantly equal to y_1 until the occurrence of the next claim, say at time $t = \tau_1 + \tau_2$, when there is a new jump of a certain length y_2, and so on.

Any function of this type is uniquely characterized by the two sequences

$$\tau_1, \tau_2, \cdots, \qquad \text{and} \qquad y_1, y_2, \cdots,$$

where the τ_i are the (essentially positive) time intervals between consecutive jumps, and the y_i are the (positive or negative) lengths of the jumps. Assuming, moreover, that there will at most be a finite number of jumps in any finite t-interval, it follows that the sum of the τ_i will be divergent:

$$\sum_1^\infty \tau_i = \infty ,$$

and that for every $t > 0$ we shall have

$$Y(t) = \sum_{\tau_1 + \cdots + \tau_n \leq t} y_n . \tag{5}$$

Any function $Y(t)$ defined in this way will be denoted as a *natural sample function* of the Y process.

Consider now a space Ω with the countable set of coördinates $\tau_1, \tau_2, \cdots, y_1, y_2, \cdots$, where the τ_i are essentially positive, while the y_i may have any real values. Each set of values of the coördinates represents a point $\omega = (\tau_1, \tau_2, \cdots, y_1, y_2, \cdots)$ of the space Ω. With each point ω such that $\sum \tau_i = \infty$, we associate the natural sample function $Y(t)$ defined by (5). There will then be a one-to-one correspondence between the set of all natural sample functions $Y(t)$ and the set of all points ω in Ω such that $\sum \tau_i = \infty$.

We now introduce a probability measure Π in the space Ω by means of the following definitions:

1) All the coördinates $\tau_1, \tau_2, \cdots, y_1, y_2 \cdots$ are mutually independent random variables.

2) The τ_i are essentially positive, and have the common probability density $e^{-\tau}$ $(\tau > 0)$.

3) The y_i have the common d.f. $P(y)$.

Since the τ_i are independent and equidistributed, the set Z of points ω such that $\sum \tau_i$ is finite is easily seen to have Π-measure zero. Neglecting the null set Z, there is a one-to-one correspondence between all points ω and all natural sample functions $Y(t)$, and accordingly the probability measure Π on Ω induces a corresponding probability measure on the set of all natural sample functions $Y(t)$. From the assumption that the τ_i have the common probability density $e^{-\tau}$, it easily follows that the probability that $Y(t)$ has exactly n jumps in a given interval of length t is given by the Poisson expression $t^n/n!\, e^{-t}$. From definitions 2 and 3, it then further follows that the random variable $Y(t)$ has, for any fixed $t > 0$, the d.f. $G(y, t)$ given by (1), and also that $Y(t)$ corresponds to a stochastic process with stationary and independent increments.

By means of the probability measure Π *defined by* 1 *to* 3 *we have thus redefined the risk process* $Y(t)$ *as a stochastic process, defined on the set of all natural sample functions*. This will be important in the study of the ruin problem (see 3.1), since it will allow us to avoid the well-known measure-theoretic difficulties connected with the definition of certain probabilities associated with a stochastic process, when the process is defined on the set of all possible functions of t.

2.3. *The risk reserve.*—According to (4), the expected value of the total amount $Y(t)$ of claims falling due in the interval $(0, t)$ is $p_1 t$. It follows that, if the company receives during each time element dt the *net risk premium* $p_1 dt$ from the totality of its policyholders, the expected value of the company's gain on the net risk business (net risk premiums minus total claims paid) will be zero during each time interval, so that the game will be a fair one.[3] In practice, however, the company will always want to have a safety margin, so that the *gross risk premium* actually paid will exceed the risk premium. Suppose that during each time element dt, the company receives the gross risk premium $(p_1 + \lambda)dt$, where $\lambda > 0$ is a given constant, which represents the *safety loading* applied by the company. We assume that all gross risk premiums are paid into a certain fund, the *risk reserve* of the company, from which all claims are paid. Assuming that the initial value of the risk reserve at time $t = 0$ is zero, the value $X(t)$ at time t will then be

$$X(t) = (p_1 + \lambda)t - Y(t)\,. \tag{6}$$

Since $X(t)$ is connected with $Y(t)$ by the trivial transformation (6), it follows that $X(t)$ is also a random function associated with a stochastic process with stationary and independent increments. The expected value of the random variable $X(t)$ is, according to (4),

$$EX(t) = \lambda t\,, \tag{7}$$

and is thus equal to the amount of safety loading that has been received up to the time t.

We now denote by $F(x, t)$ the probability of the relation $X(t) < x$ (note that we use here the sign $<$ instead of $\leqq$) according to the probability measure Π introduced in the preceding section. We then obtain from (6) and (1)

$$\begin{aligned} F(x, t) &= \Pi(X(t) < x) = \Pi(Y(t) > -x + (p_1 + \lambda)t) \\ &= 1 - \Pi(Y(t) \leqq -x + (p_1 + \lambda)t) \\ &= 1 - \sum_0^\infty \frac{t^n}{n!} e^{-t} P^{n*}(-x + (p_1 + \lambda)t)\,, \end{aligned} \tag{8}$$

and further from (6) and (2)

$$E\,e^{-sX(t)} = e^{t(p(s)-1-(p_1+\lambda)s)}\,. \tag{9}$$

To each natural sample function $Y(t)$ there corresponds, according to (6), a uniquely defined function $X(t)$, which has the same jumps as $Y(t)$, with inverted

[3] Note that we express ourselves here as if p_1 were necessarily positive. If p_1 is negative, this means that the company's insurances are predominantly of the annuity type, and the total net risk premium paid to the company by the policyholders will be replaced by a total net annuity paid by the company to the policyholders. The constant λ will still be positive, but the safety loading will take the form of a *reduction of the total annuity paid*, instead of an addition to the total net risk premium.

signs, and increases linearly throughout each interval between two consecutive jumps, the angular coefficient being $p_1 + \lambda$. These functions $X(t)$ are the natural sample functions of the $X(t)$ process, which is thus defined, in the same way as the $Y(t)$ process, by means of the probability measure induced on the space of all natural sample functions by the probability measure Π defined in 2.2.

2.4. *Asymptotic normality of $F(x, t)$.*—If it is assumed that the second order moment of $P(x)$,

$$p_2 = \int_{-\infty}^{\infty} x^2 dP(x) ,$$

exists, there is an analogue of the central limit theorem for the $X(t)$ process, which asserts that the normalized random variable

$$\frac{X(t) - \lambda t}{\sqrt{p_2 t}}$$

is asymptotically normal (0, 1). We have, in fact, for any fixed real x, as $t \to \infty$ (see Cramér [**4**], Esseen [**12**]),

$$F(\lambda t + x\sqrt{p_2 t},\, t) = \Phi(x) + O\left(\frac{1}{\sqrt{t}}\right), \tag{10}$$

where

$$\Phi(x) = \frac{1}{\sqrt{2\pi}} \int_{-\infty}^{x} e^{-1/2y^2} dy .$$

If moments of higher order of $P(x)$ exist, the relation (10) can be strengthened in the same way as the central limit theorem for sums of independent and equidistributed variables, the second member being replaced by an asymptotic expansion, containing certain derivatives of the normal function $\Phi(x)$, multiplied by powers of $t^{-1/2}$.

2.5. *Further asymptotic properties of $F(x, t)$.*—The asymptotic relation (10) holds true for any fixed real x, as $t \to \infty$, and thus furnishes information about the behavior of the distribution of $X(t)$ for values of the variable, which differ from the mean t by a quantity of the order $\pm \sqrt{t}$. If we want to study the distribution for deviations of higher orders, we should have to allow the quantity x in (10) to depend on t, and tend to $\pm\infty$ as $t \to \infty$. However, the order of approximation given by (10) is not sufficiently accurate to yield any useful information in this way.

However, if we introduce some appropriate restrictive conditions with respect to the behavior of the given d.f. $P(x)$ at infinity, it is possible to obtain the desired information about the tail values of the distribution of $X(t)$ by means of a method first used by Esscher [**11**], and then generalized by Cramér [**5**] and Feller [**13**]. We shall suppose that the integral

$$p(\sigma) = \int_{-\infty}^{\infty} e^{\sigma x} dP(x)$$

converges for all real σ in some interval $-A < \sigma < B$, where A and B are positive. (Note that this assumption will be used *only* in the present section.)

For every σ in $(-A, B)$, the function

$$\overline{P}(x) = \frac{1}{p(\sigma)} \int_{-\infty}^{x} e^{\sigma z} dP(z)$$

will then be a d.f., which for $\sigma = 0$ reduces to $P(x)$. Conversely, we have

$$P(x) = p(\sigma) \int_{-\infty}^{x} e^{-\sigma z} d\overline{P}(z) ,$$

and it is seen without difficulty that the repeated convolutions of $\overline{P}(x)$ and $P(x)$ will satisfy similar relations. In particular, we have

$$P^{n^*}(x) = (p(\sigma))^n \int_{-\infty}^{x} e^{-\sigma z} d\overline{P}^{n^*}(x) \tag{11}$$

The moments of $\overline{P}(x)$ are

$$\overline{p}_n = \int_{-\infty}^{\infty} z^n d\overline{P}(z) = \frac{1}{p(\sigma)} \int_{-\infty}^{\infty} z^n e^{\sigma z} dP(z) .$$

We further define a d.f. $\overline{F}(x, t)$ by writing, in analogy with (8), except that we take here $\lambda = 0$,

$$\overline{F}(x, t) = 1 - \sum_{0}^{\infty} \frac{t^n}{n!} e^{-t} \overline{P}^{n^*}(-x + \overline{p}_1 t) . \tag{12}$$

From (8) and (11) we then obtain

$$\begin{aligned} F(x, t) &= 1 - \sum_{0}^{\infty} \frac{t^n}{n!} e^{-t} P^{n^*}(-x + (p_1 + \lambda)t) \\ &= 1 - \sum_{0}^{\infty} \frac{t^n}{n!} e^{-t} (p(\sigma))^n \int_{-\infty}^{-x+(p_1+\lambda)t} e^{-\sigma z} d\overline{P}^{n^*}(x) . \end{aligned}$$

After some simple transformations, this can be written in either one of the two equivalent forms

$$1 - F(x, t) = e^{-\beta t} \int_{x+\gamma t}^{\infty} e^{\sigma z} d_z \overline{F}(z, tp(\sigma)) , \tag{13a}$$

or

$$F(x, t) = e^{-\beta t} \int_{-\infty}^{x+\gamma t} e^{\sigma z} d_z \overline{F}(z, tp(\sigma)) , \tag{13b}$$

where

$$\beta = 1 + \sigma \overline{p}_1 p(\sigma) - p(\sigma) = \int_{-\infty}^{\infty} (1 + (\sigma z - 1)e^{\sigma z}) dP(z) , \tag{14}$$

$$\gamma = \overline{p}_1 p(\sigma) - p_1 - \lambda = \int_{-\infty}^{\infty} z(e^{\sigma z} - 1) dP(z) - \lambda .$$

When σ increases from $-A$ to B, the quantity γ is steadily increasing, since $d\gamma/d\sigma$ is always positive. For $\sigma = 0$, we have $\gamma = -\lambda$. Hence there exists an interval, say $(-A', B')$, where A' and B' are positive, such that the equation

$$\gamma = -c \tag{15}$$

has one and only one real root σ, such that $-A < \sigma < B$, for every real c such that $\lambda - A' < c < \lambda + B'$.

We shall then have

$$\sigma \gtreqless 0 \qquad \text{according as} \qquad c \lesseqgtr \lambda .$$

Let now c be given, such that

$$\lambda - A' < c < \lambda + B' , \tag{16}$$

and take σ equal to the unique real root of (15). From (13) we then obtain, taking $x = ct$,

$$1 - F(ct, t) = e^{-\beta t} \int_0^\infty e^{\sigma z} d_z \overline{F}(z, tp(\sigma)) , \tag{17a}$$

$$F(ct, t) = e^{-\beta t} \int_{-\infty}^0 e^{\sigma z} d_z \overline{F}(z, tp(\sigma)) . \tag{17b}$$

Now $\overline{F}(x, t)$ is defined by means of $\overline{P}(x)$ in the same way as $F(x, t)$ by means of $P(x)$, except that in (12) we have taken $\lambda = 0$. Hence the normal approximation (10) is available for $\overline{F}(x, t)$, and yields

$$\overline{F}(x\sqrt{\overline{p}_2 t}, t) = \Phi(x) + O\left(\frac{1}{\sqrt{t}}\right) .$$

We now introduce this into (17a) or (17b), according as $c > \lambda$ or $c < \lambda$, and obtain after some reductions

$$\left.\begin{matrix} 1 - F(ct, t) \\ \\ F(ct, t) \end{matrix}\right\} = \left[\frac{\alpha}{\sqrt{t}} + O\left(\frac{1}{t}\right)\right] e^{-\beta t} , \tag{18}$$

where the first member should be $1 - F(ct, t)$ for $c > \lambda$, and $F(ct, t)$ for $c < \lambda$, and

$$\alpha = \frac{1}{|\sigma| \sqrt{2\pi \int_{-\infty}^{\infty} x^2 e^{\sigma x} dP(x)}} , \tag{19}$$

while β is given by (14).

Under appropriate assumptions with respect to $P(x)$, the relation (18) may, in the same way as (10), be replaced by an asymptotic expansion, the second member of (18) being replaced by an expression of the form

$$\left[\frac{\alpha}{t^{1/2}} + \cdots + \frac{\alpha_k}{t^{k/2+1}} + O\left(\frac{1}{t^{k/2+1}}\right)\right] e^{-\beta t} .$$

The relation (18) gives an asymptotic expression for the tail values of the distribution of the random variable $X(t)$, when the deviation of the variable from its mean λt is of the same order as t. It is interesting to observe that if, in the normal approximation (10), we take x equal to a positive or negative constant multiplied by $\sqrt{t}$, and neglect the error term in the second member, we get an asymptotic expression of the same form as (18), but with different values of the constants α and β. This shows that, in fact, the normal approximation breaks down when we want to consider deviations of a larger order of magnitude than $\sqrt{t}$, so that (18) yields a better asymptotic representation of the tail values of the distribution than can be obtained directly from the normal approximation (10).

In order to illustrate the above results by an example, consider the simple particular case when we have

$$P(x) = \begin{cases} 0 & x \leqq 0\,, \\ 1 - e^{-x}, & x \geqq 0\,, \end{cases}$$

so that $p_1 = 1$ and $p_2 = 2$. The equation (15) then reduces to

$$\frac{1}{(1-\sigma)^2} = 1 - (c - \lambda)$$

which gives

$$\sigma = 1 - \frac{1}{\sqrt{1-(c-\lambda)}}$$

for every given $c < \lambda + 1$. From (14) and (19) we then obtain

$$\frac{1}{\alpha} = 2\sqrt{\pi}\,(1-(c-\lambda))^{1/4}\,\left|1 - \sqrt{1-(c-\lambda)}\,\right|,$$

$$\beta = (1 - \sqrt{1-(c-\lambda)})^2\,,$$

and the asymptotic expression for the tail values will be obtained by introducing these expressions into (18). From the normal approximation (10), we should in this case obtain an asymptotic expression of the same form, but with the constants

$$\frac{1}{\alpha} = \sqrt{\pi}\,|c - \lambda|\,,$$

$$\beta = \frac{1}{4}\,(c-\lambda)^2\,,$$

which are only correct to the first order of approximation, for small values of $c - \lambda$. An analogous result can be shown to hold in the general case.

3. The ruin problem. 3.1. *Definitions.*—Consider the space Ω and the probability measure Π on this space, as introduced in 2.2. According to 2.2 and 2.3, there is a one-to-one correspondence between the points ω of the space Ω (neglecting a certain set of Π-measure zero) and the "natural sample functions" of our X process defined in 2.3:

$$X(t) = (p_1 + \lambda)\,t - \sum_{\tau_1 + \cdots + \tau_n \leqq t} y_n\,,$$

where $\tau_1, \tau_2, \cdots, y_1, y_2, \cdots$ are the coördinates of the point ω.

Let now h, u, and n be given, so that $h > 0$, $u \geqq 0$, and n is a positive integer. We then introduce the following sets of points in Ω:

$$S_{n,h} = \{\omega | X(nh) < -u\}\,,$$

$$S_h = \{\omega | X(nh) < -u \quad \text{for some} \quad n = 1, 2, \cdots\}\,, \tag{20}$$

$$S = \{\omega | X(t) \;<\; -u \quad \text{for some} \quad t > 0\}\,.$$

For any fixed t, $X(t)$ is a random variable with the distribution function $F(x, t)$, so that the set $S_{n,h}$ is certainly measurable, with Π-measure $F(-u, nh)$. From the obvious relation

$$S_h = \sum_{n=1}^{\infty} S_{n,h} ,$$

it further follows that S_h is measurable.

Consider now the set S_h for a sequence H of values of h tending to zero. Since we evidently have $S_h \subset S$ for every $h > 0$, it follows that

$$\limsup_{h \to 0} S_h \subset S ,$$

when h tends to zero on the sequence H. On the other hand, let ω be an arbitrary point of S (always neglecting the ω null-set mentioned above), and consider the corresponding sample function $X(t)$. According to the definition of S, we shall have $X(t) < -u$ for at least one value of t. However, since $X(t)$ has only a finite number of jumps in any finite interval, $X(t)$ will necessarily be $< -u$ throughout a whole t-interval. It then follows that, for every sufficiently small h, we shall be able to find an integer n such that $X(nh) < -u$, and consequently the point ω will belong to S_h. To every point ω in S, we can thus find $h_0 > 0$ such that $\omega \in S_h$ as soon as $h < h_0$. It follows that we have, as $h \to 0$ on H,

$$S \subset \liminf_{h \to 0} S_h ,$$

so that the sequence S_h must be convergent, and

$$S = \lim_{h \to 0} S_h ,$$

which shows that S is measurable.

It follows that the probabilities

$$\psi(u, h) = \Pi(S_h) = \Pi(X(nh) < -u \text{ for some } n = 1, 2, \cdots) \tag{21}$$

and

$$\psi(u) = \Pi(S) = \Pi(X(t) < -u \text{ for some } t > 0) \tag{22}$$

will be uniquely defined, and will (the above sequence H being arbitrary) satisfy the relations

$$\begin{gathered} 0 \leqq \psi(u, h) \leqq \psi(u) \leqq 1 , \\ \psi(u) = \lim_{h \to 0} \psi(u, h) . \end{gathered} \tag{23}$$

The function $\psi(u)$ will be denoted as the *ruin function* of our risk process. While $\psi(u, h)$ represents the probability that the risk reserve, starting with an initial capital u, will ultimately become ruined, when only its value in the specified time points $h, 2h, \cdots$ are considered, $\psi(u)$ represents the corresponding probability of ultimate ruin at any time $t > 0$.

The object of the investigation is to study the functions $\psi(u, h)$ and $\psi(u)$, and particularly the latter, in their dependence on the two given elements of the risk process, namely, the safety constant λ, and the d.f. $P(x)$ of the claims. In particular, it will be important to study the asymptotic behavior of the ruin function $\psi(u)$ for large u, and to find out how this behavior will react against changes of λ and $P(x)$.

3.2. *The constant R.*—With respect to the d.f. $P(x)$, we shall in this part, unless something else is explicitly stated, make only the two following assumptions:

$$\int_{-\infty}^{0} |x| dP(x) < \infty ,$$

(24)

$$\int_{0}^{\infty} e^{\sigma x} dP(x) < \infty \quad \text{for some} \quad \sigma > 0 .$$

From these relations, it follows that the mean value

$$p_1 = \int_{-\infty}^{\infty} x dP(x)$$

will exist, and that the integral defined by (3)

$$p(s) = \int_{-\infty}^{\infty} e^{sx} dP(x) ,$$

where $s = \sigma + i\tau$, will be absolutely convergent in some vertical strip $0 < \sigma \leqq r$ in the complex s-plane, and will represent a function $p(s)$, which is analytic and regular in this strip.

Consider the function

$$\pi(s) = p(s) - 1 - (p_1 + \lambda)s$$

appearing in the exponent in (9). We evidently have

$$\pi(0) = 0 , \qquad \pi'(0) = p'(0) - p_1 - \lambda = -\lambda < 0 .$$

It follows that, for sufficiently small real and positive s, the function $\pi(s)$ will be negative.

We have thus established the existence of at least one real and positive quantity r, such that the following two conditions are satisfied:

I) $p(s)$ is analytic and regular for $0 < \sigma \leqq r$,

II) $p(\sigma) - 1 - (p_1 + \lambda)\sigma < 0$ for real σ such that $0 < \sigma \leqq r$.

By R we shall denote the supremum, or the least upper bound, of all r having the above properties:

(25)
$$R = \sup r .$$

It is then clear that we shall always have $R > 0$. However, we might possibly have $R = \infty$. Now, as soon as λ and $P(x)$ are given, one and only one of the following three cases must occur:

A) $P(0) < 1$,

B) $P(0) = 1, \quad p_1 + \lambda < 0$,

C) $P(0) = 1, \quad p_1 + \lambda \geqq 0$.

A simple discussion of the above conditions I and II will show that in the cases A and B we have always $0 < R < \infty$, while in case C we have $R = \infty$.

However, if we consider the behavior of the natural sample functions $X(t)$ as defined in 2.3, it will be seen that in case C each of these functions is never decreasing, and takes the value zero at $t = 0$. It follows that, in this case, we shall have

$$\psi(u, h) = \psi(u) = 0$$

for all $u \geqq 0$, so that from the point of view of the ruin problem this case is entirely trivial.

Accordingly we shall henceforth exclude case C *from our considerations, and we may thus always consider* R *as a finite and positive quantity, uniquely determined by* λ *and* $P(x)$.

3.3. *Proof of* $\psi(u) \leqq e^{-Ru}$.—In this section we shall obtain the inequality

$$(26) \qquad 0 \leqq \psi(u, h) \leqq \psi(u) \leqq e^{-Ru}$$

for all $u \geqq 0$. The proof given here is a simplified version of a proof due to Täcklind [27].

Consider the set Z_n of all points ω in Ω defined by the relation

$$Z_n = \{\omega | X(\nu h) \geqq -u \text{ for } \nu = 1, \cdots, n-1 ; X(nh) < -u\} .$$

Z_n is certainly measurable. Further, Z_m and Z_n are disjoint for $m \neq n$, and we have

$$\sum_1^\infty Z_n = S_h ,$$

where S_h is defined by (20). Hence, writing

$$\varphi_n(u, h) = \Pi(Z_n) ,$$

we have $\varphi_n(u, h) \geqq 0$, and

$$(27) \qquad \psi(u, h) = \sum_1^\infty \varphi_n(u, h) ,$$

where $\psi(u, h) = \Pi(S_h)$ is defined by (21).

Now we have, according to (8),

$$\varphi_1(u, h) = \Pi(X(h) < -u) = F(-u, h) ,$$

and for $\nu > 1$

$$\varphi_\nu(u, h) = \int_0^\infty \varphi_{\nu-1}(v, h) d_v F(v - u, h) .$$

Hence we obtain, summing over $\nu = 2, \cdots, n$ and adding $\varphi_1 = F(-u, h)$,

$$(28) \qquad \sum_1^n \varphi_\nu(u, h) = F(-u, h) + \int_0^\infty \sum_1^{n-1} \varphi_\nu(v, h) d_v F(v - u, h) .$$

Let us now take $r > 0$ such that conditions I and II of 3.2 are satisfied. From (9) we then obtain

$$\int_{-\infty}^{\infty} e^{-rx} d_x F(x, t) = e^{t(p(r)-1-(p_1+\lambda)r)} < 1 \tag{29}$$

for all $t > 0$. Further, (28) gives

$$e^{-ru} - \sum_1^n \varphi_\nu(u, h) = g(u) + \int_0^\infty \left[e^{-rv} - \sum_1^{n-1} \varphi_\nu(v, h) \right] d_v F(v - u, h) \,, \tag{30}$$

where, using (29),

$$g(u) = e^{-ru} - F(-u, h) - \int_0^\infty e^{-rv} d_v F(v - u, h) > e^{-ru} \int_{-\infty}^\infty e^{-rv} d_v F(v, h)$$

$$- \int_{-\infty}^0 d_v F(v-u, h) - \int_0^\infty e^{-rv} d_v F(v-u, h) = \int_{-\infty}^0 (e^{-rv} - 1) d_v F(v-u, h) \geqq 0 \,.$$

Thus $g(u) > 0$ for all u, and it now follows from (30) that, if the inequality

$$\sum_1^k \varphi_\nu(u, h) \leqq e^{-ru} \qquad \text{for all} \qquad u \geqq 0 \tag{31}$$

holds for $k = n - 1$, it will also hold for $k = n$. Taking $k = 1$, we have

$$e^{-ru} - \varphi_1(u, h) = e^{-ru} - F(-u, h) \geqq g(u) > 0 \,,$$

so that (31) certainly holds for $k = 1$, and thus for all k. As $k \to \infty$, we then obtain from (27)

$$\psi(u, h) \leqq e^{-ru} .$$

For any fixed $u \geqq 0$ and $h > 0$, this inequality has thus been established for any r satisfying conditions I and II of 3.2. But then the inequality will also hold for the upper bound of these values of r, so that we have, according to (25),

$$\psi(u, h) \leqq e^{-Ru} .$$

Finally, by (23) we have $\psi(u) = \lim_{h \to 0} \psi(u, h)$, and since R is independent of h, we obtain the inequality (26).

3.4 *The fundamental integral equations.*[4]—In this section it will be shown that the

[4] The inequality (26) and the main results concerning the asymptotic behavior of $\psi(u)$ and $\psi(u, h)$ proved in part 3 have already been given in the works of F. Lundberg, though in some cases without complete proofs. For the particular case when the whole mass of the P distribution is situated on the positive part of the x-axis, the integral equation (33) for $\psi(u)$ was first given by Cramér, [2] and [3], and used to prove the asymptotic relation (56). For the case of a general P distribution, the integral equation (33) was given at a Paris conference in 1937, subsequently published in [6]. The problem was treated by Segerdahl ([24] to [26]), who used a different method and gave complete proofs of the main results, and by Täcklind [27], who worked with the integral equation (32) for $\psi(u, h)$, and used the Wiener-Hopf method to obtain the solution, $\psi(u)$ being defined as the limit of $\psi(u, h)$ as $h \to 0$, and not by its probabilistic properties. In the present paper, $\psi(u)$ is defined by (22) as a probability, and we shall apply the Wiener-Hopf method directly to the integral equation (33) satisfied by $\psi(u)$, obtaining in this way expressions which are in the main equivalent to those given by Täcklind.

functions $\psi(u, h)$ and $\psi(u)$ satisfy for all $u \geqq 0$ and $h > 0$ certain integral equations, namely

$$\psi(u, h) = F(-u, h) + \int_0^\infty \psi(v, h) d_v F(v - u, h) \tag{32}$$

and

$$\psi(u) = \int_u^\infty Q(v) dv + \int_0^\infty Q(u - v) \psi(v) dv \,, \tag{33}$$

where

$$Q(u) = \frac{\epsilon(u) - P(u)}{p_1 + \lambda} = \begin{cases} -\dfrac{P(u)}{p_1 + \lambda}, & u < 0\,, \\[2ex] \dfrac{1 - P(u)}{p_1 + \lambda}, & u \geqq 0\,. \end{cases} \tag{34}$$

(In the particular case $p_1 + \lambda = 0$, some obvious modifications will have to be made in (33).)

The equation (32) is directly obtained by making the limit passage $n \to \infty$ in (28), and using (27), so that we have only to prove (33). This will be proved by allowing h to tend to zero in (32). In doing this, we shall exclude the particular case when $p_1 + \lambda = 0$, and assume $p_1 + \lambda \neq 0$. The excluded case can be taken into account afterwards, by allowing $p_1 + \lambda$ to tend to zero in the results.

In order to perform the limit passage $h \to 0$ in (32), we first consider the behavior of each term in the second member. From (8) we obtain, as $h \to 0$,

$$F(-u, h) = h[1 - P(u + (p_1 + \lambda)h)] + O(h^2)\,,$$

where the $O(h^2)$ holds uniformly for all $u \geqq u_0 > 0$. In the same way we obtain

$$\int_0^\infty \psi(v, h) d_v F(v - u, h)$$
$$= (1 - h)\psi(u + (p_1 + \lambda)h, h) - h \int_0^\infty \psi(v, h) d_v P(u - v + (p_1 + \lambda)h) + O(h^2)$$

also uniformly for all $u \geqq u_0 > 0$.

Introducing this into (32) we obtain, replacing u by $u - (p_1 + \lambda)h$,

$$\psi(u, h) - \psi(u - (p_1 + \lambda)h, h) \tag{35}$$
$$= h\psi(u, h) - h(1 - P(u)) + h \int_0^\infty \psi(v, h) d_v P(u - v) + O(h^2)$$

uniformly in the same sense as before.

Let now $K > 0$ be a given constant, and take

$$h = \frac{K}{|p_1 + \lambda| n}\,, \tag{36}$$

where n is an integer which will ultimately be allowed to tend to infinity.

In (35) we replace u by $u + \nu K/n$, and sum the resulting equations for $\nu = 1, 2, \cdots, n$. We then obtain, multiplying by $|p_1 + \lambda|$ and taking account of the fact that, according to (35), the difference $\psi(u + K/n, h) - \psi(u, h)$ is uniformly of the order $O(h) = O(1/n)$,

$$(p_1 + \lambda)[\psi(u, h) - \psi(u + K, h)]$$

$$= -\frac{K}{n}\sum_1^n \psi\left(u + \nu\frac{K}{n}, h\right) + \frac{K}{n}\sum_1^n\left[1 - P\left(u + \nu\frac{K}{n}\right)\right]$$

$$-\frac{K}{n}\int_0^\infty \psi(v, h)d_v \sum_1^n P\left(u - v + \nu\frac{K}{n}\right) + O\left(\frac{1}{n}\right),$$

where h is given by (36), and the $O(1/n)$ holds uniformly as before. When now $n \to \infty$, the first member tends according to (23) to the limit

$$(p_1 + \lambda)[\psi(u) - \psi(u + K)].$$

A discussion of the separate terms in the second member will show that each of these terms tends to the limit obtained by replacing any Riemann sum by the corresponding integral, and writing $\psi(u)$ in the place of $\psi(u, h)$. We thus obtain in the limit

$$(p_1 + \lambda)[\psi(u) - \psi(u + K)]$$

$$= -\int_u^{u+K} \psi(v)dv + \int_u^{u+K}(1 - P(v))dv + \int_0^\infty \psi(v)(P(u-v+K) - P(u-v))dv.$$

When now $K \to \infty$ we obtain, using the inequality (26),

$$(37) \qquad (p_1+\lambda)\psi(u) = -\int_u^\infty \psi(v)dv + \int_u^\infty (1-P(v))dv + \int_0^\infty \psi(v)(1-P(u-v))dv,$$

which is easily seen to be equivalent to (33). Thus we have proved (33) for all $u > 0$. However, according to the definition (22), $\psi(u)$ is continuous to the right for all $u \geqq 0$,[5] so that (33) will also hold for $u = 0$.

3.5. *Solution of the integral equation for $\psi(u)$.*—The form of the integral equations (32) and (33) resembles the well-known Wiener-Hopf equation (see Paley-Wiener [**21**]), and in fact both these equations can be solved by the method applied by Wiener and Hopf to the equation considered by them. For the case of equation (32) this has been shown by Täcklind [**27**], who then obtained expressions for $\psi(u)$ by the limit passage $h \to 0$. Here we shall apply the Wiener-Hopf method directly to equation (33) satisfied by $\psi(u)$.

The function $\psi(u)$ has been defined by (22) for $u \geqq 0$, and equation (33) holds for all these values of u. We now define $\psi(u)$ for negative values of u by writing

$$\psi(u) = 0 \qquad \text{for} \qquad u < 0.$$

[5] From the form (37) of the integral equation, it is easily seen that $\psi(u)$ is continuous (not only continuous to the right) for all $u > 0$.

We further introduce the function $\omega(u)$ defined for all real u as follows:

$$\omega(u) = \begin{cases} 0 & \text{for} \quad u > 0 \\ \int_u^\infty Q(v)dv + \int_0^\infty Q(u-v)\psi(v)dv & \text{for} \quad u \leqq 0 \,. \end{cases}$$

By means of these definitions, we can now write equation (33) in the following extended form

$$\psi(u) + \omega(u) = \int_u^\infty Q(v)dv + \int_{-\infty}^\infty Q(u-v)\psi(v)dv \,, \tag{38}$$

which will be seen to hold for all real u.

The importance of this transformation lies in the fact that the integral in the last term is now extended over the whole real axis, and not as in (33) over only the positive half of the axis. This renders it possible to give the equation a simpler form by applying a Fourier transformation. As in part 2, we denote by $s = \sigma + i\tau$ a complex variable, and define:

$$\begin{aligned} \bar{\psi}(s) &= \int_{-\infty}^\infty e^{su}\psi(u)du = \int_0^\infty e^{su}\psi(u)du \,, \\ \bar{\omega}(s) &= \int_{-\infty}^\infty e^{su}\omega(u)du = \int_{-\infty}^0 e^{su}\omega(u)du \,. \end{aligned} \tag{39}$$

It then follows from (26) that the function $\bar{\psi}(s)$ is analytic and regular in the half-plane $\sigma < R$, where the corresponding integral is absolutely convergent. From the definition of $\omega(u)$, it further follows that $\omega(u)$ is bounded for all $u \leqq 0$, and accordingly the function $\bar{\omega}(s)$ will be analytic and regular, and the corresponding integral absolutely convergent, in the half-plane $\sigma > 0$.

We further define

$$q(s) = \int_{-\infty}^\infty e^{su}Q(u)du = -\frac{1}{s}\int_{-\infty}^\infty e^{su}\,dQ(u) = \frac{p(s)-1}{(p_1+\lambda)s} \,, \tag{40}$$

where $p(s)$ is given by (3). By condition I of 3.2, the function $q(s)$ is analytic and regular in the strip $0 < \sigma < R$, where the corresponding integrals are absolutely convergent.

We now multiply equation (38) by $e^{su}\,du$, and integrate with respect to u over $(-\infty, \infty)$, assuming $0 < \sigma < R$. All the above integrals will then be absolutely convergent, and we obtain

$$\bar{\psi}(s) + \bar{\omega}(s) = \frac{1}{s}\,q(s) + \bar{\psi}(s)q(s) \,,$$

which can be written

$$1 - \frac{p(s)}{1+(p_1+\lambda)s} = \frac{(p_1+\lambda)s}{1+(p_1+\lambda)s} \cdot \frac{1-s\bar{\omega}(s)}{1+s\bar{\psi}(s)} \,. \tag{41}$$

Let us now consider the first member of this equation. By condition II of 3.2 we have for any s in the strip $0 < \sigma < R$

$$\left|\frac{p(s)}{1+(p_1+\lambda)s}\right| \leqq \frac{p(\sigma)}{1+(p_1+\lambda)\sigma} < 1 .$$

It follows that the branch of the multivalued function

$$-\log\left(1-\frac{p(s)}{1+(p_1+\lambda)s}\right)$$

which is real and positive for real s between 0 and R will be analytic and regular throughout the strip, and may be developed in the MacLaurin series

$$-\log\left(1-\frac{p(s)}{1+(p_1+\lambda)s}\right) = \sum_1^\infty \frac{1}{n}\left(\frac{p(s)}{1+(p_1+\lambda)s}\right)^n$$

$$= \sum_1^\infty \frac{1}{n(1+(p_1+\lambda)s)^n}\int_{-\infty}^{\infty} e^{sx}d[P^{n*}(x)-1]$$

where the series will be absolutely convergent for every s in $0<\sigma<R$. We have, however,

$$\frac{1}{(1+(p_1+\lambda)s)^n} = \frac{1}{(n-1)!}\int_0^\infty y^{n-1}e^{-(1+(p_1+\lambda)s)y}dy .$$

Introducing this into the preceding relation, we obtain

$$(42) \qquad -\log\left(1-\frac{p(s)}{1+(p_1+\lambda)s}\right) = \int_{-\infty}^{\infty} e^{sx}dM(x) ,$$

where

$$(43) \qquad M(x) = \sum_1^\infty \frac{1}{n!}\int_0^\infty y^{n-1}e^{-y}[P^{n*}(x+(p_1+\lambda)y)-1]dy ,$$

provided the series in the expression for $M(x)$ can be shown to be absolutely convergent. In order to show this, let r be a real number such that $0<r<R$. We then have for any real x

$$e^{rx}(1-P^{n*}(x)) \leqq \int_x^\infty e^{ry}dP^{n*}(y) \leqq \int_{-\infty}^\infty e^{ry}dP^{n*}(y) = (p(r))^n ,$$

$$0 \leqq 1-P^{n*}(x) \leqq e^{-rx}(p(r))^n .$$

By means of this inequality, it follows that the series (43) is absolutely convergent for every x, and that we have

$$-e^{-rx}\sum_1^\infty \frac{1}{n}\left(\frac{p(r)}{1+(p_1+\lambda)r}\right)^n \leqq M(x) \leqq 0 .$$

Since r can be chosen arbitrarily in the interval $(0, R)$, it further follows that $M(x)$ is a never decreasing function of x such that $M(x) \leqq 0$ for all x and that, for any given $\epsilon > 0$,

$$|M(x)| = O(e^{\epsilon|x|}) \qquad \text{as} \quad x \to -\infty ,$$

$$|M(x)| = O(e^{-(R-\epsilon)x}) \qquad \text{as} \quad x \to +\infty .$$

Consequently the integral in the second member of (42) is absolutely convergent for $0 < \sigma < R$, and represents the function in the first member throughout this strip.

In view of a later application we observe that according to (42) we have, for fixed σ in $0 < \sigma < R$,

$$\lim_{\tau \to \pm\infty} \int_{-\infty}^{\infty} e^{(\sigma + i\tau)x} dM(x) = 0 ,$$

which shows that the never decreasing function $M(x)$ is continuous for all x. It follows that

$$\lim_{\sigma \to -\infty} \int_0^{\infty} e^{\sigma x} dM(x) = 0 . \tag{44}$$

Let us now write

$$A(s) = \exp\left[\int_0^{\infty} e^{sx} dM(x)\right] ,$$
(45)
$$B(s) = \exp\left[-\int_{-\infty}^{0} e^{sx} dM(x)\right] .$$

It will then be seen that the functions $A(s)$ and $(A(s))^{-1}$ are both regular in the half-plane $\sigma < R$, while $B(s)$ and $(B(s))^{-1}$ are regular for $\sigma > 0$. Further, given any $\epsilon > 0$, the functions $A(s)$ and $(A(s))^{-1}$ are bounded for $\sigma \leqq R - \epsilon$, while $B(s)$ and $(B(s))^{-1}$ are bounded for $\sigma \geqq \epsilon$.

According to (42), we now have for $0 < \sigma < R$

$$1 - \frac{p(s)}{1 + (p_1 + \lambda)s} = \frac{B(s)}{A(s)} , \tag{46}$$

so that (41) can be written in the form

$$(1 + (p_1 + \lambda)s)\frac{1 + s\bar{\psi}(s)}{A(s)} = (p_1 + \lambda)s\frac{1 - s\bar{\omega}(s)}{B(s)} .$$

The first member of this equation represents a function which is analytic and regular for $\sigma < R$, while the second member has the same properties for $\sigma > 0$. Since the two functions coincide in the strip $0 < \sigma < R$, each of them represents an entire function, which is analytic and regular for all finite values of s. From (39) it easily follows that $s\bar{\psi}(s)$ is bounded for $\sigma < R - \epsilon$, while $s\bar{\omega}(s)$ is bounded for $\sigma > \epsilon$, for any $\epsilon > 0$. According to the properties of $A(s)$ and $B(s)$ given above, it will then be seen that the modulus of the entire function has an upper bound of the form $c_1 + c_2|s|$, where c_1 and c_2 are constants. It follows that the entire function reduces to a polynomial of the first order, so that we have

$$(1 + (p_1 + \lambda)s)\frac{1 + s\bar{\psi}(s)}{A(s)} = (p_1 + \lambda)s\frac{1 - s\bar{\omega}(s)}{B(s)} = \alpha + \beta s , \tag{47}$$

with constant α and β. Taking $s = 0$ we obtain, since $\bar{\psi}(s)$ and $(A(s))^{-1}$ are regular at the origin,

$$\alpha = \frac{1}{A(0)} .$$

In order to find the value of β, we shall have to distinguish two cases, according as $p_1 + \lambda$ is > 0 or < 0.

If $p_1 + \lambda > 0$, the unique zero of the linear function $1 + (p_1 + \lambda)s$ is situated in the half-plane $\sigma < R$, where $\bar{\psi}(s)$ and $(A(s))^{-1}$ are both regular. It follows that $\alpha + \beta s$ has a zero at the same point, so that we have

$$\beta = \frac{p_1 + \lambda}{A(0)}$$

and (47) gives

$$\bar{\psi}(s) = \frac{1}{s}\left(\frac{A(s)}{A(0)} - 1\right). \tag{48}$$

On the other hand, if $p_1 + \lambda < 0$, we shall allow s to tend to $-\infty$ through real values in (47). According to (44) and (45) we then have $A(s) \to 1$. Further, by (39) we have

$$s\bar{\psi}(s) = -\psi(0) - \int_0^\infty e^{su} d\psi(u) ,$$

and since $\psi(u)$ is continuous for $u \geqq 0$ (see footnote 5), it follows that $s\bar{\psi}(s) \to -\psi(0)$. Now, when $p_1 + \lambda < 0$ and $u = 0$, a sample function $X(t)$ will almost surely fall below the level $-u = 0$ within an arbitrarily small time interval $(0, t_0)$. It follows that in this case we have $\psi(0) = 1$, and

$$\frac{1 + s\bar{\psi}(s)}{A(s)} \to 0 ,$$

so that we obtain

$$\beta = 0 ,$$

and accordingly

$$\bar{\psi}(s) = \frac{1}{s}\left(\frac{1}{1 + (p_1 + \lambda)s} \cdot \frac{A(s)}{A(0)} - 1\right). \tag{49}$$

Now according to (39) the function $\psi(\sigma + i\tau)/\sqrt{2\pi}$ is, for any fixed σ between 0 and R, the Fourier transform of the function $e^{\sigma u}\psi(u)$, which belongs to L_2 over $(-\infty, \infty)$, since we have defined $\psi(u) = 0$ for $u < 0$. The reciprocal Fourier formula then gives

$$\psi(u) = \frac{1}{2\pi i}\int_{\sigma - i\infty}^{\sigma + i\infty} e^{-su}\bar{\psi}(s)ds , \qquad (0 < \sigma < R) , \tag{50}$$

where the integral is to be extended over the vertical through the point $s = \sigma$ in the s plane, and will certainly exist as a limit in the mean of the integral from $\sigma - iT$ to $\sigma + iT$, as $T \to \infty$.

According as $p_1 + \lambda > 0$, or $p_1 + \lambda < 0$, we now replace $\bar{\psi}(s)$ in (50) by expression (48) or (49), and then obtain an explicit expression for the ruin function $\psi(u)$, which is the solution of the integral equation (33). Since the contribution to (50) arising from the term $-1/s$ in (48) and (49) is easily seen to be zero, the resulting expression becomes

$$\psi(u) = \frac{1}{2\pi i A(0)} \int_{\sigma - i\infty}^{\sigma + i\infty} \frac{A(s)}{s} e^{-su} ds \tag{51}$$

for $p_1 + \lambda > 0$, and

$$\psi(u) = \frac{1}{2\pi i A(0)} \int_{\sigma - i\infty}^{\sigma + i\infty} \frac{A(s)}{s(1 + (p_1 + \lambda)s)} e^{-su} ds \tag{52}$$

for $p_1 + \lambda < 0$. Introducing into the two last relations the expression

$$A(s) = \frac{1 + (p_1 + \lambda)s}{1 + (p_1 + \lambda)s - p(s)} B(s)$$

obtained from (46), and observing, in the case of (51), that a simple application of Cauchy's theorem shows that we have

$$\int_{\sigma - i\infty}^{\sigma + i\infty} \frac{B(s)}{s} e^{-su} ds = 0 ,$$

we find that the solution can in all cases be expressed by the final formula

$$\psi(u) = \frac{1}{2\pi i A(0)} \int_{\sigma - i\infty}^{\sigma + i\infty} \frac{k(s)}{s(1 + (p_1 + \lambda)s - p(s))} B(s) e^{-su} ds \tag{53}$$

where

$$k(s) = \begin{cases} p(s) , & p_1 + \lambda > 0 , \\ 1 , & p_1 + \lambda < 0 . \end{cases}$$

σ can still be given any value between 0 and R. However, the integral in (53) is easily seen to be absolutely convergent, and thus exists as an integral in the ordinary sense, and not only as a limit in the mean.

3.6. *Asymptotic properties of* $\psi(u)$.—So far, we have in this part worked only with assumptions (24) with respect to $P(x)$. We shall now introduce the following additional assumption:

$$\int_0^\infty e^{\sigma x} dP(x) < \infty \quad \text{for some} \quad \sigma = R + \alpha , \quad \text{where } \alpha > 0 . \tag{54}$$

It will then be seen from the conditions I and II of 3.2 that the point $s = R$ is a zero of the function

$$1 + (p_1 + \lambda)s - p(s) ,$$

and since the second derivative $-p''(s)$ is certainly negative at $s = R$, this will be a simple zero. Further, since we have for $0 < \sigma \leqq R$, $\tau \neq 0$,

$$|1 + (p_1 + \lambda)s - p(s)| > 1 + (p_1 + \lambda)\sigma - p(\sigma) \geqq 0 ,$$

the point $s = R$ will be the only zero in the domain $0 < \sigma \leqq R$. The domain $R < \sigma < R + \alpha$, $|\tau| > \tau_0$ will certainly be free from zeros if τ_0 is sufficiently large. It then follows that the positive quantity α in (54) can be chosen such that, for $0 < \sigma \leqq$

$R + \alpha$, the function $1 + (p_1 + \lambda)s - p(s)$ is regular and different from zero, except at the point $s = R$, which is a zero of the first order.

We can then transform the integral in (53) by applying Cauchy's theorem to the rectangle formed by the points $\sigma \pm iT$ and $R + \alpha \pm iT$, and then allowing T to tend to infinity. In this way we obtain, taking account of the residue at the pole $s = R$ of the integrand,

$$\psi(u) = Ce^{-Ru} + \frac{1}{2\pi i A(0)} \int_{R+\alpha-i\infty}^{R+\alpha+i\infty} \frac{k(s)}{s(1 + (p_1 + \lambda)s - p(s))} B(s)e^{-su} ds ,$$

where

(55) $$C = \frac{k(R)B(R)}{A(0)R(p'(R) - p_1 - \lambda)} .$$

Since the integral between the limits $R + \alpha \pm i\infty$ is still absolutely convergent, we thus have the asymptotic relation

(56) $$\psi(u) = Ce^{-Ru} + O(e^{-(R+\alpha)u})$$

as $u \to \infty$, the constant C being given by (55).

3.7. *Some particular cases.*—We shall consider the particular cases when $P(0)$ has one of the two extreme values 0 or 1. In terms of the insurance applications, the case $P(0) = 0$ signifies that all sums at risk are positive (no insurance of annuity type), while $P(0) = 1$ signifies that all sums at risk are negative (business of pure annuity type).

Case 1: $P(0) = 0$. In this case it follows from (34) that $Q(u) = 0$ for $u < 0$, and then the function $\omega(u)$ introduced in 3.5 will be

$$\omega(u) = \int_0^\infty Q(u)du = \frac{p_1}{p_1 + \lambda}$$

for all $u < 0$, so that we obtain from (39)

$$\bar{\omega}(s) = \frac{p_1}{(p_1 + \lambda)s} .$$

Then (47) gives

$$\frac{\lambda s}{B(s)} = \alpha + \beta s .$$

Now in this case we certainly have $p_1 > 0$, and a fortiori $p_1 + \lambda > 0$. Introducing the values of α and β obtained in 3.5 for the case $p_1 + \lambda > 0$, we obtain

$$\frac{\lambda s}{B(s)} = \frac{1 + (p_1 + \lambda)s}{A(0)} ,$$

or

$$\frac{B(s)}{A(0)} = \frac{\lambda s}{1 + (p_1 + \lambda)s} .$$

Hence in this case the expression (55) of the constant C in the asymptotic relation (56) takes the simple form

(57) $$C = \frac{\lambda}{p'(R) - p_1 - \lambda} .$$

CASE 2: $P(0) = 1$. In this case (34) shows that $Q(u) = 0$ for $u \geqq 0$. The integral equation (33) then reduces to

$$\psi(u) = \int_u^\infty Q(u-v)\psi(v)dv = \frac{1}{p_1 + \lambda}\int_{-\infty}^0 \psi(u-v)P(v)dv \, ,$$

which is easily seen to be satisfied by $\psi(u) = Ce^{-Ru}$ for any value of the constant C. This is the only case when the solution of the integral equation is indetermined in this sense.

Now in the case $P(0) = 1$ we have $p_1 + \lambda < 0$ (the case $p_1 + \lambda \geqq 0$ having been excluded in 3.2). However, for the case $p_1 + \lambda < 0$, we have in 3.5, in connection with the determination of the constant β in (47), made use of the condition $\psi(0) = 1$, which was not obtained from the integral equation (33), but was deduced by a direct consideration of the behavior of the sample functions $X(t)$ in our stochastic process. In the present particular case, this condition supplies the uniqueness that the integral equation fails to give, and we thus obtain the simple explicit expression

$$\psi(u) = e^{-Ru} \tag{58}$$

for the ruin function in this case.

The same expression can also be obtained from (53), where it is easily seen that, in the present case, the integrand is regular in the whole half-plane $\sigma > 0$, except at the pole $s = R$. By an application of Cauchy's theorem, we then readily find $\psi(u) = Ce^{-Ru}$, while the condition $\psi(0) = 1$ gives $C = 1$, so that we obtain (58). For further developments concerning the case $P(0) = 1$, see Saxén ([**22**] to [**23**]).

REFERENCES

[1] G. Arfwedson, "Some problems in the collective theory of risk," *Skand. Aktuarietids.*, Vol. 33 (1950), pp. 1–38.

[2] H. Cramér, "Review of F. Lundberg," [16], *Skand. Aktuarietids.*, Vol. 9 (1926), pp. 223–245.

[3] H. Cramér, "On the mathematical theory of risk," *Skand. Jubilee Volume.* Stockholm, 1930.

[4] H. Cramér, *Random Variables and Probability Distribution.* Cambridge Tracts in Math., No. 36. London, Cambridge University Press, 1937.

[5] H. Cramér, *Sur un nouveau théoreme-limite de la théorie des probabilités.* Actualités Scientifiques, No. 736. Paris, 1938.

[6] H. Cramér, "Deux conférences sur la théorie des probabilités," *Skand. Aktuarietids.*, Vol. 24 (1941), pp. 34–69.

[7] H. Cramér, "Deu Lundbergska riskteorien och teorien för stokastiska processer," *F. Lundberg Jubilee Volume.* Stockholm, 1946.

[8] A. Davidson, "Om ruinproblemet i den kollektiva riskteorien under antagande av variabel säkerhetsfaktor," *F. Lundberg Jubilee Volume.* Stockholm, 1946.

[9] J. L. Doob, *Stochastic Processes.* New York, Wiley, 1953.

[10] J. Dubourdieu, *Théorie mathématique des assurances. I, Théorie mathématique du risque dans les assurances de répartition.* Paris, Gauthier-Villars, 1952.

[11] F. Esscher, "On the probability function in the collective theory of risk," *Skand. Aktuarietids.*, Vol. 15 (1932), pp. 175–195.

[12] C. G. Esseen, "Fourier analysis of distribution functions," *Acta Math.*, Vol. 77 (1945), pp. 1–125.

[13] W. Feller, "Generalization of a probability limit theorem by Cramér," *Trans. Amer. Math. Soc.*, Vol. 54 (1943), pp. 361–372.

[14] I. N. Herstein and J. Milnor, "An axiomatic approach to measurable utility," *Econometrica*, Vol. 21 (1953), pp. 291–297.

[15] F. Lundberg, "Zur Theorie der Rückversicherung," *Verhandl. Kongr. Versicherungsmath.* Wien, 1909.

[16] F. Lundberg, *Försäkringsteknisk riskutjämning, 1–2.* Stockholm, 1926–1928.

[17] F. Lundberg, "Über die Wahrscheinlichkeitsfunktion einer Risikenmasse," *Skand. Aktuarietids.*, Vol. 13 (1930), pp. 1–83.

[18] F. Lundberg, "Some supplementary researches on the collective risk theory," *Skand. Aktuarietids.*, Vol. 15 (1932), pp. 137–158.

[19] F. Lundberg, "On the numerical application of the collective risk theory," *De Förenade Jubilee Volume.* Stockholm, 1934.

[20] J. von Neumann and O. Morgenstern, *Theory of Games and Economic Behavior*, 2d ed. Princeton University Press, 1947.

[21] R. Paley and N. Wiener, *Fourier Transforms in the Complex Domain.* Am. Math. Soc., Colloq. Pub., Vol. 19. New York, 1934.

[22] T. Saxén, "On the probability of ruin in the collective risk theory for insurance enterprises with only negative risk sums," *Skand. Aktuarietids.*, Vol. 31 (1948), pp. 199–228.

[23] T. Saxén, "Sur les mouvements aleatoires et le problème de ruine de la théorie du risque collective," *Soc. Sci. Fenn. Comment. Phys.-Math.*, Vol. 16 (1951), pp. 1–55.

[24] C. O. Segerdahl, *On Homogeneous Random Processes and Collective Risk Theory* (Thesis). Stockholm, 1939.

[25] C. O. Segerdahl, "Über einige risikotheoretische Fragestellungen," *Skand. Aktuarietids.*, Vol. 25 (1942), pp. 43–83.

[26] C. O. Segerdahl, "Some properties of the ruin function in the collective theory of risk," *Skand. Aktuarietids.*, Vol. 31 (1948), pp. 46–87.

[27] S. Täcklind, "Sur le risque de ruine dans des jeux inéquitables," *Skand. Aktuarietids.*, Vol. 25 (1942), pp. 1–42.

47.

Collective risk theory: A survey of the theory from the point of view of the theory of stochastic process

7th Jubilee Volume of Skandia Insurance Company Stockholm, 51–92 (1955)

1. Introduction

1. 1. Individual and collective risk theory. "The object of the theory of risk is to give a mathematical analysis of the random fluctuations in an insurance business, and to discuss the various means of protection against their inconvenient effects." (Cramér, **19**, p. 7.)[1] These words are taken from a paper in the 1930 Jubilee Volume of the Försäkringsaktiebolaget Skandia. — In the present paper, an attempt will be made to give a survey of certain parts of the development of the theory of risk which has taken place during the 25 years passed since 1930.

The oldest approach to the subject is to regard the total risk business of an insurance company as the resulting effect of what happens to all the individual policies issued by the company. Accordingly, this approach has been named *individual risk theory*. In this theory, the gain or loss of the company arising during a given time on one single policy is regarded as a random variable, and the mathematical development of the theory is based on a study of the probability distributions of variables of this type. The total gain or loss of the company during the same time will then be the sum of all the random variables associated with individual policies in force with the company. According to the *central limit theorem* of mathematical probability theory, this sum will be approximately normally distributed if the number of policies is large. Having recourse to appropriate data concerning the types and sums insured on all individual policies, we shall thus be able to find approximate values of various probabilities connected with the gain or loss of the company under specified conditions.

On the other hand, in *collective risk theory* as founded and developed by Dr. Filip Lundberg in a series of papers (**42—48**), no attention is paid to individual policies, the risk business of an insurance company being regarded as a whole, as a continuous game of chance between the company on one side, and the totality of policyholders on the other. In the course of this game, certain random events known as *claims* occur from time to time, and have to be settled by the company, while on the other hand the company receives a continuous flow of

[1] Numbers in bold types refer to the Bibliography at the end of the paper.

5

risk premiums from the policyholders.[1] By means of certain simplifying assumptions, it becomes possible to study the probability distributions of the fundamental random variables associated with this game, such as the total amount of claims occurring during a given time interval, the total gain of the company arising during the same interval, etc.

In the present paper, we shall only be concerned with the collective theory. Mathematically speaking, collective risk theory forms a part of the general *theory of stochastic processes*, which in recent years has had a powerful development, and has found a great number of important practical applications. The object of the present paper is to give an account of the most important results so far obtained within collective risk theory, and to show that these can be presented from a unifying point of view: the point of view of the theory of stochastic processes.

1. 2. Some fundamental problems of collective risk theory. Consider the development of the risk business of a certain company, starting from an arbitrarily chosen zero point on the time scale. Let $Y(t)$ denote the total amount of claims occurring up to a certain time $t > 0$, while $X(t)$ denotes the total gain (risk premiums received less claims paid) realized by the company up to the time t. Any amount should always be counted with its proper sign, and a loss will be counted as a negative gain.

From a mathematical point of view, $X(t)$ and $Y(t)$ are random variables, and in the first place we shall be interested in studying the probability distributions of these variables for a fixed value of t. In particular, we may want to compute the numerical values of probabilities such as the probability that $X(t)$ will be negative, the probability that $X(t)$ will fall below some given negative value $-u$, the probability that the amount of claims $Y(t)$ will exceed some given positive value a, etc. It will appear that the values of probabilities of this type depend on various factors, such as the numbers and types of policies in force, the loading of the risk premiums, the reinsurance practice adopted by the company, etc.

One of the most important problems of collective risk theory will be to investigate the probability distributions of variables like $X(t)$ and $Y(t)$, with particular regard to their dependence on the various factors just mentioned. Once we have reached a sufficiently accurate knowledge of this dependence, we shall be in a position to judge the effect on the risk business of proposed changes in loading, reinsurance policy, etc.

Further, we may consider the gain $X(t)$, or the amount of claims $Y(t)$, not

[1] In some branches of insurance, such as life annuity insurance, the situation will be reversed, so that the company is continuously paying out sums (life annuities) to the policyholders, while certain discrete sums (annuity reserves) are received by the company as the effect of random events (deaths of policyholders). If it is kept in mind throughout the paper that in general any amount paid or received by the company should be regarded as capable of being negative as well as positive, our considerations will be found to be perfectly general (cf. 3. 1).

only for a fixed value of t, but rather as a function of the real and positive variable t. The development of, say, the function $X(t)$ as t increases will characterize the development of the risk business of the company. The value assumed by $X(t)$ at any time t will be determined by certain random events, and so will the change of this value during any subsequent time interval Δt:

$$\Delta X(t) = X(t+\Delta t) - X(t).$$

According to the case, we may want to consider t as a continuous variable, capable of assuming any positive value, or as a discrete variable, which may only assume values belonging to some given sequence, say $t_1, t_2, \ldots$. (The time points t_i may, e.g., be taken as the end points of successive business periods of the company.) In the former case we shall say that $X(t)$ is a *random function* of the continuous variable t, while in the latter case the variables $X(t_1), X(t_2), \ldots$ form a *random sequence* of variables. — Similar remarks apply, of course, also to $Y(t)$.

One of the main problems that arise when we consider the random function or random sequence generated by, say, the gain $X(t)$ is the *ruin problem*. Suppose that, at time $t = 0$, the company has at its disposal a certain initial capital u, which is available for covering the losses due to random fluctuations. We may take account of this by fixing the initial value of $X(t)$ at $X(0) = u$. We then encounter the important problem of calculating the *probability of ruin*, i.e., the probability that $X(t)$ will fall below zero for *at least one* of the values of t that we choose to consider. According to the case, we may try to find the probability that ruin will take place *before a given time* T or, taking $T = \infty$, the probability of ruin *at any future time point*.

The probability of ruin will, of course, depend on the value of the initial capital u. In addition it will depend, like the probabilities considered above in connection with the random variables $X(t)$ and $Y(t)$ for a fixed value of t, on the loading of risk premiums, the reinsurance policy of the company, etc. It will be important to study the probability of ruin in its dependence of all these factors.

1.3. Summary of the present paper. Any process in nature, the development of which can be described in terms of random sequences or random functions, is known as a *stochastic process*. From the remarks made in the preceding section, it will be seen that the risk business of an insurance company forms a particular case of a stochastic process.

The mathematical theory of stochastic processes has made very considerable progress during recent years. In the present paper we shall begin by giving, in chapter 2, a very brief review of the basic concepts of this theory, in so far as they will be required in the sequel. For further details and developments concerning general probability distributions and stochastic processes, we must refer to treatises such as those by Kolmogoroff (39), Doob (30, 31), Blanc-Lapierre and Fortet (13), and Mann (51).

In chapter 3 we proceed to a study of the risk process. Certain fundamental assumptions concerning the structure of the process are introduced, and the basic random functions attached to the process are defined. It is shown that, under the assumptions just mentioned, the risk process is a stochastic process belonging to the class known as *processes with stationary and independent increments*. The probability distributions associated with the process are introduced and studied.

Chapter 4 is concerned with the probability distributions of the random variables attached to the process, for a fixed value of the time t. The asymptotic behaviour of these distributions for large values of t is investigated. Certain asymptotic formulas for the corresponding distribution functions are deduced, and their efficiency is illustrated by means of numerical examples.

The ruin problem is treated in chapter 5. The measure-theoretic tools introduced in chapter 2 are here applied to this problem, and used to give rigorous definitions of various ruin probabilities. To some readers, this procedure may seem unnecessarily complicated. So far, however, the author's attempts to gain real simplification without sacrifice of mathematical rigour have proved unsuccessful. — Certain integral equations satisfied by the most important ruin probabilities are deduced and solved. In this way, explicit analytic expressions for the ruin probabilities are obtained, and their asymptotic properties are studied.

Finally, we discuss in chapter 6 certain generalizations of the theory, some of which are of great importance for the applications to various practical insurance problems.

2. General Notions on Probability Distributions and Stochastic Processes

2. 1. Finite-dimensional spaces. In elementary probability theory, the concept of a *random variable* plays an important part. A random variable X is then defined as a quantity which, in successive observations, may assume different real values, each individual value being determined by some complex of random causes. The *probability distribution* of X is regarded as known when, with any given interval I on the real line, we are able to associate a definite number $P(I)$, the *probability* that X will take a value belonging to I.

If the probability associated with the particular interval $-\infty < X \leqq x$ is denoted by $F(x)$:

$$F(x) = P(X \leqq x),$$

we have for any $a < b$

$$P(a < X \leqq b) = F(b) - F(a), \tag{1}$$

and it follows that the probability distribution is completely known as soon as the *distribution function* (d. f.) $F(x)$ is given for all real values of x. Every d.f. is a never decreasing function of x, which is everywhere continuous to the right, and is such that $F(-\infty) = 0$, $F(+\infty) = 1$.

When $F(x)$ is given, a definite numerical probability will be associated not only with any interval I but, more generally, with any *Borel set* of points on the line. The class of all Borel sets forms a particular case of a *Borel field of sets*, i.e., a class of sets which is closed under finite and enumerable repetitions of the operations of addition, multiplication, and subtraction of sets, and which includes the particular set R, where R denotes the whole real line (*not* including the point at infinity). The Borel sets are, by definition, all sets belonging to *the smallest Borel field that includes all intervals*. In particular, any finite or enumerable sum of intervals is thus a Borel set. (Cf. Cramér, **23**.)

The probability $P(S)$ can now be regarded as a *set function*, defined for all Borel sets S. In the particular case when S is an interval $a < X \leqq b$, the value of this set function is given by (1). The set function $P(S)$ has the following three fundamental properties:

(A) $P(S) \geqq 0$ for all Borel sets S,

(B) $P(R) = 1$, where R is the whole real line,

(C) $P(S) = P(S_1)+P(S_2)+\ldots\ldots$, if $S = S_1+S_2+\ldots\ldots$
where any two of the S_i are disjoint Borel sets, so that $S_i\, S_j = 0$ for $i \neq j$.

Any set function $P(S)$ having these three properties is known as a *probability measure*, and defines a probability distribution on the line R. A probability distribution may be defined, according to the case, by its d.f. $F(x)$, or by the corresponding probability measure $P(S)$. Each of these determines the other completely.

Similarly, we may consider a *random point* or *random vector* $X = (X_1, \ldots, X_n)$ in the n-dimensional space R_n with the n real coordinates $X_1, \ldots, X_n$. The Borel sets of points in this space are all sets belonging to the smallest Borel field of sets including all n-dimensional intervals in R_n. The probability distribution of the random vector X is determined by a probability measure, which is a set function $P(S)$ defined for all Borel sets S in R_n, and satisfying the above conditions A)—C) where, of course, R has to be replaced by R_n. Alternatively, the same probability distribution may be defined by the n-dimensional distribution function $F(x_1, \ldots, x_n)$ given by the relation

$$F(x_1, \ldots, x_n) = P(X_1 \leqq x_1, \ldots, X_n \leqq x_n).$$

2. 2. General spaces. The theory of random variables and probability distributions in spaces of a finite number of dimensions is sufficient for the mathematical treatment of the majority of problems considered in classical probability theory. In many recent applications, however, we encounter problems where it is necessary to consider random elements of a more general nature, such as a random vector $X = (X_1, X_2, \ldots)$ with an infinite sequence of coordinates, or a random function $X(t)$ of a continuous variable t. We have, in fact, already seen above in 1. 2 that important problems of this type will appear in collective risk theory. Accordingly we shall have to discuss the definition and fundamental properties of probability distributions in spaces of random sequences or random functions. In the present section we shall consider the general case of a probability distribution in a completely arbitrary space, while the two following sections will be concerned with the cases of random sequences and random functions.

Let Ω denote a perfectly arbitrary space, the points ω of which may be mathematical entities of any kind. Let Σ denote a set of points ω. Just as in the case of the real line R, a Borel field $\mathfrak{B}$ of sets Σ is defined as a class of sets Σ which is closed under finite and enumerable repetitions of the operations of addition, multiplication, and subtraction of sets, and which includes the particular set Ω.

Suppose that, in the space Ω, we are given a Borel field of sets $\mathfrak{B}$ and a set function $\Pi(\Sigma)$, defined for all sets Σ belonging to $\mathfrak{B}$, and such that the conditions $A)$—$C)$ of the preceding section are satisfied (R being, of course, in this case replaced by Ω). The set function $\Pi(\Sigma)$ will then be called a *probability measure* on the sets of $\mathfrak{B}$, and we shall say that this measure defines a *probability distribution* in the space Ω, the value $\Pi(\Sigma)$ being interpreted as the probability of the event that the random point ω takes a "value" belonging to the set Σ.

We now consider a transformation

$$\omega' = g(\omega),$$

uniquely defined for every point ω of Ω, and taking ω into some point ω' of another, perfectly arbitrary space Ω'. We shall *not* suppose that this transformation will necessarily map Ω on the whole of Ω', nor that the inverse transformation is unique.

If Σ' denotes any set of points ω', the *inverse image* of Σ' will be defined as the set Σ consisting of all points ω such that the transform $\omega' = g(\omega)$ belongs to Σ'. (If, in particular, no point ω is mapped on a point ω' belonging to Σ', the inverse image of Σ' will thus be the empty set.) Any set Σ' such that its inverse image Σ belongs to the given Borel field $\mathfrak{B}$ in Ω will be called a *Π-measurable* set in Ω'. If $\mathfrak{B}'$ denotes the class of all Π-measurable sets in Ω', it is easily seen that $\mathfrak{B}'$ is also a Borel field of sets. If a set function $\Pi'(\Sigma')$ is defined for all sets Σ' of $\mathfrak{B}'$ by the relation

$$\Pi'(\Sigma') = \Pi(\Sigma),$$

$\Pi'(\Sigma')$ will be a probability measure on the sets of $\mathfrak{B}'$. The probability distribution in the image space Ω' defined in this way is known as the probability distribution *induced* by the transformation $\omega' = g(\omega)$ and the given distribution in the *reference space* Ω.

An important example of a transformation of this type is obtained by considering a real-valued function of ω:

$$X = g(\omega).$$

Suppose that, for every real k, the set of all ω such that

$$g(\omega) < k$$

belongs to the given Borel field $\mathfrak{B}$. This fact will be expressed by saying that $g(\omega)$ is a *$\mathfrak{B}$-measurable* function of ω. The transformation $X = g(\omega)$ takes every ω into a point X of the real line R, and from the measurability of $g(\omega)$ it easily follows that every interval in R has an inverse image belonging to $\mathfrak{B}$. Thus the induced Borel field $\mathfrak{B}'$ in R includes all intervals, and consequently all Borel sets in R. We thus see that $X = g(\omega)$ is a real random variable, and that the induced probability distribution in R will be the distribution of this random variable X.

Similarly, let $g_1(\omega), \ldots, g_n(\omega)$ be real-valued and $\mathfrak{B}$-measurable functions of ω. The vector $X = (X_1, \ldots, X_n)$, where $X_i = g_i(\omega)$, will then be a random vector in the space R_n, and the induced probability distribution in R_n will be the distribution of the random vector X.

2. 3. Random sequences. When a space Ω is given, a probability distribution on Ω may often be *directly defined*. Thus any never decreasing function $F(x)$ of a real variable x, which is always continuous to the right, and is such that $F(-\infty) = 0$ and $F(+\infty) = 1$, will define a probability distribution on the real line, having $F(x)$ as its d.f. Similarly, any function $F(x_1, \ldots, x_n)$ of n real variables, having the corresponding characteristic properties of an n-dimensional d.f., will define a probability distribution in the space R_n.

We shall now consider the space R_∞ consisting of all points $x = (x_1, x_2, \ldots)$ determined by an enumerably infinite sequence of coordinates, and show how a probability distribution may be directly defined in this space.

The set of all points x in R_∞ satisfying a finite system of inequalities

$$a_i < x_i < b_i \qquad (i = 1, 2, \ldots, n)$$

will be called a *finite-dimensional interval* in R_∞. In the defining inequalities, any sign $<$ may be replaced by $\leqq$, while any a_i may be $-\infty$, and any b_i may be $+\infty$. The *Borel sets* in R_∞ are all sets belonging to the smallest Borel field that includes all finite-dimensional intervals. The class of Borel sets includes all sets in R_∞ that we may have occasion to consider in connection with ordinary analytical operations. A simple case of a Borel set is the set of all points x such that $x_i \geqq k$ for all i. Since this set is the limit, as $n \to \infty$, of the n-dimensional interval defined by $x_i \geqq k$ for $i = 1, 2, \ldots, n$, it is certainly a Borel set. Hence also the complementary set, which consists of all x such that $x_i < k$ for at least one i, is a Borel set.

Suppose that, for every $n = 1, 2, \ldots$, an n-dimensional d.f. $F_n(x_1, \ldots, x_n)$ is given, and that the sequence $\{F_i\}$ satisfies the consistency condition

$$(2) \qquad \lim_{x_n \to +\infty} F_n(x_1, \ldots, x_{n-1}, x_n) = F_{n-1}(x_1, \ldots, x_{n-1})$$

An important theorem due to Kolmogoroff (39) then states[1] that the sequence $\{F_i\}$ determines a unique probability measure $P(S)$ on the Borel sets of R_∞. The probability $P(S)$ is uniquely defined for all Borel sets S. In the particular case when S is a finite-dimensional interval defined, say, by means of inequalities bearing on the n first coordinates $x_1, \ldots, x_n$, the probability $P(S)$ will be identical with the probability assigned to the interval S by the d.f. $F_n(x_1, \ldots, x_n)$.

The probability measure $P(S)$ will then define a probability distribution of

[1] For the particular case of the space $R\infty$, the general theorem proved by Kolmogoroff is easily reduced to the simplified form given here.

the *random sequence* $X_1, X_2, \ldots$, the quantity $P(S)$ being interpreted as the probability of the event that the infinite-dimensional random vector $X = (X_1, X_2, \ldots)$ takes a "value" belonging to the set S.

In the applications to be made in the sequel, we shall encounter the following important example of a directly defined probability distribution of a random sequence. Let $G_1(x), G_2(x), \ldots$ be a given sequence of one-dimensional distribution functions, and take for every $n = 1, 2, \ldots$

$$F_n(x_1, \ldots, x_n) = G_1(x_1) \ldots . G_n(x_n). \tag{3}$$

The consistency condition (2) is evidently satisfied, and it follows that the sequence $\{F_i\}$ defined by (3) determines a unique probability measure on all Borel sets of the space R_∞. It will be seen that, in the probability distribution defined by this measure, the components of the random vector $X = (X_1, X_2, \ldots)$ are *independent* random variables, every X_i having the d.f. G_i.

2. 4. Random functions. We now proceed to consider the space $\mathfrak{X}$ of all finite and real-valued functions $X(t)$ defined for all non-negative values of the real variable t. The choice of this particular range of values for t has been made with regard to the applications that will be made in the sequel, and is by no means essential.

Each "point" of the space $\mathfrak{X}$ is a function $X(t)$, which is characterized by the non-enumerable set of its values for all $t \geqq 0$. Thus we may regard each point $X(t)$ as determined by a non-enumerable set of coordinates, and accordingly speak of the space $\mathfrak{X}$ as a space of non-enumerable dimensionality.

For any finite set of t-values $t_1 < t_2 < \ldots < t_n$, the set of all functions $X(t)$ satisfying the system of inequalities

$$a_i < X(t_i) < b_i \qquad (i = 1, 2, \ldots, n)$$

will be called a *finite-dimensional interval* in $\mathfrak{X}$. The defining inequalities may be modified in the same way as in the case of the space R_∞. The *Borel sets* in $\mathfrak{X}$ are, by definition, all sets belonging to the smallest Borel field that includes all finite-dimensional intervals.

The class of Borel sets in $\mathfrak{X}$ is not, however, sufficiently general to include all sets of functions that we may want to consider in the course of ordinary analytical applications. Thus, e.g., the set of all functions $X(t)$ such that $X(t) \geqq k$ for all $t \geqq 0$ is *not* a Borel set, nor is the complementary set, which consists of all functions $X(t)$ such that $X(t) < k$ for at least one value of $t \geqq 0$. (Cf. Kolmogoroff, **39**, p. 26.) As we shall see below, this fact has important consequences for the applications to the ruin problem of collective risk theory.

A probability measure on the Borel sets of the function space $\mathfrak{X}$ may be directly defined in a similar way as in the case considered in the preceding section.

Suppose that, for any finite sequence of t-values, say $t_1, \ldots, t_n$, we are given an n-dimensional d.f.

$$F_{t_1, \ldots, t_n}(x_1, \ldots, x_n)$$

which remains invariant whenever the t_i and the x_i are subject to the same permutation. Suppose, further, that the family of all $F_{t_1}, \ldots, {}_{t_n}$ satisfies the consistency condition

$$(4) \qquad \lim_{x_n \to +\infty} F_{t_1, \ldots, t_n}(x_1, \ldots, x_n) = F_{t_1, \ldots, t_{n-1}}(x_1, \ldots, x_{n-1}).$$

By the theorem of Kolmogoroff quoted in the preceding section, the family of all $F_{t_1}, \ldots, {}_{t_n}$ then determines a unique probability measure $P(S)$ on all Borel sets S of the space $\mathfrak{X}$. In the particular case when S is a finite-dimensional interval in $\mathfrak{X}$, the probability $P(S)$ coincides with the probability that may be deduced from the relevant $F_{t_1}, \ldots, {}_{t_n}$. The probability measure $P(S)$ defines a probability distribution in the function space $\mathfrak{X}$, the probability $P(S)$ being interpreted as the probability of the event that the random function $X(t)$ will belong to the set S. This distribution has thus been uniquely determined by means of the family of its *finite-dimensional distributions* given by the $F_{t_1}, \ldots, {}_{t_n}$.

Suppose that we are concerned with the mathematical analysis of some stochastic process, the development of which may be described by a random function $X(t)$ of the time t. For any fixed value of t, the value $X(t)$ assumed by this function will then be a random variable. From the conditions of the process, it will often be possible to find the d.f. of this random variable, say $F_t(x)$. Moreover, we may be able to find also the joint distribution of the two random variables $X(t_1)$ and $X(t_2)$, and generally the joint distribution of any finite sequence of variables like $X(t_1), \ldots, X(t_n)$. Since the family of all these finite-dimensional distributions must evidently satisfy the invariance and consistency conditions stated above, they will uniquely determine the probability distribution of the random function $X(t)$ attached to the process, the corresponding probability $P(S)$ being defined for all Borel sets S of the function space $\mathfrak{X}$.

In the applications to collective risk theory, we shall see below that it is possible to proceed along the way just outlined. We shall thus be able to find the probability distribution of the random function $X(t)$, or any other function that we may choose to describe the development of the risk process. This will be a distribution defined on the Borel sets of the space $\mathfrak{X}$ consisting of all possible functions $X(t)$ defined for $t \geqq 0$. As pointed out above, we shall then encounter the difficulty that the class of Borel sets in this space is not sufficiently general to include all those sets of functions that we may find interesting. In connection with the ruin problem we shall, e.g., want to find the probability of the event that the random function $X(t)$ will never fall below some given value k, or the complementary probability that we shall have $X(t) < k$ for at least one $t \geqq 0$.

The corresponding sets of functions are not, however, Borel sets, and accordingly their probabilities are not defined. Thus it seems that the procedure indicated here will be useless for our purpose, since the probability of ruin cannot be defined by means of the finite-dimensional distributions of the process.

In order to overcome this difficulty, two different ways seem possible. In the first place, we may try to *extend the domain of definition* of the probability measure $P(S)$ to some Borel field of sets including all Borel sets, and also those particular sets of functions in which we are interested. If this were possible, our problem would be solved. It is, in fact, possible to work out a solution along these lines, but this requires fairly intricate measure-theoretic considerations, and we shall not enter upon that question here. The other possibility would be to restrict all our considerations to *functions* $X(t)$ *belonging to some appropriate subspace* $\mathfrak{X}_0$ of the space $\mathfrak{X}$, and try to define a probability measure in the restricted space $\mathfrak{X}_0$. If $\mathfrak{X}_0$ is defined as the set of all those functions $X(t)$ which seem to be a priori relevant for an adequate description of our stochastic process, we may hope to be able to define a probability measure for all those sets of *functions belonging to* $\mathfrak{X}_0$ in which we shall be interested.

The latter method is the one that we are going to use in the sequel. The requisite mathematical considerations will be given in the following section, while the application of the method to the risk process will be found in chapter 3.

2. 5. Random functions in a restricted space. Suppose that, as in 2. 2, we are given a space Ω of points ω, a Borel field $\mathfrak{B}$ of sets in Ω, and a probability measure $\Pi(\Sigma)$ defined for all sets Σ belonging to $\mathfrak{B}$. Regarding this as our basic *reference space* and *reference distribution*, we are now going to consider the probability distributions induced in certain other spaces by appropriate transformations, in accordance with 2. 2.

Let $X(t, \omega)$ be a given real-valued function of the two arguments t and ω, where t is real and non-negative, while ω is a point of Ω. Suppose that, for any fixed $t \geqq 0$, the function $X(t, \omega)$ is a $\mathfrak{B}$-measurable function of ω (cf. 2. 2). For every fixed $t \geqq 0$, the transformation

$$X = X(t, \omega)$$

will then, according to 2. 2, define a real random variable X, and the probability distribution on the real line R induced by this transformation will be the distribution of the random variable X.

Similarly, for any finite sequence of t-values, say $t_1, \ldots, t_n$, the transformation

$$X_i = X(t_i, \omega), \quad i = 1, \ldots, n,$$

will take every ω into a point $(X_1, \ldots, X_n)$ of the space R_n, and the induced probability distribution in R_n will be the joint distribution of the random variables $X_1, \ldots, X_n$.

On the other hand, for any fixed ω, the function $X(t, \omega)$ becomes a function

$$X(t) = X(t, \omega) \tag{5}$$

of the real and non-negative variable t. When ω runs through all points of the whole space Ω, we obtain in this way various functions $X(t)$, and we shall denote by $\mathfrak{X}_0$ the set of all functions $X(t)$ that may be generated according to (5), for any fixed ω in Ω.

The set $\mathfrak{X}_0$ will certainly be a subset of the set $\mathfrak{X}$ considered in the preceding section, which consists of all possible functions $X(t)$ defined for $t \geqq 0$. We shall call $\mathfrak{X}_0$ the *restricted function space* generated by the function $X(t, \omega)$. A *finite-dimensional interval* in $\mathfrak{X}_0$ is, by definition, the set of all functions in $\mathfrak{X}_0$ satisfying a finite system of inequalities

$$a_i < X(t_i) < b_i,$$

which may be modified in the same way as in the two preceding sections. The *Borel sets* in $\mathfrak{X}_0$ are all sets belonging to the smallest Borel field of sets including all finite-dimensional intervals. Obviously any Borel set in $\mathfrak{X}_0$ is the product (or intersection) of a Borel set in the general function space $\mathfrak{X}$ and the set $\mathfrak{X}_0$.

According to (5), there corresponds to every point ω of Ω a well-determined function $X(t)$ belonging to $\mathfrak{X}_0$, i.e., a "point" of the space $\mathfrak{X}_0$. Thus (5) may be regarded as a transformation which takes every point ω into a point $X(t)$ of the space $\mathfrak{X}_0$. According to 2. 2, this transformation induces a probability measure $P(S)$, which is defined for all sets S of a certain Borel field $\mathfrak{B}_0$ of sets in $\mathfrak{X}_0$. Since $X(t, \omega)$ is, for every fixed t, a $\mathfrak{B}$-measurable function of ω, it easily follows that every finite-dimensional interval in $\mathfrak{X}_0$ has an inverse image belonging to $\mathfrak{B}$. Hence the Borel field $\mathfrak{B}_0$ includes all finite-dimensional intervals, and consequently all Borel sets in $\mathfrak{X}_0$. The transformation (5) thus induces a probability distribution in $\mathfrak{X}_0$, such that the corresponding probability measure $P(S)$ will be defined at least for all Borel sets in $\mathfrak{X}_0$.

The joint distributions of the random variables $X(t_i, \omega)$ discussed above will constitute the family of *finite-dimensional distributions* corresponding to the induced distribution in $\mathfrak{X}_0$.

Let us now consider once more the situation described in the preceding section, when we are concerned with the mathematical analysis of some stochastic process, the finite-dimensional distributions of which may be deduced from the known conditions of the process. By Kolmogoroff's theorem, the family of all these distributions will determine a probability distribution in the general function space $\mathfrak{X}$. It has been pointed out, however, that certain interesting sets of functions will fall outside the domain of definition of the corresponding probability measure, so that their probabilities will not be defined.

In such a situation, it may sometimes be possible to find, according to the procedure given above in the present section, a reference distribution and a function $X(t, \omega)$ satisfying the following two conditions:

1) The restricted function space $\mathfrak{X}_0$ will contain precisely those functions $X(t)$ which seem a priori relevant for an adequate description of the process under investigation.

2) The finite-dimensional distributions of the distribution in $\mathfrak{X}_0$ induced by the transformation $X(t) = X(t, \omega)$ are identical with the given finite-dimensional distributions of the process.

When this is possible, we shall thus be in possession of a restricted function space $\mathfrak{X}_0$, and a uniquely determined probability distribution in this space, both of which seem to be naturally adapted to a study of the process. In particular we may hope that it will be possible to define, within the restricted space $\mathfrak{X}_0$, probabilities of all those sets of functions that will be of interest in connection with the process.

We shall see below that, in the case of the stochastic processes considered in connection with the collective theory of risk it is, in fact, possible to solve the existence problems by the method outlined above.

It might be asked, however, why we could not simplify the procedure by arguing in the following way. Suppose that we know the finite-dimensional distributions of the process, and also the set $\mathfrak{X}_0$ of all functions $X(t)$ that are "natural" to the process. By Kolmogoroff's theorem, we then know that the finite-dimensional distributions determine a unique probability measure $P(S)$ defined for all Borel sets S of the general function space $\mathfrak{X}$. Why not define a probability distribution in the restricted space $\mathfrak{X}_0$ simply by laying down the rule that any set $S\mathfrak{X}_0$ will have the probability $P(S)$?—The answer is that in general this definition would not be unique, since it may well occur that we shall have $S_1\mathfrak{X}_0 = S_2\mathfrak{X}_0$ for two sets S_1 and S_2 in $\mathfrak{X}$ such that the probabilities $P(S_1)$ and $P(S_2)$ are both defined but unequal. In fact, Doob (29, p. 109) has given a necessary and sufficient condition for the uniqueness of this definition, and it appears (cf. Doob and Ambrose, 32) that one form of this condition is precisely that it must be possible to find an appropriate reference distribution according to the method given above. The simplification gained by the argument discussed here would thus be only apparent.

3. The Risk Process

3.1. Fundamental assumptions. Consider the risk business of an insurance company, in any branch of insurance. Among the totality of policyholders, certain random events known as *claims* will occur from time to time. In the majority of insurance branches, a claim will usually be settled by the payment of a certain sum from the company to the policyholder involved. In return, the company collects premiums from the policyholders. For the moment, we shall only consider the *net risk premiums*, i.e., those parts of the premiums which are calculated to cover the payments of claims, without any loading for safety margin or expenses. Those parts of the premiums which, in the case of ordinary life insurance, serve to build up the reserves, will also be left out of consideration. We shall regard the net risk premiums as being continuously paid to the company, so that there will be a steady continuous flow of risk premiums from the totality of policyholders to the company, in return for the payment in the opposite direction of discrete sums in settlement of claims. This is the case of insurance with only *positive risk sums*, which includes most non-life branches, and also the ordinary types of life insurance, such as endowment insurance etc.

There are, however, certain types of insurance where the circumstances are reversed. The typical case where this situation occurs is a life annuity business. Here the company is continuously paying out annuities to the policyholders, while the random event of the death of a policyholder will instantaneously place the corresponding reserve at the free disposal of the company, thus implying a payment *from* the policyholder *to* the company or, as we shall prefer to say, the payment of a negative amount by the company to the policyholder. This is the case of insurance with only *negative risk sums*.

The cases of only positive risk sums (no annuities), or only negative risk sums (pure annuity business), constitute important particular cases, which from time to time will be given special attention in the sequel. In general, however, we shall consider an insurance business which may include insurances of both types. The amount paid by the company in settlement of a claim will then be allowed to take negative as well as positive values, and the same will hold for the risk premiums collected by the company, annuities being regarded as negative risk premiums.

In order to study the temporal development of the risk business, we introduce a time scale with an origin placed at the beginning of our investigation, and an arbitrary unit, say one year. Regarding the claims as randomly occurring events, we now make the fundamental assumption that any two disjoint intervals on the time scale are always stochastically independent in respect to the occurrence of claims.

Placing ourselves at an arbitrary point τ on the scale, let us consider the probability that there will be *just one claim* during the time interval from τ to $\tau+\Delta\tau$. According to the fundamental assumption just introduced, this probability will be independent of any information that we may possess with respect to the number and kind of claims that have occurred up to the time τ. It seems natural to assume that, for small values of $\Delta\tau$, this probability should be approximately proportional to $\Delta\tau$, and thus be of the form

$$\lambda_\tau \Delta\tau + o(\Delta\tau),$$

where $o(\Delta\tau)$ denotes a quantity which becomes small in comparison with $\Delta\tau$, as $\Delta\tau \to 0$. Since the number of policies in force with the company, and the types of the various policies, will fluctuate from time to time, the factor λ_τ will in general depend on τ. We shall assume that λ_τ is a bounded and measurable function of τ, for all $\tau \geqq 0$. We shall also assume that the probability of *more than one claim* during the interval $(\tau, \tau+\Delta\tau)$ is of the form $o(\Delta\tau)$.

We now perform a transformation of the time scale, introducing an *operational time* t, connected with the natural time τ by the relation

(6) $$t = \int_0^\tau \lambda_u \, du.$$

We then have

$$dt = \lambda_\tau \, d\tau,$$

and accordingly the probability of just one claim during the interval $(t, t+\Delta t)$ will be equal to

(7) $$\Delta t + o(\Delta t),$$

while the probability of more than one claim during the same interval will be $o(\Delta t)$.

It is well known that, under these circumstances, the occurrence of claims will constitute a stochastic process of the type known as a *Poisson process*, the probability of exactly n claims during any time interval of length t being given by the simple Poisson expression

(8) $$\frac{t^n}{n!} e^{-t}.$$

The proof of this fact is easily available in text-books, and will not be given here. Reference may be made, e.g., to Cramér (**16**, and **26**, p. 104). We observe that the mean value of the number n according to the Poisson distribution (8) is equal to t. Thus the significance of operational time as introduced by (6) is that *the time length of any period is measured by the expected number of claims during the period.*

Any claim occurring, say, at time t will be accompanied by the payment of a certain amount y from the company to some policyholder. This amount y will be regarded as a random variable which, according to the above, will in general be capable of assuming values of both signs. Our final basic assumption will be that this variable has a distribution function $P(y)$ which is independent of t, and is also independent of any information with respect to claims that may have occurred up to time t. For the moments of $P(y)$ we shall use the notation

$$(9) \qquad p_n = \int_{-\infty}^{\infty} y^n \, dP(y),$$

and we shall always assume that the first order moment p_1 is finite. The existence of any p_n with $n > 1$ will, however, not be assumed unless this is explicitly mentioned. We shall further write

$$(10) \qquad p(s) = \int_{-\infty}^{\infty} e^{sy} \, dP(y),$$

where $s = \sigma + i\tau$ is a complex variable. For $\sigma = 0$, the integral is always absolutely convergent, and

$$p(i\tau) = \int_{-\infty}^{\infty} e^{i\tau y} \, dP(y)$$

is the characteristic function of $P(y)$. At a later stage, we shall have to consider $p(s)$ also for certain values of $\sigma \neq 0$, and shall then have to introduce appropriate conditions with respect to the behaviour of $P(y)$ at infinity, so as to ensure the convergence of the integral (10) also for these values of s.

There can be no doubt that the assumptions introduced above do involve some idealization of observed facts. Nevertheless, it is believed that they are sufficiently realistic to serve as a first approximation that may be practically useful in many cases. Certain parts of the theory can be based on less restrictive assumptions, and some indications to that effect will be given in chapter 6. In particular, it will be shown that throughout chapter 4 the assumption that $P(y)$ is independent of the time may be replaced by a more general assumption.

3. 2. The amount of claims $Y(t)$. Let $Y(t)$ denote the total amount of claims paid by the company up to the time t, each amount being counted with its proper sign. As already pointed out in 1. 2, $Y(t)$ will be a random function attached to the stochastic process that we are investigating. We shall deduce the probability distribution of this random function, in an appropriate function space, by means of the general procedure outlined in 2. 5.

In the first place, we shall then have to find the finite-dimensional distributions of the process, and we now proceed to show that these are completely determined by the assumptions introduced in the preceding section. Obviously we shall always have $Y(0) = 0$. For any fixed $t > 0$, the amount $Y(t)$ will be a random variable. We denote the d.f. of this random variable by $G(y, t)$, so that we may write, in an easily understood notation,

$$G(y, t) = \text{Prob}\,[Y(t) \leqq y]. \tag{11}$$

If we suppose that there have been exactly n claims up to the time t, and that the amounts paid are $y_1, \ldots, y_n$, we have

$$Y(t) = y_1 + \ldots\ldots + y_n.$$

By assumption, the y_i are independent random variables, each having the d.f. $P(y)$. The d.f. of the sum $y_1 + \ldots + y_n$ will then be the n-fold convolution of $P(y)$ with itself, which we denote by $P^{n*}(y)$. Thus the conditional probability of the relation $Y(t) \leqq y$, relative to the hypothesis that exactly n claims have occurred, is $P^{n*}(y)$. Now the probability that exactly n claims have occurred is given by the Poisson expression (8), and since the events that exactly 0, 1, 2, ... claims have occurred are mutually exclusive, it follows that the probability of the relation $Y(t) \leqq y$ is

$$G(y, t) = \sum_0^\infty \frac{t^n}{n!} e^{-t} P^{n*}(y), \tag{12}$$

where

$$P^{0*}(y) = \varepsilon(y) = \begin{cases} 0, & y < 0, \\ 1, & y \geqq 0, \end{cases}$$

$$P^{1*}(y) = P(y),$$

and for $n > 1$

$$P^{n*}(y) = \int_{-\infty}^{\infty} P^{(n-1)*}(y-z)\,dP(z).$$

We note for later use that according to (10) and (12) we have

$$\begin{aligned} E\, e^{sY(t)} &= \int_{-\infty}^{\infty} e^{sy}\, d_y G(y, t) = \sum_0^\infty \frac{t^n}{n!} e^{-t} \int_{-\infty}^{\infty} e^{sy}\, dP^{n*}(y) \\ &= \sum_0^\infty \frac{t^n}{n!} e^{-t} [p(s)]^n = e^{t[p(s)-1]}. \end{aligned} \tag{13}$$

If, instead of the amount of claims $Y(t)$ during the interval $(0, t)$, we consider the increment

$$Y(t_0+t)-Y(t_0) \tag{14}$$

of the Y function during any interval (t_0, t_0+t) of length $t > 0$, it will be easily seen from the above deduction that this variable will have the same d.f. $G(y, t)$ as the variable $Y(t)$, for any $t_0 \geqq 0$.

Further, if (t_0, t_0+t) and (u_0, u_0+u) are any two disjoint intervals on the positive real line, such that $t > 0$ and $u > 0$, it follows from our assumptions that the increments

$$Y(t_0+t)-Y(t_0) \quad \text{and} \quad Y(u_0+u)-Y(u_0)$$

will be independent random variables. The corresponding remark applies, of course, to any finite number of disjoint intervals.

The properties of the random function $Y(t)$ thus far established will be expressed by saying that the $Y(t)$ process is a stochastic process with *stationary and independent increments* (cf. Doob, 31, p. 419), such that the increment (14) has the d.f. $G(y, t)$. It is easily seen that the joint distribution of any finite number of variables $Y(t_1), \ldots, Y(t_n)$ will be completely determined by these properties, and that the family of finite-dimensional distributions thus obtained will satisfy the conditions laid down in 2.4 for the application of Kolmogoroff's theorem.

It then follows, according to 2.4, that the finite-dimensional distributions of the $Y(t)$ process determine a unique probability distribution in the space of *all possible functions* of the real variable $t \geqq 0$. For the reasons given in 2.4 this space is, however, too general for our purpose. In the following section, we shall proceed to define a restricted function space $\mathfrak{Y}_0$, containing only those functions $Y(t)$ that are really required for an adequate description of the course of the risk process. This being done, we shall use the procedure indicated in 2.5 to define a probability distribution in this "natural" space of the process, which has the same finite-dimensional distributions as those given above.

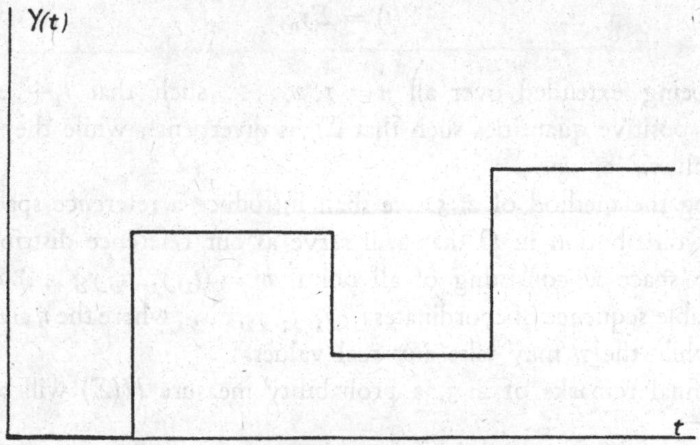

Fig. 1. Sample function of the Y(t) process.

3. 3. Probability distribution of the $Y(t)$ process. Any individual function $Y(t)$ describing the actual development of the amount of claims in one particular case, will be called a *sample function* of the $Y(t)$ process. The typical course of such a function is graphically represented in Figure 1.

As t increases from the point $t = 0$, the function $Y(t)$ will be constantly equal to zero until the first claim occurs, say at time t_1. There will then be a jump equal to the amount, say y_1, paid in settlement of this claim. The function $Y(t)$ then remains equal to y_1 until the time, say $t_1 + t_2$, when the next claim occurs. If y_2 is the amount paid on this occasion, the function will jump to the new value $y_1 + y_2$, and so on. In the case shown in Figure 1, y_1 and y_3 are positive, while y_2 is negative. For any sample function corresponding to an actual case, the number of jumps within any finite period will necessarily be finite.

Thus the "natural" sample functions of the process will consist of all functions $Y(t)$ having jumps in a certain sequence of t points, of which there are at most a finite number within any finite interval, and maintaining a constant value between any two consecutive jumps. For $t = 0$, all these functions will take the value $Y(0) = 0$.—Let $\mathfrak{Y}_0$ be the set of all functions $Y(t)$ defined for $t \geqq 0$ and satisfying these conditions.

The set $\mathfrak{Y}_0$ contains all those functions $Y(t)$ that we may possibly need for an adequate description of the temporal development of the amount of claims in an insurance business. We shall choose $\mathfrak{Y}_0$ as the natural restricted function space of the $Y(t)$ process. We now proceed to show that it is possible to define, in accordance with 2. 5, a probability distribution in the space $\mathfrak{Y}_0$, which has the finite-dimensional distributions of the $Y(t)$ process as deduced in the preceding section.

Any natural sample function $Y(t)$, i.e., any function $Y(t)$ belonging to $\mathfrak{Y}_0$, will be uniquely determined by the sequence of its discontinuity points, say $t_1, t_1 + t_2, t_1 + t_2 + t_3, \ldots$, and the sizes of the corresponding jumps, say $y_1, y_2, y_3, \ldots$. In fact, we may write for any $t > 0$

$$Y(t) = \Sigma y_n,$$

the sum being extended over all $n = 1, 2, \ldots,$ such that $t_1 + \ldots + t_n \leqq t$. The t_i are positive quantities such that Σt_i is divergent[1], while the y_i may take any real values.

Following the method of 2. 5, we shall introduce a reference space Ω, and a probability distribution in Ω that will serve as our reference distribution. We choose the space Ω consisting of all points $\omega = (t_1, y_1, t_2, y_2, \ldots)$ defined by an enumerable sequence of coordinates $t_1, y_1, t_2, y_2, \ldots$, where the t_i are essentially positive, while the y_i may take any real values.

By the final remarks of 2. 3, a probability measure $\Pi(\Sigma)$ will be uniquely

[1] It is easily seen that, in order that $Y(t)$ should have at most a finite number of discontinuities in any finite interval, it is necessary and sufficient that $\sum t_i$ should be divergent.

defined on the class $\mathfrak{B}$ of all Borel sets of this space by the following three conditions:

1) Any finite number of the coordinates t_i and y_i will be independent random variables.
2) Any t_i will have the probability density e^{-t} for all $t > 0$.
3) Any y_i will have the d.f. $P(y)$.

Let Z denote the set of all points ω such that $\sum_1^\infty t_i$ is divergent, while $Z^\star$ is the complementary set, consisting of all ω such that $\sum_1^\infty t_i$ is convergent. Z and $Z^\star$ are Borel sets, and it follows from a well-known theorem on the convergence of series of independent random variables (cf., e.g., Kolmogoroff **39**, p. 59, or Cramér **25**, p. 178) that the probability of convergence of Σt_i is zero. Thus $\Pi(Z) = 1$, while $\Pi(Z^\star) = 0$.

As in 2. 5, we shall now define a function $Y(t, \omega)$ for all $t > 0$, and all ω in Ω. We do this by writing

$$Y(t,\omega) = \begin{cases} \sum\limits_{t_1+\dots+t_n \leq t} y_n & \text{when } \omega \text{ belongs to } Z, \\ Y^\star(t) & \text{when } \omega \text{ belongs to } Z^\star. \end{cases} \tag{15}$$

Here $Y^\star(t)$ is a function belonging to $\mathfrak{Y}_0$, the same for all ω in $Z^\star$. Since $Z^\star$ is a null set, $Y^\star(t)$ may be chosen quite arbitrarily, and the choice will be of no consequence whatever in the sequel.

In order to follow the procedure of 2. 5 we first have to show that, for any fixed $t > 0$, the function $Y(t, \omega)$ is a $\mathfrak{B}$-measurable function of ω. We thus have to prove that (cf. 2. 2), for a fixed t and for every real k, the set Σ of all points ω such that $Y(t, \omega) < k$ is a Borel set in Ω. Now it is easily seen that we have

$$\Sigma = \begin{cases} \sum\limits_0^\infty \Lambda_n \Theta_n + Z^\star & \text{when } Y^\star(t) < k, \\ \sum\limits_0^\infty \Lambda_n \Theta_n & \text{when } Y^\star(t) \geq k, \end{cases}$$

where Λ_n is the set of all ω in Z satisfying the conditions

$$t_1 + \dots + t_n \leq k, \quad t_1 + \dots + t_{n+1} > k,$$

while Θ_n is the set of all ω such that

$$y_1 + \dots + y_n < k.$$

Since Λ_n, Θ_n and $Z^\star$ are all Borel sets, it follows that Σ is a Borel set.

On the other hand, for any fixed ω in Ω, (15) yields a uniquely determined function $Y(t)$ belonging to $\mathfrak{Y}_0$. According to 2.5, the transformation

$$Y(t) = Y(t, \omega)$$

then induces a uniquely determined probability distribution in the space $\mathfrak{Y}_0$, such that the probability is defined for every Borel set in $\mathfrak{Y}_0$. We shall now show that the corresponding finite-dimensional distributions are identical with the finite-dimensional distributions of the $Y(t)$ process as deduced in the preceding section.

For this purpose, we may neglect the null set Z^*, and only consider the points ω belonging to Z, so that $Y(t) = Y(t, \omega)$ will be given by the first relation (15). For any fixed t, the probability that this function $Y(t)$ will have exactly n points of discontinuity in the interval $(0, t)$ is identical with the probability that the two relations

$$t_1 + \ldots + t_n \leq t,$$
$$t_1 + \ldots + t_{n+1} > t,$$

will be jointly satisfied. A simple calculation shows that this probability is

$$\frac{t^n}{n!} e^{-t},$$

and the same argument that was used for the proof of (12) will now show that, for any fixed $t > 0$, the random variable $Y(t)$ has the d.f. $G(y, t)$ given by (12). Similarly, a simple calculation will show that the increment $Y(t_0+t) - Y(t_0)$ over an interval of length t has the same d.f. $G(y, t)$, for any $t_0 > 0$. It also follows from our assumptions concerning the variables t_i and y_i that the increments of $Y(t)$ over disjoint intervals are always independent. As in the preceding section, the properties thus established will completely determine all finite-dimensional distributions of the $Y(t)$ distribution as defined by (15).

We have thus proved that there exists a uniquely determined probability distribution in the space $\mathfrak{Y}_0$ of natural sample functions of the $Y(t)$ process, such that its finite-dimensional distributions are identical with those deduced in the preceding section from the nature of the stochastic process. This distribution, which gives a well-determined probability for every Borel set in $\mathfrak{Y}_0$, will be defined as the probability distribution of the $Y(t)$ process.

3.4. The gain $X(t)$. For the mean value of the amount of claims $Y(t)$ we obtain from (12), since each $P^{n*}(y)$ has the mean np_1,

$$E\,Y(t) = \sum_0^\infty \frac{t^n}{n!} e^{-t} np_1 = p_1 t,$$

so that the mean amount of claims during the interval $(0, t)$ is $p_1 t$. In order that the game between the company and the policyholders should be a fair one, the company should thus receive the amount $p_1 t$ as *net risk premium* for the interval. Since this result holds for any $t > 0$, it follows that the net risk premium corresponding to any time element of length dt will be $p_1 dt$. Thus there will be a continuous flow of net risk premiums collected by the company, at the rate of p_1 money units per time unit.[1]

The gain $X(t)$ realized by the company on the net risk business up to the time t will be

$$X(t) = p_1 t - Y(t), \tag{16}$$

with the mean value

$$EX(t) = p_1 t - p_1 t = 0.$$

For any fixed $t > 0$, the gain $X(t)$ is a random variable, and if we write[2]

$$F(x, t) = \text{Prob}\,[X(t) < x] \tag{17}$$

we obtain from (11) and (12)

$$\begin{aligned} F(x, t) &= \text{Prob}\,[Y(t) > p_1 t - x] \\ &= 1 - \text{Prob}\,[Y(t) \leqq p_1 t - x] \qquad (18) \\ &= 1 - \sum_0^\infty \frac{t^n}{n!} e^{-t} P^{n*}(p_1 t - x). \end{aligned}$$

The increment

$$X(t_0 + t) - X(t_0) = p_1 t - [Y(t_0 + t) - Y(t_0)]$$

will, according to 3.2, have the same d.f. $F(x, t)$ as $X(t)$, and it also follows from 3.2 that the increments of $X(t)$ over disjoint time intervals will be independent random variables.

Thus the $X(t)$ process is a stochastic process with stationary and independent increments (cf. 3.2), such that the increment $X(t_0 + t) - X(t_0)$ has the d.f. $F(x, t)$ given by (18). As in the case of the $Y(t)$ process, these properties will completely determine the family of finite-dimensional distributions of the process.

[1] The remarks made here refer, of course, to the most usual case when $p_1 > 0$. The modifications to be made when $p_1 < 0$ are evident.

[2] In all continuity points x, the function $F(x, t)$ will be identical with the d.f. of $X(t)$, while in a discontinuity point, F will be continuous to the left instead of to the right. In the sequel, we shall neglect this difference, which is entirely trivial for our purpose, and speak as if $F(x, t)$ were the d.f. of $X(t)$ for all x.

From (13) we obtain

$$Ee^{-sX(t)} = e^{-sp_1 t} E e^{sY(t)} \tag{19}$$
$$= e^{t[p(s) - 1 - p_1 s]},$$

where $p(s)$ is given by (10).

For every natural sample function $Y(t)$, as defined in 3. 3, the simple transformation (16) yields a corresponding natural sample function $X(t)$, which has the same umps as $Y(t)$, with inverted signs, and increases linearly throughout each interval between two consecutive jumps, the angular coefficient being p_1. Figure 2 shows the $X(t)$ function corresponding to the $Y(t)$ of Figure 1.

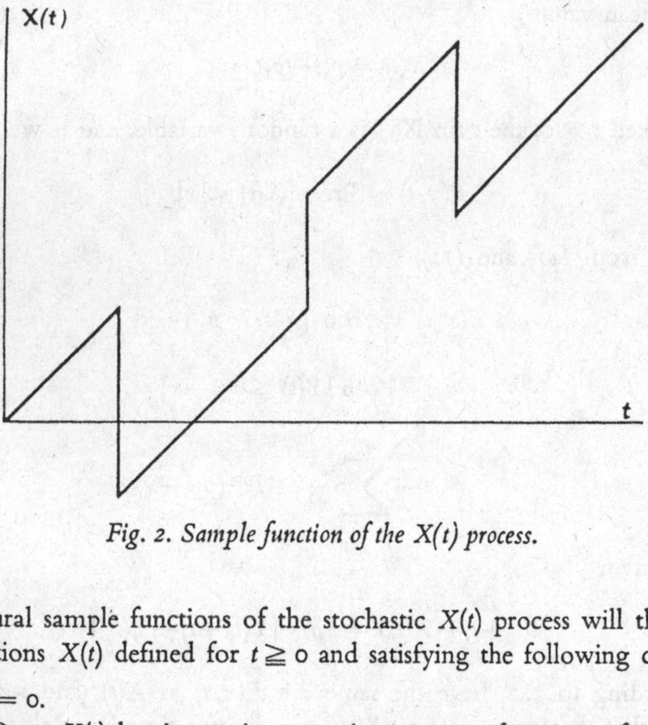

Fig. 2. Sample function of the X(t) process.

The natural sample functions of the stochastic $X(t)$ process will thus consist of all functions $X(t)$ defined for $t \geqq 0$ and satisfying the following conditions:

1) $X(0) = 0$.
2) For $t > 0$, $X(t)$ has jumps in a certain sequence of t points, of which there are at most a finite number within any finite interval. We may agree to take any $X(t)$, in accordance with the first formula (15), such that it is always continuous to the right.
3) Between two consecutive jumps, $X(t)$ is linear, with the angular coefficient p_1.

The set $\mathfrak{X}_0$ of all functions $X(t)$ satisfying these conditions is the natural restricted function space of the $X(t)$ process. The transformation (16) induces a probability distribution in the space $\mathfrak{X}_0$, corresponding to the distribution in $\mathfrak{Y}_0$ discussed in the preceding section. The probability measure $P(S)$ of this distribution in $\mathfrak{X}_0$ is defined for all Borel sets S in $\mathfrak{X}_0$.

If S is a Borel set in $\mathfrak{X}_0$, the inverse image Σ of S in the reference space Ω is defined by the transformations (15) and (16). The probability $P(S)$ is then defined by the relation (cf. 2. 2)

$$P(S) = \Pi(\Sigma),$$

where $\Pi(\Sigma)$ is the probability measure in the reference space Ω defined in the preceding section. *The probability distribution in $\mathfrak{X}_0$ thus introduced will evidently have the finite-dimensional distributions of the $X(t)$ process as deduced above, and will be defined as the probability distribution of the $X(t)$ process.*

4. The Distribution of the Gain $X(t)$ During a Fixed Period

4.1. Introductory remarks. According to 3.4 the random variable $X(t)$ has, for any fixed $t > 0$, the d.f. $F(x, t)$ given by (18), while the corresponding characteristic function (c.f.) is given by (19), for $s = -i\tau$. It follows from these expressions that the distribution of $X(t)$ is completely determined by the value of t and the d.f. $P(y)$ of risk sums.

For various practical applications, it will be important to work out methods for the numerical computation of $F(x, t)$ when t and $P(y)$ are given. It should be observed that t denotes here, as always throughout this paper, operational time as defined by (6). Accordingly, the value of t corresponding to any given period is identical with the expected number of claims during the period. In the applications, this will usually be a fairly large number, so that the problem before us is to find methods for the numerical computation of $F(x, t)$ for a given d.f. $P(y)$ and for large values of t.

We shall give some examples of practical problems where this situation comes up. Suppose, e.g., that an insurance company has at its disposal a certain sum u, which may be used to cover any losses due to random fluctuations during a certain period of, say, one year. If the expected number of claims during the year is t, the probability of a loss exceeding the available sum u will be $F(-u, t)$, and we may want to be able to compute the numerical value of this probability as accurately as possible.

For fixed values of t and u, the probability $F(-u, t)$ will only depend on the d.f. $P(y)$ of risk sums. If $P(y)$ is modified, e.g. by a change of the maximum net retention M of the company, the probability $F(-u, t)$ will also change its value, and we may be interested in finding out how this value will depend on M.

The computation of premiums for certain systems of collective reinsurance is a further example. Consider, e.g., a type of excess loss insurance under which the reinsurer covers that part of any loss on the total net risk business during a certain period t, which may exceed some agreed multiple kp_1t of the total net risk premium p_1t for the period. The net premium for this reinsurance will be

$$\int_{-\infty}^{-kp_1t} |x - kp_1t|\, d_xF(x, t). \tag{20}$$

In order to deal with problems of these types, it will be seen that we shall require accurate knowledge about the numerical values of the function $F(x, t)$, particularly for negative x and for large values of t.

The problem of finding asymptotic expressions for $F(x, t)$, valid for large values of t, was investigated already in the early works of F. Lundberg, starting with his Thesis **(42)** of 1903. Lundberg further developed his methods in a number of subsequent publications **(43—48)**. He pointed out that the $F(x, t)$ distribution becomes asymptotically normal, as t tends to infinity. We have, in fact, subject to the condition that the second moment p_2 of the $P(y)$ distribution is finite,

$$(21) \qquad F(x\sqrt{p_2 t}, t) \to \Phi(x) = \frac{1}{\sqrt{2\pi}} \int_{-\infty}^{x} e^{-\frac{u^2}{2}} du$$

for any fixed real x, as $t \to \infty$. However, as Lundberg pointed out, the approximation involved in replacing F by Φ is usually not accurate enough for practical purposes. Lundberg gave reasons for expecting certain other formulas, such a a formula of the type

$$(22) \qquad F(-kt, t) \sim \frac{\alpha}{\sqrt{t}} e^{-\beta t},$$

to give more accurate results. Here, k, α and β are positive constants, and the formula would hold asymptotically, as $t \to \infty$.

The theory of certain asymptotic expansions derived from the normal approximation (21) was developed by Cramér **(18, 19, 20)**, who gave methods for the evaluation of the corresponding errors of approximation. These were further extended by Esseen **(35)**.

A proof of (22), and an evaluation of the error of approximation involved, were given by Esscher **(34)**, who also gave other related formulas of greater accuracy, and showed by numerical illustrations that these formulas yield good results in some practical cases. The method used by Esscher was further developed by Cramér **(21)** and Feller **(36)**, who showed that it may be successfully applied to several important problems in the theory of probability. Arfwedson, in a forthcoming paper **(11)**, uses a new method to discuss the approximation formulas due to Esscher.

Extensive numerical illustrations of the use of the various approximation methods, particularly in connection with practical reinsurance problems, have been published by Hultman **(38)**, Pentikäinen **(54—56)**, Ammeter **(1—6)**, Lefèvre **(41)**, and others.

In the following section, we shall discuss the normal approximation (21), and the asymptotic expansions derived from it. Since the methods for working out exact upper limits for the errors of approximation involved in these formulas are easily available, e.g., in Cramér (**19**) and (**20**), the lengthy deductions required for this purpose will not be repeated here. Some numerical examples illustrating the accuracy of the formulas will be given in 4. 4.

In 4. 3, we shall give a deduction of the asymptotic formula (22), and of the related formulas obtained by Esscher (**34**), and some further formulas. In this connection, we shall use the method followed by Cramér (**21**). Some numerical illustrations of the use of the formulas will be given in 4. 4.

4.2. The normal and allied approximations. In this section, we shall make certain assumptions about the d.f. $P(y)$ of risk sums.

In the first place, it will be assumed that the moments p_n as defined by (9) are finite for $n \leqq 5$.

Further, we shall assume that $P(y)$ contains an absolutely continuous component, so that we may write (cf. Cramér **20**, p. 17)

$$P(y) = P_1(y) + P_2(y),$$

where P_1 and P_2 are never decreasing functions such that P_1 is absolutely continuous and not identically equal to a constant. This condition will, e.g., certainly be satisfied if we have for all y

$$P(y) = \int_{-\infty}^{y} P'(z)\, dz.$$

The characteristic function (c.f.) corresponding to the d.f. $F(x, t)$ will be denoted by

$$f(\tau, t) = E\, e^{i\tau X(t)}.$$

From (19) we obtain, taking $s = -i\tau$,

(23) $$f(\tau, t) = e^{t[p(-i\tau) - 1 + p_1 i\tau]},$$

where $p(-i\tau)$ is given by (10). From the development of $f(\tau, t)$ in MacLaurin's series for small values of τ it follows that the mean and the standard deviation of the random variable $X(t)$ are

$$E\,X(t) = 0, \qquad D\,X(t) = \sqrt{p_2 t},$$

so that the corresponding standardized variable is $X(t)/\sqrt{p_2 t}$, with the d.f.

(24) $$F(x\sqrt{p_2 t}, t).$$

We now want to study the behaviour of this d.f. as t tends to infinity, while x has any fixed real value.

The corresponding c.f. is

$$f\left(\frac{\tau}{\sqrt{p_2 t}}, t\right) = e^{t\left[p\left(-\frac{i\tau}{\sqrt{p_2 t}}\right) - 1 + \frac{p_1 i\tau}{\sqrt{p_2 t}}\right]}.$$

The expression between the brackets in the exponent in the second member is a function of $\tau/\sqrt{p_2 t}$. Regarding τ as a fixed quantity, and t as a variable tending to infinity, we develop in MacLaurin's series, and obtain

$$\log f\left(\frac{\tau}{\sqrt{p_2 t}}, t\right) = t\left[\sum_{2}^{4} \frac{p_n}{n!}\left(-\frac{i\tau}{\sqrt{p_2 t}}\right)^n + O(t^{-\frac{5}{2}})\right]$$

$$= -\frac{1}{2}\tau^2 + \frac{c_3}{3!\,t^{\frac{1}{2}}}(-i\tau)^3 + \frac{c_4}{4!\,t}(-i\tau)^4 + O(t^{-\frac{3}{2}}),$$

where we have written

(25) $$c_n = \frac{p_n}{p_2^{\frac{n}{2}}}.$$

For the c.f. itself, we then obtain by some simple calculation

$$f\left(\frac{\tau}{\sqrt{p_2 t}}, t\right) = e^{-\frac{1}{2}\tau^2}\left[1 + \frac{c_3}{3!\,t^{\frac{1}{2}}}(-i\tau)^3 + \frac{c_4}{4!\,t}(-i\tau)^4 + \frac{10\,c_3^2}{6!\,t}(-i\tau)^6\right] + O(t^{-\frac{3}{2}}).$$

We have here the first terms of an expansion of the c.f. in powers of $t^{-\frac{1}{2}}$, which may obviously be extended to further terms as long as the moments p_n are finite. The successive terms of this expansion are related to the normal d.f.

$$\Phi(x) = \frac{1}{\sqrt{2\pi}}\int_{-\infty}^{x} e^{-\frac{1}{2}u^2}\,du$$

and its derivatives by means of the relation (cf., e.g., Cramér **20**, p. 49, or **23**, p. 225)

$$\int_{-\infty}^{\infty} e^{i\tau x}\,d\Phi^{(n)}(x) = (-i\tau)^n e^{-\frac{1}{2}\tau^2}.$$

This suggests that we should have the following asymptotic expansion for the d.f. (24) of the standardized gain:

(26) $$F(x\sqrt{p_2 t}, t) = \Phi(x) + \frac{c_3}{3!\,t^{\frac{1}{2}}}\Phi^{(3)}(x) + \frac{c_4}{4!\,t}\Phi^{(4)}(x) + \frac{10 c_3^2}{6!\,t}\Phi^{(6)}(x) + O(t^{-\frac{3}{2}}).$$

It can, in fact, be shown that under the assumptions introduced above with respect to the d.f. $P(y)$ the relation (26) holds true uniformly for all real x. We thus have an asymptotic expansion, valid as $t \to +\infty$, with an error of the order $t^{-\frac{3}{2}}$. There are also other formulas giving upper limits for the error term of the expansion. The detailed proofs of these facts are, however, somewhat lengthy, and we must content ourselves here with referring the reader to our previous works on the subject (Cramér **18**, **19**, and **20**).

If the moments p_n are finite even for some $n > 5$, the expansion (26) can be extended to further terms. Thus if p_6 is finite, there will be a group of terms of the order $t^{-\frac{3}{2}}$, involving derivatives of Φ of the orders 5, 7, and 9, and giving an error term of the order t^{-2}, and so on (cf. Cramér, 18, p. 31).

On the other hand, using the methods introduced by Esseen (**35**), it can be shown that the expansion (26) holds under less restrictive conditions than those imposed above, even in certain cases when $P(y)$ contains no absolutely continuous component.

Limiting the expansion (26) to its first term, we obtain the relation

$$F(x\sqrt{p_2 t},\, t) = (\Phi x) + O(t^{-\frac{1}{2}}). \tag{27}$$

In the numerical illustrations given in 4.4, we shall use the expressions "Normal I" and "Normal II" to denote the approximation formulas obtained by neglecting the error terms in (27) and (26) respectively, thus writing:

(Normal I) $\quad F(x\sqrt{p_2 t},\, t) = \Phi(x),$

(Normal II) $\quad F(x\sqrt{p_2 t},\, t) = \Phi(x) + \dfrac{c_3}{3!\,t^{\frac{1}{2}}}\,\Phi^{(3)}(x) + \dfrac{c_4}{4!\,t}\,\Phi^{(4)}(x) + \dfrac{10 c_3^2}{6!\,t}\,\Phi^{(6)}(x).$

Both these relations are, of course, only approximately valid.

4.3. Esscher's approximation method. In addition to the assumptions concerning $P(y)$ introduced at the beginning of the preceding section, we shall now make the following further assumption. The integral (10):

$$p(s) = \int_{-\infty}^{\infty} e^{sy}\, dP(y),$$

where $s = \sigma + i\tau$, will be assumed to be absolutely convergent for all σ such that $-A < \sigma < B$, where A and B are given positive quantities. It readily follows from this assumption that the moments p_n will be finite for all values of n.

We now take h such that $-A < h < B$. The function

$$\bar{P}(y) = \frac{1}{p(h)} \int_{-\infty}^{y} e^{hz}\, dP(z) \tag{28}$$

will then be a d.f. satisfying all the conditions laid down in 4. 2 and in this section with respect to $P(y)$. (In general, of course, the constants A and B will have different values for P and $\bar{P}$.) For $h = 0$, $\bar{P}(y)$ reduces to $P(y)$. The moments of $\bar{P}(y)$

$$\bar{p}_n = \int_{-\infty}^{\infty} y^n \, d\bar{P}(y) = \frac{1}{p(h)} \int_{-\infty}^{\infty} y^n e^{hy} \, dP(y) \tag{29}$$

are finite for all values of n. The corresponding c.f., which we write here with the variable $s = \sigma + i\tau$, is

$$\int_{-\infty}^{\infty} e^{sy} \, d\bar{P}(y) = \frac{1}{p(h)} \int_{-\infty}^{\infty} e^{(s+h)y} \, dP(y) = \frac{p(s+h)}{p(h)}.$$

The n:th convolution of the d.f. $\bar{P}(y)$ with itself will be denoted by $\bar{P}^{n*}(y)$, By the well-known properties of characteristic functions, $\bar{P}^{n*}(y)$ will have the c.f.

$$\int_{-\infty}^{\infty} e^{sy} \, d\bar{P}^{n*}(y) = \left[\frac{p(s+h)}{p(h)}\right]^n. \tag{30}$$

On the other hand, the n:th convolution of $P(y)$ with itself, $P^{n*}(y)$, has the c.f. $[p(s)^n]$. Consequently, the function

$$Q_n(y) = \frac{1}{[p(h)]^n} \int_{-\infty}^{y} e^{hz} \, dP^{n*}(z)$$

is a d.f., which is readily seen to have the c.f.

$$\int_{-\infty}^{\infty} e^{sy} \, dQ_n(y) = \left[\frac{p(s+h)}{p(h)}\right]^n. \tag{31}$$

A comparison of (30) and (31) will show that $\bar{P}^{n*}(y)$ and $Q_n(y)$ are identical so that we have

$$\bar{P}^{n*}(y) = \frac{1}{[p(h)]^n} \int_{-\infty}^{y} e^{hz} \, dP^{n*}(z),$$

and conversely

$$P^{n*}(y) = [p(h)]^n \int_{-\infty}^{y} e^{-hz} \, d\bar{P}^{n*}(z). \tag{32}$$

Consider now a risk business with the d.f. $\bar{P}(y)$ of risk sums. By (18), the gain $\bar{X}(t)$ during the period t will have the d.f.

$$\bar{F}(x,t) = 1 - \sum_0^\infty \frac{t^n}{n!} e^{-t} \bar{P}^{n*}(\bar{p}_1 t - x). \tag{33}$$

On the other hand, we have by (18) and (32)

$$F(x,t) = 1 - \sum_0^\infty \frac{t^n}{n!} e^{-t} P^{n*}(p_1 t - x)$$

$$= 1 - \sum_0^\infty \frac{t^n}{n!} e^{-t} [p(h)]^n \int_{-\infty}^{p_1 t - x} e^{-hz} d\bar{P}^{n*}(z).$$

On account of the absolute convergence, we may interchange the order of summation and integration, and thus obtain

$$F(x, t) = 1 + e^{t[p(h)-1]} \int_{-\infty}^{p_1 t - x} e^{-hz} d_z \left[1 - \sum_0^\infty \frac{[tp(h)]^n}{n!} e^{-tp(h)} \bar{P}^{n*}(z) \right]$$

Introducing a new integration variable y by the substitution

$$y = tp(h)\bar{p}_1 - z,$$

we further obtain, using (33),

$$1 - F(x,t) = e^{-\beta t} \int_{x-\gamma t}^{\infty} e^{hy} d_y \bar{F}[y, tp(h)], \tag{34}$$

where, according to (10) and (29),

$$\begin{aligned} \beta &= 1 + hp(h)\bar{p}_1 - p(h) \\ &= 1 + \int_{-\infty}^{\infty} (hy - 1) e^{hy} dP(y), \end{aligned} \tag{35}$$

$$\begin{aligned} \gamma &= p_1 - p(h) p_1 \\ &= p_1 - \int_{-\infty}^{\infty} y e^{hy} dP(y). \end{aligned} \tag{36}$$

When x tends to $-\infty$, the first member in (34) tends to 1, and we obtain by subtraction

$$F(x, t) = e^{-\beta t} \int_{-\infty}^{x-\gamma t} e^{hy} d_y \bar{F}[y, tp(h)]. \tag{37}$$

By assumption, the integrals in the above formulas will be convergent for $-A < h < B$. We have

$$\frac{d\gamma}{dh} = -\int_{-\infty}^{\infty} y^2 e^{hy}\, dP(y) \leqq 0,$$

where the sign of equality holds only in the trivial case when $P(y) = \varepsilon(y)$ which implies that all risk sums are equal to zero. Excluding this trivial case from our considerations, we thus find that γ will be steadily decreasing when h increases from $-A$ to B. For $h = 0$, we have $\gamma = 0$, so that γ will decrease from some positive upper limit, say b, to some negative lower limit, say $-a$. Either a or b, or both, may be equal to $+\infty$.

For any given c such that $-a < c < b$, the equation $\gamma = c$, which may be written in explicit form:

$$(38) \qquad p_1 - \int_{-\infty}^{\infty} y\, e^{hy}\, dP(y) = c,$$

will thus have one and only one root h such that $-A < h < B$. Evidently we have

$$(39) \qquad h \gtreqless 0 \text{ according as } c \lesseqgtr 0.$$

Henceforth, we shall suppose that c has a fixed value such that $-a < c < b$, and that h is the unique root of the equation (38). We shall first suppose that $c \neq 0$, reserving the case $c = 0$ for later consideration. According to (39), c and h will then always have opposite signs.

In (34) and (37), we now take $x = ct$. Since by (38) we hawe $\gamma = c$ this gives, substituting at the same time $x\sqrt{tp(h)\bar{p}_2}$ for y,

$$(40\text{ a}) \qquad 1 - F(ct, t) = e^{-\beta t} \int_{0}^{\infty} e^{hx\sqrt{tp(h)\bar{p}_2}}\, d_x\, \bar{F}[x\sqrt{tp(h)\bar{p}_2},\, tp(h)],$$

$$(40\text{ b}) \qquad F(ct, t) = e^{-\beta t} \int_{-\infty}^{0} e^{hx\sqrt{tp(h)\bar{p}_2}}\, d_x\, \bar{F}[x\sqrt{tp(h)\bar{p}_2},\, tp(h)].$$

Both these relations hold for any c in the interval $-a < c < b$. From now on, we shall always use (40 a) when $c > 0$, and (40 b) when $c < 0$. Thus we shall have $h < 0$ in (40 a), and $h > 0$ in (40 b). Writing

$$(41) \qquad u = |h|\sqrt{tp(h)\bar{p}_2},$$

u will always be > 0, and (40 a) and (40 b) now become

$$(42\text{ a}) \qquad 1 - F(ct, t) = e^{-\beta t} \int_{0}^{\infty} e^{-ux}\, d_x \bar{F}[x\sqrt{tp(h)\bar{p}_2},\, tp(h)]$$

for $c > 0$, and

(42 b) $$F(ct, t) = e^{-\beta t} \int_{-\infty}^{0} e^{ux}\, d_x F\,[x\,\sqrt{tp(h)\bar{p}_2},\, tp(h)]$$

for $c < 0$.

We have observed above that the d.f. $\bar{P}(y)$ as defined by (28) satisfies the conditions given at the beginning of 4.2 and that, moreover, the moments $\bar{p}_n$ of $\bar{P}(y)$ are finite for all n. According to 4.2, it then follows that the corresponding d.f. $F(x, t)$ given by (33) admits an asymptotic expansion in powers of $t^{-\frac{1}{2}}$, the first terms of which will be

(43) $$\bar{F}[x\,\sqrt{tp(h)\bar{p}_2},\, tp(h)] = \\ = \Phi(x) + \frac{\bar{c}_3}{3!\,[tp(h)]^{\frac{1}{2}}}\,\Phi^{(3)}(x) + \frac{\bar{c}_4}{4!\,tp(h)}\,\Phi^{(4)}(x) + \frac{10\,\bar{c}_3^{\,2}}{6!\,tp(h)}\,\Phi^{(6)}(x) + \dots,$$

where, in analogy with (25),

$$\bar{c}_n = \frac{\bar{p}_n}{\bar{p}_2^{\frac{n}{2}}},$$

the p_n being given by (29). We now introduce the following notations

(44) $$q_n = p(h)\bar{p}_n = \int_{-\infty}^{\infty} y^n e^{hy}\, dP(y),$$
$$k_3 = \frac{hq_3}{3!\,q_2},$$
$$k_4 = \frac{h^2 q_4}{4!\,q_2},$$
$$k_6 = \frac{10\,h^2 q_3^2}{6!\,q_2^2} = \frac{1}{2}\,k_3^2.$$

The expansion (43) then becomes, after some easy reductions,

(45) $$\bar{F}[x\sqrt{tp(h)\bar{p}_2},\, tp(h)] = \Phi(x) \pm \frac{k_3}{u}\,\Phi^{(3)}(x) + \frac{k_4}{u^2}\,\Phi^{(4)}(x) + \frac{k_6}{u^2}\,\Phi^{(6)}(x) + \dots,$$

where the double sign should be identical with the sign of h. According to a remark made in 4.2, this expansion can be completed with a group of further terms, containing the factor u^3 in the denominator, and the error term will then be of the order t^{-2} or, equivalently, of the order u^{-4}, uniformly for all x.

We shall now replace $\bar{F}$ in (42 a) and (42 b) by this expansion. It will be seen that we shall then require the values of integrals of the forms

$$\int_{0}^{\infty} e^{-ux}\, d\Phi^{(n)}(x) \text{ and } \int_{-\infty}^{0} e^{ux}\, d\Phi^{(n)}(x),$$

with $u > 0$. Writing

$$E(u) = e^{\frac{u^2}{2}}[1-\Phi(u)]$$

we have the well-known asymptotic expansion

$$E(u) = \frac{1}{\sqrt{2\pi}}\left(\frac{1}{u} - \frac{1}{u^3} + \frac{1\cdot 3}{u^5} - \frac{1\cdot 3\cdot 5}{u^7} + \dots\right), \tag{46}$$

the error being, for large positive u, of the same order as the first term neglected. Further, simple calculations give

$$\int_0^\infty e^{-ux}\, d\Phi(x) = E(u),$$

$$\int_0^\infty e^{-ux}\, d\Phi^{(3)}(x) = u^3 E(u) - \frac{u^2-1}{\sqrt{2\pi}}$$

$$= \frac{1}{\sqrt{2\pi}}\left(\frac{1\cdot 3}{u^2} - \frac{1\cdot 3\cdot 5}{u^4} + \dots\right),$$

$$\int_0^\infty e^{-ux}\, d\Phi^{(4)}(x) = u^4 E(u) - \frac{u^3-u}{\sqrt{2\pi}}$$

$$= \frac{1}{\sqrt{2\pi}}\left(\frac{1\cdot 3}{u} - \frac{1\cdot 3\cdot 5}{u^3} + \dots\right),$$

$$\int_0^\infty e^{-ux}\, d\Phi^{(6)}(x) = u^6 E(u) - \frac{u^5-u^3+3u}{\sqrt{2\pi}}$$

$$= \frac{1}{\sqrt{2\pi}}\left(-\frac{1\cdot 3\cdot 5}{u} + \dots\right).$$

Finally, for $n = 0, 1, 2, \dots\dots$ we have

$$\int_{-\infty}^{0} e^{ux}\, d\Phi^{(n)}(x) = (-1)^n \int_0^\infty e^{-ux}\, d\Phi^{(n)}(x). \tag{47}$$

Using these expressions, we obtain from (42) and (45) the first terms of an asymptotic expansion of the "tail values" $1-F(ct, t)$ and $F(ct, t)$ in powers of u^{-1} or, what amounts to the same thing, in powers of $t^{-\frac{1}{2}}$. In order to simplify the writing, we define the function $G(u)$ by the relation

$$\left.\begin{array}{r} 1-F(ct, t) \\ F(ct, t) \end{array}\right\} = e^{-\beta t} G(u), \tag{48}$$

where the first member should be $1-F(ct, t)$ when $c>0$, and $F(ct, t)$ when $c<0$. Using only the first term of the expansion (45), we then obtain as a first approximation the formula

(Esscher I) $$G(u)=E(u),$$

given by Esscher **(34)**. Using all the terms of (45), the double sign of the term involving the third derivative will disappear on account of (47), and we obtain the extended formula

(Esscher II)
$$G(u)=\left(1-(k_3-k_4)u^2+k_6u^4\right)E(u)+\frac{k_3-k_4}{\sqrt{2\pi}}\left(u-\frac{1}{u}\right)-\frac{k_6}{\sqrt{2\pi}}\left(u^3-u+\frac{3}{u}\right),$$

which differs from the second approximation given by Esscher, since we have here included the terms arising from the fourth and sixth derivatives in (45).

Taking account of the order of the error term in the asymptotic expansion (45), we find that the error in Esscher I will be $O(u^{-2})=O(t^{-1})$, while Esscher II will have an error of the order $O(u^{-4})=O(t^{-2})$.

Completing Esscher I with an error term $O(u^{-2})$, and using the asymptotic expansion (46) of $E(u)$, we obtain

(49) $$G(u)=\frac{1}{u\sqrt{2\pi}}+O\left(\frac{1}{u^2}\right),$$

which gives according to (41) and (48)

(50) $$\left.\begin{matrix}1-F(ct, t)\\ F(ct, t)\end{matrix}\right\}=e^{-\beta t}\left(\frac{1}{|h|\sqrt{2\pi t p(h)\overline{p}_2}}+O\left(\frac{1}{t}\right)\right)$$

with the same convention as in the case of (48). It will be seen that this agrees with the formula (22) due to F. Lundberg. From Esscher II we obtain in the same way the extended expansion

(51) $$G(u)=\frac{1}{\sqrt{2\pi}}\left(\frac{1}{u}+\frac{K_3}{u^3}+O\left(\frac{1}{u^4}\right)\right)$$

where

$$K_3=1-3k_3+3k_4+15k_6.$$

Reintroducing the variable t, we obtain from (48) an extended expansion of the same type as (50).

Finally, a study of the limit passage $c\to 0$ will show that, in the case $c=0$, the formulas Normal I and Esscher I will always give the same results, and the same thing will be true in respect of the two formulas Normal II and Esscher II.

In order to use the above formulas for numerical computations, we have to start with a given value of c, and determine h from the equation (38). (In practice we may, of course, prefer the inverse procedure: to start with a given value of h, and determine the corresponding c from 38.) The quantities q_n and k_n are then computed from (44), using (29), while β is found from (35), and u from (41). The function $E(u)$ is computed from some table of the normal functions Φ and Φ', and can also be taken from a table of $E(u)$ given by Esscher **(34)**. Once these values have been found, approximate values of $G(u)$ may be computed from Esscher I or Esscher II, and inserted in (48). When u is not very large, it will be found that the simplified formulas (49) and (51) for $G(u)$ will give less accurate results.

Reinsurance premiums of the type (20) may be numerically computed by means of the approximation methods given above, or in 4. 2. The expression (20) can, in fact, be written in the form

$$\int_{-\infty}^{-kp_1t} |x-kp_1t|\, d_xF(x,t) = kp_1tF(-kp_1t,\, t) - t\int_{-\infty}^{-kp_1} c\,d_cF(ct,\, t)\,,$$

which is directly adapted for the use of the approximation formulas based on the Esscher method. Ammeter **(4, 5)** and Lefèvre (41) have given explicit formulas for premiums of this type.

4. 4. Numerical illustrations. In order to illustrate the efficiency of the various approximation formulas discussed in 4. 2 and 4. 3, we shall now apply them to some particular cases.

1. *Exponential distribution of risk sums.* Suppose that the d.f. $P(y)$ of risk sums is given by

$$P(y) = \begin{cases} 0, & y < 0\,, \\ 1 - e^{-y}, & y \geqq 0\,. \end{cases}$$

This implies that there are no negative risk sums, and that the positive risk sums have the probability density

$$P'(y) = e^{-y}, \quad y > 0\,.$$

The moments p_n are finite for all n, and we have

$$p_n = \int_0^\infty y^n e^{-y}\, dy = n!$$

Thus in particular the mean risk sum is $p_1 = 1$, which implies that we are choosing the mean risk sum as our unit of money.

The n:th convolution of $P(y)$ with itself is

$$P^{n*}(y) = \int_0^y \frac{z^{n-1}}{(n-1)!} e^{-z}\, dz,$$

and thus by (18) we have

$$F(x, t) = 1 - \sum_0^{\infty} \frac{t^n}{n!} e^{-t} \int_0^{t-x} \frac{z^{n-1}}{(n-1)!} e^{-z} dz$$

$$= \sum_0^{\infty} \frac{t^n}{n!} e^{-t} \int_{t-x}^{\infty} \frac{z^{n-1}}{(n-1)!} e^{-z} dz$$

$$= \sum_{n=0}^{\infty} \frac{t^n}{n!} e^{-t} \sum_{r=0}^{n-1} \frac{(t-x)^r}{r!} e^{-(t-x)}.$$

From this formula, $F(x, t)$ can be directly computed with any desired accuracy. The labour involved will, however, be fairly considerable for large t, if the computation is to be carried out by manual work.

By (25), the coefficients c_n in the expansion (26) are

$$c_n = \frac{p_n}{p_2^{\frac{n}{2}}} = 2^{-\frac{n}{2}} n!$$

so that the approximate formula Normal II becomes

$$F(x\sqrt{2t}, t) = \Phi(x) + \frac{1}{2\sqrt{2t}} \Phi^{(3)}(x) + \frac{1}{4t} \Phi^{(4)}(x) + \frac{1}{16t} \Phi^{(6)}(x).$$

In respect of the Esscher method, equation (38) becomes

$$1 - \frac{1}{(1-h)^2} = c,$$

thus giving

$$h = 1 - \frac{1}{\sqrt{1-c}},$$

where we can choose for c any real value < 1. The further quantities required for the computation of approximate values according to the formulas Esscher I and Esscher II are, according to the formulas given in the preceding section,

$$k_3 = \frac{h}{2(1-h)} = \frac{1}{2}\left(\sqrt{1-c}-1\right),$$

$$k_4 = 2\,k_3^2,$$

$$k_6 = \frac{1}{2}\,k_3^2,$$

$$\beta = 4\,k_3^2,$$

$$u = |h|\sqrt{2t(1-c)^{\frac{3}{2}}}.$$

Computations have been carried out for $t = 16$ and for $t = 400$, and the results are shown in the following tables.

TABLE I.

Values of $10^5\,F(ct,t)$ for $t = 16$. $P'(y) = e^{-y}$.

c	Exact formula	Normal I	Normal II	Esscher I	Esscher II
— 1.50	39	1	— 10	41	40
— 1.25	186	20	9	197	186
— 1.00	782	234	840	831	782
— 0.75	2 850	1 696	2 866	3 056	2 850
— 0.50	8 828	7 868	8 759	9 560	8 832
— 0.25	22 613	23 979	22 627	24 650	22 624
0.00	46 460	50 000	46 474	50 000	46 474
+ 0.25	74 615	76 021	74 625	76 875	74 630
+ 0.50	93 961	92 132	93 835	94 530	93 967
+ 0.75	99 658	98 304	99 736	99 692	99 659
+ 1.00	100 000	99 766	100 064		

TABLE II.

Absolute error in $F(ct, t)$, expressed as percentage of absolute error in Normal I. $t = 16$, $P'(y) = e^{-y}$.

c	Normal I	Normal II	Esscher I	Esscher II
— 1.50	100	129	5	3
— 1.25	100	107	7	0
— 1.00	100	11	9	0
— 0.75	100	1	18	0
— 0.50	100	7	76	0
— 0.25	100	1	149	1
0.00	100	0	100	0
+ 0.25	100	1	161	1
+ 0.50	100	7	31	0
+ 0.75	100	6	3	0

TABLE III.

Values of $10^6\, F(ct, t)$ for $t = 400$. $P'(y) = e^{-y}$.

c	Exact formula	Normal I	Normal II	Esscher I	Esscher II
— 0.30	34	11	33	36	35
— 0.15	19 492	16 949	19 498	19 787	19 495
0.00	492 944	500 000	492 948	500 000	492 948
+ 0.15	985 703	983 051	985 705	985 926	985 703
+ 0.30	999 998	999 989	999 997	999 998	999 998

TABLE IV.

Absolute error in $F(ct, t)$, expressed as percentage of absolute error in Normal I. $t = 400$, $P'(y) = e^{-y}$.

c	Normal I	Normal II	Esscher I	Esscher II
— 0.30	100	4	9	4
— 0.15	100	0	12	0
0.00	100	0	100	0
+ 0.15	100	0	8	0
+ 0.30	100	11	0	0

It will be seen that for $t = 16$ the formula Esscher II is decidedly superior to the other approximation formulas tried here. For $t = 400$, on the other hand, the formulas Normal II and Esscher II yield results of the same average order of accuracy, which is considerably better than the two I formulas.

2. *A distribution from fire insurance.* Dr. F. Esscher has kindly put at my disposal a distribution of positive risk sums, based on an experience from Swedish non-industry fire insurance, covering the years 1948—1951. Choosing the mean risk sum p_1 as our unit of money, the frequency function of this distribution is given by the formulas

$$P'(y) = Ae^{-\alpha y} + B(x+b)^{-\beta}, \qquad 0 < y < 500,$$
$$P'(y) = 0, \qquad y > 500,$$

where

$$A = 4.897954, \qquad B = 4.503, \qquad b = 6,$$
$$\alpha = 5.514588, \qquad \beta = 2.75.$$

For this distribution we have, as already noted, $p_1 = 1$. Some values of $1 - P(y)$ are given in the following table.

Through the courtesy of the Directory Committee of the Electronic Computer BESK, some approximate values of $F(ct, t)$ for this distribution, and also some values of the ruin probability $\psi(u)$ as defined in 5. 2, have been prepared

TABLE V.

Fire insurance distribution. Values of $10^6 [1-P(y)]$.

y	$10^6 [1-P(y)]$
1	88 946
5	38 680
10	20 055
50	2 197
100	687
200	182
300	67
400	22
500	0

by this computer and placed at my disposal. Mr. J. Ohlin and Mr. G. Erling have performed very valuable work in conducting and supervising the computing operations.

It would have been highly desirable to obtain at the same time also the exact values of $F(ct, t)$ for comparison with the corresponding approximate values as given below in Table VI. However, the computation of the exact values proves to be extremely laborious, and it has appeared necessary to reserve them for a later publication.

The approximate values of $F(ct, t)$ for $t = 10^3$ are given in Table VI. From these values, it is evidently not possible to draw any definite conclusions with respect to the accuracy of the various approximation formulas in this case. It should be noted, however, that in all cases given in the table (barring only the exceptional case $c = 0$), the agreement between the formulas Normal II and Esscher II is much better than the agreement between the two I formulas. Whether this may be taken as an indication that the II formulas are nearer to the exact values than the I formulas, cannot be decided until the exact values have become available.

TABLE VI.

Fire insurance distribution. Values of $10^5 F(ct, t)$ for $t = 10^3$.

h	c	$10^5 F(ct, t)$			
		Normal I	Normal II	Esscher I	Esscher II
+ 0.0070	− 0.77248	10	267	319	293
+ 0.0050	− 0.39095	2 999	4 784	5 517	4 854
+ 0.0025	− 0.13952	25 104	22 225	26 028	22 438
0	0	50 000	44 284	50 000	44 284
− 0.0200	+ 0.34791	95 292	98 864	98 623	98 519
− 0.0800	+ 0.59311	99 784	99 985	100 000	100 000

For the function $\psi(u)$, which expresses the probability of the ruin of the risk reserve of the company at some future moment, as explained in 5. 2, we have the asymptotic formula (119), which shows that for large u the probability $\psi(u)$ is approximately given by the expression

$$\psi(u) \sim Ce^{-Ru},$$

where

$$C = \frac{\lambda}{p'(R) - 1 - \lambda},$$

while R is the unique positive root of the equation

$$1 + (1+\lambda)\, R - p(R) = 0.$$

u denotes the initial value of the risk reserve, while λ is a constant representing the safety loading of the risk premiums, as defined in 5. 1.

Some exact values of the ruin probability $\psi(u)$ for $\lambda = 0.3$ have been computed by BESK, and are shown below in Table VII, together with the approximate values obtained from the above formula. For $\lambda = 0.3$ we obtain $R = 0.00736$ and $C = 0.5241$. The exact values of $\psi(u)$ have been computed from the integral equation (85).

TABLE VII.

Fire insurance distribution. Values of the ruin probability $\psi(u)$ for $\lambda = 0.3$.

u	$\psi(u)$		
	Exact	Approximate	Ratio
20	0.5039	0.4524	0.898
40	0.3985	0.3904	0.980
60	0.3280	0.3370	1.027
80	0.2757	0.2909	1.055
100	0.2346	0.2511	1.070

It is interesting to compare the values thus obtained for the ruin probability $\psi(u)$ for this extremely skew distribution of risk sums with the values of the same function for the simple exponential distribution considered above in example 1. Although both distributions have the same mean risk sum $p_1 = 1$, the exponential distribution gives according to the exact formula (137) already fur $u = 40$ a value of $\psi(u)$ slightly less than 0.0001. The safety constant λ being in both cases $= 0.3$, this value is directly comparable to the value 0.3985 given in the above table.

5. The Ruin Problem

5. 1. Introductory remarks. As before, we denote by

$$X(t) = p_1 t - Y(t)$$

the gain on the net risk business realized by the company up to the time t. Here t denotes as usual operational time (cf. 3. 1) as measured by the expected number of claims. We now assume that, during every time element dt, the company receives, besides the net risk premium $p_1 dt$, a certain *safety loading* λdt, so that the *gross risk premium* for the time element is $(p_1+\lambda)\, dt$. Here λ is assumed to be a given constant, which is essentially positive:

$$\lambda > 0,$$

while on the other hand

$$p_1 = \int_{-\infty}^{\infty} y P(y)\, dy$$

is capable of assuming values of both signs. In the case of a business where insurances with negative risk sums (such as life annuities) are predominant, p_1 will be negative, and the safety loading will then take the form of a deduction from the annuities paid by the company, instead of an addition to the premiums received by the company.

We further assume that all gross risk premiums are collected into a certain fund, the *risk reserve* of the company, from which all claims are paid (subject, of course, to obvious modifications in the case of insurances with negative risk sums). If the initial value of the risk reserve at time $t = 0$ is $u \geqq 0$, the amount of the risk reserve at time t will be

$$u + (p_1+\lambda)\, t - Y(t) = X(t) + u + \lambda t. \tag{52}$$

If, at any time t, the risk reserve takes a negative value, we shall say that the risk business is *ruined* at the time t. We then have

$$X(t) + u + \lambda t < 0, \tag{53}$$

or

$$X(t) < -u - \lambda t,$$

which means that the sample function $X(t)$ which describes the development of the risk business comes, at time t, below the straight line with the ordinate $-u-\lambda t$ (cf. fig. 3).

The *ruin problem* is the problem of finding the probability that the event of ruin will take place, subject to various specified conditions. We shall now introduce notations for certain probabilities that will be encountered in this connection.

Let $\psi(u)$ denote the probability that the risk reserve, starting from the initial value u at $t = 0$, will be ruined at some future moment, i.e., the prob-

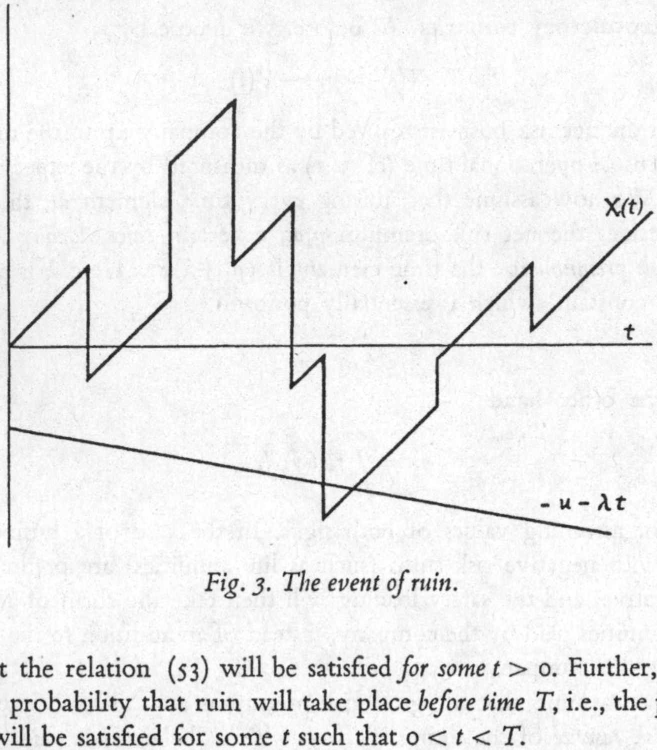

Fig. 3. The event of ruin.

ability that the relation (53) will be satisfied *for some* $t > 0$. Further, let $\psi(u, T)$ denote the probability that ruin will take place *before time* T, i.e., the probability that (53) will be satisfied for some t such that $0 < t < T$.

We may also agree to take account of the value of the risk reserve only in a sequence of specified time points, such as, e.g., the points $t = h, 2h, 3h, \ldots,$ where h is a given positive constant. This would correspond to the fact that, in practice, it is mainly the financial status of the company at the ends of successive business periods that matters. We shall denote by $\psi_h(u)$ the probability that the ruin relation (53) will be satisfied for some $t = nh$, where $n = 1, 2, 3, \ldots.$ Further, $\psi_h(u, T)$ will denote the probability that the same event takes place, with the additional condition that $nh < T$.

The functions $\psi(u)$, $\psi(u, T)$, $\psi_h(u)$, and $\psi_h(u, T)$ will be called the *ruin functions*. Their exact definitions, in terms of probability measure, will be given in the following section. It is evident that, besides on the variables explicitly given in our notation, all these functions will depend on the constant λ and the d.f. $P(y)$ of risk sums.

The functions $\psi(u)$ and $\psi_h(u)$ were introduced and studied by F. Lundberg **(45, 47, 48)**. For the case when all risk sums are positive, he obtained some fundamental results, such as the inequality

$$(54) \qquad 0 < \psi_h(u) < \psi(u) \leqq e^{-Ru},$$

and the asymptotic relations

$$(55) \qquad \psi(u) \sim Ce^{-Ru}, \qquad \psi_h(u) \sim C_h e^{-Ru},$$

valid for large values of u. Here R and C are positive constants depending only on the constant λ and the d.f. $P(y)$, while C_h depends also on h.

For the case when all risk sums are positive, Cramér **(17, 19)** showed that $\psi(u)$ satisfies an integral equation of the Volterra type, and that this equation can be solved by a complex Fourier transformation, which yields an explicit analytic expression of $\psi(u)$, and a proof of the first asymptotic relation (55). He also gave in **(22)** the integral equation satisfied by $\psi(u)$ in the general case when the risk sums may have both signs. In this case, the equation is not of the Volterra type.

Segerdahl **(59—62)** considered the general case with risk sums of both signs, and gave for this case the first rigorous proofs of the fundamental relations (54) and (55). He also made a detailed study of the particular cases of only positive, or only negative, risk sums, and proved for these cases a number of important results, some of which had been stated without complete proofs by F. Lundberg.

Täcklind **(65)** considered, for the general case, the integral equation satisfied by $\psi_h(u)$, and showed that this can be explicitly solved by means of a method due to Wiener and Hopf **(53**, p. 49**)** involving a complex Fourier transformation, combined with arguments belonging to the theory of analytic functions. From this solution, he obtained a proof of the asymptotic formula (55) for $\psi_h(u)$. He also showed that $\psi_h(u)$ tends to a definite limit as $h \to 0$, and gave an explicit expression for this limit.

Cramér **(27)** gave a probabilistic definition of the ruin function $\psi(u)$, and applied the Wiener-Hopf method directly to the integral equation satisfied by this function in the general case.

In respect of the functions $\psi(u,T)$ and $\psi_h(u,T)$ connected with the ruin problem for a finite time interval, certain preliminary results were given by F. Lundberg **(45)** and Segerdahl **(59)**, who calculated the moments of the time when ruin occurs for the first time. The problem has been thoroughly investigated by Saxén **(57, 58)** and Arfwedson **(7—10)**. The former considered mainly the case of only negative risk sums, and the latter the case of only positive risk sums. For these cases, explicit expressions for $\psi(u,T)$ are given. Arfwedson further obtains a number of results concerning the asymptotic properties of this function.

In the investigations briefly reviewed above, the problem of giving satisfactory mathematical definitions of probabilities like $\psi(u)$ and $\psi(u, T)$ has often given rise to difficulties. The corresponding question for $\psi_h(u)$ and $\psi_h(u, T)$ is much simpler, since these probabilities are attached to an enumerable sequence $X(h)$, $X(2h)$, ... of random variables (cf. 2. 3), and all relevant sets in the corresponding space will be Borel sets. When we come to probabilities like $\psi(u)$ and $\psi(u, T)$, however, we shall be operating in a space of random functions $X(t)$ of a continuous variable t, and it has been pointed out in 2. 4 that certain difficulties will arise. In most cases, these difficulties have been avoided simply by working with the limits $\lim_{h \to 0} \psi_h(u)$ and $\lim_{h \to 0} \psi_h(u, T)$, and identifying these limits, with or without an explicit definition, with the probabilities $\psi(u)$ and $\psi(u, T)$ respectively. From a mathematical point of view, such a procedure is obviously not satisfactory, unless it is proved that, e.g., the limit $\lim_{h \to 0} \psi_h(u)$ is a mathematical probability in the sense of the general definition given in 2. 2. As long as this has not been proved, it must be regarded as an open question whether the limits in question have the essential properties of mathematical probabilities.

Following Cramér (**27**), we shall in section 5. 2 of the present paper give rigorous definitions of the probabilities $\psi(u)$, $\psi(u, T)$, $\psi_h(u)$, and $\psi_h(u, T)$. It will be shown that the space $\mathfrak{X}_0$ of natural sample functions $X(t)$ can be simply referred back to a reference space Ω of an enumerable number of dimensions, and in this way we obtain definitions of the ruin probabilities, in terms of the probability distribution of the $X(t)$ process, as the measures of certain specified sets of functions. At the same time, the relations

$$\lim_{h \to 0} \psi_h(u) = \psi(u), \qquad \lim_{h \to 0} \psi_h(u, T) = \psi(u, T),$$

will be rigorously proved, instead of being taken as the definitions of the limiting probabilities.

In 5.3 we introduce certain assumptions about the d.f. $P(y)$ that will be used in the rest of this chapter. We also introduce here the Lundberg constant R, and two further important constants denoted by ϱ and γ. The basic inequalities (54) are proved in 5. 4, by a method which is a simplified version of one due to Täcklind (**65**). Both these sections are based on Cramér (**27**).

In 5. 7 and 5. 8 the fundamental integral equations satisfied by the ruin functions and by certain associated functions are deduced and discussed by means of the Wiener-Hopf method mentioned above. In 5. 9—5. 11 it is shown that the integral equation (88) satisfied by a certain auxiliary function introduced in 5. 5 leads to explicit expressions for $\psi(u)$ and $\psi(u, T)$, and also to certain results concerning the asymptotic properties of these functions. In this connection, some results due to Segerdahl **(59)** on the asymptotic properties of $\psi(u)$ are proved, and also an inequality (108) for the difference $\psi(u)-\psi(u,T)$, which is believed to be new. All these sections are concerned with the general case when there are risk sums of both signs.

The particular cases of only positive, or only negative risk sums are dealt with in 5. 12 and 5. 13. In the positive case, the asymptotic relations (55) given by F. Lundberg for $\psi(u)$ and $\psi_h(u)$ are proved. Further, the inequality (108) for $\psi(u)-\psi(u,T)$ is strengthened, and an asymptotic relation (125) for the same difference is deduced. The relation (125) has been stated (without proof) in a letter to the author by Arfwedson, who will give his proof in a forthcoming paper, together with other related results.

Finally, a certain type of exponential distributions of risk sums indicated by Täcklind **(65)** is studied in 5. 14. As a special case, this type includes the case of only positive risk sums with a simple exponential distribution, which forms one of the classical examples often discussed in works on risk theory.

5. 2. The ruin functions. We now proceed to give exact definitions of the ruin functions $\psi(u)$, $\psi(u,T)$, $\psi_h(u)$, and $\psi_h(u,T)$ discussed in the preceding section.

Consider the reference space Ω and the probability measure $\Pi(\Sigma)$ on the Borel sets of this space, as defined in 3. 3. We know by 3. 3 and 3. 4 that to every point ω in Ω, there corresponds one uniquely determined natural sample function $X(t)$, i.e., a "point" in the space $\mathfrak{X}_0$ of all natural sample functions. If S is any Borel set of functions $X(t)$ in $\mathfrak{X}_0$, and Σ is the inverse image (cf. 2. 2 and 2. 5) of S, it follows from 2. 5 that Σ will be a Borel set in Ω. The induced probability measure $P(S)$ in $\mathfrak{X}_0$ is then defined by the relation (cf. 3. 4)

$$P(S) = \Pi(\Sigma).$$

According to 3. 4, $P(S)$ is then the probability measure of the set S in the distribution corresponding to the stochastic $X(t)$ process.

If M is a given relation, or a system of relations, we denote by the symbol

$$\{X(t)\,|M\}$$

the set of all natural sample functions $X(t)$ satisfying M. Thus, if we write for any $h > 0$, $u \geqq 0$, and $n = 1, 2, \ldots$,

$$S_{n,h} = \{X(t)\,|\, X(nh)+u+\lambda nh < 0\},$$

it is seen from (53) that $S_{n,h}$ will be the set of all natural sample functions $X(t)$ such that the risk business is ruined at the particular time point $t = nh$. Obviously $S_{n,h}$ is a finite-dimensional interval, and a fortiori a Borel set, in $\mathfrak{X}_0$. It follows that the probability $P(S_{n,h})$ is well defined according to the above. We evidently have

$$P(S_{n,h}) = F(-u-\lambda nh, nh),$$

where $F(x, t)$ is the d.f. of the gain $X(t)$ as given by (18).

Further, let us write for any $h > 0$, $u \geqq 0$, $T \geqq 0$, and $n = 1, 2, \ldots$,

$$\begin{aligned} S_h(T) &= \{X(t) \mid X(nh)+u+\lambda nh < 0 \quad \text{for some } n \text{ such that } nh < T\}, \\ S_h &= \{X(t) \mid X(nh)+u+\lambda nh < 0 \quad \text{for some } n\}, \\ S(T) &= \{X(t) \mid X(t)+u+\lambda t < 0 \quad \text{for some } t \text{ such that } 0 < t < T\}, \\ S &= \{X(t) \mid X(t)+u+\lambda t < 0 \quad \text{for some } t > 0\}. \end{aligned}$$

We shall first show that all these sets are Borel sets in $\mathfrak{X}_0$. We may then assume $T > 0$ since, according to our definitions, the sets $S_h(0)$ and $S(0)$ are empty, and thus certainly Borel sets. In respect of $S_h(T)$ and S_h, it follows immediately from the relations

$$S_h(T) = \sum_{nh < T} S_{n,h}, \qquad S_h = \sum_{n=1}^{\infty} S_{n,h},$$

that they are sums of finite or enumerable sequences of Borel sets, and are thus Borel sets themselves. Further, let us consider the set $S_h(T)$ for a sequence H of values of h tending to zero. Since evidently $S_h(T)$ is a subset of $S(T)$ for every $h > 0$, which we write

$$S_h(T) \subset S(T),$$

it follows that we have

$$\limsup S_h(T) \subset S(T),$$

when h tends to zero on the sequence H. On the other hand, let $X(t)$ be any natural sample function belonging to the set $S(T)$. By the definition of $S(T)$, we have $X(t) < -u-\lambda t$ for at least one t such that $0 < t < T$. Being a natural sample function, $X(t)$ has, however, only a finite number of jumps in the interval $(0, T)$, and it follows that we shall necessarily have $X(t) < -u-\lambda t$ throughout a whole t-interval. It then follows that, for every sufficiently small h, it will be possible to find an integer n such that $nh < T$ and $X(nh) < -u-\lambda nh$, which shows that $X(t)$ belongs to $S_h(T)$. To every $X(t)$ belonging to the set $S(T)$, we can thus find $h_0 > 0$ such that $X(t)$ belongs to $S_h(T)$ as soon as $h < h_0$. It follows that we have, as h tends to zero on the sequence H,

$$S(T) \subset \liminf S_h(T).$$

Comparing the last two relations, it is seen that the sequence $S_h(T)$ is a convergent sequence of sets, and that we have

$$S(T) = \lim_{h \to 0} S_h(T),$$

which shows that $S(T)$ is a Borel set. Finally, allowing T to tend to infinity, it is evident that $S(T)$ will be monotonely increasing, and will tend to the limit S, which is thus also a Borel set.

It now follows that each of the sets $S_h(T)$, S_h, $S(T)$, and S will have a well-defined probability in the distribution of the $X(t)$ process. Obviously $S_h(T)$ is the set of all $X(t)$ such that ruin occurs at some time point $t = nh$, with $nh < T$, and similarly in respect of the three other sets.

We now define the ruin functions as mathematical probabilities in the $X(t)$ distribution by writing

$$\begin{aligned} \psi_h(u,T) &= P[S_h(T)], & \psi_h(u) &= P(S_h), \\ \psi(u,T) &= P[S(T)], & \psi(u) &= P(S). \end{aligned} \tag{56}$$

It then follows immediately from the above, observing that the sequence H was arbitrary, that we have

$$\begin{aligned} \psi(u,T) &= \lim_{h\to 0} \psi_h(u,T), \\ \psi(u) &= \lim_{h\to 0} \psi_h(u), \\ \psi_h(u) &= \lim_{T\to\infty} \psi_h(u,T), \\ \psi(u) &= \lim_{T\to\infty} \psi(u,T). \end{aligned} \tag{57}$$

It further follows directly from our definitions that all four ruin functions will be never increasing and continuous to the right with respect to u, for all $u \geqq 0$, and never decreasing and continuous to the left with respect to T, for all $T > 0$. Finally we have for all $u \geqq 0$

$$\psi_h(u, 0) = \psi(u, 0) = 0. \tag{58}$$

5.3. The constants R, γ, and ϱ. Throughout the remaining part of this chapter we shall, unless something else is explicitly mentioned, only make the following two assumptions with regard to the given d.f. $P(y)$:

$$\begin{aligned} &\int_{-\infty}^{0} |y|\, dP(y) < \infty, \\ &\int_{0}^{\infty} e^{\sigma y}\, dP(y) < \infty \quad \text{for some } \sigma > 0. \end{aligned} \tag{59}$$

It follows from these relations that the mean value p_1 will be finite, and that the integral defined by (10):

$$p(s) = \int_{-\infty}^{\infty} e^{sy}\, dP(y),$$

where $s = \sigma + i\tau$, will be absolutely convergent in some vertical strip $0 < \sigma \leqq r$ in the plane of the complex variable s, and will represent a function which is analytic and regular in this strip.

Consider the function

$$\pi(s) = 1 + (p_1 + \lambda)s - p(s), \tag{60}$$

where $\lambda > 0$ is the constant representing the safety loading on the risk premiums (cf. 5. 1). We have

$$\pi(0) = 0, \qquad \pi'(0) = p_1 + \lambda - p'(0) = \lambda > 0.$$

It follows that, for all sufficiently small real and positive s, the function $\pi(s)$ will be real and positive.

We have thus established the existence of at least one real and positive quantity r, such that the following two conditions are satisfied:

1) $p(s)$ is analytic and regular for $0 < \sigma \leqq r$,
2) $1 + (p_1 + \lambda)\, \sigma - p(\sigma) > 0$ for $0 < \sigma \leqq r$.

If these conditions are satisfied for some $r = r_0 > 0$, they will evidently be satisfied for all r such that $0 < r \leqq r_0$. We shall denote by R the supremum, or the least upper bound, of all r satisfying the conditions 1) and 2), so that

$$R = \sup r.$$

It is then clear that we shall always have $0 < R \leqq \infty$. We now proceed to prove that, in all non-trivial cases, R will be finite.

When λ and $P(y)$ are given, one and only one of the three following cases must occur:

A) $P(0) < 1$,
B) $P(0) = 1$, $p_1 + \lambda < 0$,
C) $P(0) = 1$, $p_1 + \lambda \geqq 0$.

In case A) the $P(y)$ distribution will have a positive mass on the positive y axis, so that $p(\sigma)$ will grow exponentially for large σ, provided that the corresponding integral remains convergent. When σ increases indefinitely, at least one of the conditions 1) and 2) will thus sooner or later break down, and it follows that in this case R is finite.

In case B) the condition 1) will be satisfied for all $r > 0$, but for sufficiently large σ we shall have

$$1 + (p_1 + \lambda)\,\sigma - p(\sigma) < 1 - |p_1 + \lambda|\,\sigma < 0,$$

so that condition 2) will ultimately break down, and R will be finite.

On the other hand, in case C) it is easily seen that conditions 1) and 2) will both be satisfied for all $r > 0$, so that we have in this case $R = \infty$.

However, in case C) we have $P(0) = 1$, which implies that all risk sums are negative, and further $p_1 + \lambda \geqq 0$, which implies that, during every time element dt, the company receives a non-negative gross risk premium $(p_1+\lambda)dt$. Thus in this case the company will never have anything to pay out for the risk business, and all the ruin functions are identically zero. This shows that case C) is entirely trivial and uninteresting from the point of view of the ruin problem. Accordingly we shall exclude case C) from our considerations in the sequel. This implies that, *in the particular case when all risk sums are regative*, the constant λ will always be assumed to satisfy the inequality

$$0 < \lambda < |p_1|.$$

Bearing this in mind, we may always regard R as a finite and positive quantity, uniquely determined by the constant λ and the d.f. $P(y)$.

The function $\pi(\sigma)$ defined by (60) is continuous and positive for $0 < \sigma < R$. We shall denote by γ the least upper bound of $\pi(\sigma)$ in this interval:

$$\gamma = \sup_{0<\sigma<R} [1+(p_1+\lambda)\,\sigma - p(\sigma)]. \tag{61}$$

Then γ will be finite and positive, and for every γ_0 such that $0 \leqq \gamma_0 < \gamma$ there will be an interval (α, β) with $0 \leqq \alpha < \beta \leqq R$, such that

$$1+(p_1+\lambda)\sigma - p(\sigma) > \gamma_0 \tag{62}$$

as soon as $\alpha < \sigma < \beta$. In general α and β will, of course, depend on γ_0. For $\gamma_0 = 0$, we may take $\alpha = 0$ and $\beta = R$. On the other hand, when $\gamma_0 \to \gamma$, the length of the interval (α, β) will tend to zero.

If, in particular, the derivative

$$\pi'(\sigma) = p_1 + \lambda - p'(\sigma)$$

has one single zero in an interior point $\sigma = \varrho$ of the interval $(0, R)$,[1] the function $\pi(\sigma)$ will attain its maximum at $\sigma = \varrho$, so that we have

$$\gamma = 1+(p_1+\lambda)\varrho - p(\varrho). \tag{63}$$

In this case, we can always take the interval (α, β) such that $\alpha < \varrho < \beta$. Moreover, α and β will both tend to ϱ, as $\gamma_0 \to \gamma$. (Cf. fig. 4, p. 69.)

[1] In all cases likely to be encountered in practical applications, this condition will be satisfied. However, it is easy to construct examples where this does not hold, the function $\pi(\sigma)$ being increasing all over the interval $(0,R)$.

5. 4. Proof of the inequality $\psi(u) \leqq e^{-Ru}$. Consider, in the space $\mathfrak{X}_0$, the set $W_{n,h}$ defined by the relation

$$W_{n,h} = \{X(t) \mid X(ih)+u+\lambda ih \geqq 0 \quad \text{for } i = 1, 2, \ldots, n-1; \quad X(nh)+u+\lambda nh < 0\},$$

where $h > 0$ and $u \geqq 0$. It will be seen that $W_{n,h}$ consists of all natural sample functions $X(t)$ such that the risk business is *not* ruined at the time points $t = h$, $2h, \ldots, (n-1)h$, while it is ruined at $t = nh$. Evidently $W_{n,h}$ is a finite-dimensional interval in $\mathfrak{X}_0$, and $W_{m,h}$ and $W_{n,h}$ are disjoint whenever $m \neq n$. We further have

$$\sum_{nh<T} W_{n,h} = S_h(T), \qquad \sum_{n=1}^{\infty} W_{n,h} = S_h,$$

where $S_h(T)$ and S_h are defined as in 5. 2.

Thus writing

$$\varphi_{n,h}(u) = P(W_{n,h}),$$

we obtain from (56)

$$\psi_h(u,T) = \sum_{nh<T} \varphi_{n,h}(u), \qquad \psi_h(u) = \sum_{n=1}^{\infty} \varphi_{n,h}(u). \tag{64}$$

We further have

$$\varphi_{1,h}(u) = P[X(h) < -u-\lambda h] = F(-u-\lambda h, h), \tag{65}$$

and for $n > 1$

$$\varphi_{n,h}(u) = \int_0^{\infty} \varphi_{n-1,h}(v)\, d_v F(v-u-\lambda h, h). \tag{66}$$

Summing (66) over $n = 2, \ldots, N$, and adding (65), we obtain

$$\sum_{n=1}^{N} \varphi_{n,h}(u) = F(-u-\lambda h, h) + \int_0^{\infty} \sum_{n=1}^{N-1} \varphi_{n,h}(v)\, d_v F(v-u-\lambda h, h). \tag{67}$$

Let us now take r such that $0 < r < R$. From (67) we obtain

$$e^{-ru} - \sum_{n=1}^{N} \varphi_{n,h}(u) = G + \int_0^{\infty} \left(e^{-rv} - \sum_{n=1}^{N-1} \varphi_{n,h}(v)\right) d_v F(v-u-\lambda h, h), \tag{68}$$

where

$$G = e^{-ru} - F(-u-\lambda h, h) - \int_0^{\infty} e^{-rv}\, d_v F(v-u-\lambda h, h). \tag{69}$$

However, since $0 < r < R$, the conditions 1) and 2) of 5.3 are satisfied, and it follows from (19) that we have

$$E e^{-rX(t)} = \int_{-\infty}^{\infty} e^{-rv}\, d_v F(v, t) = e^{t[p(r)-1-p_1 r]},$$

and consequently

$$\int_{-\infty}^{\infty} e^{-rv}\, d_v F(v-\lambda t, t) = e^{t[p(r)-1-(p_1+\lambda) r]} < 1$$

for all $t > 0$. Hence

$$G > e^{-ru} \int_{-\infty}^{\infty} e^{-rv}\, d_v F(v-\lambda h, h) - \int_{-\infty}^{0} d_v F(v-u-\lambda h, h) -$$

$$- \int_{0}^{\infty} e^{-rv}\, d_v F(v-u-\lambda h, h) = \int_{-\infty}^{0} (e^{-rv}-1)\, d_v F(v-u-\lambda h, h) \geqq 0.$$

It thus follows from (68) that, if the inequality

$$\sum_{n=1}^{k} \varphi_{n,h}(u) \leqq e^{-ru} \quad \text{for all } u \geqq 0 \tag{70}$$

holds for $k = N-1$, it will also hold for $k = N$. Taking $k = 1$, we have according to (69)

$$e^{-ru} - \varphi_{1,h}(u) = e^{-ru} - F(-u-\lambda h, h) \geqq G > 0,$$

so that (70) certainly holds for $k = 1$, and thus for all k. As k tends to infinity, we then obtain from (64)

$$\psi_h(u) \leqq e^{-ru}$$

for all $u \geqq 0$. For any fixed $u \geqq 0$ and $h > 0$, this inequality has thus been established for any r such that $0 < r < R$. Allowing r to tend to R, we obtain

$$\psi_h(u) \leqq e^{-Ru}.$$

Further, allowing h to tend to zero and using (57) we obtain, observing that R is independent of h,

$$\psi(u) \leqq e^{-Ru}.$$

Finally, a simple argument will show that the sets $S - S_h$ and S_h defined in 5.2 both have a positive P-measure, so that we have according to the above

$$0 < \psi_h(u) < \psi(u) \leqq e^{-Ru} \tag{71}$$

for all $u \geqq 0$, as stated above in (54). The further inequalities

$$\psi_h(u,T) \leqq \psi_h(u), \qquad \psi(u,T) \leqq \psi(u) \tag{72}$$

are evident.

5. 5. Some auxiliary functions. We shall now introduce some auxiliary functions that will be required in the sequel. We denote by

$$z = \xi + i\eta$$

a complex variable, and define for $h > 0$, $u \geqq 0$, and $\xi \leqq 0$, using (64),

$$\begin{aligned} \overline{\psi}_h(u, z) &= \sum_1^\infty e^{znh} \varphi_{n,h}(u) = \int_0^\infty e^{zT}\, d_T \psi_h(u,T), \\ \overline{\psi}(u, z) &= \int_0^\infty e^{zT}\, d_T \psi(u,T). \end{aligned} \tag{73}$$

The function $\psi_h(u,T)$ is easily seen to be continuous with respect to T at $T = 0$, so that the definition of the integral in the first formula presents no ambiguity. In the second formula, $\psi(u,T)$ will be regarded as starting from the initial value $\psi(u, 0) = 0$ given by (58), so that any jump at the point $T = 0$ will be included in the value of the integral.[1] It then follows from (57) that both integrals in (73) are absolutely convergent, and that we have (cf. Cramér, **23**, p. 74) for fixed u and z

$$\overline{\psi}(u, z) = \lim_{h \to 0} \overline{\psi}_h(u, z). \tag{74}$$

From (73) and (57) we obtain

$$\begin{aligned} \overline{\psi}_h(u, 0) &= \int_0^\infty d_T \psi_h(u,T) = \psi_h(u), \\ \overline{\psi}(u, 0) &= \int_0^\infty d_T \psi(u,T) = \psi(u), \end{aligned} \tag{75}$$

and further by means of the inequality (71)

$$\begin{aligned} |\overline{\psi}_h(u, z)| &\leqq \psi_h(u) < e^{-Ru}, \\ |\overline{\psi}(u, z)| &\leqq \psi(u) \leqq e^{-Ru}. \end{aligned} \tag{76}$$

We further define, denoting by

$$s = \sigma + i\tau$$

---

[1] It is easily seen that for $u > 0$, the function $\psi(u, T)$ will also be continuous with respect to T at $T = 0$. For $u = 0$, however, there may be a jump.

another complex variable,

$$(77)\qquad \begin{aligned}\overline{\overline{\psi}}_h(s, z) &= \int_0^\infty e^{su}\, \overline{\psi}_h(u, z)\, du,\\ \overline{\overline{\psi}}(s, z) &= \int_0^\infty e^{su}\, \overline{\psi}(u, z)\, du.\end{aligned}$$

According to (76), both integrals are absolutely convergent for $\sigma < R$. It follows that, for every fixed z such that $\xi \leqq 0$, the functions $\overline{\overline{\psi}}_h$ and $\overline{\overline{\psi}}$ of the complex variable s are analytic and regular in the half-plane $\sigma < R$. From (74) we obtain

$$(78)\qquad \overline{\overline{\psi}}(s, z) = \lim_{h \to 0} \overline{\overline{\psi}}_h(s, z).$$

Finally, a partial integration of (77) shows that we have throughout any half-plane $\sigma \leqq R - \varepsilon$, with $\varepsilon > 0$,

$$(79)\qquad |\overline{\overline{\psi}}_h(s, z)| < \frac{K}{1 + |s|}, \qquad |\overline{\overline{\psi}}(s, z)| < \frac{K}{1 + |s|},$$

where K depends only on ε.

5. 6. Further auxiliary functions. For the solution of an integral equation connected with the ruin problem, we shall require a certain auxiliary function $M(x, z)$ which we now proceed to study. We define

$$(80)\qquad M(x, z) = \sum_1^\infty \frac{1}{n!} \int_0^\infty y^{n-1} e^{-(1-z)y} \left[P^{n*}[x + (p_1 + \lambda) y] - 1 \right] dy,$$

where x is a real variable, while as in the preceding section $z = \xi + i\eta$ is a complex variable.

We shall first prove that the series in (80) is absolutely convergent for all real x, and for all z in the half-plane $\xi < \gamma$, where γ is the positive constant defined by (61). Take any γ_0 such that $0 \leqq \gamma_0 < \gamma$. According to 5. 3, we can then find an interval (α, β) with $0 \leqq \alpha < \beta \leqq R$, such that (62) holds. Take any r such that $\alpha < r < \beta$. We then have for any real x

$$e^{rx}[1 - P^{n*}(x)] \leqq \int_x^\infty e^{ry}\, dP^{n*}(y) \leqq \int_{-\infty}^\infty e^{ry}\, dP^{n*}(y) = [p(r)]^n,$$

$$0 \leqq 1 - P^{n*}(x) \leqq e^{-rx} [p(r)]^n.$$

Hence for $\xi \leqq \gamma_0$ the general term of (80) is majorated by the expression

$$\frac{[p(r)]^n}{n!} e^{-rx} \int_0^\infty y^{n-1} e^{-[1 + (p_1 + \lambda) r - \xi] y}\, dy \leqq \frac{1}{n} e^{-rx} \left(\frac{p(r)}{1 + (p_1 + \lambda) r - \gamma_0} \right)^n.$$

Hence by (62) the series (80) is absolutely convergent for all real x, and for all z such that $\xi \leqq \gamma_0$. Since γ_0 can be taken arbitrarily close to γ, the convergence holds all over the half-plane $\xi < \gamma$. It also easily follows that the convergence is uniform in every finite x-interval, and in every closed finite part of the half-plane $\xi < \gamma$. Accordingly $M(x, z)$ is, for every real x, an analytic and regular function of z in the half-plane $\xi < \gamma$. On the other hand, for every z such that $\xi < \gamma$, the function $M(x, z)$ is continuous with respect to x for all real x.

Moreover, we have for $\xi \leqq \gamma_0$

$$|M(x, z)| < Ce^{-rx}, \tag{81}$$

where C is independent of x and z, and r can be taken arbitrarily in $\alpha < r < \beta$.

Let us now take $\eta = 0$, and consider the function $M(x, \xi)$ of the real variable x, where $\xi \leqq \gamma_0$ is fixed. It is seen from (80) and (81) that $M(x, \xi)$ is a real-valued and never decreasing function of x, such that

$$M(x, \xi) \leqq 0$$

for all x, while for any $\varepsilon > 0$

$$|M(x, \xi)| < Ce^{(\alpha+\varepsilon)|x|} \quad \text{as } x \to -\infty,$$

$$|M(x, \xi)| < Ce^{-(\beta-\varepsilon)x} \quad \text{as } x \to +\infty.$$

For an arbitrary $z = \xi + i\eta$ with $\xi < \gamma$, we further have for any $\Delta x > 0$

$$|M(x+\Delta x, z) - M(x, z)| \leqq M(x+\Delta x, \xi) - M(x, \xi).$$

It follows from these relations that, for any fixed z with $\xi < \gamma$, the function $M(x, z)$ is a complex-valued function of the real variable x, which is everywhere continuous and of bounded variation in every finite interval. For $\xi \leqq \gamma_0$ the integral

$$H(s, z) = \int_{-\infty}^{\infty} e^{sx}\, d_x M(x, z),$$

where $s = \sigma + i\tau$, will be absolutely convergent for $\alpha < \sigma < \beta$. For every fixed z with $\xi \leqq \gamma_0$, the function $H(s, z)$ is thus an analytic and regular function of s in the strip $\alpha < \sigma < \beta$, which is uniformly bounded in any interior strip $\alpha + \varepsilon \leqq \leqq \sigma \leqq \beta - \varepsilon$, where $\varepsilon > 0$. Conversely, for every fixed s in the strip $\alpha < \sigma < \beta$, the function $H(s, z)$ is analytic, regular, and uniformly bounded as a function of z throughout the half-plane $\xi \leqq \gamma_0$.

On account of the absolute convergence we have, for $\xi \leqq \gamma_0$ and $\alpha < \sigma < \beta$,

$$H(s, z) = \sum_{1}^{\infty} \frac{1}{n!} \int_{0}^{\infty} y^{n-1} e^{-[1+(p_1+\lambda)s-z]y} dy \int_{-\infty}^{\infty} e^{sx} dP^{n*}(x)$$

$$= \sum_{1}^{\infty} \frac{1}{n} \left(\frac{p(s)}{1+(p_1+\lambda)s-z} \right)^n$$

$$= -\log \left(1 - \frac{p(s)}{1+(p_1+\lambda)s-z} \right),$$

the last expression representing that branch of the multi-valued logarithm, which tends to zero as s tends to infinity along a vertical in the strip $\alpha < \sigma < \beta$.

If we write

$$H^{(+)}(s, z) = \int_{0}^{\infty} e^{sx} d_x M(x, z),$$

(82) $$H^{(-)}(s, z) = \int_{-\infty}^{0} e^{sx} d_x M(x, z),$$

$$A(s, z) = e^{H^{(+)}},$$

$$B(s, z) = e^{-H^{(-)}},$$

we shall have

(83) $$1 - \frac{p(s)}{1+(p_1+\lambda)s-z} = \frac{B(s, z)}{A(s, z)},$$

and it is easily seen that, for every fixed z with $\xi \leqq \gamma_0$, both A and A^{-1} will be regular functions of s in the half-plane $\sigma < \beta$, while B and B^{-1} will be regular in $\sigma > \alpha$. Conversely A and A^{-1} will, for every fixed s with $\sigma < \beta$, be regular functions of z in the half-plane $\xi \leqq \gamma_0$, while B and B^{-1} will have the same properties for every fixed s with $\sigma > \alpha$. Moreover, if $\varepsilon > 0$ is given, A and A^{-1} will be uniformly bounded for all s and z such that $\sigma \leqq \beta - \varepsilon$ and $\xi \leqq \gamma_0$, while B and B^{-1} will be uniformly bounded for $\sigma \geqq \alpha + \varepsilon$ and $\xi \leqq \gamma_0$.

We finally observe that if, in the above developments, we take $\gamma_0 = 0$, it follows from 5. 3 that we may take $\alpha = 0$ and $\beta = R$.

5. 7. The fundamental integral equations. Allowing N to tend to infinity in (67), we obtain by means of (64)

(84) $$\psi_h(u) = F(-u-\lambda h, h) + \int_{0}^{\infty} \psi_h(v)\, d_v F(v-u-\lambda h, h).$$

Here F is to be considered as a known function, as it can be expressed by (18) in terms of the given d.f. $P(y)$. Accordingly (84) may be regarded as an integral equation for the unknown ruin function $\psi_h(u)$.

When h tends to zero in (84), it follows from (57) that the first member tends to $\psi(u)$, and it can be shown that in the limit we obtain the following integral equation for the ruin function $\psi(u)$:

$$(85) \qquad (p_1+\lambda)\,\psi(u) = \int_u^\infty Q(v)\,dv + \int_0^\infty \psi(v)\,Q(u-v)\,dv,$$

where

$$(86) \qquad Q(u) = \varepsilon(u) - P(u) = \begin{cases} -P(u), & u < 0, \\ 1-P(u), & u \geqq 0. \end{cases}$$

The integral equations (84) and (85) have been discussed by Täcklind (65) and Cramér (**27**), who have shown that they can be solved by means of a method due to Wiener and Hopf (cf. **53**, p. 49). By this method, explicit analytic expressions for $\psi_h(u)$ and $\psi(u)$ are obtained, and the asymptotic behaviour of these functions for large values of u can be studied.

This procedure can be generalized, and made to yield explicit expressions also for the ruin functions $\psi_h(u,T)$ and $\psi(u,T)$ associated with the ruin problem for a finite interval. In the present paper, we shall give a detailed account of the method as applied to $\psi(u)$ and $\psi(u,T)$, and content ourselves with some indications in respect of $\psi_h(u)$ and $\psi_h(u,T)$.

In the first place, it will be shown that the functions $\overline{\psi}_h(u,z)$ and $\overline{\psi}(u,z)$ as defined by (73) satisfy integral equations closely similar to (84) and (85), to which they reduce in the particular case $z = 0$.

From (65) and (66) we obtain for $h > 0$, $u \geqq 0$ and $\xi \leqq 0$,

$$e^{znh}\,\varphi_{n,h}(u) = \begin{cases} e^{zh}\,F(-u-\lambda h, h), & n = 1, \\ e^{zh} \int_0^\infty e^{z(n-1)h}\,\varphi_{n-1,h}(v)\,d_v F(v-u-\lambda h, h), & n > 1. \end{cases}$$

Summing over $n = 1, 2, \ldots$, we obtain according to (73)

$$(87) \qquad \overline{\psi}_h(u,z) = e^{zh}\,F(-u-\lambda h, h) + e^{zh} \int_0^\infty \overline{\psi}_h(v,z)\,d_v F(v-u-\lambda h, h).$$

This is the integral equation satisfied by $\overline{\psi}_h(u,z)$, which has thus been proved for $h > 0$, $u \geqq 0$, and for $z = \xi + i\eta$ with $\xi \leqq 0$. Taking here $z = 0$, it follows from (75) that we obtain the equation (84) for $\psi_h(u)$.

We shall now allow h to tend to zero in (87). It will be shown that in the limit we obtain, for the same values of u and z, the following integral equation for $\overline{\psi}(u,z)$:

$$(88) \quad (p_1+\lambda)\overline{\psi}(u,z) = \int_u^\infty Q(v)\,dv + z\int_u^\infty \overline{\psi}(v,z)\,dv + \int_0^\infty \overline{\psi}(v,z)\,Q(u-v)\,dv.$$

For $z = 0$, this evidently reduces to the equation (85) for $\psi(u)$.

The proof of (88) is a straightforward generalization of the proof of (85) as given by Cramér (**27**). In order to perform the limit passage $h \to 0$ in (87), we first consider the behaviour of each term in the second member. From (18) we obtain, as $h \to 0$,

$$F(-u-\lambda h, h) = h - hP[u+(p_1+\lambda)h] + O(h^2),$$

$$\int_0^\infty \bar{\psi}_h(v, z)\, d_v\, F(v-u-\lambda h, h) = (1-h)\,\bar{\psi}_h[u+(p_1+\lambda)h, z] -$$

$$-h \int_0^\infty \bar{\psi}_h(v, z)\, d_v\, P[u-v+(p_1+\lambda)h] + O(h^2),$$

where the $O(h^2)$ holds uniformly for all $u \geqq u_0 > 0$, for every fixed z with $\xi \leqq 0$. Introducing these expressions into (87) we obtain, taking account of the factor e^{zh}, and replacing u by $u-(p_1+\lambda)h$,

$$\bar{\psi}_h[u-(p_1+\lambda)h, z] - \bar{\psi}_h(u, z) = h(z-1)\,\bar{\psi}_h(u, z) + h\,[1-P(u)] -$$

$$-h \int_0^\infty \bar{\psi}_h(v, z)\, d_v\, P(u-v) + O(h^2), \tag{89}$$

uniformly in the same sense as before.

In the particular case when $p_1+\lambda = 0$, we can now directly allow h to tend to zero, and then obtain an equation which after integration with respect to u proves to be identical with (88). In the rest of the proof of (88), we may thus assume that $p_1+\lambda \neq 0$.

Let now K be a given positive constant, and take

$$h = \frac{K}{|p_1+\lambda|\,N},$$

where N is an integer which will ultimately be allowed to tend to infinity.

In (89) we replace u by $u+\frac{nK}{N}$, and sum the resulting equations for $n = 1, 2, \ldots, N$. We then obtain, multiplying by $|p_1+\lambda|$ and taking account of the fact that, according to (89), the difference $\bar{\psi}_h\left(u+\frac{K}{N}, z\right) - \bar{\psi}_h(u, z)$ is uniformly of the order $O(h) = O\left(\frac{1}{N}\right)$,

$$(p_1+\lambda)\,[\bar{\psi}_h(u, z) - \bar{\psi}_h(u+K, z)] =$$

$$= (z-1)\frac{K}{N}\sum_1^N \bar{\psi}_h\left(u+\frac{nK}{N}, z\right) + \frac{K}{N}\sum_1^N\left(1-P\left(u+\frac{nK}{N}\right)\right) -$$

62

$$-\int_0^\infty \overline{\psi}_h(v,z)\, d_v\left[\frac{K}{N}\sum_1^N P\left(u-v+\frac{nK}{N}\right)\right]+O\left(\frac{1}{N}\right).$$

When now N tends to infinity, and accordingly h tends to zero, (74) shows that the first member of the last relation tends to the limit

$$(p_1+\lambda)\,[\overline{\psi}(u,z)-\overline{\psi}(u+K,z)].$$

A discussion of the separate terms in the second member shows that each of these terms tends to the limit obtained by replacing any Riemann sum by the corresponding integral, and writing $\overline{\psi}$ in the place of $\overline{\psi}_h$. We thus obtain in the limit

$$(p_1+\lambda)[\overline{\psi}(u,z)-\overline{\psi}(u+K,z)]=(z-1)\int_u^{u+K}\overline{\psi}(v,z)\,dv+$$

$$+\int_u^{u+K}[1-P(v)]\,dv+\int_0^\infty\overline{\psi}(v,z)\,[P(u-v+K)-P(u-v)]\,dv.$$

When now K tends to infinity we obtain, using the inequality (76),

$$(p_1+\lambda)\,\overline{\psi}(u,z)=(z-1)\int_u^\infty\overline{\psi}(v,z)\,dv+$$

$$+\int_u^\infty[1-P(v)]\,dv+\int_0^\infty\overline{\psi}(v,z)\,[1-P(u-v)]\,dv,$$

which is easily seen to be equivalent to (88). Thus we have proved (88) for all $u>0$.

From (73) it follows, by means of a remark made at the end of 5.2, that $\overline{\psi}(u,z)$ is continuous to the right with respect to u, for all $u\geqq 0$. Allowing u to tend to zero in (88), it follows that (88) holds also for $u=0$.

From (88) it now follows that $\overline{\psi}(u,z)$ is continuous (and not only continuous to the right) with respect to u, for all $u>0$. Taking $z=0$ this implies, according to (75), that $\psi(u)$ is continuous for all $u>0$.

5.8. Solution of the integral equation for $\overline{\psi}(u,z)$. In the preceding section, we have proved that the functions $\overline{\psi}_h(u,z)$ and $\overline{\psi}(u,z)$ satisfy the integral equations (87) and (88) respectively. These can be solved by the Wiener-Hopf method. Taking $z=0$ in the solutions, we obtain as particular cases the solutions of the equations (84) and (85), for $\psi_h(u)$ and $\psi(u)$ respectively.

We shall now develop this procedure for the case of the integral equation (88). The equation (87) can be treated in a perfectly similar way.

The Wiener-Hopf method of solution consists in applying a complex Fourier transformation to the integral equation under investigation. In order to reach an explicit solution in this way, it will be necessary first to transform the equation (88) so as to obtain an equation holding for all real u, and not only for $u \geqq 0$, and at the same time to have the integral in the last term of (88) extended over the whole real axis, and not only over the positive half.

The function $\overline{\psi}(u, z)$ is defined by (73) for all z such that $\xi \leqq 0$, and for $u \geqq 0$. We now extend the definition to negative values of u by writing

$$\overline{\psi}(u, z) = 0 \qquad \text{for } u < 0.$$

Obviously we may then write (77) in the form

$$\text{(77 a)} \qquad \overline{\overline{\psi}}(s, z) = \int_{-\infty}^{\infty} e^{su}\, \overline{\psi}(u, z)\, du = \int_{0}^{\infty} e^{su}\, \overline{\psi}(u, z)\, du.$$

We further define, always assuming $\xi \leqq 0$,

$$\overline{\omega}(u, z) = 0 \qquad \text{for } u \geqq 0,$$

and

$$\text{(90)} \qquad \overline{\omega}(u, z) = \int_{u}^{\infty} Q(v)\, dv + z \int_{u}^{\infty} \overline{\psi}(v, z)\, dv + \int_{0}^{\infty} \overline{\psi}(v, z)\, Q(u-v)\, dv$$

for $u < 0$. Finally we write

$$\overline{\overline{\omega}}(s, z) = \int_{-\infty}^{\infty} e^{su}\, \overline{\omega}(u, z)\, du = \int_{-\infty}^{0} e^{su}\, \overline{\omega}(u, z)\, du.$$

The last integral is easily seen to be absolutely convergent for $\sigma > 0$, so that for fixed z the function $\overline{\overline{\omega}}(s, z)$ is analytic and regular in the half-plane $\sigma > 0$ of the complex variable s. By a partial integration, it is seen that we have throughout any half-plane $\sigma \geqq \varepsilon > 0$

$$\text{(91)} \qquad |\overline{\overline{\omega}}(s, z)| < \frac{K}{|s|}(1 + |z|),$$

where K depends only on ε.

By means of the above definitions, we can now write (88) in the extended form

$$\text{(92)} \qquad \begin{aligned} (p_1 + \lambda)\, \overline{\psi}(u, z) + \overline{\omega}(u, z) = \int_{u}^{\infty} Q(v)\, dv + \\ + z \int_{u}^{\infty} \overline{\psi}(v, z)\, dv + \int_{-\infty}^{\infty} \overline{\psi}(v, z)\, Q(u-v)\, dv, \end{aligned}$$

which will evidently hold for all real values of u. Moreover, the integral in the last term is now extended over the whole real axis, which does not make any difference, since by the new definition the function under the integral sign reduces to zero for negative values of v.

We now multiply (92) by $e^{su}\,du$, and integrate with respect to u from $-\infty$ to $+\infty$. By means of easy transformations we then obtain, for every s such that $0 < \sigma < R$,

$$(p_1+\lambda)\,\overline{\overline{\psi}}(s,z)+\overline{\overline{\omega}}(s,z) = \frac{p(s)-1}{s^2}+\frac{z}{s}\,\overline{\overline{\psi}}(s,z)+\frac{p(s)-1}{s}\,\overline{\overline{\psi}}(s,z),$$

which can be written in the form

$$1-\frac{p(s)}{1+(p_1+\lambda)\,s-z} = \frac{1}{1+(p_1+\lambda)\,s-z}\cdot\frac{(p_1+\lambda)\,s-z-s^2\,\overline{\overline{\omega}}(s,z)}{1+s\overline{\overline{\psi}}(s,z)}.$$

According to (83) this gives

$$(93)\qquad [1+(p_1+\lambda)\,s-z]\,\frac{1+s\overline{\overline{\psi}}(s,z)}{A(s,z)} = \frac{(p_1+\lambda)\,s-z-s^2\,\overline{\overline{\omega}}(s,z)}{B(s,z)},$$

where A and B are defined by (82). For every fixed z with $\xi \leqq 0$, it follows from the final remark of 5. 6 that A and A^{-1} will be regular functions of s for $\sigma < R$, and uniformly bounded in any interior half-plane $\sigma \leqq R-\varepsilon$, while B and B^{-1} are regular for $\sigma > 0$, and uniformly bounded for $\sigma \geqq \varepsilon$.

Thus according to the properties of $\overline{\overline{\psi}}$ and $\overline{\overline{\omega}}$ the first member of (93) represents, for every fixed z with $\xi \leqq 0$, a function of s which is analytic and regular in the half-plane $\sigma < R$, while the second member has the same properties for $\sigma > 0$. Since the two functions coincide in the strip $0 < \sigma < R$, each of them represents an entire function of s. By (79)and (91), the modulus of this entire function will be bounded by an expression of the form $c_1+c_2\,|s|$, where c_1 and c_2 are independent of s. It follows that the entire function reduces to a polynomial of the first degree, so that we have

$$(94)\qquad [1+(p_1+\lambda)\,s-z]\,\frac{1+s\overline{\overline{\psi}}\,(s,z)}{A(s,z)} = k_1+k_2\,s,$$

where k_1 and k_2 are independent of s.

Taking $s = 0$ in (94), we obtain

$$(95)\qquad k_1 = \frac{1-z}{A(0,z)}.$$

In order to determine the constant k_2, we shall have to distinguish two cases, according as $p_1+\lambda > 0$ or $p_1+\lambda \leqq 0$. It will be shown that we have

$$(96)\qquad k_2 = (p_1+\lambda)[1-\overline{\psi}(0,z)] = \begin{cases} \dfrac{p_1+\lambda}{A(0,z)}, & p_1+\lambda > 0, \\ 0\quad, & p_1+\lambda \leqq 0. \end{cases}$$

We first divide (94) by s, and allow s to tend to $-\infty$ along the real axis. We have observed in 5. 6 that $M(x, z)$ is everywhere continuous with respect to x, and it follows by (82) that $A(s, z)$ tends to 1 as $s \to -\infty$. Similarly, it follows from (77) by a partial integration, taking account of the final remark of 5. 7, that $s\bar{\bar{\psi}}(s, z)$ tends to the limit $-\bar{\psi}(0, z)$. We thus obtain the first part of (96).

If $p_1+\lambda > 0$, the unique zero of the linear factor in the first member of (94) is situated in the half-plane $\sigma < R$, where $\bar{\bar{\psi}}$ and A^{-1} are both regular. It follows that $k_1+k_2 s$ must have a zero at the same point, and this proves the expression given in (96) for the case $p_1+\lambda > 0$.

If $p_1+\lambda = 0$, it follows from the first part of (96) that we have $k_2 = 0$. If $p_1+\lambda < 0$, we have $p_1 < -\lambda$, and a consideration of the natural sample functions $X(t)$ of our stochastic process will show that $X(t)$, starting from the initial value $X(0) = 0$, and following a straight line with the angular coefficient p_1, will almost surely fall below $-\lambda t$ within an arbitrarily small time interval $(0, t_0)$. According to (53) this implies that, if we take $u = 0$, ruin will almost surely occur at once, so that we have $\psi(0, T) = 1$ for all $T > 0$. Hence by (73) we have $\bar{\psi}(0, z) = 1$, and thus the first part of (96) gives $k_2 = 0$. Thus (96) is completely proved.

From (83) and (94) we obtain

$$(97)\qquad \bar{\bar{\psi}}(s, z) = \frac{1}{s}\left(\frac{k_1+k_2 s}{1+(p_1+\lambda)s-p(s)-z}\, B(s, z)-1\right).$$

For fixed values of σ and $z = \xi + i\eta$, such that $0 < \sigma < R$ and $\xi \leqq 0$, the function of τ

$$\frac{\bar{\bar{\psi}}(\sigma+i\tau, z)}{\sqrt{2\pi}}$$

is, according to (77 a), the Fourier transform of the function $e^{\sigma u}\bar{\psi}(u, z)$, which belongs to L_2 over $(-\infty, \infty)$, since we have defined $\bar{\psi} = 0$ for $u < 0$, and the inequality (76) holds for $u \geqq 0$. The reciprocal Fourier transform then gives

$$(98)\qquad \bar{\psi}(u, z) = \frac{1}{2\pi i}\int_{\sigma-i\infty}^{\sigma+i\infty} e^{-su}\,\bar{\bar{\psi}}(s, z)\, ds,$$

where the integral is extended along any vertical in the s plane, such that $0 < \sigma < R$. The integral will certainly exist at least as a limit in the mean of the integral between $\sigma \pm iT$, as $T \to \infty$.

Assuming $u > 0$, we now introduce here the expressions (95), (96) and (97) for $\bar{\bar{\psi}}$, k_1 and k_2. Applying Cauchy's theorem to a rectangle with the corners $\sigma \pm iT$ and $K \pm iT$, and allowing first K, and then T, to tend to $+\infty$, it is easily seen that the integrals

$$\int_{\sigma-iT}^{\sigma+iT} e^{-su}\, B(s, z)\, \frac{ds}{s} \quad \text{and} \quad \int_{\sigma-iT}^{\sigma+iT} e^{-su}\, \frac{ds}{s}$$

both tend to zero as $T \to \infty$. Taking account of this fact, we obtain the following final expressions for the solution of the integral equation (88):

$$(99)\qquad \overline{\psi}(u, z) = \frac{1}{2\pi i A(0, z)} \int_{\sigma - i\infty}^{\sigma + i\infty} \frac{p(s)\, B(s, z)}{s\,[1 + (p_1 + \lambda)\, s - p(s) - z]}\, e^{-su}\, ds$$

for $p_1 + \lambda > 0$, and

$$(100)\qquad \overline{\psi}(u, z) = \frac{1 - z}{2\pi i A(0, z)} \int_{\sigma - i\infty}^{\sigma + i\infty} \frac{B(s, z)}{s\,[1 + (p_1 + \lambda)\, s - p(s) - z]}\, e^{-su}\, ds$$

for $p_1 + \lambda \leqq 0$. Both these expressions have now been proved for all $u > 0$ and for all $z = \xi + i\eta$ such that $\xi \leqq 0$. By continuity they will, however, hold also for $u = 0$. The integrals can be taken along any vertical with $0 < \sigma < R$. Since $p(s)$ and $B(s, z)$ are bounded on the line of integration, it is immediately seen that both integrals are absolutely convergent, and thus exist as ordinary absolutely convergent integrals, and not only as limits in the mean, as soon as $p_1 + \lambda \neq 0$.

5.9. Explicit expressions for $\psi(u)$ and $\psi(u, T)$. By means of the solution of the fundamental integral equation (88) given in the preceding section, we can now directly obtain explicit analytic expressions of the ruin functions $\psi(u)$ and $\psi(u, T)$ in terms of known functions.

Taking $z = 0$ in (99) and (100), we obtain according to (75) for all $u \geqq 0$

$$(101)\qquad \psi(u) = \frac{1}{2\pi i A(0, 0)} \int_{\sigma - i\infty}^{\sigma + i\infty} \frac{q(s)\, B(s, 0)}{s\,[1 + (p_1 + \lambda)\, s - p(s)]}\, e^{-su}\, ds,$$

where

$$q(s) = \begin{cases} p(s) & \text{for } p_1 + \lambda > 0, \\ 1 & \text{for } p_1 + \lambda \leqq 0. \end{cases}$$

Further, a consideration of the sample functions $X(t)$ of our stochastic process will show that, for $p_1 + \lambda \geqq 0$, the function $\psi(u, T)$ is everywhere continuous with respect to T. When $p_1 + \lambda < 0$, on the other hand, $\psi(u, T)$ may present discontinuities with respect to T. When $u > 0$, however, the particular point $T = 0$ will always be a point of continuity. On account of these remarks, we obtain from (73) by means of the inversion formula for Fourier-Stieltjes integrals (cf. Cramér **23**, p. 93) the following expression, which holds for $u > 0$ and for all continuity points T of $\psi(u, T)$,

$$(102)\qquad \psi(u, T) = \frac{1}{2\pi i} \lim_{Y \to \infty} \int_{-Y}^{Y} \frac{1 - e^{-i\eta T}}{\eta}\, \overline{\psi}(u, i\eta)\, d\eta.$$

Substituting here the expressions (99) and (100) for $\overline{\psi}(u, z)$, we obtain explicit expressions for $\psi(u,T)$.

5. 10. Asymptotic properties of $\psi(u)$. Throughout the sections 5.3—5.9, we have only made the assumptions (59) with regard to the given d.f. $P(y)$. In the present section we shall make the following additional assumption:

$$(103) \qquad \int_0^\infty e^{\sigma y}\, dP(y) < \infty \quad \text{for some } \sigma = R+\Theta \text{ with } \Theta > 0.$$

It follows from the conditions 1) and 2) of 5. 3 that the point $s = R$ will then be a zero of function

$$\pi(s) = 1+(p_1+\lambda)\, s-p(s),$$

and since the second derivative $\pi''(s) = -p''(s)$ is negative for real s, this will be a simple zero. Further, since we have for $0 < \sigma \leqq R$, $\tau \neq 0$,

$$|1+(p_1+\lambda)s-p(s)| > 1+(p_1+\lambda)\, \sigma-p(\sigma) \geqq 0,$$

the point $s = R$ will be the only zero in the strip $0 < \sigma \leqq R$. The domain $R < \sigma < R+\Theta$, $|\tau| > \tau_0$ will certainly be free from zeros if τ_0 is sufficiently large. It then follows that the positive quantity Θ in (103) can be taken such that, in the strip $0 < \sigma \leqq R+\Theta$, the function $\pi(s)$ is regular and different from zero, except at the point $s = R$, which is a zero of the first order. We shall assume that Θ has been taken in this way.

We can then transform the expression (101) for $\psi(u)$ by integrating along the contour of the rectangle with the corner points $\sigma \pm iT$ and $R+\Theta \pm iT$, and then allowing T to tend to infinity. By Cauchy's theorem we obtain, taking account of the residue at the pole $s = R$ of the integrand,

$$\psi(u) = Ce^{-Ru} + \frac{1}{2\pi i A(0,0)} \int_{R+\Theta-i\infty}^{R+\Theta+i\infty} \frac{q(s)\, B(s,0)}{s\,[1+(p_1+\lambda)\, s-p(s)]}\, e^{-su}\, ds,$$

where

$$(104) \qquad C = \frac{q(R)\, B(R,0)}{A(0,0)\; R\,[p'(R)-p_1-\lambda]}.$$

Since $q(s)$ and $B(s, 0)$ are bounded on the new line of integration, the last integral is absolutely convergent, and we thus have the asymptotic relation

$$(105) \qquad \psi(u) = Ce^{-Ru}+O(e^{-(R+\Theta)\, u})$$

as $u \to \infty$, the constant C being given by (104).

5. 11. On the difference $\psi(u)-\psi(u, T)$. As in the preceding section, we shall here assume that the d.f. $P(y)$ satisfies the conditions (59) and (103). Since the function $\pi(s)$ is zero at $s = 0$ and $s = R$, and the second derivative $\pi''(s)$ is negative for real s, it follows that, in the interval $0 < s < R$, $\pi(s)$ has a single maximum at $s = \varrho$, where ϱ is the unique root of the equation

$$\pi'(\varrho) = p_1+\lambda-p'(\varrho) = 0. \tag{106}$$

In the notation of 5. 3, we then have

$$\gamma = \pi(\varrho) = 1+(p_1+\lambda)\,\varrho-p(\varrho) > 0. \tag{107}$$

The situation is illustrated by fig. 4.

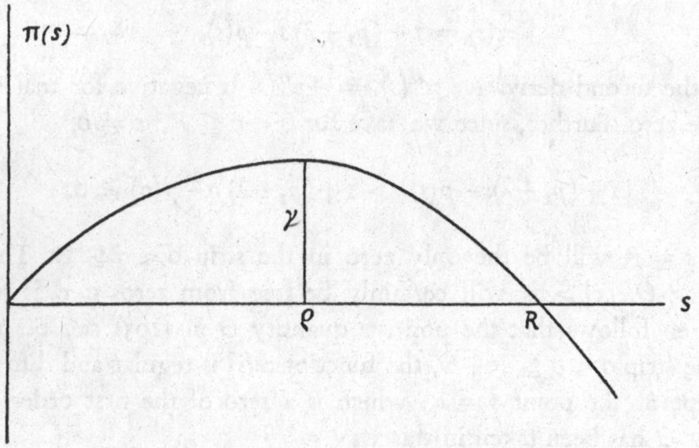

Fig. 4. The function $\pi(s)=1+(p_1+\lambda)s-p(s)$ for real and positive values of s.

For the practical applications of risk theory, it is important to know the probability $\psi(u)$ that the risk business will be ruined at some time in the future. It is also important to know whether, for a reasonably large value of T, the probability $\psi(u, T)$ that ruin will occur *before time* T is substantially smaller than $\psi(u)$. If, in a practical case, the probability of ruin within, say, 50 years can be shown to be extremely small, we may feel inclined to take it calmly if the probability of ruin at some time in a more remote future would have an appreciable value. In order to spread some light on this question, we now proceed to give an upper bound for the difference $\psi(u)-\psi(u, T)$.

We shall, in fact, prove that for any $\varepsilon > 0$ we can find C independent of u and T such that for all $u \geqq 0$, $T \geqq 0$ we have

$$0 \leqq \psi(u)-\psi(u, T) \leqq Ce^{-\varrho u-(\gamma-\varepsilon)T}, \tag{108}$$

where ϱ and γ are the quantities defined above.

For $\xi \leqq 0$ we have by (73)

$$\overline{\psi}(u, z) = \int_0^\infty e^{zT} d_T[\psi(u, T) - \psi(u)]$$

$$= \psi(u) + z \int_0^\infty e^{zT} [\psi(u) - \psi(u, T)] dT,$$

and hence

$$(109) \qquad \int_0^\infty e^{zT} [\psi(u) - \psi(u, T)] \, dT = \frac{\overline{\psi}(u, z) - \psi(u)}{z}.$$

The first member of this relation is a regular function of $z = \xi + i\eta$ in the half-plane $\xi \leqq 0$. We shall now show that the second member is regular in the larger half-plane $\xi < \gamma$. The difference $\psi(u) - \psi(u, T)$ being non-negative, it then follows from known theorems that the integral in the first member is absolutely convergent for $\xi < \gamma$.

In order to prove this, we consider the expressions (99) and (100) for $\overline{\psi}(u, z)$, specifying the line of integration by taking in both cases $\sigma = \varrho$. For every $\varepsilon > 0$, the functions $p(s)$, $B(s, z)$ and $A(0, z)^{-1}$ are then uniformly bounded for all s on the line of integration, and for all z with $\xi \leqq \gamma - \varepsilon$. In respect of $p(s)$ this is evident, while in respect of B and A^{-1} it follows from the properties of these functions established in 5. 6. On the line of integration we further have by (107) for all z with $\xi \leqq \gamma - \varepsilon$

$$|1 + (p_1 + \lambda)s - p(s) - z| \geqq 1 + (p_1 + \lambda)\varrho - p(\varrho) - \xi \geqq \varepsilon,$$

while at the same time $1 + (p_1 + \lambda)s - p(s) - z$ is for large τ of the same order of magnitude as τ. This shows that the integrals in (99) and (100), taken along the line $\sigma = \varrho$, are absolutely and uniformly convergent for all z with $\xi \leqq \gamma - \varepsilon$. Consequently $\overline{\psi}(u, z)$ is regular for $\xi \leqq \gamma - \varepsilon$ and, since ε is arbitrary, even for $\xi < \gamma$. Since $\overline{\psi}(u, 0) = \psi(u)$, it follows that the second member of (109) is regular for $\xi < \gamma$, as was to be proved.

It also follows from the above discussion of the expressions (99) and (100) that, for every $\varepsilon > 0$, we can find k independent of ξ and u such that we have for $0 \leqq \xi \leqq \gamma - \varepsilon$

$$0 < \overline{\psi}(u, \xi) < k e^{-\varrho u},$$

and hence by (109)

$$\int_0^\infty e^{(\gamma - \varepsilon)t} [\psi(u) - \psi(u, t)] \, dt < \frac{k}{\gamma - \varepsilon} e^{-\varrho u}.$$

A fortiori we have for any $T > 1$

$$\int_{T-1}^{T} e^{(\gamma-\varepsilon)t}\,[\psi(u)-\psi(u,t)]\,dt < \frac{k}{\gamma-\varepsilon}\,e^{-\varrho u},$$

and, since $\psi(u)-\psi(u,t)$ is a never increasing function of t,

$$[\psi(u)-\psi(u,T)]\int_{T-1}^{T} e^{(\gamma-\varepsilon)t}\,dt < \frac{k}{\gamma-\varepsilon}\,e^{-\varrho u},$$

$$\psi(u)-\psi(u,T) < \frac{k}{1-e^{-(\gamma-\varepsilon)}}\,e^{-\varrho u-(\gamma-\varepsilon)T} < Ce^{-\varrho u-(\gamma-\varepsilon)T},$$

where C is independent of u and T. Thus we have proved (108) for $T > 1$. That the same inequality holds also for $0 \leqq T \leqq 1$ is a trivial consequence of (71) and (72).

5. 12. The case with only positive risk sums. In addition to the assumptions (59) and (103) we shall now assume that we have

$$P(0) = 0.$$

This implies that the whole mass of the $P(y)$ distribution is situated on the positive axis, i.e., that all risk sums are positive. This is the most important case for the practical applications, and also the case that has been most thoroughly investigated from the theoretical point of view.

In this case, it follows from (10) and (103) that the function $p(s)$ is regular for all $\sigma < R+\Theta$, and that $p(s) \to 0$ as $\sigma \to -\infty$.

Some of the results given below for this case can be simply obtained by introducing the condition $P(0) = 0$ into the developments of 5. 9—5. 11. We shall, however, prefer to take a slightly different road, thus obtaining all our results for the present case by a uniform method.

In order to do so, we go back to the integral equation (87). Since in the present case it follows from (18) that we have

$$F(p_1 t, t) = 1$$

for all t, the equation (87) can now be written

$$\text{(110)}\qquad \overline{\psi}_h(u,z) = e^{zh}F(-u-\lambda h, h) + e^{zh}\int_0^{u+(p_1+\lambda)h} \overline{\psi}_h(v,z)\,d_vF(v-u-\lambda h, h),$$

where $h > 0$, $u \geqq 0$, and $z = \xi + i\eta$ with $\xi \leqq 0$.

It further follows from (18) and 5.6 that we have for $0 < r < R$ and for any real x

$$(111) \qquad F(x,t) = \sum_{0}^{\infty} \frac{t^n}{n!} e^{-t} \left[1 - P^{n*}(p_1 t - x)\right] \leqq e^{rx + t[p(r) - 1 - p_1 r]}.$$

Consequently for any $s = \sigma + i\tau$ with $\sigma < R$ we may multiply (110) by $e^{su}\, du$, and integrate with respect to u from 0 to $+\infty$. After some elementary transformations the result can be written in the form

$$(112) \qquad \left[1 - e^{-h[\pi(s) - z]}\right] \cdot \left[1 + s\,\overline{\overline{\psi}}_h(s,z)\right] = g_h(s,z),$$

where $\pi(s)$ and $\overline{\overline{\psi}}_h(s,z)$ are defined by (60) and (77) respectively, while

$$(113) \qquad \begin{aligned} g_h(s,z) = 1 - e^{-zh} F(-\lambda h, h) - e^{zh} \int_0^{(p_1+\lambda)h} e^{-su}\, d_u F(u - \lambda h, h) - \\ - s e^{zh} \int_0^{(p_1+\lambda)h} e^{-su}\, d_u F(u - \lambda h, h) \int_0^u e^{sv}\, \overline{\psi}_h(v,z)\, dv. \end{aligned}$$

It is easily seen that, for every fixed z with $\xi \leqq 0$, the first member of (112) is a regular function of s for $\sigma < R$, while the second member is regular for $\sigma > 0$.

We thus have for $0 < \sigma < R$ and $\xi \leqq 0$

$$(114) \qquad \overline{\overline{\psi}}_h(s,z) = \frac{1}{s}\left(\frac{g_h(s,z)}{1 - e^{-h[\pi(s) - z]}} - 1\right).$$

We now allow h to tend to zero, while s and z remain fixed. Using arguments similar to, but simpler than, those applied in the course of the deduction of the integral equation (88), we obtain

$$(115) \qquad \overline{\overline{\psi}}(s,z) = \lim_{h \to 0} \overline{\overline{\psi}}_h(s,z) = \frac{1}{s}\left(\frac{-z + [1 - \overline{\psi}(0,z)]\,(p_1+\lambda)s}{\pi(s) - z} - 1\right).$$

We now proceed to make various applications of (114) and (115) to the study of the ruin functions $\psi(u)$, $\psi_h(u)$ and $\psi(u,T)$. In the case of the two latter functions, we shall not give all the details of the proofs. For detailed proofs by different methods, we may refer to Segerdahl (59) in the case of $\psi_h(u)$, and to a forthcoming paper by Arfwedson in the case of $\psi(u,T)$.

I. *On the function* $\psi(u)$. The reciprocal Fourier formula (98) gives in the present case, using the remark made in the deduction of (99) and assuming $u > 0$,

$$(116) \qquad \begin{aligned} \overline{\psi}(u,z) &= \frac{1}{2\pi i} \int_{\sigma - i\infty}^{\sigma + i\infty} \frac{-z + [1 - \overline{\psi}(0,z)]\,(p_1+\lambda)s}{s\,[1 + (p_1+\lambda)\,s - p(s) - z]}\, e^{-su}\, ds \\ &= \frac{1}{2\pi i} \int_{\sigma - i\infty}^{\sigma + i\infty} \frac{-z + [1 - \overline{\psi}(0,z)]\,[p(s) + z - 1]}{s\,[1 + (p_1+\lambda)s - p(s) - z]}\, e^{-su}\, ds, \end{aligned}$$

where the integrals may be taken along any vertical with $0 < \sigma < R$. The last integral is absolutely convergent, while the first integral will certainly exist at least as a limit in the mean, as explained in the case of (98). By continuity, the last integral will represent $\psi(u, z)$ also for $u = 0$.

Taking $z = 0$ in (116), we obtain by (75)

$$(117) \qquad \psi(u) = \frac{1-\psi(0)}{2\pi i} \int_{\sigma-i\infty}^{\sigma+i\infty} \frac{p(s)-1}{s[1+(p_1+\lambda)s-p(s)]} e^{-su} ds.$$

The function under the integral is regular throughout the half-plane $\sigma \leqq R+\Theta$, except at the points $s = 0$ and $s = R$, which are poles of the first order, Θ being defined as in the proof of (105). In order to determine the constant $\psi(0)$, we take $u = 0$ in (117), which gives

$$\psi(0) = \frac{1-\psi(0)}{2\pi i} \int_{\sigma-i\infty}^{\sigma+i\infty} \frac{p(s)-1}{s[1+(p_1+\lambda)s-p(s)]} ds.$$

Applying now Cauchy's theorem to the rectangle formed by the points $\sigma \pm iT$ and $-\Sigma \pm iT$, where $\Sigma > 0$, and allowing first Σ, and then T, to tend to infinity, we find that the integral in the last relation is equal to $2\pi i$ multiplied by the residue at the pole $s = 0$, which is the only singularity inside the contour of integration. We thus obtain

$$\psi(0) = [1-\psi(0)] \frac{p_1}{\lambda},$$

$$\psi(0) = \frac{p_1}{p_1+\lambda},$$

and hence by (117)

$$(118) \qquad \psi(u) = \frac{\lambda}{2\pi i(p_1+\lambda)} \int_{\sigma-i\infty}^{\sigma+i\infty} \frac{p(s)-1}{s[1+(p_1+\lambda)s-p(s)]} e^{-su} ds.$$

The same application of Cauchy's theorem as in the proof of (105) now gives the asymptotic formula

$$(119) \qquad \psi(u) = \frac{\lambda}{p'(R)-p_1-\lambda} e^{-Ru} + O\left(e^{-(R+\Theta)u}\right),$$

as $u \to \infty$. Evidently this is the same formula as (105), with a more explicit determination of the coefficient of e^{-Ru}, which has been rendered possible by the assumption $P(0) = 0$. Some numerical applications of this formula have been given above in 4. 4, Table VII.

II. *On the function* $\psi_h(u)$. For the remaining part of this section, we shall now introduce a new assumption concerning the d.f. $P(y)$, in addition to those stated at the beginning of the section. We shall now assume, as in Chapter 4, that $P(y)$ *contains an absolutely continuous component.* Some of the results given below will hold even without this assumption.

For every $\sigma < R+\Theta$, and for every $\tau_0 > 0$, we shall then (cf. Cramér **20**, p. 25—26) be able to find $k > 0$ such that we have for $|\tau| > \tau_0$

$$(120) \qquad \begin{aligned} |p(\sigma+i\tau)| &\leqq p(\sigma)-k, \\ |p'(\sigma+i\tau)| &\leqq p'(\sigma)-k. \end{aligned}$$

According to the definition in 5.2, $\psi_h(u)$ is the probability that the risk reserve, starting at the initial value $u \geqq 0$, will be negative at any of the time points $t = h, 2h \ldots$. If we think of these time points as the ends of successive business periods of the company, h would be the expected number of claims during one business period, and could thus in general be assumed to be fairly large. It would be interesting to know if, under these circumstances, the probability $\psi_h(u)$ would be substantially smaller than the probability $\psi(u)$ of ruin at *any* future time point. We now proceed to give at least a partial answer to this question.

For $\psi(u)$ we have according to (119) an asymptotic formula of the type

$$\psi(u) \sim Ce^{-Ru}$$

for large values of u. It can be shown that, under the assumptions stated above, there is a similar formula for $\psi_h(u)$:

$$\psi_h(u) \sim C_h e^{-Ru},$$

where for large values of h we have

$$(121) \qquad C_h \sim \frac{C}{R\lambda h}.$$

We shall only give some brief indications about the proof of these relations. Taking $z = 0$ in (114), we obtain by the same argument as before

$$(122) \qquad \psi_h(u) = \bar{\psi}_h(u, 0) = \frac{1}{2\pi i} \int_{\sigma-i\infty}^{\sigma+i\infty} \frac{g_h(s, 0)}{s(1-e^{-h\pi(s)})} e^{-su} \, ds.$$

By assumption $p(s)$ is regular for $\sigma < R+\Theta$, where $\Theta > 0$. Using (120) it will then be seen that we can choose the positive quantity Θ so small that the function

$$1-e^{-h\pi(s)}$$

will be regular and different from zero for $0 < \sigma \leqq R+\Theta$, except at the point

$s = R$, which is a zero of the first order. Applying Cauchy's theorem to the integral in (122), we obtain

$$\psi_h(u) = \frac{g_h(R, 0)}{Rh\,[p'(R)-p_1-\lambda]}\, e^{-Ru} + \frac{1}{2\pi i}\int\limits_{R+\Theta-i\infty}^{R+\Theta+i\infty},$$

the function under the integral being the same as in (122). Although we are here concerned with a non-absolutely convergent integral, it is possible to prove by a method used by Täcklind (**65**, p. 15—16) that for large u the last integral is of the order $O(e^{-(R+\Theta')u})$ for any $\Theta' < \Theta$. We thus have for every fixed h

$$\psi_h(u) = C_h e^{-Ru} + O(e^{-(R+\Theta')u})$$

with

(123) $$C_h = \frac{g_h(R, 0)}{Rh(p'(R)-p_1-\lambda)}.$$

Taking $x = u-\lambda h$, $t = h$, and $r = \varrho$ in (111), we obtain

$$F(u-\lambda h, h) \leqq e^{\varrho u - h\pi(\varrho)} = e^{\varrho u - \gamma h}.$$

We further obtain for large h

$$\psi_h(u) < F(-\lambda h, h) + F(-2\lambda h, 2h) + \dots < Ke^{-\gamma h}.$$

By means of the two last inequalities, it follows without difficulty from the expression (113) of $g_h(s, z)$ that we have

$$\lim_{h\to\infty} g_h(R, 0) = 1.$$

Thus (123) gives, as $h \to \infty$,

$$C_h \sim \frac{1}{Rh[p'(R)-p_1-\lambda]} = \frac{C}{R\lambda h},$$

where C is the constant in the asymptotic expression of $\psi(u)$, as given by (119). This completes the proof of (121).

III. *On the difference* $\psi(u)-\psi(u, T)$. We still assume that $P(y)$ contains an absolutely continuous component, so that (120) holds.

In 5.11 we have proved certain results concerning the difference $\psi(u)-\psi(u,T)$, which can be considerably improved under the present conditions. In the first place, the inequality (108) can now be strengthened to

(124) $$0 \leqq \psi(u)-\psi(u,T) < Ae^{-\varrho u - \gamma T}$$

for all $u \geqq 0$, $T \geqq 0$. Moreover, we have the asymptotic relation

$$\psi(u) - \psi(u, T) \sim \frac{Bu}{T^{\frac{3}{2}}} e^{-\varrho u - \gamma T} \tag{125}$$

which holds when u and T both tend to infinity, in such a way that u is of smaller order of magnitude than $\sqrt{T}$. In these relations, A and B are constants independent of u and T.

In order to give some indications about the proofs of (124) and (125), we shall consider the relation (116). The denominator of the function under the integral in this relation will be zero whenever s is equal to a root of the equation

$$z = 1 + (p_1 + \lambda)s - p(s). \tag{126}$$

We know (cf. fig. 4, p. 67) that for real $z < \gamma$, there will be one single real root $s_0 < \varrho$ of this equation. We now proceed to consider (126) as a relation between the two complex variables $z = \xi + i\eta$ and $s = \sigma + i\tau$. It will be shown that the root $s_0 = s_0(z)$ defines an analytic function of z, which is regular and uniform in a certain domain in the plane of z.

As s runs through the vertical $\sigma = \varrho$, it follows from (120) that z will describe a path $\mathfrak{C}$ that is entirely situated to the right of the vertical $\xi = \gamma$, except at the point $s = \varrho$, where the path $\mathfrak{C}$ presents a cusp, and we have $z = \gamma$. For every $\tau_0 > 0$ we can by (120) find $k > 0$ such that the whole part of $\mathfrak{C}$ which corresponds to points $s = \varrho + i\tau$ with $|\tau| > \tau_0$ will even be situated to the right of the vertical $\xi = \gamma + k$. The ordinate of a point of $\mathfrak{C}$ is

$$\eta = (p_1 + \lambda)\tau - \int_0^\infty e^{\varrho y} \sin \tau y \, dP(y),$$

and the derivative of this with respect to τ is for $\tau \neq 0$

$$\frac{d\eta}{d\tau} = p_1 + \lambda - \int_0^\infty y e^{\varrho y} \cos \tau y \, dP(y) > p_1 + \lambda - p'(\varrho) = 0,$$

so that the path $\mathfrak{C}$ will be monotone increasing for all $\tau \neq 0$, and cannot have any double points apart from the cusp at $z = \gamma$. As $\tau \to \pm\infty$ along the vertical $\sigma = \varrho$, it follows from (126) that η will tend to $\pm\infty$ along the path $\mathfrak{C}$, while ξ will ultimately be confined between two finite and positive limits.

It then follows from known theorems on conformal representation that (126) defines a one-to-one correspondence between the half-plane $\sigma \leqq \varrho$ in the plane of s, and the domain $\mathfrak{D}$ to the left of the path $\mathfrak{C}$, including $\mathfrak{C}$ itself, in the plane of z. In the half-plane $\sigma \leqq \varrho$, the derivative of the second member of (126),

$$\pi'(s) = p_1 + \lambda - p'(s), \tag{127}$$

is different from zero, except at the point $s = \varrho$, which corresponds to $z = \gamma$. It follows that there is one branch of the inverse function[1] of (126), which we shall denote by

$$s = s_0 = s_0(z),$$

and which is regular and uniform in all points of $\mathfrak{D}$, except at the point $z = \gamma$. When z increases along the real axis from $-\infty$ to γ, the inverse function s_0 is real, and increases from $-\infty$ to ϱ. At the point $s = \varrho$, the derivative (127) is zero, while the second derivative $-p''(\varrho)$ is negative. Consequently there is an expansion

$$(128) \qquad s_0 = \varrho - \left(\frac{\gamma - z}{\frac{1}{2} p''(\varrho)} \right)^{\frac{1}{2}} + a_2(\gamma - z) + a_3(\gamma - z)^{\frac{3}{2}} + \ldots,$$

convergent for all sufficiently small values of $|z - \gamma|$.

It will now be seen from the above that to every $\eta_0 > 0$ we can find $c > 0$ and $m > 0$ with the following properties. Let $\mathfrak{D}'$ denote the half-plane $\xi \leqq \gamma + c$ in the plane of z, with a cut along the real axis from $z = \gamma$ to $z = \gamma + c$. Then the function $s_0 = s_0(z)$ will be regular and uniform in all points of $\mathfrak{D}'$, except at the point $z = \gamma$, where we have the expansion (128). Further, when s is on the vertical $\sigma = \varrho$, we have for all z in $\mathfrak{D}'$ with $|\eta| > \eta_0$

$$(129) \qquad |1 + (p_1 + \lambda)\, s - p(s) - z| > m$$

and also, trivially,

$$(130) \qquad |1 + (p_1 + \lambda)s - p(s) - z| > |(p_1 + \lambda)\tau - \eta| - p(\varrho).$$

After these preparations we now consider (116), assuming still $\xi \leqq 0$. Taking first $u = 0$ we obtain

$$\overline{\psi}(0, z) = \frac{1}{2\pi i} \int\limits_{\sigma - i\infty}^{\sigma + i\infty} \frac{-z + [1 - \overline{\psi}(0, z)]\,[p(s) + z - 1]}{s[1 + (p_1 + \lambda)s - p(s) - z]}\, ds$$

the integral being taken along any vertical with $0 < \sigma < R$. Applying Cauchy's theorem in the same way as in the case of (117) we now obtain, allowing for the residues at the poles $s = 0$ and $s = s_0$,

$$\overline{\psi}(0, z) = \overline{\psi}(0, z) + \frac{-z + [1 - \overline{\psi}(0, z)]\,[p(s_0) + z - 1]}{s_0[p_1 + \lambda - p'(s_0)]},$$

$$1 - \overline{\psi}(0, z) = \frac{z}{p(s_0) + z - 1},$$

[1] Note that in general s_0 will *not* be the only root of (126). However, an important corollary of the above argument is that, as long as z belongs to $\mathfrak{D}$, any other root of (126) will be situated in the half-plane $\sigma > \varrho$.

and hence by (116), observing that $s = s_0$ is a root of (126),

$$(131)\qquad \overline{\psi}(u, z) = \frac{z}{2\pi i(p_1+\lambda)s_0} \int_{\varrho-i\infty}^{\varrho+i\infty} \frac{p(s)-p(s_0)}{s[1+(p_1+\lambda)s-p(s)-z]} e^{-su}\, ds,$$

where we have specified the line of integration by taking $\sigma = \varrho$.[1] The point $s = s_0$ is now a regular point of the function under the integral, and the integral itself is a regular and uniform function of z in the domain $\mathfrak{D}'$, except at the point $z = \gamma$. In conjunction with the above results concerning the function s_0, this shows that $\overline{\psi}(u, z)$ is, for every fixed $u \geqq 0$, regular and uniform throughout $\mathfrak{D}'$, except at $z = \gamma$. In the vicinity of $z = \gamma$, there is an expansion in powers of $(\gamma-z)^{\frac{1}{2}}$:

$$(132)\qquad \overline{\psi}(u, z) = e^{-\varrho u}[a_0 + a_1(\gamma-z)^{\frac{1}{2}} + \ldots]$$

For the square roots, those branches should be taken which are positive for real $z < \gamma$. The coefficient of $(\gamma-z)^{\frac{n}{2}}$ is of the form

$$a_n = b_{n0} + b_{n1}u + \ldots + b_{nn}u^n + O\left(\frac{1}{1+u}\right),$$

where the b_{ij} are independent of u. In particular we have

$$(133)\qquad b_{11} = -\frac{\gamma}{(p_1+\lambda)\varrho^2} \cdot \frac{p'(\varrho)}{\left(\frac{1}{2}p''(\varrho)\right)^{\frac{3}{2}}}.$$

Further, an evaluation of the integral in (131) by means of the inequalities (129) and (130) shows that we have

$$(134)\qquad |\overline{\psi}(u, z)| < K \frac{\log(1+|\eta|)}{|\eta|} e^{-\varrho u},$$

where K is independent of u and z, for $u \geqq 0$, $\xi \leqq \gamma + c$, $|\eta| > \eta_0$.

Consider now the relation (109), which has been proved to hold for all z such that $\xi < \gamma$. It follows from (132) that, under the present conditions, the second member of (109) tends to a finite limit as $z \to \gamma$. The difference $\psi(u)$ —

---

[1] Differentiating (131) n times with respect to z, and taking then $z = 0$, we obtain an expression for the n:th moment $\int T^n\, d_T \psi(u, T)$. Applying Cauchy's theorem in the same way as in the case of $\psi(u)$ above, we find that this moment is asymptotically for large u of the form $q_n(u)\, e^{-Ru}$, where $q_n(u)$ is a polynomial of order n, in accordance with Segerdahl (59).

$-\psi(u,T)$ being non-negative, we may then conclude that the integral in the first member of (109) is convergent for $z=\gamma$, and that we have

$$\int_0^\infty e^{\gamma t}[\psi(u)-\psi(u,t)]\,dt = \frac{1}{\gamma}\,[a_0 e^{-\varrho u}-\psi(u)].$$

Hence we obtain for any $T>0$

$$[\psi(u)-\psi(u,T)]\int_0^T e^{\gamma t}\,dt < \frac{a_0}{\gamma}\,e^{-\varrho u},$$

$$\psi(u)-\psi(u,T) < \frac{a_0}{1-e^{-\gamma T}}\,e^{-\varrho u-\gamma T},$$

so that (124) is proved.

In order to prove also (125), we observe that the reciprocal Fourier formula transforms (109) into

$$\psi(u)-\psi(u,T) = \frac{1}{2\pi i}\int_{\xi-i\infty}^{\xi+i\infty}\frac{\overline{\psi}(u,z)-\psi(u)}{z}\,e^{-zT}\,dz$$

$$= \frac{1}{2\pi i}\int_{\xi-i\infty}^{\xi+i\infty}\frac{\overline{\psi}(u,z)}{z}\,e^{-zT}\,dz,$$

where we have used the remark made above in connection with (99) and (116). The integrals may be taken along any vertical in the z plane with $0<\xi<\gamma$, and the last integral is absolutely convergent, on account of (134).

By an application of Cauchy's theorem, we may now move the line of integration to the contour $\mathfrak{C}'$ of the domain $\mathfrak{D}'$ defined above. An evaluation of the integral then shows that the vertical parts of $\mathfrak{C}'$ give a contribution of the form

$$O(e^{-\varrho u-(\gamma+c)T})$$

to the value of the integral. On the other hand, the contribution due to the two edges of the cut from $z=\gamma$ to $z=\gamma+c$ is asymptotically for large u and T

$$\frac{b_{11}}{\gamma}\,u\,e^{-\varrho u-\gamma T}\cdot\frac{1}{2\pi i}\int(\gamma-z)^{\frac{1}{2}}\,e^{(\gamma-z)T}\,dz$$

the integral being extended from $z=\gamma+c$ to $z=\gamma$ along the lower edge of the cut, and back to $z=\gamma+c$ along the upper edge. The argument of $(\gamma-z)^{\frac{1}{2}}$ will be $+\frac{1}{2}\pi$ on the lower edge, and $-\frac{1}{2}\pi$ on the upper edge, so that the last expression is asymptotically equal to

$$-\frac{b_{11}}{\pi\gamma}\, u e^{-\varrho u-\gamma T} \int_0^c w^{\frac{1}{2}} e^{-wT}\, dw \sim -\frac{b_{11}}{2\gamma\sqrt{\pi}} \cdot \frac{u}{T^{\frac{3}{2}}}\, e^{-\varrho u-\gamma T}.$$

We thus finally obtain by (133)

$$\psi(u)-\psi(u,T) \sim \frac{1}{2(p_1+\lambda)\,\varrho^2\sqrt{\pi}} \cdot \frac{p'(\varrho)}{\left(\frac{1}{2}p''(\varrho)\right)^{3/2}} \cdot \frac{u}{T^{\frac{3}{2}}}\, e^{-\varrho u-\gamma T},$$

which completes the proof of (125).

5.13. The case with only negative risk sums. For the case when $P(0) = 1$, so that all risk sums are negative[1], we shall content ourselves with proving the identity

$$\psi(u) = e^{-Ru}.$$

According to 5.3, we shall in this case always have $p_1+\lambda < 0$, the case denoted as case C) in 5.3 being excluded from consideration. It then follows from (101) that we have

$$\text{(135)} \qquad \psi(u) = \frac{1}{2\pi i A(0,0)} \int_{\sigma-i\infty}^{\sigma+i\infty} \frac{B(s,0)}{s\,[1+(p_1+\lambda)s-p(s)]}\, e^{-su}\, ds,$$

the integral being taken along any vertical with $0 < \sigma < R$. Now in this case the function $p(s)$ is regular in the whole half-plane $\sigma > 0$, and tends uniformly to zero as $\sigma \to +\infty$. Consider the rectangle formed by the points $\sigma \pm iT$ and $\Sigma \pm iT$. On the contour of this rectangle, we have for all sufficiently large values of T and Σ

$$\text{(136)} \qquad |1+(p_1+\lambda)s| > |p(s)|.$$

It follows that the functions

$$1+(p_1+\lambda)s \quad \text{and} \quad 1+(p_1+\lambda)s-p(s)$$

will have the same number of zeros inside the rectangle. Consequently the point $s = R$ will be the only pole inside the rectangle for the function under the

[1] Obviously we may assume that $P(y)$ is continuous at $y = 0$, since a discontinuity at this point would correspond to the case when certain risk sums are equal to zero, which is without interest for the applications.

integral in (135). Applying Cauchy's theorem to this rectangle, and allowing first Σ, and then T, to tend to infinity, we thus obtain

$$\psi(u) = Ce^{-Ru}$$

with

$$C = \frac{B(R, 0)}{A(0, 0)\, R[p'(R) - p_1 - \lambda]}.$$

However, we have already observed in the course of the proof of (96) that, if $p_1 + \lambda < 0$, we have $\psi(0, T) = 1$ for all $T > 0$, and thus also $\psi(0) = 1$. It follows that $C = 1$, and $\psi(u) = e^{-Ru}$.

5. 14. The mixed case, with exponential distribution of positive risk sums. We shall now briefly consider the case when there may be risk sums of both signs, the negative risk sums having any distribution such that the mean value

$$\int_{-\infty}^{0} |y|\, dP(y)$$

is finite (cf. 59), while the positive risk sums have a distribution defined by

$$P'(y) = \sum_{1}^{N} k_n \beta_n e^{-\beta_n y} \qquad (y > 0),$$

$$\text{where } 0 < \beta_1 < \ldots\ldots < \beta_N, \qquad k_i > 0, \quad \text{and} \quad 1 - P(0) = \sum_{1}^{N} k_n \leqq 1.$$

Presumably many distributions of risk sums encountered in practice could be fairly well represented in this form.

In this case we have

$$p_1 = \int_{-\infty}^{0} y\, dP(y) + \sum_{1}^{N} \frac{k_n}{\beta_n}.$$

The function

$$p(s) = \int_{-\infty}^{0} e^{sy}\, dP(y) + \sum_{1}^{N} \frac{k_n \beta_n}{\beta_n - s}$$

is regular throughout the half-plane $\sigma > 0$, except at the N points $s = \beta_n$, which are poles of the first order. On the contour of the rectangle considered in the preceding section, we still have the inequality (136), if T and Σ are sufficiently large. Since the function $1+(p_1+\lambda)s$ has one zero inside the rectangle when $p_1+\lambda < 0$, and no zero when $p_1+\lambda > 0$, it follows that the function

$$\pi(s) = 1+(p_1+\lambda)s-p(s)$$

has $N+1$ zeros inside the rectangle when $p_1+\lambda < 0$, and N zeros when $p_1+\lambda > 0$. A study of the course of the function for real $s > 0$ shows that these zeros are all real. There will be one zero $R_1 = R$ between 0 and β_1, and one zero R_n in each interval $\beta_{n-1} < R_n < \beta_n$. In the case $p_1+\lambda < 0$, the $(N+1)$:st zero R_{N+1} will be $> \beta_N$.

Applying Cauchy's theorem in the same way as in the preceding section, we obtain the explicit formula

$$\psi(u) = \sum C_n e^{-R_n u},$$

the sum being extended over all zeros R_n, and C_n being given by an expression analogous to (104).

Let us now consider the particular case when

$$1-P(0) = \sum_1^N k_n = 1,$$

which implies that there are no negative risk sums. In this case we have

$$p_1 = \sum_1^N \frac{k_n}{\beta_n}.$$

Since $p_1+\lambda > 0$, there are N positive zeros $R_1, \ldots, R_N$ of $\pi(s)$. The equation $\pi(s) = 0$ now becomes

$$1+(p_1+\lambda)s-\sum_1^N \frac{k_n\beta_n}{\beta_n-s} = 0.$$

This is an algebraic equation of degree $N+1$, with the roots $0, R_1, \ldots, R_N$. From (118) we obtain in this case

$$\psi(u) = \sum_1^N \frac{\lambda}{p'(R_n)-p_1-\lambda} e^{-R_n u}.$$

In the simplest particular case $N = 1$, $k_1 = \beta_1 = 1$, we have $p_1 = 1$ and

$$\pi(s) = 1+(1+\lambda)s-\frac{1}{1-s},$$

which gives $R = \dfrac{\lambda}{1+\lambda}$, and

(137) $$\psi(u) = \frac{1}{1+\lambda} e^{-\frac{\lambda}{1+\lambda}u}.$$

In the interval $0 < s < R$, the function $\pi(s)$ attains its maximum at the point

$$s = \varrho = 1 - \frac{1}{\sqrt{1+\lambda}},$$

and we have

$$\pi(\varrho) = \gamma = (\sqrt{1+\lambda} - 1)^2.$$

These are the constants ϱ and γ which occur in the relations (124) and (125) for the difference $\psi(u) - \psi(u, T)$. Further developments and results concerning the function $\psi(u, T)$ in this case are given by Arfwedson (7—10).

6. Generalizations

6. 1. Variable distribution of risk sums. The fundamental assumptions concerning the risk process which have been introduced in 3. 1 may seem rather restrictive, and it would be important to be able to work out a theory that would be valid under more general conditions.

Let us first consider the assumption that the d.f. $P(y)$ of risk sums is independent of time. The whole theory of the probability distribution of the gain $X(t)$ during a fixed period developed in chapters 3 and 4 can be rendered independent of this assumption. Suppose that, if a claim occurs at time t, the d.f. of the sum due is $P(y, t)$, and write

$$p(s, t) = \int_{-\infty}^{\infty} e^{sy} \, d_y P(y, t),$$

$$\overline{P}(y, t) = \frac{1}{t} \int_{0}^{t} P(y, u) \, du,$$

$$\overline{p}(s, t) = \frac{1}{t} \int_{0}^{t} p(s, u) \, du.$$

Then $\overline{P}(y, t)$ will be the *average d.f. of risk sums* corresponding to the interval $(0, t)$. Subject to mild conditions of regularity we have

$$\overline{p}(s, t) = \int_{-\infty}^{\infty} e^{sy} \, d_y \overline{P}(y, t).$$

The formulas (18) and (19), which determine the probability distribution of the gain $X(t)$ during the interval $(0, t)$, will now be transformed into

$$E e^{-sX(t)} = e^{t[\overline{p}(s,t) - 1 - \overline{p}_1 s]},$$

$$F(x, t) = 1 - \sum_{0}^{\infty} \frac{t^n}{n!} e^{-t} \, \overline{P}^{n*}(\overline{p}_1 t - x),$$

where

$$\overline{p}_1 = \frac{1}{t} \int_{0}^{t} p_1 \, du.$$

The whole theory of chapter 4 will now remain valid under these more general assumptions, replacing throughout P and p by $\overline{P}$ and $\bar{p}$.

For the ruin problem, on the other hand, it hardly seems possible to work out a simple generalization on similar lines.

6. 2. Variable safety loading. In our discussion of the ruin problem in chapter 5, we have assumed (cf. 5. 1) that all gross risk premiums are collected into the risk reserve of the company, from which all claims are paid. Suppose that, in an actual case, the initial value u of the risk reserve, and the constant λ representing the safety loading have been taken such that the probability $\psi(u)$ of ruin at any future moment has a value regarded as sufficiently small. After some time, it is then very likely that the risk reserve will reach a value considerably larger than the initial value u. When this situation has arrived, it would obviously be possible to change the value of the safety loading paid into the risk reserve, adopting for the future a smaller value of λ without increasing the probability of future ruin. This would mean that a larger part of the premiums actually paid by the policyholders would be at the free disposal of the company, for bonus or otherwise.

It might then seem natural to introduce a priori a *variable safety loading*, laying down the rule that the safety loading received by the risk reserve would amount to $\lambda(x)\,dt$ during a time element dt when the value of the risk reserve is x. Here $\lambda(x)$ would be an a priori given function of x, and it seems natural to require that $\lambda(x)$ should be never increasing as x increases.

It is evident that under these circumstances the ruin problem will be much more complicated than in the case of a constant λ dealt with in chapter 5. The problem has been treated by F. Lundberg **(45)**, Laurin **(40)**, Täcklind **(65)**, and Davidson **(28)**. The results are too complicated to be quoted in any detail here, and we shall only mention one very special particular case of the general theorems proved by Davidson.

Consider the particular case discussed at the end of 5. 14 when all risk sums are positive, and have the simple exponential distribution defined by

$$P'(y) = e^{-y} \qquad (y > 0).$$

Suppose that we have

$$\lambda(x) = \lambda$$

for all $x \geqq a$, where λ and a are given constants, while $\lambda(x)$ is a given never decreasing function for $x < a$. Write

$$G(t) = e^{-\int_0^t \frac{\lambda(x)}{1+\lambda(x)}\,dx}.$$

According to Davidson **(28)** we then have the following explicit expression for the ruin function $\psi(u)$:

$$\psi(u) = 1 - \frac{\int_0^u G(t)\,dt + G(u)}{\int_0^a G(t)\,dt + \frac{1+\lambda}{\lambda}\,G(a)}.$$

If, in particular, $\lambda(x) = \lambda$ for all x, it is easily seen that this expression reduces to

$$\psi(u) = \frac{1}{1+\lambda}\,e^{-\frac{\lambda u}{1+\lambda}},$$

as deduced in 5. 14 for the case of a constant λ.

6. 3. Fluctuating basic probabilities. The assumptions of constant risk probabilities and complete independence between events in disjoint time intervals can be generalized in various ways. An important generalization consists in replacing the Poisson formula (8)

$$\frac{t^n}{n!}\,e^{-t}$$

for the probability of exactly n claims during a time interval of length t by the formula

$$(138)\qquad \frac{t^n}{n!}\left(1+\frac{t}{h}\right)^{-n-h}\left(1+\frac{1}{h}\right)\left(1+\frac{2}{h}\right)\cdots\left(1+\frac{n-1}{h}\right)$$

related to the Polya-Eggenberger urn model with contagion between drawings. Here h is a positive constant, and for $h \to \infty$ we are led back to the Poisson distribution.

The distribution (138) has been applied by O. Lundberg **(49)** to cases of sickness and disability insurance.

The same distribution has been obtained by Ammeter **(1—6)** by means of the following considerations. Suppose that the probability of n claims during a time interval of length t is given by the Poisson formula

$$\frac{(xt)^n}{n!}\,e^{-xt}$$

where the parameter x is itself a random variable, with the Pearson type III frequency function

$$\frac{h^h}{\Gamma(h)}\,x^{h-1}\,e^{-hx} \qquad (x > 0).$$

It is then easily shown that (138) will be the final expression of the probability of n claims during the interval t. In this way the basic probabilities of claims will themselves be regarded as subject to random fluctuations. Ammeter has shown that the collective risk theory can be generalized by regarding (138) as the fundamental distribution of the number of claims, instead of the Poisson distribution. In particular he obtains in this way generalizations of the Esscher approximation formulas (cf. 4. 3) for $F(ct, t)$, and also asymptotic expressions for the ruin function $\psi(u)$, generalizing formula (119) above.

Bibliography

Sk. A. = Skandinavisk Aktuarietidskrift.

M. Schw. = Mitteilungen d. Verein. Schweizerischer Versicherungsmathematiker.

1. H. AMMETER, Das Maximum des Selbstbehaltes in der Lebensversicherung unter Berücksichtigung der Rückversicherungskosten, M. Schw. 1946.
2. — A generalization of the collective theory of risk in regard to fluctuating basic probabilities, Sk. A. 1948, p. 171.
3. — Die Elemente der kollektiven Risikotheorie von festen und zufallsartig schwankenden Grundwahrscheinlichkeiten, M. Schw. 1949.
4. — The calculation of premium rates for excess of loss and stop loss reinsurance treaties, Trans. Intern. Congr. of Actuaries, Amsterdam 1953.
5. — Risikotheoretische Methoden in der Rückversicherung, Trans. Intern. Congr. of Actuaries, Madrid 1954.
6. — La théorie collective du risque et l'assurance de choses, M. Schw. 1954, p. 185.
7. G. ARFWEDSON, Some problems in the collective theory of risk, Sk. A. 1950, p. 1.
8. — A semiconvergent series with application to the collective theory of risk, Sk. A. 1952, p. 16.
9. — Research in collective risk theory. The case of equal risk sums. Sk. A. 1953, p. 1.
10. — On the collective theory of risk. Trans. Intern. Congr. of Actuaries, Madrid 1954.
11. — Forthcoming paper on collective risk theory.
12. S. BJERRESKOV, On the principles for the choice of reinsurance method and for the fixing of net retention for an insurance company, Trans. Intern. Congr. of Actuaries, Madrid 1954.
13. A. BLANC-LAPIERRE et R. FORTET, Théorie des fonctions aléatoires, Paris (Masson) 1953.
14. C. BOEHM, Versuch einer systematischen Darstellung der modernen Risikotheorie, Bl. f. Vers.-mathematik, B. 3, 1934/36.
15. E. CAMPAGNE, B. H. DE JONGH and J. N. SMIT, Contribution to the mathematical theory of the stabilization reserve and net retention in fire insurance, The Hague 1947.
16. H. CRAMÉR, Bidrag till utjämningsförsäkringens teori, Stockholm (Försäkringsinspektionen) 1919.
17. — Review of F. Lundberg (45), Sk. A. 1926, p. 223.

18. – On the composition of elementary errors, Sk. A. 1928, p. 13 and p. 141.
19. – On the mathematical theory of risk, Skandia Jubilee Volume, Stockholm 1930.
20. – Random variables and probability distributions, Cambridge Tracts in Mathematics, Cambridge Univ. Press 1937.
21. – Sur un nouveau théorème-limite de la théorie des probabilités, Actualités Scientifiques, Paris (Hermann) 1938.
22. – Deux conférences sur la théorie des probabilités, Sk. A. 1941, p. 34.
23. – Mathematical methods of statistics, Princeton Univ. Press 1946.
24. – Den Lundbergska riskteorien och teorien för stokastiska processer, F. Lundberg Jubilee Volume, Stockholm 1946, p. 25.
25. – Problems in probability theory, Ann. Math. Statistics, Vol. 18, p. 165, 1947.
26. – The elements of probability theory and some of its applications, Stockholm (Geber) and New York (Wiley) 1954.
27. – On some questions connected with mathematical risk, Univ. of California Publ. in Statistics, Vol. 2, p. 99, 1954.
28. Å. Davidson, Om ruinproblemet i den kollektiva riskteorien under antagande av variabel säkerhetsbelastning, F. Lundberg Jubilee Volume, Stockholm 1946, p. 32.
29. J. L. Doob, Stochastic processes depending on a continuous parameter, Trans. Amer. Math. Soc., Vol. 42, p. 107, 1937.
30. – Probability in function space, Bull. Amer. Math. Soc., Vol. 53, p. 15, 1947.
31. – Stochastic processes, New York (Wiley) 1953.
32. J. L. Doob and W. Ambrose, On two formulations of the theory of stochastic processes depending upon a continuous parameter, Ann. of Math., Vol. 41, p. 737, 1940.
33. J. Dubourdieu, Théorie mathématique des assurances. Premier livre: Théorie mathématique du risque dans les assurances de répartition, Paris (Gauthier-Villars) 1952.
34. F. Esscher, On the probability function in the collective theory of risk, Sk. A. 1932, p. 175.
35. C. G. Esseen, Fourier analysis of distribution functions, Acta Math., Vol. 77, p. 1, 1945.
36. W. Feller, Generalization of a probability limit theorem by Cramér, Trans. Amer. Math. Soc., Vol. 54, p. 361, 1943.
37. – An introduction to probability theory and its applications, Vol. I, New York (Wiley) 1950.
38. K. Hultman, Einige numerische Untersuchungen auf Grund der kollektiven Risikotheorie, Sk. A. 1942, p. 84 and p. 169.
39. A. Kolmogoroff, Grundbegriffe der Wahrscheinlichkeitsrechnung, Ergebn. d. Math., Vol. 2, Berlin (Springer) 1933.
40. I. Laurin, An introduction into Lundberg's theory of risk, Sk. A. 1930, p. 84.
41. J. Lefèvre, Application de la théorie collective du risque à la réassurance excess-loss, Sk. A. 1952, p. 160.
42. F. Lundberg, Approximerad framställning av sannolikhetsfunktionen. Återförsäkring av kollektivrisker. Akad. Afhandling, Uppsala (Almqvist o. Wiksell) 1903.

43. F. LUNDBERG, Zur Theorie der Rückversicherung. Trans. Intern. Congr. of Actuaries, Wien 1909.
44. — Teori för riskmassor, Stockholm (Försäkringsinspektionen) 1919.
45. — Försäkringsteknisk riskutjämning, Stockholm (F. Englund) 1926–1928.
46. — Über die Wahrscheinlichkeitsfunktion einer Risikenmasse, Sk. A. 1930, p. 1.
47. — Some supplementary researches on the collective risk theory, Sk. A. 1932, p. 137.
48. — On the numerical application of the collective risk theory, De Förenade Jubilee Volume, Stockholm 1934.
49. O. LUNDBERG, On random processes and their application to sickness and accident statistics, Thesis, Stockholm 1940.
50. — Actuarial problems appertaining to reinsurance of disability risks, Trans. Intern. Congr. of Actuaries, Madrid 1954.
51. H. B. MANN, Introduction to the theory of stochastic processes depending on a continuous parameter, Appl. Math. Series No. 24, Washington (Nat. Bur. of Standards) 1953.
52. G. OTTAVIANI, La teoria del rischio del Lundberg e il suo legame con la teoria classica del rischio, Giorn. Ist. Ital. degli Attuari 1940, p. 163.
53. R. PALEY and N. WIENER, Fourier transforms in the complex domain, Amer. Math. Soc. Colloquium Publ., Vol. 19, 1934.
54. T. PENTIKÄINEN, Einige numerische Untersuchungen über das risikotheoretische Verhalten von Sterbekassen, Sk. A. 1947, p. 75.
55. — On the net retention and solvency of insurance companies, Sk. A. 1952, p. 71.
56. — On the reinsurance of an insurance company, Trans. Intern. Congr. of Actuaries, Madrid 1954.
57. T. SAXÉN, On the probability of ruin in the collective risk theory for insurance enterprises with only negative risk sums, Sk. A. 1948, p. 199.
58. — Sur les mouvements aléatoires et le problème de ruine de la théorie de risque collective, Soc. Sc. Fennica, Comm. Phys.-Math., Vol. 16, 1951.
59. C. O. SEGERDAHL, On homogeneous random processes and collective risk theory, Thesis, Stockholm 1939.
60. — Über einige risikotheoretische Fragestellungen, Sk. A. 1942, p. 43.
61. — Some properties of the ruin function in the collective theory of risk, Sk. A. 1948, p. 46.
62. — Om maximum på egen risk inom svensk livförsäkring, Unpublished manuscript, Stockholm 1950.
63. W. SIMONSEN, Om grundlaget for den kollektive risikoteori, F. Lundberg Jubilee Volume, Stockholm 1946, p. 246.
64. A. THÉPAUT, Essai de détermination pratique du plein de conservation, Paris (Dulac) 1954.
65. S. TÄCKLIND, Sur le risque de ruine dans des jeux inéquitables, Sk. A. 1942, p. 1.
66. L. WILHELMSEN, On the stipulation of maximum net retentions in insurance companies, Trans. Intern. Congr. of Actuaries, Madrid 1954.

67. H. Wyss, Die Risikotheorie und ihre Bedeutung für die Versicherungsmathematik, M. Schw. 1953, p. 23.
68. — Die Risikotheorie als Grundlage für die Versicherungsmathematische Behandlung von Rückversicherungsproblemen, Trans. Intern. Congr. of Actuaries, Madrid 1954.

Contents

48.

Ein Satz über geordnete Mengen von Wahrscheinlichkeitsverteilungen

Teor. Verojatnost. i Primenen. **I**, 19–24 (1956)

1. Es sei A eine gegebene Menge, deren Punkte oder Elemente mit $a, b, \ldots$ bezeichnet werden. Indem wir A als unsere *Grundmenge* bezeichnen, wollen wir annehmen, dass A eine *vollständig geordnete* Menge ist, und zwar in dem folgenden Sinne.

Auf A ist eine binäre Beziehung zwischen Elementen gegeben, für die wir die Bezeichnung $\gtrsim$ wählen. Es wird angenommen, dass für jedes Paar a, b von Elementen aus A *wenigstens eine* der beiden Beziehungen

$$a \gtrsim b, \qquad b \gtrsim a$$

gilt. Es wird ferner angenommen, dass die Beziehung $\gtrsim$ transitiv ist. Au $a \gtrsim b$, $b \gtrsim c$ folgt also $a \gtrsim c$.

Mit einer Terminologie, die von gewissen Anwendungen herrührt, wollen wir die Beziehung $a \gtrsim b$ etwa so aussprechen: *a ist wenigstens so gut wie b.*

Wenn gleichzeitig $a \gtrsim a$ und $b \gtrsim a$ gilt, so schreiben wir $a \backsim b$. Es ist dann offenbar auch $b \backsim a$, und wir sagen, dass a und b *gleichwertig* sind. Aus $a \backsim b$, $b \backsim c$ folgt offenbar $a \backsim c$. Wenn $a \gtrsim b$, aber nicht $a \backsim b$ gilt, so schreiben wir $a > b$, und sagen, dass *a besser* als b ist.

Die Gesamtheit aller mit einem gegebenen a gleichwertigen Elemente heisst die *Äquivalenzklasse* von a. Um triviale Fälle auszuschliessen nehmen wir von vornherein an, dass A nicht nur eine endliche Anzahl von Äquivalenzklassen enthält.

Eine reellwertige auf A definierte Funktion $f(a)$ heisst *monoton* im Sinne der gegebenen Ordnungsbeziehung, wenn $f(a) > f(b)$ dann und nur dann gilt, wenn $a > b$. Bekanntlich lässt sich aus den oben gemachten Annahmen die Existenz einer solchen Funktion $f(a)$ nicht folgern. Wenn aber die Ordnungsbeziehungen zwischen den Elementen von A sich auf eine geeignete Menge von Wahrscheinlichkeitsverteilungen in A erweitern lassen, so kann man nach v. Neumann und Morgenstern [3] unter gewissen zusätzlichen Bedingungen die Existenz eines monotonen $f(a)$ ableiten. Wir wollen jetzt zuerst einige diesbezügliche Begriffe und Bezeichnungen einführen.

2. Es sei X_n für jedes $n = 1, 2, \ldots$ die Menge aller derjenigen Wahrscheinlichkeitsverteilungen in A, bei denen die gesamte Wahrscheinlichkeit über höchstens n Punkte aus A verteilt ist. Die Elemente von X_n nen-

2*

nen wir *n-Punktverteilungen* in A. Jedes X_n enthält dann die vorhergehenden $X_1, X_2, \ldots, X_{n-1}$ als echte Teilmengen. Wenn die Vereinigungsmenge aller X_n durch X^* bezeichnet wird, so gilt also

$$X_1 \subset X_2 \subset \ldots \subset X^*.$$

X^* ist dann die Menge aller endlichen diskreten Wahrscheinlichkeitsverteilungen in A.

Wenn x und y irgend zwei Elemente von X^* sind, so bezeichnen wir für jedes reelle λ mit $0 \leqslant \lambda \leqslant 1$ durch $\lambda x + (1-\lambda) y$ das Element von X^*, das in unmittelbar einleuchtender Weise durch lineare Verknüpfung der beiden Wahrscheinlichkeitsverteilungen x und y entsteht. Aus $x \in X_m$, $y \in X_n$ folgt dann offenbar $\lambda x + (1-\lambda) y \in X_{m+n}$.

Jedem Element a von A entspricht eine Einpunktverteilung $x_a \in X_1$, bei der die gesamte Wahrscheinlichkeit 1 im Punkte a verlegt ist. Die Ordnungsbeziehungen zwischen den a können also unmittelbar auf X_1 übertragen werden. Wenn diese Ordnungsbeziehungen sich auf die ganze Menge X^* erweitern lassen, und wenn ausserdem gewisse Regularitätsbedingungen erfüllt sind, so gibt es nach v. Neumann und Morgenstern [3] eine auf X^* definierte reellwertige und monotone Funktion $f(x)$, die ausserdem *linear* ist, d. h. die Gleichung

$$f[\lambda x + (1-\lambda) y] = \lambda f(x) + (1-\lambda) f(y) \tag{1}$$

für beliebige x, y aus X^* und für jedes reelle λ mit $0 \leqslant \lambda \leqslant 1$ befriedigt. Setzen wir dann für jedes $a \in A$

$$f(a) = f(x_a),$$

so ist leicht ersichtlich, dass für jedes $x \in X^*$

$$f(x) = \mathbf{E}_x f(a)$$

gilt, wo $\mathbf{E}_x f(a)$ die gemäss der Wahrscheinlichkeitsverteilung x berechnete mathematische Erwartung der reellen stochastischen Veränderlichen $f(a)$ bedeutet. Es ist dann $f(a)$ eine auf A definierte reellwertige Funktion mit der Eigenschaft, dass für jedes Paar von Elementen x, y aus X^* die Beziehung $x > y$ dann und nur dann gilt, wenn

$$\mathbf{E}_x f(a) > \mathbf{E}_y f(a).$$

Wenn die gegebenen Ordnungsbeziehungen als Ausdruck einer «Preferenzordnung» zwischen verschiedenen Wahrscheinlichkeitsverteilungen in A aufgefasst werden, stellt also die mathematische Erwartung der «Nutzfunktion» $f(a)$ das geeignete Kriterium für die Wahl zwischen verschiedenen Verteilungen dar. Es war ein Kriterium dieser Art, das seinerzeit Daniel Bernoulli durch die Einführung seiner «moralischen Erwartung» realisieren wollte.

3. Die Resultate von v. Neumann und Morgenstern wurden von anderen Verfassern ergänzt und verallgemeinert. (Vgl. besonders Herstein und Milnor [2], sowie Blackwell und Girshick [1].) Eine wesentliche Voraussetzung dieser Arbeiten scheint jedenfalls zu sein, dass eine vollständige

Ordnung in der ganzen Menge X^* aller endlichen diskreten Wahrscheinlichkeitsverteilungen gegeben sein muss. Wir wollen hier zeigen, dass es in dieser Hinsicht genügt, wenn man nur voraussetzt, dass *die Menge* X_2 *aller Zweipunktverteilungen geordnet ist.*

Im Folgenden sind $a, b, \ldots$ Elemente der Grundmenge A und x_a, $x_b, \ldots$ die entsprechenden Einpunktverteilungen. $x, y, z, \ldots$ sind Elemente von x_2, während $\lambda, \mu, \ldots$ reelle Zahlen im abgeschlossenen Intervall $[0, 1]$ sind. Die Menge aller λ, die eine gegebene Beziehung B erfüllen, wird durch $\{\lambda \mid B\}$ bezeichnet.

Um unsere Ausdrucksweise zu vereinfachen, wollen wir sagen, dass zwei Elemente x, y von X_2 *befreundet* sind, wenn die Gesamtzahl der Punkte, denen in x oder y eine positive Wahrscheinlichkeit zukommt, höchstens gleich zwei ist. Dann gehört offenbar auch die lineare Verknüpfung $\lambda x + (1-\lambda) y$ zu X_2. Zwei Einpunktverteilungen sind natürlich als Elemente von X_2 immer befreundet.

Satz. *Für eine gegebene vollständig geordnete Grundmenge A seien die folgenden Bedingungen* 1)—3) *erfüllt:*

1) *Die Ordnungsbeziehungen in A können auf die Menge* X_2 *aller Zweipunktverteilungen in A erweitert werden, so dass* X_2 *vollständig geordnet ist.*

2) *Es seien x und y befreundet, und* $x \gtrsim z \gtrsim y$. *Dann sind die beiden Mengen*

$$\{\lambda \mid \lambda x + (1-\lambda) y \gtrsim z\} \text{ und } \{\lambda \mid z \gtrsim \lambda x + (1-\lambda) y\} \tag{2}$$

abgeschlossen.

3) *Es seien x, y und x', y' zwei Paare von befreundeten Elementen, und es sei* $x \backsim x'$, $y \backsim y'$. *Für jedes* λ *gilt dann*

$$\lambda x + (1-\lambda) y \backsim \lambda x' + (1-\lambda) y'.$$

Dann gibt es eine reelwertige Funktion $f(a)$, *die für alle* $a \in A$ *definiert ist, so dass für irgend zwei Elemente* x, y *von* X_2 *die Beziehung* $x > y$ *dann und nur dann gilt, wenn*

$$\mathbf{E}_x f(a) > \mathbf{E}_y f(a).$$

Die Behauptung wird offenbar bewiesen sein, wenn wir die Existenz einer auf X_2 definierten, reellwertigen und monotonen Funktion $f(x)$ nachweisen können, die auf X_2 linear ist, d. h. die Gleichung (1) für beliebige befreundete x, y in X_2 und beliebige in $[0,1]$ befriedigt. Wenn wir nämlich dann $f(a) = f(x_a)$ setzen, so hat $f(a)$ die verlangten Eigenschaften.

Wir bemerken zuerst, dass keine von den Mengen (2) leer ist, da die erste Menge den Punkt $\lambda = 1$, die zweite den Punkt $\lambda = 0$ enthält. Da beide Mengen abgeschlossen sind, und ihre Vereinigungsmenge das geschlossene Intervall $[0,1]$ ist, kann die Durchschnittsmenge nicht leer sein. Unter den Voraussetzungen der Bedingung 2) gibt es also wenigstens ein λ, so dass

$$\lambda x + (1-\lambda) y \backsim z. \tag{3}$$

Für den Beweis des Satzes werden wir einige Hilfssätze brauchen.

Hilfssatz 1. *Wenn x und y befreundet sind, und* $x > y$, *so ist für* $0 < \lambda < 1$

$$x > \lambda x + (1-\lambda) y > y. \tag{4}$$

Ferner ist

$$\lambda x + (1-\lambda) y > \mu x + (1-\mu) y \tag{5}$$

dann und nur dann, wenn $\lambda > \mu$ *ist.*

Nehmen wir zuerst an, dass für irgend ein festes λ_0 mit $0 < \lambda_0 < 1$

$$\lambda_0 x + (1-\lambda_0) y \geq x > x$$

ist. Nach der obigen Bemerkung gibt es dann wenigstens ein λ, so dass

$$\lambda [\lambda_0 x + (1-\lambda_0) y] + (1-\lambda) y \backsim x$$

gilt. Offenbar muss hier $\lambda > 0$ sein, da nach Voraussetzung $x > y$ ist. Indem wir $\lambda\lambda_0 = \mu$ setzen, können wir also schliessen, dass es wenigstens ein μ mit $0 < \mu < 1$ gibt, so dass

$$\mu x + (1-\mu) y \backsim x \tag{6}$$

ist. Es sei μ_0 die untere Grenze aller μ mit dieser Eigenschaft. Nach der Bedingung 2) gilt dann (6) auch für $\mu = \mu_0$. Da (6) nicht für $\mu = 0$ gilt, ist $0 < \mu_0 < 1$. Dann ist aber nach der Bedingung 3)

$$\mu_0 [\mu_0 x + (1-\mu_0) y] + (1-\mu_0) y \backsim \mu_0 x + (1-\mu_0) y \backsim x,$$
$$\mu_0^2 x + (1-\mu_0^2) y \backsim x,$$

was einen Widerspruch enthält, da $\mu_0^2 < \mu_0$ ist. In derselben Weise schliesson wir, dass die Annahme

$$y \geq \lambda_0 x + (1-\lambda_0) y$$

zu einem Widerspruch führt, wodurch (4) bewiesen ist.

Für $\lambda > \mu > 0$ folgt sodann weiter aus (4), da $\lambda x + (1-\lambda) y$ und y befreundet sind,

$$\lambda x + (1-\lambda) y > \frac{\mu}{\lambda} [\lambda x + (1-\lambda) y] + \left(1 - \frac{\mu}{\lambda}\right) y = \mu x + (1-\mu) y.$$

Es gilt also (5) für $\lambda > \mu$, und da die umgekehrte Behauptung jetzt in trivialer Weise folgt, ist Hilfssatz 1 hiermit bewiesen.

Aus Hilfssatz 1 ergibt sich, mit Hilfe der im Zusammenhang mit (3) gemachten Bemerkung, unmittelbar der folgende

Hilfssatz 2. *Wenn x und y befreundet sind, und $x > z > y$, so gibt es ein eindeutig bestimmtes λ mit $0 < \lambda < 1$, so dass (3) gilt.*

Es seien jetzt a und b Punkte von A, so dass $a > b$, und also auch $x_a > x_b$, gilt. Die Menge aller x in X_2, für die $x_a \geq x \geq x_b$ ist, bezeichnen wir durch $I_{a,b}$. Wenn eine reellwertige Funktion $f(x)$ auf $I_{a,b}$ definiert ist und dort die oben angegebenen Monotonitäts- und Linearitätsbedingungen erfüllt, so heisst $f(x)$ auf $I_{a,b}$ monoton und linear.

Hilfssatz 3. *Es seien a, b, α, β gegebene Punkte in A, mit $a \geq \alpha > \beta \geq b$. Wenn $f(x)$ und $g(x)$ auf $I_{a,b}$ monoton und linear sind, und wenn ausserdem*

$$f(x_\alpha) = g(x_\alpha), \qquad f(x_\beta) = g(x_\beta),$$

so ist $f(x) = g(x)$ für alle x in $I_{a,b}$.

Für geeignet gewählte λ, μ, ν hat man

$$x \backsim \lambda x_\alpha + (1-\lambda) x_b,$$
$$x_\alpha \backsim \mu x_a + (1-\mu) x_b,$$
$$x_\beta \backsim \nu x_a + (1-\nu) x_b,$$

und also nach den Linearitätseigenschaften von $f(x)$

$$f(x) = \lambda f(x_a) + (1-\lambda) f(x_b),$$
$$f(x_\alpha) = \mu f(x_a) + (1-\mu) f(x_b),$$
$$f(x_\beta) = \nu f(x_a) + (1-\nu) f(x_b).$$

Da μ und ν nach Hilfssatz 1 verschieden sind, kann man $f(x)$ mit Hilfe von $f(x_\alpha)$ und $f(x_\beta)$ ausdrücken, und findet so

$$f(x) = k_1 f(x_\alpha) + k_2 f(x_\beta),$$

wo k_1 und k_2 nur von λ, μ, ν abhängen. In derselben Weise bekommt man

$$g(x) = k_1 g(x_\alpha) + k_2 g(x_\beta) =$$
$$= k_1 f(x_\alpha) + k_2 f(x_\beta) = f(x).$$

Hilfssatz 3 ist also bewiesen.

Wen nun a und b gegeben sind, so dass $a > b$, so definieren wir für alle x in $I_{a,b}$ eine reellwertige Funktion $\lambda_{a,b} = \lambda_{a,b}(x)$ durch die Beziehung

$$\lambda_{a,b} x_a + (1 - \lambda_{a,b}) x_b \backsim x.$$

Nach den Hilfssätzen 1 und 2 ist diese Definition eindeutig, und die Funktion $\lambda_{a,b}(x)$ ist auf $I_{a,b}$ monoton. Für befreundete x, y in $I_{a,b}$ hat man aber nach der Bedingung 3)

$$\mu x + (1-\mu) y \backsim \mu [\lambda_{a,b}(x) x_a + (1 - \lambda_{a,b}(x)) x_b] +$$
$$+ (1-\mu) [\lambda_{a,b}(y) x_a + (1 - \lambda_{a,b}(y)) x_b] =$$
$$= [\mu \lambda_{a,b}(x) + (1-\mu) \lambda_{a,b}(y)] x_a + [1 - \mu \lambda_{a,b}(x) - (1-\mu) \lambda_{a,b}(y)] x_b,$$

und also

$$\lambda_{a,b}[\mu x + (1-\mu) y] = \mu \lambda_{a,b}(x) + (1-\mu) \lambda_{a,b}(y),$$

so dass $\lambda_{a,b}(x)$ auch linear auf $I_{a,b}$ ist.

Um den Beweis unseres Satzes abzuschliessen nehmen wir jetzt an es seien α und β fest gegeben in A, so dass $\alpha > \beta$. Es sei ferner x beliebig in X_2. Wir wählen dann a und b, so dass x, x_α und x_β alle zu $I_{a,b}$ gehören, was natürlich immer möglich ist. Setzen wir nun

$$f(x) = \frac{\lambda_{a,b}(x) - \lambda_{a,b}(x_\beta)}{\lambda_{a,b}(x_\alpha) - \lambda_{a,b}(x_\beta)},$$

so ist $f(x)$ offenbar auf $I_{a,b}$ monoton und linear. Aus Hilfssatz 3 folgt aber, da $f(x_\alpha) = 1$, $f(x_\beta) = 0$, dass $f(x)$ von der Wahl von a und b unabhängig ist. Also ist $f(x)$ für alle x in X_2 difiniert und hat die verlangten Eigenschaften.

Jede andere Funktion $g(x)$ mit denselben Eigenschaften ist in der Form $g(x) = k_1 f(x) + k_2$ mit konstanten k_1 und k_2 darstellbar. Wenn man nämlich die Konstanten so bestimmt, dass die Gleichung für $x = x_\alpha$ und für $x = x_\beta$ stimmt, so bleibt sie nach Hilfssatz 3 für alle x gültig.

Поступила в редакцию
21.12.55

LITERATUR

[1] D. Blackwell und M. A. Girshick, Theory of Games and Statistical Decisions. New York, 1954.

[2] I. N. Herstein and J. Milnor. An axiomatic Approach to measurable Utility, Econometrica 21 (1953), S. 291.

[3] J. V. Neumann und O. Morgenstern, Theory of Games and Economic Behaviour, 2. Aufl., Princeton, 1947.

ТЕОРЕМА ОБ УПОРЯДОЧЕННЫХ МНОЖЕСТВАХ ВЕРОЯТНОСТНЫХ РАСПРЕДЕЛЕНИЙ

Г. КРАМЕР (СТОКГОЛЬМ)

(Резюме)

Пусть во множестве $A - \{a, b, c, \ldots\}$ установлено бинарное отношение порядка: для любых двух элементов имеет место хотя бы одно из соотношений:

$$a \geq b \quad (a \text{ не хуже, чем } b) \tag{1}$$
$$b \geq a \quad (b \text{ не хуже, чем } a) \tag{2}$$

Если выполнено (1), но не имеет места (2), то $a > b$ (a лучше, чем b).

Пусть $X_2 = \{x, y, z, \ldots\}$ множество всех распределений, каждое из которых сосредоточено не более чем на двух элементах множества A. Если отношение порядка для элементов множества A можно расширить на X_2, то, при некоторых простых условиях регулярности X_2 доказывается существование такой действительной функции $f(a)$, определенной на A, что для любых x, y из X_2 соотношение $x > y$ справедливо тогда и только тогда, когда

$$E_x f(a) > E_y f(a).$$

49.

On the linear prediction problem for certain stochastic processes

Ark. Mat. 4 (6), 45–53 (1959)

1. Consider an infinite sequence of complex-valued random variables

$$\ldots, x_{-2}, x_{-1}, x_0, x_1, x_2, \ldots$$

with finite second order mean values. For the sake of simplicity, we shall assume throughout that all first order mean values of the x_n reduce to zero, while there is at least one x_n having a variance different from zero:

$$E\,x_n = 0 \text{ for all } n,$$

$$E\,|\,x_n\,|^2 > 0 \text{ for some } n.$$

The *covariance function* of the x_n sequence is

$$R(m, n) = E\,(x_m \overline{x_n}).$$

We may interpret x_n as a measure of the state of some observed variable system at the time point nt, where t is a given quantity. The sequence of the x_n, with $n = \cdots, -1, 0, 1, \ldots$, will then represent the temporal development of this system, and will constitute a *stochastic process with discrete time*. In the sequel, we shall always take $t = 1$, so that the subscript n may be directly regarded as measuring time.

The *prediction problem* for a process of this kind is the problem of predicting the state of the process at some future time point, when its past development is assumed to be more or less known. In this paper we shall only be concerned with *linear least squares prediction*. Thus we shall want to find the "best possible" prediction of a certain x_n by means of linear operations acting on certain variables belonging to the past of the process, interpreting the "best possible" in the sense of minimizing the mean value of the squared error of prediction.

Consider the expression

$$\underset{c_0, \ldots, c_q}{\operatorname{Min}}\; E\,|\,x_n - c_0\,x_{n-p} - c_1\,x_{n-p-1} - \cdots - c_q\,x_{n-p-q}\,|^2 = s^2_{npq} \geqq 0,$$

where n, p and q are fixed integers with $p > 0$, $q > 0$, while the minimum has to be taken for all complex quantities $c_0, \ldots, c_q$. Then s_{npq} will be the least possible error of prediction, when x_n has to be linearly predicted in terms of x_{n-p}, x_{n-p-1},

$\ldots, x_{n-p-q}$. Obviously s_{npq} will never increase when q increases, while n and p remain fixed. The limit

$$\lim_{q\to\infty} s_{npq} = \sigma_{np} \geqq 0$$

will thus always exist, and will be called the *prediction error* for x_n, when prediction is based on all the variables of the process up to and including x_{n-p}. We easily find that

$$0 \leqq \sigma_{n1} \leqq \sigma_{n2} \leqq \cdots.$$

The same thing may be expressed a little differently, if we look at the question from the point of view of Hilbert space geometry. Consider the Hilbert space H_{n-p} spanned by all the variables $x_{n-p}, x_{n-p-1}, \ldots$. The elements of this space are random variables, which are either finite linear combinations of $x_{n-p}, x_{n-p-1}, \ldots$, or limits in the mean of sequences of such combinations. The inner product and the norm are defined by the usual expressions

$$(y, z) = E(y\bar{z}), \quad \|y\| = (E|y^2|)^{\frac{1}{2}}.$$

By $P_{n-p}(x_n)$ we denote the projection of x_n on H_{n-p}. Then $z = P_{n-p}(x_n)$ is the uniquely determined element of H_{n-p} which minimizes the distance $\|x_n - z\|$. We shall call $P_{n-p}(x_n)$ the *best possible linear prediction* of x_n in terms of $x_{n-p}, x_{n-p-1}, \ldots$. The corresponding error of prediction will be

$$\sigma_{np} = \|x_n - P_{n-p}(x_n)\|.$$

If $\sigma_{np} = 0$ for all n and p, exact linear prediction is always possible. In this case every x_n can be exactly represented in terms of variables belonging to an arbitrarily remote past of the process. A process of this kind will be called a *deterministic* process.

On the other hand, every process such that $\sigma_{np} > 0$ for at least one pair of values of n and p, will be called *non-deterministic*. Every non-deterministic process may be represented as the sum of a deterministic component and a linear combination of certain *innovations*, which represent the "new" random impulses entering into the process at certain moments. In fact, it can be shown (cf. Cramér, 2) that for every non-deterministic process, there exists a uniquely determined, finite or infinite sequence of integers

$$\cdots < r_{-1} < r_0 < r_1 < \cdots$$

such that

$$x_n = \sum_{r_k \leqq n} c_{nr_k} \xi_{r_k} + y_n. \tag{1}$$

Here y_n is the deterministic component of the process, while the ξ_{r_k} are random variables such that

$$E\,\xi_{r_k} = 0, \quad E(\xi_{r_j}\bar{\xi}_{r_k}) = \delta_{jk},$$
$$E(x_n\,\bar{\xi}_{r_k}) = c_{nr_k}, \quad E(y_n\,\bar{\xi}_{r_k}) = 0.$$

Further $\sum_{r_k \leqq n} |c_{nr_k}|^2$ is convergent, so that the series in the expression for x_n converges in the mean. If n is a member of the r_k sequence, the quantity c_{nn} is real and positive. Finally, the prediction error σ_{np} is given by the expression

$$\sigma_{np}^2 = \sum_{n-p<r_k\leq n} |c_{nr_k}|^2.$$

It follows that $\sigma_{n1}>0$ when and only when n is a member of the r_k sequence. Thus it will be seen that every r_k is a *point of indetermination*, where the process receives an *innovation* proportional to ξ_{r_k}. On the other hand, when n is different from all r_k, we have $\sigma_{n1}=0$, so that x_n can be exactly predicted in terms of the preceding variables $x_{n-1}, x_{n-2}, \ldots$, and there is no innovation corresponding to the time point n. Thus in the expression (1) of x_n, the non-deterministic component is a linear combination of the innovations received by the process in all its points of indetermination preceding or coinciding with the time point n.

2. The linear prediction problem has been thoroughly studied for the important class of *stationary* processes, which are characterized by the fact that the covariance function $R(m,n)=E(x_m\overline{x_n})$ only depends on the time difference $m-n$. (We are not here concerned with the so called strictly stationary processes, which satisfy more stringent conditions.) For this class of processes, the sequence r_k considered in the preceding section contains every integer n, and the representation (1) reduces to a decomposition theorem due to H. Wold, 1. With respect to the general theory of stationary processes, and in particular the prediction problem, we may refer e.g. to the books by Wiener, 1, and Doob, 1.

One of the most important properties of the class of stationary processes is that they admit a *spectral representation* by means of a stochastic integral of Fourier type. Loève (1) has introduced a more general class of processes, which he calls *harmonizable* processes, and which possess spectral representations of a similar kind. In this paper, we shall consider the prediction problem for a group of harmonizable processes satisfying certain regularity conditions.

Consider a stochastic process such that x_n is given by the stochastic integral

$$x_n = \int_0^{2\pi} e^{inu}\,dz(u), \tag{2}$$

where $z(u)$ denotes, for every u in the interval $0\leq u\leq 2\pi$, a random variable such that

$$E\,z(u)=0, \quad E\,(z(u)\,\overline{z(v)})=F(u,v),$$

where $F(u,v)$—in general complex-valued—is of bounded variation over the square C: $0\leq u, v\leq 2\pi$. Obviously

$$F(v,u)=\overline{F(u,v)},$$

while $F(u,u)$ is real and non-negative. Under these conditions, the stochastic integral (2) can be defined as a limit in the mean of certain Riemann sums, and the variable x_n determines a harmonizable stochastic process. (Cf. also Cramér 1.)

In the sequel, we shall assume that the function $F(u,v)$ satisfies two additional conditions, which we denote by (A) and (B). Thus we shall assume:

$$F(u,v)=\int_0^u\int_0^v f(s,t)\,ds\,dt, \tag{3}$$

where

(*A*) $f(s,t)$ belongs to L^2 over the square C.

(*B*) $f(s,t)$ is bounded in the vicinity of the diagonal $s=t$ of C, so that there exist positive constants h and M such that $|f(s,t)|<M$ for $|s-t|<h$.

The covariance function of the x_n process is then given by

$$R(m,n)=E(x_m\overline{x_n})=\int_0^{2\pi}\int_0^{2\pi} e^{i(mu-nv)}\,dF(u,v)$$
$$=\int_0^{2\pi}\int_0^{2\pi} e^{i(mu-nv)} f(u,v)\,du\,dv. \tag{4}$$

More generally, for any $g(u)$ and $h(u)$ belonging to L^2 over $(0,2\pi)$, the random variables

$$\xi=\int_0^{2\pi} g(u)\,dz(u),\quad \eta=\int_0^{2\pi} h(u)\,dz(u)$$

are well defined, and we have

$$E(\xi\bar{\eta})=\int_0^{2\pi}\int_0^{2\pi} g(u)\,\overline{h(v)}\,f(u,v)\,du\,dv. \tag{5}$$

The function $F(u,v)$ is called the *spectral function* of the x_n process, and is said to define the *spectral distribution* of the process, which is a distribution of complex-valued "mass" over the square C, such that every surface element $du\,dv$ carries the mass $f(u,v)\,du\,dv$. The function $f(u,v)$ is the *spectral density* of the process, while $z(u)$ defines the corresponding *spectral process*.

I have elsewhere (Cramér, 2) given a sufficient condition that a harmonizable process will be deterministic. In the present note, I shall be concerned with an x_n process as defined by (2), and such that the corresponding spectral density $f(u,v)$ satisfies the conditions (*A*) and (*B*). It is proposed to find, for this process, the conditions under which an arbitrarily given sequence of integers r_k will constitute the complete set of points of indetermination of the process.

3. Suppose that we are given an x_n process as defined by (2), with a spectral function $F(u,v)$ satisfying (3), and a spectral density $f(u,v)$ satisfying the conditions (*A*) and (*B*). Let there further be given an increasing sequence of integers:

$$\cdots<r_{-1}<r_0<r_1<\cdots,$$

which may be finite or infinite, in one or both directions.

In order that the given r_k sequence will constitute the complete set of points of indetermination of the given x_n process, the following condition is necessary and sufficient. The spectral function $F(u,v)$ should admit a development of the form:

$$F(u,v)=\sum_{r_k}\varphi_k(u)\,\overline{\varphi_k(v)}+G(u,v), \tag{6}$$

where $G(u,v)$ is the spectral function of a deterministic harmonizable process, while the series in the second member converges absolutely for all u and v, and we have

$$\varphi_k(u) = \sum_{n=r_k}^{\infty} \alpha_{n r_k} e^{-inu} + \beta_k u + \gamma_k, \tag{7}$$

the one-sided trigonometric series occurring here being absolutely convergent for all u, while the α, β and γ are constants such that $\alpha_{r_k r_k} \neq 0$ for every $r_k \neq 0$, $\beta_k = 0$ for $r_k > 0$, and the series

$$\sum_{r_k \leq n} |\alpha_{n r_k}|^2,$$

extended over all $r_k \leq n$, converges for every fixed n.

We shall first show that the condition is necessary. Since by hypothesis r_k is a point of indetermination of the given x_n process, the innovation

$$c_{r_k r_k} \xi_{r_k} = x_{r_k} - P_{r_k - 1}(x_{r_k})$$

is not identically zero. As in the preceding section, we may take $c_{r_k r_k}$ real and positive, and such that $E|\xi_{r_k}|^2 = 1$. As before we write

$$c_{n r_k} = E(x_n \bar{\xi}_{r_k}),$$

observing that we have

$$c_{n r_k} = 0 \text{ for } n < r_k, \tag{8}$$

as the innovation associated with ξ_{r_k} does not enter into the process until the time point r_k, and is thus uncorrelated with every x_n preceding this time point.

We now define a random variable w by writing

$$w = \int_0^{2\pi} u \, dz(u),$$

where $z(u)$ is the spectral process appearing in (2). The stochastic integral is defined in the way indicated in the preceding section. We further write

$$d_k = E(w \bar{\xi}_{r_k}) \tag{9}$$

and

$$\varphi_k(u) = -\frac{1}{2\pi i} \sum_{n=r_k}^{\infty}{}' \frac{c_{n r_k}}{n} e^{-inu} + \frac{1}{2\pi} c_{0 r_k}(u+\pi) - \frac{1}{2\pi} d_k, \tag{10}$$

where the accent on the summation sign indicates that, if the value $n=0$ falls between the limits of summation, the corresponding term should be omitted. Clearly this is an expression of the form postulated in (7). In order to show that the trigonometric series appearing here is absolutely convergent, we observe that

$$|c_{n r_k}|^2 \leq E|x_n|^2 E|\xi_{r_k}|^2 = E|x_n|^2 = \int_0^{2\pi}\int_0^{2\pi} e^{in(u-v)} f(u,v)\, du\, dv.$$

Thus by condition (A) the quantities $|c_{n r_k}|^2$ are, for a fixed r_k, the Fourier coefficients of a function in L^2, so that the series $\sum_n |c_{n r_k}|^4$ is convergent. Hence by Hölder's inequality it is easily shown that $\sum_n{}' \frac{c_{n r_k}}{n}$ is absolutely convergent. It has already been remarked in connection with (1) that the series $\sum_{r_k \leq n} |c_{nr_k}|^2$ is convergent for every n. We finally observe that, in consequence of (8), the lower limit of summation in (10) may be replaced by $-\infty$, since the new terms thus introduced all reduce to zero.

We now consider the expression

$$G_K(u, v) = F(u, v) - \sum_{k=-K}^{K} \varphi_k(u) \overline{\varphi_k(v)}, \tag{11}$$

with appropriate modification in case the r_k sequence is finite in either direction. We shall first show that $G_K(u, v)$ is, for every K, the covariance function of some random variable $z_K(u)$ defined for every u in $(0, 2\pi)$, so that

$$G_K(u, v) = E(z_K(u) \overline{z_K(v)})$$

for all u and v in $(0, 2\pi)$. In order to show that $G_K(u, v)$ is a covariance, it is sufficient to show (cf. Loève, 1) that

$$\int_0^{2\pi}\int_0^{2\pi} q(u) \overline{q(v)}\, G_K(u, v)\, du\, dv \geqq 0 \tag{12}$$

for any continuous $q(u)$. Obviously it is even sufficient to show that (12) holds for any trigonometric polynomial $q(u)$. Taking

$$q(u) = \sum_{n=A}^{B} \lambda_n e^{inu},$$

$$Q(u) = \sum_{n=A}^{B}{}' \frac{\lambda_n}{in} (e^{inu} - 1) + \lambda_0 (u - 2\pi),$$

we have

$$Q'(u) = q(u), \quad Q(2\pi) = 0.$$

By (3) we have

$$F(u, 0) = F(0, v) = 0,$$

so that we obtain by partial integration, using (4), (5) and (9),

$$\int_0^{2\pi}\int_0^{2\pi} q(u) \overline{q(v)}\, F(u, v)\, du\, dv = \int_0^{2\pi}\int_0^{2\pi} Q(u) \overline{Q(v)}\, f(u, v)\, du\, dv =$$

$$= E \left| \sum_{A}^{B}{}' \frac{\lambda_n}{in} (x_n - x_0) + \lambda_0 (w - 2\pi x_0) \right|^2. \tag{13}$$

On the other hand

$$\int_0^{2\pi}\int_0^{2\pi} q(u) \overline{q(v)}\, \varphi_k(u) \overline{\varphi_k(v)}\, du\, dv = \left| \int_0^{2\pi} q(u) \varphi_k(u)\, du \right|^2, \tag{14}$$

and by some simple calculation we find

$$\int_0^{2\pi} q(u) \varphi_k(u)\, du = -\sum_{n=A}^{B}{}' \frac{\lambda_n}{in} (c_{nr_k} - c_{0r_k}) - \lambda_0 (d_k - 2\pi c_{0r_k}) =$$

$$= -E\left\{ \left(\sum_{A}^{B}{}' \frac{\lambda_n}{in} (x_n - x_0) + \lambda_0 (w - 2\pi x_0) \right) \bar{\xi}_{r_k} \right\}. \tag{15}$$

Writing

$$X=\sum_A^{B}{}' \frac{\lambda_n}{i\,n}(x_n-x_0)+\lambda_0(w-2\pi x_0),$$

we obtain from (11), (13), (14) and (15)

$$\int_0^{2\pi}\int_0^{2\pi} q(u)\overline{q(v)}\,G_K(u,v)\,du\,dv=E|X|^2-\sum_k|E(X\,\xi_{rk})|^2.$$

The ξ_{rk} being orthogonal random variables, (12) now follows directly from Bessel's inequality.

Thus $G_K(u,v)$ as defined by (11) is always a covariance function. It follows that $G_K(u,u)\geqq 0$, so that, allowing K to tend to infinity, the sum over all k

$$\sum_k|\varphi_k(u)|^2$$

is convergent, and consequently by the Schwarz inequality

$$\sum_k\varphi_k(u)\overline{\varphi_k(v)}$$

is absolutely convergent. Thus $G_K(u,v)$ tends to a limit $G(u,v)$ as $K\to\infty$.

The limit of a sequence of covariance functions being itself a covariance function (cf. Loève 1), we have now proved that $F(u,v)$ admits a development of the form (6), where $G(u,v)$ is a covariance function, while $\varphi_k(u)$ has the form (7), the stated convergence conditions being satisfied. It remains to show that $G(u,v)$ is the spectral function of a deterministic harmonizable process.

In order that a covariance function $G(u,v)$ defined in the square C: $0\leqq u, v \leqq 2\pi$, should be the spectral function of some harmonizable process, it is necessary and sufficient that $G(u,v)$ should be of bounded variation over C. We shall first show that this property holds here.

Since $G(u,v)$ is a covariance function, there exists for every u in $(0,2\pi)$ a random variable $Z(u)$ such that

$$G(u,v)=F(u,v)-\sum_k\varphi_k(u)\overline{\varphi_k(v)}=E(Z(u)\overline{Z(v)}). \tag{16}$$

For an arbitrary sub-interval $(u,u+h)$ of $(0,2\pi)$ we then have, taking differences in the obvious way,

$$\Delta_2 G(u,u)=\Delta_2 F(u,u)-\sum_k|\Delta\varphi_k(u)|^2=E|\Delta Z(u)|^2,$$

and thus

$$E|\Delta Z(u)|^2\leqq\Delta_2 F(u,u). \tag{17}$$

Let now

$$0=u_0<u_1<\cdots<u_m=2\pi,$$
$$0=v_0<v_1<\cdots<v_n=2\pi,$$

be two arbitrary sub-divisions of the interval $(0,2\pi)$. All differences occurring in the sequel will be understood to be related in the obvious way to the sub-intervals in these divisions.

From (16) we obtain

$$\Delta_2 G(u_r, v_s) = E(\Delta Z(u_r) \overline{\Delta Z(v_s)}),$$

and further by the Schwarz inequality, using (17),

$$\begin{aligned}|\Delta_2 G(u_r, v_s)| &\leq E\{|\Delta Z(u_r)| \cdot |\Delta Z(v_s)|\} \\ &\leq \sqrt{E|\Delta Z(u_r)|^2 \cdot E|\Delta Z(v_s)|^2} \\ &\leq \sqrt{\Delta_2 F(u_r, u_r) \cdot \Delta_2 F(v_s, v_s)}.\end{aligned}$$

Summing over all the sub-intervals in both variables, we thus obtain by means of the condition (B), assuming that all sub-intervals are sufficiently small,

$$\sum_{r,s} |\Delta_2 G(u_r, v_s)| \leq \sum_r \sqrt{\Delta_2 F(u_r, u_r)} \cdot \sum_s \sqrt{\Delta_2 F(v_s, v_s)} \leq 4\pi^2 M,$$

which shows that $G(u, v)$ is of bounded variation over C.

Comparing now the general representation (1) of a non-deterministic x_n process, and the development (6) of the spectral function $F(u, v)$ in the present case, it is easily verified that the term $\sum \varphi_k(u) \overline{\varphi_k(v)}$ in (6), where $\varphi_k(u)$ is determined by (10), is the spectral function of the sum $\sum c_{nr_k} \xi_{r_k}$ in (1), while the remaining term $G(u, v)$ is the spectral function of the deterministic component y_n of the x_n process. We have thus completed the proof that the given condition is necessary.

The proof that the condition is also sufficient is now very simple. We first observe that the predictionary properties (prediction being always understood in the sense of linear least squares prediction) of a stochastic process are entirely determined by the covariance function of the process. Thus if we are given an x_n process with the covariance function $R(m, n)$, and if we can show that $R(m, n)$ is the covariance function of *some* stochastic process having the given sequence of the r_k for its points of indetermination, it follows that the given x_n process will have precisely the same points of indetermination. For a harmonizable process, the covariance function is uniquely determined by the spectral function $F(u, v)$, so that the same remark applies here to $F(u, v)$.

Suppose now that we are given an r_k sequence, and an x_n process with a spectral function $F(u, v)$ satisfying all our conditions. In particular, $F(u, v)$ will then be given by the development (6), where $\varphi_k(u)$ is given by (7). According to the remark just made, we shall then only have to show that $F(u, v)$ is the spectral function of some stochastic process having the given r_k for its points of indetermination.

Consider a stochastic process x_n^* represented in the form (1), where we take

$$\begin{aligned} c_{nr_k} &= -2n\pi i \alpha_{nr_k} \quad \text{for } r_k \leq n,\ n \neq 0, \\ c_{0r_k} &= 2\pi \beta_k \qquad\qquad \text{for } r_k \leq 0, \end{aligned}$$

while the ξ_{r_k} are orthogonal random variables, and also orthogonal to the y_n, which are the variables of a deterministic process with the spectral function $G(u, v)$ appearing in (6). Then it will be immediately seen that the x_n^* process has the given r_k sequence for its points of indetermination, so that the proof is hereby completed.

We finally remark that the development of the difference $F(u,v)-G(u,v)$, which follows from (6), is formally analogous to the well-known development of the kernel $F(u,v)-G(u,v)$ in terms of its characteristic functions. However, it is easily shown by means of examples that the $\varphi_k(u)$ appearing in (6) are not necessarily identical with the characteristic functions of the corresponding kernel.

REFERENCES

CRAMÉR, H., 1. A contribution to the theory of stochastic processes. Proceedings of the Second Berkeley Symposium on Mathematical Statistics and Probability. Berkeley 1951.

2. Remarques sur le problème de prédiction pour certaines classes de processus stochastiques. To appear in the Publications de l'Institut de Statistique de l'Université de Paris.

DOOB, J. L., Stochastic processes. New York (Wiley) 1953.

LOÈVE, M., Fonctions aléatoires du second ordre. Supplément to P. LÉVY, Processus stochastiques et mouvement brownien. Paris. (Gauthier-Villars) 1948.

WIENER, N., Extrapolation, interpolation and smoothing of stationary time series. New York (Wiley) 1949.

WOLD, H., A study in the analysis of stationary time series. Thesis, Stockholm. Uppsala 1938.

Tryckt den 28 augusti 1959

Uppsala 1959. Almqvist & Wiksells Boktryckeri AB

50.

On some classes of nonstationary stochastic processes

Proc. Fourth Berkeley Symp. Math. Statist. Prob., Vol. II, 57–78
University of California Press, Berkeley and Los Angeles 1961
(Reprinted in Random Processes)

1. Introduction

This paper will be concerned with stochastic processes with finite second-order moments. We start from a given probability space $(\Omega, \mathfrak{F}, \mathfrak{P})$, where Ω is a space of points ω, while $\mathfrak{F}$ is a Borel field of sets in Ω, and $\mathfrak{P}$ is a probability measure defined on sets of $\mathfrak{F}$.

Any $\mathfrak{F}$-measurable complex-valued function $X = x(\omega)$ defined for all $\omega \in \Omega$ will be denoted as a *random variable*. We shall always assume that

$$(1) \qquad \begin{aligned} Ex &= \int_\Omega x(\omega)\, d\mathfrak{P} = 0, \\ E|x|^2 &= \int_\Omega |x(\omega)|^2\, d\mathfrak{P} < \infty. \end{aligned}$$

Two random variables which are equal except on a null set with respect to $\mathfrak{P}$ will be regarded as identical, and equations containing random variables are always to be understood in this sense.

A family of random variables $x(t) = x(t, \omega)$, defined for all t belonging to some given set T, will be called a *stochastic process* defined on T. With respect to T, we shall consider only two cases:

(i) T is the set of all integers $n = 0, \pm 1, \pm 2, \cdots$,

(ii) T is the set of all real numbers t.

With the usual terminology borrowed from the applications, we shall in these cases talk respectively of a stochastic process with *discrete time*, or with *continuous time*. In the first case, where we are concerned with a sequence of random variables, we shall usually write x_n in place of $x(n)$.

With due modifications, the majority of our considerations may be extended to cases where T is some other set of real numbers.

We shall also consider *finite-dimensional vector-valued stochastic processes*, writing

$$(2) \qquad \mathbf{x}(t) = \{x^{(1)}(t), x^{(2)}(t), \cdots, x^{(q)}(t)\},$$

where $\mathbf{x}(t)$ is a q-dimensional column vector, while the components $x^{(1)}(t), \cdots, x^{(q)}(t)$ are stochastic processes in the above sense.

57

The *covariance functions* of the $\mathbf{x}(t)$ process are

$$R_{jk}(t, u) = E\{x^{(j)}(t)\bar{x}^{(k)}(u)\} = \bar{R}_{kj}(u, t), \tag{3}$$

where $j, k = 1, 2, \cdots, q$. These are all finite, since we consider only variables with finite second-order moments. In the particular case $q = 1$, we are concerned with one single stochastic process $x(t)$, and there is only one covariance function

$$R(t, u) = E\{x(t)\bar{x}(u)\}. \tag{4}$$

2. Stationary stochastic processes

In the important particular case when all R_{jk} are functions of the difference $t - u$, so that we have

$$R_{jk}(t, u) = r_{jk}(t - u), \tag{5}$$

the $\mathbf{x}(t)$ process is known as a *stationary stochastic process*. We shall not here be concerned with the so-called *strictly stationary* processes, which satisfy more restrictive conditions.

The class of stationary processes possesses useful and interesting properties, which have been thoroughly studied. The present paper is the outcome of an attempt to generalize some of these properties to certain classes of nonstationary processes. In particular, various properties related to the problem of *linear least squares prediction* will be considered. For the sake of brevity, we shall in the sequel always use the word "prediction" in the sense of "linear least squares prediction."

In the first part of the paper, the general case of a vector-valued stochastic process with finite second-order moments will be studied. For such a process, there exists a uniquely defined decomposition into a *deterministic* and a *purely nondeterministic* component, which are mutually orthogonal.

In the case of a vector process with discrete time, the properties of this decomposition are a straightforward generalization of the well-known Wold decomposition [12] for stationary one-variable processes with discrete time. This case will be treated in detail in section 4 of the present paper.

For a process with continuous time, on the other hand, the properties of the nondeterministic component are somewhat more complicated than in the case of a stationary process. This case will be dealt with in section 5. We shall here give only the main lines of the argument, and state our main results. Complete proofs will be given in a forthcoming publication in the *Arkiv för Matematik*.

The decomposition into a deterministic and a purely nondeterministic component forms the basis of a *time-domain analysis* of a given stochastic process, generalizing the well-known properties of stationary processes.

For the class of stationary processes, "the chief advantage of turning from the time-domain analysis of stochastic processes to the *frequency-domain* or *spectral analysis* is the possibility of using the powerful methods of harmonic analysis" (Wiener and Masani [11], p. 140). This can be done for stationary processes, since there exist for these processes *spectral representations* in the form of Fourier-Stieltjes integrals, both for the process variables themselves and for the associated covariance functions.

In the second part of the present paper, we shall consider certain classes of nonstationary processes admitting spectral representations of a similar kind. Our results in this part of the paper are of a much less definite character than those given in the first part. In fact, only some highly preliminary results concerned with the spectral analysis of processes will be given here. Also we shall here consider only one-variable processes with discrete time, although most of our results may be generalized to the vector case, and also to processes with continuous time.

Two classes of stochastic processes admitting spectral representations will be considered, each including the stationary processes as a particular case.

For a stationary process with discrete time, it is well known that there exists a representation in the form of a stochastic Fourier-Stieltjes integral

$$x_n = \int_0^{2\pi} e^{inu}\, dz(u), \tag{6}$$

where $z(u)$ is a stochastic process with *orthogonal increments.* If we consider a process x_n representable in the same form, but without requiring that $z(u)$ should necessarily have orthogonal increments, we shall be led to a class of stochastic processes first introduced by Loève [7], [8], and called by him *harmonizable processes.* Obviously this class contains the class of stationary processes. The harmonizable processes will be considered in section 6 of the present paper.

Finally, in section 7 we shall consider a different kind of generalization of the concept of a stationary process. With respect to a stationary process with discrete time, it is well known that there exists a *unitary shift operator* U, which takes every x_n into the immediately following variable x_{n+1}. The properties of this operator are intimately connected with the properties of the stationary process. The more general class of processes obtained when it is only assumed that the shift operator is *normal* has been studied by Getoor [5]. In section 7 we shall give some very preliminary results concerning the spectral analysis of this class of processes, restricting ourselves to the case of processes with discrete time.

PART I. THE GENERAL VECTOR-VALUED PROCESS WITH FINITE SECOND-ORDER MOMENTS

3. Notation, deterministic and nondeterministic processes

All random variables $x, y, \cdots$ defined on the given probability space $(\Omega, \mathfrak{F}, \mathfrak{P})$, and satisfying (1), form a Hilbert space $\mathfrak{H}$, if the inner product and the norm are defined by the usual expressions

$$(x, y) = E(x\bar{y}), \qquad ||x||^2 = E|x|^2. \tag{7}$$

Whenever we use the term *convergence* with respect to a sequence of random variables, it will be understood that we refer to convergence in the topology induced by this norm, that is, convergence in quadratic mean.

We now consider a vector-valued stochastic process $\mathbf{x}(t) = \{x^{(1)}(t), \cdots, x^{(q)}(t)\}$, where the $x^{(j)}(t)$ are complex-valued stochastic processes, defined for all $t \in T$. Every random variable $x^{(j)}(t)$ is assumed to satisfy (1), and is thus an element of $\mathfrak{H}$. In order to avoid trivial difficulties we shall always suppose that, for every $j = 1, \cdots, q$,

$$E|x^{(j)}(t)|^2 > 0 \tag{8}$$

for at least one $t \in T$. We shall say that two processes $\mathbf{x}(t)$ and $\mathbf{y}(t)$ of this type are *orthogonal*, in symbols $\mathbf{x}(t) \perp \mathbf{y}(t)$, if

$$E\{x^{(j)}(t)\overline{y}^{(k)}(u)\} = 0, \tag{9}$$

for $j, k = 1, \cdots, q$ and all $t, u \in T$.

Let $\mathfrak{H}(\mathbf{x}, t)$ denote the subspace of $\mathfrak{H}$ spanned by the random variables $x^{(j)}(u)$ for $j = 1, \cdots, q$ and all u such that $u \in T$ and $u \leqq t$. We shall write this

$$\mathfrak{H}(\mathbf{x}, t) = \mathfrak{S}\{x^{(j)}(u), j = 1, \cdots, q, u \in T, u \leqq t\}, \tag{10}$$

where $\mathfrak{S}$ stands for "span." Instead of $\mathfrak{H}(\mathbf{x}, +\infty)$, we shall write simply $\mathfrak{H}(\mathbf{x})$.

As t decreases, the set $\mathfrak{H}(\mathbf{x}, t)$ can never increase. It follows that, when $t \to -\infty$, the set $\mathfrak{H}(\mathbf{x}, t)$ must tend to a limiting set, which we denote by $\mathfrak{H}(\mathbf{x}, -\infty)$. We thus have for any $t_1 < t_2$

$$\mathfrak{H}(\mathbf{x}, -\infty) \subset \mathfrak{H}(\mathbf{x}, t_1) \subset \mathfrak{H}(\mathbf{x}, t_2) \subset \mathfrak{H}(\mathbf{x}) \subset \mathfrak{H}. \tag{11}$$

It can be said that $\mathfrak{H}(\mathbf{x}, t)$ contains all the information available when we know the development of all the component processes $x^{(j)}(u)$ up to and including the point t. In the terminology used by Wiener and Masani ([11], p. 135), we can say that $\mathfrak{H}(\mathbf{x}, t)$ represents the *past and present* of $\mathbf{x}(t)$, while $\mathfrak{H}(\mathbf{x}, -\infty)$ corresponds to the *remote past* of the process.

If $\mathfrak{M}$ is any subspace of $\mathfrak{H}$, that is, any closed linear manifold in $\mathfrak{H}$, we denote by $P_{\mathfrak{M}}y$ the projection on $\mathfrak{M}$ of an arbitrary element y of $\mathfrak{H}$. When $\mathfrak{M} = \mathfrak{H}(\mathbf{x}, t)$, we write simply P_t instead of $P_{\mathfrak{M}}$.

Similarly, if $y^{(1)}, \cdots, y^{(q)}$ are any elements of $\mathfrak{H}$, and $\mathbf{y}$ denotes the column vector

$$\mathbf{y} = (y^{(1)}, \cdots, y^{(q)}), \tag{12}$$

we shall write

$$P_{\mathfrak{M}}\mathbf{y} = (P_{\mathfrak{M}}y^{(1)}, \cdots, P_{\mathfrak{M}}y^{(q)}), \tag{13}$$

replacing $P_{\mathfrak{M}}$ by P_t in the particular case when $\mathfrak{M} = \mathfrak{H}(\mathbf{x}, t)$.

The projection $P_{t-h}x^{(j)}(t)$ is, among all elements y of the subspace $\mathfrak{H}(\mathbf{x}, t - h)$, that which minimizes the norm $||x^{(j)}(t) - y||$. Accordingly $P_{t-h}x^{(j)}(t)$ is, from the point of view of linear least squares prediction, the best possible prediction of $x^{(j)}(t)$ in terms of all variables $x^{(1)}(u), \cdots, x^{(q)}(u)$ with $u \leqq t - h$. The norm

(14) $$\sigma_{th}^{(j)} = \|x^{(j)}(t) - P_{t-h}x^{(j)}(t)\|$$

is the corresponding *error of prediction*. For $0 < h < k$, we obviously have

(15) $$0 \leqq \sigma_{th}^{(j)} \leqq \sigma_{tk}^{(j)}.$$

Going back to relation (11), we shall now consider two extreme possibilities with respect to the subspace $\mathfrak{H}(\mathbf{x}, -\infty)$, namely

(16) $$\begin{aligned} &(A) \qquad \mathfrak{H}(\mathbf{x}, -\infty) = \mathfrak{H}(\mathbf{x}), \\ &(B) \qquad \mathfrak{H}(\mathbf{x}, -\infty) = 0. \end{aligned}$$

In case (A), it follows from (11) that $\mathfrak{H}(\mathbf{x}, t) = \mathfrak{H}(\mathbf{x}, -\infty)$ for all t. Thus, in particular, $x^{(j)}(t) \in \mathfrak{H}(\mathbf{x}, -\infty)$ for all j and t, and it follows that the prediction error $\sigma_{th}^{(j)}$ reduces to zero for $j = 1, \cdots, q$, for all $t \in T$ and all $h > 0$. Hence for every $t \in T$ the components $x^{(j)}(t)$ of $\mathbf{x}(t)$ all can be exactly predicted by means of the information provided by the arbitrarily remote past of the process. In this case we shall say that the $\mathbf{x}(t)$ process is *deterministic*. Every $\mathbf{x}(t)$ process not satisfying condition (A) will be called *nondeterministic*.

On the other hand, in case (B) we can say that the information provided by the remote past of the $\mathbf{x}(t)$ process is, in the limit, of no value for the prediction of the component variables $x^{(j)}(t)$ at any given time t. Thus every piece of information contained in the process at the instant t must have entered the process as an *innovation* at some definite instant $u \leqq t$ in the past or present. Accordingly, a process satisfying condition (B) will be called a *purely nondeterministic* process. (Wiener and Masani [11] use the term *regular* process.) For such a process, the irrelevance of the remote past for prediction purposes may be expressed by the relation

(17) $$\lim_{h \to \infty} \sigma_{th}^{(j)} = \|x^{(j)}(t)\|,$$

which holds for $j = 1, \cdots, q$ and every $t \in T$.

4. The discrete case

When T is the set of all integers $n = 0, \pm 1, \cdots$, we are concerned with a discrete vector process

(18) $$\mathbf{x}_n = (x_n^{(1)}, \cdots, x_n^{(q)}),$$

where it is assumed only that each component $x_n^{(j)}$ is a complex-valued stochastic process with discrete time parameter n, zero mean-values, and finite second-order moments. Using the notations introduced above, we have

(19) $$P_{n-1}\mathbf{x}_n = (P_{n-1}x_n^{(1)}, \cdots, P_{n-1}x_n^{(q)}).$$

Writing

(20) $$\xi_n = \mathbf{x}_n - P_{n-1}\mathbf{x}_n, \qquad \xi_n^{(j)} = x_n^{(j)} - P_{n-1}x_n^{(j)}$$

it follows that

(21) $$\xi_n = (\xi_n^{(1)}, \cdots, \xi_n^{(q)}).$$

The sequence of vector-valued random variables ξ_n defines a vector-valued stochastic process with discrete time parameter n. The ξ_n process will be called the *innovation process* corresponding to the given $\mathbf{x}_n$ process. This name may be justified by the following remarks.

If, for a certain value of n, we have

$$\xi_n = (0, \cdots, 0), \tag{22}$$

this signifies that every component $\xi_n^{(j)}$ is zero, that is, every $x_n^{(j)}$ is contained in the subspace $\mathfrak{H}(\mathbf{x}, n-1)$. From the prediction point of view this means that every component $x_n^{(j)}$ of $\mathbf{x}_n$ can be predicted exactly by means of the information available when we know the development of the $\mathbf{x}$ process up to and including the instant $n-1$. Obviously this can be expressed by saying that no innovation enters the $\mathbf{x}$ process at the instant n or, equivalently, that the innovation received by the process at this instant reduces to zero.

Suppose, on the other hand, that for a certain n, the vector variable ξ_n has at least one component $\xi_n^{(j)}$ such that $E|\xi_n^{(j)}|^2 > 0$. Then $\xi_n^{(j)} = x_n^{(j)} - P_{n-1}x_n^{(j)}$ does not reduce to zero, so that $x_n^{(j)}$ cannot be predicted exactly in terms of the variables $x_m^{(1)}, \cdots, x_m^{(q)}$ with $m \leqq n-1$. The variable $\xi_n^{(j)}$ then represents the innovation received by the component $x_n^{(j)}$ at the instant n, and consequently $\xi_n = (\xi_n^{(1)}, \cdots, \xi_n^{(q)})$, which is not identically zero, is the *innovation* entering into the vector process $\mathbf{x}$ at the instant n.

The set of all those values of n, for which the innovation ξ_n does not reduce to zero, may be said to form the *innovation spectrum* of the $\mathbf{x}_n$ process. This set contains precisely all those time points where a new impulse, or an innovation, enters into the process.

The innovation spectrum may be empty, finite, or infinite; and it will be readily seen that, to any given set of integers n, we can construct an $\mathbf{x}_n$ process having this set for its innovation spectrum. For a deterministic process, the innovation spectrum is evidently empty, while for any nondeterministic process it must contain at least one value of n. For a nondeterministic stationary process, the innovation spectrum includes all integers n.

Let now n be a given integer, and consider the set of q random variables $\xi_n^{(1)}, \cdots, \xi_n^{(q)}$, with covariance matrix

$$\mathbf{R}_n = \{E(\xi_n^{(j)}\bar{\xi}_n^{(k)})\}, \qquad j, k = 1, \cdots, q. \tag{23}$$

The rank r_n of $\mathbf{R}_n$ is equal to the maximum number of linearly independent variables among $\xi_n^{(1)}, \cdots, \xi_n^{(q)}$. Thus $0 \leqq r_n \leqq q$, and $r_n > 0$ if, and only if, n belongs to the innovation spectrum of the $\mathbf{x}$ process.

Let $\mathfrak{k}(\mathbf{x}, n)$ denote the r_n-dimensional space spanned by the variables $\xi_n^{(1)}, \cdots, \xi_n^{(q)}$. Since $\xi_n^{(j)} \in \mathfrak{H}(\mathbf{x}, n)$, it is seen that $\mathfrak{k}(\mathbf{x}, n)$ is a subspace of $\mathfrak{H}(\mathbf{x}, n)$.

It follows from (20) that $\xi_n^{(j)}$ is always orthogonal to $\mathfrak{H}(\mathbf{x}, n-1)$, and consequently $\mathfrak{k}(\mathbf{x}, n) \perp \mathfrak{H}(\mathbf{x}, n-1)$. It also follows from (20) that any two variables $\xi_m^{(j)}$ and $\xi_n^{(k)}$ with $m \neq n$ are orthogonal, so that $\mathfrak{k}(\mathbf{x}, m) \perp \mathfrak{k}(\mathbf{x}, n)$ when $m \neq n$.

If we orthogonalize the set of variables $\xi_n^{(1)}, \cdots, \xi_n^{(q)}$, we shall obtain a set of

r_n variables, say $\eta_n^{(1)}, \cdots, \eta_n^{(r_n)}$, forming a complete orthonormal system in $\mathfrak{k}(\mathbf{x}, n)$, and in addition $q - r_n$ zero variables. If n does not belong to the innovation spectrum, $r_n = 0$, and the space $\mathfrak{k}(\mathbf{x}, n)$ reduces to zero.

The vector sum of the orthogonal family of subspaces $\mathfrak{k}(\mathbf{x}, m)$ with $m \leqq n$,

$$\mathfrak{K}(\mathbf{x}, n) = \mathfrak{k}(\mathbf{x}, n) + \mathfrak{k}(\mathbf{x}, n - 1) + \cdots \tag{24}$$

is the space spanned by all $\xi_m^{(j)}$ with $j = 1, \cdots, q$ and $m \leqq n$. Obviously this is a subspace of $\mathfrak{H}(\mathbf{x}, n)$. It will be seen that the set of all variables $\eta_m^{(j)}$, where $j = 1, \cdots, r_m$ and $m \leqq n$, forms a complete orthonormal system in $\mathfrak{K}(\mathbf{x}, n)$. This remark will be used later.

We now proceed to the proof of the following lemma which, in the stationary case, corresponds to part (b) of lemma 6.10 of Wiener and Masani [11].

LEMMA 1. *The space $\mathfrak{K}(\mathbf{x}, n)$ is, within $\mathfrak{H}(\mathbf{x}, n)$, the orthogonal complement of $\mathfrak{H}(\mathbf{x}, -\infty)$. In symbols*

$$\begin{aligned} &\mathfrak{K}(\mathbf{x}, n) \perp \mathfrak{H}(\mathbf{x}, -\infty), \\ &\mathfrak{H}(\mathbf{x}, n) = \mathfrak{K}(\mathbf{x}, n) + \mathfrak{H}(\mathbf{x}, -\infty). \end{aligned} \tag{25}$$

We have already seen that $\mathfrak{k}(\mathbf{x}, m)$ is orthogonal to $\mathfrak{H}(\mathbf{x}, m - 1)$, and a fortiori orthogonal to $\mathfrak{H}(\mathbf{x}, -\infty)$. Consequently the vector sum $\mathfrak{K}(\mathbf{x}, n)$ is orthogonal to $\mathfrak{H}(\mathbf{x}, -\infty)$.

Further, since $\mathfrak{K}(\mathbf{x}, n)$ and $\mathfrak{H}(\mathbf{x}, -\infty)$ are both subspaces of $\mathfrak{H}(\mathbf{x}, n)$, we have

$$\mathfrak{K}(\mathbf{x}, n) + \mathfrak{H}(\mathbf{x}, -\infty) \subset \mathfrak{H}(\mathbf{x}, n). \tag{26}$$

On the other hand we have $x_n^{(j)} = \xi_n^{(j)} + P_{n-1}x_n^{(j)}$, so that every element of $\mathfrak{H}(\mathbf{x}, n)$ is the limit of a convergent sequence of variables, each of which is the sum of a linear combination of $\xi_n^{(1)}, \cdots, \xi_n^{(q)}$ and an element of $\mathfrak{H}(\mathbf{x}, n - 1)$. Since $\mathfrak{k}(\mathbf{x}, n)$ and $\mathfrak{H}(\mathbf{x}, n - 1)$ are orthogonal, it follows that every element of $\mathfrak{H}(\mathbf{x}, n)$ is the sum of one element of $\mathfrak{k}(\mathbf{x}, n)$ and one of $\mathfrak{H}(\mathbf{x}, n - 1)$, so that

$$\mathfrak{H}(\mathbf{x}, n) \subset \mathfrak{k}(\mathbf{x}, n) + \mathfrak{H}(\mathbf{x}, n - 1) \subset \mathfrak{K}(\mathbf{x}, n) + \mathfrak{H}(\mathbf{x}, n - 1). \tag{27}$$

By repeated application of this relation we obtain

$$\mathfrak{H}(\mathbf{x}, n) \subset \mathfrak{K}(\mathbf{x}, n) + \mathfrak{H}(\mathbf{x}, n - p) \tag{28}$$

for every $p > 0$, and finally, as $p \to \infty$,

$$\mathfrak{H}(\mathbf{x}, n) \subset \mathfrak{K}(\mathbf{x}, n) + \mathfrak{H}(\mathbf{x}, -\infty). \tag{29}$$

This, together with (26), completes the proof of the lemma.

We can now prove the analogue of the Wold decomposition for the $\mathbf{x}_n$ process, thus generalizing theorem 6.11 of Wiener and Masani [11].

THEOREM 1. *For any given $\mathbf{x}_n$ process, there is a uniquely determined decomposition*

$$\mathbf{x}_n = \mathbf{u}_n + \mathbf{v}_n \tag{30}$$

having properties (a) *and* (b).

(a) $\mathbf{u}_n = (u_n^{(1)}, \cdots, u_n^{(q)})$ and $\mathbf{v}_n = (v_n^{(1)}, \cdots, v_n^{(q)})$, *where all $u_n^{(j)}$ and $v_n^{(j)}$ belong to $\mathfrak{H}(\mathbf{x}, n)$.*

(b) *The* $\mathbf{u}_n$ *and* $\mathbf{v}_n$ *processes are orthogonal, and* $\mathbf{u}_n$ *is purely nondeterministic, while* $\mathbf{v}_n$ *is deterministic. The nondeterministic component* $\mathbf{u}_n$ *has, in addition, property* (c).

(c) $\mathbf{u}_n$ *can be expressed as a linear combination of those innovations* $\boldsymbol{\xi}_p$ *of the* $\mathbf{x}_n$ *process that have entered into the process before or at the instant* n,

$$\mathbf{u}_n = \sum_{p=-\infty}^{n} \mathbf{A}_{np}\boldsymbol{\xi}_p, \tag{31}$$

where the $\mathbf{A}_{np} = \{a_{np}^{(jk)}\}$ *are* $q \times q$ *matrices, such that the development formally obtained for any component* $u_n^{(j)}$,

$$u_n^{(j)} = \sum_{p=-\infty}^{n} \sum_{k=1}^{q} a_{np}^{(jk)}\xi_p^{(k)}, \tag{32}$$

is convergent in the topology of $\mathfrak{H}$. *Thus, writing*

$$c_{np}^{(j)} = \left\| \sum_{k=1}^{q} a_{np}^{(jk)}\xi_p^{(k)} \right\| = \left\{ E\left| \sum_{k=1}^{q} a_{np}^{(jk)}\xi_p^{(k)} \right|^2 \right\}^{1/2}, \tag{33}$$

we have

$$\sum_{p=-\infty}^{n} (c_{np}^{(j)})^2 < \infty \tag{34}$$

for all n *and for* $j = 1, \cdots, q$. *The coefficients* $a_{np}^{(jk)}$ *are uniquely determined if, and only if, the rank* r_p *has the maximum value* q, *while the* $c_{np}^{(j)}$ *are uniquely determined for all* n, p, *and* j.

Proof. For all n and for $j = 1, \cdots, q$ we take for $u_n^{(j)}$ and $v_n^{(j)}$ the projections of $x_n^{(j)}$ on the subspaces $\mathfrak{K}(\mathbf{x}, n)$ and $\mathfrak{H}(\mathbf{x}, -\infty)$ respectively. It then follows from lemma 1 that $u_n^{(j)}$ and $v_n^{(j)}$ belong to $\mathfrak{H}(\mathbf{x}, n)$, and that we have

$$\begin{aligned} x_n^{(j)} &= u_n^{(j)} + v_n^{(j)}, \\ \mathbf{x}_n &= \mathbf{u}_n + \mathbf{v}_n, \end{aligned} \tag{35}$$

where

$$\mathbf{u}_n = (u_n^{(1)}, \cdots, u_n^{(q)}), \qquad \mathbf{v}_n = (v_n^{(1)}, \cdots, v_n^{(q)}). \tag{36}$$

Since $u_m^{(j)}$ belongs to $\mathfrak{K}(\mathbf{x}, m)$, while $v_n^{(k)}$ belongs to $\mathfrak{H}(\mathbf{x}, -\infty)$, it further follows from lemma 1 that $u_m^{(j)}$ and $v_n^{(k)}$ are always orthogonal. Thus the $\mathbf{u}_n$ and $\mathbf{v}_n$ processes are orthogonal, according to the definition given in section 3. If, in accordance with section 3, we define

$$\begin{aligned} \mathfrak{H}(\mathbf{u}, n) &= \mathfrak{S}(u_m^{(j)}, j = 1, \cdots, q, m \leqq n), \\ \mathfrak{H}(\mathbf{v}, n) &= \mathfrak{S}(v_m^{(j)}, j = 1, \cdots, q, m \leqq n), \end{aligned} \tag{37}$$

we thus find that $\mathfrak{H}(\mathbf{u}, m)$ and $\mathfrak{H}(\mathbf{v}, n)$ are orthogonal for all m and n.

Since all $u_m^{(j)}$ and $v_m^{(j)}$ with $m \leqq n$ belong to $\mathfrak{H}(\mathbf{x}, n)$, we have

$$\mathfrak{H}(\mathbf{u}, n) + \mathfrak{H}(\mathbf{v}, n) \subset \mathfrak{H}(\mathbf{x}, n). \tag{38}$$

On the other hand, it follows from (35) and from the orthogonality of the $\mathbf{u}_n$ and $\mathbf{v}_n$ processes that

$$\mathfrak{H}(\mathbf{x}, n) \subset \mathfrak{H}(\mathbf{u}, n) + \mathfrak{H}(\mathbf{v}, n). \tag{39}$$

Hence by lemma 1, $\mathfrak{H}(\mathbf{u}, n) + \mathfrak{H}(\mathbf{v}, n) = \mathfrak{H}(\mathbf{x}, n) = \mathfrak{K}(\mathbf{x}, n) + \mathfrak{H}(\mathbf{x}, -\infty)$. By the definition of the $u_n^{(j)}$ and $v_n^{(j)}$ we have, however, $\mathfrak{H}(\mathbf{u}, n) \subset \mathfrak{K}(\mathbf{x}, n)$ and $\mathfrak{H}(\mathbf{v}, n) \subset \mathfrak{H}(\mathbf{x}, -\infty)$, and thus obtain

$$\begin{aligned} \mathfrak{H}(\mathbf{u}, n) &= \mathfrak{K}(\mathbf{x}, n), \\ \mathfrak{H}(\mathbf{v}, n) &= \mathfrak{H}(\mathbf{x}, -\infty). \end{aligned} \tag{40}$$

From lemma 1 we then obtain $\mathfrak{H}(\mathbf{u}, -\infty) = \mathfrak{K}(\mathbf{x}, -\infty) = 0$, so that the $\mathbf{u}_n$ process is purely nondeterministic. On the other hand $\mathfrak{H}(\mathbf{v}, -\infty) = \mathfrak{H}(\mathbf{x}, -\infty) = \mathfrak{H}(\mathbf{v}, n)$ for every n, and so the $\mathbf{v}_n$ process is deterministic.

The properties (a) and (b) of the decomposition considered here are thus established, and we shall now prove that this is the only decomposition of the given $\mathbf{x}_n$ process that has these properties. Suppose, in fact, that $\mathbf{u}_n$ and $\mathbf{v}_n$ are any processes satisfying (35), and having the properties (a) and (b) stated in the theorem. Then (38) and (39) will still hold, so that we obtain as before

$$\mathfrak{H}(\mathbf{x}, n) = \mathfrak{H}(\mathbf{u}, n) + \mathfrak{H}(\mathbf{v}, n) \tag{41}$$

for all n. It is readily seen that, owing to the orthogonality of $\mathfrak{H}(\mathbf{u}, n)$ and $\mathfrak{H}(\mathbf{v}, n)$, this holds even for $n = -\infty$, and we obtain

$$\mathfrak{H}(\mathbf{x}, -\infty) = \mathfrak{H}(\mathbf{u}, -\infty) + \mathfrak{H}(\mathbf{v}, -\infty). \tag{42}$$

However, on account of property (b), we have $\mathfrak{H}(\mathbf{u}, -\infty) = 0$, and thus

$$\mathfrak{H}(\mathbf{x}, -\infty) = \mathfrak{H}(\mathbf{v}, -\infty) = \mathfrak{H}(\mathbf{v}, n) \tag{43}$$

for all n. Hence by (41) and lemma 1 we find that relations (40) will still hold. In the decomposition $x_n^{(j)} = u_n^{(j)} + v_n^{(j)}$, the first component must then be the projection of $x_n^{(j)}$ on $\mathfrak{K}(\mathbf{x}, n)$, and the second the projection on $\mathfrak{H}(\mathbf{x}, -\infty)$, so that the decomposition is unique.

It finally remains to prove property (c). In order to do this, we use the remark made above about the completeness of the orthonormal system $\eta_m^{(j)}$ in the space $\mathfrak{K}(\mathbf{x}, n)$. The corresponding Fourier development of the element $u_n^{(j)} \in \mathfrak{K}(\mathbf{x}, n)$ will have the form

$$u_n^{(j)} = \sum_{p=-\infty}^{n} \sum_{k=1}^{r_p} b_{np}^{(jk)} \eta_p^{(k)} \tag{44}$$

with

$$\sum_{p=-\infty}^{n} \sum_{k=1}^{r_p} |b_{np}^{(jk)}|^2 < \infty. \tag{45}$$

For any fixed p, the orthogonal variables $\eta_p^{(1)}, \cdots, \eta_p^{(r_p)}$ are certain linear combinations of the innovation components $\xi_p^{(1)}, \cdots, \xi_p^{(q)}$. The coefficients appearing in these linear combinations will be uniquely determined if, and only if, the rank r_p has its maximum value q. Replacing the $\eta_p^{(k)}$ in the above development of $u_n^{(j)}$ by their expressions in terms of the $\xi_p^{(k)}$, we obtain the development given under (c), and we find that

$$(c_{np}^{(j)})^2 = \sum_{k=1}^{r_p} |b_{np}^{(jk)}|^2. \tag{46}$$

The proof of theorem 1 is thus completed.

We can now immediately state the following result, which gives the application of theorem 1 to the prediction problem for the $\mathbf{x}_n$ process.

THEOREM 2. *Let h be any positive integer. With the notation of theorem* 1, *the best prediction of the component $x_n^{(j)}$ in terms of all variables $x_p^{(1)}, \cdots, x_p^{(q)}$ with $p \leqq n - h$ will be*

$$P_{n-h}x_n^{(j)} = \sum_{p=-\infty}^{n-h} \sum_{k=1}^{q} a_{np}^{(jk)} \xi_p^{(k)} + v_n^{(j)}, \tag{47}$$

with the corresponding error of prediction

$$\sigma_{nh}^{(j)} = ||x_n^{(j)} - P_{n-h}x_n^{(j)}|| = \left\{ \sum_{p=n-h+1}^{n} (c_{np}^{(j)})^2 \right\}^{1/2}. \tag{48}$$

This follows directly from theorem 1 and from the definition (14) of the error of prediction, if we observe that $v_n^{(j)}$, as well as all $\xi_p^{(k)}$ with $p \leqq n - h$, belong to $\mathfrak{H}(\mathbf{x}, n - h)$, while all $\xi_p^{(k)}$ with $p > n - h$ are orthogonal to this space.

It should be noted that the coefficients $a_{np}^{(jk)}$ in the expression for the best prediction of $\mathbf{x}_n^{(j)}$ given in theorem 2 depend on certain covariances of the $\mathbf{x}$ process up to the time n. Accordingly, any statistical estimation of this prediction by means of theorem 2 must be based on information concerning the covariance structure of the process up to the time n, either from a priori knowledge (as in the case when the process is assumed to be stationary), or from previous statistical experience.

5. The continuous case

REMARK. I am indebted to Professor K. Itô for the observation that there are interesting points of contact of this section and a work by T. Hida on "Canonical representations of Gaussian processes," which will shortly appear in the *Memoirs of the College of Science, University of Kyoto*.

We now consider a q-dimensional stochastic vector process

$$\mathbf{x}(t) = \{x^{(1)}(t), \cdots, x^{(q)}(t)\}, \tag{49}$$

the parameter set T being the set of all real numbers t. Each component $x^{(j)}(t)$ is a complex-valued stochastic process with continuous time t, and the covariance functions $R_{jk}(t, u)$ defined by (3) are all assumed to be finite.

When t increases from $-\infty$ to $+\infty$, the point $x^{(j)}(t)$ describes a curve in the Hilbert space $\mathfrak{H}(\mathbf{x})$, and the $\mathbf{x}(t)$ process is made up by the set of q curves corresponding to the components $x^{(1)}(t), \cdots, x^{(q)}(t)$. The subspace $\mathfrak{H}(\mathbf{x}, t_0)$ is spanned by the arcs of these curves that belong to the domain $t \leqq t_0$. The properties of the family of all subspaces $\mathfrak{H}(\mathbf{x}, t)$, where t ranges from $-\infty$ to $+\infty$, will play an important part in the sequel.

The following theorem is directly analogous to the first part of theorem 1, and can be proved along similar lines, so that we may content ourselves here with stating the theorem.

THEOREM 3. *There is a unique decomposition of the* $\mathbf{x}(t)$ *process,*

$$\mathbf{x}(t) = \mathbf{u}(t) + \mathbf{v}(t), \tag{50}$$

having properties (a) *and* (b).

(a) $\mathbf{u}(t) = \{u^{(1)}(t), \cdots, u^{(q)}(t)\}$ *and* $\mathbf{v}(t) = \{v^{(1)}(t), \cdots, v^{(q)}(t)\}$, *where all* $u^{(j)}(t)$ *and* $v^{(j)}(t)$ *belong to* $\mathfrak{H}(\mathbf{x}, t)$.

(b) *The* $\mathbf{u}(t)$ *and* $\mathbf{v}(t)$ *processes are orthogonal, and* $\mathbf{u}(t)$ *is purely nondeterministic, while* $\mathbf{v}(t)$ *is deterministic.*

The second part of theorem 1 is concerned with the representation of the nondeterministic component of a given process with *discrete* time as a linear function of the innovations associated with the past and present of the process. For the nondeterministic component of a process with *continuous* time there exists, in fact, an analogous representation. However, the circumstances are somewhat more complicated than in the discrete case, and we shall here only give some preliminary discussion and state our main results, reserving complete proofs for a forthcoming publication.

For our present purpose, it will be sufficient to deal with the purely nondeterministic component $\mathbf{u}(t)$ of the given $\mathbf{x}(t)$ process, and we may then as well assume that $\mathbf{x}(t)$ itself is purely nondeterministic, that is, the deterministic component $\mathbf{v}(t)$ is identically zero. Further, we shall find it convenient to introduce a certain regularity condition relating to the behavior of the $\mathbf{x}(t)$ process in points of discontinuity. Thus it will be assumed throughout the rest of the present section that we are dealing with a vector process $\mathbf{x}(t)$ satisfying the following two conditions.

(C_1) $\mathbf{x}(t)$ is purely nondeterministic, that is, $\mathfrak{H}(\mathbf{x}, -\infty) = 0$.

(C_2) The limits $x^{(j)}(t-0)$ and $x^{(j)}(t+0)$ exist (as always in the $\mathfrak{H}$ topology) for $j = 1, \cdots, q$ and for every real t.

We shall then write

$$\mathbf{x}(t-0) = \{x^{(1)}(t-0), \cdots, x^{(q)}(t-0)\}, \tag{51}$$

and similarly for $\mathbf{x}(t+0)$.

It follows without difficulty from condition (C_2) that *the space* $\mathfrak{H}(\mathbf{x})$ *is separable*, and that the set of *points of discontinuity of* $\mathbf{x}(t)$, that is, the set of all t such that at least one of the relations

$$\mathbf{x}(t-0) = \mathbf{x}(t) = \mathbf{x}(t+0) \tag{52}$$

is not satisfied, is at most enumerable.

Let us now consider the family of subspaces $\mathfrak{H}(\mathbf{x}, t)$ of the space $\mathfrak{H}(\mathbf{x})$. As t increases from $-\infty$ to $+\infty$, the $\mathfrak{H}(\mathbf{x}, t)$ form a never decreasing set of subspaces, with $\mathfrak{H}(\mathbf{x}, -\infty) = 0$ and $\mathfrak{H}(\mathbf{x}, +\infty) = \mathfrak{H}(\mathbf{x})$. The limits $\mathfrak{H}(\mathbf{x}, t \pm 0)$ will exist for all t. If $(t, t + h)$ and $(u, u + k)$ are disjoint intervals, the orthogonal complements

$$\mathfrak{H}(\mathbf{x}, t + h) - \mathfrak{H}(\mathbf{x}, t) \quad \text{and} \quad \mathfrak{H}(\mathbf{x}, u + k) - \mathfrak{H}(\mathbf{x}, u) \tag{53}$$

are mutually orthogonal.

The set of all t such that for any $h > 0$ we have

$$\mathfrak{H}(\mathbf{x}, t + h) - \mathfrak{H}(\mathbf{x}, t - h) \neq 0 \tag{54}$$

will be called the *innovation spectrum* of the $\mathbf{x}(t)$ process. A point t such that at least one of the relations

$$\mathfrak{H}(\mathbf{x}, t - 0) = \mathfrak{H}(\mathbf{x}, t) = \mathfrak{H}(\mathbf{x}, t + 0) \tag{55}$$

is not satisfied, is a *point of discontinuity* of the innovation spectrum. The space $\mathfrak{H}(\mathbf{x})$ being separable, it follows immediately that the set of all discontinuity points is at most enumerable.

A discontinuity point of the innovation spectrum will not necessarily be a discontinuity point of the process, nor conversely. We shall make some remarks concerning the relations between these two kinds of discontinuities.

Let us first consider the case of a left discontinuity of the innovation spectrum, that is, a point t such that

$$\mathfrak{M}(t) = \mathfrak{H}(\mathbf{x}, t) - \mathfrak{H}(\mathbf{x}, t - 0) \neq 0. \tag{56}$$

Then it is easily shown that

$$\mathbf{x}(t) - \mathbf{x}(t - 0) \neq 0, \tag{57}$$

so that t is also a left discontinuity point of the process. Further, if $y^{(j)}$ denotes the projection of $x^{(j)}(t) - x^{(j)}(t - 0)$ on $\mathfrak{M}(t)$, we have $y^{(j)} \neq 0$ for at least one j, and the subspace $\mathfrak{M}(t)$ is spanned by the variables $y^{(1)}, \cdots, y^{(q)}$, and has thus at most q dimensions.

Thus in particular (56) implies (57). The converse statement is however not true: a left discontinuity of the process may, in fact, be a continuity point of the innovation spectrum.

Proceeding now to the case of a right discontinuity, it can be shown that neither of the two relations

$$\mathfrak{N}(t) = \mathfrak{H}(\mathbf{x}, t + 0) - \mathfrak{H}(\mathbf{x}, t) \neq 0 \tag{58}$$

and

$$\mathbf{x}(t + 0) - \mathbf{x}(t) \neq 0 \tag{59}$$

implies the other. In fact, it can be shown by examples that (58) may be satisfied even in a continuity point of the process, while on the other hand (59) may be satisfied even in a continuity point of the innovation spectrum.

The only implication that exists between the relations (56), (57), (58), and (59) is thus that (56) implies (57).

As in section 3, we now denote by $P_t z$ the projection of any point $z \in \mathfrak{H}(\mathbf{x})$ on the subspace $\mathfrak{H}(\mathbf{x}, t)$. When t increases from $-\infty$ to $+\infty$, the P_t form a never decreasing set of projections, with $P_{-\infty} = 0$ and $P_{+\infty} = I$. For $h > 0$, the difference $P_{t+h} - P_t$ is the projection on $\mathfrak{H}(\mathbf{x}, t+h) - \mathfrak{H}(\mathbf{x}, t)$. The limits $P_{t\pm 0}$ exist for every t, and are the projections on $\mathfrak{H}(\mathbf{x}, t \pm 0)$ respectively.

For an arbitrary random variable z in $\mathfrak{H}(\mathbf{x})$, we now define a stochastic process by writing for all real t

$$z(t) = P_t z. \tag{60}$$

It then follows from the above that $z(t)$ defines a complex-valued stochastic process with *orthogonal increments*, such that

$$\begin{aligned} z(-\infty) &= 0, & z(+\infty) &= z, \\ Ez(t) &= 0, & E|z(t)|^2 &= F(t, z), \end{aligned} \tag{61}$$

where $F(t, z)$ is, for any fixed z, a real, never decreasing and bounded function of t, such that

$$F(-\infty, z) = 0, \qquad F(+\infty, z) = E|z|^2. \tag{62}$$

The points of increase of $z(t)$, that is, the points t such that for any $h > 0$

$$E|z(t+h) - z(t-h)|^2 = F(t+h, z) - F(t-h, z) > 0, \tag{63}$$

form a subset of the innovation spectrum of $\mathbf{x}(t)$. Similarly, the left (right) discontinuities of $z(t)$ form a subset of the left (right) discontinuities of the innovation spectrum. Any increment $z(t+h) - z(t)$ belongs to the subspace $\mathfrak{H}(\mathbf{x}, t+h) - \mathfrak{H}(\mathbf{x}, t)$, and may thus be regarded as a part of the innovation received by the $\mathbf{x}(t)$ process during the interval $(t, t+h)$.

We now denote by $\mathfrak{L}(z)$ the subspace of $\mathfrak{H}(\mathbf{x})$ spanned by all the variables $z(u)$ for $-\infty < u < +\infty$, and by $\mathfrak{L}^*(z)$ the set of all random variables y representable in the form

$$y = \int_{-\infty}^{\infty} g(u)\, dz(u) \tag{64}$$

with

$$E|y|^2 = \int_{-\infty}^{\infty} |g(u)|^2\, dF(u, z) < \infty. \tag{65}$$

If no u is at the same time a left and a right discontinuity of $z(u)$, then $\mathfrak{L}(z)$ and $\mathfrak{L}^*(z)$ are identical (see Doob [4], pp. 425–429). The variable y given by (64 will belong to $\mathfrak{H}(\mathbf{x}, t)$ if, and only if, we have $g(u) = 0$ for almost all $u > t$, "almost all" referring to the $F(u, z)$-measure on the u-axis.

By means of the theory of spectral multiplicity in Hilbert space (see, for example, Stone [10], chapter VII, and Halmos [6]), we can now show that with any $\mathbf{x}(t)$ process satisfying (C_1) and (C_2) it is possible to associate a number N, which may be a finite positive integer or equal to $+\infty$, such that we can find N random variables $z_1, \cdots, z_N$ belonging to $\mathfrak{H}(\mathbf{x})$, with the properties

(a) $\mathfrak{L}(z_n) = \mathfrak{L}^*(z_n), \qquad n = 1, \cdots, N.$

(b) $\mathfrak{L}(z_m) \perp \mathfrak{L}(z_n), \qquad m \neq n.$

(c) $$\mathfrak{H}(\mathbf{x}) = \mathfrak{L}(z_1) + \cdots + \mathfrak{L}(z_N).$$

(d) N is the smallest number having the properties (a), (b), and (c).

In particular, since any component variable $x^{(j)}(t)$ of $\mathbf{x}(t)$ evidently belongs to $\mathfrak{H}(\mathbf{x}, t)$, we have the expression

$$x^{(j)}(t) = \sum_{k=1}^{N} \int_{-\infty}^{t} g_k^{(j)}(t, u)\, dz_k(u) \tag{66}$$

for $j = 1, \cdots, q$. If $N = \infty$, the series appearing here will converge in quadratic mean, so that

$$\sum_{k=1}^{N} \int_{-\infty}^{t} |g_k^{(j)}(t, u)|^2\, dF(u, z_k) < \infty. \tag{67}$$

If now we define a column vector

$$\mathbf{z}(u) = \{z_1(u), \cdots, z_N(u)\} \tag{68}$$

and a $q \times N$ matrix

$$\mathrm{G}(t, u) = \{g_k^{(j)}(t, u)\}, \qquad j = 1, \cdots, q;\ k = 1, \cdots, N, \tag{69}$$

we finally obtain the required expression for the vector variable $\mathbf{x}(t)$ in terms of past and present innovations of the process, as stated in the following theorem.

THEOREM 4. *The vector variable* $\mathbf{x}(t)$ *of any stochastic process satisfying conditions* (C_1) *and* (C_2) *can be expressed in the form*

$$\mathbf{x}(t) = \int_{-\infty}^{t} \mathrm{G}(t, u)\, d\mathbf{z}(u), \tag{70}$$

where $\mathbf{z}(u)$ *is an N-dimensional vector process with orthogonal increments, while* $\mathrm{G}(t, u)$ *is a* $q \times N$ *matrix, in accordance with* (68) *and* (69). *The development* (66) *formally obtained for the component* $x^{(j)}(t)$ *is then convergent as shown by* (67).

It will be seen that this is directly analogous to the last part of theorem 1, except that certain sums have been replaced by integrals, and that the q-dimensional random vector $\boldsymbol{\xi}_p$ has been replaced by the N-dimensional vector $\mathbf{z}(u)$. It is the fact that the multiplicity N may have any integral value from 1 to ∞ that introduces additional complication into the continuous case. It is possible to construct examples corresponding to any given value of N, even when it is required that $\mathbf{x}(t)$ be everywhere continuous (or even differentiable in quadratic mean). We finally state the following theorem, which is the continuous analogy of theorem 2.

THEOREM 5. *Let* $h > 0$ *be given. For any* $\mathbf{x}(t)$ *process satisfying conditions* (C_1) *and* (C_2), *the best prediction of the component* $x^{(j)}(t)$ *in terms of all variables* $x^{(1)}(u), \cdots, x^{(q)}(u)$ *with* $u \leq t - h$ *will be*

$$P_{t-h}x^{(j)}(t) = \sum_{k=1}^{N} \int_{-\infty}^{t-h} g_k^{(j)}(t, u)\, dz_k(u), \tag{71}$$

with the corresponding error of prediction

$$\sigma_{th}^{(j)} = \|x^{(j)}(t) - P_{t-h}x^{(j)}(t)\| \tag{72}$$

$$= \left\{\sum_{k=1}^{N} \int_{t-h}^{t} |g_k^{(j)}(t, u)|^2 \, dF(u, z_k)\right\}^{1/2}$$

PART II. ON TWO CLASSES OF PROCESSES ADMITTING SPECTRAL REPRESENTATIONS

6. Harmonizable processes

We shall now consider a one-dimensional process with discrete time, such that x_n is given by a stochastic Fourier-Stieltjes integral

$$x_n = \int_0^{2\pi} e^{inu} \, dz(u), \tag{73}$$

where $z(u)$ denotes, for $0 \leqq u \leqq 2\pi$, a complex-valued random variable satisfying the conditions

$$Ez(u) = 0, \qquad E\{z(u)\bar{z}(v)\} = F(u, v). \tag{74}$$

It will be assumed that the complex-valued covariance function $F(u, v)$ is of bounded variation over the square C defined by $0 \leqq u, v \leqq 2\pi$, in the sense that, for every subdivision of C in a finite number of rectangles, we have

$$\sum |\Delta_2 F| < K, \tag{75}$$

the sum being extended over all the rectangles, and the constant K being independent of the subdivision.

The integral (73) will then exist as a limit in quadratic mean of certain Riemann sums. Processes of this type have been introduced by Loève [7], [8], and have been called by him *harmonizable processes.* The covariance function corresponding to the process defined by (73) is

$$R(m, n) = E(x_m \bar{x}_n) = \int_0^{2\pi} \int_0^{2\pi} e^{i(mu-nv)} \, dF(u, v). \tag{76}$$

Conversely, if the covariance function of a certain x_n process is given by (76), where $F(u, v)$ is a covariance function satisfying (75) it is known (Loève [8], Cramér [1]) that there exists a process $z(u)$ satisfying (74), and such that x_n is given by the integral (73).

Without changing the value of the integral (73), we can always suppose that $z(u)$ is everywhere continuous to the right in quadratic mean, so that $z(u + 0) = z(u)$. The function $F(u, v)$ then defines a *complex mass distribution* over C, such that the mass carried by any rectangle $h < u \leqq h + \Delta h$, $k < v \leqq k + \Delta k$ is equal to the second-order difference $\Delta_2 F(u, v)$ corresponding to this rectangle.

It follows from the Hermite-symmetric properties of covariances that the masses carried by two sets of points symmetrically situated with respect to the diagonal $u = v$ of the square C are always complex conjugates. If a point set belonging to the diagonal $u = v$ carries a mass different from zero, this mass will be real and positive.

The function $F(u, v)$ will be called the *spectral function* of the x_n process, while the distribution defined by F is the *spectral distribution* of the process.

In the particular case when the whole spectral mass is situated on the diagonal $u = v$, it follows from the general symbolic relation

$$E\{dz(u)\, d\bar{z}(v)\} = d_{u,v}F(u, v) \tag{77}$$

that the $z(u)$ process has orthogonal increments, and so in this case the x_n process is *stationary*.

In the general case, $F(u, v)$ may be represented as a sum of three components, each of which is a covariance of bounded variation over C,

$$F = F_1 + F_2 + F_3. \tag{78}$$

Here F_1 is absolutely continuous, with a *spectral density* $f_1(u, v)$ such that

$$F_1(u, v) = \int_0^u \int_0^v f_1(s, t)\, ds\, dt. \tag{79}$$

On the other hand, the F_2 and F_3 distributions both have their total masses concentrated in sets of two-dimensional Lebesgue measure zero. For F_2 this set is at most enumerable, each point carrying a mass different from zero, while the F_3 set is nonenumerable, and each single point carries the mass zero. In the stationary case, the F_1 component is absent, while the F_2 and F_3 components have their total masses situated on the diagonal $u = v$.

A sufficient condition that the harmonizable x_n process given by (73) will be deterministic can be obtained in the following way. The x_n process will be deterministic if, and only if, for every n and every $h > 0$ we can find a finite number of constants $c_0, c_1, \cdots, c_r$ such that the quantity

$$W = E|x_n - c_0x_{n-h} - c_1x_{n-h-1} - \cdots - c_rx_{n-h-r}|^2 \geqq 0 \tag{80}$$

will be arbitrarily small. Writing

$$g(u) = e^{inu} - c_0e^{i(n-h)u} - \cdots - c_re^{i(n-h-r)u}, \tag{81}$$

it follows from (76) that we have

$$W = \int_0^{2\pi} \int_0^{2\pi} g(u)\bar{g}(v)\, dF(u, v) \tag{82}$$

and hence by the Schwarz inequality

$$W^2 \leqq \int_0^{2\pi} \int_0^{2\pi} |g(u)|^2\, |dF(u, v)| \int_0^{2\pi} \int_0^{2\pi} |g(v)|^2\, |dF(u, v)|. \tag{83}$$

By the symmetry of the spectral distribution, the two factors in the last member are equal, so that we obtain

$$W \leqq \int_0^{2\pi} |g(u)|^2\, dG(u), \tag{84}$$

where

(85) $$G(u) = \int_0^u \int_0^{2\pi} |dF(s, t)|.$$

Now $G(u)$, being a never decreasing and bounded function of u, has almost everywhere in $(0, 2\pi)$ a nonnegative derivative $G'(u)$, and the integral

(86) $$\int_0^{2\pi} \log G'(u)\, du$$

will be finite or equal to $-\infty$. In particular, if $G'(u) = 0$ on a set of positive measure, the integral will certainly have the value $-\infty$.

If the integral (86) has the value $-\infty$, it follows from well-known theorems in the prediction theory for stationary processes that the coefficients c_j can be chosen so as to make the second member of (80) as small as we please. Thus we have the following result (Cramér [2]).

THEOREM 6. *If we have*

(87) $$\int_0^{2\pi} \log G'(u)\, du = -\infty$$

the x_n process is deterministic.

In particular, if the F_1 and F_3 components in (78) are absent, so that the whole mass of the F distribution is concentrated in isolated points, it will be seen that $G'(u) = 0$ almost everywhere, so that (87) will certainly hold, and the x_n process will be deterministic.

Consider now, on the other hand, a process x_n with a spectral function F having an absolutely continuous component F_1 not identically zero. Moreover, let us suppose that the spectral density $f_1(u, v)$ corresponding to F_1 belongs to L_2 over the square C. For such a process, we shall give a sufficient condition that it is nondeterministic. Let

(88) $$f_1(u, v) = \sum_{p=1}^{\infty} \mu_p \varphi_p(u) \bar{\varphi}_p(v)$$

be the expansion of $f_1(u, v)$ in terms of its eigenvalues μ_p and eigenfunctions $\varphi_p(u)$. The μ_p are real and positive, the $\varphi_p(u)$ are a set of orthonormal functions in $(0, 2\pi)$, and the series converges in quadratic mean over C. We then have

THEOREM 7. *Suppose that, in the expansion* (88), *there is a p such that the Fourier series of the eigenfunction $\varphi_p(u)$*

(89) $$\varphi_p(u) \sim \sum_{q=-\infty}^{\infty} b_{pq} e^{-iqu}, \qquad \sum_{q=-\infty}^{\infty} |b_{pq}|^2 < \infty$$

is "one-sided" in the sense that for a certain m it satisfies the conditions

(90) $$b_{pq} = 0 \text{ for } q < m, \qquad b_{pm} \neq 0.$$

Then the x_n process is nondeterministic, and the point $n = m$ belongs to its innovation spectrum.

Taking $n = m$ and $h = 1$ in the expressions (80) and (81) for W and $g(u)$ we have, in fact (see Riesz and Nagy [9], p. 240),

(91) $$W = \int_0^{2\pi}\int_0^{2\pi} g(u)\bar{g}(v)\, dF(u, v) \geqq \int_0^{2\pi}\int_0^{2\pi} g(u)\bar{g}(v) f_1(u, v)\, du\, dv$$
$$= \sum_{p=1}^{\infty} \mu_p \left|\int_0^{2\pi} g(u)\varphi_p(u)\, du\right|^2$$

and thus by hypothesis

(92) $$W \geqq \mu_p \left|\int_0^{2\pi} g(u)\varphi_p(u)\, du\right|^2 = 4\pi^2\mu_p|b_{pm}|^2$$

independently of the choice of the coefficients c_j. This obviously signifies that x_m cannot be predicted exactly in terms of the variables $x_{m-1}, x_{m-2}, \cdots$, so that the prediction error

(93) $$||x_m - P_{m-1}x_m||$$

is positive, and m belongs to the innovation spectrum of the x_n process, which is thus nondeterministic.

It follows from well-known theorems that, when the conditions of theorem 7 are satisfied, we have

(94) $$\int_0^{2\pi} \log|\varphi_p(u)|\, du > -\infty.$$

The converse of this statement is, however, not true; (94) may be satisfied even in a case when $\varphi_p(u)$ does not have a one-sided Fourier expansion. A simple example is obtained by taking

(95) $$2\pi f_1(u, v) = \varphi(u)\bar{\varphi}(v)$$

with

(96) $$\varphi(u) = \begin{cases} e^{iu}, & 0 \leqq u \leqq \pi, \\ e^{-iu}, & \pi < u \leqq 2\pi. \end{cases}$$

It can also be shown by examples that there are nondeterministic processes with a spectral density $f_1(u, v)$ belonging to L_2 that do not satisfy the conditions of theorem 7 for any value of p.

By imposing a further restrictive condition on the behavior of the spectral density it is possible, however, to obtain a criterion which is both necessary and sufficient in order that a given harmonizable process be nondeterministic, and even have an a priori given set of integers as its innovation spectrum (Cramér [3]). Thus, in particular, it follows that, any set of integers being given, there always exists a harmonizable process having this set as its innovation spectrum.

7. Processes with normal shift operator

For a stationary process with discrete time, we have the integral representation

(97) $$x_n = \int_0^{2\pi} e^{inu}\, dz(u),$$

where $z(u)$ has orthogonal increments. In the preceding section we considered the generalization obtained by dropping the assumption that $z(u)$ has orthogonal increments, and we saw that this leads to the class of harmonizable processes.

We now consider a different kind of generalization of (97), which leads to a different class of processes. To this effect, we now regard the integration variable u in (97) as a *complex* variable, and suppose that the integration is extended over a certain domain D in the plane of u, and that $z(u)$ is defined for all u belonging to D.

After an appropriate change of variables, the integral corresponding to (97) then takes the form

$$x_n = \int_D w^n \, dz(\rho, \lambda), \tag{98}$$

where $w = \rho \exp(i\lambda)$, while $z(\rho, \lambda)$ is a random variable satisfying the conditions

$$Ez(\rho, \lambda) = 0, \qquad E|z(\rho, \lambda)|^2 < K \tag{99}$$

for all ρ, λ such that w belongs to D. As in the stationary case, we still suppose that $z(\rho, \lambda)$ has orthogonal increments, so that we have in the usual symbolism

$$\begin{aligned} E\{dz(\rho_1, \lambda_1)\, \overline{dz}(\rho_2, \lambda_2)\} &= 0, \qquad w_1 \neq w_2, \\ E|dz(\rho, \lambda)|^2 &= dF(\rho, \lambda), \end{aligned} \tag{100}$$

where $F(\rho, \lambda)$ is a nonnegative and never decreasing function of ρ and λ, which is bounded throughout D. The integral (98) can then be defined in the same way as before, and we obtain

$$R(m, n) = E(x_m \bar{x}_n) = \int_D w^m \overline{w^n} \, dF(\rho, \lambda). \tag{101}$$

The function $F(\rho, \lambda)$ will be called the *spectral function* of the x_n process, and defines the *spectral distribution* of the process, which is a distribution of real and positive mass over the domain D.

We now introduce the further assumption that the domain D is entirely situated within the ring

$$\rho_1 \geqq \rho \geqq \rho_0 > 0. \tag{102}$$

We observe that this includes the particular case of a stationary process, when the domain D reduces to the unit circle.

In the present case we obtain from (101) for any complex constants c_j and any positive integer Q

$$0 \leqq E\left|\sum_{j=-Q}^{Q} c_j x_j\right|^2 = \sum_{j,k=-Q}^{Q} c_j \bar{c}_k R(j, k) = \int_D \left|\sum_{-Q}^{Q} c_j w^j\right|^2 dF, \tag{103}$$

and hence

$$\begin{aligned} E\left|\sum_{-Q}^{Q} c_j x_{j+1}\right|^2 &\leqq \rho_1^2 E\left|\sum_{-Q}^{Q} c_j x_j\right|^2, \\ E\left|\sum_{-Q}^{Q} c_j x_{j-1}\right|^2 &\leqq \frac{1}{\rho_0^2} E\left|\sum_{-Q}^{Q} c_j x_j\right|^2. \end{aligned} \tag{104}$$

According to Getoor [5], these inequalities imply that there is a *shift operator* N uniquely defined and bounded throughout the Hilbert space $\mathfrak{H}(x)$ of the x_n process, and such that

$$(105) \qquad N^m x_n = x_{m+n}$$

for all $m, n = 0, \pm 1, \cdots$. It also follows from the work of Getoor that in our case N is a *normal* operator in $\mathfrak{H}(x)$. We have, in fact,

$$(106) \qquad \begin{aligned}(Nx_m, x_n) &= E(x_{m+1}\bar{x}_n) = \int_D w^{m+1}\overline{w^n}\, dF \\ &= \int_D w^m(\overline{\bar{w}w^n})\, dF = E(x_m\bar{y}_n),\end{aligned}$$

where

$$(107) \qquad y_n = N^* x_n = \int_D \bar{w}w^n\, dz.$$

It follows that

$$(108) \qquad NN^* x_n = N^* N x_n = \int_D |w|^2 w^n\, dz.$$

Thus N commutes with its adjoint N^*, and consequently N is normal. In the particular case of a stationary process, when the spectral mass is wholly situated on the unit circle, it is well known that N is even a unitary operator.

By an argument quite similar to that used for the deduction of the inequalities (104), we obtain for $n > 0$

$$(109) \qquad \begin{aligned}\rho_0^{2n} E\left|x_0 - \sum_{j=1}^{Q} c_j x_{-j}\right|^2 &\leqq E\left|x_n - \sum_{j=1}^{Q} c_j x_{n-j}\right|^2 \\ &\leqq \rho_1^{2n} E\left|x_0 - \sum_{j=1}^{Q} c_j x_{-j}\right|^2.\end{aligned}$$

The coefficients c_j being arbitrary, this shows that we have for the prediction errors σ_{nh}, where h is any positive integer,

$$(110) \qquad \rho_0^{2n}\sigma_{0h}^2 \leqq \sigma_{nh}^2 \leqq \rho_1^{2n}\sigma_{0h}^2.$$

For $n < 0$ we obtain in the same way

$$(111) \qquad \rho_1^{2n}\sigma_{0h}^2 \leqq \sigma_{nh}^2 \leqq \rho_0^{2n}\sigma_{0h}^2.$$

From these inequalities, we obtain directly the following theorem.

THEOREM 8. *If the x_n process defined by* (98) *and* (102) *is nondeterministic, we have $\sigma_{nh} > 0$ for all n and all $h > 0$. In particular, the innovation spectrum of the process then contains all $n = 0, \pm 1, \cdots$.*

Suppose now that the spectral distribution defined by $F(\rho, \lambda)$ has a nonvanishing absolutely continuous component. We may then write

$$(112) \qquad \begin{aligned}J = E\left|x_0 - \sum_1^Q c_j x_{-j}\right|^2 &= \int_D \left|1 - \sum_1^Q \frac{c_j}{w^j}\right|^2 dF \\ &\geqq \int_{\rho_0}^{\rho_1}\int_0^{2\pi}\left|1 - \sum_1^Q \frac{c_j}{w^j}\right|^2 f(\rho, \lambda)\, d\rho\, d\lambda,\end{aligned}$$

where $f(\rho, \lambda)$ is nonnegative and integrable. From the inequalities between arithmetic and geometric means we further obtain

$$(113)\quad \frac{1}{2\pi(\rho_1-\rho_0)} J \geqq \exp\left\{\frac{1}{2\pi(\rho_1-\rho_0)}\int_{\rho_0}^{\rho_1}\int_0^{2\pi}\left[\log f(\rho,\lambda) + 2\log\left|1-\sum_1^Q \frac{c_j}{w^j}\right|\right] d\rho\, d\lambda\right\}$$
$$= G(f)\exp\left\{\frac{1}{\pi(\rho_1-\rho_0)}\int_{\rho_0}^{\rho_1}\int_0^{2\pi}\log\left|1-\sum_1^Q \frac{c_j}{w^j}\right| d\rho\, d\lambda\right\},$$

where

$$(114)\quad G(f) = \exp\left\{\frac{1}{2\pi(\rho_1-\rho_0)}\int_{\rho_0}^{\rho_1}\int_0^{2\pi}\log f(\rho,\lambda)\, d\rho\, d\lambda\right\}.$$

From Jensen's theorem we obtain, however,

$$(115)\quad \exp\left\{\frac{1}{\pi(\rho_1-\rho_0)}\int_{\rho_0}^{\rho_1}\int_0^{2\pi}\log\left|1-\sum_1^Q \frac{c_j}{w^j}\right| d\rho\, d\lambda\right\}$$
$$= \exp\left\{\frac{2}{\rho_1-\rho_0}\int_{\rho_0}^{\rho_1}\log\frac{|w_1\cdots w_k|}{\rho^k}\, d\rho\right\} \geqq 1,$$

where $w_1, \cdots, w_k$ are the zeros of $1-\sum_1^Q c_j w^{-j}$ outside the circle $|w| = \rho$. Consequently

$$(116)\quad \frac{1}{2\pi(\rho_1-\rho_0)} J \geqq G(f),$$

and it follows that, if $G(f) > 0$, then the x_n process is nondeterministic, and we have

$$(117)\quad \sigma_{n1}^2 \geqq 2\pi(\rho_1-\rho_0)G(f).$$

In the particular case when there is an expansion

$$(118)\quad \log f(\rho,\lambda) = \sum_{-\infty}^{\infty}\frac{b_j}{\rho^{|j|}}c^{-ij\lambda}, \qquad b_{-j} = \bar{b}_j,$$

absolutely convergent for $\rho \geqq \rho_0$, it can even be proved that the sign of equality holds in (117).

Finally, we may observe that it is also possible to give a sufficient condition for a deterministic process, corresponding at least partly to theorem 6. In fact, it can be shown that if the spectral distribution defined by $F(\rho, \lambda)$ is discrete, and if the set of points carrying a positive mass has at most a finite number of limiting points, then the x_n process is deterministic.

REFERENCES

[1] H. Cramér, "A contribution to the theory of stochastic processes," *Proceedings of the Second Berkeley Symposium on Mathematical Statistics and Probability*, Berkeley and Los Angeles, University of California Press, 1951, pp. 329–339.

[2] ———, "Remarques sur le problème de prédiction pour certaines classes de processus stochastiques," *Colloque sur le Calcul des Probabilités*, Paris, 1959, pp. 103–112.
[3] ———, "On the linear prediction problem for certain stochastic processes," *Ark. Mat.*, Vol. 4 (1959), pp. 45–53.
[4] J. L. Doob, *Stochastic Processes*, New York, Wiley, 1953.
[5] R. K. Getoor, "The shift operator for non-stationary stochastic processes," *Duke Math. J.*, Vol. 23 (1956), pp. 175–187.
[6] P. R. Halmos, *Introduction to Hilbert Space and the Theory of Spectral Multiplicity*, New York, Chelsea, 1951.
[7] M. Loève, "Fonctions aléatoires du second ordre," appendix to P. Lévy, *Processus Stochastiques et Mouvement Brownien*, Paris, Gauthier-Villars, 1948.
[8] ———, *Probability Theory*, New York, Van Nostrand, 1955.
[9] F. Riesz and B. Sz.-Nagy, *Leçons d'Analyse Fonctionnelle*, Paris, 1955 (3rd ed.).
[10] M. H. Stone, *Linear Transformations in Hilbert Space and their Applications to Analysis*, New York, The American Mathematical Society, 1932.
[11] N. Wiener and P. Masani, "The prediction theory of multivariate stochastic processes, I," *Acta Math.*, Vol. 98 (1957), pp. 111–150.
[12] H. Wold, *A Study in the Analysis of Stationary Time Series*, Stockholm, Almqvist & Wiksell, 1954.

51.

On the structure of purely non-deterministic stochastic processes

Ark. Mat. 4 (19), 249–266 (1961)
(Reprinted in Random Processes)

1. The purpose of this paper is to give the proofs of some results recently communicated in a lecture at the Fourth Berkeley Symposium on Mathematical Statistics and Probability [1].

Although these results were expressed in the language of mathematical probability, they may equally well be regarded as concerned with the properties of certain curves in Hilbert space. In this first section of the Introduction we shall briefly state some results of the paper in Hilbert space language, and then in the following sections recur to the "mixed" language which seems convenient when probability questions are treated with the methods of Hilbert space geometry.

Let $\mathfrak{H}$ be a complex Hilbert space, and let, for every real τ, a set of q elements $x_1(\tau), x_2(\tau), \ldots, x_q(\tau)$ of $\mathfrak{H}$ be given. As τ runs through all real values, each element $x_j(\tau)$ describes a "curve" C_j in the space $\mathfrak{H}$. Let $C_j(t)$ denote the "arc" of C_j corresponding to values of $\tau \leqslant t$, and denote by $\mathfrak{H}(\mathbf{x}, t)$ the smallest subspace of $\mathfrak{H}$ containing the arcs $C_1(t), \ldots, C_q(t)$.

As t increases, the $\mathfrak{H}(\mathbf{x}, t)$ form a never decreasing family of subspaces, and the limiting spaces $\mathfrak{H}(\mathbf{x}, +\infty)$ and $\mathfrak{H}(\mathbf{x}, -\infty)$ will exist. It will be assumed that the following two conditions are satisfied:

(A) The strong limits $x_j(t \pm 0)$ exist for $j = 1, \ldots, q$ and for all real t.

(B) The space $\mathfrak{H}(\mathbf{x}, -\infty)$ contains only the zero element of $\mathfrak{H}$.

The projection of an arbitrary element z of $\mathfrak{H}(\mathbf{x}, +\infty)$ on the subspace $\mathfrak{H}(\mathbf{x}, t)$ will be denoted by $P_t z$.

We propose to show that the $x_j(t)$ can be simultaneously and linearly expressed in terms of certain mutually orthogonal elements. For this purpose we shall use considerations closely related to the theory of spectral multiplicity of self-adjoint transformations in a separable Hilbert space (cf. e.g. [7], Chapter VII).

It will be shown that it is possible to find a sequence $z_1, \ldots, z_N$ of elements of $\mathfrak{H}(\mathbf{x}, +\infty)$ such that we have for every $j = 1, \ldots, q$ and for all real t

$$x_j(t) = \sum_{n=1}^{N} \int_{-\infty}^{t} g_{jn}(t, \lambda)\, d z_n(\lambda), \tag{1}$$

where $z_n(\lambda) = P_\lambda z_n$. Here N may be a finite integer or equal to $\aleph_0$, the g_{jn} are complex-valued functions of the real variables t, λ, and the integrals are appropriately defined. Two increments $\Delta z_m(\lambda)$ and $\Delta z_n(\mu)$ are always orthogonal if $m \neq n$, while for $m = n$ they are orthogonal if they correspond to disjoint intervals.

We shall study the properties of the expansion (1), and in particular it will be shown that we have for any $u < t$

$$P_u x_j(t) = \sum_{n=1}^{N} \int_{-\infty}^{u} g_{jn}(t, \lambda)\, d z_n(\lambda). \tag{2}$$

When certain additional conditions are imposed, N is the smallest cardinal number such that a representation of the form (1) holds.

If the elements of $\mathfrak{H}$ are interpreted as random variables, the set of curves $C_1, \ldots, C_q$ will correspond to q simultaneously considered stochastic processes with a continuous time parameter. The above results then yield a representation of such a set of processes in terms of past and present "innovations", as well as an explicit expression for the linear least squares prediction, as will be shown in the sequel.

2. Consider a random variable x defined on a once for all given probability space, and satisfying the relations

$$E x = 0, \quad E|x|^2 < \infty. \tag{3}$$

The set of all random variables defined on the given probability space and satisfying (3) forms a Hilbert space $\mathfrak{H}$, if the inner product and the norm are defined in the usual way:

$$(x, y) = E(x\bar{y}), \quad \|x\|^2 = E|x|^2.$$

If two random variables x, y belonging to $\mathfrak{H}$ are such that

$$\|x - y\|^2 = E|x - y|^2 = 0,$$

they will be considered as identical, and we shall write

$$x = y.$$

In the sequel, any equation between random variables should be interpreted in this sense.

Whenever we are dealing with the *convergence* of a sequence of random variables, it will always be understood that we are concerned with convergence in the topology induced by the norm in $\mathfrak{H}$, which in probabilistic terminology corresponds to convergence in quadratic mean.

A family of complex-valued random variables $x(t)$, where t is a real parameter ranging from $-\infty$ to $+\infty$, and $x(t) \in \mathfrak{H}$ for every t, will be called a *one-dimensional stochastic process* with continuous time parameter t. Further, if

$x_1(t), \ldots, x_q(t)$ are q random variables, each of which is associated with a process of this type, the (column) vector

$$\mathbf{x}(t) = (x_1(t), \ldots, x_q(t)) \tag{4}$$

defines a *q-dimensional stochastic vector process* with continuous time t.

For every fixed t, each component $x_j(t)$ of the vector process (4) is a point in the Hilbert space $\mathfrak{H}$. As t increases from $-\infty$ to $+\infty$, this point describes a curve C_j in $\mathfrak{H}$, and so we are led to consider the set of curves $C_1 \ldots, C_q$ mentioned in the preceding section. The subspace $\mathfrak{H}(\mathbf{x}, t)$ is, from the present point of view, the subspace spanned by the random variables $x_1(\tau), \ldots, x_q(\tau)$ for all $\tau \leqslant t$, and we shall write this

$$\mathfrak{H}(\mathbf{x}, t) = \mathfrak{S}\{x_1(\tau), \ldots, x_q(\tau); \quad \tau \leqslant t\}.$$

$\mathfrak{H}(\mathbf{x}, t)$ may be regarded as the set of all random variables that can be obtained by means of linear operations acting on the components of $\mathbf{x}(\tau)$ for all $\tau \leqslant t$. We evidently have

$$\mathfrak{H}(\mathbf{x}, t_1) \subset \mathfrak{H}(\mathbf{x}, t_2)$$

whenever $t_1 < t_2$. It follows that the limiting spaces $\mathfrak{H}(\mathbf{x}, +\infty)$ and $\mathfrak{H}(\mathbf{x}, -\infty)$ exist, and also that the limiting spaces $\mathfrak{H}(\mathbf{x}, t \pm 0)$ exist for all t.

Following Wiener & Masani [8], we shall say that the space $\mathfrak{H}(\mathbf{x}, t)$ represents the *past and present* of the vector process (4), as seen from the point of view of the instant t. The limiting space $\mathfrak{H}(\mathbf{x}, -\infty)$ will be called the *remote past* of the process, while the space $\mathfrak{H}(\mathbf{x}, +\infty)$ will be briefly denoted by $\mathfrak{H}(\mathbf{x})$, and called the *space of the* $\mathbf{x}(t)$ *process*. We then have for any t

$$\mathfrak{H}(\mathbf{x}, -\infty) \subset \mathfrak{H}(\mathbf{x}, t) \subset \mathfrak{H}(\mathbf{x}, +\infty) = \mathfrak{H}(\mathbf{x}) \subset \mathfrak{H}.$$

In the particular case of a vector process $\mathbf{x}(t)$ satisfying

$$\mathfrak{H}(\mathbf{x}, -\infty) = \mathfrak{H}(\mathbf{x}), \tag{5}$$

it will be seen that complete information concerning the process is already contained in the remote past. Accordingly such a process will be called a *deterministic* process.

Any process not satisfying (5) will be called *non-deterministic*. In the extreme case when we have

$$\mathfrak{H}(\mathbf{x}, -\infty) = 0, \tag{6}$$

the remote past does not contain any information at all, and the process is said to be *purely non-deterministic*.

It is known [1] that any vector process (4) can be represented as the sum of a deterministic and a purely non-deterministic component, which are mutually orthogonal. In the present paper, we shall mainly be concerned with the structure of the latter component, and we may then as well assume that the given $\mathbf{x}(t)$ process itself is purely non-deterministic, i.e. that $\mathfrak{H}(\mathbf{x}, -\infty) = 0$.

Before introducing this assumption we shall, however, in sections 4–5 study the properties of the general $\mathbf{x}(t)$ process, without imposing more than the mildly restrictive condition (A), relating to the behaviour of the process in its points of discontinuity.

In section 6 we shall then, besides condition (A), introduce condition (B) which states that the $\mathbf{x}(t)$ process is purely non-deterministic. Throughout the rest of the paper, we shall then be concerned with processes satisfying both conditions (A) and (B).

In the particular case of a *stationary* one-dimensional process $x(t)$ satisfying (A) and (B), it is known that $x(t)$ can be linearly represented in terms of past innovations by an expression of the form [1]

$$x(t) = \int_{-\infty}^{t} g(t-\lambda)\, dz(\lambda),. \tag{7}$$

where $z(\lambda)$ is a process *with orthogonal increments,* which may be called an *innovation process* of $x(t)$.

In sections 8–9, we shall be concerned with representations of a similar kind, but generalized in two directions: the assumption of stationarity will be dropped, and a vector process $\mathbf{x}(t)$ will be considered instead of the one-dimensional $x(t)$. It will be shown that, for any $\mathbf{x}(t)$ process satisfying (A) and (B), we have a representation of the form (1) indicated in section 1 above. The $z_n(\lambda)$ occurring in (1) will now be one-dimensional stochastic processes with orthogonal increments. Accordingly we may say that, in the general case, we are concerned with an innovation process $(z_1(\lambda), \ldots, z_N(\lambda))$ which is multi-dimensional, and possibly even infinite-dimensional.

3. In a series of papers, P. Lévy (cf. [4–6], and further references there given) has investigated the properties of stochastic processes representable, in the notation of the present paper, in the form

$$x(t) = \int_{0}^{t} g(t,\lambda)\, dz(\lambda),$$

where $z(\lambda)$ is a *normal* (i.e Gaussian or Laplacian) process with independent increments. His investigations have been continued in a recent paper by T. Hida [3], who has also considered the more general representation

$$x(t) = \sum_{n=1}^{N} \int_{0}^{t} g_n(t,\lambda)\, dz_n(\lambda)$$

derived from considerations of spectral multiplicity, as well as the case when the lower limits of the integrals are $-\infty$ instead of zero.

---

[1] Cf. e.g. Doob [2], p. 588, and the references there given. The corresponding representation for a one-dimensional stationary process with *discrete* time parameter was first found by Wold [9], and was generalized to the vector case by Wiener & Masani [8].

In connection with my Berkeley lecture [1], Professor K. Itô kindly drew my attention to the work of Mr. Hida, which was then in the course of being printed. Evidently there are interesting points of contact between Mr. Hida's line of investigation and the one pursued in the present paper.

Discontinuities and innovations

4. Consider a stochastic vector process as defined in section 2,

$$\mathbf{x}(t) = (x_1(t), \ldots, x_q(t)),$$

and let us suppose that the following condition is satisfied:

(A) *The limits* $x_j(t-0)$ *and* $x_j(t+0)$ *exist for* $j = 1, \ldots, q$ *and for all real* t.

We shall then write

$$\mathbf{x}(t-0) = (x_1(t-0), \ldots, x_q(t-0))$$

and similarly for $\mathbf{x}(t+0)$. Any point t such that at least one of the relations

$$\mathbf{x}(t-0) = \mathbf{x}(t) = \mathbf{x}(t+0)$$

is not satisfied, is a *discontinuity point* of the $\mathbf{x}$ process. The point t is a *left* or *right* discontinuity, or both, according as $\mathbf{x}(t-0) \neq \mathbf{x}(t)$, or $\mathbf{x}(t) \neq \mathbf{x}(t+0)$, or both. We shall now prove the following Lemma.

Lemma 1. *For any* $\mathbf{x}(t)$ *process satisfying* (A), *we have*

(*a*) *For* $j = 1, \ldots, q$, *the functions* $E|x_j(t)|^2$ *are bounded throughout every finite* t*-interval.*
(*b*) *The set of all discontinuity points of the* $\mathbf{x}$ *process is at most enumerable.*
(*c*) *The Hilbert space* $\mathfrak{H}(\mathbf{x})$ *is separable.*

In order to prove (*a*), let us suppose that the non-negative function $E|x_j(t)|^2$ were not bounded in a certain finite interval I. Then it would be possible to find a sequence of points $\{t_n\}$ in I, and converging monotonely to a limit t^*, such that $E|x_j(t_n)|^2 \to \infty$. Clearly this is not compatible with condition (A), so that our hypothesis must be wrong, and point (*a*) is proved.

Point (*b*) will be proved if we can show that, for each j, the one-dimensional process $x_j(t)$ has at most an enumerable set of discontinuity points. A discontinuity point of $x_j(t)$ is then, of course, a point t such that at least one of the relations

$$x_j(t-0) = x_j(t) = x_j(t+0)$$

is not satisfied. Clearly it will be enough to show, e.g., that the inequality

$$x_j(t-0) \neq x_j(t) \tag{8}$$

cannot be satisfied in more than an enumerable set of points.

According to our conventions, (8) is equivalent to the relation

$$s(t) = E\,|\,x_j(t-0) - x_j(t)\,|^2 > 0.$$

We are going to show first that, given any positive number h and any finite interval I, there is at most a finite number of points t in I such that $s(t) > h$. This being shown, the desired result follows immediately by allowing first h to tend to zero, and then I to tend the whole real axis.

Suppose, in fact, that I contains an infinite set of points t with $s(t) > h$. We can then find an infinite sequence of points $\{t_n\}$ in I, converging to a limit t^*, and such that

$$s(t_n) = E\,|\,x_j(t_n - 0) - x_j(t_n)\,|^2 > h \tag{9}$$

for all n. Evidently we can even find a *monotone* sequence $\{t_n\}$ having these properties. Let us suppose, e.g., that t_n converges *decreasingly* to t^*. (The increasing case can, of course, be treated in the same way.) On account of condition (A) we can then find a sequence of positive numbers $\{\varepsilon_n\}$ tending to zero, such that

$$t_n - \varepsilon_n > t_{n+1},$$

$$E\,|\,x_j(t_n - \varepsilon_n) - x_j(t_n - 0)\,|^2 < \frac{c^2 h^2}{16\,K} \tag{10}$$

for all n, where c and K are constants such that $0 < c < 1$ and $E\,|\,x_j(t)\,|^2 < K$ throughout I. The existence of such a constant K follows from point (a) of the Lemma, which has already been established.

Now for any random variables u and v we have

$$\begin{aligned} E\,|\,u+v\,|^2 &= E\,|\,u\,|^2 + E\,|\,v\,|^2 + E(u\bar{v}) + E(\bar{u}v) \\ &\geqslant E\,|\,u\,|^2 + E\,|\,v\,|^2 - 2\sqrt{E\,|\,u\,|^2\,E\,|\,v\,|^2}. \end{aligned}$$

Taking here

$$u = x_j(t_n - 0) - x_j(t_n),$$
$$v = x_j(t_n - \varepsilon_n) - x_j(t_n - 0),$$

we obtain from (9) and (10)

$$E\,|\,x_j(t_n - \varepsilon_n) - x_j(t_n)\,|^2 \geqslant h - 2\sqrt{4\,K \cdot \frac{c^2 h^2}{16\,K}} = (1-c)\,h \tag{11}$$

for all n. On the other hand, the sequences $\{t_n - \varepsilon_n\}$ and $\{t_n\}$ both converge decreasingly to t^*. Consequently by condition (A) we have (convergence, as usual, in the $\mathfrak{H}$ topology, i.e. in quadratic mean)

$$x_j(t_n - \varepsilon_n) \to x_j(t^* + 0),$$

$$x_j(t_n) \to x_j(t^* + 0),$$

and thus $$x_j(t_n - \varepsilon_n) - x_j(t_n) \to 0,$$

as n tends to infinity. However, this is incompatible with (11), so that our hypothesis must be wrong, and thus point (*b*) of the Lemma is proved.

The last part of the Lemma now follows immediately, if we consider an enumerable set $\{t_n\}$ including all discontinuity points of $\mathbf{x}(t)$ as well as an everywhere dense set of continuity points. The set of all finite linear combinations of the random variables $x_j(t_n)$ for $j = 1, \ldots, q$ and $n = 1, 2, \ldots$, with coefficients whose real and imaginary parts are both rational, is then an enumerable set dense in $\mathfrak{H}(\mathbf{x})$.

5. Consider now the family of subspaces $\mathfrak{H}(\mathbf{x}, t)$ associated with an $\mathbf{x}(t)$ process satisfying condition (A) of the preceding section.

As observed in section 2, $\mathfrak{H}(\mathbf{x}. t)$ never decreases as t increases. If $\mathfrak{H}(\mathbf{x}, t)$ effectively increases when t ranges over some interval $t_1 < t \leqslant t_2$, so that we have

$$\mathfrak{H}(\mathbf{x}, t_1) \neq \mathfrak{H}(\mathbf{x}, t_2),$$

this means that some new information has entered into the process during that interval. Accordingly we shall then say that the process has received an *innovation* during the interval $t_1 < t \leqslant t_2$, and we shall regard this innovation as being represented by the orthogonal complement

$$\mathfrak{H}(\mathbf{x}, t_2) - \mathfrak{H}(\mathbf{x}, t_1). \tag{12}$$

In fact, if this complement reduces to the zero element, the two spaces are identical, and no innovation has entered during the interval, while in the opposite case (12) is the set of all differences between an element of $\mathfrak{H}(\mathbf{x}, t_2)$ and its projection on $\mathfrak{H}(\mathbf{x}, t_1)$.

If, for a certain value of t, we have

$$\mathfrak{H}(\mathbf{x}, t-h) \neq \mathfrak{H}(\mathbf{x}, t+h)$$

for any $h > 0$, this means that there is a non-vanishing innovation associated with any interval containing t as an interior point. The set of all points t having this property will be called the *innovation spectrum* of the $\mathbf{x}(t)$ process.

$\mathfrak{H}(\mathbf{x}, t)$ being never decreasing as t increases, it follows that the limiting spaces $\mathfrak{H}(\mathbf{x}, t \pm 0)$ will exist for every t. Any point such that at least one of the relations

$$\mathfrak{H}(\mathbf{x}, t-0) = \mathfrak{H}(\mathbf{x}, t) = \mathfrak{H}(\mathbf{x}, t+0)$$

is not satisfied, will certainly belong to the innovation spectrum, and will be called a *discontinuity point* of that spectrum. As in section 4, the terms *left* and *right* discontinuities will be used in the obvious sense.

The set of all discontinuity points constitutes the *discontinuous part* of the innovation spectrum. Let $t_1, t_2, \ldots$, be the points of this set (it will be shown below that the set is at most enumerable), and form the vector sum

$$\mathfrak{G}(\mathbf{x}, t) = \sum_{t_k \leqslant t} \mathfrak{H}(\mathbf{x}, t_k + 0) - \mathfrak{H}(\mathbf{x}, t_k - 0)$$

and the orthogonal complement $\mathfrak{F}(\mathbf{x}, t) = \mathfrak{H}(\mathbf{x}, t) - \mathfrak{G}(\mathbf{x}, t)$. The set of all points t such that $\mathfrak{F}(\mathbf{x}, t-h) \neq \mathfrak{F}(\mathbf{x}, t+h)$ for any $h > 0$ constitutes the *continuous part* of the innovation spectrum of $\mathbf{x}(t)$. If the discontinuous part is not a closed set, its limiting points will belong to the innovation spectrum, without necessarily belonging to any of the two parts here defined.

When t is a discontinuity point of the innovation spectrum, the number of dimensions of the subspace $\mathfrak{H}(\mathbf{x}, t) - \mathfrak{H}(\mathbf{x}, t-0)$ will be called the *left multiplicity* of the point t. Similarly the *right multiplicity* of t is the number of dimensions of $\mathfrak{H}(\mathbf{x}, t+0) - \mathfrak{H}(\mathbf{x}, t)$. If t is not a left (right) discontinuity, the left (right) multiplicity of t is of course equal to zero. We shall now prove the following Lemma.

Lemma 2. *For any* $\mathbf{x}(t)$ *process satisfying* (A), *we have*

(a) *The set of discontinuity points of the innovation spectrum is at most enumerable.*

(b) *In a left discontinuity point, the left multiplicity is at most equal to* q.

(c) *In a right discontinuity point, the right multiplicity may be any finite integer, or equal to* $\aleph_0$.

(d) *A left discontinuity of the innovation spectrum is always at the same time a left discontinuity of the process. On the other hand, a right discontinuity of the innovation spectrum is not necessarily a discontinuity of the process.*

We first observe that the two subspaces $\mathfrak{H}(\mathbf{x}, t_2) - \mathfrak{H}(\mathbf{x}, t_1)$ and $\mathfrak{H}(\mathbf{x}, u_2) - \mathfrak{H}(\mathbf{x}, u_1)$ corresponding to disjoint time intervals are always orthogonal. It follows that the subspaces

$$\mathfrak{H}(\mathbf{x}, t+0) - \mathfrak{H}(\mathbf{x}, t-0)$$

corresponding to different discontinuity points are orthogonal. By Lemma 1, the space $\mathfrak{H}(\mathbf{x})$ is separable, and cannot include more than an enumerable set of mutually orthogonal subspaces. Hence follows the truth of point (a) of the Lemma.

Point (b) of the Lemma asserts that the orthogonal complement $\mathfrak{H}(\mathbf{x}, t) - \mathfrak{H}(\mathbf{x}, t-0)$ has at most q dimensions. By definition, the space $\mathfrak{H}(\mathbf{x}, t)$ is spanned by the variables $x_j(\tau)$ with $j = 1, \ldots, q$ and $\tau \leq t$. Writing

$$x_j(t) = y_j + P_{t-0}\, x_j(t), \quad (j = 1, \ldots, q), \tag{13}$$

where P_{t-0} denotes the projection on $\mathfrak{H}(\mathbf{x}, t-0)$, it will be seen that every element of $\mathfrak{H}(\mathbf{x}, t)$ is the sum of an element of $\mathfrak{H}(\mathbf{x}, t-0)$ and a linear combination of the variables $y_1, \ldots, y_q$, which belong to $\mathfrak{H}(\mathbf{x}, t)$ and are orthogonal to $\mathfrak{H}(\mathbf{x}, t-0)$. Consequently the orthogonal complement $\mathfrak{H}(\mathbf{x}, t) - \mathfrak{H}(\mathbf{x}, t-0)$ is identical with the space spanned by $y_1, \ldots, y_q$, and thus has at most q dimensions. It will also be seen that, choosing the variables $y_1, \ldots, y_q$ in an appropriate way, we can construct examples of processes having, in a given point, a left discontinuity of the innovation spectrum with any given multiplicity not exceeding q. Thus (b) is proved.

We shall now prove (c) by constructing an example of an $\mathbf{x}(t)$ process, the innovation spectrum of which has, in a given point, a right discontinuity of infinite multiplicity. It will then be easily seen how the example may be modified in order to produce a discontinuity of any given finite multiplicity.

Let $z_1, z_2, \ldots$ be an infinite orthonormal sequence of random variables belonging to $\mathfrak{H}$. Denoting by $p_1, p_2, \ldots$ the successive prime numbers ($p_1 = 2$, $p_2 = 3, \ldots$), we define a process $\mathbf{x}(t) = (x_1(t), \ldots, x_q(t))$ by taking

$$x_2(t) = \cdots = x_q(t) = 0 \quad \text{for all } t,$$

$$x_1(t) = 0 \quad \text{for} \quad t \leqslant 0,$$

$$x_1(t) = t\, z_n \quad \text{for} \quad t = p_n^{-k}, \quad k = 1, 2, \ldots$$

$$x_1(t) = t\, z_1 \quad \text{for} \quad t > \tfrac{1}{2}.$$

For values of t in the interval $0 < t < \frac{1}{2}$, which are not of the form p_n^{-k}, we define $x_1(t)$ by linear interpolation:

$$x_1(t) = \frac{(t_2 - t)\, x_1(t_1) + (t - t_1)\, x_1(t_2)}{t_2 - t_1},$$

where $t_1 = p_{n_1}^{-k_1}$ and $t_2 = p_{n_2}^{-k_2}$ are the nearest values below and above t, for which $x_1(t)$ has been defined above. Some easy calculation shows that we have for all $t > 0$

$$E\,|x_1(t)|^2 \leqslant t^2.$$

This shows that the point $t = 0$ is a continuity point for the vector process $\mathbf{x}(t)$. Since every other real t is evidently also a continuity point, the $\mathbf{x}(t)$ process is everywhere continuous, and *a fortiori* satisfies condition (A).

On the other hand, we obviously have $\mathfrak{H}(\mathbf{x}, t) = 0$ for $t \leqslant 0$, while for every $t > 0$ the space $\mathfrak{H}(\mathbf{x}, t)$ will include the infinite orthonormal sequence $z_1, z_2, \ldots$. The point $t = 0$ will thus be a right discontinuity of the innovation spectrum of $\mathbf{x}(t)$, with an infinite right multiplicity. A case with any given finite multiplicity n is obtained if the variables $z_{n+1}, z_{n+2}, \ldots$ are replaced by zero. We have thus proved point (c) of the Lemma.

We observe that, by some further elaboration of the example given above, we may construct a process having the same multiplicity properties in each point of an everywhere dense set of values of t. We can even arrange the example so that the mean square derivative of $\mathbf{x}(t)$ exists for every t.

The last part of the Lemma follows simply from the above proofs of (b) and (c). If t is a left discontinuity of the innovation spectrum, at least one of the variables y_j occurring in (13) must be different from zero. Suppose, e.g., that $y_1 \neq 0$. Then

$$\| x_1(t) - x_1(t-0) \| \geqslant \| x_1(t) - P_{t-0}\, x_1(t) \| = \| y_1 \| > 0,$$

and so t is a left discontinuity of $x_1(t)$, and consequently also of $\mathbf{x}(t)$. On the other hand, the process $\mathbf{x}(t)$ constructed in the proof of (c) provides an example of a process having in $t = 0$ a right discontinuity point of the innovation spectrum (even of infinite multiplicity), which is nevertheless a continuity point of the process. This completes the proof of Lemma 2.

Innovation processes

6. From now on, it will be assumed that we are dealing with a vector process $\mathbf{x}(t)$ satisfying not only condition (A) of section 4, but also the following condition:

(B) $\mathbf{x}(t)$ *is a purely non-deterministic process, i.e. we have* $\mathfrak{H}(\mathbf{x}, -\infty) = 0$.

In the Hilbert space $\mathfrak{H}(\mathbf{x}) = \mathfrak{H}(\mathbf{x}, +\infty)$ of the process, we shall in general denote by $P_{\mathfrak{M}}$ the projection operator whose range is the subspace $\mathfrak{M}$. However, when $\mathfrak{M}$ is the particular subspace $\mathfrak{H}(\mathbf{x}, t)$, we write simply P_t instead of $P_{\mathfrak{H}(\mathbf{x}, t)}$.

As t increases from $-\infty$ to $+\infty$, the P_t form a never decreasing family of projections, with

$$P_{-\infty} = 0, \quad P_{+\infty} = I.$$

P_{t+0} is the projection on $\mathfrak{H}(\mathbf{x}, t+0)$, and similarly for P_{t-0}. The difference $P_{t_2} - P_{t_1}$, where $t_1 < t_2$, denotes the projection on the orthogonal complement $\mathfrak{H}(\mathbf{x}, t_2) - \mathfrak{H}(\mathbf{x}, t_1)$. It follows, in particular, that the projections $P_{t_2} - P_{t_1}$ and $P_{u_2} - P_{u_1}$ will be mutually orthogonal, as soon as the corresponding time intervals are disjoint.

Further, the points t of the innovation spectrum are characterized by the property

$$P_{t+h} - P_{t-h} > 0$$

for any $h > 0$, while the discontinuity points of that spectrum are characterized by the relation

$$P_{t+0} - P_{t-0} > 0.$$

Consider now any element z of the Hilbert space $\mathfrak{H}(\mathbf{x})$, and let us define a stochastic process $z(\lambda)$ by writing for any real λ

$$z(\lambda) = P_\lambda z. \tag{14}$$

It then follows from the above that $z(\lambda)$ is a process *with orthogonal increments*, such that

$$z(-\infty) = 0, \quad z(+\infty) = z.$$

We have $E z(\lambda) = 0$, and if we write

$$E|z(\lambda)|^2 = F_z(\lambda),$$

$F_z(\lambda)$ will be a never decreasing function of the real variable λ, such that

$$F_z(-\infty) = 0, \quad F_z(+\infty) = E|z|^2.$$

The points of increase of $z(\lambda)$, i.e. the points λ such that the increment $z(\lambda+h) - z(\lambda-h)$ does not reduce to zero for any $h > 0$, are identical with the points of increase of $F_z(\lambda)$, and form a subset of the innovation spectrum of the $\mathbf{x}(t)$ process. Similarly the left (right) discontinuities of $z(\lambda)$ are identical with the left (right) discontinuities of $F_z(\lambda)$, and form a subset of the set of all left (right) discontinuities of the innovation spectrum of $\mathbf{x}(t)$. Any increment $dz(\lambda)$ belongs to the subspace $d_\lambda \mathfrak{H}(\mathbf{x}, \lambda)$, and is thus built up by a certain part of the elements which enter as innovations into the $\mathbf{x}(t)$ process between the time points $t=\lambda$ and $t=\lambda+d\lambda$.

On account of these facts, we shall denote the (one-dimensional) $z(\lambda)$ process as a *partial innovation process* associated with the given vector process $\mathbf{x}(t)$.

7. For any $t \leqslant +\infty$, we shall denote by $\mathfrak{H}(z, t)$ the Hilbert space spanned by the random variables $z(\lambda)$ for all $\lambda \leqslant t$:

$$\mathfrak{H}(z, t) = \mathfrak{S}\{z(\lambda);\quad \lambda \leqslant t\}.$$

It follows from (14) that $z(\lambda)$ is always an element of $\mathfrak{H}(\mathbf{x}, \lambda)$, and consequently $\mathfrak{H}(z, t)$, is a subspace of $\mathfrak{H}(\mathbf{x}, t)$:

$$\mathfrak{H}(z, t) \subset \mathfrak{H}(\mathbf{x}, t).$$

Instead of $\mathfrak{H}(z, +\infty)$ we shall write briefly $\mathfrak{H}(z)$. Evidently $\mathfrak{H}(z, t)$ is the projection of $\mathfrak{H}(z)$ on $\mathfrak{H}(\mathbf{x}, t)$.

If no λ is at the same time a left and a right discontinuity of $z(\lambda)$, then $\mathfrak{H}(z, t)$ is, for every $t \leqslant +\infty$, identical[1] with the set $\mathfrak{H}^*(z, t)$ of all random variables y representable in the form

$$y = \int_{-\infty}^{t} g(\lambda)\, dz(\lambda) \tag{15}$$

with an F_z-measurable g such that

$$E|y|^2 = \int_{-\infty}^{t} |g(\lambda)|^2 \, dF_z(\lambda) < \infty.$$

On the other hand, if in a certain point $\lambda < t$ the left and the right jumps of $z(\lambda)$, say u and v respectively, are both different from zero, all random variables $Au + Bv$ with constant A and B will belong to $\mathfrak{H}(z, t)$ while, with the usual definition of the integral (15), the discontinuity at λ will only provide the variables $A(u+v)$ as elements of $\mathfrak{H}^*(z, t)$.

We shall require the following Lemma, which is only a restatement of familiar facts concerning Hilbert space.

Lemma 3. *If y and z are elements of $\mathfrak{H}(\mathbf{x})$ such that $y \perp \mathfrak{H}(z)$, then $\mathfrak{H}(y) \perp \mathfrak{H}(z)$.*

Since $\mathfrak{H}(z)$ is the space spanned by all $z(\lambda) = P_\lambda z$, the relation $y \perp \mathfrak{H}(z)$ is equivalent to $y \perp P_\lambda z$ for all real λ. By the same argument, the assertion of Lemma 3 is equivalent to the relation

$$P_\lambda y \perp P_\mu z$$

---

[1] Cf., e.g., [2], p. 425–428. The integral in (15) should be so defined that, if the upper limit t is a discontinuity of $z(\lambda)$, a left jump of $z(\lambda)$ is included in the value of the integral, but not a right jump.

for all λ and μ. Now $y = P_\lambda y + w$, where $w \perp \mathfrak{H}(\mathbf{x}, \lambda)$, and thus in particular $w \perp P_\lambda z$. Hence for any λ

$$P_\lambda y = y - w \perp P_\lambda z.$$

Suppose now first $\lambda > \mu$. Then $P_\mu P_\lambda = P_\mu$, and thus

$$P_\lambda y = P_\mu y + w',$$

where $w' \perp \mathfrak{H}(\mathbf{x}, \mu)$. Hence in particular $w' \perp P_\mu z$. Since we have just proved that $P_\mu y \perp P_\mu z$, it follows that $P_\lambda y \perp P_\mu z$.

On the other hand, if $\lambda < \mu$ we have $P_\mu z = P_\lambda z + w''$, where $w'' \perp \mathfrak{H}(\mathbf{x}, \lambda)$, and thus $w'' \perp P_\lambda y$. Hence we obtain as before $P_\mu z \perp P_\lambda y$, and so Lemma 3 is proved.

Representation of $\mathbf{x}(t)$

8. We begin by proving the following Lemma, assuming as before that we are dealing with a given vector process $\mathbf{x}(t)$ satisfying conditions (A) and (B).

Lemma 4. *It is possible to find a finite or infinite sequence $z_1, z_2, \ldots$ of non-vanishing elements of $\mathfrak{H}(\mathbf{x})$ such that we have for every $t \leqslant +\infty$*

$$\mathfrak{H}(z_j, t) \perp \mathfrak{H}(z_k, t) \quad \text{for} \quad j \neq k, \tag{16}$$

$$\mathfrak{H}(\mathbf{x}, t) = \mathfrak{H}(z_1, t) + \mathfrak{H}(z_2, t) + \ldots, \tag{17}$$

where the second member of (17) *denotes the vector sum of the mutually orthogonal spaces involved.*

We first observe that it is sufficient to show that we can find a z_n sequence such that (16) and (17) hold for $t = +\infty$, since their validity for any finite t then easily follows.

By Lemma 1, the space, $\mathfrak{H}(\mathbf{x})$ is separable. Neglecting trivial cases, we may assume that $\mathfrak{H}(\mathbf{x})$ is infinite-dimensional. Thus a complete orthonormal system in $\mathfrak{H}(\mathbf{x})$ will form an infinite sequence, say $z_1^*, z_2^*, \ldots$. Starting from the sequence of the z_n^*, we shall now construct an infinite sequence $z_1, z_2, \ldots$ satisfying (16) and (17) for $t = +\infty$. Discarding any z_n which reduces to zero, we then obtain a finite or enumerable sequence of non-vanishing elements having the same properties, and so the Lemma will be proved.

We define the z_n sequence by the relations

$$\begin{aligned} z_1 &= z_1^*, \\ z_2 &= z_2^* - P_{\mathfrak{M}_1} z_2^*, \\ &\cdots\cdots \\ z_n &= z_n^* - P_{\mathfrak{M}_{n-1}} z_n^*, \\ &\cdots\cdots \end{aligned} \tag{18}$$

where $\mathfrak{M}_n$ denotes the vector sum

$$\mathfrak{M}_n = \mathfrak{H}(z_1) + \cdots + \mathfrak{H}(z_n).$$

Then for any $n > 1$ we have $z_n \perp \mathfrak{M}_{n-1}$, and consequently $z_n \perp \mathfrak{H}(z_j)$ for $j = 1, \ldots, n-1$. By Lemma 3 it then follows that we have $\mathfrak{H}(z_j) \perp \mathfrak{H}(z_k)$ for $j \neq k$, so that (16) is satisfied.

We further have

$$z_n^* = z_n + P_{\mathfrak{M}_{n-1}} z_n^*,$$

and since $z_n \in \mathfrak{H}(z_n)$, this shows that $z_n^* \in \mathfrak{H}(z_n) + \mathfrak{M}_{n-1} = \mathfrak{M}_n$. Now $\mathfrak{M}_n$ is a subset of the infinite vector sum $\mathfrak{H}(z_1) + \mathfrak{H}(z_2) + \ldots$. This sum, which is a subspace of $\mathfrak{H}(\mathbf{x})$, will thus contain the complete orthonormal system $z_1^*, z_2^*, \ldots$, and will consequently be identical with $\mathfrak{H}(\mathbf{x})$, so that (17) is also satisfied, and the Lemma is proved.

9. The sequence $z_1, z_2, \ldots$ considered in Lemma 4 is not uniquely determined, and we now proceed to show that it can be chosen in a way which will suit our purpose.

By Lemma 2, the discontinuities of the innovation spectrum of $\mathbf{x}(t)$ form an at most enumerable set. Let them be denoted by $\lambda_1. \lambda_2, \ldots$, and consider the subspaces

$$\mathfrak{U}_k = \mathfrak{H}(\mathbf{x}, \lambda_k) - \mathfrak{H}(\mathbf{x}, \lambda_k - 0),$$

$$\mathfrak{B}_k = \mathfrak{H}(\mathbf{x}, \lambda_k + 0) - \mathfrak{H}(\mathbf{x}, \lambda_k),$$

for $k = 1, 2, \ldots$. If h_k and j_k are the numbers of dimensions of $\mathfrak{U}_k$ and $\mathfrak{B}_k$ respectively, then by Lemma 2 we have $0 \leqslant h_k \leqslant q$, while j_k may be any finite non-negative integer or $\aleph_0$. According to the terminology of section 5, h_k is the left multiplicity of the point λ_k, while j_k is the right multiplicity. The number

$$N' = \sup_k \,(h_k + j_k)$$

will be called the *multiplicity of the discontinuous part* of the innovation spectrum. Let

$$u_{k1}, \ldots, u_{kh_k},$$

$$v_{k1}, \ldots, v_{kj_k}$$

be complete orthonormal systems in $\mathfrak{U}_k$ and $\mathfrak{B}_k$ respectively. If $\mathfrak{U}_k$ or $\mathfrak{B}_k$ reduces to zero, the corresponding system does of course not occur; however, when λ_k is a discontinuity, h_k and j_k cannot both be equal to zero, so that at least one of the u and v systems must contain a non-vanishing number of terms.

If z denotes any of the u_{kh} or v_{kj}, it will be seen that we have $P_\lambda z = z$ for $\lambda > \lambda_k$, and $P_\lambda z = 0$ for $\lambda < \lambda_k$. It follows that the space $\mathfrak{H}(z)$ will then be one-dimensional, and consist of all constant multiples of z. If y is any variable in $\mathfrak{H}(\mathbf{x})$ such that $y \perp z$, it then follows from Lemma 3 that $\mathfrak{H}(y) \perp \mathfrak{H}(z)$.

Suppose now that one of the orthonormal variables from which we started our proof of Lemma 4, say z_n^*, is identical with one of the u_{kh} or v_{kj}. By means of the remark just made, it then follows from the relations (18) that z_n^*

is orthogonal to $z_1, \ldots, z_{n-1}$, and thus also orthogonal to $\mathfrak{M}_{n-1}$. Consequently by (18) we obtain $z_n = z_n^*$.

Choosing the orthonormal system $z_1^*, z_2^*, \ldots$, so that all the variables u_{kh} and v_{kj} $(k=1, 2, \ldots, h=1, \ldots, h_k, j=1, \ldots, j_k)$ occur in it, we thus see that the same variables will also occur in the sequence $z_1, z_2, \ldots$ constructed according to (18), and satisfying the conditions of Lemma 4. Besides the u_{kh} and v_{kj}, there may be other elements in the z_n sequence. Let $w_1, w_2, \ldots$ be those elements, if any, which are different from all the u_{kh} and v_{kj}. We are going to show that, if

$$w(\lambda) = P_\lambda w$$

s the partial innovation process corresponding to any of the w_n, then $w(\lambda)$ *has no discontinuities.*

In fact, by a remark made in section 6, any discontinuity of $w(\lambda)$ would be a discontinuity of the innovation spectrum of $\mathbf{x}(t)$, i.e. equal to one of the λ_k. The corresponding jump of $w(\lambda)$, say w^*, would then be an element of the space $\mathfrak{H}(\mathbf{x}, \lambda_k+0) - \mathfrak{H}(\mathbf{x}, \lambda_k-0)$. At the same time, w^* would belong to the space $\mathfrak{H}(w)$, and by Lemma 4 would thus be orthogonal to all the u_{kh} and v_{kj}. However, the latter variables form a complete orthonormal system in $\mathfrak{H}(\mathbf{x}, \lambda_k+0) - \mathfrak{H}(\mathbf{x}, \lambda_k-0)$ so that we must have $w^*=0$, and it follows that $w(\lambda)$ has no discontinuities.

Let now the sequence of non-vanishing elements $z_1, z_2, \ldots$ considered in Lemma 4 be chosen in all possible ways that are consistent with the requirement that all the u_{kh} and v_{kj} should occur in it. Let, in each case, M denote the cardinal number of the corresponding sequence $w_1, w_2, \ldots$, formed by those elements which are different from all the u_{kh} and v_{kj}. The numbers M will then have a non-negative lower bound:

$$N'' = \inf M,$$

which we shall call the *multiplicity of the continuous part* of the innovation spectrum. Finally

$$N = \max(N', N'') \tag{19}$$

will be called the *spectral multiplicity* of the $\mathbf{x}(t)$ process. As soon as $\mathbf{x}(t)$ is not identically zero, N will be a finite positive integer, or equal to $\aleph_0$.

It follows from the definition of the multiplicity N'' of the continuous part that it is possible to find a sequence $z_1, z_2, \ldots$ satisfying the above requirements, and such that the corresponding set $w_1, w_2, \ldots$ will have precisely the cardinal number N''. We shall then say that these w_n form a *minimal w sequence.*[1]

In the sequel, $w_1, w_2, \ldots$ will denote the elements of a fixed minimal w sequence. We now propose to construct, by means of this given w sequence, a particular sequence $z_1, z_2, \ldots$ satisfying the conditions of Lemma 4, which will then be used for the proof of our representation theorem for $\mathbf{x}(t)$.

---

[1] By an adaptation of the proofs of theorem 7.5 and 7.6 of Stone [7] to the case considered here, it can be shown that a minimal w sequence can be chosen in such a way that the set of all points of increase of $w_1(\lambda) = P_\lambda w_1$ is identical with the continuous part of the innovation spectrum of $\mathbf{x}(t)$, and includes the corresponding set of any $w_n(\lambda)$ with $n > 1$ as a subset. As this property is not indispensable for the proof of the representation theorem given below, we shall restrict ourselves here to this remark.

We first observe that, owing to the way in which the w_j have been chosen, we have by Lemma 4 for every $t \leqslant +\infty$

$$\mathfrak{H}(\mathbf{x}, t) = \sum_{k,h} \mathfrak{H}(u_{kh}, t) + \sum_{k,j} \mathfrak{H}(v_{kj}, t) + \sum_{j} \mathfrak{H}(w_j, t), \tag{20}$$

where the sums denote vector addition, and all the $\mathfrak{H}$ spaces appearing in the second member are mutually orthogonal.

For any discontinuity point λ_k we now arrange the corresponding variables u_{kh} and v_{kj} into a single sequence

$$s_{k1}, s_{k2}, \ldots,$$

where the number of terms will be $h_k + j_k$. The new sequence $z_1, z_2, \ldots$ which we have in view is then defined by taking

$$z_n = w_n + \sum_k s_{kn}, \tag{21}$$

where the w_n are the elements of our fixed minimal w sequence. The summation is extended over all discontinuity points λ_k, and we take $s_{kn} = 0$ whenever $n > h_k + j_k$, and $w_n = 0$ whenever $n > N''$. The number of non-vanishing terms in the z_n sequence defined in this way will evidently be equal to the spectral multiplicity N of the $\mathbf{x}(t)$ process as defined by (19).

According to a remark made above, any space $\mathfrak{H}(s_{kn})$ is one-dimensional, and consists of all constant multiples of the variable s_{kn}. Hence it easily follows that we have for $t \leqslant +\infty$

$$\mathfrak{H}(z_n, t) = \mathfrak{H}(w_n, t) + \sum_k \mathfrak{H}(s_{kn}, t),$$

and further according to (20)

$$\mathfrak{H}(\mathbf{x}, t) = \mathfrak{H}(z_1, t) + \mathfrak{H}(z_2, t) + \ldots,$$
$$\mathfrak{H}(z_j, t) \perp \mathfrak{H}(z_k, t) \quad \text{for} \quad j \neq k, \tag{22}$$

so that the z_n defined by (21) satisfy the conditions of Lemma 4.

We observe that it follows from (21) that no $z_n(\lambda)$ can have a left and a right discontinuity in the same point λ. In fact, any discontinuity of $z_n(\lambda)$ will be either a left discontinuity with a jump u_{kh}, or a right discontinuity with a jump v_{kj}. By a remark made in section 6, the space $\mathfrak{H}(z_n, t)$ will then be identical with the set of all random variables y representable in the form (15).

In our given q-dimensional (column) vector process

$$\mathbf{x}(t) = (x_1(t), \ldots, x_q(t))$$

every component $x_j(t)$ is a random variable belonging to $\mathfrak{H}(\mathbf{x}, t)$. It then follows from (15) and (22) that we have for all real t

$$x_j(t)=\sum_{n=1}^{N}\int_{-\infty}^{t} g_{jn}(t,\lambda)\,d z_n(\lambda). \tag{23}$$

If the multiplicity N is infinite, the series in the second member will converge in the usual sense, so that we have

$$\sum_{n=1}^{N}\int_{-\infty}^{t} |g_{jn}(t,\lambda)|^2\,d F_{z_n}(\lambda)<\infty. \tag{24}$$

Introducing the $q\times N$ order matrix function

$$\mathbf{G}(t,\lambda)=\{g_{jn}(t,\lambda)\}, \tag{25}$$

$(j=1,\ldots,q;\,n=1,\ldots,N)$, and the N-dimensional (column) vector process

$$\mathbf{z}(\lambda)=(z_1(\lambda),\ldots,z_N(\lambda)), \tag{26}$$

it is seen that (23) may be written

$$\mathbf{x}(t)=\int_{-\infty}^{t}\mathbf{G}(t,\lambda)\,d\,\mathbf{z}(\lambda). \tag{27}$$

The vector process $\mathbf{z}(\lambda)$ defined by (26) has *orthogonal increments*, in the sense that two increments $\Delta z_j(\lambda)$ and $\Delta z_k(\mu)$ are always orthogonal if $j\neq k$, while for $j=k$ they are orthogonal if the corresponding time intervals are disjoint.

If we denote by $\mathfrak{H}(\mathbf{z},t)$ the Hilbert space spanned by the variables $z_1(\lambda),\ldots,z_N(\lambda)$ for all $\lambda\leqslant t$:

$$\mathfrak{H}(\mathbf{z},t)=\mathfrak{S}\{z_1(\lambda),\ldots,z_N(\lambda);\ \lambda\leqslant t\},$$

it follows from (22) that we have for every $t\leqslant+\infty$

$$\mathfrak{H}(\mathbf{x},t)=\mathfrak{H}(\mathbf{z},t). \tag{28}$$

According to the representation formula (27) and the property expressed by (28), it seems appropriate to call $\mathbf{z}(\lambda)$ a *total innovation process* associated with the given $\mathbf{x}(t)$. While $\mathbf{z}(\lambda)$ is not uniquely determined, its dimensionality N is uniquely determined by (19) as the spectral multiplicity of $\mathbf{x}(t)$. It is also seen that N is the smallest cardinal number for which there exists a representation of the form (27), with the properties specified by (22)–(28).

Summing up our results, we now have the following representation theorem.

Theorem 1. *Any stochastic vector process* $\mathbf{x}(t)$ *satisfying conditions* (A) *and* (B) *can be represented in the form* (27), *where* $\mathbf{G}(t,\lambda)$ *and* $\mathbf{z}(\lambda)$ *are defined by* (25) *and* (26). N *is the spectral multiplicity of the* $\mathbf{x}(t)$ *process. If* N *is infinite, the expansions* (23) *formally obtained for the components* $x_j(t)$ *are convergent in quadratic mean, as shown by* (24). $z_1,\ldots,z_N$ *are random variables in* $\mathfrak{H}(\mathbf{x})$ *satisfying* (22), *and such that no* $z_j(\lambda)=P_\lambda z_j$ *has a left and a right discontinuity in the same point* λ. *The vector process* $\mathbf{z}(\lambda)$ *has orthogonal increments and satisfies* (28).

No representation with these properties holds for any smaller value of N.

If $\mathbf{x}(t)=(x_1(t), \ldots, x_q(t))$ is a vector process satisfying (A) and (B), each component $x_j(t)$, regarded as a one-dimensional process, has a certain spectral multiplicity N_j, and thus by Theorem 1 may be represented in the following form

$$x_j(t)=\sum_{n=1}^{N_j}\int_{-\infty}^{t} g_{jn}(t,\lambda)\,d z_{jn}(\lambda).$$

It can then be shown (although the proof is slightly more involved than may possibly be expected) that the spectral multiplicity N of $\mathbf{x}(t)$ satisfies the inequality

$$N \leqslant \sum_{j=1}^{q} N_j. \tag{29}$$

Consider, in particular, the case of a *stationary* vector process $\mathbf{x}(t)$, i.e. a process such that every second order covariance moment of the components is a function of the corresponding time difference:

$$E\left(x_j(t)\,\overline{x_k(u)}\right)=R_{jk}(t-u).$$

This process will satisfy (A) and (B) if and only if (*a*) the functions $R_{11}(t), \ldots, R_{qq}(t)$ are continuous at $t=0$, and (*b*) each component $x_j(t)$ is a purely non-deterministic stationary process. When these conditions are satisfied, each $x_j(t)$ has a representation of the form (7), and accordingly the spectral multiplicity of $x_j(t)$ is equal to one. It then follows from (29) that we have in this case $N \leqslant q$.

On the other hand, the process $\mathbf{x}(t)$ constructed in connection with the proof of Lemma 2, point (*c*), evidently provides an example of a vector process satisfying (A) and (B), and having an infinite spectral multiplicity N. As already observed, this example may be easily modified so as to yield a process with any given finite multiplicity.

If, in the relation (27), all components of the vectors on both sides are projected on the space $H(\mathbf{x}, u)$, where $u<t$, we finally obtain the following theorem, denoting by $P_u \mathbf{x}(t)$ the vector with the components $P_u x_j(t)$, for $j=1, \ldots, q$.

Theorem 2. *The best linear (least squares) prediction of* $\mathbf{x}(t)$ *in terms of all variables* $x_j(\tau)$ *with* $j=1, \ldots, q$ *and* $\tau \leqslant u$ *is given by the expression*

$$P_u \mathbf{x}(t)=\int_{-\infty}^{u} \mathbf{G}(t,\lambda)\,d\mathbf{z}(\lambda).$$

The square of the corresponding error of prediction for any component $x_j(t)$ *is*

$$E\left|x_j(t)-P_u x_j(t)\right|^2=\sum_{n=1}^{N}\int_{u}^{t}\left|g_{jn}(t,\lambda)\right|^2 d F_{z_n}(\lambda).$$

REFERENCES

1. CRAMÉR, H., On some classes of non-stationary stochastic processes. To appear in: Proc. Fourth Berkeley Symp. on Mathematical Statistics and Probability.
2. DOOB, J. L., Stochastic processes. Wiley, New York, 1953.
3. HIDA, T., Canonical representations of Gaussian processes and their applications. Mem. Coll. Sci. Kyoto, A33 (1960).
4. LÉVY, P., A special problem of Brownian motion, and a general theory of Gaussian random functions. Proc. Third Berkeley Symp. on Mathematical Statistics and Probability, *2*, 133 (1956).
5. ——, Sur une classe de courbes de l'espace de Hilbert et sur une équation intégrale non linéaire. Ann. Ec. Norm. Sup. *73*, 121 (1956).
6. ——, Fonctions aléatoires à corrélation linéaire. Illinois J. Math. *1*, 217 (1957).
7. STONE, M. H., Linear transformations in Hilbert space. American Math. Soc. colloquium publication. New York, 1932.
8. WIENER, N., and MASANI, P., The prediction theory of multivariate stochastic processes. Acta Math. *98*, 111 (1957); and *99*, 93 (1958).
9. WOLD, H., A Study in the Analysis of Stationary Time Series. Thesis, University of Stockholm. Uppsala, 1938.

Tryckt den 22 februari 1961

Uppsala 1961. Almqvist & Wiksells Boktryckeri AB.

52.

Model building with the aid of stochastic processes

Bull. Inst. Internat. Statist. **39**, 2, 3–30 (1961)
(Reprinted 1964 in Technometrics **6**, 133–159)

I. Introduction.

1. In building a mathematical model of some group of observed phenomena, we establish a correspondence between the main features of our observations on the one side, and the abstract mathematical concepts used as building-stones of our model on the other side. In order that the model should serve a useful purpose, this correspondence must be so chosen that it yields an adequate idealized *description* of the available observations in terms of the mathematical concepts used.

If, in the course of repeated observations, we have been able to verify that a proposed model gives a sufficiently accurate and permanent agreement with observed facts, we may feel justified in using the model for an *analysis* of the structure of the phenomena under investigation, for *predictions* bearing on the results of future observations, and for practical *decision-making.*

In all cases when an element of *randomness* appears in the phenomena, the mathematical model will include some *probabilistic* concepts. Certain quantities will be regarded as random variables, the probability distributions of which must be more or less specified in the course of building the model. The process of verifying the accuracy of the model, and of using it for analysis, prediction, and decision-making, will then assume *statistical* character.

The classical theory of random variables in spaces of a finite number of dimensions is able to deal adequately with a large number of situations, where probabilistic models are required. During recent times, however, an increasing number of important cases have arisen in widely different fields of application, where it is necessary to consider models involving the use of *random variables in infinite-dimensional spaces.* The object of this paper is to give a brief survey of some typical cases of model building in situations of this kind.

The mathematical framework used in this connection is the modern theory of *stochastic processes.* In the following sections of this Introduction, we shall give a very brief sketch of the historical background of this theory, and then proceed to recall some fundamental definitions and properties associated with the main classes of stochastic processes that will be used in the sequel.

2. During the nineteenth century, the molecular theory of the structure of matter attained a dominant role in physics. Each portion of matter was conceived as a sys-

1

tem, consisting of a very large number of small particles, subject to the laws of Newtonian mechanics. Theoretically, the development in time of such a system from a given initial state would be uniquely determined by the differential equations of the movement, if certain geometrical and physical properties of the particles were regarded as known. However, a deterministic model of this kind would be completely useless in practice, on account of the large number of particles, and our lack of knowledge concerning their properties.

Accordingly, we find already at an early stage that probabilistic models are introduced in the kinetic theory of matter. In the works of Clausius, Maxwell and Boltzmann, from about 1860 on, models of this kind were systematically used in kinetic gas theory. A little later, the foundations of a general theory of statistical models in mechanics were laid by Gibbs (1902). The temporal development of a mechanical system possessing a very large number of degrees of freedom is here regarded as governed by probability laws, about which more or less definite hypotheses are made.

It soon became clear that this was an extremely fertile point of view. Not only in various parts of physics, but in a large number of other applied fields, we are interested in studying the temporal development of some system, which may be regarded as subject to randomly varying influences. Let us suppose that we observe a certain variable quantity x associated with a system of this kind, and that at the time t this quantity assumes the value $x(t)$. As t varies through a certain range of values, we thus obtain a function $x(t)$ associated with the temporal development of the system.

In a probabilistic model of this phenomenon, we may regard $x(t)$, for each fixed value of t, as a random variable, and propose to study the probability relations between the variables $x(t)$ for various values of t. From another point of view, we may consider the various possible « realizations » of the development of the system, i.e. the various possible functions $x(t)$, as random elements, the probabilistic properties of which have to be studied.

For an idealized mathematical picture, it will in most cases be desirable to consider a quantity like $x(t)$ for an infinity of values of t. Obviously this implies the necessity of studying random variables and probability distributions in spaces of infinitely many dimensions.

A general theory of infinite-dimensional probability distributions did not exist until the beginning of the 1930's. Already from the beginning of the century, however, various important particular cases were considered, in connection with the construction of special probability models designed for use in widely different applied fields. We shall briefly mention a number of these pioneering works, all published before 1930.

3. In a series of works beginning in 1900, Bachelier introduced and developed a probabilistic model for stock market operations. His methods were largely heuristic, but the model was mathematically equivalent to the well-known model encountered in 1906 by Einstein and Smoluchovski, in connection with the theory of the Brownian movement. A rigorous mathematical treatment of this model was given in 1923 by Wiener.

In the Einstein-Smoluchovski-Wiener model of the Brownian movement, the fundamental random variable $x(t)$ will represent *e.g.* the abscissa at time t of a small

2

particle immersed in a liquid, and subject to a tremendous number of irregular shocks from the molecules of the liquid. In order that a function $x(t)$ of the time t should be acceptable as a mathematical idealization of the path described by a physical particle, it seems essential to require that it should be a continuous function of t, however irregular. Thus, it was an important result when Wiener was able to show that, in his version of the Brownian movement model $x(t)$ is, with a probability equal to one, everywhere continuous as a function of t.

A radically different type of probabilistic model for the temporal development of a physical system presented itself in connection with the study of sequences of random events occurring successively in time. In such a case, if we denote by $x(t)$ the number of events occurred from the beginning of our observations up to the instant t, it is obvious that $x(t)$ will be an essentially discontinuous function of t, increasing by unit steps in those time points where an event occurs. In the case when the events are incoming calls at a telephone exchange, it was shown by Erlang in 1909 that $x(t)$ may, under reasonable assumptions, be expected to have a Poisson distribution. This result became the starting point for the construction of important probabilistic models for telephone traffic and other types of queueing problems. On the other hand, a famous experiment published by Rutherford and Geiger in 1910 showed that the same Poisson model is applicable to radioactive disintegrations, where $x(t)$ may denote e.g. the number of α particles radiated from a portion of radium during t seconds.

The Poisson model is at least approximately applicable to many other types of random events, e.g. to the occurrence of claims in an insurance company. A simple model for the total risk business of such a company may be built on the assumption that claims occur according to the Poisson model, while the amount due under any claim is a random variable with a given distribution. The theory of this model was developed by F. Lundberg in a series of works from 1909 on, and by Cramér from 1919.

In the course of the study of time series appearing e.g. in geophysical, economic, and demographic statistics, intricate problems connected with the interaction between random and causal factors were encountered. Problems of correlation in and between time series, and of the detection of periodic components in time series, were investigated from the point of view of probabilistic models in a series of papers by Yule from 1921 on. Important contributions to this subject were also given during the same epoch by Slutsky.

The foundations of a statistical theory of turbulence were laid by Taylor in the 1920's. This case differs from all those mentioned above in that we are here concerned with a random variation not only in time, but also in space. In fact, a typical random variable encountered in this theory will be a component, say $u(x, y, z, t)$, of the velocity of the turbulent liquid at the point x, y, z, t of the space-time « field ». A probabilistic model of turbulence will have to be built on certain hypotheses concerning the joint distributions of random variables of this type. Models for spacial random variation were also considered at an early stage in connection with varions ecological distribution problems.

About 1930, time was evidently ripe for the construction of a general theory. A great number of particular cases involving random variation in time or space had been considered, but mathematical rigour was often lacking, and the need was felt for a

3

general structure into which each particular case could be fitted in its proper place.

The foundations of such a general mathematical theory of random variation were laid from 1931 on by the works of Kolmogoroff, Khintchine, Lévy, Feller, and Doob, followed by a large number of other authors.

4. The name *stochastic process* was applied, in the first instance, to any observable process developing in time under the influence of causes, which are conceived as randomly variable. In the majority of practical applications, the state of such a system is observed either for a sequence of discrete and equidistant time points, or for a continuously variable time t. Accordingly, we shall talk of a stochastic process with *discrete* or *continuous* time. The former case will usually present itself in connection with repeated statistical observations or measurements, while the latter case occurs e.g. when we are concerned with a curve drawn by some continuously registering instrument.

The state of such a system at any time t will, in general, be represented by a certain number of quantities, such as $x(t)$ etc. For the purpose of mathematical model building these quantities should, for any fixed t, be regarded as random variables. On the other hand, a particular *realization* of the process will be expressed by the numerical course, in an individual observed case, of the functions $x(t)$ etc., throughout the range of values of t considered.

Gradually, the name of stochastic process came to be applied not only to the physical process itself, but also to the mathematical model set up for its study. Thus e.g. the collection of random variables $x(t)$ encountered in the study of the Brownian movement, with their joint probability distributions for all considered values of t, will constitute a stochastic process, forming the mathematical counterpart of the physical process represented by the moving particle. The same mathematical process may be characterized by an appropriate probability distribution in the space of all functions $x(t)$ that we consider as possible alternatives for describing the path of the particle.

From a purely mathematical point of view, a stochastic process may thus be defined as a *family of random variables* $x(t)$ depending on a parameter t, or alternatively as a *probability distribution in a certain space of functions* $x(t)$. We shall not go deeper into the questions connected with the relations between these two modes of definition, but shall only make some general comments that will be useful in the sequel.

Obviously it is by no means necessary for the mathematical theory that the parameter t should represent time. In theoretical developments t may be regarded as an element of some arbitrarily given parameter set T, and in practical applications t may sometimes represent time, sometimes a spatial variable, and sometimes a point in space-time. Also the random variable $x(t)$ may, according to the case, be a real-valued or a complex-valued variable, or a vector variable in some specified space.

In order that a stochastic process should be appropriately defined, we must require that the probability distribution of $x(t)$ should be known for every fixed t in the parameter set T. Further, the joint distribution of any pair $x(t_1)$, $x(t_2)$ must be known, and generally the joint distribution of any finite collection of random variables $x(t_1)$... $x(t_n)$, where all the t_j are points of T. All these probability distributions constitute

4

the *family of finite-dimensional distributions* associated with the stochastic $x(t)$ process. The members of this family must satisfy certain evident consistency conditions. Thus if the $x(t)$ are real-valued random variables, and if we consider the two-dimensional distribution of the pair $x(t_1)$, $x(t_2)$, the marginal distribution corresponding to $x(t_1)$ should obviously be identical with the distribution given for the single variable $x(t_1)$. Analogous relations must hold for all finite-dimensional distributions.

When the consistency conditions are satisfied, the knowledge of the family of finite-dimensional distributions will often be sufficient for the solution of problems raised by practical applications. In particular this will generally be the case when we are concerned with questions of correlation between the various $x(t)$, or their correlation with other random variables that may be involved. On the other hand, when we are investigating properties of continuity, derivability etc. of $x(t)$ as a function of t, it will usually be necessary to specify the space of *sample functions* $x(t)$, in which we want to operate. This specification must be made with due regard to the particular circumstances in the problem actually at hand. In most ordinary applications, this can be made without great difficulties, as we shall see in some examples.

5. Suppose that the family of finite-dimensional distributions of a certain real-valued stochastic $x(t)$ process is given, and satisfies the necessary consistency conditions. Let $t > t_1 > \ldots > t_n$, where t and the t_j are positive or negative integers in the case of a process with discrete time, but may have any real values in the case of a process with continuous time.

Let $a, a_1 \ldots, a_n$ be given real numbers, and consider the conditional probability of the relation $x(t) \leq a$, relative to the hypothesis that $x(t_j) = a_j$ for $j = 1, \ldots n$.

Suppose that this conditional probability is independent of the choice of $a_2 \ldots, a_n$. We then have, in the usual symbolism,

$$\mathrm{P}\{x(t) \leq a \mid x(t_j) = a_j, j = 1, \ldots, n\} = \mathrm{P}\{x(t) \leq a \mid x(t_1) = a_1\}.$$

This means that, given the state of the process at time t_1, no additional information concerning the development of the process before the instant t_1 will have any influence on the conditional distribution of the process at time $t > t_1$. The class of stochastic processes satisfying this condition is the important class of *Markov processes.*

A Markov process with discrete time is usually called a *Markov chain.* In this case we write x_n in the place of $x(t)$, and suppose that n may take all positive and negative integral values.

An important particular case of a Markov process with continuous time is obtained if we suppose that the increments of $x(t)$ corresponding to non-overlapping time intervals are always mutually independent random variables. A stochastic process satisfying this condition is known as a process with *independent increments*, and it is readily seen that such a process will satisfy the characteristic property of Markov processes given above. If, in addition, the probability distribution of the increment

$$\Delta x(t) = x(t + \Delta t) - x(t)$$

only depends on the length Δt of the time interval, but is independent of the location of the interval on the time axis, we talk of a process with *stationary and independent increments.*

5

Processes with independent increments have important applications in statistical model building, and some typical examples will be given in sections 7 - 10 below. More general Markov processes, and some related types of non-Markov processes, will then be discussed in sections 11-14.

6. A further class of stochastic processes with important applications in model building is the class of *stationary processes*. We shall here only be concerned with the class of processes usually known as *second order stationary*, or *weakly stationary*, but for the sake of brevity we shall simply call them *stationary*.

For the definition of a stationary process, we shall suppose that the fundamental random variable $x(t)$ of the process is complex-valued, and has for every t the mean value zero and a finite variance :

$$E\{x(t)\}=0, \quad E\{|x(t)|^2\}<\infty.$$

(When, in an applied problem, we want to consider the particular case of a real-valued process, this can be simply performed by imposing a certain symmetry condition, as will be shown later.) We then define the covariance moment of $x(t)$ and $x(u)$ by the expectation

$$E\{x(t)\overline{x(u)}\},$$

where the bar indicates the conjugate complex value. This moment will always be finite, since we have by the Schwarz inequality

$$|E\{x(t)\overline{x(u)}\}|^2 \leqq E\{|x(t)|^2\} E\{|x(u)|^2\}.$$

Suppose now that the covariance moment is invariant under a translation in time, so that we have

$$E\{x(t+h)\overline{x(u+h)}\}=E\{x(t)\overline{x(u)}\}.$$

for any h (arbitrary real in the case of a continuous time process, integer in the discrete case). A process satisfying this condition will be called a *stationary* process.

Taking $h=-u$, it will be seen that the covariance moment of a stationary process is a function of the time diffreence $t-u$, so that we may write

$$E\{x(t)\overline{x(u)}\}=r(t-u), \tag{1}$$

where $r(t)$ is known as the *covariance function* of the *process*.

In the case of a process with continuous time, $-\infty<t<\infty$, a necessary and sufficient condition that a given function $r(t)$ should be the covariance function of a stationary process is that $r(t)$ should have a *spectral representation*

$$r(t)=\int_{-\infty}^{\infty} e^{it\lambda}\, dF(\lambda), \tag{2}$$

where the *spectral function* $F(\lambda)$ is real, never decreasing and bounded. This representation is due to Khintchine [34].

In the discrete time case, we write x_n and r_n in the place of $x(t)$ and $r(t)$ and the spectral representation now assumes the form

$$r_n = \int_{-\pi}^{\pi} e^{in\lambda}\, dF(\lambda), \tag{3}$$

where F (λ) is still real, never decreasing and bounded. (Cf. Wold [55].)

There is also a similar spectral representation for the process variable $x(t)$ or r_n itself. We have, in fact, in the two cases referred to above (Cramer [12], [13]).

$$\begin{aligned} x(t) &= \int_{-\infty}^{\infty} e^{it\lambda}\, dz(\lambda), \\ x_n &= \int_{-\pi}^{\pi} e^{in\lambda}\, dz(\lambda). \end{aligned} \tag{4}$$

Here $z(\lambda)$, the *spectral process* variable, is a complex-valued random variable with mean value $E\{z(\lambda)\} = 0$ for all λ, attached to a stochastic process with *orthogonal increments,* satisfying

$$E\left\{ \Delta z(\lambda)\, \overline{\Delta z(\mu)} \right\} = 0$$

whenever $(\lambda, \lambda + \Delta\lambda)$ and $(\mu, \mu + \Delta\mu)$ are disjoint intervals, and

$$E\left\{ \left| \Delta z(\lambda) \right|^2 \right\} = \Delta F(\lambda),$$

where F (λ) is the spectral function introduced above. The integrals for $x(t)$ and x_n are defined as limits in quadratic mean of the corresponding Riemann sums.

The spectral properties of stationary processes play an important part in the applications.

II. *Processes with independent increments*

7. *The Wiener process.* Consider the idealized case of a particle moving on a straight line under the only influence of random molecular shocks, no outer forces being present. A simple way of introducing a mathematical model of the Brownian movement of this particle consists in regarding that movement as a limiting case of an elementary discrete « random walk ».

Let $x(t)$ denote the abscissa of the particle at time t, and suppose that the position of the particle at $t = t_0$ is known : $x(t_0) = x_0$. Suppose further that, at every instant $t = t_0 + nh$ $(n = 1, 2, \ldots)$, the particle receives a shock resulting in a displacement, which may take one of the values $+\sigma\sqrt{h}$ or $-\sigma\sqrt{h}$, each with a probability equal to 1/2. Here h and σ are given positive quantities. The displacement due to each particular shock is assumed to be stochastically independent of the effects of all previous shocks, and also independent of the initial value x_0.

The probability that, as a result of the first n shocks, there will be precisely s displacements equal to $+\sigma\sqrt{h}$ and $n - s$ equal to $-\sigma\sqrt{h}$ is then

$$\binom{n}{s} 2^{-n},$$

and we shall have

$$x(t_0+nh)-x(t_0)=s\sigma\sqrt{h}-(n-s)\,\sigma\sqrt{h}=2\sigma\sqrt{h}\left(s-\frac{n}{2}\right).$$

It then follows from the elementary theory of the game of heads or tails that the increment $x(t_0+nh)-x(t_0)$ will have zero mean value and a variance equal to $\sigma^2 nh$. Moreover, in the limit as n tends to infinity, the random variable

$$\frac{x(t_0+nh)-x(t_0)}{\sigma\sqrt{nh}}$$

will be asymptotically normally distributed, with zero mean and unit variance, and will be stochastically independent of $x(t_0)$.

Let us now allow h to tend to zero as n tends to infinity, in such a way that

$$nh \longrightarrow t-t_0.$$

This means that the time interval between consecutive shocks, as well as the displacement caused by each particular shock, will tend to zero. It then seems plausible that in the limit we shall obtain a random variable $x(t)$, depending continuously on t in a way that will serve as a mathematical idealization of the path described by the physical particle.

According to the above, the increment $x(t)-x(t_0)$ obtained by this limit process will be normally distributed with zero mean and variance $\sigma^2(t-t_0)$, and will be independent of $x(t_0)$. In the same way we find that, given any two disjoint intervals on the time axis, say $(t, t+\Delta t)$ and $(u, u+\Delta u)$, the corresponding increments

$$\Delta x(t)=x(t+\Delta t)-x(t) \quad \text{and} \quad \Delta x(u)=x(u+\Delta u)-x(u)$$

will be mutually independent random variables, each normally distributed, with zero means and variances $\sigma^2\Delta t$ and $\sigma^2\Delta u$ respectively. We have thus arrived at a stochastic process with *stationary and independent increments*, as defined above in section 5.

If we suppose that the particle starts from the origin at time $t=0$, we have $x(0)=0$, and it will be seen that, for every t on the positive time axis, $x(t)$ is a normally distributed random variable, with zero mean and variance $\sigma^2 t$. The probability density of $x(t)$,

$$f(x,t)=\frac{1}{\sqrt{2\pi\sigma^2 t}}\,e^{-\frac{x^2}{2\sigma^2 t}} \tag{5}$$

satisfies the differential equation

$$\frac{\partial f}{\partial t}=\frac{1}{2}\sigma^2\frac{\partial^2 f}{\partial x^2}, \tag{6}$$

and it is known that (5) is the only solution of (6) satisfying for $t>0$ the conditions

$$f(x,t)\geqq 0,$$

$$\int_{-\infty}^{\infty} f(x,t)\,dx=1,$$

$$\lim_{t\to 0} f(x,t)=0 \quad \text{for all} \quad x\neq 0.$$

8

The two first of these state that $f(x, t)$ is a probability density in x, while the third expresses the initial condition $x(0) = 0$.

Einstein showed by physical arguments that the differential equation (6) will hold, at least in a first approximation, and then deduced the properties of the probability distribution of $x(t)$. According to Einstein, the constant σ^2 has the value

$$\sigma^2 = \frac{kT}{3\pi r\eta} \tag{7}$$

where k is the Boltzmann constant, T the absolute temperature and η the viscosity coefficient of the liquid; while the particle is assumed to be spherical and of radius r.

The probability density (5) and the expression (7) for σ^2 have been tested by observation, and found to give a satisfactory agreement for values of t which are large in comparison with the intervals between successive shocks. In particular, Perrin has shown that (7) yields satisfactory values of the Boltzmann constant k (cf. Chandrasekhar [7]).

The limiting process by which we have here deduced the properties of the $x(t)$ process is of course only heuristic. A rigorous mathematical theory of the process was first given by Wiener [53], cf. also [54]. It was shown by Wiener that it is possible to define, as mentioned at the end of section 4 above, a space of sample functions $x(t)$ appropriate for the study of the Brownian movement considered here. We now proceed to indicate very briefly how this can be done.

Let $y_0, y_1, y_2, \ldots$ and $z_1, z_2, \ldots$ be real random variables, which are all mutually independent and normally distributed with zero means and unit variances. By these properties, a probability measure P is uniquely defined for all Borel sets of the space Ω where the y_j and z_k are coordinates. Evidently Ω has an enumerable infinity of dimensions, and every system of values of the y_j and z_k determines a point ω in Ω.

The series

$$\sum_1^{\infty} \left(y_n \cos 2n\pi t + z_n \sin 2n\pi t \right)$$

is almost certainly divergent, i.e. the series diverges for every ω, with the exception at most of a set of points ω of P-measure zero. On the other hand, the series obtained by term-by-term integration is readily seen to be convergent in quadratic mean, in respect of P-measure. Writing

$$x_N(t) = \sigma y_0 t + \frac{\sigma}{\pi\sqrt{2}} \sum_{n=1}^{2^N} \frac{y_n \sin 2n\pi t + z_n(1 - \cos 2n\pi t)}{n}$$

it is even possible to show that, with a probability equal to one, $x_N(t)$ converges uniformly for $0 \leqq t \leqq 1$ to a limit $x(t)$ as N tends to infinity. To every point ω, with the exception at most of a set of P-measure zero, there thus corresponds a function

$$x(t) = \lim_{N \to \infty} x_N(t)$$

uniquely determined and continuous for $0 \leqq t \leqq 1$. Moreover $x_N(t)$, being a finite linear combination of the normally distributed random variables y_j and z_k, is itself for every fixed t a normally distributed random variable, and this also applies to the limit $x(t)$. It is, in fact, easily shown that $x(t)$ is normally distributed with zero mean and variance $\sigma^2 t$, and generally that $x(t)$ has precisely the same distributional properties as those heuristically deduced above. In particular, the family of random variables $x(t)$ defines a stochastic process with independent increments, such that any increment $\Delta x(t) = x(t + \Delta t) - x(t)$ is normally distributed, with zero mean and variance $\sigma^2 \Delta t$.

The sample space in which we are now operating contains, in a precisely defined sense, almost only continuous functions $x(t)$. As already pointed out, this is a desirable property, since we want to use $x(t)$ as a model function for describing the path of a physical particle. On account of the continuity of the $x(t)$ functions, it is now possible to solve various problems of e.g. the following type : what is the probability that the abscissa of the moving particle, starting from the initial position $x(0) = 0$, will at least once during the time interval $0 < t \leqq T$ exceed a given positive quantity M? The answer to this question is contained in the formula

$$P\{x(t) > M \text{ for some } t \text{ in } 0 < t \leqq T\} = 2P\{x(T) > M\},$$

which can be proved by simple symmetry arguments. Taking derivatives with respect to T, we are led to an expression for the « first-passage » probability that $x(t)$, starting from $x(0) = 0$, will attain the « barrier » M for *the first time* during the infinitesimal time element $(t, t + dt)$. The expression obtained is

$$\frac{M}{\sigma\sqrt{2\pi t^3}} e^{-\frac{M^2}{2\sigma^2 t}} dt. \tag{8}$$

Even though the sample functions $x(t)$ are almost certainly continuous, it can be shown that almost no $x(t)$ has a derivative $x'(t)$ for any value of t. In other words, the particles in this mathematical model of the Brownian movement have no well-defined velocities. This shows that the model can only be accepted as a first approximation, and we shall in fact indicate in section 14 how a somewhat more realistic model can be built. However, it will then be necessary to give up the simple property that the $x(t)$ process has independent increments.

The one-dimensional Brownian movement discussed here is, of course, only a simple particular case of the really interesting physical problem. For the mathematical theory of the corresponding model in any number of dimensions, we refer to Lévy [39].

8. *The Poisson process.* Random events, occurring successively in time at irregular intervals, are encountered in many applied fields. In section 3 we have already mentioned the cases of telephone calls, radio-active disintegrations, and claims in an insurance company.

In order to obtain a model for such cases that may serve as a first approximation, and as a starting point for various generalizations, we assume that the following three conditions are satisfied :

a. The probability $p_n(t)$ that exactly n events will occur in a time interval of length t depends on n and t, but not on the position of the interval on the time scale;

b. The numbers of events occurring in disjoint time intervals are mutually independent random variables;

c. The probability of more than one event in a small time interval of length dt is of a smaller order of magnitude than dt. In the usual symbolism, this probability is thus $o(dt)$ as dt tends to zero.

If we start observations at time 0, and denote by $x(t)$ the number of events observed up to time t, it follows from a and b that, just as in the Wiener process, $x(t)$ will be a random variable associated with a stochastic process with *stationary and independent increments*. However, in the present case $x(t)$ can only assume the values 0, 1, 2..., and will thus be an essentially discontinuous function of t. We propose to show that $x(t)$, as well as any increment $x(u+t) - x(u)$, will have a distribution of the Poisson type.

By condition a, we have for any non-negative t and u, and for $n = 0, 1, 2...$

$$p_n(t) = \mathrm{P}\{x(u+t) - x(u) = n\},$$

where as usual P denotes the probability of the relation written between the brackets. Condition b gives for any t and u

$$p_0(t+u) = p_0(t)p_0(u).$$

The only solution of this equation such that $0 < p_0(t) \leqq 1$ is

$$p_0(t) = e^{-\lambda t}$$

where λ is a non-negative constant. Excluding the trivial case $p_0(t) = 1$ for all t, we may assume $\lambda > 0$. (The further trivial solution $p_0(t) = 0$ is clearly incompatible with b and c.) For small dt this gives

$$p_0(dt) = 1 - \lambda dt + o(dt),$$

while from c we obtain

$$p_1(dt) = 1 - p_0(dt) + o(dt) = \lambda dt + o(dt). \tag{9}$$

Further, b and c give for $n \geqq 1$

$$p_n(t+dt) = (1 - \lambda dt)p_n(t) + \lambda p_{n-1}(t)dt + o(dt).$$

Dividing by dt, and allowing dt to tend to zero, we see that the derivative $p_n'(t)$ exists and satisfies the equation

$$p'_n(t) = \lambda\,[p_{n-1}(t) - p_n(t)]. \tag{10}$$

This system of equations is easily solved, and the unique solution satisfying the initial conditions $p_n(0) = 0$ for all $n \geqq 1$ is

$$p_n(t) = \frac{(\lambda t)^n}{n!} e^{-\lambda t}.$$

Thus $x(t)$, as well as any increment $x(u+t) - x(u)$, has a Poisson distribution with the parameter λt.

In the literature, the conditions a—c for the Poisson process are usually given in a somewhat modified form, condition a being replaced by the explicit postulation of the validity of the relation (9). The form given here is found e.g. in Khintchine [35]. The equations (10), and the deduction of the Poisson distribution from them, were given (probably not for the first time) by Cramer [8]. In section 10 we shall briefly mention the more general type of process obtained by omitting condition *c*, and requiring only that *a* and *b* should be satisfied.

The space of « natural » sample functions $x(t)$ for the Poisson process can be specified by means of enumerably many random parameters, as shown above for the Wiener process. Starting as before with $x(0) = 0$, let the first event happen at time t_1, the second at time $t_1 + t_2$, and so forth, so that generally the n : th event occurs at time $t_1 + t_2 + \ldots + t_n$. Any system of values of the t_j will then uniquely determine a realization $x(t)$ of the process. In fact, $x(t)$ will be the step functiontaking the value $x(t) = 0$ for $0 \leqq t < t_1$, and the value $x(t) = n$ if $t_1 + \ldots + t_n \leqq t$ while $t_1 + \ldots + t_{n+1} > t$. The t_j will be independent random variables taking only positive values, and it is easily shown that all the t_j will have the same probability density $\lambda e^{-\lambda t}$.

The t_j, which are the intervals between consecutive events, can be directly observed, and their distribution tested by standard methods. The same remark applies, of course, to the Poisson distribution of increments $x(u + t) - x(u)$ of a given length t. It is well known that such tests have been repeatedly made, and that the agreement, *e.g.* in respect of radioactive disintegrations and telephone calls, has often been satisfactory.

9. *Generalization. The risk process.* An important generalization of the Poisson process is obtained in the following way. With the n : *th* event in a Poisson process we associate a certain quantity w_n, where $w_1, w_2, \ldots$ are independent values of a random variable w with a given distribution function $G(w)$. Taking as before $x(0) = 0$, we then define any increment $x(u + t) - x(u)$ as the sum of all the w_j associated with events having occurred in the time interval $(u, u + t)$. Evidently the $x(t)$ process obtained in this way has independent increments, and the Poisson process corresponds to the particular case when $w_j = 1$ for all j.

The process thus defined may serve as a first approximation model for the risk business of an insurance company. We then regard the events as claims from the policyholders of the company, while $w_1, w_2, \ldots$ denote the amounts due under the successive claims. We shall consider the w_j as essentially positive, so that the distribution function $G(w)$ reduces to zero for $w < 0$.

The random variable $x(t)$ will now denote the total amount due under all claims up to time t. Choosing the time unit so that the constant λ of the Poisson process becomes equal to one, we then have the following expression for the distribution function of $x(t)$

$$F(y,t) = P\{x(t) \leqq y\} = \sum_0^\infty \frac{t^n}{n!} e^{-t} G_n(y),$$

where G_n is the n : *th* repeated convolution of the *d.f.* G with itself. Writing

$$g_1 = \int_0^\infty w dG(w) , \qquad g_2 = \int_0^\infty w^2 dG(w) ,$$

we have

$$E\,x(t) = g_1 t , \qquad E\,[x(t) - g_1 t]^2 = g_2 t.$$

12

For the applications it is important to be able to compute $F(y, t)$ numerically for large values of y and t. The function $G(w)$ can be directly observed, and estimates for its moments g_1 and g_2 computed. An approximate solution of the computation problem for $F(y, t)$ is then obtained by means of the theorem that, for large t, the variable

$$\frac{x(t) - g_1 t}{\sqrt{g_2 t}}$$

is asymptotically normal, with zero mean and unit variance. However, the approximation provided by the normal distribution is in most cases not sufficiently accurate, and more elaborate approximation methods will have to be used.

A further problem of practical importance is the *ruin problem.* Let us suppose that, during every time element dt, the company collects from its policyholders the *risk premium* amount $(g_1 + h)\, dt$, where $h > 0$ is a given constant representing the *safety loading* applied by the company. The claims are paid from the risk premium income. If at any moment the difference between accumulated claims and collected risk premiums

$$x(t) - (g_1 + h)t$$

takes a positive value, this represents a deficit that will have to be covered from the available funds of the company.

Suppose now that the company is prepared to meet a maximum deficit amounting to a given value u, but will be ruined if at any moment the deficit exceeds u. The probability that the company will be ruined at some future moment is then according to our model

$$\psi(u) = P\{ x(t) - (g_1 + h)t > u \text{ for some } t > 0 \}.$$

Writing $H(w) = 1 - G(w)$ for $w > 0$, it is then possible to show that $\Psi(u)$ satisfies the integral equation

$$(g_1 + h)\, \psi(u) = \int_u^\infty H(w)\, dw + \int_0^u H(u - w)\, \psi(w)\, dw$$

which can be explicitly solved, and gives the asymptotic relation

$$\psi(u) \sim A e^{-Ru}$$

for large u, where A and R are constants depending only on $G(w)$ and h. Under certain additional conditions, R is the smallest positive root of the equation

$$\int_0^\infty e^{Rw}\, dG(w) = 1 + (g_1 + h)\, R.$$

It is also possible to obtain some corresponding results for the probability that ruin will take place *before a given time T*, but in this case the situation is more intricate, and the results so far obtained are less definite.

After the pioneering work of F. Lundberg mentioned in section 3, the theory of this model and its various generalizations was developed by Cramér [8, 9, 11, 14, 16] and other authors. For references and an account of recent results, cf. Cramér [16] and Segerdahl [49,50]. Certain generalizations of the model will be mentioned in section 11.

13

The same model has been applied also in other fields. Thus Pólya and H.A. Einstein [19,48] have applied it to the transport of stones by rivers. Feller [23] p. 270, indicates applications to the damages caused by lightning, and to certain problems in animal ecology.

10. *Multiple events. Multi-dimensional processes.* An interesting particular case of the risk process studied in the preceding section is obtained if we assume that G (w) is the distribution function of a variable w only capable of assuming positive integral values. Let q_k be the probability of the value $w = k$, for $k = 1,2...$ The random variable $x(t)$, which is the sum of the independent values $w_1, w_2,...$ associated with all events occurred during $(0, t)$, can then be represented in the form

$$x(t) = x_1(t) + 2x_2(t) + 3x_3(t) + \dots. \tag{11}$$

Here $x_k(t)$ is the number of those events during $(0, t)$, for which the associated w assumes the value k. It is then easily shown that $x_1(t), x_2(t),...$ are mutually independent random variables, and that $x_k(t)$ forms a Poisson process with the parameter q_k. The series (11) converges with probability one. If, in addition, we suppose that $\Sigma k^2 q_k$ converges, (11) will also converge in quadratic mean. The distribution function F (y, t) of $x(t)$ can, in the present case, be represented as the convolution of a sequence of distribution functions corresponding to the successive terms of the expansion (11).

The expansion (11) may thus be regarded as a representation of the $x(t)$ process as the sum of independent Poisson processes of single, double, triple,... events. It may be remarked that this is the most general type of process satisfying conditions *a* and *b* (but not *c*) of section 8. This type of processes may be used as models *e.g.* in the statistical treatment of car accidents, where the multiplicity of an event would correspond to the number of cars involved in an accident.

Of greater importance for model building purposes are, however, the *spatial* processes obtained when the parameter t does not represent time, but is conceived as a point or a vector in some multi-dimensional euclidean space T. The simple Poisson process can be extended to this case by a straightforward generalization of conditions *a-c* of section 8. Instead of the length of a time interval we then have to consider the volume or measure of a variable set in T, and everything goes through as in the case of a temporal process in section 8. The risk process of section 9 can also be extended to this case. Similarly, if condition *c* is removed, we obtain a generalization of the processes of multiple events as given by (11) to the case of a multi-dimensional parameter t. Processes of this type have important applications as models for random distributions in space, such as the distribution of galaxies (cf. Neyman and Scott [43,44] and certain distributions encountered in forest surveys and other ecological investigations. For a survey of problems and methods in the latter type of applications we refer to Matérn [58].

III. Further Markov and related processes

The Wiener and the Poisson processes are typical representatives of the two most important classes of Markov processes used in model building : the purely continuous and the purely discontinuous processes. We shall use both of them as starting points for various generalizations. Owing to limitations of space, it will only be possible to discuss a small number of models of each of the two kinds, and we shall not even have occasion to mention the general forms of Markov processes and their properties.

It will be practical to invert the order followed above, and consider first some genralizations of the Poisson process, proceeding then to the continuous cases.

11. *Random events.* Consider, as in the Poisson process, a stream of random events starting at time $t = 0$, and denote as before by $p_n(t)$ the probability that exactly n events have occurred within time t. We propose to generalize the Poisson process in two different ways.

A. Suppose that the probability relations associated with a certain time element $(t, t + dt)$ may depend on the number n of the events that have occurred up to time t. More precisely, we make the two following assumptions characterizing a *pure birth process* :

a. The probability of one event during $(t, t + dt)$ is $\lambda_n dt + o\,(dt)$;
b. The probability of more than one event during $(t, t + dt)$ is $o\,(dt)$.

We then obtain as before, in generalization of (10),

$$p'_0(t) = -\lambda_0 p_0(t) \; ,$$

(12)

$$p'_n(t) = \lambda_{n-1}\, p_{n-1}(t) - \lambda_n p_n(t) \; .$$

There is a unique solution satisfying the initial conditions $p_0\,(0) = 1$, $p_n\,(0) = 0$ for $n > 0$. In particular we have $p_0\,(t) = e^{-\lambda_0 t}$. We always have $p_n\,(t) \geqq 0$ for all n and t, and if $\Sigma \frac{1}{\lambda_n}$ is divergent, we also have $\sum_0^{\infty} p_n\,(t) = 1$, so that the $p_n\,(t)$ satisfy all necessary conditions for a probability distribution. On the other hand, if $\Sigma \frac{1}{\lambda_n}$ converges, it can be shown (cf. Feller [23], that for some t we shall have $\sum_0^{\infty} p_n\,(t) < 1$, which should be taken as an indication that there is a positive probability of the occurrence of an infinite number of events within a finite time. We shall give three examples of the application of this model.

1. Consider a population of individuals, each of which has a probability λdt of giving birth to a new individual during any time element dt, the different individuals multiplying independently of one another. If there are n_0 individuals at time $t = 0$, and if the size of the population at time t is $n_0 + n$, we have to take $\lambda_n = (n_0 + n)\lambda$ The solution of (12) is then

$$p_n(t) = \binom{-n_0}{n} (e^{-\lambda t} - 1)^n \, e^{-n_0 \lambda t}$$

The mean value of the population size $n_0 + n$ at time t is then

$$\mathrm{E}\,\{(n_0 + n)\} = \sum_0^{\infty} (n_0 + n)\, p_n\,(t) = n_0 e^{\lambda t}$$

This model has been applied i.a. to certain cases of biological evolution, but evi-

15

dently it is strongly oversimplified, and cannot be expected to give more than a crude approximation. We shall make below some remarks on the possibility of introducing more realistic models.

2. Consider the case of the equations (12) with $n_o = 1$ and

$$\lambda_n = \lambda n\,(N - n),$$

where λ and N are given constants. We then obtain a stochastic process corresponding to the well-known (deterministic) logistic scheme of growth studied by Verhulst and Pearl. This process may be used to describe the development of a biological population, for which there is a definite maximum size N, or as a model for the spread of an epidemic or a rumour within a group of N individuals. (Cf. e.g. Feller [21], Bartlett [2], [3], D. G. Kendall [29], [31].) Consider e.g. the time T when the number of events occurred attains its maximum N (i.e. the time when all the members of the group have been infected, or have heard the rumour). It can be shown that the mean value of T is

$$E\{T\} = \frac{2}{\lambda}\sum_{1}^{N-1}\frac{1}{n} \sim \frac{2}{\lambda}\log N$$

3. We finally consider an example where the λ_n in (12) depend also on the time t. Taking

$$\lambda_n = \frac{n+h}{t+h}, \tag{13}$$

where h is a positive constant, the solution of (12) will be

$$p_n(t) = \binom{-h}{n}\left(-\frac{t}{t+h}\right)^n\left(\frac{h}{t+h}\right)^h. \tag{14}$$

Thus we obtain here, as in example 1, a negative binomial distribution, also known as a limiting form of the Pólya distribution. It is well known that this distribution has an interesting connection with the Poisson distribution. We have, in fact,

$$p_n(t) = \int_0^\infty \frac{(\lambda\, t)^n}{n\,!}\, e^{-\lambda t}\cdot\frac{h^h}{\Gamma(h)}\lambda^{h-1}\, e^{-h\lambda}\, d\lambda,$$

and this shows that we may regard the stochastic « birth » process specified by (13) and (14) as the result of a random selection of one from a population of Poisson processes, the parameters λ being distributed with the density

$$\frac{h^h}{\Gamma(h)}\lambda^{h-1}\, e^{-h\lambda}.$$

This process has important applications in the theory of insurance risk, where its connection with the Poisson process opens a way to take account of a heterogeneity in risk conditions among the individuals in a given group of policy-holders. It can be generalized in the same way as shown in section 9 for the Poisson process, so as to allow for random fluctuations of the amounts due under successive claims. We may refer to works of O. Lundberg [40], Ammeter [1] and Philipson [47].

B. A different generalization of the Poisson process can be obtained in the following way. Suppose, as in section 8, that the first event in our stream occurs at time t_1,

16

while t_n for $n > 1$ denotes the time interval between the events numbered $n-1$ and n. We assume that the t_n are independent and equidistributed random variables with the common distribution function $G(t)$. The Poisson process then corresponds to the particular case when $G(t) = 1 - e^{-\lambda t}$, $G'(t) = \lambda e^{-\lambda t}$.

Denoting as before by $x(t)$ the number of random events observed up to time t, we then obtain for every $G(t)$ with $G(0) = 0$ a well-defined stochastic process. *Except in the Poisson case, this process is not a Markov process.* In fact, the probability that an event will occur between t and $t + dt$ is

$$\frac{G'(u)}{1 - G(u)} dt, \tag{15}$$

where u is the time elapsed since the occurrence of the last interval before t. In order that the process should be a Markov process, (15) must be independent of u, and it is easily shown that this condition is only satisfied for the negative exponential distribution corresponding to the Poisson process.

It thus appears that the negative exponential distribution is the only distribution endowed with a « lack of memory », in the sense that the probability of an event occurring within $(t, t + dt)$ is equal to $\lambda\, dt$, where λ is a constant independent of the time elapsed since the occurrence of the last event before t.

In some applied fields, the negative exponential distribution does not seem to agree well with the observations, and it has been suggested that a distribution of the χ^2 type, with an even number $2k$ of degrees of freedom, might be preferable. In this case, each t_n could be regarded as the sum of k independent variables, each with a negative exponential distribution. When we are concerned with the growth of certain biological populations, where the t_n are related to the intervals between the « birth » of a micro-organism and its own fission, this assumption may be supported by theoretical arguments (cf. Kendall, [30]). The case of a general $G(t)$ has been studied e.g. by Bellman and Harris [4].

12. *The birth and death process.* As in the preceding section, we now consider a stochastic process such that the fundamental variable $x(t)$ can only assume non-negative integral values. We assume, however, that there are two different kinds of events, both of which may occur at any moment t when we have $x(t) = n > 0$. The first kind of event will be called a «birth», and the second kind a «death». If a birth occurs within $(t, t + dt)$, we have $x(t + dt) = n + 1$, and we suppose as before that the corresponding probability is $\lambda_n dt + o(dt)$. On the other hand, a death yields the value $x(t + dt) = n - 1$, the corresponding probability being $\mu_n dt + o(dt)$. The occurrence of more than one event within $(t, t + dt)$ is still supposed to be of the form $o(dt)$. When $x(t) = 0$ no death is possible ($\mu_0 = 0$), while a birth may or may not be possible, according as $\lambda_0 > 0$ or $\lambda_0 = 0$. As before, we write $p_n(t) = P\{x(t) = n\}$ and then obtain the following equations first given by Feller [21].

$$\begin{aligned} p'_0(t) &= -\lambda_0 p_0(t) + \mu_1 p_1(t), \\ p'_n(t) &= -(\lambda_n + \mu_n) p_n(t) + \lambda_{n-1} p_{n-1}(t) + \mu_{n+1} p_{n+1}(t), \quad (n > 0), \end{aligned} \tag{16}$$

Under fairly general conditions, it can be shown that this system has a unique solution

17

such that $p_n(t) \geqq 0$ and $\Sigma p_n(t) = 1$ for all n and t, and that the limits

$$p_n = \lim_{t \to \infty} p_n(t)$$

exist for all n. We then have $\Sigma p_n \leqq 1$, and if the sign of equality holds here, the p_n may be considered as the probabilities corresponding to a limiting state of « statistical equilibrium » of the process. On the other hand, if $\Sigma p_n < 1$, the limiting « steady state » distribution has a positive probability situated at infinity.

We shall give three examples of the application of this type of process for construction of models.

1. If $x(t) = n$ is the size at time t of a population of mutually independent individuals, each of which can die or give birth to a new individual, we may take $\lambda_n = n\lambda$, $\mu_n = n\mu$. This implies, of course, that any age-dependence of birth and death rates is neglected. Hence, in particular, the model is too crude to be applied to human populations. A more general model, where λ and μ depend on t, has been studied by D.G. Kendall [28], who has also investigated the random fluctuations of the age distribution of the population.

For constant λ and μ, the solution of (16) is, in the case of a population descending from one single individual ($n_0 = 1$)

$$p_0(t) = \mu z,$$

$$p_n(t) = (1 - \mu z)(1 - \lambda z)(\lambda z)^{n-1}, \quad (n > 0),$$

where

$$z = \frac{e^{(\lambda - \mu)t} - 1}{\lambda e^{(\lambda - \mu)t} - \mu}.$$

For an arbitrary $n_0 \geqq 1$, the $p_n(t)$ distribution is the n_0 times repeated convolution of the above distribution with itself. In particular $p_0(t) = (\mu z)^{n_0}$, so that the probability of ultimate extinction of the population is

$$p_0 = \lim_{t \to \infty} p_0(t) = \begin{cases} \left(\dfrac{\mu}{\lambda}\right)^{n_0} & \text{for } \lambda > \mu, \\ 1 & \text{for } \lambda \leqq \mu. \end{cases}$$

For any fixed $n \geqq 1$, it follows from (16) that $p_n = \lim_{t \to \infty} p_n(t) = 0$, so that we have $\Sigma p_n = p_0$. If $\lambda > \mu$, we have $p_0 < 1$, and thus the limiting « steady state » distribution has the probability p_0 at zero and $1 - p_0$ at infinity. When $\lambda \leqq \mu$, on the other hand, $p_0 = 1$, and the limiting distribution has its whole mass concentrated at zero. In any case, the population may either die out or increase beyond all limits, while the probability of a population size remaining between fixed non-zero limits is always zero. The mean value of the population size at time t is

$$\mathrm{E}\{x(t)\} = n_0 e^{(\lambda - \mu)t}.$$

2. The birth and death process may serve as a model in certain telephone traffic problems. Suppose that there are N trunklines available, and that incoming calls form a Poisson process with parameter λ, while the duration of a conversation (the « holding

18

time ») has a negative exponential distribution with parameter μ. It then follows from the remarks made at the end of the preceding section that, if a certain line is busy at time t, the probability that it will become free during $(t, t + dt)$ is μdt, independently of the time elapsed since the beginning of the conversation. We assume that there is no waiting line, so that a call made at an instant when all N lines are busy will not be satisfied, and will have to be repeated later. Denoting by $p_n(t)$ the probability that exactly n lines are busy at time t, where $0 \leqq n \leqq N$, we have to take $\lambda_n = \lambda$, $\mu_n = n\mu$ in (16) for $n < N$, while for $n = N$

$$p'_N(t) = -N\mu p_N(t) + \lambda p_{N-1}(t).$$

Replacing in the differential equations all $p'_n(t)$ by zero and all $p_n(t)$ by p_n, we obtain a system of linear equations for the limiting probabilities $p_n = \lim_{t \to \infty} p_n(t)$. The solution is given by the « Erlang loss formula » (cf. Brockmeyer, etc., [6] also Cramér [15], p. 107).

$$p_n = \frac{\frac{\alpha^n}{n!}}{1 + \frac{\alpha}{1!} + \dots \frac{\alpha^N}{N!}},$$

where $\alpha = \frac{\lambda}{\mu}$. In particular, for $n = N$ we obtain the probability p_N that, in the limiting state of statistical equilibrium, all N lines will be busy, so that an incoming call will be « lost ». As $N \to \infty$, the p_n: s evidently tend to the probabilities of a Poisson distribution with parameter α.

3. If we modify the preceding example by assuming that a call incoming when all lines are busy will join a waiting line common for all N trunklines, and will be satisfied in order of arrival, when lines are becoming free, we obtain a model applicable to certain « queueing problems ». Instead of trunklines we may have « counters », and instead of lengths of telephone conversations, there will be « service times » for successive « customers ».

In this case, we shall denote by $x(t)$ the total number of customers *engaged in waiting or being served* at time t. Writing as before $p_n(t) = P\{x(t) = n\}$, the equations (16) will still hold with $\lambda_n = \lambda$ for all n, $\mu_n = n\mu$ for $n \leqq N$ and $\mu_n = N\mu$ for $n > N$. The limits $p_n = \lim_{t \to \infty} p_n(t)$ can be found as before. If $\alpha = \frac{\lambda}{\mu} \geqq N$, we have $p_n = 0$ for all n, so that the waiting line will almost certainly grow beyond all limits as $t \to \infty$. On the other hand, if $\alpha < N$, the steady state probabilities p_n can be found from the expressions

$$\frac{p_n}{p_0} = \begin{cases} \frac{\alpha^n}{n!} & \text{for } n \leqq N, \\ \frac{\alpha^N}{N!}\left(\frac{\alpha}{N}\right)^{n-N} & \text{for } n > N, \end{cases}$$

where p_0 is determined by the condition $\Sigma p_n = 1$. In the limiting distribution, the probability that a newcoming customer will have to wait is

$$p_N + p_{N+1} + \dots = p_N \frac{N}{N-\alpha},$$

while the mean number of customers in the waiting line is

$$p_{N+1} + 2p_{N+2} + \dots = p_N \frac{N\alpha}{(N-\alpha)^2}.$$

The distribution of the waiting time T for a newcoming customers has a discrete probability

$$1 - p_N \frac{N}{N-\alpha}$$

at $T = 0$, and a continuous part with the density

$$\mu N p_N e^{-\mu (N-\alpha) T}$$

for $T > 0$. The mean value of T is

$$E\{T\} = p_N \frac{N}{\mu (N-\alpha)^2}.$$

There is an extensive literature on queueing problems, where probabilistic models are studied under various conditions. For a recent survey of the field, we refer to Khintchine [35].

13. *Generalizations.* The birth and death process has been generalized in various directions, some of which have already been indicated. A further generalization can be obtained by introducing an « immigration rate » (cf. D. G. Kendall [30]). Suppose that, in example 1 of the preceding section, there is in every time element dt a probability νdt of a new individual entering into the population from outside. If we have $\lambda < \mu$ and $n_0 = 0$, there will then be a steady state distribution of the negative binomial type, with

$$p_n = \binom{-\beta}{n} (-\alpha)^n (1-\alpha)^\beta,$$

where $\alpha = \frac{\lambda}{\mu}$ and $\beta = \frac{\nu}{\lambda}$. In some recent works (cf. D. G. Kendall [32], Neyman [41, 42] and further references there given) this type of process has been used as one of the elements of a stochastic model for the genesis of cancer. It is then assumed that the population just described consists of « first order » mutations of cells, and that these in their order are able to produce « second order » mutations, which are subject to a birth and death process with $\lambda > \mu$, so that their number will, with a certain probability, ultimately tend to grow exponentially.

14. *Diffusion. Brownian movement.* The Poisson process, as well as its various generalizations discussed in sections 9-13, have the characteristic property that the associated random variable $x(t)$ changes its value only in discrete jumps. The sample functions are step functions, and the probability that a jump will occur within a small

20

time element dt is proportional to dt, the only alternative being that there is no change in the value of $x(t)$.

The Wiener process discussed in section 7 gives an example of a radically different type. Here it is certain that *some*-change will occur in any time element dt, but the change will in general be small when dt is small, and the sample functions of the process are almost certainly continuous. The heuristic argument employed in section 7 to represent the Wiener process as a limiting case of a simple random walk is capable of being generalized, and leads to more general Markov processes of the continuous variation type. These processes provide probabilistic models for important physical processes, particularly in the theory of diffusion.

The foundations of a general theory of Markov processes, including both continuous and discontinuous cases, were laid by Kolmogoroff [36], who introduced the celebrated Kolmogoroff equations, of which equations (6) and (16) of the present paper are simple particular cases. The general theory was further developed by Feller [20], [22], Doob [18], and several other authors. For the application of continuous processes to diffusion theory we may refer to Khintchine [33], Chandrasekhar [7] and Wang-Uhlenbeck [51].

Owing to limitations of space, we can here only treat one simple example, and we shall choose the process known as the Ornstein-Uhlenbeck process, which attempts to give a more realistic picture of the Brownian movement than the classical Wiener process (cf. Ornstein and Uhlenbeck [45], Wang and Uhlenbeck [51], Doob [17]).

Consider, as in section 7, the one-dimensional Brownian movement of a free particle, which at time $t = 0$ has the initial position $x(0) = 0$, and velocity $v(0) = v_0$. The equation of motion for this particle (the Langevin equation) is

$$m \frac{dv}{dt} + \beta v(t) - y(t) = 0,$$

where m is the mass of particle, while

$$\beta = b \pi r \eta$$

is a friction coefficient (r and η have the same significance as in section 7), and $y(t)$ represents the random molecular impacts. It seems to agree with current physical theory, if we assume that the impulses corresponding to disjoint time intervals are independent and normally distributed, each having zero mean and a variance proportional to the lenght of the interval. The random variable $y(t)$ will then be extremely irregular in its dependence on t, and it seems preferable to rewrite the equation in the form (cf. Doob [17]).

$$m\, dv(t) + \beta v(t)\, dt - dz(t) = 0, \tag{17}$$

where $z(t)$ is associated with a stochastic process with stationary, independent and normally distributed increments, i.e. a Wiener process. By formal calculation we obtain the expression

$$v(t) = e^{-\frac{\beta t}{m}} \left(v_0 + \frac{1}{m} \int_0^t e^{\frac{\beta u}{m}} dz(u) \right). \tag{18}$$

It is easily shown that this really represents a solution of (17), in the sense that the first member of (17), after introducing the expression (18), becomes a random variable with both mean and variance of the form $o\,(dt)$, as dt tends to zero.

The velocity $v\,(t)$ as given by (18) is a normally distributed random variable associated with a Markov process, the mean and covariance moment of which are

$$\mathrm{E}\left\{ v\,(t) \right\} = v_0\, e^{-\frac{\beta t}{m}},$$

$$\mathrm{E}\left\{ \left(v\,(t) - v_0\, e^{-\frac{\beta t}{m}}\right)\left(v\,(u) - v_0\, e^{-\frac{\beta u}{m}}\right) \right\} = \frac{\rho^2}{2\,m\,\beta}\left(e^{-\frac{\beta\,|\,t-u\,|}{m}} - e^{-\frac{\beta\,(t+u)}{m}}\right),$$

where $\mathrm{E}\left\{ |dz\,(t)|^2 \right\} = \rho^2 dt$. As t tends to infinity, the influence of the initial velocity v_0 disappears, and the moments of the limiting (normal) $v\,(t)$ distribution are $\mathrm{E}\{v(t)\} = 0$ and

$$\mathrm{E}\left\{ v(t)\, v(u) \right\} = \frac{\rho^2}{2\,m\,\beta}\, e^{-\frac{\beta\,|\,t-u\,|}{m}}.$$

In the limit we thus obtain a process which is at the same time Markov, normal and stationary, the covariance moment being a function of the time difference $t-u$. It can be shown (Doob, l.c.) that the sample functions of this process are almost certainly continuous, but have no derivatives. For the displacement $x\,(t)$ we have

$$x(t) = \int_0^t v\,(u)\, du.$$

$x\,(t)$ is normally distributed, and has almost certainly everywhere a derivative $x'\,(t) = v\,(t)$, but no second derivative. In this model, the velocity of the moving particle is thus well defined, but not its acceleration.

In the limiting distribution for large t we have $\mathrm{E}\{x\,(t)\} = 0$ and for any $h > 0$

$$\mathrm{E}\left\{ \left(x(t+h) - x(t) \right)^2 \right\} = \frac{m\,\rho^2}{\beta^3}\left(e^{-\frac{\beta h}{m}} - 1 + \frac{\beta h}{m}\right).$$

The second member of the last expression is aymptotically $\frac{\rho^2}{2m\beta}\, h^2$ for small h, and $\frac{\rho^2}{\beta^2}\, h$ for large h. Taking

$$\rho^2 = 12\,\pi\, r\, \eta\, k\, \mathrm{T},$$

we obtain in accordance with (7)

$$\frac{\rho^2}{\beta^2} = \sigma^2 = \frac{k\,\mathrm{T}}{3\,\pi\, r\, \eta}.$$

IV. Stationary and related processes

When dealing with statistical time series, it often seems reasonable to assume that the underlying chance mechanism is more or less invariant under a translation in time. Similarly, in the case of a spatial random distribution, invariance under a translation in space can sometimes be assumed. In situations of this type, we are led to consider models based on stationary processes (cf. section 6 above).

For surveys of the extensive literature on stationary processes and their applications, we refer to works by Doob [18], Blanc-Lapierre and Fortet [9], Grenander and Rosenblatt [24], Grenander and Szegö [25], Jaglom [26]. In the present paper, it is only possible to mention very briefly some typical cases of the use of this class of stochastic processes for model building. (One such case has already been discussed in the preceding section.)

15. *Discrete parameter.* Consider a process with the spectral representation (cf. section 6)

$$x_n = \int_{-\pi}^{\pi} e^{in\lambda}\, dz(\lambda), \quad \mathrm{E}\left\{ | dz(\lambda) |^2 \right\} = d\,\mathrm{F}(\lambda). \tag{19}$$

We shall suppose that the spectral function F is absolutely continuous, and that the *spectral density* $f(\lambda) = \mathrm{F}'(\lambda)$ is almost everywhere > 0, and satisfies the condition

$$\int_{-\pi}^{\pi} \log f(\lambda)\, d\lambda > -\infty. \tag{20}$$

It is then possible to show that there exists a sequence of mutually orthogonal random variables z_n $(-\infty < n < +\infty)$ and a sequence of constants $c_0, c_1, \ldots$ such that

$$\mathrm{E}\{z_n\} = 0, \quad \mathrm{E}\{z_m \overline{z_n}\} = \delta_{mn},$$

$$\sum_0^\infty |c_n|^2 = \int_{-\pi}^{\pi} f(\lambda)\, d\lambda,$$

$$x_n = \sum_0^\infty c_k z_{n-k}, \tag{21}$$

where the last series converges in quadratic mean. The z_n can be regarded as mutually orthogonal (uncorrelated) « impulses » or « innovations », each z_n entering into the process at the instant corresponding to the subscript n. Then (21) shows that x_n can for any n be linearly expressed in terms of past and present innovations, the weights c_k being independent of n.

We remark in passing that if the weights c_k are allowed to depend also on n, a spectral representation of the form (19) will still hold under certain conditions, but the spectral process $z(\lambda)$ will then in general not have orthogonal increments. We then obtain an instance of a more general class of stochastic processes known as *harmonizable processes.*

In order to show an example of the use of (21) for model building, we denote by $g(z) = g_0 + g_1 z + \ldots + g_p z^p$ a polynomial in the complex variable z such that $g(z) \neq 0$ for $|z| < 1$, and consider the case when the spectral density $f(\lambda)$ is given by

$$2\pi f(\lambda) = |g(e^{-i\lambda})|^{-2} \tag{22}$$

In this case the c_k will be the coefficients of the power series expansion of $[g(z)]^{-1}$, and it then follows from the representation (21) that we have the following *stochastic difference equation* for x_n

$$g_0 x_n + g_1 x_{n-1} + \ldots + g_p x_{n-p} = z_n.$$

In certain applications, e.g. to econometric problems (cf. Wold [55-57]), an equation of this type is given, the z_n being regarded as orthonormal random «disturbances». The solution with respect to x_n then takes the form of a spectral representation (19), with a spectral density given by (22), and this can be used as a model for the x_n process.

If the disturbances are supposed to be orthogonal but not necessarily normalized, we are again under certain conditions led to the more general harmonizable type of spectral representation.

16. *Continuous parameter.* We suppose that a spectral density $f(\lambda) = F'(\lambda)$ exists and satisfies the condition

$$\int_{-\infty}^{\infty} \frac{\log f(\lambda)}{1+\lambda^2}\, d\lambda > -\infty.$$

(We observe that, in particular, this condition is satisfied for the stationary Markov process encountered in section 14, where we have the covariance function $E\{x(s)\, \overline{x(t)}\} = e^{-k|s-t|}$ and the spectral density $f(\lambda) = \dfrac{k}{\pi(k^2+\lambda^2)}$).

The representation (21) can then be generalized to the continuous case. We have here

$$x(t) = \int_{-\infty}^{t} c(t-u)\, dw(u), \qquad \int_0^{\infty} |c(u)|^2\, du < \infty, \tag{23}$$

where the *innovation process* $w(u)$ has orthogonal increments such that

$$E\{dw(u)\} = 0. \quad E\{|dw(u)|^2\} = du$$

The covariance function of the $x(t)$ process is

$$E\left\{x(s)\, \overline{x(t)}\right\} = \int_{-\infty}^{\mathrm{Min}\,(s,t)} c(s-u)\, \overline{c(t-u)}\, du = \frac{1}{2\pi} \int_{-\infty}^{\infty} e^{i(s-t)\lambda} |\gamma(\lambda)|^2\, d\lambda,$$

where $\gamma(\lambda)$ is the Fourier-Plancherel transform

$$\gamma(\lambda) = \int_0^{\infty} e^{-i\lambda u} c(u)\, du,$$

the integral being defined as a limit in quadratic mean. It follows that the spectral density $f(\lambda)$ of $x(t)$ is

$$f(\lambda) = \frac{1}{2\pi} |\gamma(\lambda)|^2.$$

The representation (23) can be used as a model e.g. in various noise and filtering problems of radio engineering. As a simple example we may consider the so called *shot noise.* Suppose that the arrivals of electrons at the anode of a vacuum tube constitute a Poisson process with parameter μ. If $n(t)$ is the number of arrivals up to time t, and if $b(t)$ is the intensity of current at time t due to an electron arrived at time 0, the total intensity of current at time t is

$$x(t) = \int_{-\infty}^{t} b(t-u)\, dn(u) = x_1(t) + x_2(t),$$

24

where we have supposed that the effects of the electrons are simply additive, while

$$x_1(t) = \mu \int_0^{\infty} b(t)\, dt,$$

$$x_2(t) = \int_{-\infty}^{t} c(t-u)\, dw(u),$$

$$c(t) = \mu\, b(t), \quad w(u) \equiv \frac{n(u) - \mu u}{\mu}.$$

Thus the current is composed of a constant, « direct current » component $x_1(t)$ and a « shot noise » component $x_2(t)$ due to the random fluctuations in the arrivals of electrons. The latter component forms a stationary process with a representation of the form (23).

The concept of a stationary stochastic process can be straightforwardly generalized to the case when the parameter t is a point in some multi-dimensional euclidean space. In generalization of section 6, we then have to lay down the condition that the first and second order moments of $x(t)$ should be invariant under a translation in t. Spectral representations analogous to those given in section 6 hold also in this case. Moreover, under appropriate conditions, the representation (23) can be generalized to this case. The $w(u)$ process may, e.g., be a multi-dimensional Poisson process (cf. section 10), which generates a set of randomly located « centers » spreading their influence over the neigh-bourhood in accordance with the weight function $c(t-u)$. For a general survey of the theory and applications of this class of processes we refer to Matérn [58]. Cf. also e.g. Wittle [52] and, for the applications to turbulence problems, the theory of ocean waves, etc,, Grenander and Rosenblatt [24].

RÉFÉRENCES

[1] AMMETER, H. A generalization of the collective theory of risk in regard to fluctuating basic probabilities. *Skandinavisk Aktuarietidskrift*, 1948, 171-198.

[2] BARTLETT, M. S. Some evolutionary stochastic processes. *Journal of the Royal Statistical Society* (*B*), vol. 11 (1949), 211-229.

[3] BARTLETT, M. S. An introduction to stochastic processes. Cambridge University Press, 1955.

[4] BELLMAN, R. and HARRIS, T. E. On the theory of age-dependent stochastic branching processes. *Proceedings, National Academy of Sciences*, vol. 37 (1948), 601-604.

[5] BLANC-LAPIERRE, A. and FORTET, R. *Théorie des fonctions aléatoires.* Paris, Masson et Cie, 1953.

[6] BROCKMEYER, E. HALSTROM, H. L. and JENSEN, A. *The life and works of A. K. Erlang.* Copenhaguen, Academy of Technical Sciences, 1948.

[7] CHANDRASEKHAR, S. Stochastic problems in physics and astronomy. *Reviews of Modern Physics*, vol. 15 (1943), 1-89.

[8] CRAMÉR, H. *Bidrag till utjämningsförsäkringens teori.* Stockholm, Royal Insurance Board, 1919.

[9] CRAMÉR, H. *On the mathematical theory of risk.* Skandia Jubilee Volume, Stockholm 1930, 7-84.

[10] CRAMÉR, H. *Random variables and probability distributions.* Cambridge University Press 1937 (2 *nd* éd. 1961).

[11] CRAMÉR, H. Deux conférences sur la théorie des probabilités. *Skandinavisk Aktuarietidskrift*, 1941, 34-69.

[12] CRAMÉR, H. On harmonic analysis in certain functional spaces. *Arkiv för Matematik, Astronomi och Fysik*, vol. 28 B (1942), Nr 12.

[13] CRAMÉR, H. *On the theory of stochastic processes.* Tenth Congress of Scandinavian Mathematicians, Copenhaguen 1947, 28-39.

[14] CRAMÉR, H. On some questions connected with mathematical risk. *University of California Publications in Statistics*, vol. 2 (1954), 99-123.

[15] CRAMÉR, H. *The elements of probability theory.* New York, Wiley, 1955.

[16] CRAMÉR, H. *Collective risk theory.* Skandia Centennial Volume, Stockholm 1955, 1-92.

[17] DOOB, J. L. The Brownian movement and stochastic equations. *Annals of Mathematics*, vol. 43 (1942), 351-369.

[18] DOOB, J. L. *Stochastic processes.* New York, Wiley, 1953.

[19] EINSTEIN, H. A. Der Geschiebebetrieb als Wahrscheinlichkeitsproblem. *Dissertation,* Technische Hochschule Zürich, 1937.

[20] FELLER, W. Zur Theorie der stochastischen Prozesse. *Mathematische Annalen,* vol. 113 (1936), 113-160.

[21] FELLER, W. Die Grundlagen der Volterraschen Theorie des Kampfes ums Dasein in wahrscheinlichkeitstheoretischer Behandlung. *Acta Biotheoretica,* vol. 5 (1939), 11-40.

[22] FELLER, W. On the integrodifferential equations of purely discontinuous Markoff processes. *Transactions, American Mathematical Society,* vol. 48 (1940), 488-515.

[23] FELLER, W. *An introduction to probability theory and its applications.* New York Wiley, 2 *nd,* éd. 1957.

[24] GRENANDER, U. and ROSENBLATT, M. *Statistical analysis of stationary time series.* New York, Wiley, 1957.

[25] GRENANDER, U. and SZEGÖ, G. *Toeplitz forms and their applications.* University of California Press, 1958.

[26] JAGLOM, A. M. *Einführung in die Theorie stationärer Zufallsfunktionen.* Berlin, Akademie-Verlag, 1959.

[27] KAC, M. Random walk and the theory of Brownian motion. *American Mathematical Monthly,* vol. 54 (1947), 369-391.

[28] KENDALL, G. D. On the generalized birth and death process. *Annals of Mathematical Statistics,* vol. 19 (1948), 1-15.

[29] KENDALL, D. G. Stochastic processes and population growth. *Journal of the Royal Statistical Society (B),* vol. 11 (1949), 230-282.

[30] KENDALL, D. G. Les processus stochastiques de croissance en biologie. *Annales de l'Institut Henri-Poincaré,* vol. 13 (1952), 43-108.

[31] KENDALL, D. G. La propagation d'une épidémie ou d'un bruit dans une population limitée. *Publications de l'Institut statistique de l'Université de Paris,* vol. 6 (1947), 307-311.

[32] KENDALL, D. G. Birth and death processes, and the theory of carcinogenesis. *Biometrika,* vol. 47 (1960), 13-21.

[33] KHINTCHINE, A. *Asymptotische Gesetze der Wahrscheinlichkeitsrechnung.* Berlin, Springer, 1933.

[34] KHINTCHINE, A. Korrelationstheorie der stationären stochastischen Prozesse. *Mathematische Annalen,* vol. 109, (1934), 604-615.

[35] KHINTCHINE, A. *Mathematical methods in the theory of queueing.* London, Griffin, 1960.

[36] KOLMOGOROFF, A. Über die analytischen Methoden in der Wahrscheinlichkeitsrechnung. *Mathematische Annalen,* vol. 104 (1931), 415-458.

[37] LÉVY, P. Sur les intégrales dont les éléments sont des variables aléatoires indépendantes. *Annali di Pisa,* vol. 3 (1934), 337-366.

[38] LÉVY, P. *Théorie de l'addition des variables aléatoires.* Paris, Gauthier-Villars, 1937 (2 *nd,* éd. 1954).

[39] LÉVY, P. *Processus stochastiques et mouvement brownien.* Paris, Gauthier-Villars, 1948.

[40] LUNDBERG, O. *On random processes and their application to sickness and accident statistics.* Thesis, Stockholm, 1940.

[41] NEYMAN, J. *A two-stage mutation theory of carcinogenesis.* International Institute of Statistics, Tokyo Session 1960.

[42] NEYMAN, J. Indeterminism in science and new demands on statisticians. *Journal of the American Statistical Association,* vol. 55 (1960), 625-639.

[43] NEYMAN, J. and SCOTT, E. L. A theory of the spatial distribution of galaxies. *The Astrophysical Journal,* vol. 116 (1952), 144-163.

[44] NEYMAN, J. and SCOTT, E. L. Statistical approach to problems of cosmology, *Journal of the Royal Statistical Society (B),* vol. 20 (1958), 1-43.

[45] ORNSTEIN, L. S. and UHLENBECK, G. E. On the theory of the Brownian motion. *Physical Review,* vol. 36 (1930), 823-841.

[46] PALM, C. *Intensitätsschwankungen im Fernsprechverkehr.* Ericson Technics, Stockholm, 1943.

[47] PHILIPSON, C. Note on the application of compound Poisson processes to sickness and accident statistics. *Astin Bulletin,* vol. 1 (1960), 224-237.

[48] POLYA, G. *Zur Kinematik der Geschiebebewegung.* Versuchsanstalt für Wasserbau, Technische Hochschule Zurich, 1937.

[49] SEGERDAHL, C. O. *On homogeneous random processes and collective risk theory.* Thesis, Stockholm, 1939.

[50] SEGERDAHL, C. O. Asurvey of results in the collective theory of risk. *Probability and Statistics, The Harald Cramèr Volume,* NewYork, Wiley, 1959.

[51] WANG, M. C. and UHLENBECK, G. E. On the theory of Brownian motion II. *Reviews of Modern Physics,* vol. 17 (1945), 323-350.

[52] WHITTLE, P. On stationary processes in the plane. *Biometrika,* vol. 41 (1954), 434-449.

[53] WIENER, N. Differential-space. *Journal of Mathematics and Physics,* vol. 2 (1923), 131-174.

[54] WIENER, N. and PALEY, R. *Fourier transforms in the complex domain.* American Mathematical Society, New York, 1934.

[55] WOLD, H. *A study in the analysis of stationary* time *series.* Thesis, Stockholm 1938 (2 *nd,* éd. 1954).

[56] WOLD, H. *Demand analysis.* New York, Wiley, 1932.

[57] Wold, H. Ends and means in econometric model building. *Probability and Statistics The Harald Cramér Volume*, New York, Wiley, 1959.

[58] Matérn, B. Spatial variation. *Communications of the Swedish Forestry Research Institute*, vol. 49 (1960), 1-144.

RÉSUMÉ

Dans l'introduction de ce travail, on donne une esquisse du développement historique qui, en partant de divers cas isolés d'une variation aléatoire dans le temps ou l'espace, a conduit à la formation d'une théorie générale des processus stochastiques.

Ensuite, quelques exemples caractéristiques de la construction de modèles statistiques à l'aide des processus stochastiques sont étudiés.

Ainsi, dans les paragraphes 7-10 on traite quelques modèles basés sur des processus à accroissements indépendants, à savoir les processus dits de Wiener et de Poisson, ainsi que quelques généralisations importantes de ces processus, qui ont été appliquées par exemple à la théorie du risque d'une compagnie d'assurance.

Dans les paragraphes 11-13, on donne des exemples de l'application des processus de Markoff discontinus plus généraux. Ici on étudie divers cas des événements aléatoires, ayant des applications biologiques, téléphoniques, etc. Le paragraphe 14 contient un aperçu du modèle du mouvement Brownien dû à Ornstein et Uhlenbeck, qui se sert d'un processus continu à la fois Markoff et stationnaire.

Finalement, les paragraphes 15-16 sont consacrés à quelques applications des processus stationnaires, en général non Markoff, à la construction de modèles statistiques.

IMPRIMERIE NATIONALE.

J. H. 110431.

53.

On the maximum of a normal stationary stochastic process

Bull. Amer. Math. Soc. **68** (5), 512–516 (1962)

Communicated by W. Feller, May 1, 1962

1. Let $x(t)$ with $-\infty < t < +\infty$ be the variables of a real, separable, normal and stationary stochastic process, such that $E[x(t)] = 0$ and $E[x^2(t)] = 1$. Let the covariance function of the process be

$$r(t) = E[x(t)x(0)] = \int_0^\infty \cos \lambda t f(\lambda) d\lambda,$$

and assume that the spectral density $f(\lambda)$ is of bounded variation in $(-\infty, \infty)$ and satisfies the condition

$$\int_0^\infty \lambda^2 (\log(1+\lambda))^a f(\lambda) d\lambda < \infty$$

for some $a > 1$.

Then it is known (Hunt [5], Belayev [1]) that the sample functions $x(t)$ will almost certainly be everywhere continuous and have continuous first derivatives $x'(t)$. Consequently for every fixed $t > 0$ the maximum

$$\max_{0 \leqq u \leqq t} x(u)$$

will be a random variable defined but for equivalence.

For the sake of typographical convenience, we write in the sequel simply $\max x(u)$, omitting the subscript $0 \leqq u \leqq t$, and similarly in respect of $\min x(u)$.

The object of this note is to prove the relation

$$\text{(1)} \qquad \lim_{t\to\infty} P\left[\left| \max x(u) - (2 \log t)^{1/2} \right| < \frac{\log \log t}{(\log t)^{1/2}} \right] = 1.$$

The notation $P[\cdots]$ denotes here, as throughout the sequel, the probability of the relation between the brackets.

A similar relation was recently given for the case of a normal stationary sequence x_n with $n = 0, \pm 1, \cdots$ by Berman [2].

2. We shall first prove that

---

[1] Research work done (Tech. Report No. 1) partially under Contract NASw-334, National Aeronautics and Space Administration.

(2) $$P\left[\max x(u) \leqq (2 \log t)^{1/2} - \frac{\log \log t}{(\log t)^{1/2}}\right] \to 0$$

as $t \to \infty$.

Let $c > 0$ be given, and define a random variable $y(u)$ by writing for any real u

$$y(u) = \begin{cases} 1 & \text{if } x(u) > c, \\ 0 & \text{if } x(u) \leqq c. \end{cases}$$

Then $y(u)$ will define a stationary process such that

$$E[y(u)] = P[x(u) > c] = \int_c^\infty \phi(x)dx.$$

$$E[y(u)y(v)] = P[x(u) > c,\, x(v) > c]$$
$$= \int_c^\infty \int_c^\infty \phi(x, y; r)dxdy,$$

where

$$\phi(x) = \frac{1}{(2\pi)^{1/2}} \exp\left(-\frac{x^2}{2}\right),$$

$$\phi(x, y; r) = \frac{1}{2\pi(1 - r^2)^{1/2}} \exp\left(-\frac{x^2 - 2rxy + y^2}{2(1 - r^2)}\right),$$

$$r = r(u - v).$$

It follows (cf. e.g. Loève [6, pp. 472, 520]) that the integral

$$z(t) = \int_0^t y(u)du$$

is defined both in quadratic mean and as a sample function integral, and that the two integrals coincide, but for equivalence. Then $z(t)$ will, with probability 1, be equal to the Lebesgue measure of the set of points u in $[0, t]$ such that $x(u) > c$. Thus $z(t) \geqq 0$ with probability 1, and

(3) $$P[z(t) = 0] = P[\max x(u) \leqq c].$$

For all sufficiently large c we have (Loève, l.c.)

(4) $$E[z(t)] = t \int_c^\infty \phi(x)dx > \frac{t}{3c} \exp\left(-\frac{c^2}{2}\right),$$

and further

$$E[z^2(t)] = \int_0^t \int_0^t dudv \int_c^\infty \int_c^\infty \phi(x, y; r)dxdy$$

with $r = r(u-v)$.

For any fixed r in $(-1, 1)$ we have the identity

$$\int_c^\infty \int_c^\infty \phi(x, y; r)dxdy$$

$$= \left(\int_c^\infty \phi(x)dx \right)^2 + \frac{1}{2\pi} \int_0^r \exp\left(-\frac{c^2}{1+w} \right) \frac{dw}{(1-w^2)^{1/2}}.$$

(For $r=0$ the identity is obvious, and some calculation will show that the derivatives of both sides with respect to r are equal.)

It then follows that the variance of $z(t)$ is

$$\begin{aligned} \operatorname{Var}[z(t)] &= \frac{1}{2\pi} \int_0^t \int_0^t dudv \int_0^{r(u-v)} \exp\left(-\frac{c^2}{1+w} \right) \frac{dw}{(1-w^2)^{1/2}} \\ &< \frac{1}{\pi^2} \int_0^t \int_0^t |r(u-v)| \exp\left(-\frac{c^2}{1+|r(u-v)|} \right) dudv. \end{aligned} \tag{5}$$

From our assumptions concerning the spectral density $f(\lambda)$, it follows that there exist positive constants k and m such that

$$|r(t)| < \frac{k}{|t|} \qquad \text{for all } t,$$

$$|r(t)| \leqq 1 - m^2t^2 \qquad \text{for } |t| \leqq 2k.$$

(The latter inequality is easily proved by means of Cramér [**4**, Lemma 1].)

Dividing the domain of integration in (5) into two parts, defined respectively by $|u-v| > 2k$ and $|u-v| \leqq 2k$, and using in each part the appropriate inequality for $|r(u-v)|$, we obtain from (5) by some straightforward estimation

$$\operatorname{Var}[z(t)] < 2kt \log t \exp\left(-\frac{2c^2}{3} \right) + \frac{2\pi^{1/2}}{m} \cdot \frac{t}{c} \exp\left(-\frac{c^2}{2} \right). \tag{6}$$

Now the Tchebychev inequality gives

$$P[z(t) = 0] \leqq \frac{\operatorname{Var}[z(t)]}{E^2[z(t)]}.$$

Taking

$$c = (2 \log t)^{1/2} - \frac{\log \log t}{(\log t)^{1/2}},$$

we then obtain from (3), (4) and (6)

$$P[\max x(u) \leqq c] < A((\log t)^2 t^{-1/8} + (\log t)^{1/2 - 2^{1/2}})$$

where A is independent of t. Since the second member obviously tends to zero as $t \to \infty$, (2) is proved.

3. It now remains to prove that

$$P\left[\max x(u) \geqq (2 \log t)^{1/2} + \frac{\log \log t}{(\log t)^{1/2}}\right] \to 0 \tag{7}$$

as $t \to \infty$. For any $c > 0$ we evidently have

$$\begin{aligned} P[\max x(u) \geqq c] &= P[\min x(u) \leqq c \leqq \max x(u)] + P[\min x(u) > c] \\ &= P_1 + P_2. \end{aligned}$$

P_1 is, for any continuous sample function $x(u)$, the probability of at least one "crossing" with the level c within $[0, t]$, i.e., the probability that there is at least one point u in $[0, t]$ such that $x(u) = c$. Let N denote the total number of such points, and write $p_n = P[N = n]$ for $n = 0, 1, \cdots$. Then

$$P_1 = p_1 + p_2 + \cdots \leqq p_1 + 2p_2 + \cdots = E[N]. \tag{8}$$

However, it is known (Bulinskaya [3]) that under the present conditions

$$E[N] = \frac{(\lambda_2)^{1/2}}{\pi} t \exp\left(-\frac{c^2}{2}\right), \tag{9}$$

where λ_2 denotes the second moment of $f(\lambda)$. Further

$$\begin{aligned} P_2 &= P[\min x(u) > c] \leqq P[x(0) > c] \\ &= \int_c^\infty \phi(x) dx < \frac{1}{c(2\pi)^{1/2}} \exp\left(-\frac{c^2}{2}\right). \end{aligned} \tag{10}$$

Taking now

$$c = (2 \log t)^{1/2} + \frac{\log \log t}{(\log t)^{1/2}},$$

it follows from (8), (9) and (10) that P_1 and P_2 both tend to zero as $t \to \infty$, so that (7) is proved. Finally, the result (1) follows from (2) and (7).

References

1. Yu. K. Belayev, *Continuity and Hölder's conditions for sample functions of stationary Gaussian processes*, Proc. Fourth Berkeley Symp., Vol. 2, pp. 23–33, 1961.

2. S. M. Berman, *A law of large numbers for the maximum in a stationary Gaussian sequence*, Ann. Math. Statist. **33** (1962), 93–97.

3. E. V. Bulinskaya, *On the mean number of crossings of a level by a stationary Gaussian process*, Teor. Verojatnost. i Primenen. **6** (1961), 474–477.

4. H. Cramér, *Random variables and probability distributions*, Cambridge Tracts in Mathematics, Vol. 36, Cambridge University Press, Cambridge, 1937. 2nd ed. to appear in 1962.

5. G. A. Hunt, *Random Fourier transforms*, Trans. Amer. Math. Soc. **71** (1951), 38–69.

6. M. Loève, *Probability theory*, 2nd ed., Van Nostrand, Princeton, N. J., 1960.

Research Triangle Institute

54.

Décompositions orthogonales de certains processus stochastiques

Ann. Fac. Sci. Clermont **8**, 15–21 (1962)

1 - Soient x, y,... des variables complexes aléatoires définies sur un espace de probabilité fixe. Nous supposerons que toute variable a une valeur moyenne égale à zéro, et un moment fini de second ordre :

$$E\,x = 0, \qquad E|x|^2 < \infty.$$

Deux variables x et y telles que $E|x - y|^2 = 0$ sont considérées comme identiques.

L'ensemble de toutes les variables aléatoires satisfaisant à ces conditions constitue un espace H de Hilbert, en définissant le produit scalaire et la norme par les relations :

$$(x,y) = E\,x\bar{y}, \quad \|x\|^2 = E|x|^2.$$

La convergence en norme d'une suite d'éléments $x_1, x_2, \ldots$ coïncide alors avec la convergence en moyenne quadratique (m. q.) des variables aléatoires $x_1, x_2, \ldots$

Deux éléments x,y sont orthogonaux, $x \perp y$, si l'on a :

$$(x,y) = E\,x\bar{y} = 0.$$

Considérons un processus stochastique défini par une famille de variables aléatoires x(t), où t est un paramètre variant d'une manière continue sur tout l'axe réel $(-\infty, \infty)$. Dans un grand nombre d'applications importantes t signifie le temps, et nous emploierons souvent une terminologie qui se réfère à ce cas. Chaque x(t) représentant un point de H, le processus x(t) définit une "courbe" C dans cet espace. Le plus petit sous-espace $H(x) \subset H$ renfermant toute la courbe C sera appelé l'espace du processus x(t). C'est le sous-espace de H engendré par toutes les variables x(t) avec $-\infty < t < +\infty$.

D'autre part, considérons l'arc C(t) de la courbe C formé par tous les points x(u) avec $u \leq t$, et désignons par H(x,t) le plus petit sous-espace de H renfermant C(t). On voit que H(x,t) est le sous-espace engendré par tous les x(u) avec $u \leq t$. Tout élément de H(x,t) est donc, soit une combinaison linéaire d'un nombre fini de variables $x(u_1), \ldots, x(u_k)$ avec $u_j \leq t$, soit la limite en m. q. d'une suite convergente de telles combinaisons. Dans un certain sens on peut donc dire que l'espace H(x,t) renferme la totalité de l'information contenue dans le passé et le présent du processus x(t), du point de vue de l'instant t. Evidemment l'espace H(x,t) n'est jamais décroissant quand t croît, et l'union de tous les H(x,t) pour $-\infty < t < +\infty$ coïncide avec H(x). En désignant par $H(x,-\infty)$ l'intersection de tous les H(x,t), on a donc :

$$H(x,-\infty) \subseteq H(x,t) \subseteq H(x).$$

Le sous-espace $H(x,-\infty)$ renferme l'information contenue dans le passé infiniment éloigné du processus x(t). Considérons les deux cas extrêmes :

(A) $$H(x,-\infty) = 0,$$

(B) $$H(x,-\infty) = H(x),$$

où nous avons désigné par O le sous-espace formé par le seul élément zéro de H.

Dans le cas (A), il est évident que le passé infiniment éloigné du processus x(t) ne nous apprend rien sur le développement futur du processus. Toute information contenue dans ce processus doit donc être entrée comme une innovation à un certain moment. Dans ce cas nous dirons que x(t) est un processus parfaitement non-déterministe.

Dans le cas (B), au contraire, toute information relativement au processus est déjà contenue dans son passé infiniment éloigné, et nous dirons alors que x(t) est un processus déterministe.

Dans le cas général, on aura $O \subset H(x,-\infty) \subset H(x)$, et nous dirons que x(t) est non-déterministe, sans être parfaitement non-déterministe.

Deux processus x(t) et y(t) sont dits orthogonaux l'un à l'autre, si nous avons $x(s) \perp y(t)$ pour tous s et t réels.

2 - Soit x(t) un processus donné tel que pour tout t :

$$Ex(t) = 0, \qquad E|x(t)|^2 < \infty. \tag{1}$$

Nous dirons que la relation :

$$x(t) = u(t) + v(t) \tag{2}$$

définit une décomposition orthogonale du processus x(t) si les processus u(t) et v(t) sont orthogonaux l'un à l'autre, et si l'on a pour tout t :

$$u(t) \in H(x,t), \quad v(t) \in H(x,t).$$

THEOREME 1 - Pour tout processus x(t) satisfaisant à (1), il y a une décomposition orthogonale (2) uniquement déterminée telle que u(t) est parfaitement non-déterministe, tandis que v(t) est déterministe.

Dans mes travaux antérieurs sur ce sujet (1), j'ai énoncé ce théorème sans donner une démonstration détaillée. Voici comment on peut le démontrer.

Désignons par $P_t x$ la projection orthogonale d'un élément arbitraire de H sur le sous-espace H(x,t), et posons :

$$v(t) = P_{-\infty} x(t), \qquad u(t) = x(t) - v(t).$$

D'abord il est évident que les deux composantes u(t) et v(t) appartiennent à H(x,t). En effet, v(t) appartient à $H(x,-\infty)$, qui est un sous-espace de H(x,t), tandis que $u(t) = x(t) - v(t) \in H(x,t)$.

D'autre part, v(t) étant la projection de x(t) sur $H(x,-\infty)$, et $x(t) = u(t) + v(t)$, on voit que u(t) doit être orthogonal à $H(x,-\infty)$ pour tout t, et il s'ensuit que $u(s) \perp v(t)$ pour tout s et t. Les processus u(t) et v(t) sont donc orthogonaux l'un à l'autre.

En désignant par H(u,t) et H(v,t) les sous-espaces de H qui, pour les processus u(t) et v(t), correspondent à H(x,t), il suit des remarques précédentes qu'on a pour tout t :

$$\begin{aligned} &H(u,t) \subseteqq H(x,t), \\ &H(u,t) \perp H(x,-\infty), \\ &H(v,t) \subseteqq H(x,-\infty). \end{aligned} \tag{3}$$

En faisant tendre t vers $-\infty$ dans les deux premières de ces relations, on voit tout de suite qu'on a :

$$H(u,-\infty) = 0 ;$$

u(t) est donc un processus parfaitement non-déterministe.

Or, la relation $x(t) = u(t) + v(t)$ montre que le sous-espace H(x,t) est contenu dans la somme vectorielle des sous-espaces mutuellement orthogonaux H(u,t) et H(v,t), qui sont eux-mêmes des sous-espaces de H(x,t) ; donc :

$$H(x,t) = H(u,t) + H(v,t). \tag{4}$$

---

(1) Voir : Colloque sur le Calcul des Probabilités, Paris 1959, p. 103-112 ; Proceedings of the Fourth Berkeley Symposium 1961, Vol. II, p. 57-78 ; Arkiv för Matematik, Stockholm 1961, Vol. 4, Nr 19.

Il s'ensuit en particulier :

$$H(x,-\infty) \subsetneqq H(u,t) + H(v,t)$$

En combinant ceci avec les deux dernières relations (3), on obtient :

$$H(v,t) = H(x,-\infty)$$

pour tout t, ce qui donne :

$$H(v,-\infty) = H(v) = H(x,-\infty). \tag{5}$$

v(t) est donc un processus déterministe. De (4) et (5) on déduit aussi :

$$H(x,t) = H(u,t) + H(x,-\infty),$$

ce qui montre que H(u,t) est le complément orthogonal de $H(x,-\infty)$ par rapport à H(x,t).

Il reste à démontrer que la décomposition est unique. Supposons donc que :

$$x(t) = u_1(t) + v_1(t)$$

est une décomposition orthogonale satisfaisant aux conditions du théorème, mais différente de celle qu'on vient de considérer. Posons :

$$\begin{aligned} u_1(t) &= u(t) + z(t), \\ v_1(t) &= v(t) - z(t), \end{aligned} \tag{6}$$

u(t) et v(t) étant définis comme ci-dessus. Alors on a, puisque les composantes $u_1(t)$ et $v_1(t)$ doivent appartenir à H(x,t),

$$H(x,t) \subsetneqq H(u_1,t) + H(v_1,t) \subsetneqq H(x,t),$$

$$H(x,t) = H(u_1,t) + H(v_1,t) = H(u_1,t) + H(v_1),$$

$$H(v_1) = H(x,-\infty) = H(v).$$

On voit maintenant que $H(u_1,t)$ est le complément orthogonal de $H(x,-\infty)$ par rapport à H(x,t), c'est-à-dire identique à H(u,t). Il suit donc de (6) que nous avons :

$$z(t) \in H(u,t),$$

$$z(t) \in H(v).$$

Or, les sous-espaces H(u,t) et H(v) sont orthogonaux l'un à l'autre. Donc z(t) doit se réduire à zéro pour tout t, et il s'ensuit que :

$$u_1(t) = u(t),$$

$$v_1(t) = v(t).$$

La démonstration du théorème est donc achevée.

3 - Considérons maintenant un processus x(t) qui est parfaitement non-déterministe, de manière que x(t) coïncide avec sa composante non-déterministe u(t), tandis que la composante déterministe v(t) se réduit à zéro. Supposons encore, pour éliminer des cas trop irréguliers, que les limites en m.q.

$$\begin{aligned} x(t+0) &= \text{l.i.m.}\ x(t+h), \\ x(t-0) &= \text{l.i.m.}\ x(t-h) \end{aligned} \tag{7}$$

existent pour tout t, quand h tend vers zéro par des valeurs positives.

Dans mon travail dans l'Arkiv f. Matematik cité plus haut j'ai montré que, sous ces conditions, x(t) admet une décomposition orthogonale dans un sens généralisé, que je vais maintenant rappeler.

Le sous-espace H(x,t) défini ci-dessus n'est jamais décroissant quand t croît et l'on a, x(t) étant parfaitement non-déterministe, $H(x,-\infty) = 0$. Les compléments orthogonaux :

$$\Delta H(x,t) = H(x,t+\Delta t) - H(x,t),$$
$$\Delta H(x,u) = H(x,u+\Delta u) - H(x,u)$$

sont orthogonaux l'un à l'autre si les intervalles $(t, t+\Delta t]$ et $(u, u+\Delta u]$ sont disjoints. On voit que le sous-espace $\Delta H(x,t)$ est engendré par les innovations reçues par le processus x(t) pendant l'intervalle de temps $(t, t+\Delta t]$. Si, pour tout $h > 0$, on a :

$$H(x,t+h) - H(x,t-h) \supset 0, \tag{8}$$

le processus reçoit donc effectivement des innovations dans tout intervalle $(t-h, t+h]$, et l'ensemble de toutes les valeurs de t telles que (8) est satisfaite s'appelle le <u>spectre d'innovations</u> du processus x(t).

Si, pour une certaine valeur de t appartenant au spectre d'innovations, le premier membre de (8) ne tend pas vers zéro avec h, nous dirons que t est un <u>point de discontinuité</u> du spectre ; dans le cas contraire, c'est un <u>point de continuité</u>. J'ai démontré (l. c.) que, si les limites (7) existent pour tout t, l'espace H(x) est séparable, et l'ensemble des points de discontinuité du spectre d'innovations est au plus dénombrable.

Soit maintenant z une variable aléatoire quelconque dans H(x), et posons :

$$z(t) = P_t\, z, \tag{9}$$

P_t désignant, comme plus haut, la projection sur le sous-espace H(x,t). Il suit alors des propriétés de la famille H(x,t) que z(t) est un processus stochastique <u>à accroissements orthogonaux</u>. Dans mon travail cité, j'ai démontré le théorème suivant.

THEOREME 2 - <u>Soit x(t) un processus parfaitement non-déterministe tel que les limites (7) existent pour tout t. Alors il existe un nombre N uniquement déterminé, qui s'appelle la multiplicité spectrale du processus x(t), et qui peut être un entier positif fini, ou bien égal à $+\infty$, ayant les propriétés suivantes. On peut trouver N éléments $z_1, \ldots, z_N$ de H(x) tels que les processus correspondants $z_1(t), \ldots, z_N(t)$ définis par (9) sont orthogonaux l'un à l'autre, et que l'on a pour tout t</u> :

$$x(t) = \sum_{n=1}^{N} \int_{-\infty}^{t} g_n(t,u)\, d z_n(u), \tag{10}$$

$$H(x,t) = \sum_{n=1}^{N} H(z_n, t), \tag{11}$$

<u>où les g_n sont des fonctions non aléatoires telles que</u> :

$$\sum_{n=1}^{N} \int_{-\infty}^{t} |g_n(t,u)|^2\, E|\, d z_n(u)|^2 < \infty .$$

<u>Enfin, N est le plus petit nombre ayant ces propriétés.</u>

On voit que (10) donne une décomposition du processus donné dans une infinité d'éléments :

$$g_n(t,u)\, d z_n(u), \tag{12}$$

qui sont tous orthogonaux l'un à l'autre, et qui appartiennent tous à H(x,t). C'est donc une généralisation de la simple décomposition orthogonale $x(t) = u(t) + v(t)$ considérée auparavant.

Chaque élément (12) peut être considéré comme l'effet à l'instant t d'une innovation $dz_n(u)$ entrée dans l'intervalle (u, u + du). Donc (10) donne une représentation linéaire de x(t) au moyen de toutes les innovations appartenant au passé et au présent du processus, du point de vue de l'instant t. Le processus vectoriel :

$$\{z_1(t), \ldots, z_N(t)\}$$

peut être regardé comme le processus d'innovations lié au processus donné x(t). On voit que, dans le cas général, le processus d'innovations sera multi-dimensionnel, sa dimensionnalité étant égale à la multiplicité spectrale N de x(t).

4 - Le théorème 2 donne une généralisation de certaines propriétés bien connues des processus stationnaires à une classe plus générale de processus stochastiques. Pour les processus stationnaires, on sait que la multiplicité est toujours égale à l'unité : N = 1. Il est donc bien naturel de poser la question s'il existent des processus x(t) satisfaisant aux conditions du théorème 2, ayant une multiplicité N donnée, finie ou infinie. Dans mon travail cité, j'ai répondu à cette question par l'affirmative, en construisant un exemple d'un tel processus.

Cependant, dans cet exemple le spectre d'innovations est purement discontinu, et ne contient en effet qu'un seul point. Il serait d'un intérêt bien plus grand de connaître un exemple d'un processus x(t) à spectre d'innovations partout continu, ayant une multiplicité N donnée.

L'exemple suivant d'un processus x(t) avec ces propriétés a été construit et discuté au cours d'une correspondance entre M. A. N. Kolmogorov et moi. Soit :

$$z_1(t),\ z_2(t), \ldots$$

une suite infinie de processus orthogonaux l'un à l'autre et avec des accroissements orthogonaux, et supposons que l'on a pour tous n et t :

$$E\, z_n(t) = 0, \qquad E\,|z_n(t)|^2 = \int_{-\infty}^{t} f(u)\, du,$$

où f(u) est partout continue, bornée et positive, avec un moment fini de second ordre.

Soit d'autre part $A_1, A_2, \ldots$ une suite d'ensembles disjoints de nombres réels, tout A_n étant de mesure positive dans chaque intervalle Δ :

$$\text{Mes}\,(\Delta \cap A_n) > 0;$$

désignons par $b_n(s)$ la fonction caractéristique de l'ensemble A_n, et posons pour $u \leq t$

$$g_n(t,u) = \int_u^t b_n(s)\, ds.$$

On trouve alors sans difficulté que la série :

$$x(t) = \sum_1^{\infty} \int_{-\infty}^{t} g_n(t,u)\, dz_n(u) = \sum_1^{\infty} \int_{-\infty}^{t} b_n(u)\, z_n(u)\, du \tag{13}$$

converge en m. q. pour tout t, et représente un processus x(t) satisfaisant aux conditions du théorème 2. On a évidemment :

$$H(x,t) \subseteq \sum_1^{\infty} H(z_n, t). \tag{14}$$

D'autre part, on a :

$$\int_{-\infty}^{t} b_n(u)\, z_n(u)\, du = \int_{-\infty}^{t} b_n(u)\, dx(u),$$

$$z_n(t) = \lim_{h \to 0} \frac{\int_{t-h}^{t} b_n(u)\, z_n(u)\, du}{\text{Mes}\,([t-h, t] \cap A_n)},$$

19

la dernière relation ayant lieu dans chaque point de continuité de la fonction $z_n(t)$, c'est-à-dire avec une probabilité égale à l'unité pour chaque t. On tire de là :

$$z_n(t) \in H(x,t),$$
$$H(z_n,t) \subseteqq H(x,t),$$

et d'après (14), pour tout t,

$$H(x,t) = \sum_1^\infty H(z_n,t)$$

Pour les compléments orthogonaux correspondant à un intervalle $(t, t+\Delta t]$, on obtient :

$$\Delta H(x,t) = \sum_1^\infty \Delta H(z_n,t),$$

et on voit que le spectre d'innovations de x(t) est partout continu, et contient tout l'axe réel.

Nous allons montrer explicitement que x(t) a une multiplicité $N = \infty$. Supposons, en effet, que N a une valeur finie, et que :

$$x(t) = \sum_{n=1}^{N} \int_{-\infty}^{t} h_n(t,u)\, dy_n(u) \tag{15}$$

est pour x(t) l'analogue de la représentation (10) du théorème 2. Alors les $y_n(u)$ sont des processus orthogonaux l'un à l'autre, et à accroissements orthogonaux, tels que

$$H(x,u) = \sum_{n=1}^{N} H(y_n,u) = \sum_{n=1}^{\infty} H(z_n,u),$$

et pour un intervalle $(u, u+\Delta u)$:

$$\Delta H(x,u) = \sum_{n=1}^{N} \Delta H(y_n,u) = \sum_{n=1}^{\infty} \Delta H(z_n,u).$$

On peut donc écrire :

$$E\,|\,dy_n(u)|^2 = dJ_n(u),$$

$$E\, dy_m(u)\, d\overline{z_n(v)} = \begin{cases} 0 & \text{pour } u \neq v, \\ d\,K_{mn}(u) & \text{pour } u = v, \end{cases}$$

J_n étant non décroissante, tandis que K_{mn} est à variation bornée dans tout intervalle fini. On a alors pour $u < u+\Delta u < t < t+\Delta t$:

$$\sum_{m=1}^{N} \int_{u}^{u+\Delta u} \Delta_t h_m(t,u)\, dy_m(u) = \sum_{m=1}^{\infty} \int_{t}^{t+\Delta t} b_m(s)\, ds. \int_{u}^{u+\Delta u} dz_m(u), \tag{16}$$

où nous avons écrit :

$$\Delta_t h_m(t,u) = h_m(t+\Delta t, u) - h_m(t,u).$$

Il s'ensuit :

$$\sum_{m=1}^{N} \int_{u}^{u+\Delta u} |\Delta_t h_m(t,u)|^2\, dJ_m(u) = \int_{u}^{u+\Delta u} f(u)\, du. \sum_{m=1}^{\infty} \left(\int_{t}^{t+\Delta t} b_m(s) ds \right)^2.$$

Cette identité ayant lieu pour tout intervalle $(u, u+\Delta u)$ tel que $u < u+\Delta u < t$, on voit que la contribution de la composante non absolument continue de $J_m(u)$ - s'il y en a une - à la valeur du premier membre doit se réduire à zéro, de manière qu'on aura, presque partout en u :

$$\sum_{m=1}^{N} |\Delta_t h_m(t,u)|^2 J_m'(u) = f(u) \sum_{m=1}^{\infty} \left(\int_{t}^{t+\Delta t} b_m(s)\, ds \right)^2.$$

En multipliant (16) par $\int_u^{u+\Delta u} \overline{d z_n(u)}$, on obtient :

$$\sum_{m=1}^{N} \int_u^{u+\Delta u} \Delta_t h_m(t,u)\, d K_{mn}(u) = \int_t^{t+\Delta t} b_n(s)\, ds . \int_u^{u+\Delta u} f(u)\, du.$$

et par un raisonnement analogue à celui de tout à l'heure :

$$\sum_{m=1}^{K} \Delta_t h_m(t,u)\; K'_{mn}(u) = f(u) \int_t^{t+\Delta t} b_n(s)\, ds \tag{17}$$

presque partout en u. En donnant N + 1 intervalles disjoints $(t_j, t_j + \Delta t_j)$, on peut toujours supposer les ensembles $A_1, A_2, \ldots$ choisis tels que la matrice :

$$\left\{ \int_{t_j}^{t_j+\Delta t_j} b_n(s)\, ds \right\} \qquad (n = 1,2,\ldots;\ j = 1,2,\ldots, N+1)$$

aura le rang N + 1. Or, la matrice dont les éléments sont donnés par le premier membre de l'identité (17), en remplaçant successivement t par $t_1, \ldots, t_{N+1}$ et en prenant n = 1,2,..., sera au plus du rang N. Il y a donc ici une contradiction, et la représentation (15) ne peut pas avoir lieu. Le processus x(t) a donc une multiplicité infinie.

En modifiant cet exemple d'une manière évidente, on peut construire un processus ayant une multiplicité finie N donnée d'avance.

DISCUSSION

M. FORTET - A quelles décompositions corrélatives de la fonction de covariance correspondent les décompositions indiquées pour la fonction aléatoire ?

M. CRAMER - Dans le cas stationnaire les composantes non déterministes et déterministes correspondent à des composantes déterminées du spectre de la fonction de covariance.

Pour les classes de processus harmonisable il y a correspondance incomplète entre les composantes du spectre de la fonction de covariance et cette décomposition.

M. FORTET - Quel est le mode de convergence utilisé lorsque N = + ∞ ?

M. CRAMER - Si . N + ∞ la convergence est définie en moyenne quadratique, la condition de convergence est :

$$\sum_{1}^{\infty} \int_{-\infty}^{t} |g_n(t,u)|^2\, d F_n(u) < \infty$$

Lorsque F_n est définie par :

$$E\, |dz_n(u)|^2 = dF\, n(u)$$

M. ITO - Can you use the same method to construct an example which needs a countable number of orthogonal processes ?

M. CRAMER - Yes, by the same method. I treated a general example in my paper.

M. BASS - Dans la représentation :

$$X(t) = \sum_{n=1}^{N} \int_{-\infty}^{t} g_n(t,u)\, d Z_n(u)$$

Les propriétés d'orthogonalité de $Z_n(u)$ relativement à u et à n présentent de l'analogie. Y-a-il de l'intérêt à remplacer n par une variable continue, et $Z_n(u)$ par un accroissement correspondant à cette variable ?

55.

On the approximation to a stable probability distribution

Studies in Mathematical Analysis and Related Topics
(Essays in Honor of Georg Polya). Ed. by G. Szegö et al.,
Stanford University Press, Stanford, California, 70–76 (1962)

1. The name "Central Limit Theorem of Probability Theory," which is now in general use, seems to have made its first appearance in George Pólya's paper [5], written in 1919.

The various classical forms of this theorem, due to de Moivre, Laplace, Chebyshev, Lyapunov, and Markov, give conditions under which the probability distribution of a sum of n independent random variables becomes asymptotically normal as n tends to infinity. During the last thirty years this theorem has served as a starting point for a great number of important investigations, sharpening and generalizing the classical results in various directions. An admirable survey of these modern investigations is given in [4].

Throughout all these works, the method of characteristic functions, or Fourier-Stieltjes transforms, has played a fundamental part. If $F(x)$ is a distribution function (d.f.) with the characteristic function (c.f.) $f(t)$, we have

$$f(t) = \int_{-\infty}^{\infty} e^{itx}\, dF(x)\,, \tag{1}$$

and conversely $F(x)$ may be expressed in terms of $f(t)$ by a reciprocal integral formula of Fourier type. The latter integral is, however, in general not absolutely convergent, and is accordingly not very well suited as an instrument for the study of the asymptotic properties of $F(x)$. A modified formula, containing an absolutely convergent integral, was used by Lyapunov for the proof of his form of the classical limit theorem.

In [5] Pólya points out that a fruitful approach to the central limit theorem consists in making use of a simple modification of the "discontinuous factor" due to Dirichlet, in the form of an absolutely convergent integral currently used in the analytic theory of numbers. In modern terminology, Pólya is here concerned with the relation between a d.f. and its c.f., and he observes that the *integral* of the d.f. is expressed by an absolutely convergent integral, which may be advantageously used for the proof of probability limit theorems.

In various modified forms, this method has been systematically used for improving and generalizing the classical limit theorems of probability theory. Thus, for example, the theory of asymptotic expansions of certain distribution functions approximating the normal d.f., as given in my Cambridge Tract [3], is based on certain generalized forms of the absolutely convergent integral expressions used by Pólya. In the present paper I propose to show how the same method may be used for studying the asymptotic properties of the probability distribution of a sum of independent and equally distributed random variables, in the case when the limiting distribution is of a non-normal stable type.

In order to avoid trivial complications, I shall consider only some simple particular cases, but it will be evident that the method is capable of being straightforwardly applied also to more general cases.

2. Consider an infinite sequence of independent and equally distributed random variables $X_1, X_2, \cdots$, and let $F(x) = P(X_j \leqq x)$ denote the common d.f. of the X_j, while $f(t)$ is the corresponding c.f. given by (1). If we write $S_n = X_1 + \cdots + X_n$, there will under certain known conditions [4, chap. 7] exist constants A_n and B_n such that the d.f.

$$F_n(x) = P\left(\frac{S_n}{B_n} - A_n \leqq x\right) \tag{2}$$

tends for every x to a limiting d.f. $G(x)$ as n tends to infinity. $G(x)$ will then be a *stable* d.f., and we shall here consider only the case when $G(x)$ is non-normal, with the characteristic exponent α $(0 < \alpha < 2)$, while it is possible to take $B_n = n^{1/\alpha}$ in (2). In this case the given d.f. $F(x)$ is said to belong to the domain of normal attraction of the stable d.f. $G(x)$ [4, p. 181].

In order to simplify the writing, we shall assume that the given d.f. $F(x)$ is symmetric, so that we have in all continuity points of $F(x)$

$$F(-x) = 1 - F(x)\,. \tag{3}$$

By a theorem due to Gnedenko [4, pp. 181–82], a necessary and sufficient condition that $F(x)$ will belong to the domain of normal attraction of some non-normal stable d.f. with characteristic exponent α $(0 < \alpha < 2)$ is then that we have, as $x \to +\infty$,

$$1 - F(x) = \frac{p}{x^\alpha} + o\left(\frac{1}{x^\alpha}\right), \tag{4}$$

where $p > 0$ is a constant. If this condition is satisfied, we can take $A_n = 0$ and $B_n = n^{1/\alpha}$ in (2), and the limiting d.f. will be a stable and symmetric d.f. $G_\alpha(x)$ with the c.f.

$$g_\alpha(t) = \exp\{-P\,|\,t\,|^\alpha\}\,, \tag{5}$$

where

$$P = 2p\int_0^\infty \frac{\sin x}{x^\alpha}dx > 0 . \tag{6}$$

The d.f. $F_n(x)$ given by (2) will then be the n-fold composition ("Faltung") of $F(n^{1/\alpha}x)$ with itself. Denoting by $f_n(t)$ the c.f. corresponding to $F_n(x)$, we thus have for all x and t, as $n \to +\infty$,

$$\begin{aligned} F_n(x) &= [F(n^{1/\alpha}x)]^{n*} \to G_\alpha(x) , \\ f_n(t) &= [f(t/n^{1/\alpha})]^n \to g_\alpha(t) . \end{aligned} \tag{7}$$

We next consider the asymptotic behavior of the difference $F_n(x) - G_\alpha(x)$. It will be desirable to find an upper bound for the modulus of this difference, and also, if possible, some kind of asymptotic expansion valid for large n.

In the case when the limiting distribution is normal, these approximation problems have been thoroughly studied by means of the method of characteristic functions (see, for example, [3, chap. 7, and 4, chap. 8]). For the stable distributions, the corresponding problems have been investigated by Bergström [1, 2], who used a totally different method. If I am not mistaken, the Fourier-Stieltjes transform (c.f.) methods have so far not yielded any definite results in this field. It will be shown below that these methods may be applied to the problem and will, under appropriate conditions, lead to simple and explicit results, which seem to be only partly contained in the work of Bergström.

3. In order to obtain definite asymptotic results, it will be necessary to introduce somewhat more restrictive assumptions about $F(x)$ than those expressed by (3) and (4). I shall here consider only the simple case in which $F(x)$ satisfies, in addition to the symmetry relation (3), the following three conditions:

As $x \to +\infty$ we have

$$1 - F(x) = \frac{p}{x^\alpha} + \frac{q}{x^\beta} + r(x) , \tag{i}$$

where $r(x) = o(x^{-\beta})$ $(0 < \alpha < \beta < 2, p > 0)$, and q is any real constant. Further

(ii) if $\beta \leq 1$, we assume that $r(x)$ is monotonic for all sufficiently large $x > 0$.

Finally we assume

$$\limsup_{|t| \to \infty} |f(t)| < 1 . \tag{iii}$$

As before, $G_\alpha(x)$ will denote the stable d.f. corresponding to the c.f. $g_\alpha(t)$ given by (5) and (6). We shall write

$$\begin{aligned} G_{\alpha\beta}(x) &= -\frac{1}{2\pi i}\int_{-\infty}^{\infty} \frac{|t|^\beta}{t} \exp\{-P|t|^\alpha - itx\}\,dt \\ &= \frac{1}{\pi}\int_0^\infty t^{\beta-1} \exp\{-Pt^\alpha\} \sin tx\,dt , \end{aligned} \tag{8}$$

and

$$Q = 2q\int_0^\infty \frac{\sin x}{x^\beta}\,dx\,.$$

THEOREM. *Let $F(x)$ be a given d.f. satisfying the symmetry relation* (3) *and the conditions* (i)-(iii). *For $F_n(x)$ given by* (7), *we then have, as $n \to \infty$:*

(A) *if $\beta < 2\alpha$,*

$$F_n(x) = G_\alpha(x) - \frac{Q}{n^{(\beta/\alpha)-1}}G_{\alpha\beta}(x) + o\left(\frac{1}{n^{(\beta/\alpha)-1}}\right);$$

(B) *if $2\alpha < \beta < 2$,*

$$F_n(x) = G_\alpha(x) - \frac{P^2}{2n}G_{\alpha,2\alpha}(x) + o\left(\frac{1}{n}\right).$$

The constants implied by the o's may depend on α and β but are independent of x and n. In the limiting case when $\beta = 2\alpha$, the last relation holds with P^2 replaced by $P^2 + 2Q$.

As the following proof will show, the conditions of the theorem could easily be generalized in various directions. We might, for example, omit the symmetry condition, replace the factor $x^{-\beta}$ by $x^{-\alpha}(\log x)^{-\gamma}$ with $\gamma > 0$, bring in more terms of the asymptotic expansions, and so on.

PROOF. Without restricting the generality, we may assume that $x = 0$ is a point of continuity for $F(x)$. The symmetry relation (3) then gives $F(0) = \frac{1}{2}$ and further, as $t \to 0$,

$$\begin{aligned} f(t) &= 1 - 2t\int_0^\infty [1 - F(x)]\sin tx\,dx \\ &= 1 - 2t\int_m^\infty [1 - F(x)]\sin tx\,dx + O(t^2)\,, \end{aligned}$$

where $m > 0$ has been taken so large that, $\varepsilon > 0$ being given, we have $|r(x)| < \varepsilon x^{-\beta}$ for $x > m$ and moreover, in the case $\beta \leqq 1$, the function $r(x)$ is monotonic for $x > m$. From condition (i) we now obtain

$$\begin{aligned} f(t) &= 1 - 2pt\int_m^\infty \frac{\sin tx}{x^\alpha}dx - 2qt\int_m^\infty \frac{\sin tx}{x^\beta}dx \\ &\qquad - 2t\int_m^\infty r(x)\sin tx\,dx + O(t^2)\,. \end{aligned}$$

In the two first integrals the lower limit may be replaced by 0, since the error thus committed is of the form $O(t^2)$. This gives us

$$f(t) = 1 - P|t|^\alpha - Q|t|^\beta - 2\int_{m|t|}^\infty r\left(\frac{x}{|t|}\right)\sin x\,dx + O(t^2)\,.$$

Assuming t to be so small that $m|t| < 1$, we have

$$\int_{m|t|}^\infty r\left(\frac{x}{|t|}\right)\sin x\,dx < \varepsilon|t|^\beta\int_0^1 x^{1-\beta}\,dx + J\,,$$

where for $1 < \beta < 2$

$$J = \left| \int_1^\infty r\left(\frac{x}{|t|}\right) \sin x \, dx \right| < \varepsilon |t|^\beta \int_1^\infty x^{-\beta} dx \, ,$$

while for $0 < \beta \leqq 1$, by the second mean value theorem,

$$J \leqq 2 \left| r\left(\frac{1}{|t|}\right) \right| < 2\varepsilon \, |t|^\beta \, .$$

Since $\varepsilon > 0$ is arbitrary, it thus follows that we have in all cases, as $t \to 0$,

$$f(t) = 1 - P|t|^\alpha - Q\,|t|^\beta + o(|t|^\beta) \, . \tag{9}$$

In this relation we now replace t by $n^{-1/\alpha} t$, and assume

$$|t| \leqq n^{(1/\alpha) - 1/3} \, . \tag{10}$$

As n tends to infinity we then obtain, uniformly for all t satisfying (10),

$$f\left(\frac{t}{n^{1/\alpha}}\right) = 1 - P\frac{|t|^\alpha}{n} - Q\,\frac{|t|^\beta}{n^{\beta/\alpha}} + o\left(\frac{|t|^\beta}{n^{\beta/\alpha}}\right) .$$

It now follows, by an argument analogous to the proof of Lemma 2 in [3], that for the c.f. $f_n(t)$ given by (7) we have

$$\log f_n(t) = -P|t|^\alpha - Q\frac{|t|^\beta}{n^{(\beta/\alpha)-1}} - P^2\frac{|t|^{2\alpha}}{2n} + o\left(\frac{|t|^\beta}{n^{(\beta/\alpha)-1}} + \frac{|t|^{2\alpha}}{n}\right).$$

We now obtain for $\beta < 2\alpha$

$$\log f_n(t) = -P|t|^\alpha - Q\frac{|t|^\beta}{n^{(\beta/\alpha)-1}} + o\left(\frac{|t|^\beta}{n^{(\beta/\alpha)-1}}\right),$$

and for $2\alpha < \beta < 2$

$$\log f_n(t) = -P|t|^\alpha - P^2\frac{|t|^{2\alpha}}{2n} + o\left(\frac{|t|^{2\alpha}}{n}\right).$$

In the limiting case in which $\beta = 2\alpha$, we finally have

$$\log f_n(t) = -P|t|^\alpha - (P^2 + 2Q)\frac{|t|^{2\alpha}}{2n} + o\left(\frac{|t|^{2\alpha}}{n}\right).$$

All these relations hold as $n \to \infty$, uniformly for all t satisfying (10).

The rest of the proof will be given only for the case $\beta < 2\alpha$, the other cases being quite similar. A further development of the argument in accordance with the above-quoted proof from [3] gives

$$\begin{aligned} f_n(t) = {} & \exp\{-P|t|^\alpha\} - \frac{Q}{n^{(\beta/\alpha)-1}}|t|^\beta \exp\{-P|t|^\alpha\} \\ & + o\left(\frac{|t|^\beta}{n^{(\beta/\alpha)-1}} \exp\{-P|t|^\alpha\}\right) + O\left(\frac{|t|^{2\beta}}{n^{2[(\beta/\alpha)-1]}} \exp\{-\tfrac{1}{2}P|t|^\alpha\}\right) \end{aligned} \tag{11}$$

as $n \to \infty$, uniformly for all t satisfying (10).

Further, it follows from condition (iii) and known properties of characteristic functions that we have $|f(t)| \leqq k < 1$ for $|t| \geqq 1$. Hence by Lemma 1 of [3] we obtain

$$|f(t)| \leqq 1 - \frac{1-k^2}{8} t^2 \exp\left\{-\frac{1-k^2}{8} t^2\right\}$$

for $|t| \leqq 1$. Thus for $n^{(1/\alpha)-1/3} < |t| < n^{1/\alpha}$ we have

$$|f_n(t)| = \left|f\left(\frac{t}{n^{1/\alpha}}\right)\right|^n \leqq \exp\left\{-\frac{1-k^2}{8} n^{1/3}\right\}. \tag{12}$$

Finally, for $|t| \geqq n^{1/\alpha}$ we have

$$|f_n(t)| = \left|f\left(\frac{t}{n^{1/\alpha}}\right)\right|^n \leqq k^n. \tag{13}$$

Writing

$$R_n(x) = F_n(x) - G_\alpha(x) + \frac{Q}{n^{(\beta/\alpha)-1}} G_{\alpha\beta}(x),$$

$$r_n(t) = f_n(t) - \exp\{-P|t|^\alpha\} + \frac{Q}{n^{(\beta/\alpha)-1}} |t|^\beta \exp\{-P|t|^\alpha\},$$

we have, according to Theorem 12 of [3, 2d ed.],

$$\int_x^{x+h} (y-x)^{\omega-1} R_n(y)\, dy = -\frac{1}{2\pi i}\int_{-\infty}^{\infty} \frac{r_n(t)}{t} e^{-itx}\, dt \int_0^h u^{\omega-1} e^{-itu}\, du$$

for $0 < \omega < \alpha$, any real x, and any $h > 0$. (The proof is given in [3] under somewhat more restrictive assumptions concerning $R_n(x)$, and for $0 < \varepsilon < 1$, but it is easily seen that the same proof holds under the present conditions if ω is confined to the interval $0 < \omega < \alpha$.)

Using the relations (11), (12), and (13) and denoting by C an absolute constant, we have

$$\left|\int_x^{x+h} (y-x)^{\omega-1} R_n(y)\, dy\right| \leqq \frac{C}{\omega}\int_0^{\infty} \frac{|r_n(t)|}{t^{\omega+1}}\, dt$$
$$= o\left(\frac{1}{\omega n^{(\beta/\alpha)-1}}\right) + O\left(\frac{k^n}{\omega^2 n^{\omega/\alpha}}\right),$$

the constants implied by the o and O being independent of x, h, and ω. By simple estimations it now follows, as in the proof of Theorem 25 in [3, 2d ed.], that for any given $\varepsilon > 0$ we can find n_0 and M independent of x, h, and ω such that for all $n > n_0$ we have

$$|R_n(x)| < M\left(h + \frac{k^n}{\omega h^\omega}\right) + \varepsilon \frac{1}{h^\omega n^{(\beta/\alpha)-1}}.$$

Taking here $h = 1/n$ and $\omega = 1/\log n$, and observing that $k < 1$ is a fixed positive constant and that $\varepsilon > 0$ is arbitrarily small, we obtain

$$R_n(x) = o\left(\frac{1}{n^{(\beta/\alpha)-1}}\right).$$

We have thus proved our Theorem for the case $\beta < 2\alpha$, and the other cases can be proved similarly.

University of Stockholm

REFERENCES

[1] BERGSTRÖM, H., On Asymptotic Expansions of Probability Functions, *Skand. Aktuarietidskr.*, **34** (1951), 1-34.

[2] BERGSTRÖM, H., On Distribution Functions with a Limiting Stable Distribution Function, *Ark. Mat.*, **2** (25) (1953).

[3] CRAMÉR, H., *Random Variables and Probability Distributions* (Cambridge Tracts in Mathematics, and Mathematical Physics, No. 36). Cambridge: Cambridge Univ. Press, 1937 (second edition to appear in 1962).

[4] GNEDENKO, B. V., and A. N. KOLMOGOROV, Limit Distributions for Sums of Independent Random Variables. Cambridge, Mass.: Addison-Wesley, 1954 (translation by K. L. Chung from the Russian original published in 1949).

[5] PÓLYA, G., Über den zentralen Grenzwertsatz der Wahscheinlichkeitsrechnung und das Momentenproblem, *Math. Z.*, **8** (1920), 171-81.

56.

On asymptotic expansions for sums of independent random variables with a limiting stable distribution

Sankhyā Ser. **A 25**, 13–24 (1963)

SUMMARY. **The sum of n independent and identically distributed random variables will, under certain known conditions and after appropriate normalization, converge in distribution. The limiting distribution will then belong to the class of stable distributions. In the particular case when the limiting distribution is normal it is well known that, under certain additional conditions, the distribution function $F_n(x)$ of the normalized sum admits an asymptotic expansion in powers of $n^{-1/2}$, valid as n tends to infinity. The present paper considers the analogous question for the case of a non-normal stable limiting distribution. Sufficient conditions for the validity of an asymptotic expansion for $F_n(x)$ in this case are given, and explicit expressions for the terms of the expansion are deduced. As in the normal case, the terms are functions of x, multiplied with certain negative powers of n.**

1. Introduction

Let $x_1, x_2, \ldots,$ be independent and identically distributed random variables, with a given common distribution function (d.f.) $F(x)$, and the characteristic function (c.f.)

$$f(t) = \int_{-\infty}^{\infty} e^{itx}\, dF(x).$$

If it is possible to find constants $B_n > 0$ and A_n such that the d.f. of the normalized sum

$$\frac{x_1+x_2+\ldots+x_n-A_n}{B_n} \qquad \ldots \quad (1.1)$$

tends to a limiting d.f. $G(x)$ as n tends to infinity, we say that $F(x)$ belongs to the *domain of attraction* of $G(x)$.

Denoting by $F_n(x)$ and $f_n(t)$ the d.f. and c.f. of the random variable (1.1) we then have, for any continuity point x of $G(x)$, and for all real t

$$F_n(x) = F^{n*}(A_n+B_nx) \rightarrow G(x),$$

$$f_n(t) = e^{-itA_n/B_n}\left[f\left(\frac{t}{B_n}\right)\right]^n \rightarrow g(t), \qquad \ldots \quad (1.2)$$

where $F^{n*}(x)$ denotes the n times repeated convolution of $F(x)$ with itself, while $g(t)$ is the c.f. corresponding to the d.f. $G(x)$.

It is well known that, in order that $G(x)$ should possess a domain of attraction in this sense, it is necessary and sufficient that $G(x)$ should be the d.f. of a *stable* distribution. The c.f. $g(t)$ is then given by the expression

$$\log g(t) = hit - c|t|^{\alpha}[1+\mu i\omega(t, \alpha)], \qquad \ldots \quad (1.3)$$

where h is any real constant, while c, α and μ are constants such that $c \geqslant 0$, $0 < \alpha \leqslant 2$, $-1 \leqslant \mu \leqslant 1$. Finally,

$$\omega(t, \alpha) = \begin{cases} \operatorname{sgn} t \,.\, tg\dfrac{\alpha\pi}{2} & \text{when } \alpha \neq 1, \\ \operatorname{sgn} t \,.\, \dfrac{2}{\pi}\log|t| & \text{when } \alpha = 1. \end{cases} \qquad \dots \quad (1.4)$$

The case $c = 0$ is trivial, and will be excluded from our considerations, so that we may throughout assume $c > 0$.

When the limiting relations (1.2) hold, we can evidently always, by an appropriate modification of A_n and B_n obtain a limiting distribution such that, in the expression (1.3) for the c.f., we have $h = 0$ and $c = 1$.

The parameter α is called the *characteristic exponent* of the stable distribution. For $\alpha = 2$ the distribution is normal; for $0 < \alpha < 2$ we have non-normal stable distributions.

In order that $F(x)$ should belong to the domain of attraction of a non-normal stable law with the characteristic exponent α, it is necessary and sufficient that we should have, as $x \to +\infty$,

$$F(-x) \sim \frac{h_1(x)}{x^{\alpha}}, \qquad 1 - F(x) \sim \frac{h_2(x)}{x^{\alpha}}, \qquad \dots \quad (1.5)$$

where for any constant $k > 0$ $\qquad \dfrac{h_i(kx)}{h_i(x)} \to 1, \qquad (i = 1, 2),$

while the ratio $h_1(x)/h_2(x)$ tends to a constant limit (cf. Gnedenko and Kolmogorov (1949, p. 175); also Dynkin (1955)). We may, e.g., take $h_i(x) = c_i(\log x)^q$.

Particularly important is the case when the $h_i(x)$ can be taken as constants, so that

$$F(-x) \sim \frac{c_1}{x^{\alpha}}, \qquad 1 - F(x) \sim \frac{c_2}{x^{\alpha}}, \qquad \dots \quad (1.6)$$

where $c_1 \geqslant 0$, $c_2 \geqslant 0$, and $c_1 + c_2 > 0$. The normalizing constants B_n in (1.1) can then be determined by the formula

$$B_n = bn^{1/\alpha}, \qquad \dots \quad (1.7)$$

where $b > 0$ is a constant. In this case, $F(x)$ is said to belong to the *domain of normal attraction* of the limiting stable law (cf. Gnedenko and Kolmogorov (1949) p. 191).

In two papers (Cramér, 1925, 1928) and in my Cambridge Tract (Cramér, 1937) I proved the existence of an asymptotic expansion for $F_n(x)$, as $n \to \infty$ in the case of a *normal* limiting distribution. I also considered the case when the x_i in (1.1) are not identically distributed. Moreover, I gave (Cramér, 1928) a similar asymptotic expansion for the probability density $F_n'(x)$.

All these results, for the normal limit law, have since been generalized and improved by various authors [Esseen (1944); Gnedenko and Kolmogorov (1949); Petrov (1959) and others]. For the case of a non-normal stable limit law, on the other hand, comparatively little seems to be known. Some results have been given in two papers by Bergström (1951 and 1953), and I have recently published a preliminary note on the subject in the Anniversary Volume dedicated to G. Pólya (Cramér, 1962).

In the present paper, the problem of asymptotic expansions for $F_n(x)$ will be considered for the case of normal attraction to a non-normal stable law. It will be shown that, when appropriate conditions are imposed on the given d.f. $F(x)$, there exists for the d.f. $F_n(x)$ given by (1.2) an asymptotic expansion as $n \to \infty$, the successive terms of which tend to zero as certain, in general fractional, powers of $1/n$. Similar expansions can be shown to exist in the case of non-identically distributed variables and also, under appropriate conditions, for the probability density $F_n'(x)$.

In the case of non-normal attraction, the situation is in general quite different. When e.g., the functions $h_i(x)$ in (1.5) are of the form $c_i(\log x)^q$ with a non-integral q, there is, under similar conditions as in the normal attraction case, an asymptotic expansion for $F_n(x)$. However, the successive terms of this expansion only tend to zero as powers of $\frac{1}{\log n}$. This case will not be further dealt with in the present paper.

2. Theorems on asymptotic expansions

Throughout the following, α, β and γ will be real constants satisfying the relations

$$0 < \alpha < 2, \qquad \alpha < \beta < \gamma. \qquad \dots \quad (2.1)$$

Further, c_1 and c_2 are non-negative constants, at least one of which differs from zero, while d_1 and d_2 are any real constants. In particular, we may have $d_1 = d_2 = 0$. However, if c_1 or c_2 is equal to zero, we assume that the corresponding d is non-negative.

Let $F(x)$ be a given d.f. satisfying the following conditions (A) and (B) :

(A) *As* $x \to +\infty$, *we have*

$$F(-x) = \frac{c_1}{x^\alpha} + \frac{d_1}{x^\beta} + r_1(x),$$

$$1-F(x) = \frac{c_2}{x^\alpha} + \frac{d_2}{x^\beta} + r_2(x)$$

$$r_i(x) = O\left(\frac{1}{x^\gamma}\right), \qquad (i = 1, 2).$$

(B) *In the case when* $0 < \gamma \leqslant 1$, *the functions*

$$r_1(x) \pm r_2(x)$$

are assumed to be monotone for all sufficiently large $x > 0$.

It will be seen that condition (A) implies considerably stronger assumptions about the infinitary behaviour of $F(x)$ than do the relations (1.6). Some assumptions of this kind will be necessary in order to get more precise information concerning the asymptotic behaviour of $F_n(x)$ for large n than the one provided by the limit relation (1.2). However, the method used below will be applicable even if condition (A) is modified in various ways, e.g. by the introduction of further terms in the asymptotic expressions of $F(-x)$ and $1-F(x)$.

For our Theorem 2, we shall also use the following condition (C), bearing on the c.f. $f(t)$ corresponding to $F(x)$:

(C) $\qquad \limsup |f(t)| < 1 \quad \text{as } |t| \to \infty.$

It is well known that, in particular, this condition is always satisfied when the d.f. $F(x)$ contains an absolutely continuous component.

We shall denote by $G_\alpha(x)$ the stable d.f. corresponding to the c.f. $g_\alpha(t)$ obtained from (1.3) by taking $h = 0$ and $c = 1$, so that

$$g_\alpha(t) = \exp\,[-|t|^\alpha(1+\mu i\omega(t, \alpha)], \qquad \text{... (2.2)}$$

where we now take
$$\mu = \frac{c_1-c_2}{c_1+c_2}$$

while $\omega(t, \alpha)$ is given by (1.4).

Further, we shall use the notation

$$k = k_1\alpha+k_2(\beta-\alpha)+k_3(2-\alpha)+k_4, \qquad \text{... (2.3)}$$

where k_1, k_2, k_3 and k_4 are non-negative integers, at least one of which differs from zero. It follows, in particular, that we always have $k > 0$.

If $P_k(t)$ is a polynomial in t of degree $k_1+k_2+k_3+k_4-1$, with complex coefficients depending on the k_i, we write

$$G_\alpha(x; P_k) = -\frac{1}{\pi}\,\mathcal{I}_m\Big[\int_0^\infty t^{k+\alpha-1}P_k(t^\alpha)g_\alpha(t)e^{-itx}dt\Big]. \qquad \text{... (2.4)}$$

It will be shown below in Lemma 1 that $G_\alpha(x; P_k)$ is a function of the real variable x which is everywhere continuous, has derivatives of all orders, tends to zero as $x\to\pm\infty$, and is of bounded variation over the whole real axis.

We shall now state our two main theorems.

Theorem 1 : *Suppose that α, β and γ are not integers. If the given d.f. $F(x)$ satisfies conditions* (A) *and* (B) *it is possible to choose the normalizing constants A_n and $B_n = bn^{1/\alpha}$ in* (1.1), *and the polynomials P_k such that, as $n\to\infty$,*

$$F_n(x) = G_\alpha(x)+\sum_{0<k<\lambda} G_\alpha(x; P_k)n^{-k/\alpha}+O(n^{-\lambda/\alpha}), \qquad \text{... (2.5)}$$

uniformly for all real x, where $\lambda = \min(1, \gamma-\alpha)$. The summation is extended over all k given by (2.3), *which satisfy the inequality $0 < k < \lambda$.*

Theorem 2 : *Suppose that α, β and γ are not integers. If $\gamma-\alpha > 1$, and the given d.f. $F(x)$ satisfies conditions* (A) *and* (C) *the summation in the second member of* (2.5) *may be extended over all k such that $0 < k < \gamma-\alpha$, and the remainder term will be of the order*

$$O(n^{-(\gamma-\alpha)/\alpha}).$$

Remark (1) : If α or β (or both) is an integer, powers of log t will occur under the integral sign in (2.4), and the terms of the asymptotic expansion in (2.5) will be multiplied by polynomials in log n. If γ is an integer, the majorants of the remainder terms in Theorems 1 and 2 must be multiplied by log n. Since the explicit formulas for these cases are somewhat cumbersome, we shall restrict ourselves to this remark.

Remark (2) : Note that, in the case of Theorem 2, we necessarily have $\gamma > 1$, so that condition (B) does not apply.

Remark (3) : In particular cases, some of the $G_\alpha(x; P_k)$ may be identically zero. An extreme example of this phenomenon is obtained by taking $F(x) = G_\alpha(x)$. For appropriately chosen normalizing constants, $F_n(x)$ will then be identical with $G_\alpha(x)$.

3. Some lemmas

Before we can proceed to the proofs of the main theorems, we shall have to state and prove a number of lemmas.

Lemma 1 : *The function $G_\alpha(x; P_k)$ defined by* (2.4) *is an everywhere continuous real-valued function of the real variable x, which has derivatives of all orders, tends to zero as $x \to \pm\infty$, and is of bounded variation over the whole real axis. Moreover, we have for* $t > 0$

$$\int_{-\infty}^{\infty} e^{itx}\, d_x G_\alpha(x;\, P_k) = t^{k+\alpha} P_k(t^\alpha)\, g_\alpha(t). \qquad \ldots \quad (3.1)$$

For $t < 0$, the integral evidently takes the complex conjugate of its value for $-t > 0$.

It will be seen from the expression (2.2) of $g_\alpha(t)$ that the integral in the second member of (2.4), as well as all the integrals obtained by repeated differentiation with respect to x, are absolutely and uniformly convergent for all real x, and tend to zero as $x \to \pm\infty$. It thus only remains to show that $G_\alpha(x; P_k)$ is of bounded variation, and that we have the relation (3.1).

In order to prove the bounded variation property, it is sufficient to show that, for any $q > \alpha$, the integral

$$J(x) = \int_0^{\infty} t^{q-1} g_\alpha(t) e^{-itx}\, dt$$

is of bounded variation over $(-\infty, \infty)$. The derivative $J'(x)$ can be calculated by formal differentiation under the integral sign, and we have to show that $J'(x)$ is absolutely integrable over $(-\infty, \infty)$. The proof will be different according as $q \leqslant 1$, or $q > 1$.

If $q \leqslant 1$, we must have $0 < \alpha < 1$. For $x > 0$ we then obtain

$$J'(x) = -i \int_0^{\infty} t^q g_\alpha(t) e^{-itx} dt = -i\, x^{-(q+1)} \int_0^{\infty} t^q \exp\Big[-t^\alpha x^{-\alpha}\Big(1 + \mu i t g \frac{\alpha\pi}{2}\Big) - it\Big]\, dt.$$

A method used in a similar case by Skorohod (1954) can then be applied to show that the last integral is bounded as $x \to +\infty$. Since $|J'(-x)| = |J'(x)|$, we thus have

$$J'(x) = O(|x|^{-(q+1)})$$

as $|x| \to \infty$, which shows that $|J'(x)|$ is integrable.

For $q > 1$, a repeated partial integration shows that

$$J'(x) = -i\int_0^\infty H(t,x)g''_\alpha(t)dt,$$

where
$$H(t,x) = \int_0^t du \int_0^u v^q e^{-ivx}dv.$$

By elementary calculations we obtain

$$|H(t,x)| < K\frac{t^q}{x^2},$$

and it then follows from the expression (2.2) for $g_\alpha(t)$ that in this case $J'(x) = O(x^{-2})$ so that $J'(x)$ is integrable.

Thus $G_\alpha(x; P_k)$ is of bounded variation over the whole real x-axis, so that the Stieltjes integral in (3.1) exists. It is then easily seen that (3.1) is the Fourier-Stieltjes transform which is reciprocal to (2.4). Since both integrals are absolutely convergent, (3.1) is certainly valid, and the proof of Lemma 1 is completed.

We now proceed to two lemmas concerning the properties of the c.f. $f_n(t)$ defined by (1.2), for large values of n.

Lemma 2 : *Suppose that α, β and γ are not integers, and that $F(x)$ satisfies conditions* (A) *and* (B). *Then it is possible to choose the normalizing constants A_n and $B_n = bn^{1/\alpha}$ in* (1.1), *and the polynomials P_k such that, uniformly for $0 < t \leqslant n^{1/\alpha-1/3}$, we have*

$$f_n(t) = e^{-Ct^\alpha}[1+\sum_{0<k<\gamma-\alpha} t^{\alpha+k}P_k(t^\alpha)n^{-k/\alpha}]+O[t^\gamma(1+t^\delta)e^{-\frac{1}{2}t^\alpha}n^{-(\gamma-\alpha)/\alpha}],$$

where
$$C = 1+\frac{c_1-c_2}{c_1+c_2}\, itg\,\frac{\alpha\pi}{2}, \qquad \dots (3.2)$$

while δ is a positive constant only dependent on α, β and γ. The summation index k is given by (2.3), *and the summation is extended over all non-negative integers $k_1, \dots, k_4$ such that $0 < k < \gamma-\alpha$. For $-n^{1/\alpha-1/3} \leqslant t < 0$ we have, of course, $f_n(t) = \overline{f_n(-t)}$.*

In order to prove this lemma, we shall first deduce an expansion of the c.f. $f(t)$ for small positive values of t with an error term of the order $O(t^\gamma)$. We use in the sequel the notation $Q(it)$ to denote an unspecified polynomial in the argument it, with real coefficients, of degree $< \gamma-1$, not necessarily the same in different formulas. If $\gamma < 1$, the polynomial Q will be identically zero.

By partial integration we obtain

$$f(t) = 1-t\int_0^\infty (1-F(x)+F(-x))\sin tx\,dx+it\int_0^\infty(1-F(x)-F(-x))\cos tx dx.$$

The integrals from 0 to 1 can be developed in convergent power series. In the integrals from 1 to ∞, we replace $F(-x)$ and $1-F(x)$ by their expressions according to condition (A) and then obtain, using the known properties of integrals of the form $\int x^{\gamma} \cos x\, dx$ and $\int x^{\gamma} \sin x\, dx$, assuming $t > 0$,

$$f(t) = 1 - Lt^{\alpha} - Mt^{\beta} - R(t) - itQ(it) - O(t^{\gamma}),$$

where
$$L = \left[c_1 + c_2 + i(c_1 - c_2) tg \frac{\alpha\pi}{2} \right] \int_0^{\infty} \frac{\sin x}{x^{\alpha}}\, dx,$$

$$M = \rho(d_1 + d_2) + i\sigma(d_1 - d_2),$$

$$R(t) = t \int_1^{\infty} (r_1(x) + r_2(x)) \sin tx\, dx + it \int_1^{\infty} (r_1(x) - r_2(x)) \cos tx\, dx.$$

Here ρ and σ are real constants only depending on β. We now show that $R(t)$ is, for small $t > 0$, of the form $itQ(it) + O(t^{\gamma})$. This will be explicitly shown for the cosine term; the sine term can be dealt with in the same way.

Consider first the case $0 < \gamma < 1$. Writing $r(x) = r_1(x) - r_2(x)$, we have, denoting by K an unspecified constant,

$$it \int_1^{1/t} r(x) \cos tx\, dx \Bigg| = \Bigg| \int_t^1 r\left(\frac{x}{t}\right) \cos x dx \Bigg| < Kt^{\gamma} \int_0^1 \frac{dx}{x^{\gamma}},$$

and by the second mean value theorem, assuming t so small that $r(x)$ is monotone for $x > 1/t$,

$$\Bigg| it \int_{1/t}^{\infty} r(x) \cos tx\, dx = \int_1^{\infty} r\left(\frac{x}{t}\right) \cos x\, dx = \Bigg| r\left(\frac{1}{t}\right) \int_1^{z} \cos dx \Bigg| < Kt^{\gamma},$$

so that our assertion is proved for this case.

Now suppose $\gamma > 1$, and let $2h - 1 < \gamma < 2h + 1$, where h is a positive integer. Then

$$it \int_1^{\infty} r(x) \cos tx\, dx = it \int_1^{\infty} r(x) \left(\cos tx - 1 + \ldots \pm \frac{(tx)^{2h-2}}{(2h-2)!} \right) dx + itQ(it),$$

where Q is an even polynomial of degree $2h - 2 < \gamma - 1$. Majorating the cosine difference in two different ways, we find that the first term in the second member is majorated by

$$Kt^{\gamma} \int_t^{\infty} \cos x - 1 + \ldots \pm \frac{x^{2h-2}}{(2h-2)!} \Bigg| \frac{dx}{x^{\gamma}} < Kt^{\gamma} \left(\int_0^1 x^{2h-\gamma}\, dx + \int_1^{\infty} x^{2h-2-\gamma}\, dx \right).$$

Since $2h - \gamma > -1$, while $2h - 2 - \gamma < -1$, both integrals are convergent, and our assertion is proved. We thus finally have, as $t > 0$ tends to zero,

$$f(t) = 1 - Lt^{\alpha} - Mt^{\beta} - itQ(it) - O(t^{\gamma}). \qquad \ldots \quad (3.3)$$

We now choose the normalizing constants B_n in (1.1) so that

$$B_n = bn^{1/\alpha},$$

where b is the positive root of the equation

$$b^{\alpha} = (c_1 + c_2) \int_0^{\infty} \frac{\sin x}{x^{\alpha}}\, dx.$$

In (3.3) we replace t by t/B_n, and write as an abbreviation

$$\tau = \frac{t}{n^{1/\alpha}}. \quad \dots \quad (3.4)$$

It will then be seen that as long as t remains confined to the interval $0 < t \leqslant n^{1/\alpha-1/3}$, we shall have $0 < \tau \leqslant n^{-1/3}$, so that τ tends to zero as $n \to \infty$, uniformly for all t in the interval considered. Hence we obtain, taking account of the value of the constant L,

$$f\left(\frac{t}{B_n}\right) = f\left(\frac{\tau}{b}\right) = 1 - C\tau^\alpha - D\tau^\beta - i\tau Q(i\tau) + O(\tau^\gamma), \quad \dots \quad (3.5)$$

where C is given by (3.2) while D, as well as the coefficients of the polynomial Q, is independent of n and t. The error term is $O(\tau^\gamma)$, uniformly for all t in the interval considered.

We now choose the normalizing constants A_n in (1.1) so that

$$A_n = -nbq,$$

where q is the constant term of the polynomial $Q(i\tau)$ appearing in (3.5). From (1.2) we then obtain

$$f_n(t) = e^{nqi\tau}\left[f\left(\frac{\tau}{b}\right)\right]^n$$

and further by means of (3.5)

$$\log f_n(t) = n \log\left[e^{qi\tau} f\left(\frac{\tau}{b}\right)\right] = n \log [1 - Z + O(\tau^\gamma)],$$

where $$Z = (C\tau^\alpha + D\tau^\beta)e^{qi\tau} + i\tau Q_0(i\tau), \quad \dots \quad (3.6)$$

$Q_0(i\tau)$ denoting a polynomial in $i\tau$, with the same properties as before, but *without a constant term*. Since $\alpha < \beta$ and $\alpha < 2$, it follows that $Z = O(\tau^\alpha)$, so that $Z \to 0$ as $n \to \infty$. Consequently

$$\log f_n(t) = n \log (1 - Z) + O(n\tau^\gamma) = -n \sum_{\nu=1}^{p} \frac{1}{\nu} Z^\nu + O(n\tau^\gamma),$$

the integer p being taken so that $p\alpha < \gamma \leqslant (p+1)\,\alpha$. Expanding the power Z^ν by means of (3.6), we obtain a linear aggregate of powers of τ, each term having an exponent of the form

$$j = j_1\alpha + j_2\beta + 2j_3 + j_4 = (j_1 + j_2 + j_3 - 1)\alpha + j_2(\beta - \alpha) + j_3(2 - \alpha) + j_4 + \alpha,$$

where the j_i are non-negative integers such that $j_1 + j_2 + j_3 = \nu$. It will be seen that we always have $j > \alpha$, except only in the case when $\nu = j_1 = 1$, $j_2 = j_3 = j_4 = 0$. Thus we may write, replacing j by $k + \alpha$ and using (3.4),

$$\log f_n(t) = -nC\tau^\alpha + n \sum_k D_k \tau^{k+\alpha} + O(n\tau^\gamma) = -Ct^\alpha + t^\alpha \sum_k D_k \tau^k + O(t^\alpha \tau^{\gamma-\alpha}), \quad \dots \quad (3.7)$$

where k is given by

$$k = k_1\alpha+k_2(\beta-\alpha)+k_3(2-\alpha)+k_4, \qquad \dots \quad (3.8)$$

the k_i running through all non-negative integers such that

$$0 < k < \gamma-\alpha, \qquad \text{and} \qquad k_1+1 \geqslant k_2+k_3.$$

(The second inequality is another way of writing the evident inequality $j_1+j_2+j_3 \geqslant j_2+j_3$.) In particular, the powers τ^α, $\tau^{\beta-\alpha}$, $\tau^{2-\alpha}$ and τ will always appear in the sum in the last member of (3.7), in so far as their exponents are $<\gamma-\alpha$. This remark will be used in a moment.

Taking

$$\lambda = \min(\alpha, \beta-\alpha, 2-\alpha) > 0,$$

(3.7) may be written

$$\log f_n(t) = -Ct^\alpha+U+V,$$

where

$$U = t^\alpha \sum_k D_k\tau^k = O(t^\alpha\tau^\lambda),$$

$$V = O(t^\alpha\tau^{\gamma-\alpha}).$$

It follows in particular that for all sufficiently large n, and for all t in the interval considered, we have

$$|U| < \frac{1}{4}t^\alpha \qquad \text{and} \qquad |V| < \frac{1}{4}t^\alpha.$$

Further

$$f_n(t) = e^{-Ct^\alpha}.\, e^u.\, e^v = e^{-Ct^\alpha}\left[\sum_{\nu=0}^{q}\frac{u^\nu}{\nu!}+O(U^{q+1}e^{1/4t^\alpha})\right][1+O(Ve^{1/4t^\alpha})].$$

Taking here the integer q such that $q\lambda < \gamma-\alpha \leqslant (q+1)\lambda$, we have, observing that according to (3.2) the real part of C is equal to unity,

$$f_n(t) = e^{-Ct^\alpha}\left(1+\sum_{\nu=1}^{q}\frac{u^\nu}{\nu!}\right)+O[(t^\alpha+t^{(q+1)\alpha})\tau^{\gamma-\alpha}e^{-\frac{1}{2}t^\alpha}]. \qquad \dots \quad (3.9)$$

Now we have for $\nu = 1, \dots, q$

$$U^\nu = t^{\nu\alpha}[\sum_k D_{k\nu}\tau^k+O(\tau^{\gamma-\alpha})] \qquad \dots \quad (3.10)$$

where the $D_{k\nu}$ are constants, while k is given by (3.8) the k_i being now certain non-negative integers satisfying

$$0 < k < \gamma-\alpha. \qquad \dots \quad (3.11)$$

Let $k_1, \dots, k_4$ be any given set of non-negative integers satisfying (3.11). It then easily follows that we must have

$$k_1+k_2+k_3+k_4 \leqslant q,$$

so that a term in $U^{k_1+k_2+k_3+k_4}$ will certainly appear in the sum in the second member of (3.9). According to a previous remark, the expansion (3.10) of this power will include a term in

$$\tau^k = (\tau^\alpha)^{k_1}(\tau^{\beta-\alpha})^{k_2}(\tau^{2-\alpha})^{k_3}\tau^{k_4}.$$

On the other hand, it is obvious that no U^ν with $\nu > k_1+\ldots+k_4$ can include the power τ^k with the given values of the k_i. It follows that we may write

$$\sum_{\nu=1}^{q} \frac{U^\nu}{\nu\,!} = t^\alpha \sum_k \tau^k P_k(t^\alpha)+O[(t^\alpha+t^{(q+1)\alpha})\tau^{\gamma-\alpha}], \qquad \ldots \quad (3.12)$$

where the sum in the second member is extended over all k given by (3.8) and satisfying (3.11), while $P_k(t^\alpha)$ is a polynomial in t^α of degree $k_1+\ldots+k_4-1$. Finally, we obtain from (3.4), (3.9) and (3.12)

$$f_n(t) = e^{-Ct^\alpha}[1+ \sum_k t^{\alpha+k}P_k(t^\alpha)n^{-k/\alpha}]+O[t^\gamma(1+t^{q\alpha})e^{-\frac{1}{2}t^\alpha}\, n^{-(\gamma-\alpha)/\alpha}\,]$$

uniformly for $0 < t \leqslant n^{1/\alpha-1/3}$, so that Lemma 2 is proved.

Lemma 3 : *If $F(x)$ satisfies condition* (A) *there are positive constants p and q such that*

$$|f_n(t)| < e^{-pn^{1/3}}$$

for
$$n^{1/\alpha-1/3} < |t| \leqslant q\, n^{1/\alpha}.$$

When $F(x)$ satisfies condition (A), it is obvious that the modulus of the c.f. $f(t)$ cannot reduce to a constant. Hence we must have $|f(t)| < 1$ for all $t \neq 0$ in some neighbourhood of the origin. Thus we can find $g > 0$ such that $|f(t)| \leqslant h < 1$ for $g \leqslant |t| \leqslant 2g$. It then follows from Lemma 1 of my Cambridge Tract (Cramér, 1937) that

$$|f(t)| \leqslant 1-\frac{1-h^2}{8g^2}\, t^2 \leqslant \exp\left(-\frac{1-h^2}{8g^2}\, t^2\right)$$

for $|t| \leqslant g$. (In the work quoted, this is proved under the assumption that $|f(t)| \leqslant h < 1$ holds for all $|t| \geqslant g$. The only property used in the proof is, however, that this inequality holds for $g \leqslant |t| \leqslant 2g$.)

Further by (1.2), taking $B_n = bn^{1/\alpha}$,

$$|f_n(t)| = |f\left(\frac{t}{bn^{1/\alpha}}\right)|^n,$$

and thus for $n^{1/\alpha-1/3} < |t| \leqslant gbn^{1/\alpha}$

$$|f_n(t)| \leqslant \exp\left[-\frac{1-h^2}{8b^2g^2}\, n\left(\frac{t}{n^{1/\alpha}}\right)^2\right] < \exp\left(-\frac{1-h^2}{8b^2g^2}\, n^{1/3}\right).$$

Taking $p = \dfrac{1-h^2}{8b^2g^2}$, $q = gb$, Lemma 3 is proved.

We finally state the following lemma which is contained in one due to Esseen (1944).

Lemma 4: *Let $F(x)$ be any d.f. with the c.f. $f(t)$. Let $G(x)$ be a real function of bounded variation over the whole real axis, such that $G(-\infty) = 0$, $G(+\infty) = 1$, while the derivative $G'(x)$ exists everywhere and satisfies $|G'(x)| < K$ for some constant K. Write*

$$g(t) = \int_{-\infty}^{\infty} e^{itx}\, dG(x).$$

Suppose that for some positive constants T and ϵ we have

$$\int_{-T}^{T} \left| \frac{f(t)-g(t)}{t} \right| dt = \epsilon.$$

Then there are positive constants A and B independent of T and ϵ such that for all real x

$$|F(x)-G(x)| < A\epsilon + \frac{B}{T}.$$

4. Proofs of the theorems

In Lemma 4, we now replace $F(x)$ by $F_n(x)$, and $G(x)$ by

$$G_\alpha(x)+ \sum_k G_\alpha(x;\ P_k)n^{-k/\alpha}, \qquad \dots \quad (4.1)$$

the sum being extended over all k given by (2.3) or (3.8), and satisfying

$$0 < k < \lambda = \min(1, \gamma-\alpha).$$

Then $f(t)$ will be replaced by $f_n(t)$, while according to Lemma 1 we have to replace $g(t)$ by

$$e^{-Ct^\alpha}\,[1+ \sum_k t^{\alpha+k}\, P_k(t^\alpha)n^{-k/\alpha}\,]$$

for $t > 0$, C being given by (3.2), and by the complex conjugate value of $g(-t)$ for $t < 0$.

Further, $G'(x)$ will exist everywhere, and satisfy the inequality $|G'(x)| < K$, where as before K denotes an unspecified constant, independent of x and n.

From Lemma 2 we then obtain

$$\int_{|t|\leqslant n^{1/\alpha-1/3}} \left| \frac{f(t)-g(t)}{t} \right| dt < K\, n^{-\lambda/\alpha}, \qquad \dots \quad (4.2)$$

while Lemma 3 gives, taking account of the behaviour of $g(t)$ for large $|t|$,

$$\int_{n^{1/\alpha}-1/3<|t|\leqslant qn^{1/\alpha}} \left| \frac{f(t)-g(t)}{t} \right| dt < K \log n\, e^{-pn^{1/3}} < K\, n^{-(\gamma-\alpha)/\alpha}.$$

Thus we may take $T = qn^{1/\alpha}$ and $\epsilon = Kn^{-\lambda/\alpha}$ in Lemma 4, which now yields

$$|F_n(x)-G_\alpha(x)- \sum_k G_\alpha(x;\ P_k)n^{-k/\alpha}\,| < An^{-\lambda/\alpha} + Bn^{-1/\alpha} < Kn^{-\lambda/\alpha} \qquad \dots \quad (4.3)$$

so that Theorem 1 is proved.

In order to prove Theorem 2, we extend the summation in (4.1) over all k satisfying

$$0 < k < \gamma-\alpha.$$

By Lemma 2, the second member of (4.2) may then be replaced by an expression of the form $Kn^{-(\gamma-\alpha)/\alpha}$. We further observe that, if $F(x)$ satisfies condition (C) there is a constant $r < 1$ such that for $|t| > qn^{1/\alpha}$ we have

$$|f_n(t)| < r^n$$

and hence

$$\int_{qn^{1/\alpha}<|t|\leqslant n^{(\gamma-\alpha)/\alpha}} \left|\frac{f(t)-g(t)}{t}\right| dt < K(r^n \log n + e^{-n^{1/2}}) < Kn^{-(\gamma-\alpha)/\alpha}.$$

Thus in this case we may take $T = n^{(\gamma-\alpha)/\alpha}$ and $\epsilon = Kn^{-(\gamma-\alpha)/\alpha}$ in Lemma 4. The last member of (4.3) will then be replaced by $Kn^{-(\gamma-\alpha)/\alpha}$, and the proof of Theorem 2 is completed.

References

BERGSTRÖM, H. (1951): On asymptotic expansions of probability functions. *Skandinavisk Aktuarietidskrift*, 1-33.

——— (1953): On distribution functions with a limiting stable distribution function. *Arkiv för Matematik*, 2, Number 25.

CRAMÉR, H. (1925): On some classes of series used in mathematical statistics. Sixth Scandinavian Mathematical Congress, Copenhaguen, 399-425.

——— (1928): On the composition of elementary errors. *Skandinavisk Aktuarietidskrift*, 13-74 and 141-180.

——— (1937): Random variables and probability distributions. *Cambridge Tracts in Mathematics*, No. 36, Cambridge, second edition 1962.

———(1962): On the approximation to a stable probability distribution. *Essays in honor of George Pólya*, Stanford University Press.

DYNKIN, E. B. (1955): Some limit theorems for sums of independent random variables with infinite mathematical expectations. *Izvestija Akad. Nauk SSSR, Ser. Math.*, 19, 247-266. *Selected Translation in Math. Stat. and Prob.*, 1 (1961), 171-190.

ESSEEN, C. G. (1944): Fourier analysis of distribution functions. *Acta Mathematica*, 77, 1-125.

GNEDENKO, B. V. and KOLMOGOROV, A. N. (1954): *Limit distributions for sums of independent random variables.* Cambridge, Mass., Translation from the Russian original, published in 1949.

PETROV, V. V. (1959): Asymptotic expansions for distributions of sums of independent random variables. *Teorija Verojatnost. i Primenen*, 4, 220-224.

SKOROHOD, A. V. (1954): On a theorem concerning stable distributions. *Uspehi Mat. Nauk* (N.S.), 9, 2, 189-190. *Selected Translations in Math. Stat. and Prob.*, 1 (1961), 169-170.

Paper received: November, 1962.

57.

Stochastic processes as curves in Hilbert space

Teor. Verojatnost i Primenen. 9 (2), 169–179 (1964)
(Reprinted in Random Processes)

1. In this paper the theory of spectral multiplicity in a separable Hilbert space will be applied to the study of stochastic processes $x(t)$, where $x(t)$ is a complex-valued random variable with a finite second-order moment, while the parameter t may take any real values.

For an account of multiplicity theory we may refer to Chapter 7 of Stone's book [13] which deals with the case of a separable space. The treatment of the subject found e.g. in the books by Halmos [6] and Nakano [10] is mainly concerned with the more general and considerably more intricate case of a non-separable space.

Our considerations will apply to certain classes of curves in a purely abstract Hilbert space, and it is only a question of terminology when, throughout this paper, we confine ourselves to that particular realization of Hilbert space which has proved useful in probability theory.

We finally observe that all our statements may be directly generalized to the case of vector processes of the form $\{x_1(t), \cdots, x_n(t)\}$.

2. In Sections 2—4 we shall now introduce some basic definitions and some auxiliary concepts.

Consider the set of all complex-valued random variables x defined on a fixed probability space and satisfying the relations

$$\mathbf{E}\{x\} = 0, \qquad \mathbf{E}\{|x|^2\} < \infty.$$

Two variables x and y will be regarded as identical if

$$\mathbf{E}\{|x-y|^2\} = 0.$$

The set of all these variables forms a Hilbert space H, if the inner product is defined in the usual way:

$$(x, y) = \mathbf{E}\{x\bar{y}\}.$$

Convergence of sequences of random variables will always be understood as strong convergence in the topology thus introduced, i.e. as convergence in quadratic mean according to probability terminology.

If for every real t a random variable $x(t) \in H$ is given, the set of variables $x(t)$ may be regarded as a stochastic process with continuous time t or, alternatively, as a curve C in the Hilbert space H. It is well known that various properties of stochastic processes have been studied by regarding them as curves in H (cf., e.g., [7], [8], [11], [12], [2], [3]). We shall give in this paper some further applications of this point of view.

We define certain subspaces of H associated with the $x(t)$ curve or process by writing

$$H(x) = S\{x(u), -\infty < u < \infty\},$$
$$H(x, t) = S\{x(u), u \leqq t\},$$
$$H(x, -\infty) = \bigcap_t H(x, t).$$

Here we denote by $S\{\cdots\}$ the subspace of H spanned by the random variables indicated between the brackets.

Evidently $H(x)$ is the smallest subspace of H which contains the whole curve C generated by the $x(t)$ process, while $H(x, t)$ is the smallest subspace containing the "arc" of C formed by all points $x(u)$ with $u \leqq t$. If the parameter t is interpreted as time, $H(x, t)$ will perresent the "past and present" of the $x(t)$ process from the point of view of the instant t, while $H(x, -\infty)$ represents the "infinitely remote past" of the process.

For the processes $y(t), z(t), \cdots$ we use in the sequel the analogous notations $H(y, t)$, $H(z, t), \cdots$ for subspaces defined in the corresponding way.

In the sequel we shall only consider stochastic processes $x(t) \in H$ which are assumed to satisfy the following conditions (A) and (B):

(A) The subspace $H(x, -\infty)$ contains only the zero element of H.

(B) For all t the limits $x(t \pm 0)$ exist and $x(t-0) = x(t)$.

The condition (A) implies that $x(t)$ is a *regular*, or *purely non-deterministic* process (cf. [3]). From (B) it follows, as shown in [3], that the space $H(x)$ is separable, and that the $x(t)$ curve has at most an enumerable number of discontinuities. The condition $x(t-0) = x(t)$ is not essential, and is introduced here only in order to avoid some trivial complications.

As t increases through real values, the subspaces $H(x, t)$ will obviously form a never decreasing family. For a fixed finite t the union of all $H(x, u)$ with $u < t$ may not be a closed set, but if we define $H(x, t-0)$ as the closure of this union, it is easily proved that we always have $H(x, t-0) = H(x, t)$. Similarly, the union of the $H(x, u)$ for all real u may not be closed, but if $H(x, +\infty)$ is defined as its closure, we shall have $H(x, +\infty) = H(x)$.

Suppose that a certain time point t is such that for all $h > 0$ we have $H(x, t-h) \neq H(x, t+h)$. Every time interval $t-h < u < t+h$ will then contain at least one $x(u)$ not included in $H(x, t-h)$. This may be expressed by saying that the process receives a new impluse, or an *innovation*, during the interval $(t-h, t+h)$ for every $h > 0$. The set of all time points t with this property will be called the *innovation spectrum* of the $x(t)$ process.

3. Suppose that we are given a stochastic process $x(t)$ satisfying the conditions (A) and (B). By P_t we shall denote the projection operator in the Hilbert space $H(x)$ with range $H(x, t)$. It then follows from the properties of the $H(x, t)$ family given above that we have

$$\begin{aligned} &P_t \leqq P_u \text{ for } t < u, \\ (1) \qquad &P_{t-0} = P_t \text{ for all } t, \\ &P_{-\infty} = 0, \ P_{+\infty} = 1, \end{aligned}$$

where 0 and 1 denote respectively the zero and the identity operator in $H(x)$.

It follows that the P_t form a *spectral family* of projections or a *resolution of the identity* according to Hilbert space terminology. As we have seen, the projections P_t are, for all real t, uniquely determined by the given $x(t)$ process. We shall return to the properties of the P_t family in Section 5.

For any random variable $z \in H(x)$ with $\mathbf{E}\{|z|^2\} = 1$ we now define a stochastic process by writing

$$z(t) = P_t z.$$

It is then readily seen that $z(t)$ will be a process with orthogonal increments satisfying (A) and (B). Writing

$$F_z(t) = \mathbf{E}\{|z(t)|^2\},$$

it follows that $F_z(t)$ is, for any fixed z, a distribution function of t such that $F_z(t-0) = F_z(t)$ for all t.

The points of increase of $z(t)$ coincide with the points of increase of $F_z(t)$ and form a subset of the innovation spectrum of $x(t)$, and, accordingly, we shall denote $z(t)$ as a *partial innovation process* associated with $x(t)$. The space $H(z, t)$ is spanned by the random variables $z(u)$ with $u \leq t$, and it is known (cf., e.g., [5], p. 425—428) that $H(z, t)$ is identical with the set of all random variables of the form

$$\int_{-\infty}^{t} g(t, u)dz(u),$$

where $g(t, u)$ is a non-random function such that the integral

$$\int_{-\infty}^{t} |g(t, u)|^2 dF_z(u)$$

is convergent.

4. Consider now the class Q of all distribution functions $F(t)$ determined such as to be continuous to the left for all t. We introduce a partial ordering in Q by saying that F_1 is *superior* to F_2, and writing $F_1 > F_2$, whenever F_2 is absolutely continuous with respect to F_1. If $F_1 > F_2$ and $F_2 > F_1$, we say that F_1 and F_2 are *equivalent*.

The set of all distribution functions equivalent to a given $F(t)$ forms an *equivalence class* R. A partial ordering is introduced in the set of all equivalence classes in the obvious way by writing $R_1 > R_2$ when the corresponding relation holds for any $F_1 \in R_1$ and $F_2 \in R_2$. A point t is called a point of increase for the equivalence class R whenever t is a point of increase for any $F \in R$.

In the sequel we shall be concerned with never increasing sequences of equivalence classes:

$$R_1 > R_2 > \cdots > R_N. \tag{2}$$

The number N of elements in a sequence of this form, which may be finite or infinite, will be called the *total multiplicity* of the sequence. We shall also define a *multiplicity function* $N(t)$ of the sequence (2) by writing for any real t

$N(t) =$ the number of those R_n in (2) for which t is a point of increase.

$N(t)$, like N, may be finite or infinite, and the total multiplicity N will obviously satisfy the relation

$$N = \sup N(t),$$

where t runs through all real values.

If, in particular, we have $N(t) = 0$ for all t in some closed interval $[a, b]$, all functions $F(t)$ belonging to any of the equivalence classes in the sequence (2) will be constant throughout $[a, b]$.

5. To any $x(t)$ satisfying the conditions (A) and (B) there corresponds according to Section 3 a uniquely determined spectral family of projections P_t satisfying (1). It then follows from the theory of spectral multiplicity in a separable Hilbert space ([13], Chapter 7) that to the same $x(t)$ there corresponds a uniquely determined, never increasing sequence (2) of equivalence classes, having the following properties:

If N is the total multiplicity of the sequence (2), it is possible to find N orthonormal random variables $z_1, \cdots, z_N \in H(x)$ such that the corresponding processes with orthogonal increments defined in Section 3 satisfy the relations

$$\begin{aligned} &F_{z_n}(t) \in R_n, \\ (3) \qquad &H(z_m, t) \perp H(z_n, t), \qquad m \neq n, \\ &H(x, t) = \sum_1^N H(z_n, t), \end{aligned}$$

where the last sum denotes the vector sum of the mutually orthogonal subspaces involved.

Now $x(t)$ is always an element of $H(x, t)$ and from Section 3 we then obtain the following theorem, previously given in somewhat less precise form in [2] and [3].

Theorem 1. *To any stochastic process $x(t)$ satisfying* (A) *and* (B) *there corresponds a uniquely determined sequence* (2) *of equivalence classes such that $x(t)$ can be represented in the form*

$$x(t) = \sum_1^N \int_{-\infty}^t g_n(t, u) dz_n(u), \tag{4}$$

where the $z_n(u)$ are mutually orthogonal processes with orthogonal increments satisfying (3). *The $g_n(t, u)$ are non-random functions such that*

$$\sum_1^N \int_{-\infty}^t |g_n(t, u)|^2 dF_{z_n}(u) < \infty.$$

The number N, which is called the total spectral multiplicity of the $x(t)$ process, is the uniquely determined number of elements in (2) *and may be finite or infinite. No representation of the form* (4) *with these properties exists for any smaller value of N.*

The sequence (2) corresponding to a given $x(t)$ process will be said to determine the *spectral type* of the process.

The relation (4) gives a linear representation of $x(t)$ in terms of past and present innovation elements $dz_n(u)$. The *total innovation process* associated with $x(t)$ is an N-dimensional vector process $\{z_1(t), \cdots, z_N(t)\}$ where, as before, N may be finite or infinite.

It is interesting to compare this with the situation in the case of a regular process with *discrete* time ([2], Theorem 1) where a similar representation always holds with $N = 1$.

Also in the particular case of a *stationary* process with continuous time, satisfying (A) and (B), it follows from well-known theorems that we have $N = N(t) = 1$ for all t, and that the only element in the corresponding sequence (2) may be represented by any absolutely continuous distribution function $F(t)$ having an everywhere positive density function.

6. The best linear least squares prediction of $x(t+h)$ in terms of all $x(u)$ with $u \leqq t$ is obtained from (4) in the form

$$P_t x(t+h) = \sum_1^N \int_{-\infty}^t g_n(t+h, u) dz_n(u).$$

The error involved in this prediction is

$$x(t+h) - P_t x(t+h) = \sum_1^N \int_t^{t+h} g_n(t+h, u) dz_n(u). \tag{5}$$

Now consider the multiplicity function $N(t)$ associated with the sequence (2), as defined in Section 3. Suppose that in the closed interval $t \leqq u \leqq t+h$ we have $N(u) \leqq N_1 < N$. Then all terms with $n > N_1$ in the second member of (5) will reduce to zero, so that the innovation entering into the process during $[t, t+h]$ will only be of dimensionality N_1. Speaking somewhat loosely, we may say that the multiplicity function $N(t)$ determines for every t the dimensionality of the innovation element $\{dz_1(t), \cdots\}$.

If, in particular, $N(u) = 0$ for $t \leqq u \leqq t+h$, it follows that the process does not receive any innovation at all during this interval. Accordingly in this case the whole second member of (5) reduces to zero, so that exact prediction is possible over the interval considered.

7. We now introduce the correlation function of the $x(t)$ process:

$$r(s, t) = \mathbf{E}\{x(s)\overline{x(t)}\}.$$

As before we assume that all stochastic processes considered satisfy the conditions (A) and (B). We proceed to prove the following theorem, which shows that the spectral type of a process is uniquely determined by the correlation function.

Theorem 2. *Let $x(t)$ and $y(t)$ be two processes satisfying* (A) *and* (*B*) *and having the same correlation function $r(s, t)$. The sequences of equivalence classes, which correspond to $x(t)$ and $y(t)$ in the way described in Theorem* 1, *are then identical.*

$x(t)$ and $y(t)$ define two curves situated, respectively, in the spaces $H(x)$ and $H(y)$. We now define a transformation V from the x-curve to the y-curve by writing

$$Vx(t) = y(t),$$

and extend this definition by linearity to the linear manifold in $H(x)$ determined by all points $x(t)$. It is readily seen that this definition is unique, and that the transformation is isometric. It follows, in fact, from the equality of the correlation functions that any linear relation $\sum c_n x(t_n) = 0$ implies and is implied by the corresponding relation $\sum c_n y(t_n) = 0$, which shows that the transformation is unique, while the isometry follows from the identity

$$r(s, t) = (x(s), x(t)) = (Vx(s), Vx(t)).$$

The transformation can now be extended to an isometric transformation V defined in the whole space $H(x)$. If we consider the restriction of V to $H(x, t)$, it is immediately seen that we have for all t

$$VH(x, t) = H(y, t).$$

Denoting by $P_t^{(x)}$ and $P_t^{(y)}$ the spectral families of projections corresponding, respectively, to $x(t)$ and $y(t)$, we then obtain

$$VP_t^{(x)}V^{-1} = P_t^{(y)}.$$

Thus the two spectral families are isometrically equivalent, and the assertion of the theorem now follows directly from Hilbert space theory. In the particular case when $H(x) = H(y)$, the transformation V will be unitary.

On the other hand, two processes with isometrically equivalent spectral families do not necessarily have the same correlation function. In other words, *the correlation function is not uniquely determined by the spectral type.*

In order to see this, it is enough to consider the two processes $x(t)$ and $y(t) = f(t)x(t)$, where $f(t)$ is a non-random function such that $0 < m < |f(t)| < M$ for all t. It is clear that $H(x, t) = H(y, t)$ for all t, while the correlation functions differ by the factor $f(s)\overline{f(t)}$.

8. In this section it will be shown that we can always find a stochastic process possessing any given spectral type. We shall even prove the more precise statement contained in the following theorem.

Theorem 3. *Suppose that a sequence of equivalence classes of the form* (2) *is given. Then there exists a harmonizable process*

$$x(t) = \int_{-\infty}^{\infty} e^{it\lambda} dy(\lambda)$$

which has the spectral type defined by the given sequence.

Comparing this statement with the final remark in Section 5, it will be seen how restricted the class of stationary processes is in comparison with the class of harmonizable processes.

In order to prove the theorem we denote by $A_1, A_2, \cdots$ a sequence of disjoint sets of real points such that the measure of every A_n is positive in any non-vanishing interval.[1] If $\alpha_n(v)$ is the characteristic function of A_n, we thus have

$$\int_a^b \alpha_n(v)dv > 0$$

for all n and for any real $a < b$.

We further take in each equivalence class R_n appearing in the given sequence (2) a distribution function $F_n(t) \in R_n$. Obviously we can choose the functions $F_1, \cdots, F_N$ so that the integrals

$$k_n^2 = \int_{-\infty}^{\infty} e^{t^2} dF_n(t) \tag{6}$$

converge for all n. We then have $1 \leqq k_n < \infty$. Assuming that the basic probability field is not too restricted, we can then find N mutually orthogonal stochastic processes $z_1(t), \cdots, z_N(t)$ with orthogonal increments such that

$$F_{z_n}(t) = \mathbf{E}\{|z_n(t)|^2\} = F_n(t).$$

We now introduce the following definition:

$$g_n(t, u) = \begin{cases} \dfrac{1}{nk_n} e^{-t} \displaystyle\int_u^t (t-v)\alpha_n(v)dv, & u < t, \\ 0, & u \geqq t, \end{cases} \tag{7}$$

and

$$x_n(t) = \int_{-\infty}^t g_n(t, u)dz_n(u), \tag{8}$$

$$x(t) = \sum_1^N x_n(t).$$

We then have for $u < t$

$$0 < g_n(t, u) < \frac{1}{nk_n} e^{-t}(t-u)^2,$$

and hence by (6),

$$\begin{aligned} \mathbf{E}\{|x_n(t)|^2\} &< \frac{1}{n^2 k_n^2} e^{-2t} \int_{-\infty}^{\infty} (t-u)^4 dF_n(u) \\ &\leqq \frac{8}{n^2 k_n^2} e^{-2t} \int_{-\infty}^{\infty} (t^4+u^4) dF_n(u) \leqq \frac{8(t^4+k_n^2)}{n^2 k_n^2}, \end{aligned}$$

so that the series for $x(t)$ converges in quadratic mean if $N = \infty$. (We note that the $x_n(t)$, like the $z_n(t)$, are mutually orthogonal.)

---

[1] The use of the sets A_n for the construction of processes with given multiplicity properties goes back to a correspondence between Professor Kolmogorov and the present author (cf. [4]). A simple way of constructing the A_n is the following: Let $1 < n_1 < n_2 < \cdots$ be positive integers such that $\sum_1^\infty 1/n_k$ converges. Almost every real x then has a unique expansion $x = r_0 + \sum_1^\infty r_k/(n_1 \cdots n_k)$, where the r_k are integers and $0 \leqq r_k < n_k$ for $k \geqq 1$. If A_n is the set of those x for which the number of zeros among the r_k with $k \geqq 1$ is finite and of the form $2^n(2p+1)$ where p is a non-negative integer, then the sequence $A_1, A_2, \cdots$ has the required properties.

We now proceed to prove a) that the $x(t)$ process defined by (8) has the given spectral type, and b) that it is harmonizable.

It follows from the construction of the $z_n(t)$ and from (8) that we have

$$F_{z_n}(t) \in R_n,$$
$$H(z_m, t) \perp H(z_n, t), \qquad m \neq n,$$
$$H(x, t) \subset \sum_1^N H(z_n, t).$$

If we can show that the sign of equality holds in the last relation, the relations (3) will be satisfied and it then follows that the $x(t)$ process defined by (8) has the given spectral type. In order to prove this it is sufficient to show that we have

$$z_n(t) \in H(x, t)$$

for all n and t.

We have

$$e^t x(t) = \sum_1^N \int_{-\infty}^t g_n(t, u) dz_n(u) = \sum_1^N \frac{1}{nk_n} \int_{-\infty}^t (t-u)\alpha_n(u) z_n(u) du.$$

It is shown without difficulty that the derivative in q.m. of this random function exists for all t and has the expression

$$\frac{d}{dt}(e^t x(t)) = \sum_1^N \frac{1}{nk_n} \int_{-\infty}^t \alpha_n(u) z_n(u) du, \tag{9}$$

where the last sum converges in q.m. We now want to show that for almost all t (Lebesgue measure) we may differentiate once more in q. m., and so obtain

$$\frac{d^2}{dt^2}(e^t x(t)) = \sum_1^N \frac{1}{nk_n} \alpha_n(t) z_n(t). \tag{10}$$

In order to prove this we must show that the random variable

$$W = \sum_1^N \frac{1}{nk_n} \left(\frac{1}{h} \int_t^{t+h} \alpha_n(u) z_n(u) du - \alpha_n(t) z_n(t) \right)$$

converges to zero in q. m. for almost all t as $h \to 0$. We have $W = W_1 + W_2$, where

$$W_1 = \sum_1^N \frac{1}{nk_n h} \int_t^{t+h} \alpha_n(u) (z_n(u) - z_n(t)) du,$$

$$W_2 = \sum_1^N \frac{1}{nk_n} z_n(t) \left(\frac{1}{h} \int_t^{t+h} \alpha_n(u) du - \alpha_n(t) \right).$$

Now both W_1 and W_2 are sums of mutually orthogonal random variables and we have

$$\mathbf{E}|W_1|^2 = \sum_1^N \frac{1}{n^2 k_n^2 h^2} \int_t^{t+h} \int_t^{t+h} \alpha_n(u)\alpha_n(v)[F_n(\min(u, v)) - F_n(t)]\,du\,dv$$
$$\leq \sum_1^N \frac{2}{n^2 h^2} \int_t^{t+h} (t+h-u)[F_n(u)-F_n(t)]du \leq 2\sum_1^N \frac{F_n(t+h)-F_n(t)}{n^2}$$

and

$$\mathbf{E}|W_2|^2 = \sum_1^N \frac{1}{n^2 k_n^2} F_n(t) \left[\frac{1}{h}\int_t^{t+h} \alpha_n(u)du - \alpha_n(t)\right]^2$$
$$\leq \sum_1^N \frac{1}{n^2}\left[\frac{1}{h}\int_t^{t+h} \alpha_n(u)du - \alpha_n(t)\right]^2.$$

However, all the F_n are continuous almost everywhere, and it follows that W_1 tends to zero in q. m. for almost all t. On the other hand, the metric density of any A_n exists almost everywhere and is equal to $\alpha_n(t)$ so that W_2 tends to zero in q. m. almost everywhere. Thus we have shown that (10) holds for almost all t.

Let now m be a given integer, $1 \leq m \leq N$. The sets A_n being disjoint, it then follows from (10) that for almost all $t \in A_m$

$$\frac{d^2}{dt^2}(e^t x(t)) = \frac{1}{mk_m} z_m(t).$$

The first member of the last relation is evidently an element of $H(x, t)$, so that we have $z_m(t) \in H(x, t)$ for almost all $t \in A_m$. Now A_m is of positive measure in every non-vanishing interval, while $z_m(t)$ is by definition everywhere continuous to the left in q. m. Thus $z_m(t) \in H(x, t)$ for all t and all $m = 1, \cdots, N$, and according to the above this proves that $x(t)$ has the given spectral type.

In order to prove also that $x(t)$ is harmonizable we introduce the Fourier transform $h_n(\lambda, u)$ of $g_n(t, u)$ with respect to t. From (7) we obtain

$$h_n(\lambda, u) = \int_{-\infty}^{\infty} g_n(t, u)e^{-it\lambda}dt = \frac{1}{nk_n}\int_u^{\infty} e^{-t(1+i\lambda)}dt \int_u^t (t-v)\alpha_n(v)dv.$$

The double integral is absolutely convergent and we have

$$h_n(\lambda, u) = \frac{1}{nk_n}\int_u^{\infty} \alpha_n(v)dv \int_v^{\infty} (t-v)e^{-t(1+i\lambda)}dt$$
$$= \frac{1}{nk_n(1+i\lambda)^2}\int_u^{\infty} \alpha_n(v)e^{-v(1+i\lambda)}dv,$$

$$|h_n(\lambda, u)| < \frac{e^{-u}}{nk_n(1+\lambda^2)}. \tag{11}$$

Thus $h_n(\lambda, u)$ is, for any fixed u, absolutely integrable with respect to λ. On the other hand, it follows from (7) that $g_n(t, u)$ is everywhere continuous, so that we have the inverse Fourier formula

$$g_n(t, u) = \frac{1}{2\pi}\int_{-\infty}^{\infty} h_n(\lambda, u)e^{it\lambda}d\lambda. \tag{12}$$

Now the correlation functions of $x_n(t)$ and $x(t)$ are, by (7) and (8),

$$r_n(s, t) = \mathbf{E}\{x_n(s)\overline{x_n(t)}\} = \int_{-\infty}^{\infty} g_n(s, u)\overline{g_n(t, u)}dF_n(u),$$

$$r(s, t) = \mathbf{E}\{x(s)\overline{x(t)}\} = \sum_1^N r^n(s, t).$$

Replacing the g_n by their expressions according to (12) we obtain

$$r_n(s, t) = \frac{1}{(2\pi)^2}\int_{-\infty}^{\infty}\int_{-\infty}^{\infty} e^{i(s\lambda - t\mu)}\,d\lambda\,d\mu\int_{-\infty}^{\infty} h_n(\lambda, u)\overline{h_n(\mu, u)}dF_n(u),$$

the inversion of the order of integration being justified by absolute convergence according to (6) and (11).

If we write

$$c_n(\lambda, \mu) = \frac{1}{(2\pi)^2}\int_{-\infty}^{\infty} h_n(\lambda, u)\overline{h_n(\mu, u)}dF_n(u),$$

$$C_n(\lambda, \mu) = \int_{-\infty}^{\lambda}\int_{-\infty}^{\mu} c_n(\rho, \sigma)d\rho\,d\sigma,$$

it follows from well-known criteria (cf., e.g., [9], p. 466—469) that $C_n(\lambda, \mu)$ is a correlation function. Further $C_n(\lambda, \mu)$ is of bounded variation over the whole (λ, μ)-plane, its variation being bounded by the expression

$$\int_{-\infty}^{\infty}\int_{-\infty}^{\infty} |c_n(\lambda,\mu)|\,d\lambda\,d\mu < \frac{1}{(2\pi)^2 n^2 k_n^2}\int_{-\infty}^{\infty}\frac{d\lambda}{1+\lambda^2}\int_{-\infty}^{\infty}\frac{d\mu}{1+\mu^2}\int_{-\infty}^{\infty} e^{-2u}dF_n(u)$$

$$< \frac{1}{4n^2k_n^2}\int_{-\infty}^{\infty} e^{1+u^2}dF_n(u) < \frac{1}{n^2}$$

obtained from (6) and (11).

It now follows that we have

$$r_n(s, t) = \int_{-\infty}^{\infty}\int_{-\infty}^{\infty} e^{i(s\lambda - t\mu)}\,d_{\lambda,\mu}C_n(\lambda, \mu),$$

$$r(s, t) = \int_{-\infty}^{\infty}\int_{-\infty}^{\infty} e^{i(s\lambda - t\mu)}\,d_{\lambda,\mu}C(\lambda, \mu),$$

where

$$C(\lambda, \mu) = \sum_1^N C_n(\lambda, \mu)$$

is a correlation function which, according to the above, is of bounded variation over the whole (λ, μ)-plane. Hence we may conclude (cf. [1], and [9], p. 476) that $x(t)$ is a harmonizable process

$$x(t) = \int_{-\infty}^{\infty} e^{it\lambda}dy(\lambda),$$

where $y(\lambda)$ has the correlation function $C(\lambda, \mu)$. We note that $x(t)$ is a regular process and is everywhere continuous in quadratic mean. The proof is completed.

Received by the editors
November 27, 1963

REFERENCES

[1] H. CRAMÉR, *A contribution to the theory of stochastic processes*, Proc. Second Berkeley Sympos. Math. Statist. and Prob., 1951, pp. 329–339.

[2] H. CRAMÉR, *On some classes of non-stationary stochastic processes*, Proc. 4-th Berkeley Sympos. Math. Statist. and Prob., II, 1961, pp. 57–77.

[3] H. CRAMÉR, *On the structure of purely non-deterministic stochastic processes*, Arkiv Math., 4, 1961, pp. 249–266.

[4] H. CRAMÉR, *Décompositions orthogonales de certains procès stochastiques*, Ann. Fac. Sciences Clermont, 11, 1962, pp. 15–21.

[5] J. L. DOOB, *Stochastic Processes*, Wiley, N.Y., 1953.

[6] P. R. HALMOS, *Introduction to Hilbert Space and the Theory of Spectral Multiplicity*, Chelsea, N.Y., 2-nd ed. 1957.

[7] A. N. KOLMOGOROV, *Curves in Hilbert space which are invariant with respect to a one-parameter group of motions*, DAN SSSR, 26, 1940, pp. 6–9. (In Russian.)

[8] A. N. KOLMOGOROV, *Wiener's spiral and some other interesting curves in Hilbert space*, DAN SSSR, 26, 1940, pp. 115–118. (In Russian.)

[9] M. LOÈVE, *Probability Theory*, 3-rd ed., Van Nostrand, Princeton, N.J., 1963.

[10] H. NAKANO, *Spectral Theory in the Hilbert Space*, Jap. Soc. for the Promotion of Science, Tokyo, 1953.

[11] J. v. NEUMANN AND I. J. SCHOENBERG, *Fourier integrals and metric geometry*, Trans. Amer. Math. Soc., 50, 1941, pp. 226–251.

[12] M. S. PINSKER, *Theory of curves in Hilbert space with stationary n-th increments*, Izv. AN SSSR, Ser. Mat., 19, 1955, pp. 319–344. (In Russian.)

[13] M. H. STONE, *Linear Transformations in Hilbert Space*, American Math. Soc., N.Y., 1932.

STOCHASTIC PROCESSES AS CURVES IN HILBERT SPACE

HARALD CRAMÉR (STOCKHOLM)

(*Summary*)

Regular complex-valued random processes $x(t)$ with finite moments of second order are studied by methods of Hilbert space geometry. A representation formula (4) is given for the process $x(t)$ in terms of "past and present innovations". The number N is called the complete spectral multiplicity of the process $x(t)$ and is the smallest number for which such a representation exists. It is shown that the multiplicity of $x(t)$ is uniquely determined by the corresponding correlation function and that one can always find a harmonizing process $x(t)$ which has the multiplicity prescribed in advance.

58.

with M. R. Leadbetter

The moments of the number of crossings of a level by a stationary normal process

Ann. Math. Statist. **36**, 1656–1663 (1965)

0. Summary. In this paper we consider the number N of upcrossings of a level u by a stationary normal process $\xi(t)$ in $0 \leqq t \leqq T$. A formula is obtained for the factorial moment $M_k = \mathcal{E}\{N(N-1)\cdots(N-k+1)\}$ of any desired order k. The main condition assumed in the derivation is that $\xi(t)$ have, with probability one, a continuous sample derivative $\xi'(t)$ in the interval $[0, T]$. This condition involves hardly any restriction since an example shows that even a slight relaxation of it causes all moments of order greater than one to become infinite. The moments of the number of downcrossings or total number of crossings can be obtained analogously.

1. Introduction. The problem of obtaining the mean number of crossings, (or equivalently upcrossings) of a given level, by a stationary normal process in a given time, has received a good deal of attention in the literature. In fact, a complete solution to this problem has now been given by Ylvisaker [8]. However, moments of order greater than one of the number of crossings of a level have received less attention. The variance was obtained by Steinberg et al. [6], using somewhat heuristic arguments. Rozanov and Volkonski [7] point out in a footnote that the formula given in [6] for the variance is valid under certain precise conditions, of which the main one is that the covariance function of the process have a finite sixth derivative at the origin. Finally in this connection, the variance has been obtained by Leadbetter and Cryer [4] under conditions which assume just a little more than the existence of a second derivative of the covariance function.

There is virtually no literature available in connection with moments of the number of crossings of a level, of higher order than the second. (A partial result is indicated by Ivanov at the end of his paper [3].) It will be our purpose here to obtain explicit expressions for such moments. From conversations with Dr. Yu. K. Belayev, we know that he has investigated certain problems related to the asymptotic behaviour of the moments of the number of crossings. The explicit expression for these moments which we prove seems, however, to be new.

Our treatment in the sequel will be in terms of upcrossings and we shall obtain explicit expressions for factorial moments of arbitrary orders, under conditions which are very close to the necessary ones. Corresponding formulae for moments of all orders for the downcrossings, or total number of crossings, follow similarly.

2. Moments of the number of upcrossings. We shall, throughout, consider a

---

Received 8 March 1965.

[1] Research was supported in part by the Office of Naval Research.

real valued stationary normal process $\{\xi(t): 0 \leqq t \leqq T\}$ having (for convenience) zero mean, spectrum $F(\lambda)$ possessing a continuous component, and covariance function $r(\tau) = \int_{-\infty}^{\infty} e^{i\lambda\tau}\, dF(\lambda)$. We shall further assume that $\xi(t)$ has, with probability one, a continuous sample derivative $\xi'(t)$ on the interval $[0, T]$. Sufficient conditions for this latter property in terms of the behaviour of the covariance function are given, for example, in [1]. Write N for the number of *upcrossings* of the level u by $\xi(t)$ in $0 \leqq t \leqq T$; that is N is the number of points t in that interval for which $\xi(t) = u$, $\xi'(t) > 0$. Then the following result holds.

THEOREM. *If $\{\xi(t): 0 \leqq t \leqq T\}$ is a normal stationary process, as described, possessing, with probability one, a continuous sample derivative, and k is any positive integer, then*

$$(1) \quad M_k = \mathcal{E}\{N(N-1)\cdots(N-k+1)\}$$
$$= \int_0^T \cdots \int_0^T dt_1 \cdots dt_k \int_0^\infty \cdots \int_0^\infty y_1 \cdots y_k p_t(u, \mathbf{y})\, dy_1 \cdots dy_k$$

in which $p_t(u, \mathbf{y}) = p_t(u, \cdots, u, y_1, \cdots, y_k)$, $p_t(x_1, \cdots, x_k, y_1, \cdots, y_k)$ denoting the joint density for the random variables $\xi(t_1) \cdots \xi(t_k)$, $\xi'(t_1) \cdots \xi'(t_k)$.

We note here that it follows from the appendix that, when all t_i are different, p_t is the density corresponding to a non singular joint distribution since $F(\lambda)$ is assumed to have a continuous component.

Before proceeding to the proof we note that the theorem can easily be modified to refer to "downcrossings" or the total number of crossings of the level u in time T. The discussion will be given here in terms of upcrossings, however.

The following proof is divided into two parts A and B. In Part A it is shown that M_k does not exceed the expression on the right hand side of (1), whereas in Part B the reverse inequality is proved. The techniques are straightforward but quite different in each part. It is a perhaps somewhat surprising feature, however, that in both parts use can be made of Fatou's lemma to give the essential inversions of limiting operations with integrations, in order that inequalities in the desired (opposite) directions may be obtained.

PROOF OF THE THEOREM.

Part A. Write $\xi(t) = \xi(t, \omega)$ to exhibit explicit dependence on the "sample point" $\omega \in \Omega$. Let S denote the set of all ω such that $\xi'(t)$ is continuous in the interval $I = [0, T]$, while the equation $\xi(t) = u$ has at most a finite number of roots t in I, and while further $\xi(0) \neq u \neq \xi(T)$ and $\xi'(t) \neq 0$ whenever $\xi(t) = u$. According to Bulinskaya [2], Theorem 1, we then have

$$(2) \qquad P(S) = 1.$$

Write now $(N)_k = N(N-1)\cdots(N-k+1)$ for $k = 1, 2, \cdots$ and define the functions $\delta_n(x)$, $\sigma(x)$ by

$$\delta_n(x) = n, \qquad |x| \leqq 1/(2n)$$
$$= 0, \qquad \text{otherwise}$$

and

$$\sigma(x) = x, \qquad x > 0,$$
$$= 0, \qquad \text{otherwise}.$$

Let $D(\epsilon)$ denote the domain in the k-dimensional space R^k with coordinates $t_1 \cdots t_k$ defined by the inequalities

$$0 < t_i < T \qquad \text{for } i = 1 \cdots k,$$

$$|t_i - t_j| > \epsilon \qquad \text{for } i \neq j.$$

Define also the random variable $J_k(n, \epsilon, \omega)$ by the relation

$$(3) \quad J_k(n, \epsilon, \omega) = \int \cdots \int_{D(\epsilon)} \prod_{i=1}^{k} \{\delta_n[\xi(t_i) - u]\sigma[\xi'(t_i)]\} \, dt_1 \cdots dt_k .$$

We shall now proceed to prove that

$$(4) \qquad \mathcal{E}\{(N)_k\} \leqq \lim_{\epsilon \to 0} \lim_{n \to \infty} \mathcal{E}\{J_k(n, \epsilon, \omega)\}.$$

In order to prove the validity of (4) we define a subset $S(h)$ of S consisting of all $\omega \varepsilon S$ for which the following two conditions are satisfied:

(a) the distance between any two upcrossings in I by $\xi(t)$ of the level u is greater than $2h$,

(b) for any zero $t = t_0$ of the derivative $\xi'(t)$ in I, we have $|\xi(t_0) - u| > h$.

According to the definition of S every $\omega \varepsilon S$ must also belong to $S(h)$ for some $h > 0$ so that

$$(5) \qquad S(h) \uparrow S \quad \text{as} \quad h \downarrow 0.$$

Take now any fixed $\omega \varepsilon S(h)$, and let $t = r_1, \cdots, r_N$ be all the upcrossings of the corresponding $\xi(t) = \xi(t, \omega)$ in I. Consider the k-dimensional interval I^k in the space R^k, and let $A_{j_1, \cdots, j_k}$ denote the point in I^k with coordinates $t_1 = r_{j_1}, \cdots, t_k = r_{j_k}$, where each j_i may assume the values $1, 2, \cdots, N$. Clearly there are N^k different points A, and among these there are exactly $(N)_k$ points A' such that no two of the j_i are equal. Since $\omega \varepsilon S(h)$, these points A' will all be situated in the domain $D(2h)$, while the remaining $N^k - (N)_k$ points A will fall outside $D(2h)$, and even outside $D(\epsilon)$, for any $\epsilon > 0$.

Considering, still the same fixed $\omega \varepsilon S(h)$ we now take n and ϵ such that $0 < n^{-1} < \epsilon < h$ and consider the integral $J_k(n, \epsilon, \omega)$ defined by (3). The contribution to $J_k(n, \epsilon, \omega)$ arising from small disjoint k dimensional blocks about each point A' is easily seen to be just $(N)_k$ for all sufficiently large n (i.e. a unit contribution from each such block). The contribution from the remaining region is zero for all sufficiently large n. (This can be seen clearly from a picture by taking $k = 2$ and writing down the integrals involved.) Hence for any fixed $\epsilon < h$, we can always find n_0 so large that, for all $n > n_0$ we have

$$J_k(n, \epsilon, \omega) = (N)_k ,$$

and hence also

$$(N)_k = \lim_{n \to \infty} J_k(n, \epsilon, \omega).$$

Since this holds for any $\epsilon < h$, while the first member is independent of ϵ, it follows that

$$(6) \qquad (N)_k = \lim_{\epsilon \to 0} \lim_{n \to \infty} J_k(n, \epsilon, \omega)$$

for any fixed $\omega \varepsilon S(h)$. But h can be chosen arbitrarily small and since $S(h) \uparrow S$ as $h \downarrow 0$ it follows that (6) holds for any $\omega \varepsilon S$, i.e. with probability one. Finally an application of Fatou's lemma to the ϵ and n-limits yields the result (4). Thus from (4) we obtain

$$\mathcal{E}\{(N)_k\} \leqq \lim_{\epsilon\to 0} \lim_{n\to\infty} \int \cdots \int_{D(\epsilon)} dt_1 \cdots dt_k$$

$$[n^k \int_{u-(2n)^{-1}}^{u+(2n)^{-1}} \cdots \int_{u-(2n)^{-1}}^{u+(2n)^{-1}} dx_1 \cdots dx_k \int_0^\infty \cdots \int_0^\infty y_1 \cdots y_k \tag{7}$$

$$\cdot p_t(x_1 \cdots x_k , y_1 \cdots y_k)\, dy_1 \cdots dy_k].$$

The entire expression in square brackets on the right hand side of (7) clearly converges to $\int_0^\infty \cdots \int_0^\infty y_1 \cdots y_k\, p_t(u, \mathbf{y})\, dy_1 \cdots dy_k$. Further, it can be readily shown that this expression is bounded for all $t_1 \cdots t_k$ in the region $D(\epsilon)$ (using the fact that the determinant of the covariance matrix of $\xi(t_1) \cdots \xi(t_k)$, $\xi'(t_1) \cdots \xi'(t_k)$ is bounded away from zero). Hence by dominated convergence

$$\mathcal{E}\{(N)_k\} \tag{8}$$

$$\leqq \lim_{\epsilon\to 0} \int \cdots \int_{D(\epsilon)} dt_1 \cdots dt_k \int_0^\infty \cdots \int_0^\infty y_1 \cdots y_k\, p_t(u, \mathbf{y})\, dy_1 \cdots dy_k .$$

Finally by monotone convergence it follows that

$$\mathcal{E}\{(N)_k\} \leqq \int_0^T \cdots \int_0^T dt_1 \cdots dt_k \int_0^\infty \cdots \int_0^\infty y_1 \cdots y_k\, p_t(u, \mathbf{y})\, dy_1 \cdots dy_k . \tag{9}$$

Part B. In order to prove the reverse inequality to (9) we adopt a different procedure (due to Ylvisaker [8]) for counting the number of upcrossings by $\xi(t)$ in $0 \leqq t \leqq T$. First, however, we note that if χ_i, $i = 1, 2, \cdots$ are each either zero or one, and $M = \sum_{i=1}^m \chi_i$, then, for any integer $k \leqq m$,

$$(M)_k = M(M-1) \cdots (M-k+1) = \sum{}' \chi_{i_1} \cdots \chi_{i_k} \tag{10}$$

where $\sum'$ denotes summation over all possible ordered sets of distinct integers $i_1 \cdots i_k$. For M is just the number of non zero χ_i, and the right hand side of (10) therefore represents the number of ordered sets of distinct integers $i_1 \cdots i_k$ such that each corresponding χ_i is non zero, taken out of a total of M possible integers i for which $\chi_i \neq 0$. But this number is simply $M(M-1) \cdots (M-k+1)$ as required.

Write now $\xi_i = \xi(Ti/2^n)$, $i = 0, 1, \cdots, 2^n$, $n = 1, 2, \cdots$. Let $\chi_i = 1$ if $\xi_i < u < \xi_{i+1}$, and $\chi_i = 0$ otherwise. Then if $N_n = \sum_{i=1}^{2^n} \chi_i$ we have $N_n \uparrow N$ a.s. (*A* detailed proof of this latter statement is given by Ylvisaker [8].) Hence by monotone convergence,

$$\mathcal{E}\{(N_n)_k\} \to \mathcal{E}\{(N)_k\} \quad \text{as} \quad n \to \infty. \tag{11}$$

Now from (10) we have with $m = 2^n$,

$$\mathcal{E}\{(N_n)_k\} = \sum{}'_{i_1 \cdots i_k} P\{\chi_{i_1} = \chi_{i_2} = \cdots = \chi_{i_k} = 1\}. \tag{12}$$

We note that no terms for which $|i_r - i_s| = 1$ for any r, s appear since we cannot have $\chi_{i_r} = \chi_{i_r+1} = 1$. Write $\eta_i = 2^n(\xi_{i+1} - \xi_i)/T$. Then

$$P\{\chi_{i_1} = \chi_{i_2} = \cdots = \chi_{i_k} = 1\} = P\{u - 2^{-n}T\eta_{i_r} < \xi_{i_r} < u, r = 1 \cdots k\}$$
$$= \int_0^\infty \cdots \int_0^\infty dy_1 \cdots dy_k \int_{u-2^{-n}Ty_1}^u \cdots$$
$$\int_{u-2^{-n}Ty_k}^u p_{n,\mathbf{i}_j}(x_1 \cdots x_k, y_1 \cdots y_k)\, dx_1 \cdots dx_k,$$

where $p_{n,\mathbf{i}_j}$ is the joint density for the distribution of $\xi_{i_1} \cdots \xi_{i_k}, \eta_{i_1} \cdots \eta_{i_k}$. (That this distribution is non singular follows from the Appendix.) By a change of the x-variables in this expression we thus obtain from (12)

$$(13)\quad \mathcal{E}\{(N_n)_k\} = 2^{-kn}T^k \sum{}' \int_0^\infty \cdots \int_0^\infty dy_1 \cdots dy_k \int_{-y_1}^0 \cdots$$
$$\cdot \int_{-y_k}^0 p_{n,\mathbf{i}_j}(u + 2^{-n}Tx_1, \cdots, u + 2^{-n}Tx_k, y_1, \cdots, y_k)\, dx_1 \cdots dx_k.$$

Let A_n be the subset of I^k for which no two of $(t_1 \cdots t_k)$ are contained in the same or adjacent intervals of the form $(rT/2^n, (r+1)T/2^n)$. Write

$$\Psi_{nt}(x_1 \cdots x_k, y_1 \cdots y_k) = p_{n\mathbf{i}_j}(x_1 \cdots x_k, y_1 \cdots y_k)$$

for all $(t_1 \cdots t_k)$ in A_n such that t_r lies in the interval $(i_rT/2^n, (i_r+1)T/2^n)$ for each r, and $\Psi_{nt}(x_1 \cdots x_k, y_1 \cdots y_k) = 0$ for $\mathbf{t} \not\in A_n$. Then (using the remark following Equation (12)), (13) may be rewritten as

$$(14)\quad \int_0^T \cdots \int_0^T dt_1 \cdots dt_k \int_0^\infty \cdots \int_0^\infty dy_1 \cdots dy_k \int_{-y_1}^0 \cdots$$
$$\cdot \int_{-y_k}^0 \Psi_{nt}(u + 2^{-n}Tx_1, \cdots, u + 2^{-n}Tx_k, y_1 \cdots y_k)\, dx_1 \cdots dx_k.$$

Let now $(t_1 \cdots t_k)$ be a fixed point in A_n. Then Ψ_{nt} is a $2k$-dimensional normal density function. Suppose that $i_rT/2^n \leq t_r < (i_r+1)T/2^n$, $r = 1, \cdots, k$. Then corresponding to the point t_r we have the random variables $\xi(i_rT/2^n)$, $\eta(i_rT/2^n)$, yielding the following typical members of the covariance matrix for Ψ_{nt} for example:

$$\operatorname{var}(\xi_{i_1}) = r(0), \qquad \text{writing } \xi_i \text{ for } \xi(iT/2^n),$$
$$\operatorname{cov}(\xi_{i_1}, \xi_{i_2}) = r_p, \qquad \text{writing } r_i \text{ for } r(iT/2^n),\ p = i_1 - i_2,$$
$$\operatorname{cov}(\xi_{i_1}, \eta_{i_1}) = 2^n(r_1 - r_0)/T,$$
$$\operatorname{cov}(\xi_{i_1}, \eta_{i_2}) = 2^n(r_{p+1} - r_p)/T,$$
$$\operatorname{var}\eta_{i_1} = 2^{2n+1}(r_0 - r_1)/T^2,$$
$$\operatorname{cov}(\eta_{i_1}, \eta_{i_2}) = -2^{2n}[r_{p+1} - 2r_p + r_{p-1}]/T^2.$$

For the fixed t_1, t_2 considered i_1, i_2, p depend on n. It is an easy exercise to show that if $\tau = t_2 - t_1$ the above elements converge (in the order given), as $n \to \infty$, to $r(0)$, $r(\tau)$, $r'(0)$, $r'(\tau)$, $-r''(0)$, $-r''(\tau)$, respectively. Similar conclusions hold for the elements corresponding to any pair t_i, t_j. Further, the set A_n converges to almost the whole space I^k. But this means that the integrand in (14) must converge a.e. to $p_t(u, \mathbf{y})$ as $n \to \infty$ and hence, by Fatou's lemma,

$$(15)\quad \mathcal{E}\{(N)_k\} \geq \int_0^T \cdots \int_0^T dt_1 \cdots dt_k \int_0^\infty \cdots \int_0^\infty y_1 \cdots y_k p_t(u, \mathbf{y})\, dy_1 \cdots dy_k.$$

Combining (9) and (15) we obtain the desired equality and hence the truth of the theorem follows.

3. A case when $M_{k_i} = +\infty$. Formula (1) was obtained under the condition that $\xi(t)$ have a continuous sample derivative, with probability one. However, this assumption was used in Part A of the proof, but not at all in Part B. Hence if the right hand side of (1) is infinite, the equation is true with both sides infinite. We now give an example of a case where the integral on the right of (1) is infinite, and hence the corresponding moment is infinite.

For this example we take a covariance function of the form

$$r(\tau) = 1 - \lambda_2\tau^2/2 - \tau^2/\log|\tau| + o(\tau^2/\log|\tau|). \tag{16}$$

That this can be done follows from Theorem 5 of Pitman [5]. In fact we can choose $F(\lambda)$ so that $1 - F(\lambda) = 1/(2\lambda^2 \log^2\lambda)$ for all sufficiently large λ, to give the desired form (16).

Consider now the case $k = 2$, and $u = 0$. Then one can show by some calculation that

$$\int_0^\infty \int_0^\infty y_1y_2p_t(0, \mathbf{y})dy_1\, dy_2 \sim K|\mathbf{\Lambda}|^{\frac{1}{2}}/(1 - r^2(\tau))$$

where K denotes some constant and $\mathbf{\Lambda}$ is the covariance matrix for $\xi(t_1)$, $\xi(t_2)$, $\xi'(t_1)$, $\xi'(t_2)$, $\tau = t_2 - t_1$. But straightforward calculation shows that

$$|\mathbf{\Lambda}| \sim \lambda_2\tau^2/\log^2|\tau| \quad \text{as} \quad \tau \to 0,$$

and hence

$$\int_0^\infty \int_0^\infty y_1y_2p_t(0, \mathbf{y})dy_1\, dy_2 \sim K/(|\tau| \log|\tau|) \quad \text{as} \quad \tau \to 0.$$

It follows from this that the right hand side of (1) is infinite, in this case.

Finally we note a sufficient condition for $\xi(t)$ to possess a continuous sample derivative, with probability one, is that

$$r(\tau) = 1 - \lambda_2\tau^2/2 + O\{\tau^2/|\log|\tau||^a\}$$

for some $a > 1$. This follows from the work of Belayev [1]. In our case $r(\tau)$ given by (16) just fails to satisfy this requirement. Hence it appears that the requirements that ξ have a continuous derivative and that the right hand side of (1) be finite, which are *sufficient* for M_k to be finite and given by (1), are also very close to being *necessary* for this to be the case.

APPENDIX

It was stated, in writing down certain density functions that if $t_1 \cdots t_k$ are distinct time points, then

(i) the joint distribution of $\xi(t_1) \cdots \xi(t_k)$, $\xi'(t_1) \cdots \xi'(t_k)$ is non singular, and

(ii) the joint distribution of $\xi(t_1) \cdots \xi(t_k)$ is non singular.

We shall now prove (i), and hence (ii) will also follow.

Let (as assumed throughout) $F(\lambda)$ have a continuous component and write

$\mathbf{\Lambda} = [\Lambda_{ij}]$ for the covariance matrix of $\xi(t_1) \cdots \xi(t_k)$, $\xi'(t_1) \cdots \xi'(t_k)$. Let $\mathbf{A} = [A_{ij}]$ denote the covariance matrix of $\xi(t_1) \cdots \xi(t_k)$, $\mathbf{B} = [B_{ij}]$ that for $\xi'(t_1) \cdots \xi'(t_k)$, and $\mathbf{C}$ the matrix of "cross" covariances, $C_{ij} = \operatorname{cov}\,(\xi(t_i), \xi'(t_j))$. Then

$$\mathbf{\Lambda} = \begin{bmatrix} \mathbf{A} & \mathbf{C} \\ \mathbf{C}' & \mathbf{B} \end{bmatrix}.$$

Let $\boldsymbol{\theta}$ denote the column vector with elements $\{\theta_1 \cdots \theta_k, \phi_1 \cdots \phi_k\}$, where θ_i, ϕ_i are complex numbers which are not all zero. Then we have

$$A_{jl} = \int \exp\,[i(t_j - t_l)\lambda]\, dF(\lambda),$$

$$C_{jl} = -\int i\lambda \exp\,[i(t_j - t_l)\lambda]\, dF(\lambda),$$

$$B_{jl} = \int \lambda^2 \exp\,[i(t_j - t_l)\lambda]\, dF(\lambda).$$

From this we see that

$$\boldsymbol{\theta}^* \mathbf{\Lambda} \boldsymbol{\theta} = \int [|\textstyle\sum_j \theta_j e^{it_j\lambda}|^2 + \lambda^2 |\sum_j \phi_j e^{it_j\lambda}|^2 - i\lambda \sum_j \theta_j e^{i\lambda t_j} \sum_l \phi_l^* e^{-i\lambda t_l}$$
$$+ i\lambda \textstyle\sum_j \theta_j^* e^{-i\lambda t_j} \sum_l \phi_l e^{i\lambda t_l}]\, dF(\lambda),$$

in which a * denotes the complex conjugate of a scalar, and the conjugate transpose of a vector. Thus

$$\boldsymbol{\theta}^* \mathbf{\Lambda} \boldsymbol{\theta} = \int |\textstyle\sum_{j=1}^k \theta_j e^{it_j\lambda} + i\lambda \sum_{j=1}^k \phi_j e^{it_j\lambda}|^2\, dF(\lambda).$$

Now since the t_j are distinct and θ_j, ϕ_j not all zero it follows that $\sum_{j=1}^k (\theta_j e^{it_j\lambda} + i\lambda\phi_j e^{it_j\lambda})$ is a non constant regular function of λ, viewed as a complex variable, and hence cannot vanish for more than a finite number of values of λ in any bounded region. Hence we must have $\boldsymbol{\theta}^* \mathbf{\Lambda} \boldsymbol{\theta} > 0$ since $F(\lambda)$ is not purely discrete. Thus $\mathbf{\Lambda}$ is a strictly positive definite matrix and the distribution thus defined is non singular.

The condition that $F(\lambda)$ have a continuous component is a convenient one. However it is (as pointed out by the referee) worth noting that the above proof does apply to spectra which are purely discrete, provided that the set of jumps has at least one finite limiting point.

Finally we note here that the above argument can be easily generalized to include an arbitrary number of derivatives. That is if $F(\lambda)$ has a continuous component and is such that $\xi(t)$ has n sample derivatives $\xi(t)\xi'(t) \cdots \xi^{(n)}(t)$, then for any distinct $t_1, t_2, \cdots, t_k$, the joint distribution of $\xi(t_1) \cdots \xi(t_k) \cdots \xi^{(n)}(t_1) \cdots \xi^{(n)}(t_k)$ is non singular.

REFERENCES

[1] Belayev, Yu. K. (1960). Continuity and Hölder's conditions for sample functions of stationary Gaussian processes. *Proc. Fourth Berkeley Symp. Math. Statist. Prob.* **2** 23–33. Univ. of California Press.

[2] Bulinskaya, E. C. (1961). On the mean number of crossings of a level by a stationary Gaussian process. *Teor. Veroyatnost. i Primenen* **6** 474–477.

[3] IVANOV, V. A. (1960). On the average number of crossings of a level by sample functions of a stochastic process. *Teor. Veroyatnost. i Primenen* **5** 319–323.
[4] LEADBETTER, M. R. and CRYER, J. D. (1965). The variance of the number of zeros of a stationary normal processes. *Bull. Amer. Math. Soc.* **71** 561–563.
[5] PITMAN, E. J. G. (1960). Some theorems on characteristic functions of probability distributions. *Proc. Fourth Berkeley Symp. Math. Statist. Prob.* **2** 393–402.
[6] STEINBERG, H., SCHULTHEISS, P. M., WOGRIN, C. A., and ZWEIG, F. (1955). Short-time frequency measurements of narrow-band random signals by means of zero counting process. *J. Appl. Phys.* **26** 195–201.
[7] VOLKONSKII, V. A. and ROZANOV, YU. (1961). Some limit theorems for random functions II. *Teor. Veroyatnost. i Primenen* **6** 202–215.
[8] YLVISAKER, N. D. (1965). The expected number of zeros of a stationary Gaussian process. *Ann. Math. Statist.* **36** 1043–1046.

59.

A limit theorem for the maximum values of certain stochastic processes

Teor. Verojatnost. i Primenen. 10, 137–139 (1965)

In their paper [1], Volkonskij and Rozanov prove an important theorem (Theorem 3.1) concerning the number of intersections of the trajectories of a normal (Gaussian) stationary stochastic process with a given level. The purpose of the present note is to deduce from their theorem an asymptotic expression for the maximum in a large time interval of a normal stationary process, with discrete as well as with continuous time.

We first lay down the following definitions:

$\xi_0, \xi_1, \xi_2, \ldots$ are normal and independent random variables;

$x_0, x_1, x_2, \ldots$ are the random variables of a normal and stationary stochastic process with discrete time;

$x(t)$ with $0 \leqslant t < \infty$ are the random variables of a normal and stationary stochastic process with continuous time.

All these random variables are assumed to be real, with zero means and unit variances.

In respect of the stationary processes x_n and $x(t)$, we shall assume that they satisfy the *strong mixing condition* required in [1], Theorem 3.1. In respect of the continuous time process $x(t)$, we shall finally assume that there a spectral density $f(\lambda)$ such that

$$\mathrm{E}\{x(t)x(0)\} = \int_0^\infty \cos t\lambda\, f(\lambda)\, d\lambda,$$

$$\lambda_n = \int_0^\infty \lambda^n f(\lambda)\, d\lambda < \infty \quad \text{for } 0 \leqslant n \leqslant 4.$$

The sample functions $x(t)$ will then, with a probability 1, be everywhere continuous and have continuous derivatives $x(t)$.

A well-known classical result asserts that, if a random variable u is defined by writing

$$\max_{0\leqslant n\leqslant T} \xi_n = \sqrt{2\log T} - \frac{\log\log T + \log 4\pi}{2\sqrt{2\log T}} + \frac{u}{\sqrt{2\log T}},$$

then we shall have for any real z

$$\lim_{T\to\infty} \mathrm{P}\{u \leqslant z\} = e^{-e^{-z}}.$$

We now introduce two further random variables v and w by means of the relations

$$\max_{0\leqslant n\leqslant T} x_n = \sqrt{2\log T} - \frac{\log\log T + \log 4\pi}{2\sqrt{2\log T}} + \frac{v}{\sqrt{2\log T}},$$

$$\max_{0\leqslant t\leqslant T} x(t) = \sqrt{2\log T} - \frac{\log (2\pi/\sqrt{\lambda_2})}{\sqrt{2\log T}} + \frac{w}{\sqrt{2\log T}}.$$

Theorem. *Under the conditions stated above with respect to x_n and $x(t)$, the variables v and w will have the same limiting distribution as u:*

$$\lim_{T\to\infty} \mathrm{P}\{v \leqslant z\} = \lim_{T\to\infty} \mathrm{P}\{w \leqslant z\} = e^{-e^{-z}}. \tag{1}$$

This is a stronger form of earlier results given (under somewhat wider conditions) by Berman [2] and Cramér [3].

From Theorem 3.1 of [1] it follows in particular that we have, under the given conditions, for any real y and any non-negative integer k

$$\lim_{c\to\infty} \mathrm{P}\left\{N_c\left(\frac{y}{\mu_c}\right) = k\right\} = \frac{y^k}{k!}\, e^{-y}. \tag{2}$$

$N_c(T)$ denotes here, in the case of the $x(t)$ process with continuous time, the number of upcrossings of the sample curve $x(t)$ with the level c in the time interval $0 \leqslant t \leqslant T$, while (cf. [4] and [5])

$$\mu_c = \mathrm{E}\{N_c(1)\} = \frac{\sqrt{\lambda_2}}{2\pi}\, e^{-\frac{c^2}{2}}. \tag{3}$$

On the other hand, in the case of a discrete time process, $N_c(T)$ denotes the number of upcrossings of the broken line joining the points (n, x_n) with the level c in the time interval $0 \leqslant n \leqslant T$, while

$$\mu_c = \mathrm{E}\{N_c(1)\} \sim \frac{1}{c\sqrt{2\pi}}\, e^{-\frac{c^2}{2}} \tag{4}$$

as $c \to \infty$.

Taking in (2) $k = 0$, $y = e^{-z}$, and writing

$$T = \frac{e^{-z}}{\mu_c}, \tag{5}$$

it will be seen that, when z is fixed, T and c tend to infinity at the same time. We then obtain from (2)

$$\lim_{T\to\infty} \mathrm{P}\{N_c(T) = 0\} = e^{-e^{-z}},$$

which holds both for the case of continuous and discrete time.

Let us now consider the continuous case. We then have

$$\mathrm{P}\{N_c(T) = 0\} = \mathrm{P}\{\max_{0\leqslant t\leqslant T} x(t) \leqslant c\} + \mathrm{P}\{\min_{0\leqslant t\leqslant T} x(t) \geqslant c\},$$

where the last term is no greater than $\mathrm{P}\{x(0) \geqslant c\}$, and thus tends to zero as $c \to \infty$. It follows that we have

$$\lim_{T\to\infty} \mathrm{P}\{\max_{0\leqslant t\leqslant T} x(t) \leqslant c\} = e^{-e^{-z}}. \tag{6}$$

From (3) and (5) we obtain

$$c^2 = 2(\log T - A + z) \quad \text{with } A = \log\frac{2\pi}{\sqrt{\lambda_2}},$$

and consequently

$$c = \sqrt{2 \log T} - \frac{A}{\sqrt{2 \log T}} + \frac{z}{\sqrt{2 \log T}} O\left((\log T)^{-\frac{3}{2}}\right).$$

The second member of (6) being a continuous function of z it now follows that we have

$$\lim_{T \to \infty} P\left\{\max_{0 \leq t \leq T} x(t) \leq \sqrt{2 \log T} - \frac{A}{\sqrt{2 \log T}} + \frac{z}{\sqrt{2 \log T}}\right\} = e^{-e^{-z}}.$$

This proves the limiting relation (1) for the continuous case. The discrete case is proved in the same way, using the expression (4) for μ_c instead of (3).

Additional Note (August 17, 1964). After the above paper was completed and sent to the Editor, it came to my knowledge that the discrete time case of my theorem had been previously found by D. M. Chibisov [6] and, under somewhat different conditions, by S. M. Berman [7].

Поступила в редакцию
3.9.64

BIBLIOGRAPHY

[1] В. А. Волконский, Ю. А. Розанов, Некоторые предельные теоремы для случайных функций, I; II, Теория вероят. и ее примен., IV, 2 (1959), 186—207; VI, 2 (1961), 202—215.

[2] S. M. Berman, A law of large numbers for the maximum in a stationary Gaussian sequence, Ann. Math. Statist., 33, 1 (1962), 93—97.

[3] H. Cramér, On the maximum of a normal stationary stochastic process, Bull. Amer. Math. Soc., 68, 5 (1962), 512—516.

[4] В. А. Иванов, О среднем числе пересечений некоторого уровня выборочными функциями вероятностного процесса, Теория вероят. и ее примен., V, 3 (1960), 352—357.

[5] Е. В. Булинская, О среднем числе пересечений некоторого уровня стационарным гауссовским процессом, Теория вероят. и ее примен., VI, 4 (1961), 474—478.

[6] Д. М. Чибисов, Замечание о работе Н. В. Смирнова «Предельные законы распределения для членов вариационного ряда», Теория вероят. и ее примен., VII, 3 (1962), 357—358.

[7] S. M. Berman, Limit theorems for the maximum term in stationary sequences, Ann. Math. Statist., 35, 2 (1964), 502—516.

ПРЕДЕЛЬНАЯ ТЕОРЕМА ДЛЯ МАКСИМУМОВ НЕКОТОРЫХ СЛУЧАЙНЫХ ПРОЦЕССОВ

Г. КРАМЕР (ДЬЮРСХОЛЬМ, ШВЕЦИЯ)

(Резюме)

В предположении, что выполнено условие сильного перемешивания, показано, что максимум гауссовского стационарного процесса имеет (после надлежащей нормировки) то же самое предельное распределение, как и максимум последовательности независимых нормальных случайных величин. Это справедливо как для процессов с непрерывным временем, так и для процессов с дискретным временем. Нормировка для одного случая несколько отличается от нормировки для другого.

60.

On the intersections between the trajectories of a normal stationary stochastic process and a high level

Ark. Mat. **6** (20), 337–349 (1966)

1. Introduction

Let $\xi(t)$ be a real-valued, normal and stationary stochastic process with a continuous time parameter t, varying over $(-\infty, \infty)$. Suppose that $E\xi(t)=0$ for all t, and that the covariance function

$$r(t-u)=E\xi(t)\xi(u)$$

satisfies the following two conditions:

$$r(t)=1-\frac{1}{2!}\lambda_2 t^2+\frac{1}{4!}\lambda_4 t^4+o(t^4) \tag{1}$$

as $t\to 0$, and

$$r(t)=O(|t|^{-\alpha}) \tag{2}$$

for some $\alpha>0$, as $t\to\pm\infty$. We may, of course, assume $\alpha<1$.

It follows[1] from (1) that there exists an equivalent version of $\xi(t)$ having, with probability one, a continuous sample function derivative $\xi'(t)$, and it will be supposed that $\xi(t)$ has, if required, been replaced by this equivalent version.

Let $u>0$ be given, and consider the intersections between the trajectories $\eta=\xi(t)$ of the ξ process and the horizontal line, or "level", $\eta=u$ during some finite time interval, say $0<t<T$. It follows from a theorem due to Bulinskaja [2] that, with probability one, there are only a finite number of such intersections, and also that, with probability one, there is no point of tangency between the trajectory $\eta=\xi(t)$ and the level $\eta=u$ during the interval $(0,T)$. Thus, with probability one, every point with $\xi(t)=u$ can be classified as an "upcrossing" or a "downcrossing" of the level u, according as $\xi'(t)$ is positive or negative.

The present paper will be concerned with the upcrossings, and their asymptotic distribution in time, as the level u becomes large. It will be obvious that the case of a large negative u, as well as the corresponding problem for the downcrossings, can be treated in the same way.

The upcrossings may be regarded as a *stationary stream of random events* (cf.,

[1] With respect to the general theory of the normal stationary process and its sample functions we refer to the forthcoming book [4] by Cramér and Leadbetter.

e. g. Khintchine's book [5]), and it is well known that the simplest case of such a stream occurs when the successive events form a Poisson process. However, a necessary condition for this case is (Khintchine, l. c., pp. 11–12) that the numbers of events occurring in any two disjoint time intervals should be independent random variables, and it is readily seen that this condition cannot be expected to be satisfied by the stream of upcrossings. On the other hand, it seems intuitively plausible that the independence condition should be at least approximately satisfied when the level u becomes very large, provided that values of $\xi(t)$ lying far apart on the time scale can be supposed to be only weakly dependent.

Accordingly it may be supposed that, subject to appropriate conditions on $\xi(t)$, the stream of upcrossings will tend to form a Poisson process as the level u tends to infinity.

That this is actually so was first proved by Volkonskij and Rozanov in their remarkable joint paper [8]. They assumed, in addition to the condition (1) above, that $\xi(t)$ satisfies the so-called *condition of strong mixing*. This is a fairly restrictive condition, as can be seen e. g. from the analysis of the strong mixing concept given by Kolmogorov and Rozanov [6]. Moreover, in an actual case it will not always be easy to decide whether the condition is satisfied or not. On the other hand, various interesting properties of the $\xi(t)$ trajectories follow as corollaries from the asymptotic Poisson character of the stream of upcrossings (cf. Cramér [3], Cramér and Leadbetter [4]), so that it seems highly desirable to prove the latter property under simpler and less restrictive conditions.

It will be shown in the present paper that it is possible to replace the condition of strong mixing by the condition (2) above. This is considerably more general, and also simpler to deal with in most applications.

I am indebted to Dr. Yu. K. Belajev for stimulating conversations about the problem treated in this paper.

2. The main theorem

The number of upcrossings of $\xi(t)$ with the level u during the time interval (s,t) will be denoted by $N(s,t)$. This is a random variable which, by the above remarks, is finite with probability one. When $s=0$, we write simply $N(t)$ instead of $N(0,t)$. From the stationarity and the condition (1) it follows (Bulinskaja [2], Cramér and Leadbetter [4], Ch. 10) that the mean value of $N(s,t)$ is, for $s<t$,

$$EN(s,t)=EN(t-s)=\mu(t-s), \tag{3}$$

where
$$\mu=EN(1)=\frac{\sqrt{\lambda_2}}{2\pi}e^{-u^2/2}. \tag{4}$$

The quantity μ will play an important part in the sequel. We note that μ is a function of the level u, and that μ tends to zero as u tends to infinity. From (3) we obtain for any $\tau>0$

$$EN(\tau/\mu)=\tau.$$

For the study of the stream of upcrossings, it will then seem natural to choose $1/\mu$ as a scaling unit of time, thus replacing t by τ/μ. We might expect, e.g., that the probability distribution of $N(\tau/\mu)$ will tend to some limiting form when μ tends to zero while τ remains fixed. This is in fact the case, as shown by the following theorem, first proved by Volkonskij and Rozanov under more restrictive conditions, as mentioned above.

Theorem. *Suppose that the normal and stationary process $\xi(t)$ satisfies the conditions (1) and (2). Let $(a_1,b_1),\ldots,(a_j,b_j)$ be disjoint time intervals depending on u in such a way that, for $i=1,\ldots,j$,*

$$b_i - a_i = \tau_i/\mu,$$

the integer j and the positive numbers $\tau_1,\ldots\tau_j$ being independent of u. Let $k_1,\ldots,k_j$ be non--negative integers independent of u. Then

$$\lim_{u\to\infty} P\{N(a_i,b_i)=k_i \text{ for } i=1,\ldots,j\} = \prod_{i=1}^{j} \frac{\tau_i^{k_i}}{k_i!} e^{-\tau_i}.$$

Thus, when time is measured in units of $1/\mu$, the stream of upcrossings will asymptotically behave as a Poisson process as the level u tends to infinity.

We shall first give the proof for the case $j=1$, when there is one single interval (a,b) of length $b-a=\tau/\mu$. Owing to the stationarity it is sufficient to consider the interval $(0,\tau/\mu)$. Writing

$$T=\tau/\mu, \quad P_k = P\{N(T)=k\}, \tag{5}$$

we then have to prove the relation

$$\lim_{u\to\infty} P_k = \frac{\tau^k}{k!} e^{-\tau} \tag{6}$$

for any given $\tau>0$ and non-negative integer k, both independent of u. Once this has been achieved, the proof of the general case will follow in a comparatively simple way.

The proof of (6) is rather long, and will be broken up in a series of lemmas. In the following section we shall introduce some notations that will be used in the course of the proof. The lemmas will be given in sections 4 and 5, while section 6 contains the proof of the case $j=1$ of the theorem, and section 7 the proof of the general case.

3. Notations

The level u will be regarded as a variable tending to infinity, and we must now introduce various functions of u. It will be practical to define them as functions of μ, where μ is the function of u given by (4). Writing as usual $[x]$ for the greatest integer $\leqslant x$, we define

$$\left.\begin{aligned} m_1 &= [\mu^{-1}], \quad m_2 = [\mu^{\beta-1}], \\ M &= [T/\mu^{\beta}]. \quad n = [M/(m_1+m_2)]+1. \end{aligned}\right\} \tag{7}$$

Here β is any number satisfying the relation

$$0<(k+4)\beta<\alpha<1, \tag{8}$$

where α is the constant occurring in the condition (2), while k is the integer occurring in (6). We further write

$$q=T/M, \quad t_1=m_1 q, \quad t_2=m_2 q, \tag{9}$$

and divide the interval $(0,T)$ on the time axis into subintervals, alternatively of length t_1 and t_2, starting from the origin. We shall refer to these subintervals as t_1- and t_2-intervals respectively, the former being regarded as closed and the latter as open. Each t_i-interval $(i=1,2)$ consists of m_i subintervals of length q. The whole interval $(0,T)$, which consists of M intervals of length q, is covered by n pairs of t_1- and t_2-intervals, the nth pair being possibly incomplete. Any two distinct t_1-intervals are separated by an interval of length at least equal to t_2. An important use will be made of this remark in the proof of Lemma 5 below.

The quantities defined by (7) and (9) are all functions of u. It will be practical to express their order of magnitude for large u in terms of μ. The following relations are easily obtained from (5), (7) and (9):

$$\left.\begin{aligned} q\sim\mu^{\beta}, \quad & n\sim\tau\mu^{-\beta}, \\ t_1\sim\mu^{\beta-1}, \quad & t_2\sim\mu^{2\beta-1}. \end{aligned}\right\} \tag{10}$$

We now define a stochastic process $\xi_q(t)$ by taking

$$\xi_q(\nu q)=\xi(\nu q)$$

for all integers ν, and determining ξ_q by linear interpolation in the interval between two consequtive νq. To any sample function of the $\xi(t)$ process will then correspond a sample function of $\xi_q(t)$, which is graphically represented by the broken line joining the points $[\nu q, \xi(\nu q)]$. For the number of upcrossings of this broken line with the u level we use the notations $N_q(s,t)$ and $N_q(t)$, corresponding to $N(s,t)$ and $N(t)$. The probability corresponding to P_k as defined by (5) is

$$P_k^{(q)}=P\{N_q(T)=k\}. \tag{11}$$

4. Lemmas 1–3

Throughout the rest of the paper we assume that the $\xi(t)$ process satisfies the conditions of the above theorem.

Lemma 1. *If T and q are given by* (5) *and* (9), *τ and k being fixed as before, we have*

$$\lim_{u\to\infty}(P_k-P_k^{(q)})=0.$$

Evidently $N_q(T)\leqslant N(T)$. We shall prove that the non-negative and integer-valued random variable $N(T)-N_q(T)$ converges in first order mean to zero, as $u \to \infty$. It then follows that the probability that $N(T)-N_q(T)$ takes any value different from zero will tend to zero, and so the lemma will be proved.

By (3) and (5), the mean value of the number $N(T)$ of upcrossings of $\xi(t)$ in $(0,T)$ is for every u

$$EN(T)=T\mu=\tau.$$

It will now be proved that the mean value $EN_q(T)$ tends to the limit τ as $u \to \infty$, so that we have

$$\lim_{u\to\infty} E\{N(T)-N_q(T)\}=0.$$

Since $N(T)-N_q(T)$ is non-negative, this implies convergence in first order mean to zero, so that by the above remark the lemma will be proved.

Consider first the number $N_q(q)$ of upcrossings of $\xi_q(t)$ in the interval $(0,q)$. This number is one, if $\xi(0)<u<\xi(q)$, and otherwise zero, so that

$$EN_q(q)=P\{\xi(0)<u<\xi(q)\}.$$

Now $\xi(0)$ and $\xi(q)$ have a joint normal density function, with unit variances and correlation coefficient $r=r(q)$. By (1) and (10) we have

$$r(q)=1-\tfrac{1}{2}\lambda_2 q^2+O(q^4). \tag{12}$$

For the probability that $\xi(0)<u$ and $\xi(q)>u$ we obtain by a standard transformation

$$EN_q(q)=\frac{1}{\sqrt{2\pi}}\int_u^\infty e^{-x^2/2}\Phi\left(\frac{u-rx}{\sqrt{1-r^2}}\right)dx,$$

where as usual

$$\Phi(x)=\frac{1}{\sqrt{2\pi}}\int_{-\infty}^x e^{-t^2/2}\,dt.$$

By some straightforward evaluation of the integral from $u+(1-r)^{\frac{1}{4}}$ to infinity we obtain, using (12), and denoting by K an unspecified positive constant,

$$EN_q(q)=\frac{1}{\sqrt{2\pi}}\int_u^{u+(1-r)^{\frac{1}{4}}} e^{-x^2/2}\Phi\left(\frac{u-rx}{\sqrt{1-r^2}}\right)dx+O[\exp(-e^{Ku^2})].$$

For the first term in the second member we obtain, using again (12), the expression

$$\frac{1}{\sqrt{2\pi}}e^{-u^2/2}(1+O(q^{\frac{1}{2}}u))\int_u^{u+(1-r)^{\frac{1}{4}}}\Phi\left(\frac{u-rx}{\sqrt{1-r^2}}\right)dx$$

$$=\frac{\sqrt{1-r^2}}{r\sqrt{2\pi}}e^{-u^2/2}(1+O(q^{\frac{1}{2}}u))\left(\int_{-\infty}^0\Phi(y)\,dy+O(qu)\right)=\frac{\sqrt{\lambda_2}}{2\pi}qe^{-u^2/2}(1+O(q^{\frac{1}{2}}u)),$$

so that $$EN_q(q)=q\mu+o(q\mu).$$

If ν denotes an integer, which may tend to infinity with u, it then follows from the stationarity that we have

$$EN_q(\nu q) = \nu q\mu + o(\nu q\mu). \tag{13}$$

In particular, taking $\nu = M$ we obtain from (9) and (5)

$$EN_q(T) = T\mu + o(T\mu) \to \tau.$$

According to the above remarks, this proves the lemma.

We now consider the number $N_q(t_1)$ of ξ_q upcrossings in an interval of length $t_1 = m_1 q$, observing that by (10) we have $t_1 \sim \mu^{\beta-1}$.

Lemma 2. *We have*

$$\lim_{u\to\infty} \frac{E\{N_q(t_1)[N_q(t_1)-1]\}}{EN_q(t_1)} = 0.$$

By (3) we have $EN(t_1) = t_1\mu$, while (13) gives for $\nu = m_1$

$$EN_q(t_1) = t_1\mu + o(t_1\mu) \sim EN(t_1).$$

Further, since $N_q(t_1) \leqslant N(t_1)$,

$$E\{N_q(t)[N_q(t_1)-1]\} \leqslant E\{N(t_1)[N(t_1)-1\}.$$

The truth of the lemma will then follow from the corresponding relation with N_q replaced by N. Now this latter relation is identical with the relation proved by Volkonskij and Rozanov [8] in their Lemma 3.4. It is proved by them without any mixing hypothesis, assuming only that $\xi(t)$ is regular (or purely non-deterministic) and that $r(t)$ has a fourth order derivative at $t = 0$. Their proof is valid without any modification whatever, if these conditions are replaced by our conditions (1) and (2). Thus we may refer to their paper for the proof of this lemma. We note that the proof is based on the important work of S. O. Rice [7].

Lemma 3. *As $u \to \infty$, we have*

$$P\{N_q(t_1) = 0\} = 1 - q + o(q),$$
$$P\{N_q(t_1) = 1\} = q + o(q),$$
$$P\{N_q(t_1) > 1\} = o(q).$$

For any random variable ν taking only non-negative integral values we have, writing $\pi_i = P\{\nu = i\}$ and assuming $E\nu^2 < \infty$,

$$E\nu = \pi_1 + 2\pi_2 + 3\pi_3 + \dots,$$
$$E\nu(\nu-1) = 2\pi_2 + 6\pi_3 + \dots,$$

and consequently

$$E\nu - E\nu(\nu-1) \leqslant \pi_1 \leqslant 1 - \pi_0 \leqslant E\nu. \tag{14}$$

Taking $\nu = N_q(t_1)$, and observing that by (10) we have $EN_q(t_1) \sim t_1\mu \sim q$, the truth of the lemma follows directly from Lemma 2.

5. Lemmas 4–5

For each $r = 1, 2, \ldots, n$, we now define the following events, i.e. the sets of all $\xi(t)$ sample functions satisfying the conditions written between the brackets:

$$c_r = \{\text{exactly one } \xi_q \text{ upcrossing in the } r\text{th } t_1\text{-interval}\},$$
$$d_r = \{\text{at least one } \xi_q \text{ upcrossing in the } r\text{th } t_1\text{-interval}\},$$
$$e_r = \{\xi(\nu q) > u \text{ for at least one } \nu q \text{ in the } r\text{th } t_1\text{-interval}\}.$$

Further, let C_k denote the event that c_r occurs in exactly k of the t_1-intervals in $(0, T)$, while the complementary event c_r^* occurs in the $n-k$ others. D_k and E_k are defined in the corresponding way, using respectively d_r and e_r instead of c_r.

Lemma 4. *For the probability* $P_k^{(q)}$ *defined by* (11) *we have*

$$\lim_{u\to\infty} [P_k^{(q)} - P\{E_k\}] = 0.$$

We shall prove that each of the differences $P_k^{(q)} - P\{C_k\}$, $P\{C_k\} - P\{D_k\}$ and $P\{D_k\} - P\{E_k\}$ tends to zero as $u \to \infty$.

By (13) and (14) the probability of at least one ξ_q upcrossing in an interval of length t_2 is at most $EN_q(t_2) = t_2\mu + o(t_2\mu)$. Thus the probability of at least one ξ_q upcrossing in at least one of the n t_2-intervals in $(0, T)$ is by (10)

$$O(nt_2\mu) = O(\mu^\beta),$$

and thus tends to zero as $u \to \infty$. It follows that we have

$$P_k^{(q)} - P\{\text{total number of } \xi_q \text{ upcrossings in all } n\ t_1\text{-intervals} = k\} \to 0. \quad (15)$$

On the other hand, by the stationarity of $\xi(t)$, Lemma 3 remains true if $N_q(t_1)$ is replaced by the number of ξ_q upcrossings in any particular t_1-interval. Since the interval $(0, T)$ contains n of these intervals, it follows from (10) that the probability of more than one ξ_q upcrossing in at least one of the t_1-intervals is $o(nq) = o(1)$, and thus tends to zero as $u \to \infty$.

From (15) and the last remark, it now readily follows that the differences $P_k^{(q)} - P\{C_k\}$ and $P\{C_k\} - P\{D_k\}$ both tend to zero as $u \to \infty$. It thus only remains to show that this is true also for $P\{D_k\} - P\{E_k\}$.

By the definitions of the events D_k and E_k we have

$$\left.\begin{aligned} P\{D_k\} &= \sum P\{d_{r_1} \ldots d_{r_k} d_{s_1}^* \ldots d_{s_{n-k}}^*\}, \\ P\{E_k\} &= \sum P\{e_{r_1} \ldots e_{r_k} e_{s_1}^* \ldots e_{s_{n-k}}^*\}, \end{aligned}\right\} \quad (16)$$

the summations being extended over all $\binom{n}{k}$ groups of k different subscripts $r_1, \ldots, r_k$ selected among the numbers $1, \ldots, n$, while in each case $s_1, \ldots, s_{n-k}$ are the remaining $n-k$ subscripts.

Let $\nu_r q$ denote the left endpoint of the rth t_1-interval, and denote by g_r the event

$$g_r = \{\xi(\nu_r q) \geq u\}$$

of probability

$$P\{g_r\} = O\left(\frac{1}{u} e^{-u^2/2}\right).$$

Then for every $r = 1, \ldots, n$

$$d_r \subset e_r \quad \text{and} \quad e_r - d_r \subset g_r,$$

so that
$$P\{e_r - d_r\} = O\left(\frac{1}{u} e^{-u^2/2}\right). \tag{17}$$

Similarly
$$e_r^* \subset d_r^*, \quad \text{and}$$
$$d_r^* - e_r^* = e_r - d_r \subset g_r,$$
$$P\{d_r^* - e_r^*\} = O\left(\frac{1}{u} e^{-u^2/2}\right).$$

By a simple recursive argument (16) then yields, using (8) and (10),

$$P\{D_k\} - P\{E_k\} = O\left(\frac{n^{k+1}}{u} e^{-u^2/2}\right) = O\left[\exp\left(-\frac{1-(k+1)\beta}{2} u^2\right)\right] \to 0,$$

which proves the lemma.

By definition, E_k is composed of certain events e_r and e_s^*. Each of these is associated with one particular t_1-interval, and it has been observed above that any two t_1-intervals are separated by an interval which is of length $\geqslant t_2$, and thus tends to infinity with u. By means of the condition (2) it will now be shown that the component events of E_k are asymptotically independent, as $u \to \infty$. Moreover, owing to stationarity, the probability

$$p = P\{e_r\} \tag{18}$$

is independent of r, so that by (16) the asymptotic independence will be expressed by the following lemma.

Lemma 5. *The probability p being defined by* (18) *we have, as $u \to \infty$,*

$$P\{E_k\} - \binom{n}{k} p^k (1-p)^{n-k} \to 0.$$

In order to prove this lemma, we consider the points νq on the time axis for all integers ν such that νq belongs to one of the t_1-intervals in $(0, T)$. Each t_1-interval, which we regard as closed, contains $m_1 + 1$ points νq, and there are $n-1$ complete and one possibly incomplete such interval in $(0, T)$. If L is the total number of points νq in all t_1-intervals, we thus have

$$(n-1)(m_1+1) < L \leqslant n(m_1+1).$$

Let $\eta_1, \ldots, \eta_L$ be the random variables $\xi(\nu q)$ corresponding to all these L points

νq, ordered according to increasing ν. The η_i sequence will consist of n groups, each corresponding to one particular t_1-interval.

Further, let $f_1(y_1, \ldots, y_L)$ be the L-dimensional normal probability density of $\eta_1, \ldots, \eta_L$, and let Λ_1 be the corresponding covariance matrix. (Our reasons for using the subscript 1 here and in the sequel will presently appear.) From (16) we obtain

$$P\{E_k\} = \int_{E_k} f_1 dy = \sum \int_{e_{r_1} \ldots e^*_{s_1} \ldots} f_1 dy, \tag{19}$$

where the abbreviated notation should be easily understood, the summation being extended as explained after (16).

Let us now consider one particular term of the sum in the last member of (19), say the term where the group of subscripts $r_1, \ldots, r_k$ coincides with the integers $1, \ldots, k$. It will be readily seen that any other term can be treated in the same way as we propose to do with this one. This term is

$$F(1) = \int_G f_1 dy,$$

where G denotes the set

$$G = e_1 \ldots e_k e^*_{k+1} \ldots e^*_n.$$

$F(1)$ may be regarded as a function of the covariances which are elements of the matrix Λ_1. Let us consider in particular the dependence of $F(1)$ on those covariances $\varrho_{ij} = E\eta_i\eta_j$ which correspond to variables η_i and η_j *belonging to different t_1-intervals*. If all covariances ϱ_{ij} having this character are replaced by $\lambda_{ij} = h\varrho_{ij}$, with $0 \leqslant h \leqslant 1$, while all other elements of Λ_1 remain unchanged, the resulting matrix will be

$$\Lambda_h = h\Lambda_1 + (1-h)\Lambda_0, \tag{20}$$

while the density function f_1 will be replaced by a certain function f_h. Evidently f_0, corresponding to the covariance matrix Λ_0, will be the normal density function that would apply if the groups of variables η_i belonging to different t_1-intervals were all mutually independent, while the joint distribution within each group were the same as before.

Thus Λ_1 and Λ_0 are both positive definite, and it then follows from (20) that the same is true for Λ_h, so that f_h is always a normal probability density. Writing

$$F(h) = \int_G f_h dy,$$

it follows from the remarks just made that we have

$$F(0) = \int_{e_1} f_0 dy \ldots \int_{e_k} f_0 dy \int_{e^*_{k+1}} f_0 dy \ldots \int_{e^*_n} f_0 dy = P\{e_1\} \ldots P\{e_k\} P\{e^*_{k+1}\} \ldots P\{e^*_n\}.$$

By stationarity this reduces to

$$F(0) = p^k (1-p)^{n-k}, \tag{21}$$

where p is given by (18).

We shall now evaluate the difference $F(1)-F(0)$ by a development of a method used by S. M. Berman [1]. We note that for any normal density function $f(x_1, \ldots, x_n)$ with zero means and covariances r_{ij} we have

$$\frac{\partial f}{\partial r_{ij}} = \frac{\partial^2 f}{\partial x_i \partial x_j}.$$

In our case, $f_h(y_1, \ldots, y_L)$ is a normal density, depending on h through the covariances $\lambda_{ij} = h\varrho_{ij}$. Hence

$$F'(h) = \int_G \frac{df_h}{dh} dy = \sum \varrho_{ij} \int_G \frac{\partial f_h}{\partial \lambda_{ij}} dy = \sum \varrho_{ij} \int_G \frac{\partial^2 f_h}{\partial y_i \partial y_j} dy, \tag{22}$$

the summation being extended over all i, j such that η_i and η_j belong to different t_1-intervals. With respect to the integral over the set $G = e_1 \ldots e_k e^*_{k+1} \ldots e^*_n$ occurring in the last sum in (22), we have to distinguish three different cases.

Case A. When η_i and η_j both belong to t_1-intervals of subscripts $> k$, say to the t_1-intervals of subscripts $k+1$ and $k+2$ respectively, integration with respect to y_i and y_j has to be performed over e^*_{k+1} and e^*_{k+2} respectively. By definition of the sets e_r, both y_i and y_j thus have to be integrated over $(-\infty, 0)$, and so we obtain by direct integration with respect to y_i and y_j

$$\int_G \frac{\partial^2 f_h}{\partial y_i \partial y_j} dy = \int_G f_h(y_i = y_j = u)\, dy'. \tag{23}$$

The notation used in the last integral is to be understood so that we have to take $y_i = y_j = u$ in f_h, and then integrate with respect to all y's expect y_i and y_j. As $f_h > 0$ always, we have

$$0 < \int_G \frac{\partial^2 f_h}{\partial y_i \partial y_j} dy < \int_{-\infty}^{\infty} \ldots \int_{-\infty}^{\infty} f_h(y_i = y_j = u)\, dy'.$$

The last integral, where all the y's except y_i and y_j are integrated out, yields the joint density function of the random variables corresponding to η_i and η_j in the normal distribution with covariance matrix Λ_h, for the values $y_i = y_j = u$, so that

$$0 < \int_G \frac{\partial^2 f_h}{\partial y_i \partial y_j} dy < \frac{1}{2\pi(1-h^2\varrho_{ij}^2)^{\frac{1}{2}}} \exp\left[-u^2/(1+h|\varrho_{ij}|)\right]. \tag{24}$$

Case B. Let now η_i and η_j both belong to t_1-intervals of subscripts $\leqslant k$, say to those of subscripts 1 and 2 respectively. Then integration with respect to each of the groups of variables to which y_i and y_j belong has to be performed over e_1 and e_2 respectively. By definition, e_1 is the set of all points in the y space such that at least one of the y's associated with the first t_1-interval exceeds u, and correspondingly for e_2. Some reflection will then show that the integration indicated in the first member of (23) can still be carried out directly, and yields the same result, with the only difference that in the second member of (23) the integration has to be performed over a set G', obtained from G by replacing e_1 and e_2 by e_1^* and e_2^* respectively. It follows that the inequality (24) still holds.

Case C. Finally we have the case when η_i and η_j belong to t_1-intervals of different kinds, say to the first and the $(k+1)$st respectively. As before the integration in the first member of (23) can be carried out directly. In this case, however, we obtain the relation (23) with a changed sign of the second member, and e_1 replaced by e_1^* in the expression of the domain of integration. In this case we thus obtain the inequality (24) with changed inequality signs.

Thus in all three cases we have the inequality

$$\left|\int_G \frac{\partial^2 f_h}{\partial y_i \partial y_j} dy\right| < \frac{1}{2\pi(1-h^2\varrho_{ij}^2)^{\frac{1}{2}}} \exp\left[-u^2/(1+h|\varrho_{ij}|)\right]. \qquad (25)$$

Now ϱ_{ij} is the covariance between the variables $\eta_i = \xi(\nu_i q)$ and $\eta_j = \xi(\nu_j q)$, where the points $\nu_i q$ and $\nu_j q$ belong to different t_1-intervals, and are thus separated by an interval of length at least equal to t_2. By the condition (2) we then have

$$|\varrho_{ij}| = |r(\nu_i q - \nu_j q)| < K t_2^{-\alpha},$$

where as usual K denotes an unspecified positive constant. Further, there are less than $L^2 \leqslant n^2(m_1+1)^2$ covariances ϱ_{ij}. Owing to stationarity some of the ϱ_{ij} are equal, but it is easily seen that this does not affect our argument. It then follows from (22) and (25), using (7) and (10), that we have

$$|F'(h)| < K n^2 m_1^2 t_2^{-\alpha} e^{-u^2} < K\mu^{\alpha-4\beta},$$

$$|F(1)-F(0)| = \left|\int_0^1 F'(h)\,dh\right| < K\mu^{\alpha-4\beta}.$$

This holds for any of the $\binom{n}{k}$ terms in the last member of (19), and $F(0)$ will in all cases be given by (21), so that we finally obtain

$$\left|P\{E_k\} - \binom{n}{k} p^k (1-p)^{n-k}\right| < K\binom{n}{k}\mu^{\alpha-4\beta} < K\mu^{\alpha-(k+4)\beta}.$$

By (8) we have $(k+4)\beta < \alpha$, so that the last member tends to zero as $u \to \infty$, and the lemma is proved.

6. Proof of the case j=1 of the theorem

By (18), $p = P\{e_r\}$ is defined as the probability that at least one of the random variables $\xi(0), \xi(q), \xi(2q), \ldots, \xi(m_1 q)$ takes a value exceeding u. According to (17), this differs from the probability $P\{d_r\}$ of at least one ξ_q upcrossing in the first t_1-interval by a quantity of the order

$$O\left(\frac{1}{u} e^{-u^2/2}\right).$$

By Lemma 3, the latter probability is

$$P\{d_r\} = q + o(q).$$

Thus we obtain from Lemma 5, observing that by (10) we have $1/u\,e^{-u^2/2} = o(q)$,

$$P\{E_k\} - \binom{n}{k}[q + o(q)]^k [1 - q + o(q)]^{n-k} \to 0.$$

By (10) we have $nq \to \tau$, and thus

$$\lim_{u\to\infty} P\{E_k\} = \frac{\tau^k}{k!} e^{-\tau}.$$

Lemmas 1 and 4 then finally give the relation (6) that was to be proved:

$$\lim_{u\to\infty} P_k = \frac{\tau^k}{k!} e^{-\tau}.$$

Thus we have proved the simplest case of the theorem, when $j = 1$, so that there is only one interval.

7. Proof of the general case

The generalization to the case of an arbitrary number $j > 1$ of intervals is now simple.

For any $\varepsilon > 0$, it follows from the result just proved that, for every $i = 1, \dots, j$, the random variable

$$N(a_i + \varepsilon/\mu, b_i - \varepsilon/\mu),$$

where $b_i - a_i = \tau_i/\mu$, will be asymptotically Poisson distributed with parameter $\tau_i - 2\varepsilon$. In the same way as in the proof of Lemma 5 it is shown that these j variables are asymptotically independent, so that we have

$$P\{N(a_i + \varepsilon/\mu, b_i - \varepsilon/\mu) = k_i \text{ for } i = 1, \dots, j\} \to \prod_{i=1}^{j} \frac{(\tau_i - 2\varepsilon)^{k_i}}{k_i!}\, e^{-(\tau_i - 2\varepsilon)}. \tag{26}$$

From the asymptotic Poisson distributions of the variables

$$N(a_i, a_i + \varepsilon/\mu) \quad \text{and} \quad N(b_i - \varepsilon/\mu, b_i)$$

it further follows that, with a probability exceeding $1 - 2j\varepsilon$, these variables will ultimately be zero for all $i = 1, \dots, j$. Since j is fixed, and $\varepsilon > 0$ is arbitrarily small, the truth of the theorem then follows from (26).

REFERENCES

1. BERMAN, S. M., Limit theorems for the maximum term in stationary sequences. Ann. Math. Stat. *35*, 502 (1964).
2. BULINSKAJA, E. V., On the mean number of crossings of a level by a stationary Gaussian process. Teor. Verojatnost. i Primenen. *6*, 474 (1961).
3. CRAMÉR, H., A limit theorem for the maximum values of certain stochastic processes. Teor. Verojatnost. i Primenen. *10*, 137 (1965).
4. CRAMÉR, H., and LEADBETTER, M. R., Stationary and Related Stochastic Processes. To be published by Wiley and Sons, New York.
5. KHINTCHINE, A. Y., Mathematical Methods in the Theory of Queueing. Griffin and Co., London 1960.
6. KOLMOGOROV, A. N., and ROZANOV, YU. A., On strong mixing conditions for stationary Gaussian processes. Teor. Verojatnost. i Primenen. *5*, 222 (1960).
7. RICE, S. O., Distribution of the duration of fades in radio transmission. Bell Syst. Techn. Journ. *37*, 581 (1958).
8. VOLKONSKIJ, V. A., and ROZANOV, YU. A., Some limit theorems for random functions. Teor. Verojatnost. i Primenen. *4*, 186 (1959) and *6*, 202 (1961).

Tryckt den 4 januari 1966

Uppsala 1966. Almqvist & Wiksells Boktryckeri AB

61.

On stochastic processes whose trajectories have no discontinuities of the second kind

Ann. Mat. Pura Appl. **71**, 85–92 (1966)

In memory of Guido Castelnuovo, in the recurrence of the first centenary of his birth.

Summary. - *It is shown that, under the condition* (1) *below, the trajectories of the stochastic process* $\xi(t)$ *can, after replacing the* $\xi(t)$ *process by an equivalent version* $\eta(t)$, *at most have discontinuities of the first kind, i.e. simple jumps.*

Let $\xi(t)$ denote a real-valued stochastic process, where the real parameter t varies over some finite interval. In order to simplify the writing, we shall assume that t is confined to the closed interval [0, 1]. We shall say that two processes $\xi(t)$ and $\eta(t)$ are *equivalent*, if

$$P\{\xi(t)=\eta(t)\}=1$$

for every t in [0, 1].

When a process $\xi(t)$ is given, it is often important to know conditions which imply the existence of an equivalent process $\eta(t)$ such that, with probability one, the *trajectories* or *sample functions* $\eta(t)$ possess some simple analytic property, such as continuity, differentiability etc., throughout the whole t-interval considered. One of the earliest known conditions of this type is contained in a theorem due to KOLMOGOROV and published by SLUTSKY [8] in the Gior. dell'Istituto Italiano degli Attuari in 1937. This theorem gives a sufficient condition for sample function continuity which has later been improved by other authors [1, 2, 4–7], who have also given conditions for properties such as differentiability, existence of discontinuities of specified types, etc.

In the present paper we shall prove a theorem giving conditions for the existence of an equivalent $\eta(t)$ process, almost all sample functions of which have at most discontinuities of the first kind. We shall say that a bounded function $f(t)$ has a discontinuity of the first kind at $t=a$ if the limits $f(a+0)$ and $f(a-0)$ exist and are unequal. If every t in [0, 1] is a point of continuity or a discontinuity of the first kind for $f(t)$, we say that $f(t)$ has *at most discontinuities of the first kind* in [0, 1].

We shall first state and prove our main Theorem, and then proceed to give a simple Corollary convenient for certain applications.

Theorem. - *Let $\xi(t)$ be a stochastic process defined for $t \in [0, 1]$ Suppose that, for $t_1, t_2, t_3 \in [0, 1]$, $t_1 < t_2 < t_3$, $t_3 - t_1 = h$, we have.*

$$P\{\,|[\xi(t_3) - \xi(t_2)] \cdot [\xi(t_2) - \xi(t_1)]| \geqq g^2(h)\} \leqq q(h), \tag{1}$$

where g and q are even functions of h, never increasing as $h \downarrow 0$, and such that

$$\sum_1^\infty g(2^{-n}) < \infty \quad \text{and} \quad \sum_1^\infty 2^n q(2^{-n}) < \infty.$$

Then there exists a process $\eta(t)$ equivalent to $\xi(t)$ and such that, with probability one, the sample functions (trajectories) $\eta(t)$ have at most discontinuities of the first kind in $[0, 1]$

If, in particular, the $\xi(t)$ process is separable, it then follows from known theorems that the conclusion of the theorem holds for $\xi(t)$ itself, without having recourse to an equivalent $\eta(t)$.

The proof will be based on some Lemmas concerning ordinary (non-random) functions $x(t)$ defined in $[0, 1]$. Let X denote the space of all finite-valued functions $x(t)$, and define the following sets in X:

$A_{n,r}$ is the set of all $x(t)$ such that $|x(t_{n,r+1}) - x(t_{n,r})| < g(2^{-n})$, where g is the function of the Theorem, $t_{n,r} = r/2^n$, $n = 1, 2, \ldots$ and $r = 0, 1, \ldots, 2^n - 1$.

$B_{n,r} = A_{n,r-1} + A_{n,r}$ for $r = 1, 2, \ldots, 2^n - 1$.

$$C_n = \prod_{j=n}^\infty \prod_{r=1}^{2^j - 1} B_{j,r}, \qquad C = \sum_1^\infty C_n = \lim_{n\to\infty} C_n.$$

Further, define

$D =$ the set of all dyadic numbers $t_{n,r} = r/2^n$ in $[0, 1]$.

When $x(t)$ is any function defined in $[0, 1]$, we denote by $x_D(t)$ the restriction of $x(t)$ to the set D of dyadic numbers.

Lemma 1. - *Suppose $x(t) \in A_{n,r} C_{n+1}$. For any dyadic number $t_{m,s}$ such that $t_{n,r} \leqq t_{m,s} \leqq t_{n,r+1}$ we then have*

$$|x(t_{m,s}) - x(t_{n,r})| < \sum_n^\infty g(2^{-j}),$$

$$|x(t_{m,s}) - x(t_{n,r+1})| < \sum_n^\infty g(2^{-j}).$$

Bisect first the interval $[t_{n,r}, t_{n,r+1}]$ by the point $t_{n+1,2r+1}$. Since $x(t) \in C_{n+1} \subset B_{n+1,2r+1}$, it follows that at least one of the inequalities

$$|x(t_{n+1,2r+1}) - x(t_{n,r})| < g(2^{-n-1}),$$

(2)

$$|x(t_{n+1,2r+1}) - x(t_{n,r+1})| < g(2^{-n-1}).$$

must be satisfied. Since, moreover, $x(t) \in A_{n,r}$, it further follows that the first members of the last two inequalities must both be less than

$$g(2^{-n}) + g(2^{-n-1}).$$

By bisection of the two partial intervals $[t_{n,r}, t_{n+1,2r+1}]$ and $[t_{n+1,2r+1}, t_{n,r+1}]$ we obtain in the same way

$$|x(t_{n+2,s}) - x(t_{n+r})| < g(2^{-n}) + g(2^{-n-1}) + g(2^{-n-2}),$$

$$|x(t_{n+2,s}) - x(t_{n,r+1})| < g(2^{-n}) + g(2^{-n-1}) + g(2^{-n-2}),$$

for $s = 2^2r, 2^2r+1, \ldots, 2^2r+2^2$. By repeated bisection we obtain in the same way for any $k = 1, 2, \ldots$

$$|x(t_{n+k,s}) - x(t_{n,r})| < g(2^{-n}) + \ldots + g(2^{-n-k}),$$

$$|x(t_{n+k,s}) - x(t_{n,r+1})| < g(2^{-n}) + \ldots + g(2^{-n-k}),$$

for $s = 2^kr, 2^kr+1, \ldots, 2^kr+2^k$. As this process can be continued indefinitely, the Lemma is proved.

LEMMA 2. - *Suppose* $x(t) \in B_{n,r} C_{n+1}$. *Then* $t \in [t_{n,r-1}, t_{n,r+1}]$ *exists such that, for any dyadic* $t_{m,s}$,

$$|x(t_{m,s}) - x(t_{n,r-1})| < \sum_n^\infty g(2^{-j}) \text{ for } t_{n,r-1} \leqq t_{m,s} < t,$$

$$|x(t_{m,s}) - x(t_{n,r+1})| < \sum_n^\infty g(2^{-j}) \text{ for } t < t_{m,s} \leqq t_{n,r+1}.$$

(*t may be dyadic or not.*)

We shall say that any dyadic $t_{m,s} \in [t_{n,r-1}, t_{n,r+1}]$ belongs to the *lower class* if the first inequality of the Lemma is satisfied, and to the *upper class* if the second inequality is satisfied. We must then show that t (dyadic or not) can be found so that all $t_{m,s}$ to the left of t belong to the lower class, while all $t_{m,s}$ to the right of t belong to the upper class.

Since $x(t) \in B_{n,r}$, at least one of the inequalities

$$(3) \qquad \begin{aligned} |x(t_{n,r}) - x(t_{n,r-1})| &< g(2^{-n}), \\ |x(t_{n,r}) - x(t_{n,r+1})| &< g(2^{-n}), \end{aligned}$$

must be satisfied. If both are satisfied, we have $x(t) \in A_{n,r-1} A_{n,r}$, and it follows from Lemma 1 that the Lemma is true if we take $t = t_{n,r}$. Suppose, then, that the first inequality (3) is satisfied, but not the second. Then count all dyadics in the interval $[t_{n,r-1}, t_{n,r}]$ to the *lower class*. Bisect the remaining interval $[t_{n,r}, t_{n,r+1}]$ by the point $t_{n+1,2r+1}$. As in the proof of the preceding Lemma, it follows from $x(t) \in C_{n+1} \subset B_{n+1,2r+1}$ that at least one of the inequalities (2) must be satisfied. If both are satisfied, it follows from Lemma 1 that the Lemma is true if we take $t = t_{n+1,2r+1}$. Suppose this time that the second inequality (2) is true, but not the first, and then count all dyadics in the interval $[t_{n+1,2r+1}, t_{n,r+1}]$ to the *upper class*. Bisect the remaining interval $[t_{n,r}, t_{n+1,2r+1}]$ by the point $t_{n+2,4r+1}$, and proceed in the same way. By this process, we obtain for every $k = 1, 2, \ldots$ two intervals $[t_{n,r-1}, t_{n+k,s}]$ and $[t_{n+k,s+1}, t_{n,r+1}]$ such that all dyadics in the former interval belong to the lower class, while all those in the latter interval belong to the upper class. The $t_{n+k,s}$ will form a never decreasing sequence, and the $t_{n+k,s+1}$ a never increasing one. As k increases, both sequences will tend to a common limit, which we denote by t, and this proves the Lemma.

Lemma 3. - *Suppose $x(t) \in C$. Then the restriction $x_D(t)$ of $x(t)$ to the set D of all dyadics has at most discontinuities of the first kind.*
(We say that $x_D(t)$ has a discontinuity of the first kind at $t = t_0$ if the left and right limiting values when t tends to t_0 *on dyadic numbers*, $x_D(t_0 - 0)$ and $x_D(t_0 + 0)$, exist and are unequal.)

Suppose that $x_D(t)$ has at $t = t_0$ a discontinuity which is not of the first kind. Then at least one of the limits $x_D(t_0 \pm 0)$ does not exist. Suppose, e.g., that $x_D(t_0 - 0)$ does not exist. Then $k > 0$ exist such that, for every $\varepsilon > 0$, the interval $(t_0 - \varepsilon, t_0)$ contains a sequence of dyadic points $t_1 < t_2 < \ldots$ such that $|x_D(t_{j+1}) - x_D(t_j)| > k$ for all j.

Now by hypotesis $x(t) \in C = \lim C_n$, and since C_n is never decreasing, this implies that $x(t) \in C_n$ for all sufficiently large n. Then we can take n, r and ε such that

$$x(t) \in C_n \subset B_{n,r} C_{n+1},$$

$$[t_{n,r-1}, t_{n,r+1}] \supset (t_0 - \varepsilon, t_0),$$

$$2 \sum_n^{\infty} g(2^{-j}) < k.$$

Then by Lemma 2 there must be two points t_j and t_{j+1} belonging to the same (upper or lower) class in $[t_{n,r-1}, t_{n,r+1}]$, so that have, t_j and t_{j+1} being dyadics,

$$|x_D(t_{j+1}) - x_D(t_j)| < 2 \sum_n^{\infty} g(2^{-\nu}) < k.$$

This is a contradiction, and thus Lemma 3 is proved.

Moreover, it follows from Lemma 2 that, under the conditions of Lemma 3, $x_D(t)$ is bounded on the set D.

LEMMA 4. - *Suppose $x(t) \in C$, and let $y(t)$ be a function defined for all $t \in [0, 1]$, such that for every t at least one of the relations*

$$y(t) = x_D(t+0), \qquad y(t) = x_D(t-0), \tag{4}$$

holds. Then $y(t)$ has at most discontinuities of the first kind.

By definition, for any t at least one of the relations

$$y(t) = \lim_{u \uparrow t} x_D(u), \qquad y(t) = \lim_{u \downarrow t} x_D(u), \tag{5}$$

will hold, where in both cases u tends to t throughdyadic numbers. We have to prove that the limits

$$\lim_{t \uparrow t_0} y(t) \quad \text{and} \quad \lim_{t \downarrow t_0} y(t)$$

both exist for every t_0, without restrictions on the values of t. Let us prove, e.g., that the first limit exists. Consider an increasing sequence $t_1 < t_2 < \ldots \to t_0$. For every t_j we can find a dyadic u_j such that

$$|u_j - t_j| < \frac{1}{2} \operatorname{Min}(t_j - t_{j-1}, t_{j+1} - t_j),$$

$$|x_D(u_j) - y(t_j)| < \varepsilon_j,$$

where $\varepsilon_1, \varepsilon_2, \ldots$ is a sequence of positive numbers tending to zero. (Of course, if the first relation (5) is satisfied for $t = t_j$, we take $u_j < t_j$, while in the opposite case we take $u_j > t_j$.)

The u_j will now form an increasing sequence of dyadic numbers tending to t_0, so that by Lemma 3 the limit

$$\lim_{j \to \infty} x_D(u_j) = x_D(t_0 - 0)$$

exists. It then follows from the last inequality that $y(t_j)$ tends to the same limit, so that the Lemma is proved.

For the proof of the Theorem, we shall further require the following two probabilistic Lemmas.

LEMMA 5. - *Let* $\xi(t)$ *be a stochastic process satisfying the conditions of the Theorem. Consider the sample functions* $\xi(t)$ *of the process as points in the function space X. Then*

$$P\{\xi(t) \in C\} = 1.$$

By definition of the set $B_{n,r}$, the complementary set $B^*_{n,r}$ is the set of all functions $\xi(t)$ such that both inequalities

$$|\xi(t_{n,r+1}) - \xi(t_{n,r})| \geqq g(2^{-n}),$$

$$|\xi(t_{n,r}) - \xi(t_{n,r-1})| \geqq g(2^{-n}),$$

are satisfied, and consequently also the inequality

$$|(\xi(t_{n,r+1}) - \xi(t_{n,r}))\cdot(\xi(t_{n,r}) - \xi(t_{n,r-1}))| \geqq g^2(2^{-n}).$$

By the inequality (1) of the Theorem we then have

$$P(B^*_{n,r}) \leqq q(2^{-n}).$$

Further

$$C^*_n = \sum_{j=n}^{\infty} \sum_{r=1}^{2^j-1} B^*_{j,r},$$

$$P(C^*_n) \leqq \sum_{j=n}^{\infty} \sum_{r=1}^{2^j-1} P(B^*_{j,r}) \leqq \sum_{j=n}^{\infty} 2^j q(2^{-j}),$$

$$C^* = \prod_1^{\infty} C^*_n = \lim_{n\to\infty} C^*_n,$$

$$P(C^*) = \lim_{n\to\infty} P(C^*_n) = 0,$$

$$P(C) = 1.$$

and the Lemma is proved.

For the last Lemma, and the subsequent proof of the Theorem, we shall explicitly consider the given stochastic process as a family of random variables $\xi(t) = \xi(t, \omega)$, where ω is the elementary event of the basic probability space. For a fixed ω we obtain a sample function $\xi(t)$ of the process, and we denote by $\xi_D(t) = \xi_D(t, \omega)$ the restriction of this function to the t set D of dyadic numbers.

For every sample function $\xi(t) \in C$, the limits $\xi_D(t \pm 0)$ exist for all t by Lemma 3. For every sample function $\xi(t)$ not belonging to C, we define $\xi_D(t-0) = \xi_D(t+0) = 0$ for all t.

It then follows from Lemma 5 that, for every fixed t, the limits $\xi_D(t \pm 0) = \xi_D(t \pm 0, \omega)$ exist as limits with probability 1 of random variables; hence they are themselves random variables.

LEMMA 6. - *Let $\xi(t)$ be a stochastic process satisfing the conditions of the Theorem. For every fixed $t \in [0, 1]$, the probability that the random variable $\xi(t)$ differs from both variables $\xi_D(t \pm 0)$ is zero.*

In the inequality (1) of the Theorem, we take $t_1 = t_{n, r-1}$, $t_2 = t$, and $t_3 = t_{n, r+1}$, choosing r so that $t_1 < t_2 < t_3$. We then have $t_3 - t_1 = h = 2^{-n+1}$. As n tends to infinity, we then have with probability 1

$$\xi(t_1) \to \xi(t_D - 0) \quad \text{and} \quad \xi(t_3) \to \xi_D(t+0),$$

and it now follows from (1) that

$$P\{(\xi(t) - \xi_D(t-0)) \cdot (\xi(t) - \xi_D(t+0) \neq 0\} = 0,$$

so that the Lemma is proved.

PROOF OF THE THEOREM. - For every fixed $t \in [0, 1]$, we now denote by U_t, V_t, and W_t the ω sets (events) defined respectively by the relations

$$
\begin{aligned}
&U_t\colon \xi(t) = \xi_D(t-0), \\
(6) \qquad &V_t\colon \xi(t) = \xi_D(t+0) \neq \xi_D(t-0), \\
&W_t\colon \xi(t) \neq \xi_D(t-0) \quad \text{and} \quad \xi(t) \neq \xi_D(t+0).
\end{aligned}
$$

Then U_t, V_t and W_t are disjoint sets, and we have by Lemma 6

$$(7) \qquad P(U_t + V_t) = 1, \qquad P(W_t) = 0.$$

For every t and ω we now define

$$\eta(t, \omega) = \begin{cases} \xi_D(t-0, \omega) & \text{for} \quad \omega \in U_t, \\ \xi_D(t+0, \omega) & \text{for} \quad \omega \in V_t + W_t. \end{cases}$$

It then follows from (6) and (7) that we have for the random variable $\eta(t) = \eta(t, \omega)$

$$P\{\xi(t) = \eta(t)\} = 1 \qquad \text{for every} \qquad t \in [0, 1],$$

so that the processes $\xi(t)$ and $\eta(t)$ are equivalent.

On the other hand, consider any fixed sample function $\xi(t) \in C$. We then have by definition for every $t \in [0, 1]$

$$\eta(t) = \xi_D(t-0) \quad \text{or} \quad \eta(t) = \xi_D(t+0),$$

and it follows from Lemma 4 that $\eta(t)$ has at most discontinuities of the first kind. However, by Lemma 5 the sample functions $\xi(t)$ not belonging to C, and thus also the corresponding functions $\eta(t)$, form an ω null set, so that the Theorem is proved.

By a straightforward application of the TCHEBYCHEV inequality, we obtain the following simple Corollary, which is included in a Theorem proved by CHENTSOV [2].

COROLLARY. - *If, with the notations of the Theorem, we have*

$$E|[\xi(t_3) - \xi(t_2)]\cdot[\xi(t_2) - \xi(t_1)]|^p \leqq Kh^{1+r},$$

where p, r and K are positive constants the conclusion of the Theorem holds.

In particular, the conditions of the Corollary are satisfied for any process $\xi(t)$ with independent increments such that for all t, $t+h \in [0, 1]$ we have

$$E[\xi(t+h) - \xi(t)]^2 < A\,|h|$$

with constant A. This follows by taking $p = 2$ and $r = 1$. This applies, in particular, to the POISSON process.

BIBLIOGRAPHY

[1] YU. K. BELAJEV, *Continuity and Hölder's conditions for sample functions of stationary Gaussian processes*, Proc. Fourth Berkeley Symp, 2 (1961), 23-33.

[2] N. N. CHENTSOV, *Weak convergence of stochastic processes whose trajectories have no discontinuities of the second kind and the heuristic approach to the Kolmogorov-Smirnoff tests*, Teor. Verojatnost. i Primenen., 1 (1956), 140-144.

[3] H. CRAMÉR, *Mathematical Methods of Statistics*, Princeton Univ. Press., 1946.

[4] R. L. DOBRUSCHIN, *The continuity condition for sample functions of a martingale*, Teor. Verojatnost. i Primenen, 3 (1958), 92-93.

[5] E. B. DYNKIN, *Criteria of continuity and of absence of discontinuities of the second kind for trajectories of a Markov random process*, Izv. Akad. Nauk SSSR, 16 (1952), 563-572.

[6] G. A. HUNT, *Random Fourier Transforms*, Trans. Amer. Math. Soc., 71 (1951), 38-69.

[7] M. LOÈVE, *Probability Theory*. 2nd Ed., Van Nostrand, Princeton, 1960.

[8] E. SLUTSKY, *Qualche proposizione relativa alla teoria delle funzioni aleatorie*, Giornale dell'Ist. Italiano degli Attuari, 8 (1937), 193-199.

62.

On extreme values of certain stochastic processes

Research Papers in Statistics
Festschrift for J. Neyman, ed. by F. N. David
Wiley, London New York Sydney, 73–78 (1966)

Dedicated to Professor Jerzy Neyman

1 Let $x(t)$ be the real-valued random variables of a separable and normal (Gaussian) stochastic process, defined for all real values of the parameter t. Questions relative to the extreme values of $x(t)$ in some finite t-interval, say $0 \leqslant t \leqslant T$, are important in various fields of application.

Under appropriate continuity conditions, the maximum

$$\max_{0 \leqslant t \leqslant T} x(t)$$

will be a well-defined random variable. For the particular case when $x(t)$ is stationary, fairly precise results are known with respect to the asymptotic behaviour of the probability distribution of this maximum value, as T tends to infinity. (Cf. Cramér (1962, 1965) and, for the case of a process with discrete time, Berman (1962, 1964).)

In this paper, we propose to deal with a more general class of normal stochastic processes, which contains the normal stationary processes as a subclass. It will be shown that for this class it is possible to reach at least some corresponding results, although considerably less precise than in the stationary case.

Let $x(t)$ be real, separable, and normal, but not necessarily stationary. Suppose that $x(t)$ has zero mean value for all t, and covariance function $r(t, u)$:

$$Ex(t) = 0, \qquad Ex(t)x(u) = r(t, u)$$

Suppose further that

$$\sigma^2(t) = Ex^2(t) = r(t, t) > 0$$

for all t, and write

$$\rho(t, u) = \frac{r(t, u)}{\sigma(t)\sigma(u)}$$

The mixed derivative

$$r_{11}(t, u) = \frac{\partial^2 r}{\partial t\, \partial u}$$

is assumed to exist everywhere, and to be continuous in every diagonal point $t = u$. For the first order partial derivatives we use the corresponding notations r_{10} and r_{01}.

It then follows [cf. Loève, 1963 : p. 520] that, with probability one, the sample functions of $x(t)$ are continuous in every finite interval. If sufficiently strong regularity conditions are imposed on $r_{11}(t, u)$, it even follows [Loève, *loc. cit.*] that, with probability one, there will exist an everywhere continuous sample derivative $x'(t)$. We shall here assume that this is the case. Then $Ex'(t) = 0$, and we introduce the notations

$$\gamma^2(t) = Ex'^2(t) = r_{11}(t, t)$$

$$\mu(t) = \frac{Ex'(t)x(t)}{\gamma(t)\sigma(t)} = \frac{r_{10}(t, t)}{\gamma(t)\sigma(t)}$$

$$\omega(t) = \frac{\gamma(t)}{\sigma(t)} \sqrt{1 - \mu^2(t)}$$

Under the conditions thus laid down, the maximum

$$M(T) = \max_{0 \leqslant t \leqslant T} \frac{x(t)}{\sigma(t)}$$

will be a well-defined random variable. We propose to study the asymptotic behaviour of the probability distribution of $M(T)$, as T tends to infinity.

If $N_c(T)$ denotes the total number of crossings in the interval $0 \leqslant t \leqslant T$ between the normalized trajectory of the process

$$y = \frac{x(t)}{\sigma(t)}$$

and the level $y = c$, it follows from the work of Leadbetter and Cryer (1964) that we have in the present notations

$$EN_c(T) = \frac{1}{\pi} e^{-c^2/2} \int_0^T \omega(t)\, dt \tag{1}$$

In the particular case of a stationary process, assuming without restriction of generality $Ex^2(t) = 1$, we have

$$\sigma(t) = 1, \quad \gamma(t) = \sqrt{\lambda_2}, \quad \mu(t) = 0, \quad \omega(t) = \sqrt{\lambda_2}$$

where λ_2 is the second-order spectral moment. The expression (1) then reduces to the well-known formula

$$EN_c(T) = \frac{\sqrt{\lambda_2}}{\pi} Te^{-c^2/2}$$

proved to hold under very general conditions by Bulinskaja (1961).

Returning to the general case, suppose that we have for some $\alpha > 0$

$$\omega(t) = O(t^{\alpha-1}) \tag{2}$$

as $t \to \infty$, and take

$$c = (2\alpha \log T)^{1/2} + h(T)/(\log T)^{1/2}$$

with any $h(T) \to \infty$, arbitrarily slowly, as $T \to \infty$. (In the stationary case, we may evidently take $\alpha = 1$.) We then have

$$EN_c(T) = O[T^\alpha \exp(-\alpha \log T - \tfrac{1}{2}\sqrt{2\alpha}\, h(T))] \to 0$$

as $T \to \infty$. It follows that

$$P\{M(T) > c\} = P\{N_c(T) \geqslant 1\} + o(1)$$
$$\leqslant EN_c(T) + o(1) \to 0$$

and hence

$$P\{M(T) > \sqrt{2\alpha \log T} + h(T)/\sqrt{\log T}\} \to 0$$

or

$$P\{M(T) \leqslant \sqrt{2\alpha \log T} + h(T)/\sqrt{\log T}\} \to 1 \tag{3}$$

as $T \to \infty$. Thus we have obtained an *upper limit* for the maximum value $M(T)$, valid with a probability tending to one as $T \to \infty$.

2 It seems considerably more difficult to obtain a *lower limit* for $M(T)$, and we shall here only indicate some results in this direction, which, however, can so far only be proved under rather more restrictive conditions.

We shall now have to replace condition (2) by the more restrictive condition

$$kt^{\alpha-1} < \omega(t) < Kt^{\alpha-1} \tag{4}$$

for all sufficiently large t, with $0 < k < K$, and $0 < \alpha \leqslant 1$. (We have already noted that $\alpha = 1$ in the stationary case.) Writing

$$y(t) = \begin{cases} 1 & \text{for} \quad x(t) > c\sigma(t) \\ 0 & \text{for} \quad x(t) \leqslant c\sigma(t) \end{cases}$$

$$z(T) = \int_0^T y(t)\, dt$$

it will be seen that $z(T)$ is the measure of the set of points t in $(0, T)$ such that $x(t) > c\sigma(t)$. This set is a finite set of intervals, and by its measure we simply understand the total length of these intervals. We then have [cf. Cramér, 1962]

$$Ez(T) = \frac{1}{\sqrt{2\pi}} T \int_c^\infty e^{-t^2/2}\, dt$$

$$\operatorname{Var} z(T) = \frac{1}{2\pi} \int_0^T \int_0^T dt\, du \int_0^{\rho(t,u)} e^{-c^2/(1+v)}(1 - v^2)^{-1/2}\, dv$$

and

$$P\{M(T) \leqslant c\} = P\{z(T) = 0\} \leqslant \frac{\operatorname{Var} z(T)}{[Ez(T)]^2}$$

Hence we obtain without difficulty

$$P\{M(T) \leqslant c\} < \frac{c^2}{T^2} \int_0^T \int_0^T |\rho(t, u)| \exp \frac{c^2|\rho(t, u)|}{1 + |\rho(t, u)|}\, dt\, du \tag{5}$$

We now want to show that, if c is appropriately fixed as a function of T, the second member of the last inequality tends to zero as $T \to \infty$.

In order to evaluate the integral, we divide the domain of integration in the two parts defined respectively by the inequalities $|t - u| \geqslant \varepsilon$ and $|t - u| < \varepsilon$, for some $\varepsilon > 0$ to be fixed later. The first part is easily evaluated, if we suppose that for every $\varepsilon > 0$ we can find $A > 1$ and $B > 0$ such that for $|t - u| \geqslant \varepsilon$ we have

$$|\rho(t, u)| < (A + B|t - u|)^{-1}$$

and we shall not consider this part further.

The critical part of the integral is the part with $|t - u| < \varepsilon$, which lies along the diagonal of the square of integration. In every diagonal point (t, t), the correlation function ρ takes the value

$\rho(t, t) = 1$. When the variable point (u, v) tends to the diagonal point (t, t), it can be shown that we have

$$1 - \rho(u, v) \sim \tfrac{1}{2}\omega^2(t)(u - v)^2$$

Let us now suppose that this holds uniformly, in the sense that $\varepsilon > 0$ can be found such that for $|u - t| < \varepsilon$, $|v - t| < \varepsilon$ and for all sufficiently large t, we have

$$1 - \rho(u, v) > m\omega^2(t)(u - v)^2$$

with a constant $m > 0$. Then it can be shown that the second member of (5) is bounded by a constant multiple of

$$\frac{c}{T^2} e^{c^2/2} \int_0^T \frac{dt}{\omega(t)} + \frac{c^2 \log T}{T} \exp \frac{c^2}{2A}$$

the first term arising from the diagonal part of the domain of integration. Now we obtain from (4)

$$\int_0^T \frac{dt}{\omega(t)} = O(T^{2-\alpha})$$

and taking

$$c = \sqrt{2\alpha \log T} - \frac{\log \log T}{\sqrt{2\alpha \log T}}$$

it finally follows from (5) that we have, as $T \to \infty$,

$$P\left\{M(T) \leqslant \sqrt{2\alpha \log T} - \frac{\log \log T}{\sqrt{2\alpha \log T}}\right\} \to 0$$

or

$$P\left\{M(T) > \sqrt{2\alpha \log T} - \frac{\log \log T}{\sqrt{2\alpha \log T}}\right\} \to 1 \tag{6}$$

The two relations (3) and (6) together give a fairly precise idea of the probable order of magnitude of the maximum value $M(T)$ for large values of T. Unfortunately, it has so far only been possible to prove relation (6) under rather more restrictive conditions than relation (3).

As an example we may consider the case when we have

$$x(t) = \int_0^t y(u)\, du$$

where $y(u)$ is normal and stationary, with mean $Ey(u) = 0$ and variance $Ey^2(u) = 1$, while the integral may be defined as an integral in quadratic mean. It is then easily shown that the conditions for the validity of the upper limit relation (3) for the maximum $M(T)$ will be satisfied if we take $\alpha = \frac{1}{2}$. On the other hand, the methods used in the second part of this paper do not seem to enable us to prove the validity of the lower limit relation (6).

REFERENCES

Berman, S. M. (1962). A law of large numbers for the maximum in a stationary Gaussian sequence. *Ann. Math. Statist.*, **33**, 93–97.

Berman, S. M. (1964). Limit theorems for the maximum term in stationary sequences. *Ann. Math. Statist.*, **35**, 502–16.

Bulinskaja, E. V. (1961). On the mean number of crossings of a level by a stationary Gaussian process. *Teor. Veroj. i Primenen*, **6**, 474–77.

Cramér, H. (1962). On the maximum of a normal stationary stochastic process. *Bull. Amer. Math. Soc.*, **68**, 512–16.

Cramér, H. (1965). A limit theorem for the maximum values of certain stochastic processes. *Teor. Veroj. i Primenen*, **10**, 137–39.

Leadbetter, M. R. and Cryer, J. D. (1964). On the mean number of curve crossings by non-stationary normal processes. *Res. Triangle Inst.*, Techn. Report 1.

Loève, M. (1963). *Probability theory*, 3rd ed. Van Nostrand, Princeton, N.J.

63.

A contribution to the multiplicity theory of stochastic processes

Proc. Fifth Berkeley Symp. Math. Stat. Prob., Vol. II, Part 1
University of California Press, Berkeley and Los Angeles, 215–221 (1967)
(Reprinted in Random Processes)

1. Introduction

In a paper [1] read before the Fourth Berkeley Symposium in 1960, I communicated the elements of a theory of spectral multiplicity for stochastic processes. A related theory was given about the same time by Hida [5]. Since then, I have developed the theory in some subsequent papers [2]–[4], the most recent of which contains the text of a lecture given at the Seventh All-Soviet Conference of Probability and Mathematical Statistics in Tbilisi 1963. Further important work in the field has been made by Kallianpur and Mandrekar [6]–[8].

Many interesting problems arising in connection with this theory are still unsolved. The object of this paper is to offer a small contribution to the investigation of one of these problems.

We shall begin by giving in section 2 a brief survey of the results of multiplicity theory so far known for the simplest case of one-dimensional processes. For proofs and further developments we refer to the papers quoted above. A major unsolved problem will be discussed in section 3, whereas section 4 is concerned with some aspects of the well-known particular class of stationary processes, which are relevant for our purpose. Finally, section 5 is concerned with the construction of a class of examples which may be useful in the further study of the problem stated in section 3.

2. Spectral multiplicity of stochastic processes

Consider a stochastic process $x(t)$, where $x(t)$ is a complex-valued random variable defined on a fixed probability space, while t is a real-valued parameter. In general we shall allow t to take any real values, and shall only occasionally consider the case when t is restricted to the integers. We shall always assume that the relations

$$Ex(t) = 0, \qquad E|x(t)|^2 < \infty \tag{2.1}$$

are satisfied for all t.

We denote by $H(x)$ the Hilbert space spanned in a well-known way by the random variables $x(t)$ for all t, while $H(x, t)$ is the subspace of $H(x)$ spanned only by the $x(u)$ with $u \leq t$. The tail space $H(x, -\infty)$ may be regarded as representing the "infinitely remote past" of the process. If $H(x, -\infty)$ only contains the zero element of $H(x)$, the $x(t)$ process is said to be purely nondeterministic.

All processes $x(t)$ considered in the sequel will be assumed to satisfy the following conditions (A) and (B):

(A) the process is purely nondeterministic;

(B) the limits in quadratic mean $x(t + 0)$ and $x(t - 0) = x(t)$ exist for every t. Under these conditions the space $H(x)$ will be separable. We note that, in the case of a parameter t taking only integral values, condition (B) is irrelevant.

Let us now for a moment consider the case when t is restricted to the integers, so that we are concerned with a sequence of random variables x_n, with $n = 0, \pm 1, \cdots$. Then there exists a sequence of mutually orthogonal random variables z_n with

$$\begin{aligned} Ez_n &= 0, \qquad E|z_n|^2 = 1 \text{ or } 0 \text{ for every } n, \\ Ez_m\overline{z_n} &= 0 \qquad \text{for} \quad m \neq n, \end{aligned} \tag{2.2}$$

such that

$$x_n = \sum_{k=-\infty}^{n} c_{nk} z_k, \tag{2.3}$$

where the series

$$\sum_{k=-\infty}^{n} |c_{nk}|^2 \tag{2.4}$$

converges for every n, so that the expression for x_n converges in quadratic mean. The variable z_n may then be regarded as a (normalized) *innovation* entering into the process at time $t = n$.

By analogy, we might expect to have in the case of a continuous paramter t a representation of the form

$$x(t) = \int_{-\infty}^{t} g(t, u)\, dz(u) \tag{2.5}$$

where $z(u)$ would be a process with orthogonal increments, the increment $dz(u)$ representing the *innovation element* entering into the $x(t)$ process during the time element $(u, u + du)$.

However, in general this is not true. The situation in the continuous case turns out to be more complicated than in the discrete case. In general the innovation associated with a given time element must be regarded as a multi-dimensional or even infinite-dimensional random variable, so that the representation (2.5) is definitely too simple.

In order to present the representation formula which in the general case takes the place of (2.5), we must first consider the class C of all real-valued and never decreasing, not identically constant functions $F(t)$ which are continuous to the left for all t. A subclass D of C is called an *equivalence class* if any two functions F_1 and F_2 in D are mutually absolutely continuous. If D_1 and D_2 are equivalence classes, D_1 is said to be *superior* to D_2, and we write $D_1 > D_2$, if any $F_2 \in D_2$ is absolutely continuous relative to any $F_1 \in D_1$. Evidently, the relation $D_1 > D_2$ does not exclude the case that the two classes are identical.

Consider now a finite or infinite never increasing sequence of equivalence classes

$$D_1 > D_2 > \cdots > D_N, \tag{2.6}$$

where N may have any of the values $1, 2, \cdots, \infty$. Then N will be called the *total multiplicity* of the sequence. Further, let $N(t)$ for every t denote the number of those classes in (2.6) for which t is a point of increase of the corresponding functions F. Then $N(t)$ is called the *multiplicity function* of the sequence (2.6). Like N, $N(t)$ may be finite or infinite, and we have

$$N = \sup N(t), \tag{2.7}$$

where t runs through all real values.

The fundamental proposition of multiplicity theory for stochastic processes is the following. To any $x(t)$ stochastic process satisfying conditions (A) and (B), there is a uniquely determined sequence of the form (2.6) such that the following properties hold. For every $n = 1, 2, \cdots, N$, there is a process $z_n(t)$ of orthogonal increments, such that

$$\begin{aligned} &Ez_n(t) = 0, \qquad E|z_n(t)|^2 = F_n(t) \in D_n, \\ &Ez_m(t)\overline{z_n(u)} = 0 \qquad \text{for } m \neq n \text{ and all } t, u, \\ &H(x, t) = \sum_1^N H(z_n, t) \qquad \text{for all } t, \end{aligned} \tag{2.8}$$

where the last sum denotes the vector sum of the orthogonal subspaces $H(z_n, t)$. We then have for every t the representation

$$x(t) = \sum_1^N \int_{-\infty}^t g_n(t, u)\, dz_n(u), \tag{2.9}$$

where the g_n are nonrandom functions such that

$$\sum_1^N \int_{-\infty}^t |g_n(t, u)|^2\, dF_n(u) < \infty. \tag{2.10}$$

It is important to observe that the D_n sequence (2.6) is uniquely determined by the $x(t)$ process. Thus, in particular, the multiplicity function $N(t)$ and the total multiplicity N are also uniquely determined by $x(t)$. Accordingly, we shall say that the D_n sequence, as well as $N(t)$ and N, are spectral multiplicity characteristics of the stochastic process $x(t)$.

On the other hand, the $g_n(t, u)$ and $z_n(u)$ occurring in the representation (2.9) are not uniquely determined by the $x(t)$ process. Thus, for a given $x(t)$ we may have different representations of the form (2.9), all satisfying the relations (2.8). However, the D_n sequence (2.6), as well as the multiplicity characteristics $N(t)$ and N, will be identical for all these representations.

According to the representation (2.9), we may say that the multiplicity function $N(t)$ determines the dimensionality of the innovation element $[dz_1(u), dz_2(u), \cdots]$ entering into the process during the time element $(u, u + du)$.

It has been shown that the multiplicity characteristics of a given stochastic process $x(t)$ are uniquely determined by the covariance function of the process

$$r(t, u) = Ex(t)\overline{x(u)}. \tag{2.11}$$

We finally remark that the multiplicity theory as outlined above can be directly generalized to stochastic vector processes of a very general kind. We shall, however, not deal with these generalizations in the present paper.

3. Processes of total multiplicity $N = 1$

According to the above, we know that any given stochastic process $x(t)$ satisfying (A) and (B) has multiplicity characteristics which are uniquely determined by the process, and even by the covariance function $r(t, u)$ of the process.

On the other hand, so far we know very little about those properties of the process, or of the corresponding covariance function, which determine the actual values of multiplicity characteristics like $N(t)$ and N.

In the discrete case it follows from the above that, by analogy, it can be said that the total multiplicity is always $N = 1$. In the continuous case, the important class of (second-order) stationary processes has even $N(t) = 1$ for all t, and consequently $N = 1$, as follows from well-known properties of these processes to be presently recalled.

In view of these examples, it might well be asked if there exist any stochastic processes with a total multiplicity exceeding unity. The answer to this question is that such processes do, in fact, exist. It can even be shown that, as soon as we proceed from the class of stationary processes to the more general class of harmonizable processes introduced by Loève, any prescribed multiplicity properties may occur. In fact, it has been shown in [4] that, given any D_n sequence (2.6), there exists a harmonizable process $x(t)$ associated with this given D_n sequence. However, the example of such a process given in [4] is of a very special kind, and the corresponding representation (2.9) contains functions $g_n(t, u)$ having rather pathological properties, not likely to occur in applications to any physical problems.

Accordingly, it seems to be a problem of some interest to study more closely those properties of a stochastic process which determine the actual values of the multiplicity characteristics. In particular, it would be interesting to be able to define some fairly general class of processes having total multiplicity $N = 1$.

A natural approach to this last problem might be to start from the class of stationary processes, which always have $N = 1$, and then try to generalize the definition, still keeping sufficiently near the property of stationarity to conserve the multiplicity characteristic $N = 1$. We propose to give in the sequel an example of a generalization of this type. In order to do this, we must first recall some of the relevant properties of stationary processes.

4. Stationary processes

Let $x(t)$ be a (second-order) stationary process, satisfying (A) and (B). It then follows that the covariance function

$$r(t) = Ex(t+h)\overline{x(h)} \tag{4.1}$$

is everywhere continuous, and has the spectral representation

$$r(t) = \int_{-\infty}^{\infty} e^{it\lambda} f(\lambda)\, d\lambda \tag{4.2}$$

with a spectral density $f(\lambda) > 0$ for almost all λ (Lebesgue measure), such that $f(\lambda) \in L_1(-\infty, \infty)$, and

$$\int_{-\infty}^{\infty} \frac{\log f(\lambda)}{1+\lambda^2}\, d\lambda > -\infty. \tag{4.3}$$

The random variable $x(t)$ has the corresponding spectral representation

$$x(t) = \int_{-\infty}^{\infty} e^{it\lambda}\, dw(\lambda), \tag{4.4}$$

where $w(\lambda)$ is a process with orthogonal increments such that

$$E\, dw(\lambda) = 0, \qquad E|\, dw(\lambda)\,|^2 = f(\lambda)\, d\lambda. \tag{4.5}$$

Further, there exists a complex-valued function $h(\lambda) \in L_2(-\infty, \infty)$ and a process $z(t)$ of orthogonal increments such that

$$E\, dz(t) = 0, \qquad E|\, dz(t)\,|^2 = dt, \qquad |h(\lambda)|^2 = f(\lambda), \tag{4.6}$$

while the Fourier transform $g(t)$ of $h(\lambda)$ reduces to zero for $t < 0$, and we have the representation

$$x(t) = \int_{-\infty}^{t} g(t-u)\, dz(u) \tag{4.7}$$

with

$$H(x, t) = H(z, t) \tag{4.8}$$

for all t. The functions $h(\lambda)$ and $g(t)$ are uniquely determined, up to a constant factor of absolute value 1. Comparing this with the general representation formula (2.9), it is seen that the stationary process $x(t)$ has the multiplicity characteristics $N = 1$ and $N(t) = 1$ for all t.

5. A class of harmonizable processes with $N = 1$

We shall now define a class of harmonizable processes containing the stationary process $x(t)$ given by (4.4) or (4.7) as a particular case, and such that the multiplicity characteristics are the same as for $x(t)$, that is $N = 1$ and $N(t) = 1$ for all t.

Let $Q(\rho)$ be a never-decreasing function of the real variable ρ such that Q has a jump of size 1 at $\rho = 0$, whereas $Q(-\infty) = 0$, $Q(+\infty) < 2$, and

$$Q(\rho) + Q(-\rho) = Q(+\infty) \tag{5.1}$$

in all continuity points ρ of Q. The Fourier-Stieltjes transform of Q will then be real and positive, so that we may define an everywhere positive, continuous and bounded function $q(u)$ by the relation

$$[q(u)]^2 = \int_{-\infty}^{\infty} e^{-i\rho u}\, dQ(\rho). \tag{5.2}$$

We now define a stochastic process $X(t)$ by writing

$$X(t) = \int_{-\infty}^{t} g(t-u)q(u)\, dz(u), \tag{5.3}$$

where $g(t)$ and $z(u)$ are the same as in (4.7). As $q(u)$ is bounded, and $g(t) \in L_2(0, \infty)$, the integral in (5.3) exists as a quadratic mean integral. When Q is identically constant except for the jump at $\rho = 0$, it is seen that $X(t)$ reduces to the stationary process $x(t)$ given by (4.7).

We shall now first show that $X(t)$ has the required multiplicity characteristics. According to (2.8) and (2.9), we have to show that $H(X, t) = H(z, t)$ for all t. As it evidently follows from (5.3) that $H(X, t) \subset H(z, t)$, it will be sufficient to show that the opposite inclusion relation is also true. If, for some t, this were not so, there would be a nonzero element in $H(z, t)$ orthogonal to $X(u)$ for all $u \leq t$. Now every nonzero element in $H(z, t)$ is of the form

$$\int_{-\infty}^{t} m(v)\, dz(v), \tag{5.4}$$

with a quadratically integrable $m(v)$ not almost everywhere equal to zero. If this is orthogonal to $X(u)$ for $u \leq t$, we have

$$\int_{-\infty}^{u} g(u-v)q(v)\overline{m(v)}\, dv = 0 \tag{5.5}$$

for all $u \leq t$. However, since $q(v)$ is bounded and positive, it would follow that there is a nonzero element in $H(z, t)$ orthogonal to $x(u)$ for all $u \leq t$, in contradiction with the relation (4.8). Thus our assertion is proved.

We now proceed to prove that $X(t)$ as defined by (5.3) is a harmonizable process, and to deduce an expression for its spectral distribution. From (5.3) we obtain for the covariance function $R(s, t)$ of $X(t)$ the expression

$$\begin{aligned} R(s, t) = EX(s)\overline{X(t)} &= \int_{-\infty}^{\infty} g(s-u)\overline{g(t-u)}\, [q(u)]^2\, du \\ &= \int_{-\infty}^{\infty} g(s-u)\, \overline{g(t-u)}\, du \int_{-\infty}^{\infty} e^{-i\rho u}\, dQ(\rho). \end{aligned} \tag{5.6}$$

As $g(t) = 0$ for $t < 0$, and $g(t) \in L_2(0, \infty)$, it follows that the double integral is absolutely convergent, so that

$$R(s, t) = \int_{-\infty}^{\infty} dQ(\rho) \int_{-\infty}^{\infty} e^{-i\rho u} g(s-u)\overline{g(t-u)}\, du. \tag{5.7}$$

By the Parseval formula, this gives

$$R(s, t) = \int_{-\infty}^{\infty} \int_{-\infty}^{\infty} e^{i[s\lambda - t(\lambda+\rho)]} h(\lambda)\overline{h(\lambda+\rho)}\, d\lambda\, dQ(\rho). \tag{5.8}$$

Substituting here μ for $\lambda + \rho$, it will be seen that this is the expression of a harmonizable covariance function. The corresponding spectral mass is distributed over the (λ, μ)-plane so that the infinitesimal strip between the lines $\mu = \lambda + \rho$ and $\mu = \lambda + \rho + d\rho$ contains the mass $dQ(\rho)$, whereas the distribution within the strip has the relative density $h(\lambda)\overline{h(\mu)}$. Again we see that, in the particular case when $Q(\rho)$ is identically constant except for the jump at $\rho = 0$, the whole spectral mass is situated on the diagonal $\lambda = \mu$, so that we have the covariance function of a stationary process with spectral density $|h(\lambda)|^2 = f(\lambda)$. As soon as $Q(\rho)$ has some variation outside the point $\rho = 0$, we have the two-dimensional spectral distribution of a harmonizable covariance.

Thus the covariance function of the $X(t)$ process, given by (5.3), is harmonizable, and it then follows from known properties of harmonizable processes that $X(t)$ itself is harmonizable; that is, we have

$$X(t) = \int_{-\infty}^{\infty} e^{it\lambda}\, dZ(\lambda), \tag{5.9}$$

where the covariance function $EZ(\lambda)\overline{Z(\mu)}$ is obtained from the expression (5.8) with $\mu = \lambda + \rho$. At the same time, we have seen that the harmonizable process $X(t)$ has the multiplicity characteristics $N = 1$ and $N(t) = 1$ for all t.

REFERENCES

[1] H. Cramér, "On some classes of non-stationary stochastic processes," *Proceedings of the Fourth Berkeley Symposium on Mathematical Statistics and Probability*, Berkeley and Los Angeles, University of California Press, 1960, Vol. II, pp. 57–78.

[2] ———, "On the structure of purely non-deterministic stochastic processes," *Ark. Mat.*, Vol. 4 (1961), pp. 249–266.

[3] ———, "Décompositions orthogonales de certains processus stochastiques," *Ann. Fac. Sci., Clermont*, Vol. 8 (1962), pp. 15–21.

[4] ———, "Stochastic processes as curves in Hilbert space," *Teor. Verojatnost. i Primenen.*, Vol. 9 (1964), pp. 195–204.

[5] T. Hida, "Canonical representations of Gaussian processes and their applications," *Mem. Coll. Sci., Kyoto, Ser. A*, Vol. 33 (1960), pp. 109–155.

[6] G. Kallianpur and V. Mandrekar, "On the connection between multiplicity theory and Hanner's time domain analysis of weakly stationary stochastic processes," Department of Statistics, University of Minnesota, Technical Report 49 (1964).

[7] ———, "Multiplicity and representation theory of purely non-deterministic stochastic processes," Department of Statistics, University of Minnesota, Technical Report 51 (1964).

[8] ———, "Semi-groups of isometries and the representation and multiplicity of weakly stationary stochastic processes," to appear in *Ark. Mat.* (1966).

64.

Historical review of Filip Lundberg's works on risk theory

Skand. Aktuarietidskr. **52**, Supplement (3–4), 6–12 (1969)

1. Filip Lundberg's works on risk theory were all written at a time when no general theory of stochastic processes existed, and when collective reinsurance methods, in the present day sense of the word, were entirely unknown to insurance companies. In both respects his ideas were far ahead of his time, and his works deserve to be generally recognized as pioneering works of fundamental importance.

In this introductory chapter, an attempt will be made to review very briefly some of his leading ideas, particularly in his earlier works, and to indicate their proper place in the historical development. We propose to deal mainly with the mathematical aspect of the works, and then add some remarks concerning the applications to insurance practice.

2. During the first ten years of the present century several works were published which, to a modern reader, appear as forerunners of the general theory of stochastic processes created during the 1930's. All these works deal with the temporal development of some variable system exposed to random influences.

Bachelier, in a paper of 1900, analyzed the fluctuations of stock market prices, and was led to an important particular case of stochastic process which, in 1905, was encountered by Einstein in his famous study of the Brownian movement. This process is designed to give a mathematical description of the continuous, although extremely irregular, path of a particle suspended in a fluid and subject to random molecular shocks.

Between these works of Bachelier and Einstein, the Uppsala University Thesis of Filip Lundberg appeared in 1903. This Thesis has always had a firmly established reputation for being impossible to understand. However, when one looks at the Thesis to-day, in the light of the development that has taken place since its publication, much of the darkness seems to disappear, and one cannot help being struck by his ability to deal intuitively with concepts and methods that would have to wait another thirty years before being put on a rigorous foundation.

In the first part of the Thesis, Lundberg studies the essentially discontinuous variations of the accumulated amount of claims incurred by an insurance company. He is thus led to a stochastic process of a type entirely different from the

Skand. AktuarTidskr. Suppl. 1969

continuous Brownian movement process of Bachelier and Einstein. The "Poisson process", well known to-day for its numerous applications, comes out as a very special case of the process studied by Lundberg, arising when all sums at risk are equal. It is explicitly pointed out in the Thesis that the Poisson formula associated with this special case is simply obtained from his general theory.

Several years after Lundberg, in 1909, the Poisson process was introduced by Erlang in the study of telephone traffic problems, and it has since proved to be an indispensable tool in various branches of communication engineering. About the same time, the Poisson process was used by Rutherford and Geiger in connection with their experiments with the disintegration of radioactive material.

It was not until the 1930's that the foundation of a general and rigorous theory of stochastic processes were laid by Kolmogorov, Khintchine, Lévy, Feller, and other authors. In this connection it became clear that the Brownian movement process and the Lundberg risk process, introduced many years earlier by the pioneers, form the main building stones of an important general class of stochastic process, known as processes with independent increments.

3. The risk process studied by Lundberg in his Thesis of 1903, his Congress paper of 1909, and the paper on "Risk Masses" of 1919, is concerned with the analysis of the case when claims arise according to a Poisson process, while the amounts due under the successive claims vary in a random way, and are mutually independent. The three mentioned papers cover the general case when the distribution of the sums at risk may change with time, while the great work "Försäkringsteknisk riskutjämning" of 1926–28 deals with the important particular case of a stationary distribution of risk sums.

The chief tool used by Lundberg for the analysis of this process is a functional equation which, to a modern eye, reveals itself as a particular case of the famous "forward equation" introduced in 1931 by Kolmogorov for a general class of continuous processes, and in 1940 by Feller for the corresponding class of discontinuous processes, which includes the Lundberg process as a special case. We shall give some comments on the deduction and the use of this fundamental equation. It is hardly necessary to point out that the methods used by Lundberg do not satisfy modern standards of rigour; the remarkable thing is that his intuitive methods led him to correct and valuable results.

Let $Y(t)$ denote the accumulated amount of claims due up to the (operational) time t, while $G(y, t)$ is the distribution function (d.f.) of the random variable $Y(t)$. Further, let $P(y, t)$ denote the d.f. of the amount due under a claim arising at time t. The "forward equation" of the $Y(t)$ process is obtained by considering the events that may happen during an infinitesimal time element $(t, t+dt)$:

$$G(y,t+dt) = (1-dt)\,G(y,t) + dt\int G(y-z,t)\,dP(z,t),$$

where the integration with respect to z has to be extended over $(-\infty, \infty)$ if risk sums of both signs are possible. This gives, subject to appropriate regularity conditions,

$$\frac{\partial G(y,t)}{\partial t} = \int_{-\infty}^{\infty} [G(y-z,t) - G(y,t)]\, dP(z,t). \tag{1}$$

Apart from the change in notations, this is identical with the basic equation given on p. 4 of Lundberg's Thesis, and on p. 10 of the "Risk Masses" paper. Together with other relations of a similar type, this equation is basic to most of his work on risk theory.

If the general forward equation for a discontinuous Markov process given on p. 490 of Feller's 1940 paper is specialized by adapting it to the conditions of the Lundberg $Y(t)$ process, it will be easily seen that we arrive precisely at the equation (1).

It is interesting to find that, already in the early papers (see e.g. Thesis, p. 6), it is quite clear to Lundberg that the basic relation (1) will hold even for an arbitrary initial distribution given by a d.f. $G(y, 0)$. Thus he considers a stochastic process as generated by the initial distribution $G(y, 0)$ and the transition probabilities determined by the equation (1). It will be seen that this comes very close to modern approaches in the theory of Markov processes.—In practical insurance applications $G(y, 0)$ will, of course, be the d.f. that has the whole mass 1 placed at the point $y=0$. However, Lundberg often uses processes with more general initial distributions in order to obtain results pertaining to the special conditions encountered in practice.

For the particular case when the risk sum distribution $P(y, t)$ is independent of t, Lundberg points out (Risk Masses, p. 14) that the equation (1) is satisfied by taking

$$G(y,t) = \sum_{0}^{\infty} \frac{t^n}{n!} e^{-t} P^{n*}(y), \tag{2}$$

where P^{n*} denotes the nth repeated convolution of the d.f. P with itself. Lundberg interprets this solution in the obvious way, pointing out that the nth term of the sum is the product of the Poisson probability that exactly n claims will occur up to the time t, and the probability that the total amount of the n claims will not exceed y. In the further particular case when all the sums at risk are equal to 1, we are led to the simple Poisson process, as already mentioned in the Thesis.

With respect to the explicit expression (2), Lundberg makes the characteristic comment that it does not represent any considerable progress as compared with (1), as it is often more advantageous to deal directly with (1) in the investigation of the risk process.

4. If the risk sum distribution $P(y)$ is independent of t, with the moments

$$p_n = \int_0^\infty y^n dP(y)$$

(we now consider the case when all risk sums are positive), the net risk premium corresponding to the (operational) time element dt will be $p_1 dt$. Lundberg supposes that this net risk premium, increased by the security loading $\lambda p_1 dt$, is fed into a risk reserve, from which all claims are paid. If $X(t)$ is the amount of the risk reserve at time t, while the initial amount is $X(0)=u$, we shall have

$$X(t) = u + (1+\lambda)p_1 t - Y(t).$$

Denoting the d.f. of the random variable $X(t)$ by $F(x, t)$, the forward equation now takes the form

$$\frac{\partial F}{\partial t} + (1+\lambda)p_1 \frac{\partial F}{\partial x} = \int_0^\infty [F(x+z, t) - F(x, t)]\, dP(z),$$

which is found in the Lundberg Congress paper of 1909, p. 908, and in the 1926 paper, p. 22. Lundberg even gives a more general version of this relation, obtained when the security loading λ is supposed to depend on x.

The mean and the variance of $X(t)$ are

$$EX(t) = u + \lambda p_1 t, \quad D^2X(t) = p_2 t.$$

For large values of t, the normalized risk reserve

$$\frac{X(t) - (u + \lambda p_1 t)}{\sqrt{p_2 t}}$$

is approximately normally distributed, with zero mean and unit variance. This means that, for every fixed x, the probability

$$Pr\{X(t) < u + \lambda p_1 t + x\sqrt{p_2 t}\} = F(u + \lambda p_1 t + x\sqrt{p_2 t}, t)$$

tends, as $t \to \infty$, to the limit

$$\Phi(x) = \frac{1}{\sqrt{2\pi}} \int_{-\infty}^x e^{-(t^2/2)}\, dt.$$

This result is obtained by Lundberg already in his Thesis, by means of the basic equation (1). Throughout his earlier papers, he spends a considerable amount of work in trying to estimate the magnitude of the error involved in this and related approximations.

Gradually, however, he seems to lose confidence in approximation methods founded on the normal distribution, as he does not find their numerical results sufficiently accurate. He then tries a different approach which leads, heuristically, to an approximation formula of the type

$$F(u + kt, t) \sim \frac{a}{\sqrt{t}} e^{-\beta t},$$

and a similar formula for the positive tail of the F distribution. Here k is a real constant $<\lambda p_1$, while α and β are positive. The approach thus indicated by Lundberg was used in 1932 by F. Esscher to work out a complete proof of the approximation formulae.—It is well known that the main problems of finding numerical approximations to the functions appearing in this connection have quite recently found very satisfactory solutions.

5. The temporal development of the risk reserve $X(t)$ is extensively studied in the Congress paper of 1909, and in subsequent works, particularly the 1926–1928 paper. In this connection, an important part is played by the ruin problem: What is the probability that $X(t)$, starting from the given initial value $X(0)=u$, will fall down to zero at any moment before a given (operational) time T?—This problem is studied, in the first place for the simplest case when $T=\infty$, by a penetrating analysis of the trajectories, or sample functions, of the "random walk" associated with the stochastic process under consideration. His methods had to be largely heuristic; however, most of the results obtained in this way have later been justified by rigorous methods.

In this brief historic introduction there is no room for a further discussion of methods and results in this part of Lundberg's work. In view of certain misconceptions that have appeared it is, however, necessary to point out that Lundberg repeatedly emphasizes the practical importance of some arrangement which automatically prevents the risk reserve from growing unduly. This point is, in fact, extensively discussed in the papers of 1909, 1919 and 1926–28. One possible arrangement proposed to this end is to work with a security factor $\lambda=\lambda(x)$ which is a decreasing function of the risk reserve $X(t)=x$. Another possibility is to dispose, at predetermined epochs, of part of the risk reserve for bonus distribution. By either method, the growth of the risk reserve may be efficiently controlled. What Lundberg does in this connection is really to work with a rather refined case of what has much later come to be known as a random walk with two barriers.

From certain quarters, the Lundberg theory has been declared to be unrealistic because, it is asserted, no limit is imposed on the growth of the risk reserve. In view of what has been said above, it would seem that these critics have not read the author they are criticizing. For a non-Scandinavian author there is, of course the excuse that most of Lundberg's works are written in Swedish.

6. The main object of risk theory, as developed by Lundberg, is to provide a theoretical basis for practical reinsurance arrangements. The word reinsurance is understood by him in a very general sense, as including any technical arrangement designed to smooth out the random fluctuations in the annual financial results of the risk business of a certain company. This can be effected within the company itself, e.g. by appropriate rules for the administration of a risk reserve, or by setting up a joint risk reserve for several companies, or by an agreement with an outside reinsurance company.

Already in the second part of the 1903 Thesis, Lundberg discusses some questions belonging to this order of ideas, and his views are further developed in all his subsequent papers, from the 1909 Congress paper on. In this introduction we can only make some very brief remarks on this subject, which will be treated in later chapters.

From the beginning, Lundberg strongly emphasizes the necessity of judging any proposed reinsurance method from the point of view of its efficiency. Its smoothing power has to be weighed against the cost of reinsurance, and the possible decrease of business volume that may be involved. In the Thesis, the problem is stated in a very simple form, which later was to be replaced by more sophisticated approaches. We shall finish up this introduction by some indications about the discussion of this point in the Thesis.

The total amount of claims incurred by a company during a certain business period is a random variable, say Y, with the d.f. $G(y)$. Let it be agreed that a reinsurer (a risk reserve or an outside company) should contribute the amount $w(Y)$, which is a steadily increasing function of Y, satisfying the relation

$$\int_0^\infty w(y)\,dG(y) = 0. \tag{3}$$

The net amount of claims incurred by the company will then be $Y-w(Y)$. If the result of the risk business is favourable, Y is small, and it follows from (3) that $w(Y)$ must be negative, which means that the company has to pay a reinsurance premium amounting to $|w(Y)|$. On the other hand, if Y is large, $w(Y)$ will be positive, and the effect of the reinsurance will be to reduce the company's expenditure for claims.

Now suppose that the relation (3), which expresses the "fairness" of the game, has been calculated according to a loaded table of risk rates (mortality rates, or whatever the case may be). The use of the "true" risk rates would then lead to another d.f. $G^*(y)$ for the amount of claims. We shall then have $G^*(y) > G(y)$, and the substitution of $G^*(y)$ instead of $G(y)$ in (3) will give a negative value to the integral. This means that the reinsurer will make an average gain of the amount

$$\left|\int_0^\infty w(y)\,dG^*(y)\right|.$$

Using the normal approximations for $G(y)$ and $G^*(y)$, Lundberg shows how the function $w(Y)$ can be determined so as to minimize the expected gain of the reinsurer, while still providing an efficient smoothing of risk fluctuations.

It is clear that already in this simple approach, taken from the earliest publication of Lundberg, we can discern important analogies with the ideas which lie behind modern collective reinsurance methods. These analogies become even more apparent in the 1909 Congress paper, where the importance of analyzing the financial effect of any proposed reinsurance method is particularly emphasized. The subject will be taken up for full discussion in later chapters of the present work, and so it will be convenient for the historical introduction to stop here.

65.

On streams of random events

Skand. Aktuarietidskr. **52**, Supplement (3–4), 13–23 (1969)

1. Introduction

In various applied fields, we encounter streams of events occurring successively in time in a more or less random way. Instances of such *random events* are the claims arising in an insurance company, the calls on a telephone line, the arrival of customers at a counter, etc. In this paper we shall deal with some of the mathematical models set up for the probabilistic analysis of streams of this character, with particular reference to the streams of insurance claims.

The theory of these models constitutes an important part of the general theory of *stochastic processes*. In order to introduce the theory in a rigorous way, it will be necessary to recall some of the fundamental concepts and results of the general theory, which will also be required in subsequent papers in this book. These basic parts of the theory will be briefly indicated in Section 2 of this paper, where also the immediate application to streams of random events will be made. For proofs and further details we may refer to Chapters 1–4 of the recent book by Cramér and Leadbetter: *Stationary and Related Stochastic Processes* (New York, Wiley 1967), and to Cramér: Collective Risk Theory (Stockholm, Skandia 1955).

The simplest case of a stream of random events is the classical *Poisson process*, which will be discussed in Section 3. It is well known that this process has found important applications in risk theory, as well as in telephone traffic and many other fields. It also forms the point of depart for generalizations in various directions. Some of these generalizations, which have shown themselves useful in risk theory, will be discussed in Section 4, while some further generalizations will be briefly indicated in Section 5.

2. Stochastic processes

Without going deeper into the measure-theoretic foundations of mathematical probability theory, a *random variable* X may be defined as an ordinary real variable endowed with a *probability distribution*. This distribution is uniquely characterized by its distribution function (d.f.) $F(x)$, which for every real x expresses the probability that the variable X takes a value $\leqslant x$. In symbols we thus have

$$F(x) = \Pr\{X \leqslant x\}.$$

(We shall always use the notation Pr{——} for the probability that the relation or relations within the brackets are satisfied.) The d.f. $F(x)$ is a never decreasing function of the real variable x, which is always continuous to the right, and such that $F(-\infty)=0$ and $F(+\infty)=1$. By the additive properties of probability we have for any $a<b$

$$F(b)-F(a)=\Pr\{a<X\leqslant b\}.$$

Moreover, the d.f. $F(x)$ determines uniquely the probability that X takes a value belonging to any Borel set B on the real axis.

In an actual experiment or observation the variable X will take a definite observed value, or sample value, say X_0. The d.f. $F(x)$ then represents the probability, before the experiment, that X_0 will be $\leqslant x$.

When several random variables, say $X_1, ..., X_n$, are considered simultaneously, their *joint probability distribution* is determined by the n-dimensional d.f.

$$F(x_1, ..., x_n)=\Pr\{X_1\leqslant x_1, ..., X_n\leqslant x_n\}.$$

By the same argument as in the one-dimensional case, the d.f. $F(x_1, ..., x_n)$ uniquely determines the probability that the variable point with coordinates $X_1, ..., X_n$ belongs to any given Borel set in n-dimensional space.

Suppose now that we are studying the temporal development of some system which is subject to randomly varying influences, such as the risk reserve of an insurance company. Let us say that we start our observations at time $t=0$, and then follow the development in time of some quantity associated with the system, say the amount of the risk reserve. For every time point $t>0$ the value of this quantity will be a random variable, say $X(t)$. Thus we are concerned with a *family of random variables depending on a parameter t*. We shall say that such a family of random variables constitutes a *stochastic process*.

In an actual observation of the development of the system, say during the interval $0<t<T$, every random variable $X(t)$ will take a definite sample value $X_0(t)$. The result of our observations will thus be a collection of sample values $X_0(t)$, or a *sample function* $X_0(t)$ observed for $0<t<T$. The sample function $X_0(t)$ is often also called a *realization* of the stochastic process under observation. In an applied problem, we are usually interested in computing the probability, before any observations, that a realization or sample function will satisfy certain given conditions. Thus, e.g., when discussing the ruin problem we may want to find the probability that $X(t)$ will never become negative during the interval $0<t<T$. In order to be able to define and compute probabilities of this kind, we shall first have to consider the joint distributions of finite groups of the variables $X(t)$.

Let $t_1<t_2<...<t_n$ be an arbitrary finite set of time points, all >0. The random variables $X(t_1), ..., X(t_n)$ will have a joint probability distribution, characterized by a certain n-dimensional d.f., say $F(x_1, ..., x_n; t_1, ..., t_n)$, where the t_i are parameters. The collection of all these distributions, for $n=1, 2, ...$, and for all possible choices of the t_i, is known as the *family of finite-dimensional distributions* of the $X(t)$ process. This is one of the basic concepts of the theory.

Many of the essential properties of a stochastic process can be expressed as properties of the finite-dimensional distributions. In particular such properties form the basis of a classification of stochastic processes.

It is obvious that any family of finite-dimensional distributions must satisfy the following *consistency condition*: If we allow one of the variables in the d.f. $F(x_1, \ldots, x_n; t_1, \ldots, t_n)$, say x_k, to tend to $+\infty$, the restriction imposed on the random variable $X(t_k)$ will fall away in the limit, so that the result of the limit passage must be the member of the F family which corresponds to all the x_i and t_i with the exception of x_k and t_k. (The *symmetry condition* usually required has no significance here, as we have supposed the t_i sequence to be increasing.)

Conversely, if a family of finite-dimensional distributions is given, such that the consistency condition is satisfied, it follows from a celebrated theorem of Kolmogorov that there always exists a stochastic process $X(t)$ associated with this particular family of distributions. For any finite set $t_1, \ldots, t_n$, the given finite-dimensional distributions determine the probability that the random variables $X(t_1), \ldots, X(t_n)$ will take values belonging to any Borel set in n-dimensional space. Moreover, the totality of finite-dimensional distributions determine uniquely the probability that the whole sample function $X(t)$ will satisfy given conditions, as long as these are sufficiently simple.

At this point, however, certain difficulties arise. It turns out that probabilities of the kind mentioned above, such as the probability that $X(t)$ will never become negative, are in general not sufficiently simple to be uniquely determined by the finite-dimensional distributions. Still, as soon as we restrict ourselves to stochastic processes associated with streams of random events, or with risk theory, this difficulty can be overcome. In fact, in all such cases we may legitimately assume that, with a probability equal to one, all our sample functions have at most a finite number of simple discontinuities in any finite t-interval. A sample function will then be uniquely determined by its values in a countable set of t points, say the set of all rational t, and the relevant probabilities will be uniquely determined by the finite-dimensional distributions. In the sequel we shall only consider such cases, and shall be able to regard all probabilities dealt with as uniquely determined by the finite-dimensional distributions of the process under discussion.

We now turn to the application of Kolmogorov's theorem to a stochastic process naturally associated with a stream of random events. Assuming as above that observations start at time $t=0$, we denote by $X(t)$ the number of events occurring in the semiclosed interval $(0, t]$. A sample function of this process will be a never decreasing function of t, increasing only by steps of unit height in every time point where an event occurs. The number of events occurring in an arbitrary semiclosed interval $(s, t]$ with $0 \leqslant s < t$, will then be

$$N(s, t) = X(t) - X(s).$$

Let $(s_1, t_1], \ldots, (s_n, t_n]$ be any finite set of disjoint intervals, and suppose that the joint distributions of the random variables $N(s_1, t_1), \ldots, N(s_n, t_n)$ are given.

As soon as these distributions satisfy simple consistency conditions analogous to those mentioned above, it is easily seen that they will define a family of consistent finite-dimensional distributions for the $X(t)$ process, so that we may regard all the relevant probabilities associated with the stream of events under consideration as uniquely defined.

If the joint distribution of the variables $N(s_i+h, t_i+h)$, for $i=1, 2, ..., n$, is always independent of h, we shall say that we are concerned with a *stationary* stream of random events. For any stationary stream, the probability that exactly n events occur in the interval $(s, s+t]$ is independent of s and equal to the corresponding probability for the interval $(0, t]$. For a stationary stream it can be shown that, as t tends to zero, the expression

$$\frac{1}{t} Pr\{N(0,t) \geqslant 1\}$$

tends to a definite positive limit λ, which may be finite or infinite. The constant λ is called the *intensity* of the stream. If, moreover, we have

$$\frac{1}{t} Pr\{N(0,t) \geqslant 2\} \rightarrow 0$$

as $t \rightarrow 0$, the stationary stream is said to be *regular*. Obviously this implies that the probability of more than one event in a small time interval will be small compared to the length of the interval.

Now suppose that we are concerned with a stationary and regular stream, such that the mean number of events in a time interval of unit length

$$\mu = EN(0, 1)$$

is finite. Then it can be shown that we have the important relation

$$\lambda = \mu,$$

so that the intensity of the stream is finite and coincides with the mean number of events in a unit interval.

3. The Poisson process

The simplest probabilistic model for a stream of random events is the Poisson process, studied by Filip Lundberg in his 1903 Thesis on risk theory, and 1909 by Erlang in his work on telephone engineering problems. Except for the value of a numerical parameter, the Poisson process is uniquely determined by the following three conditions: (A) Stationarity, (B) Regularity, and (C) Absence of after-effects.

Stationary and regularity have been defined in the preceding Section. By the absence of after-effects we mean that the random variables $N(s, t)$ corresponding to disjoint time intervals are always independent. We now proceed to show that the well-known expressions for the probabilities associated with a Poisson process can be deduced from the three conditions.

From the conditions A and C it evidently follows that the joint distributions of any finite set of variables $N(s, t)$ corresponding to disjoint intervals $(s, t]$ will be uniquely determined by the function

$$p_n(t) = \Pr\{N(0, t) = n\}$$

for $n = 1, 2, \ldots$ and all $t > 0$. By the same conditions we have

$$p_0(t+u) = p_0(t)p_0(u).$$

It is well known that the only solution to this equation such that $0 < p_0(t) \leqslant 1$ for all t is

$$p_0(t) = e^{-\lambda t},$$

where λ is a non-negative constant. Excluding the trivial case $p_0(t) = 1$ for all t, we may assume $\lambda > 0$. For small t this gives

$$p_0(t) = 1 - \lambda t + o(t),$$

while from B we obtain

$$\begin{aligned} p_1(t) &= 1 - p_0(t) + o(t) \\ &= \lambda t + o(t). \end{aligned}$$

Further, B and C give for $n \geqslant 1$

$$p_n(t+h) = (1 - \lambda h)p_n(t) + \lambda h p_{n-1}(t) + o(h).$$

Dividing by h and allowing h to tend to zero, we see that the derivative $p_n(t)$ exists and satisfies the equation

$$p_n(t) = \lambda[p_{n-1}(t) - p_n(t)].$$

Substituting here $p_n(t) = e^{-\lambda t} q_n(t)$, it follows that we have

$$q_n'(t) = \lambda q_{n-1}(t)$$

with the initial conditions $q_0(t) = 1$ for all t, and $q_n(0) = 0$ for $n \geqslant 1$. It is easily seen that this gives

$$q_n(t) = \frac{(\lambda t)^n}{n!}$$

and thus finally

$$p_n(t) = \frac{(\lambda t)^n}{n!} e^{-\lambda t},$$

which is the well-known expression for the Poisson distribution. The constant λ is, by the definition in the preceding section, the intensity of the Poisson stream of events, and evidently this coincides with the mean value of the number of events in a time interval of unit length:

$$EN(0,1) = \sum_0^\infty n p_n(1) = \lambda.$$

It is also easily verified that the joint distributions of the variables $N(s, t)$ defined by this expression of $p_n(t)$ satisfy the necessary consistency conditions and thus determine the family of finite-dimensional distributions of a well defined stochastic process.

The random function $X(t) = N(0, t)$ associated with the Poisson process expresses the number of events from the beginning of our observations until time t. Evidently $X(0) = 0$, and $X(t)$ increases by unit steps at every time point where an event occurs. The lengths of the time intervals between consecutive events are mutually independent random variables, each with a probability density equal to $\lambda e^{-\lambda t}$. This probability distribution has the interesting property that, if we place ourselves at an arbitrary time point, the time t until the next event occurs will have the same probability density $\lambda e^{-\lambda t}$, independently of the time already passed since the immediately preceding event. This may be expressed as a "lack of memory" property, which turns out to be characteristic for the Poisson process.

4. Generalizations, I

In recent years an extensive theory of general types of streams of random events has developed. In this brief survey it is impossible even to indicate more than a few scattered results of this theory, and we shall have to restrict ourselves to some particular cases, which may be naturally regarded as generalizations in various directions of the simple Poisson process. In the present section we shall consider some of these generalizations which have been or could possibly be applied to the study of streams of insurance claims.

The most important of these generalizations, which was introduced already at an early stage, consists in the replacement of ordinary ("absolute") time by an *operational time*, by which a non-stationary case is reduced to stationarity.

According to the Poisson process, the mean number of events in an infinitesimal time element $d\tau$ is $\lambda d\tau$, where λ is the constant intensity of the process. When applying the theory e.g. to the stream of claims in an insurance company, where the business volume is strongly variable, it is of course entirely unrealistic to regard λ as a constant. If we replace λ by a non-random function of time, say $\lambda(\tau)$, the mean number of claims during the time interval $(0, \tau]$ will be

$$t = \int_0^\tau \lambda(u)\, du,$$

and we now introduce t as an operational time instead of the absolute time τ. The three postulates for the Poisson process given in the preceding section will often be found reasonable as applied to operational time, and the result will then be that the probability of exactly n claims in the *operational time interval* $(0, t]$ is given by the Poisson expression with $\lambda = 1$:

$$p_n(t) = \frac{t^n}{n!} e^{-t}.$$

By changing the unit of operational time we shall, of course, come back to the expression for $p_n(t)$ given in the preceding section, and in the sequel we shall use the Poisson probability in this more general form.

Even after the introduction of operational time, it will often be found that the postulates leading to the Poisson process imply an oversimplification too strong to yield realistic results. We must then ask ourselves if something might be gained by relaxing one or more of the conditions A–C given in the preceding section.

If we suppose that condition A can be taken care of by the introduction of an appropriately defined operational time, we come next to condition B. If we give up this condition, we shall be concerned with a stream where *multiple events* may occur, and we shall be naturally led to the expression

$$X(t) = X_1(t) + 2X_2(t) + 3X_3(t) + \dots$$

for the total number of events in the (operational) time interval $(0, t]$. Here $X_1(t), X_2(t), \dots$ are the random variables associated with mutually independent Poisson processes of simple events, each satisfying the conditions A–C. If $X_k(t)$ has the Poisson parameter λ_k, the probability of a k-uple event in the infinitesimal time element dt is $\lambda_k dt$, and the mean total number of events will be $\Sigma k \lambda_k dt$. Under appropriate convergence conditions these expressions will be valid, and an expression for the probability of a total number of n events in a time interval of length t is easily deduced. Streams of this type will occur in certain insurance branches, where claims may occur in "bunches", such as in car accident insurance, where the multiplicity of an event will correspond to the number of cars involved in an accident. A multi-dimensional generalization of this type of streams has been applied with great success to the study of the distribution of galaxies in cosmical space.

Turning now to the question of replacing the independence condition C by some less restrictive condition, we are faced with a multitude of widely different possibilities. Some of these, which are of great importance for various applications outside the insurance field, will be briefly mentioned in the following section. Here we shall only indicate two forms of generalization with applications to the study of streams of insurance claims.

It is, in fact, evident that a condition of independence between subsequent time intervals can never be expected to be strictly satisfied under practical insurance conditions. In order to work out a realistic theory of insurance risk, the replacement of the independence condition by something less restrictive, but more realistic, must be regarded as a task of fundamental importance.

Taking the simple Poisson process as our point of departure, a natural possibility would seem to be to introduce some kind of *random variation* of the parameter λ, thereby allowing for the lack of homogeneity and the influence of random factors on the development of the process. The first idea that comes to the mind would be to regard λ as a random variable, the value of which is fixed by a separate random experiment before the beginning of our observations. If λ is a random variable with the probability density $f(\lambda)$, the probability of n claims in a time interval of length t would then amount to

$$p_n(t) = \int_0^\infty \frac{(\lambda t)^n}{n!} e^{-\lambda t} f(\lambda)\, d\lambda$$

A well-known case which leads to a simple explicit expression is the case when $f(\lambda)$ corresponds to a gamma distribution:

$$f(\lambda) = \frac{h^h}{\Gamma(h)} \lambda^{h-1} e^{-h\lambda}$$

where h is a constant parameter. This choise of $f(\lambda)$ yields

$$p_n(t) = \binom{-h}{n} \left(-\frac{t}{t+h}\right)^n \left(\frac{h}{t+h}\right)^h,$$

so that we obtain for the number of claims n a negative binomial distribution, also known as a limiting form of the Pólya distribution. It should be noted that in this case, the parameter λ will vary about a mean value equal to one.

However, this might seem to be a somewhat primitive method to introduce some form of randomness into the intensity of the stream of claims. In fact, as we have already observed, a fixed value of λ would be determined by a random experiment before the start of the stream, and the whole stream of claims would then proceed as a Poisson stream with a constant value of the parameter, fixed in advance by this experiment. It might seem more in agreement with real conditions to allow λ to vary in a random way as time proceeds, or in other words to regard $\lambda = \lambda(t)$ as a random function of t.

According to this approach, the stream of claims should be represented as a Poisson process with a parameter $\lambda(t)$ which is, for every $t > 0$, the random variable of a stochastic process. The probability of a claim in the infinitesimal time element dt would then be $\lambda(t)dt$, and in this way we should allow both for the possibility of random variations in the intensity of the stream, and for the correlation which may be expected to exist between the conditions in adjacent time points.

Suppose, e.g., that the variation of $\lambda(t)$ could be satisfactorily represented by a stationary stochastic process, with a mean value $E\lambda(t) = 1$, and a covariance function

$$r(t-u) = E[\lambda(t) - 1][\lambda(u) - 1].$$

Writing

$$z = \int_0^t \lambda(u)\,du,$$

we shall then have $Ez = t$, and under fairly general conditions the standard deviation of z will be of the same order of magnitude as $\sqrt{t}$. The probability $p_n(t)$ of exactly n claims within the time interval $(0, t]$ will now be

$$p_n(t) = E\frac{z^n}{n!}e^{-z}.$$

5. Generalizations, II

From the great variety of generalized streams of events encountered in many applications outside the insurance field, we choose as a typical example the so called *birth and death process*, which has important applications to the study of the growth of biological populations, to telephone traffic, queueing problems, etc.

In this case we are still concerned with a random variable $X(t)$ which can only assume non-negative integral values. We assume, however, that we consider a stream where there are two different kinds of events, "births" and "deaths". A birth may occur at any time $t>0$, and increases the value of $X(t)$ by one. A death can only occur at a time when $X(t)$ has a positive value, and decreases the value of $X(t)$ by one. If, at a certain moment t, we have $X(t)=n$, the probability of an event during the infinitesimal time element between t and $t+dt$ is assumed to be $\lambda_n dt$ for a birth, and $\mu_n dt$ for a death, where $\mu_0=0$. We assume that the process is regular, so that the occurrence of more than one event in the time element dt is small compared to dt. Introducing the probability

$$p_n(t) = \Pr\{X(t) = n\},$$

we then obtain, by similar arguments as in the case of the Poisson process,

$$p_0'(t) = -\lambda_0 p_0(t) + \mu_1 p_1(t),$$

and for $n>0$

$$p_n'(t) = -(\lambda_n+\mu_n)p_n(t) + \lambda_{n-1}p_{n-1}(t) + \mu_{n+1}p_{n+1}(t).$$

Under fairly general conditions, it can be shown that this system has a unique solution such that $p_n(t) \geqslant 0$ and $\Sigma p_n(t) = 1$ for all t, and that the limits

$$p_n = \lim_{t\to\infty} p_n(t)$$

exist for all n. We then have $\Sigma p_n \leqslant 1$, and if the sign of equality holds here, the system will be said to tend to a limiting state of statistical equilibrium as t tends to infinity. On the other hand, if $\Sigma p_n < 1$, there will be a positive probability that the size $X(t)$ of the "population" tends to infinity. We shall give three examples of different applications of this process.

1. Let $X(t)=n$ denote the size at time t of a biological population descending from one single individual, so that $X(0)=1$. We assume that the population consists of mutually independent individuals, each of which has the same constant probability to die or give birth to a new individual. We can then take $\lambda_n = n\lambda$, $\mu_n = n\mu$, where λ and μ are constants. The solution of the fundamental equations is in this case

$$p_0(t) = \mu z,$$

$$p_n(t) = (1-\mu z)(1-\lambda z)(\lambda z)^{n-1}, \quad (n>0),$$

where

$$z = \frac{e^{(\lambda-\mu)t}-1}{\lambda e^{(\lambda-\mu)t}-\mu},$$

and $z = t/(1+\lambda t)$ in the limiting case $\lambda = \mu$.

It follows that the limiting probabilities as $t \to \infty$ are

$$p_0 = \begin{cases} \dfrac{\mu}{\lambda} & \text{for } \lambda > \mu, \\ 1 & \text{for } \lambda \leqslant \mu, \end{cases}$$

and $p_n = 0$ for every $n > 0$. Thus if $\lambda \leqslant \mu$ we have the probability one that the population will sooner or later die out, while in the opposite case $\lambda > \mu$ there is a positive probability of extinction, and another positive probability that the population will increase beyond all limits. In either case, the probability of a population size remaining between fixed non-zero limits is equal to zero.

2. The birth and death process may serve as a model in certain telephone traffic problems. Suppose that there are N trunklines available at a certain station, and that incoming calls form a simple Poisson process with parameter λ, while the duration of a conversation has a negative exponential distribution with probability density $\mu e^{-\mu t}$. It then follows from the remark at the end of section 3 that, if a certain line is busy at time t, the probability that it will become free during the following time element dt is μdt, independently of the time elapsed since the beginning of the conversation. We assume that there is no waiting line, so that a call made at an instant when all N lines are busy will not be satisfied, and will have to be repeated later. Denoting by $p_n(t)$ the probability that exactly n lines are busy at time t, where $0 \leqslant n \leqslant N$, we may take $\lambda_n = \lambda$, $\mu_n = n\mu$ in the above equations for the birth and death process, for $n < N$, while for $n = N$ we obtain

$$p_N'(t) = -N\mu p_N(t) + \lambda p_{N-1}(t).$$

The limiting probabilities $p_n = \lim_{t\to\infty} p_n(t)$ are in this case given by the famous *Erlang loss formula*

$$p_n = \frac{\dfrac{\alpha^n}{n!}}{1 + \dfrac{\alpha}{1!} + \ldots + \dfrac{\alpha^N}{N!}}$$

with $\alpha = \lambda/\mu$. In particular, for $n = N$ we obtain the probability p_N that, in the limiting state of statistical equilibrium, all N lines will be busy, so that an incoming call will be lost. As $N \to \infty$, the p_n evidently tend to the probabilities of a Poisson distribution with parameter α.

3. If we modify the preceding example by assuming that a call incoming when all N lines are busy will join a waiting line common for all N trunklines, and will be satisfied in order of arrival, when lines are becoming free, we obtain a model applicable to certain queueing problems. Instead of trunklines we may have "counters", and instead of lengths of telephone conversations, there will be "service times" for the successive customers.

In this case, we denote by $X(t)$ the total number of customers engaged in waiting or being served at time t, and we write as before $p_n(t) = \Pr\{X(t) = n\}$. If $\alpha = \lambda/\mu \geqslant N$, the limiting probability p_n will be equal to zero for all n, so that the waiting line will almost certainly grow beyond all limits as $t \to \infty$. On the other hand, if $\alpha < N$, the limiting probabilities, corresponding to the state of statistical equilibrium, are given by the expressions

$$p_n = \begin{cases} \dfrac{\alpha^n}{n!}\, p_0 & \text{for } n \leqslant N, \\[2ex] \left(\dfrac{\alpha}{N}\right)^{n-N} \dfrac{\alpha^N}{N!}\, p_0 & \text{for } n > N, \end{cases}$$

where p_0 is determined by the condition $\Sigma p_n = 1$. In particular, the limiting probability that a newcoming customer will have to wait is

$$p_N + p_{N+1} + \ldots = \frac{N p_0}{N - \alpha} \cdot \frac{\alpha^N}{N!}.$$

66.

Structural and statistical problems for a class of stochastic processes

The first Samuel Stanley Wilks Lecture
Princeton University Press, Princeton, 3–30 (1974)
(Reprinted in Random Processes)

1. Introduction

It is an honor and a privilege to have been asked to give the first S. S. Wilks Memorial Lecture.

Sam Wilks entered the field of statistical science at a time when it still had all the freshness of youth. Even during his early years he was able to make outstanding contributions to its development. Later, while always pursuing his scientific research, he became engaged in administrative work to an extent that rendered him the well-deserved name "A Statesman of Statistics." He organized the teaching of statistics in this university and elsewhere, always making it clear that this teaching should be built on rigorous mathematical foundations. And, last but not least, Sam was a friend of his many friends, always ready to go out of his way to help a friend, in every possible way.

The early statistical works of Sam Wilks belong to the classical theory of statistical inference. Accordingly, they are concerned with the probability distributions of groups of random variables, the number of which may be very large, but is always assumed to be finite. It is well known that, in many fields of applications, we nowadays encounter important problems that involve an infinite number of random variables, and so cannot be solved within the frames of the classical theory. Some of these problems, which come up in the theory of stochastic processes, may be regarded as straightforward generalizations of inference problems treated by Wilks and other authors. It seems to me that it will be in the spirit of Sam Wilks to choose some of these generalized problems as the subject of a Wilks Memorial Lecture.

A systematic attempt to generalize classical statistical inference theory to stochastic processes first appeared in the 1950 Stockholm University thesis by Ulf Grenander [8]. Since then, a vigorous development has taken place, based on works by a large number of authors.

A group of probabilistic problems that seems to have been of fundamental importance throughout this development is concerned with finding convenient *types of representation* for the classes of stochastic processes under consideration, so as to enable us to investigate the properties of the processes, and, in particular, to work out methods of statistical inference. As examples of such types of representation, we may think of the spectral representation for stationary processes, the Karhunen-Loève representation in terms of eigenfunctions and eigenvalues of a symmetrical kernel, etc.

In the sequel we shall be concerned with one particular mode of representation, which is applicable to a large class of stochastic processes. After having introduced the general representation formula and discussed its properties, we shall consider the possibility of its application to some problems of statistical inference.

Let us first present a simple argument that suggests the type of representation formula in question. Consider the real-valued random variable $x(t)$ of a stochastic process with a continuous time parameter t. In various applications it seems natural to regard the value assumed by $x(t)$ at a given instant t as the accumulated effect of a stream of random *impulses* or *innovations* acting throughout the past. Suppose, e.g., that $x(t)$ represents the intensity at time t of an electric current generated by a stream of electrons arriving at an anode at randomly distributed time points. Let an electron arriving at time u bring an impulse $dz(u)$, while the effect at time t of a unit impulse acting at time u is measured by a *response function* $g(t, u)$. Assuming the effects to be simply additive, we should then be heuristically led to try a representation formula for $x(t)$ of the type

$$x(t) = \int_{-\infty}^{t} g(t, u)\, dz(u), \tag{1.1}$$

with some appropriate definition of the stochastic integral in the second member.

4

Similar arguments will apply to various other applied problems, and thus the following question arises in a natural way: *Is it possible to define some general class of stochastic processes that will admit a representation more or less similar to the formula* (1.1)?

It will appear that this is, in fact, possible if we modify (1.1) by replacing the second member of that formula by the sum of a certain number of terms of the same form. In some cases the number of terms will be equal to one, so that (1.1) will apply without modification.

It turns out that the most adequate approach to our question is supplied by the geometry of Hilbert space. In the following section we shall briefly recall some known facts from Hilbert space theory, and then show how this theory leads to a complete answer to our question.

2. The Representation Formula

We shall consider real-valued random variables $x = x(\omega)$ defined on a fixed probability space $(\Omega, \mathbf{F}, P)$, and satisfying the conditions

$$Ex = 0, \qquad Ex^2 < \infty.$$

It is well known that the set of all such random variables forms a real Hilbert space H, if addition and multiplication by real numbers are defined in the obvious way, while the inner product of two elements x and y is defined by

$$(x, y) = Exy.$$

The norm of an element x is then

$$\|x\| = (Ex^2)^{1/2}.$$

Two elements x and y are regarded as identical if the norm of their difference is zero. We then have $E(x - y)^2 = 0$, and thus

$$P(x = y) = 1.$$

All equalities between random variables occurring in the sequel should be understood in this sense, so that the two members are equal with probability one.

Convergence in norm in the Hilbert space H is identical with convergence in quadratic mean (q.m.) in probabilistic terminology.

A stochastic process is a family of real-valued random variables

$x(t) = x(t, \omega)$. We shall not be concerned in this paper with the more general cases of complex or vector-valued processes. The parameter t, which will be interpreted as time, is supposed to be confined to a finite or infinite real interval $T = (A, B)$, where A may be finite or $-\infty$, while B may be finite or $+\infty$. We shall always suppose that for every $t \in T$ we have

$$(2.1) \qquad Ex(t) = 0, \qquad Ex^2(t) < \infty,$$

so that $x(t)$ is a point in the Hilbert space H, while the family of all $x(t)$ may be regarded as defining a *curve* in H.

We shall further also impose the important condition that the q.m. limits

$$(2.2a) \qquad x(t-0) = \lim_{h \uparrow 0} x(t+h), \qquad x(t+0) = \lim_{h \downarrow 0} x(t+h)$$

exist for all $t \in T$, and that we have

$$(2.2b) \qquad x(t-0) = x(t),$$

so that $x(t)$ is everywhere continuous to the left in q.m.

When a stochastic process satisfying (2.1) and (2.2) is given, we define the *Hilbert space $H(x)$ of the process* as the subspace of H spanned by the random variables $x(t)$ for all $t \in T$, i.e., the closure in norm of all finite linear combinations

$$(2.3) \qquad c_1x(t_1) + c_2x(t_2) + \cdots + c_nx(t_n),$$

where the c_r and t_r are real constants, and $t_r \in T$. It can be shown [3, p. 253] that under the conditions (2.1) and (2.2) the Hilbert space $H(x)$ is *separable.*

We further define a family of subspaces $H(x, t)$ of $H(x)$, such that $H(x, t)$ is the closure in norm of all those linear combinations (2.3), where the t_r are restricted by the conditions $t_r \in T$ and $t_r \leqq t$. If the lower extreme A of T is finite, we define $H(x, t)$ for $t \leqq A$ as the space containing only the zero element, i.e., the random variable that is almost everywhere equal to zero.

Obviously the space $H(x, t)$ will never decrease as t increases. Consequently the limiting space $H(x, -\infty)$ will exist, whether A is finite or not, and will be identical with the intersection of all the $H(x, t)$. We then have for $t < u$

$$H(x, -\infty) \subset H(x, t) \subset H(x, u) \subset H(x).$$

Using the terminology of Wiener and Masani [18], we may say that $H(x, t)$ represents the *past and present* of the $x(t)$ process, from the point of view of the instant t, while $H(x, -\infty)$ represents the *remote past* of the process. With respect to $H(x, -\infty)$ there are two extreme particular cases that present a special interest.

If $H(x, -\infty) = H(x)$, we may say that already the remote past contains all available information concerning the process. As soon as the remote past is known, the future development of the process can be accurately predicted. In this case we say that the $x(t)$ process is *deterministic*.

The opposite extreme case occurs when $H(x, -\infty)$ contains only the zero element of $H(x)$, i.e., the random variable that is almost everywhere equal to zero. We have already noted above that this case is always present when A is finite. The remote past is in this case useless for prediction purposes. Speaking somewhat loosely, we may say that any piece of information present in the process, at the instant t for example, must have entered as a new impulse at some definite instant in the past. The $x(t)$ process is then said to be *purely nondeterministic* (some authors use the term *linearly regular*).

Any $x(t)$ satisfying (2.1) and (2.2) can be built up in a simple way by components belonging to these two extreme types. In fact, $x(t)$ can be uniquely represented in the form

$$x(t) = u(t) + v(t),$$

where $u(t)$ and $v(t)$ both belong to $H(x, t)$, while $u(t)$ is deterministic and $v(t)$ purely nondeterministic. Moreover, for any s, t the elements $u(s)$ and $v(t)$ are orthogonal or, in probabilistic terminology, the random variables $u(s)$ and $v(t)$ are uncorrelated.

From the point of view of the applications, deterministic processes do not seem to have very great interest, and we shall in the sequel be concerned only with the opposite extreme case of purely nondeterministic processes.

Consider now the never decreasing family $H(x, t)$ of subspaces of $H(x)$, and let P_t denote the projection operator on $H(x)$ with range $H(x, t)$. The family P_t of projections, where $t \in T$, then has the following properties

$$P_t \leqq P_u \qquad \text{for } t < u,$$
$$P_{t-0} = P_t,$$
$$P_A = 0, \qquad P_B = 1,$$

where 0 and 1 denote respectively the zero and the identity operator in $H(x)$. It follows that, in Hilbert space terminology, the P_t form a *resolution of the identity* or a *spectral family of projections.** We note that the P_t are uniquely determined by the given $x(t)$ process.

For any random variable z in $H(x)$, we now define a stochastic process $z(t)$ by writing

$$z(t) = P_t z. \tag{2.4}$$

It then follows from the properties of the P_t family that $z(t)$ will be a *process of orthogonal increments* defined for $t \in T$, and such that

$$E\, dz(t) = 0, \qquad E(dz(t))^2 = dF(t),$$

where $F(t)$ is a never decreasing function that is everywhere continuous to the left.

The subspaces $H(z)$ and $H(z, t)$ are, by definition, spanned by the random variables $z(t)$ in the same way as $H(x)$ and $H(x,t)$ are spanned by the $x(t)$. The space $H(z)$ is known as a *cyclic* subspace of $H(x)$, and consists of all random variables of the form

$$\int_A^B h(u)\, dz(u), \tag{2.5}$$

where $h(u)$ is a nonrandom F-measurable function such that

$$\int_A^B h^2(u)\, dF(u) < \infty,$$

the stochastic integral (2.5) being defined as an integral in q.m. Similarly, $H(z, t)$ is the set of all random variables obtained by replacing in (2.5) the upper limit by t, with a corresponding integrability condition over (A, t).

Consider now the class of all real-valued and never decreasing, not identically constant functions $F(t)$ defined for $t \in T$ and everywhere continuous to the left. We introduce a partial ordering in this

* For the elementary theory of Hilbert space, with applications to random variables, we may refer to [5, Ch. 5]. Those parts of the advanced theory that will be used here are developed in [17, Chs. 5–7], and in [15]. With respect to the condition $P_{t-0} = P_t$, which is usually replaced by $P_{t+0} = P_t$, see a remark in [12, p. 19].

class by saying that F_1 is *superior* to F_2, and writing $F_1 > F_2$, whenever F_2 is absolutely continuous with respect to F_1. If F_1 and F_2 are mutually absolutely continuous, we say that they are *equivalent.* The set of all F equivalent to a given F_0 forms the *equivalence class* of F_0. In the set of all equivalence classes C, a partial ordering is introduced in the obvious way by writing $C_1 > C_2$ when the corresponding relation holds for any $F_1 \in C_1$ and $F_2 \in C_2$. Evidently the relation $C_1 > C_2$ does not exclude that the classes may be identical.

We can now proceed to the formal statement of the representation theorem that will provide the basis of this paper. It has been shown above how a given $x(t)$ process uniquely determines the corresponding family of subspaces $H(x, t)$ and the associated spectral family of projections P_t. According to the theory of self-adjoint linear operators in a separable Hilbert space, the projections P_t, in their turn, uniquely determine a number N, which may be a positive integer, or equal to infinity, such that the following theorem holds.

Theorem 2.1. Let $x(t)$ be a purely nondeterministic stochastic process defined for $t \in T = (A, B)$, and satisfying the conditions (2.1) *and* (2.2). *Then there exists a number N uniquely determined by the $x(t)$ process, which will be called the multiplicity of the process and may have one of the values* $1, 2, \ldots, \infty$. *The random variable $x(t)$ can then be represented by the expansion*

$$x(t) = \sum_{1}^{N} \int_{A}^{t} g_n(t, u)\, dz_n(u), \tag{2.6}$$

which holds for all $t \in T$ and satisfies the following conditions $Q_1, \ldots, Q_4$:

(Q_1) *The $z_n(u)$ are mutually orthogonal processes of orthogonal increments such that*

$$E\, dz_n(u) = 0, \qquad E(dz_n(u))^2 = dF_n(u),$$

where F_n is never decreasing and everywhere continuous to the left.

(Q_2) *The $g_n(t, u)$ are nonrandom functions such that*

$$Ex^2(t) = \sum_{1}^{N} \int_{A}^{t} g_n^2(t, u)\, dF_n(u) < \infty.$$

(Q_3) $F_1 > F_2 > \cdots > F_N$.

(Q_4) *For any $t \in T$ the Hilbert space $H(x, t)$ is the vector sum of the mutually orthogonal corresponding subspaces of the z_n processes:*

$$H(x, t) = H(z_1, t) \oplus \cdots \oplus H(z_N, t).$$

The expansion (2.6) *is a canonical representation of $x(t)$ in the sense that no expansion of the same form, satisfying $Q_1, \ldots, Q_4$, exists for any smaller value of N.*

The expansion (2.6) gives a linear representation of $x(t)$ in terms of past and present N-dimensional *innovation elements* $[dz_1(u), \ldots, dz_N(u)]$. The *response function* $g_n(t, u)$ corresponds to the component $dz_n(u)$ of the innovation element. The quantity $g_n(t, t)$ is the corresponding *instantaneous response function.*

It should be observed that, while the multiplicity N is uniquely determined by the $x(t)$ process, the innovation processes $z_n(u)$ and the response functions $g_n(t, u)$ are *not* uniquely determined. It is, e.g., obvious that a positive and continuous factor $h(u)$ may be transferred from $g_n(t, u)$ to $dz_n(u)$ or vice versa. This remark will be used later.

However, the sequence of equivalence classes $C_1, \ldots, C_N$ of the functions $F_1, \ldots, F_N$ appearing in (Q_3) is, in fact, uniquely determined by the $x(t)$ process. Thus any expansion of $x(t)$ in the form (2.6), satisfying $Q_1, \ldots, Q_4$, will have functions $F_1, \ldots, F_N$ belonging to the same equivalence classes.

The representation (2.6) gives immediately an explicit expression for the best *linear least squares prediction* of $x(t)$ in terms of the development of the x process up to the instant $s < t$. In fact, this best prediction is

$$P_s x(t) = \sum_1^N \int_A^s g_n(t, u)\, dz_n(u),$$

the error of prediction being

$$x(t) - P_s x(t) = \sum_1^N \int_s^t g_n(t, u)\, dz_n(u). \tag{2.7}$$

The canonical representation (2.6) was first given by Hida [10] for the case of a normal process, and by Cramér [2, 3, 4] for the general case considered here, and also for finite-dimensional vector proc-

10

esses. Generalizations to higher vector processes have been investigated by Kallianpur and Mandrekar, e.g. in [11].

3. The Multiplicity Problem

Any $x(t)$ process considered in the sequel will be supposed to be defined for $t \in T = (A, B)$, to be purely nondeterministic and satisfy (2.1) and (2.2). Generally this will not be explicitly mentioned, but should always be understood.

For such a process we have a canonical representation of the form (2.6), and a problem that now immediately arises is *how to determine the multiplicity N corresponding to a given* $x(t)$. No complete solution of this *multiplicity problem* is so far known. In this and the two following sections, we propose to discuss the problem and present some preliminary results.

The corresponding problem for a process with discrete time is easily solved. In this case there is always a representation corresponding to (2.6) with $N = 1$, the integral being replaced by a sum of mutually orthogonal terms [2]. However, as soon as we pass to the case of continuous time, the situation is entirely different.

For the important case of a *stationary* process $x(t)$, defined over $T = (-\infty, \infty)$, we still have $N = 1$. In fact, it is well known that such a process admits a representation

$$x(t) = \int_{-\infty}^{t} g(t-u)\, dz(u), \tag{3.1}$$

where $z(u)$ is a process with orthogonal increments such that $E(dz(u))^2 = du$, while $H(x, t) = H(z, t)$. Obviously this is a canonical representation with $N = 1$ and $F(u) = u$.

On the other hand it is known [4, p. 174] that, in the more general class of *harmonizable* processes, it is possible to construct explicit examples of processes with any given finite or infinite multiplicity N, and even with any given sequence of equivalence classes $C_1, \ldots, C_N$ such that $C_1 > \cdots > C_N$.

From the point of view of statistical applications, it seems important to be able to define some fairly general class of $x(t)$ processes having multiplicity $N = 1$, since in this case the expression (2.6) will evidently be easier to deal with than in a case with $N > 1$. In this connection it should be observed that every example so far explicitly given of a process with $N > 1$ contains response functions $g_n(t, u)$

showing an extremely complicated behavior. Thus we might hope that, by imposing some reasonable regularity conditions on the $g_n(t, u)$ and the $z_n(u)$ occurring in (2.6), we should obtain a class of processes with $N = 1$.

This can in fact be done, as will be shown in Section 5. In the present section, we shall prove a lemma on multiplicity that will be used in the sequel.

Suppose that a stochastic process $x(t)$ is given by the expression

$$x(t) = \sum_{1}^{N} \int_{A}^{t} g_n(t, u)\, dz_n(u), \tag{3.2}$$

where the g_n and the z_n are known. This expression has the same form as (2.6), and we shall want to know if it may serve as a canonical representation of $x(t)$, in the sense of Theorem 2.1. Usually it will then be comparatively easy to find out whether the conditions Q_1, Q_2 and Q_3 of Theorem 2.1 are satisfied or not. Suppose now that these conditions are in fact satisfied. In order that (3.2) should be a canonical representation, it is then necessary and sufficient that the remaining condition Q_4 should also be satisfied. In the lemma given below, this condition will be expressed in a form that, in some cases, permits an easy application.

By (3.2) $x(t)$ is expressed as a sum of N terms, the nth term being an element of the Hilbert space $H(z_n, t)$. It follows that we have for all $t \in T$

$$H(x, t) \subset H(z_1, t) \oplus \cdots \oplus H(z_N, t).$$

In order that the two members of this relation should be identical, as required by the condition Q_4, it is then necessary and sufficient that the space in the second member should not contain any element orthogonal to $H(x, t)$. Now, by the remarks in connection with (2.5), any element y belonging to the vector sum of the mutually orthogonal $H(z_n, t)$ spaces has the form

$$y = \sum_{1}^{N} \int_{A}^{t} h_n(u)\, dz_n(u),$$

where $A < t < B$. If y is not almost everywhere equal to zero, we have

$$0 < Ey^2 = \sum_{1}^{N} \int_{A}^{t} h_n^2(u)\, dF_n(u) < \infty. \tag{3.3}$$

Further, if y is orthogonal to $x(s)$ for $A < s \leqq t$,

$$\text{(3.4)} \qquad Eyx(s) = \sum_1^N \int_A^s h_n(u) g_n(s, u)\, dF_n(u) = 0.$$

Now the space $H(x, t)$ is spanned by the $x(s)$ for all s such that $A < s \leqq t$, and so we obtain the following lemma.

Lemma 3.1. Suppose that the expansion (3.2) *is known to satisfy conditions Q_1, Q_2, and Q_3 of Theorem 2.1. Then* (3.2) *is a canonical representation of $x(t)$ if and only if, for every $t \in T$, it is impossible to find functions $h_1(u), \ldots, h_N(u)$ satisfying* (3.3), *and such that* (3.4) *holds for all s in $(A, t]$.*

4. Regularity Conditions

Let us consider one single term in the expansion (3.2), say

$$\int_A^t g(t, u)\, dz(u),$$

where $z(u)$ is a process of orthogonal increments, satisfying the conditions Q_1 and Q_2 of Theorem 2.1. We shall say briefly that this term satisfies the *regularity conditions R*, if

(R_1) $g(t, u)$ and $\dfrac{\partial g(t, u)}{\partial t}$ are bounded and continuous for $u, t \in T$ and $u \leqq t$.

(R_2) $g(t, t) = 1$ for $t \in T$.

(R_3) $F(u)$ as defined by $E(dz(u))^2 = dF(u)$ is absolutely continuous and not identically constant, with a derivative $f(u) = F'(u)$ having at most a finite number of discontinuity points in any finite subinterval of T.

It will be convenient to add a few remarks with respect to the condition R_2. As pointed out in connection with Theorem 2.1, the canonical expansion (2.6) represents $x(t)$ as the accumulated effect of a stream of N-dimensional random impulses or innovations $[dz_1(u), \ldots, dz_N(u)]$ acting throughout the past of the process. The response function $g_n(t, u)$ measures the effect at time t of a unit impulse in the nth component acting at time u, and $g_n(t, t)$ is the corresponding instantaneous response function associated with the instant t. In an applied problem it may often seem natural to assume that we always have $g_n(t, t) > 0$. If this is so, the canonical representation (2.6) may be transformed by writing

$$\overline{g_n}(t, u) = \frac{g_n(t, u)}{g_n(u, u)},$$

$$d\overline{z_n}(u) = g_n(u, u)\, dz_n(u),$$

so that we obtain

$$x(t) = \sum_1^N \int_A^t \overline{g_n}(t, u)\, d\overline{z_n}(u),$$

which is obviously a canonical representation of $x(t)$ with $\overline{g_n}(t, t) = 1$. Thus it will be seen that, if we are willing to assume $g_n(t, t) > 0$ for all n and t, the condition R_2 will imply no further restriction of generality.

5. Classes of Processes with Multiplicity $N = 1$

We now proceed to prove the following theorem.

Theorem 5.1. Let X be the class of all $x(t)$ processes admitting a canonical representation (2.6) *such that each term in the second member satisfies the conditions R of Section* 4. *Then every $x(t) \in X$ has multiplicity $N = 1$.*

Let $x(t)$ be a process belonging to the class X, and suppose that $x(t)$ has multiplicity $N > 1$. We have to show that this is impossible.

By hypothesis, $x(t)$ has a canonical representation with $N > 1$, every term of which satisfies the regularity conditions R. From the regularity condition R_3 and the condition Q_3 of Theorem 2.1, it easily follows that we can find a finite subinterval of T, say $T_1 = (A_1, B_1)$, such that the derivatives $f_1(u)$ and $f_2(u)$ are bounded away from zero for all $u \in T_1$. Take any point t in T_1. It then follows from Lemma 3.1 that our theorem will be proved if we can find functions $h_1(u)$, $\ldots$, $h_N(u)$ satisfying (3.3), and such that (3.4) is satisfied for $A < s \leqq t$.

We now take $h_n(u) = 0$ for all u when $n > 2$, and $h_1(u) = h_2(u) = 0$ for $u \leqq A_1$ and $u \geqq B_1$. The relation (3.4) is then certainly satisfied for $A < s \leqq A_1$, while for $A_1 < s \leqq t$ it becomes

$$\sum_1^2 \int_{A_1}^s h_n(u) g_n(s, u) f_n(u)\, du = 0. \tag{5.1}$$

14

By the regularity conditions R_1 and R_2, this relation may be differentiated with respect to s, and gives

$$\sum_1^2 \left[h_n(s) f_n(s) + \int_{A_1}^{s} h_n(u) f_n(u) \frac{\partial g_n(s, u)}{\partial s} \, du \right] = 0. \tag{5.2}$$

Consider now the two integral equations of Volterra type obtained by equating the nth term of the first member of (5.2) to $(-1)^n$, where $n = 1$ and 2. Each of these can be solved by the classical iteration method, regarding $h_n(s)f_n(s)$ as the unknown function. The uniquely determined solutions are bounded and continuous for $A_1 < s \leqq t$, and are not almost everywhere equal to zero. These solutions satisfy (5.2), and, by integration, we obtain (5.1). Since by hypothesis $f_1(u)$ and $f_2(u)$ are bounded away from zero for $A_1 < u < B_1$, it follows that (3.3) is also satisfied, and the proof is completed.

Consider now the case of an $x(t)$ process given by the expression (3.2) with $N = 1$:

$$x(t) = \int_A^t g(t, u) \, dz(u), \tag{5.3}$$

where the second member is known to satisfy the regularity conditions R. It cannot be directly inferred from the preceding theorem that (5.3) is a canonical representation of $x(t)$ in the sense of Theorem 2.1. In fact, it might be possible that $x(t)$ would have muliplicity $N > 1$, with a canonical representation the terms of which do not satisfy the conditions R.

However, in the case when the lower extreme A of T is finite, we can actually prove that (5.3) is a canonical representation. On the other hand, when $A = -\infty$, we are only able to prove this under an additional condition, as shown by the following theorem.

Theorem 5.2. Let $x(t)$ be given for $t \in T = (A, B)$ by (5.3), *with a second member satisfying the conditions Q_1 and Q_2 of Theorem* 2.1, *as well as the regularity conditions R of Section* 4. *If A is finite, $x(t)$ has multiplicity $N = 1$, and* (5.3) *is a canonical representation. When $A = -\infty$, the same conclusion holds under the additional condition that*

$$\int_{-\infty}^{t} \left| \frac{\partial g(t, u)}{\partial t} \right| du < \infty \tag{5.4}$$

for all $t \in T$.

By Lemma 3.1 this theorem will be proved if we can show that, for any $t \in T$, it is impossible to find a function $h(u)$ such that

$$0 < \int_A^t h^2(u) f(u)\, du < \infty \tag{5.5}$$

and

$$\int_A^s h(u) g(s, u) f(u)\, du = 0 \tag{5.6}$$

for all s in $(A, t]$. As in the preceding proof, (5.6) may be differentiated, and gives

$$h(s)f(s) + \int_A^s h(u) f(u) \frac{\partial g(s, u)}{\partial s}\, du = 0.$$

This is a homogeneous integral equation of the Volterra type, and it follows from the classical theory that, in the case when A is finite, the only solution is $h(s)f(s) = 0$ for all s in $(A, t]$. If $A = -\infty$, it is easily shown that, under the assumption (5.4), the classical proof will hold without modification also in this case. Obviously the solution $h(u)f(u) = 0$ cannot satisfy (5.5), and so the proof is completed.

REMARK. We observe that Theorem 5.2 does not hold without the regularity conditions R. Consider, in fact, the case when $T = (A, B)$ is a finite interval, and $x(t)$ is given by (5.3), with a $z(u)$ satisfying R_3, and

$$g(t, u) = \frac{2(u - A)}{t - A} - 1.$$

Then R_2 is satisfied, but not R_1, as $\frac{\partial g}{\partial t} = -\frac{2(u - A)}{(t - A)^2}$ is not bounded as $t \to A$ and $A < u < t$. Taking $h(u)f(u) = 1$, and assuming that $f(u)$ is bounded away from zero over (A, B), it will be seen that (5.5) and (5.6) are both satisfied. Thus, in this case, (5.3) is not a canonical representation, and the multiplicity of $x(t)$ remains unknown.

6. Some Properties of Processes with $N = 1$

Let $x(t)$ be a process of multiplicity $N = 1$, with a canonical representation

$$x(t) = \int_A^t g(t, u)\, dz(u) \tag{6.1}$$

16

satisfying the conditions of Theorem 5.2. We shall discuss some properties of $x(t)$.

UNIQUENESS. It will be shown that, under the conditions stated, the representation (6.1) of $x(t)$ is unique in the sense that, for any two representations of $x(t)$ satisfying the same conditions, and characterized by the subscripts 1 and 2 respectively, we have

$$g_1(t, u) = g_2(t, u),$$

$$z_1(t) - z_1(u) = z_2(t) - z_2(u),$$

for any $t, u \in T$ and $u \leqq t$. We note that the last relation signifies that the random variables in the two members are equivalent, i.e., equal with probability one.

For the proof of this assertion we use the expression (2.7) for the error involved in a least squares prediction. Since this error depends only on the process $x(t)$ itself, it must be the same for both representations, so that we have

$$(6.2)\quad x(t+h) - P_t x(t+h) = z_n(t+h) - z_n(t) - \int_t^{t+h} [1 - g_n(t+h, u)]\, dz_n(u)$$

for $n = 1$ and 2, and for any $h > 0$. Now the variance of $z_n(t+h) - z_n(t)$ is $F_n(t+h) - F_n(t)$ while, by conditions R_1 and R_2, the variance of the last term in (6.2) is for small h small relative to $F_n(t+h) - F_n(t)$. Allowing h to tend to zero it then follows that we have $f_1(t) = f_2(t)$. For any $u < t$ we now divide the interval $[u, t)$ in n subintervals of equal length, and apply (6.2) to each subinterval. Summing over the subintervals and allowing n to tend to ∞, we then obtain by the same argument as before $z_1(t) - z_1(u) = z_2(t) - z_2(u)$. Finally, for $u < u + h < t$ we have by (2.7) for $n = 1$ and 2

$$P_{u+h}x(t) - P_u x(t) = g_n(t, u+h)[z_n(u+h) - z_n(u)] - \int_u^{u+h} [g_n(t, u+h) - g_n(t, v)]\, dz_n(v),$$

and an argument of the same type as above will then show that $g_1(t, u) = g_2(t, u)$.

THE COVARIANCE FUNCTION. Since $Ex(t) = 0$ we have for the covariance function

(6.3) $$r(s, t) = Ex(s)x(t) = \int_A^{\min(s,t)} g(s, u)g(t, u)f(u)\, du,$$

which shows that $r(s, t)$ is everywhere continuous in the interval $T \times T$ in the (s, t)-plane. Moreover, by the regularity conditions R_1 and R_2, it follows from (6.3) that $r(s, t)$ has partial derivatives $\frac{\partial r}{\partial s}$ and $\frac{\partial r}{\partial t}$, which are continuous at every point with $s \neq t$. On the diagonal $s = t$ these derivatives are discontinuous; we have, e.g.,

(6.4)
$$\lim_{s \uparrow t} \frac{r(s, t) - r(t, t)}{s - t} = \int_A^t g(t, u) \frac{\partial g(t, u)}{\partial t} f(u)\, du + f(t),$$

$$\lim_{s \downarrow t} \frac{r(s, t) - r(t, t)}{s - t} = \int_A^t g(t, u) \frac{\partial g(t, u)}{\partial t} f(u)\, du.$$

Thus at the point $s = t$, there is a jump of the height $f(t)$ in the partial derivatives. We shall see later that this fact has interesting applications in the case of normal processes.

SOME PARTICULAR CASES. In the case when $g(t, u) = p(t)q(u)$ is the product of one function of t and one function of u, it follows from the condition R_2 that $p(t)q(t) = 1$, so that

$$g(t, u) = \frac{p(t)}{p(u)}.$$

For the covariance function we then have

$$r(s, t) = p(s)p(t) \int_A^{\min(s,t)} \frac{f(u)}{p^2(u)}\, du.$$

Hence for $s < t < u$ the covariance function satisfies the relation

$$r(s, u)r(t, t) = r(s, t)r(t, u).$$

For the correlation coefficient

$$\rho(s, t) = \frac{r(s, t)}{[r(s, s)r(t, t)]^{1/2}}$$

the corresponding relation is

$$\rho(s, u) = \rho(s, t)\rho(t, u).$$

This is the characteristic relation of a *Markov process in the wide sense* according to the terminology of Doob [6, p. 233]. If, in addition, $x(t)$ is a normal process, it is even *strict sense Markov*. In the particular case when $A = -\infty$ and

$$p(t) = e^{-ct}, \qquad f(u) = 2c,$$

we have

$$r(s, t) = \rho(s, t) = e^{-c|s-t|},$$

so that $x(t)$ is the well-known stationary Markov process.

When $g(t, u) = 1$ for all t and u, and $z(A) = 0$, (6.1) gives $x(t) = z(t)$, so that $x(t)$ is now a *process of orthogonal increments*. In the special case when $x(t) = z(t)$ is normal, A is finite, and $f(u) = 1$, we have the Wiener (or Brownian movement) process.

Finally, when $g(t, u) = g(t - u)$, $f(u) = 1$, and $A = -\infty$, we have the case of a *stationary process*.

7. Normal Processes, Equivalence

From now on we shall restrict ourselves to *normal* (or Gaussian) processes, i.e., processes such that all their finite-dimensional distributions are normal. As far as statistical applications are concerned, the assumption of normality will often imply an oversimplification. Nevertheless it has been found useful and, in any case, the investigation of normal processes should be regarded as an important starting point for more realistic research work.

Consider a normal process $x(t)$ given by the canonical representation (6.1), and thus having multiplicity $N = 1$. As before, we assume that the second member of (6.1) satisfies the conditions of Theorem 5.2. Any finite number of elements of the Hilbert space $H(x)$ then have a normal joint distribution, and since $H(x) = H(z)$ it follows that the $z(u)$ process is also normal. Conversely, if $z(u)$ is known to be normal, (6.1) shows that $x(t)$ is also normal.

We shall begin by choosing our basic probability space in a way that will be convenient for the sequel. Consider the measurable space $(W, \mathbf{B})$, where W is the space consisting of all finite and real-valued functions $w(\cdot)$ defined on the real line, while $\mathbf{B}$ is the smallest σ-field of sets in W that includes every set formed by all functions $w(\cdot)$ satisfying a finite number of inequalities of the form

$$a_r < w(t_r) \leqq b_r$$

for $r = 1, 2, \ldots, n$.

Let P be a probability measure defined on the sets of $\mathbf{B}$, and consider the probability space $(W, \mathbf{B}, P)$. A point ω of the space W is a function $\omega = w(\cdot)$, and a random variable on $(W, \mathbf{B}, P)$ is a measurable functional of $w(\cdot)$.

Consider the family $x(t) = x(t, \omega)$ of random variables on this probability space defined by the relation

$$x(t, \omega) = x(t, w(\cdot)) = w(t).$$

This $x(t)$ family defines a stochastic process, any finite-dimensional distribution of which, say for the parameter values $t = t_1, t_2, \ldots, t_n$, is given by probabilities of the form

$$P(a_r < x(t_r, \omega) \leqq b_r) = P(a_r < w(t_r) \leqq b_r)$$

for $r = 1, 2, \ldots, n$. If P is such that all these distributions are normal, $x(t)$ is a normal stochastic process, and we say that P is a *normal probability measure* on $(W, \mathbf{B})$.

Any two probability measures P_0 and P_1 on $(W, \mathbf{B})$ are said to be *perpendicular*, if there exists a set of functions $w(\cdot)$ belonging to $\mathbf{B}$, say the set $S \in \mathbf{B}$, such that

$$P_0(S) = 0, \qquad P_1(S) = 1.$$

Thus, in this case, P_0 and P_1 have their total probability masses concentrated on two disjoint sets. It is then possible to test the hypothesis that the $x(t)$ distribution is given by P_0 against the alternative hypothesis P_1 with complete certainty by choosing S as our critical region. In practical applications this should be regarded as a singular limiting case that will hardly ever be more than approximately realized.

On the other hand, P_0 and P_1 are called *equivalent*, and we write $P_0 \sim P_1$, if the two measures are mutually absolutely continuous so that, for any set $S \in \mathbf{B}$, the relations $P_0(S) = 0$ and $P_1(S) = 0$ always imply one another. In this case the so called *Radon-Nikodym derivative*

$$p(\omega) = \frac{dP_1(\omega)}{dP_0(\omega)}$$

exists, and we have for any set $S \in \mathbf{B}$

$$P_1(S) = \int_S p(\omega)\, dP_0(\omega).$$

The derivative $p(\omega)$ corresponds to the *likelihood ratio* in the classical theory of statistical inference, and can be used, in the same way as the likelihood ratio, for the construction of statistical tests by means of the basic *Neyman-Pearson Lemma.*

Hajek [9] and Feldman [7] proved the important theorem that *two normal probability measures are either perpendicular or equivalent.* No intermediate case can occur. A fundamental problem in the theory of normal stochastic processes will thus be to find out when two given normal probability measures are equivalent, and when they are perpendicular. During recent years a large number of works dealing with this problem have been published. We refer, e.g., to papers by Parzen [13, 14], Yaglom [19], and Rozanov [16]. In our subsequent discussion of equivalence problems for normal processes of multiplicity $N = 1$, we shall follow mainly the last-mentioned author.

The finite-dimensional distributions of a normal process $x(t)$ are uniquely determined by the *mean* and *covariance functions*

$$m(t) = Ex(t), \qquad r(s, t) = E(x(s) - m(s))(x(t) - m(t)).$$

If $P(m, r)$ denotes the normal probability measure associated with $m(t)$ and $r(s, t)$, it can be proved without difficulty that we have

$$P(m_0, r_0) \sim P(m_1, r_1)$$

when and only when

$$P(m_0, r_0) \sim P(m_1, r_0) \sim P(m_1, r_1).$$

In order to find conditions for the equivalence of two normal probability measures it is thus sufficient to consider two particular cases, viz.: (a) the case of different means but the same covariance; and (b) the case of the same mean but different covariances. We shall consider these two cases separately.

CASE A. DIFFERENT MEANS, BUT THE SAME COVARIANCE. As we can always add a nonrandom function to the given stochastic $x(t)$ process, we may consider two normal probability measures P_0 and P_1 such that, denoting by E_0 and E_1 the corresponding expectations,

$$E_0 x(t) = 0, \qquad E_0 x(s) x(t) = r(s, t),$$

$$E_1 x(t) = m(t), \qquad E_1(x(s) - m(s))(x(t) - m(t)) = r(s, t).$$

Let $H_0(x)$ be the Hilbert space spanned by the random variables $x(t)$, with the inner product

$$(y, z) = E_0 yz.$$

The mean

$$E_1 y = M(y)$$

is finite for all finite linear combinations y of the $x(t_i)$, and is a linear functional in $H_0(x)$ such that

$$M(x(t)) = m(t).$$

A necessary and sufficient condition for the equivalence of P_0 and P_1 is that $M(y)$ is a continuous linear functional in $H_0(x)$ [16, p. 31].

If this condition is satisfied, it is known from Hilbert space theory that there is a uniquely determined element $y \in H_0(x)$ such that

$$M(v) = (v, y) = E_0 yv \tag{7.1}$$

for all $v \in H_0(x)$. In particular, taking $v = x(t)$, we obtain

$$m(t) = E_0 yx(t). \tag{7.2}$$

Since the variables $x(t)$ span $H_0(x)$, the relation (7.2) implies (7.1), so that (7.2) is a necessary and sufficient equivalence condition.

We now apply this equivalence criterion to the case when P_0 is the probability measure corresponding to the normal $x(t)$ process given by the canonical expression (6.1):

$$x(t) = \int_A^t g(t, u)\, dz(u),$$

while P_1 is associated with the process $m(t) + x(t)$. Since in this case $H_0(x) = H_0(z)$, a necessary and sufficient condition for the equivalence of P_0 and P_1 is that $m(t)$ can be represented in the form

$$m(t) = E_0 yx(t) = \int_A^t h(u)\, g(t, u)\, f(u)\, du \tag{7.3}$$

for some $y \in H_0(z)$ given by

$$y = \int_A^B h(u)\, dz(u). \tag{7.4}$$

Then y is a normally distributed random variable with $E_0 y = 0$ and

$$E_0 y^2 = \int_A^B h^2(u)\, f(u)\, du < \infty. \tag{7.5}$$

The set of all functions $m(t)$ such that P_0 and P_1 are equivalent is thus identical with the set of all functions represented by (7.3) for some $y \in H_0(x) = H_0(z)$, i.e., for some $h(u)$ satisfying (7.5). This set of $m(t)$ functions, with an inner product defined by $(m_1, m_2) = (y_1, y_2) = E_0 y_1 y_2$, is known as the *reproducing kernel Hilbert space* of the $x(t)$ process. The usefulness of this concept has been brought out by the works of Parzen quoted above.

The Radon-Nikodym derivative is in this case

$$p(\omega) = \frac{dP_1(\omega)}{dP_0(\omega)} = e^{y - 1/2 E_0 y^2},$$

where y is given by (7.4). The most powerful test of the hypothesis P_0 against the alternative P_1 is obtained by taking as critical region the set of all $\omega = w(\cdot)$ such that $p(\omega) > c$, the constant c being determined by

$$P_0(p(\omega) > c) = P_0(y > \tfrac{1}{2} E_0 y^2 + \log c) = \alpha,$$

y being normally distributed with zero mean and variance given by (7.5), while α is the desired level of the test.

As before we assume that the canonical expression (6.1) satisfies the conditions of Theorem 5.2, so that we may differentiate (7.3) and obtain a Volterra integral equation

$$m'(t) = h(t)\, f(t) + \int_A^t h(u)\, f(u) \frac{\partial g(t, u)}{\partial t}\, du.$$

From this equation an explicit expression for $h(t)$ may be obtained by the iteration method.

In the particular case of a *normal Markov process* with $g(t, u) = p(t)/p(u)$ discussed in Section 6, the expression (7.3) for $m(t)$ becomes

$$m(t) = p(t) \int_A^t \frac{h(u)\, f(u)}{p(u)}\, du,$$

which gives

$$h(t) = \frac{p(t)}{f(t)} \cdot \frac{d}{dt}\left(\frac{m(t)}{p(t)}\right). \tag{7.6}$$

It is thus necessary and sufficient for equivalence that the function $h(t)$ given by (7.6) satisfies the convergence condition (7.5).

For $p(t) = 1$ we have $g(t, u) = 1$, and $x(t)$ becomes a normal process with independent increments. Suppose that we observe this in the interval $0 \leqq t \leqq T$. From (7.6) we then obtain $h(t) = m'(t)/f(t)$, and the condition (7.5) now becomes

$$\int_0^T \frac{(m'(t))^2}{f(t)} \, dt < \infty.$$

Taking here $f(t) = 1$, we have the particular case of the Wiener process.

CASE B. THE SAME MEAN, BUT DIFFERENT COVARIANCES. We now consider two probability measures P_0 and P_1, and a normal process $x(t)$ with the canonical representation

$$x(t) = \int_A^t g(t, u) \, dz(u),$$

which, as before, is assumed to satisfy the conditions of Theorem 5.2. We suppose that, according to the probability measure P_i for $i = 0$ and 1,

$$g(t, u) = g_i(t, u),$$

$$E_i x(t) = E_i z(t) = 0,$$

$$E_i x(s)x(t) = r_i(s, t) = \int_A^{\min(s,t)} g_i(s, u) g_i(t, u) f_i(u) \, du.$$

We denote by $H_i(x) = H_i(z)$ the Hilbert space spanned by the random variables $x(t)$, with the inner product

$$(y, z)_i = E_i yz.$$

The direct product space

$$H_0(x) \otimes H_1(x) = H_0(z) \otimes H_1(z)$$

consists of all functions

$$y(\omega_0, \omega_1) = \int_A^B \int_A^B h(u, v) \, dz(u, \omega_0) \, dz(v, \omega_1),$$

where ω_i is a point in the probability space $(W, \mathbf{B}, P_i)$, and $h(u, v)$ satisfies

24

$$(7.7) \qquad \int_A^B \int_A^B h^2(u, v) f_0(u) f_1(v)\, du\, dv < \infty.$$

Then [16, p. 54] P_0 *and* P_1 *are equivalent if and only if we have, for all* s, t *and for some* $y \in H_0(z) \otimes H_1(z)$,

$$(7.8) \qquad \begin{aligned} r_1(s, t) - r_0(s, t) &= E_0 E_1 \left[y(\omega_0, \omega_1)\, x(s, \omega_0) x(t, \omega_1) \right] \\ &= \int_A^s \int_A^t h(u, v) g_0(s, u) g_1(t, v) f_0(u) f_1(v)\, du\, dv. \end{aligned}$$

Under the present conditions it follows from (7.8) that the difference $r_1(s, t) - r_0(s, t)$ has everywhere continuous partial derivatives of the first order. Hence the discontinuities of the partial derivatives of r_1 and r_0 at the diagonal $s = t$ must cancel in the difference, so that by (6.4) we obtain as a necessary condition for equivalence that

$$(7.9) \qquad f_1(t) = f_0(t)$$

for all t.

In the case when P_0 and P_1 are both associated with normal Markov processes such that $g_i(t, u) = p_i(t)/p_i(u)$, the necessary and sufficient condition for equivalence becomes by (7.8) and (7.9)

$$r_1(s, t) - r_0(s, t) = p_0(s) p_1(t) \int_A^s \int_A^t h(u, v) \frac{f(u) f(v)}{p_0(u) p_1(v)}\, du\, dv,$$

where $f = f_0 = f_1$. This gives

$$(7.10) \qquad h(s, t) = \frac{p_0(s) p_1(t)}{f(s) f(t)} \cdot \frac{\partial^2}{\partial s \partial t} \left(\frac{r_1(s, t) - r_0(s, t)}{p_0(s) p_1(t)} \right).$$

Thus P_0 and P_1 are equivalent if and only if $h(s, t)$ as given by (7.10) satisfies the convergence condition (7.7).

Consider the particular case when P_0 is associated with a Wiener process with $f_0(t) = 1$ observed in the interval $0 \leqq t \leqq T$, while P_1 corresponds to a normal Markov process with $g_1(t, u) = p_1(t)/p_1 u)$ observed in the same interval. By (7.9) a necessary condition for equivalence is that we have $f_1(t) = f_0(t) = 1$. If this condition is satisfied, (7.10) gives, after some calculation,

$$(7.11) \qquad h(s, t) = \frac{p_1'(\max(s, t))}{p_1(t)},$$

Now it follows from the regularity condition R_1 that $h(s, t)$ as given by (7.11) is bounded for all $s, t \in (0, T)$. Thus the convergence condition (7.7) is satisfied, and the condition $f_1(t) = f_0(t) = 1$ is in this case both necessary and sufficient for equivalence.

In this last case the Radon-Nikodym derivative is

$$p(\omega) = \frac{dP_1(\omega)}{dP_0(\omega)} = Ke^{-1/2y(\omega,\omega)},$$

where ω is a point in the probability space $(W, \mathbf{B}, P_0)$, while

$$y(\omega, \omega) = \int_0^T \int_0^T h(u, v)\, dz(u, \omega)\, dz(v, \omega),$$

$h(u, v)$ is given by (7.11), and K is a constant determined so as to render the integral of $p(\omega)\, dP_0(\omega)$ over the whole space equal to 1.

8. Normal Processes, Estimation

In this final section we shall briefly discuss some problems of statistical estimation for a normal process given by the canonical representation (6.1), subject to the same conditions as before. By (6.1) the diagonal $s = t$ of the (s, t)-plane is a line of discontinuity for the first order partial derivatives of the covariance function $r(s, t)$, the jump at the point $s = t$ being $f(t)$, in the sense specified by (6.4).

Let now $a < b$ be fixed constants, and consider the random variable S_n given by the expression

$$S_n = \sum_{a\cdot 2^n}^{b\cdot 2^n} [x(k/2^n) - x((k-1)/2^n)]^2.$$

For any given n an observed value of S_n can be computed as soon as a realization of the $x(t)$ process is known for $a \leqq t \leqq b$. It follows from a theorem due to Baxter [1] that, as $n \to \infty$,

$$S_n \to \int_a^b f(t)\, dt,$$

where the convergence takes place both in q.m. and with probability one.

Thus from one single observed realization of $x(t)$, we may compute the quadratic variation S_n for some large value of n, and the value obtained will give us an estimate of $\int_a^b f(t)\,dt$ with a variance tending to zero as $n \to \infty$.

In order to find an estimate of a parametric function connected with $g(t, u)$, we take $s < t$ and write for $k = 0, 1, 2, \ldots$

$$s_{k,n} = s - k \cdot 2^{-n}.$$

Consider the linear regression of the random variable $x(t)$ on the variables $x(s_{k,n})$ for $k = 0, 1, \ldots, m$, where $m = n \cdot 2^n$. The regression polynomial, say

$$x_{s,n}(t) = c_{0,n}x(s_{0,n}) + \cdots + c_{m,n}x(s_{m,n})$$

is the projection of $x(t)$ on the finite-dimensional Hilbert space spanned by the random variables $x(s_{0,n}), \ldots, x(s_{m,n})$. It is easily proved that, under the present conditions, $x_{s,n}(t)$ converges in q.m. as $n \to \infty$ to the best linear prediction

$$x_s(t) = P_s x(t)$$

of $x(t)$ in terms of the whole development of the process up to the instant s. Thus the residual

$$x(t) - x_{s,n}(t)$$

converges in q.m. to the prediction error

$$x(t) - P_s x(t)$$

given by (2.7). Hence we obtain from (2.7), as $n \to \infty$,

$$E[x(t) - x_{s,n}(t)]^2 \to \int_s^t g^2(t, u) f(u)\,du. \tag{8.1}$$

For a given large value of n, a sample value of the residual variance in the first member of (8.1) can be computed if a sufficient number of realizations of the $x(t)$ process are available. This will then serve as an estimate of the parametric function in the second member.

REFERENCES

1. Baxter, G., A strong limit theorem for Gaussian processes, *Proc. Am. Math. Soc.*, 7 (1956), 522–527.
2. Cramér, H., On some classes of non-stationary stochastic processes, *Proc. 4th Berkeley Symp. Math. Stat. and Prob.*, II (1961), 57–77.
3. ———, On the structure of purely non-deterministic stochastic processes, *Arkiv Mat.*, 4 (1961), 249–266.
4. ———, Stochastic processes as curves in Hilbert space, *Teoriya Veroj. i ee Primen.*, 9 (1964), 169–179.
5. ———, and Leadbetter, M. R., *Stationary and related stochastic processes,* New York: Wiley, 1967.
6. Doob, J. L., *Stochastic processes,* New York: Wiley, 1953.
7. Feldman, J., Equivalence and perpendicularity of Gaussian processes, *Pacif. J. Math.*, 8 (1958), 699–708.
8. Grenander, U., Stochastic processes and statistical inference, *Arkiv Mat.*, 1 (1950), 195–277.
9. Hajek, J., On a property of normal distributions of any stochastic process, *Czech. Math. J.*, 8 (1958), 610–618, and *Selected Transl. Math. Stat. Prob., 1* (1961), 245–252.
10. Hida, T., Canonical representations of Gaussian processes and their applications, *Mem. Coll. Sc. Kyoto Ser. A,* 32 (1960), 109–155.
11. Kallianpur, G., and Mandrekar, V., Multiplicity and representation theory of purely non-deterministic stochastic processes, *Teorija Veroj. i ee Primen.*, 10 (1965), 553–581.
12. Nagy, B. v. Sz., Spektraldarstellung linearer Transformationen des Hilbertschen Raumes. *Ergeb. Mathematik,* 5, Berlin, 1942.
13. Parzen, E., An approach to time series analysis, *Ann. Math. Stat.*, 32 (1961), 951–989.
14. ———, Statistical inference on time series by RKHS methods, *Techn. Report 14,* Stanford Univ., Statistics Dept., 1970.
15. Plessner, A. I., and Rokhlin, V. A., Spectral theory of linear operators, *Uspekhi Mat. Nauk,* 1 (1946), 71–191 (Russian).

16. Rozanov, Yu. A., Infinite-dimensional Gaussian distributions, *Trudy Ord. Lenina, Mat. Inst. V. A. Steklova,* 108 (1968), 1–136 (Russian).
17. Stone, M. H., Linear transformations in Hilbert space and their applications to analysis, *Am. Math. Soc. Colloquium Publications,* 25, New York, 1932.
18. Wiener, N., and Masani, P., The prediction theory of multivariate stochastic processes, *Acta Math.,* 98 (1957), 111–150, and 99 (1958), 93–137.
19. Yaglom, A. M., On the equivalence and perpendicularity of two Gaussian probability measures in function space, *Proc. Symp. Time Series Analysis,* New York: Wiley, 1963, 327–346.

67.

with M. R. Leadbetter and R. J. Serfling

On distribution function–moment relationships in a stationary point process

Z. Wahrscheinlichkeitstheorie und verw. Gebiete **18**, 1–8 (1971)

1. Introduction

We shall, throughout, consider a stationary point process, and denote by $N(s, t)$ the number of events in the (semiclosed) interval $(s, t]$. It will further be assumed that the mean number of events per unit time $EN(0, 1)=\lambda$ is finite, and that there is zero probability of the occurrence of multiple events (hence the process is regular (orderly) in the sense that $Pr\{N(0, t)>1\}=o(t)$ as $t\downarrow 0$). Write

$$G_n(t)=Pr\{N(0, t)\geqq n\}, \qquad n=0, 1, 2, \ldots. \tag{1.1}$$

This is the probability that the n-th event after time zero occurs no later than time t, i.e., $G_n(t)$ is the distribution function for the time to the n-th event after time zero (which is a well defined random variable).

We now modify $G_n(t)$ to define a corresponding probability, but now conditioned by the occurrence of an event "at" time zero in the following precise sense:

$$F_n(t)=\lim_{\delta\downarrow 0} Pr\{N(0, t)\geqq n \mid N(-\delta, 0)\geqq 1\} \tag{1.2}$$

and we shall write

$$F_n(t)=Pr\{N(0, t)\geqq n \mid \text{Event at } 0\} \tag{1.3}$$

it being understood that the right hand side of (3) is defined as the limit occurring in Eq. (1.2); i.e., the zero probability condition that an event occurs precisely at time zero, is replaced by the positive probability condition $N(-\delta, 0)\geqq 1$, and the limit is taken as $\delta\downarrow 0$.

The limit in (1.2) does in fact exist and moreover $F_n(t)$ is a distribution function for each n. Further, we may interpret $F_n(t)$ as the distribution function for the interval from time zero to the n-th event after time zero, given an event occurred "at" time zero. Alternatively "conditional probabilities" such as (1.3) may be defined (and manipulated) from the general theory of Palm distributions ([4]). Here, however, we use the analytical approach described (cf. [1]).

Corresponding to (1.1) and (1.2), we define

$$v_n(t)=G_n(t)-G_{n+1}(t)=Pr\{N(0, t)=n\}, \tag{1.4}$$

$$u_n(t)=F_n(t)-F_{n+1}(t)=Pr\{N(0, t)=n \mid \text{Event at } 0\}. \tag{1.5}$$

Then for fixed t, $v_n(t)$ is the probability distribution of the non-negative integer valued random variable $N(0, t)$, whereas $u_n(t)$ is the distribution of $N(0, t)$ conditional on the occurrence of an event "at" time zero.

---

* Research supported by the Office of Naval Research under Contract No. N00014-67-A-0321-0002.
** Research supported by the U.S. Department of Transportation under Contract No. FH-11-6890.

1 Z. Wahrscheinlichkeitstheorie verw. Geb., Bd. 18

We next define two types of factorial moments, viz.,

$$\beta_k(t) = \sum_{n=k}^{\infty} n(n-1)\dots(n-k+1)\, v_n(t) = k! \sum_{n=k}^{\infty} \binom{n}{k} v_n(t), \tag{1.6}$$

$$\alpha_k(t) = \sum_{n=k}^{\infty} n(n-1)\dots(n-k+1)\, u_n(t) = k! \sum_{n=k}^{\infty} \binom{n}{k} u_n(t). \tag{1.7}$$

It is apparent at once (writing $N(0, t) = N$), that

$$\beta_k(t) = E N(N-1)\dots(N-k+1). \tag{1.8}$$

That is $\beta_k(t)$ is the k-th factorial moment of $N(0, t)$. It will be shown in Section 3 that the corresponding quantities $\alpha_k(t)$ satisfy

$$\alpha_k(t) = \lim_{\delta \downarrow 0} E\{N(N-1)\dots(N-k+1) \,|\, N(-\delta, 0) \geqq 1\}. \tag{1.9}$$

Anticipating this (intuitively obvious) fact, we shall refer to the $\alpha_k(t)$ defined by (1.8) as the *conditional factorial moments* of $N(0, t)$.

One of the main purposes of this paper is to relate the distribution functions $G_n(t)$, $F_n(t)$ to the corresponding factorial moment sequences $\beta_k(t)$, $\alpha_k(t)$. The practical importance of this (which will be discussed further later) lies in the fact that in a number of interesting cases there is no known way of calculating the distribution functions directly, whereas expressions for the factorial moments are known. This applies particularly when the events are zeros of a stationary stochastic process – a case discussed in [3].

The complexity of the expressions for the factorial moments makes it desirable to calculate as few of them as possible in some practical cases (cf. [3]). In Section 2, we obtain inequalities giving upper and lower bounds for the distribution functions based on essentially as few or as many of the moments as we please. From these inequalities exact series expressions for the distribution functions (involving an infinite number of the moments) will be given, together with necessary and sufficient conditions for their validity. Further, the conditional and unconditional moments will be there related to each other, yielding as a corollary, a series for $F_n(t)$ given in [2] under slightly more restrictive conditions.

In Section 3, we prove some further results relative to the conditional distribution functions and moments, in particular showing that (1.9) holds.

2. Bounds and Series for $F_n(t)$, $G_n(t)$

To assist in obtaining the desired inequalities for $F_n(t)$ and $G_n(t)$, we first prove two lemmas. The first of these is rather obvious but is stated for repeated reference.

Lemma 2.1. *Let* $\{p_j\}$ *be a probability distribution on* $\{0, 1, 2 \dots\}$, $P_j = \sum_{r=j}^{\infty} p_r$, *and* γ_j *the corresponding j-th factorial moment* $\sum_{m=j}^{\infty} m(m-1)\dots(m-j+1)\, p_m$. *Then* $\gamma_j/j! = \sum_{s=j}^{\infty} \binom{s-1}{j-1} P_s \,(\leqq \infty)$.

This is proved simply by substituting $P_s=\sum_{r=s}^{\infty} p_r$ and changing the order of summation of the non-negative quantities involved.

The next lemma contains the main calculation involved in obtaining the inequalities for F_n, G_n.

Lemma 2.2. *Let n, k be positive integers and (with the notation of Lemma 2.1) suppose that the factorial moments γ_j are finite for $1\leqq j\leqq n+k$. Write*

$$S(=S_{n,k})=\sum_{j=n}^{n+k}(-)^{j-n}\binom{j-1}{n-1}\gamma_j/j!.$$

Then $P_n=S+(-)^{k+1}T$, where $T\geqq 0$.

$$\left(\textit{Specifically, } T=\sum_{j=n+k+1}^{\infty}\binom{j-1}{n-1}\binom{j-n-1}{k}P_j.\right)$$

Proof. By Lemma 2.1,

$$S=\sum_{j=n}^{n+k}(-)^{j-n}\binom{j-1}{n-1}\sum_{s=j}^{\infty}\binom{s-1}{j-1}P_s$$

$$=\sum_{s=n}^{\infty}A_s P_s \tag{2.1}$$

where

$$A_s=\sum_{j=n}^{\min(s,n+k)}(-)^{j-n}\binom{j-1}{n-1}\binom{s-1}{j-1}.$$

For $s\leqq n+k$, we have

$$A_s=\binom{s-1}{n-1}\sum_{j=n}^{s}(-)^{j-n}\binom{s-n}{j-n}$$

$$=\binom{s-1}{n-1}\sum_{r=0}^{s-n}(-)^r\binom{s-n}{r}$$

which is one if $s=n$ and zero if $s>n$. Hence

$$\sum_{s=n}^{n+k}A_s P_s=P_n. \tag{2.2}$$

On the other hand, for $s>n+k$,

$$A_s=\binom{s-1}{n-1}\sum_{r=0}^{k}(-)^r\binom{s-n}{r}=\binom{s-1}{n-1}(-)^k\binom{s-n-1}{k}$$

by virtue of the known result $\sum_{r=0}^{k}(-)^r\binom{s}{r}=(-)^k\binom{s-1}{k}$ (which is easily shown by induction on k). Hence $\sum_{s=n+k+1}^{\infty}A_s P_s=(-)^k T$ and the desired result follows by using (2.1) and (2.2).

1*

We now use this lemma, identifying p_j with $v_j(t)$ (or $u_j(t)$), P_j with $G_j(t)$ (or $F_j(t)$), and γ_j with $\beta_j(t)$ (or $\alpha_j(t)$) (cf., Eqs. (1.4) to (1.7)). This leads at once to the inequalities for $F_n(t)$, $G_n(t)$ contained in the following result.

Theorem 2.3. *For given positive integers n, k, suppose the (conditional) factorial moments $\alpha_j(t)$ are finite for $1 \leqq j \leqq n+k$. Then, if k is even,*

$$F_n(t) \leqq \sum_{j=n}^{n+k} (-)^{j-n} \binom{j-1}{n-1} \alpha_j(t)/j! \tag{2.3}$$

whereas the inequality is reversed if k is odd. Corresponding inequalities hold for $G_n(t)$ (if $\beta_j(t) < \infty$ for $1 \leqq j \leqq n+k$) by simply replacing $\alpha_j(t)$ by $\beta_j(t)$.

The inequalities of the type (2.3), and the reverse inequalities for odd k values, thus provide a sequence of upper and lower bounds for $F_n(t)$, each based on just a finite number of the moments $\alpha_j(t)$. We may also obtain an exact series expression for $F_n(t)$ (and similarly for $G_n(t)$) under the conditions of the next theorem. This exact expression does, however, involve an infinite number of the moments.

Theorem 2.4. *Let $\alpha_j(t)$ be finite for $j = 1, 2 \ldots$. Then*

$$F_n(t) = \sum_{j=n}^{\infty} (-)^{j-n} \binom{j-1}{n-1} \alpha_j(t)/j! \quad (n = 1, 2 \ldots) \tag{2.4}$$

if and only if the j-th term of the series tends to zero as $j \to \infty$. That is, if

$$\binom{j-1}{n-1} \alpha_j(t)/j! \to 0 \quad \text{as } j \to \infty,$$

the series converges (not necessarily absolutely) to $F_n(t)$, and conversely. The theorem is true for $G_n(t)$ if $\beta_j(t)$ is substituted for $\alpha_j(t)$ in the statement.

Proof. If S_k denotes the partial sum $\sum_{j=n}^{n+k} (-)^{j-n} \binom{j-1}{n-1} \alpha_j(t)/j!$, it follows from Theorem 2.3 that the S_k alternately over- and underestimate $F_n(t)$. Thus $|S_k - F_n(t)| \leqq |S_{k+1} - S_k|$. Hence if the terms of the series tend to zero, it follows that $|S_{k+1} - S_k| \to 0$ and hence $S_k \to F_n(t)$, as required.

Conversely, if the series converges (to $F_n(t)$), the terms, of course, tend to zero.

In our next result, we relate the unconditional and conditional factorial moments to each other. First we note that if (for some fixed integer $r > 0$), we have $EN^r(0, t) < \infty$ for some $t > 0$ then it follows simply by Minkowski's inequality and stationarity that $EN^r(0, t) < \infty$ for *all* $t > 0$.

Theorem 2.5.

$$\beta_{k+1}(t)/(k+1)! = \lambda \int_0^t \frac{\alpha_k(u)}{k!} du \leqq \infty, \qquad k = 1, 2 \ldots.$$

Hence, if for some k, $EN^{k+1}(0, t) < \infty$ for some $t > 0$ (and hence for all $t > 0$), then $\beta_{k+1}(t)$ is absolutely continuous with density $\lambda(k+1)\alpha_k(t)$. Further, under this finiteness restriction, $\beta_{k+1}(t)$ has, for all $t \geqq 0$, the right hand derivative $D^+ \beta_{k+1}(t) = \lambda(k+1)\alpha_k(t)$.

Proof. From Lemma 2.1, we have

$$\beta_{k+1}(t)/(k+1)! = \sum_{n=k+1}^{\infty} \binom{n-1}{k} G_n(t).$$

Now a simple transformation of "Palm's Formulae" (cf., [1, Section 3.1]) shows that $G_n(t) = \lambda \int_0^t u_{n-1}(s)\,ds$ and hence

$$\beta_{k+1}(t)/(k+1)! = \lambda \int_0^t \left\{ \sum_{n=k}^{\infty} \binom{n}{k} u_n(s) \right\} ds,$$

where the sum and integral have been exchanged since the integrands are all non-negative. The first statement of the theorem now follows from the definition (1.7) of $\alpha_k(t)$.

If $EN^{k+1}(t) < \infty$ for $t > 0$, then $\beta_{k+1}(t) < \infty$ and the (non-decreasing) function $\alpha_k(t)$ must be finite for all t. It thus follows from the relation $\alpha_k(t)/k! = \sum_{n=k}^{\infty} \binom{n-1}{k-1} F_n(t)$ (Lemma 2.1) and the right-continuity of $F_n(t)$, that $\alpha_k(t)$ is continuous on the right. Thus the integral expression for $\beta_{k+1}(t)$ just obtained may be differentiated on the right at t to yield the final conclusion of the theorem.

By using this result, we may now write $F_n(t)$ in terms of the *unconditional* moments $\beta_k(t)$ (which may be simpler to write down in practice than the $\alpha_k(t)$). Specifically, if $\beta_k(t) < \infty$ for all $k = 1, 2 \ldots$, then $\alpha_k(t) < \infty$ for $k = 1, 2 \ldots$ and (2.4) may be rewritten as

$$F_n(t) = \lambda^{-1} \sum_{k=n+1}^{\infty} (-)^{k-n-1} \binom{k-2}{n-1} \beta_k'(t)/k! \tag{2.5}$$

where β_k' is written for $D^+ \beta_k$.

Eq. (2.5) thus expresses $F_n(t)$ in terms of the naturally occurring quantities $\beta_k(t)$ (this is the relation obtained in [2] under slightly more restrictive conditions).

3. The Conditional Moments

In this section, we prove two theorems which are independently of interest and from which Eq. (1.9) will follow. Special cases concern the "renewal function" $H(t) = \sum_{n=1}^{\infty} F_n(t)$.

Theorem 3.1. *Let t_i be the random variable which is the position of the i-th event after the origin. Let $t > 0$, $h > 0$. Then, for each positive integer k,*

$$\alpha_k(t) = (\lambda h)^{-1} E\left\{ \sum_{t_i \in (0,h)} k! \binom{N(t_i, t_i + t)}{k} \right\}. \tag{3.1}$$

Proof. Write $\lambda_{i,n} = 1$ if $N(t_i, t_i + t) \geqq n$, $\lambda_{i,n} = 0$ otherwise, and $N_{n,t} = \sum_{t_i \in (0,h)} \lambda_{i,n}$. That is $N_{n,t}$ is the number of $t_i \in (0, h)$ such that $N(t_i, t_i + t) \geqq n$. It follows from

[1, Section 5, Lemma] that $F_n(t)=(\lambda h)^{-1} EN_{n,t}$ and hence, since $N_{n,t}\geqq 0$,

$$\begin{aligned}\lambda h \alpha_k(t)/k! &= \lambda h \sum_{n=k}^{\infty}\binom{n-1}{k-1} F_n(t) = E\left\{\sum_{n=k}^{\infty}\binom{n-1}{k-1} N_{n,t}\right\}\\ &= E\left\{\sum_{t_i\in(0,h)}\sum_{n=k}^{\infty}\binom{n-1}{k-1}\lambda_{i,n}\right\}\\ &= E\left\{\sum_{t_i\in(0,h)}\sum_{n=k}^{N(t_i,t_i+t)}\binom{n-1}{k-1}\right\}\\ &= E\left\{\sum_{t_i\in(0,h)}\binom{N(t_i,t_i+t)}{k}\right\}\end{aligned}$$

since $\sum_{n=k}^{m}\binom{n-1}{k-1}=\binom{m}{k}$. Thus the required result follows.

Corollary. *The "renewal function"* $H(t)=\sum_{n=1}^{\infty} F_n(t)$ *satisfies*

$$H(t)=(\lambda h)^{-1} E\{\sum_{t_i\in(0,h)} N(t_i, t_i+t)\}.$$

Theorem 3.2. *Let k be a positive integer. Then for $h>0$, $t>0$,*

$$\lim_{\delta\downarrow 0} E\left\{\binom{N(0,t)}{k}\Big| N(-\delta,0)\geqq 1\right\}=(\lambda h)^{-1} E\left\{\sum_{t_i\in(0,h)}\binom{N(t_i,t_i+t)}{k}\right\}. \tag{3.2}$$

If $EN^{k+1}(0,t)<\infty$ for some $t>0$ (and hence all $t>0$), then both sides of (3.2) *are finite. Otherwise, both sides are infinite.*

Proof. Let $\chi_{j,m}=1$ if there is at least one event in the interval $((j-1)h/m, jh/m]$, $\chi_{j,m}=0$ otherwise, $j=0,1\ldots m$. Now with probability one for fixed i and for m sufficiently large (depending on the "sample point ω"), we have $N(t_i,t_i+t)=N(jh/m, jh/m+t)$ if $j\,(=j(t_i,m))$ is that integer such that $t_i\in((j-1)h/m, jh/m]$. (For m may be chosen large enough so that there are no events in $(t_i, jh/m]$ or $(t_i+t, jh/m+t]$.) Considering all of the (finite number of) t_i in this way, we see that

$$\sum_{j=1}^{m}\chi_{j,m}\binom{N(jh/m, jh/m+t)}{k}\to\sum_{t_i\in(0,h)}\binom{N(t_i,t_i+t)}{k} \tag{3.3}$$

with probability one, as $m\to\infty$.

Assume now that $EN^{k+1}(0,t)<\infty$ for $t>0$. Then since $N(jh/m, jh/m+t)\leqq N(0,h+t)$ and $\sum_{j=1}^{m}\chi_{j,m}\leqq N(0,h)$, it follows that the left hand side of (3.3) is dominated by $N(0,h)\binom{N(0,h+t)}{k}$ which in turn does not exceed $N^{k+1}(0,h+t)$. But $EN^{k+1}(0,h+t)<\infty$ and thus from (3.3), by dominated convergence,

$$\begin{aligned}E\left\{\sum_{t_i\in(0,h)}\binom{N(t_i,t_i+t)}{k}\right\}&=\lim_{m\to\infty}\sum_{j=1}^{m}E\left\{\chi_{j,m}\binom{N(jh/m, jh/m+t)}{k}\right\}\\ &=\lim_{m\to\infty} mE\left\{\chi_{0,m}\binom{N(0,t)}{k}\right\}.\end{aligned} \tag{3.4}$$

Now for $\delta>0$, write $\chi_\delta=1$ if $N(-\delta,0)\geqq 1$, $\chi_\delta=0$ otherwise, and let m be the integer part $[h\delta^{-1}]$ of h/δ. Then $h(m+1)^{-1}\leqq\delta\leqq hm^{-1}$ and thus $\chi_{0,m+1}\leqq\chi_\delta\leqq\chi_{0,m}$. Hence

$$\frac{h}{\delta(m+1)}(m+1)\,E\left\{\chi_{0,m+1}\binom{N(0,t)}{k}\right\}\leqq\frac{h}{\delta}E\left\{\chi_\delta\binom{N(0,t)}{k}\right\}$$
$$\leqq\frac{h}{\delta m}mE\left\{\chi_{0,m}\binom{N(0,t)}{k}\right\}.$$

But the two extreme terms of this inequality each converge to the left hand side of (3.4) ($\delta m\to h$) and hence so does the middle term. That is

$$E\left\{\sum_{t_i\in(0,h)}\binom{N(t_i,t_i+t)}{k}\right\}=\lim_{\delta\downarrow 0}\left[\frac{h}{\delta}E\left\{\binom{N(0,t)}{k}\middle|\chi_\delta=1\right\}Pr(\chi_\delta=1)\right]$$

and the conclusion (3.2) follows since $Pr(\chi_\delta=1)=Pr\{N(-\delta,0)\geqq 1\}\sim\lambda\delta$ as $\delta\downarrow 0$. It is clear from the proof that both sides of (3.2) are finite in this case.

In the case where $EN^{k+1}(0,t)=\infty$ for $t>0$, we show that both sides of (3.2) are infinite whenever $h<t$. It will follow that this is then also true for all $h\geqq t$ since the left hand side is independent of h and the expectation on the right hand side is non-decreasing in h.

Assume, then, that $EN^{k+1}(0,t)=\infty$, and $h<t$. Then, writing $N(0,h)=N$, we have $N(t_i,t_i+t)\geqq N(t_i,h)=N-i$ when $t_i\in(0,h)$, and hence

$$\sum_{t_i\in(0,h)}\binom{N(t_i,t_i+t)}{k}\geqq\sum_{i=1}^{N-k}\binom{N-i}{k}=\binom{N}{k+1}$$

which has infinite expectation since $EN^{k+1}=\infty$. Thus the right hand side of (3.2) is infinite and, by Fatou's lemma, so is the limit of the mean of the left hand side of (3.3). The same argument used in the case above (when $EN^{k+1}(0,t)$ was finite) now shows that the left hand side of (3.2) is infinite, as required.

By combining Theorems 3.1 and 3.2, we immediately obtain Eq. (1.9). Stating this formally, we have the following theorem.

Theorem 3.3. *The conditional factorial moments* $\alpha_k(t)$ *defined by* (1.7) *are also given by* (1.9), *viz.*,

$$\alpha_k(t)=\lim_{\delta\downarrow 0}E\left\{k!\binom{N(0,t)}{k}\middle|N(-\delta,0)\geqq 1\right\},\quad k=1,2\ldots.$$

$\alpha_k(t)$ *is finite for all* $t>0$ *if and only if* $EN^{k+1}(0,t)<\infty$ *for all* $t>0$. *In fact, if either* $\alpha_k(t)$ *or* $EN^{k+1}(0,t)$ *is finite for some* $t>0$, *both are finite for all* $t>0$.

Stated in another way, this theorem shows that if either $\alpha_k(t)$ or $\beta_{k+1}(t)$ is finite for some $t>0$, then both $\alpha_k(t)$ and $\beta_{k+1}(t)$ are finite for all $t>0$.

A particular case again concerns the "renewal function" $H(t)=\sum_1^\infty F_n(t)$ (i.e., $k=1$). For this case, we thus have

$$H(t)=\lim_{\delta\downarrow 0}E\{N(0,t)|N(-\delta,0)\geqq 1\}$$

$H(t)$ being finite for all $t>0$ if and only if $EN^2(0,t)<\infty$ for some (hence all) $t>0$.

References

1. Leadbetter, M. R.: On streams of events and mixtures of streams. J. roy. statist. Soc. Ser. B **28**, 218–227 (1966).
2. – On the distribution of the times between events in a stationary stream of events. J. roy. statist. Soc. Ber. B **31**, 295–302 (1969).
3. Longuet Higgins, M. S.: The distribution of intervals between zeros of a stationary random function. Philos. Trans. roy. Soc. London, Ser. A **254**, 557–599 (1962).
4. Matthes, K.: Stationäre zufällige Punktfolgen I. Jber. Deutsch. Math. Verein. **66**, 66–79 (1963).

H. Cramér
Skärviksväken 33
Djursholm
Sweden

M. R. Leadbetter
University of North Carolina
Chapel Hill, N.C. 27514
USA

R. J. Serfling
Florida State University
Dept. of Statistics
Tallahassee, Florida 32306
USA

(Received January 6, 1970)

68.

On the history of certain expansions used in mathematical statistics

Biometrika **59**, 205–207 (1972)
(Reprinted in Studies in the History of Statistics and Probability, Vol. II
ed. by M. G. Kendall and R. L. Plackett, Griffin, London 1977)

SUMMARY

This note is concerned with the history of the expansions known to probabilitists and statisticians as Charlier's A-series and Edgeworth's series. It has been pointed out (Gnedenko & Kolmogorov, 1968, Chapter 8) that both these types of expansions appear already in the mathematical work of Tchebychev, so that the names usually attached to them are historically incorrect. In the sequel we shall, however, denote them by the names familiar to statisticians. Both expansions are closely connected with the attempts to give a refinement of the classical central limit theorem in probability. The properties which are relevant in this connexion are the asymptotic properties of the sum of a small number of terms. It will be stated here that the first valid proof of these properties was given in two papers by the present author (1925, 1928).

Some key words: History of expansions for distributions; Central limit theorem; Edgeworth series; Charlier series.

In a recent paper, Särndal (1971) gives a historical account of the work connected with the hypothesis of elementary errors, which has been performed by what he calls the Scandinavian school in statistical theory. The work centring about Charlier's A-series occupies an important place in this account. Särndal's presentation of this work does, however, call for some complementary remarks.

Consider the following simple case of the classical central limit theorem. Suppose that we are given a sequence of mutually independent and equidistributed random variables $z_1, z_2, \ldots$ with a common distribution of the continuous type, having zero mean, standard deviation σ and central moments $\mu_3, \mu_4, \ldots$, which need not all be finite. Consider the standardized sum variable

$$y_n = \frac{z_1 + \ldots + z_n}{n^{\frac{1}{2}}\sigma},$$

and let $f_n(x)$ be the density function of y_n. The central limit theorem then asserts that, as $n \to \infty$, under appropriate conditions, $f_n(x)$ tends to $\phi(x)$, the normal $N(0, 1)$ density function.

In order to improve the approximation to $f_n(x)$ supplied by the normal density function, Charlier (1905) introduced his A-series, which gives for $f_n(x)$ an expansion with $\phi(x)$ as its leading term, followed by terms containing the successive derivatives of ϕ, from $\phi^{(3)}$ on:

$$f_n(x) = \phi(x) + \frac{c_3}{3!}\phi^{(3)}(x) + \frac{c_4}{4!}\phi^{(4)}(x) + \ldots. \tag{1a}$$

The coefficients c_i depend on the moments of the common distribution of the z_i, the expressions of c_3 and c_4 being

$$c_3 = -\frac{\mu_3}{n^{\frac{1}{2}}\sigma^3}, \quad c_4 = \frac{1}{n}\left(\frac{\mu_4}{\sigma^4} - 3\right). \tag{2}$$

Thus with increasing n both c_3 and c_4 tend to zero, the order of smallness being $n^{-\frac{1}{2}}$ for c_3, and n^{-1} for c_4. But the simple rule suggested by these relations does not hold, as shown, for example, by c_6, which is of the same order n^{-1} as c_4.

Charlier (1905) asserts that the sum of a small number of terms of the A-series (1 *a*) gives for large n a better approximation to $f_n(x)$ than does the leading term $\phi(x)$ alone. Särndal (1971, p. 378) writes in this connexion: 'Realizing that the normal curve itself was only a first approximation to the distribution of a sum of errors, Charlier (1905) used Laplacean analytic techniques to show how the approximation could be improved by adding correction terms, the importance of which ought to be smaller the larger the number of error sources.' He further (p. 380) briefly indicates the method of proof used by Charlier

(1905), and then states that Charlier thus arrives at this expansion 'by extending to a finer degree of approximation Laplace's analytical treatment of the distribution of a sum of random variables'. However, Särndal does not mention the fact that Charlier's proof of his statement was entirely false, as it depended essentially on an illegitimate use of what we now call the inversion formula for characteristic functions.

In the course of his discussion of Charlier's 1905 paper, Särndal (1971, p. 380) gives the correct form of the inversion formula which expresses $f_n(x)$ in terms of the corresponding characteristic function. As is well known, this formula contains an integral extended between the limits $\pm\infty$. In Charlier's 1905 paper this integral appears, however, with the limits $\pm\pi$. In a later paper, not quoted by Särndal, Charlier (1914) admits in a brief footnote on p. 2 that the correct limits of the integral should be $\pm\infty$. But he does not mention that his whole proof of 1905 was essentially based on the use of the incorrect limits $\pm\pi$ and becomes invalid when the correct infinite limits are introduced. In fact, the assertion which Charlier claimed to have proved in 1905 is true only in a modified form, and under more restrictive conditions than those given by him. No correct proof of any statement of this character was ever published by Charlier or by any of his students.

Thus the problem of what may be denoted as the asymptotic properties of the A-series remained open for a number of years, until it was taken up and at least partially solved in two papers by the present author (Cramér, 1925, 1928). An account of some of the results obtained in these papers may be found in the book of Gnedenko & Kolmogorov (1968, Chapter 8), as well as in a Cambridge Tract (Cramér, 1970, Chapter 7).

The first paper (Cramér, 1925) is a preliminary publication, while more explicit results were given by Cramér (1928). It is there proved that, under appropriate conditions, asymptotic properties similar to those asserted by Charlier do, in fact, hold for the A-series, but that the corresponding relations appear in a considerably simpler form when the density function $f_n(x)$ is replaced by the corresponding distribution function $F_n(x)$, and the A-series is replaced by the expansion known as Edgeworth's series. The A-series for $F_n(x)$ is obtained by formal integration of (1*a*), which gives

$$F_n(x) = \Phi(x) + \frac{c_3}{3!}\Phi^{(3)}(x) + \frac{c_4}{4!}\Phi^{(4)}(x) + \ldots, \tag{1b}$$

where $\Phi(x)$ is the normal $N(0,1)$ distribution function. The series introduced by Edgeworth (1905) assumes for the distribution function $F_n(x)$ the form

$$F_n(x) = \Phi(x) + P_1(x) + P_2(x) + \ldots. \tag{3}$$

Here the $P_i(x)$ are linear aggregates of the derivatives of $\Phi(x)$, the expressions for P_1 and P_2 being

$$P_1(x) = \frac{c_3}{3!}\Phi^{(3)}(x),$$

$$P_2(x) = \frac{c_4}{4!}\Phi^{(4)}(x) + \frac{1}{2}\left(\frac{c_3}{3!}\right)^2\Phi^{(6)}(x).$$

It thus follows from (2) that P_1 is of the order $n^{-\frac{1}{2}}$, while P_2 is of the order n^{-1}. Edgeworth (1905, 1907) showed that generally, if the requisite moments are finite, P_i is of the order $n^{-\frac{1}{2}i}$.

The asymptotic properties of the Edgeworth expansion (3) for $F_n(x)$ were for the first time given by the present author (1925, 1928). As shown in these papers, the expansion (3) does really have the asymptotic character that Charlier tried to prove for the A-series. It is, in fact, proved (Cramér, 1928, p. 59) that, under fairly general conditions, any partial sum of (3) gives an approximation to $F_n(x)$ with an error of the same order of magnitude for large n as the first neglected term. The Edgeworth expansion for the density function $f_n(x)$, which is obtained by formal differentiation of (3), has similar asymptotic properties, but only under rather more restrictive conditions (Cramér, 1928, p. 63; Gnedenko & Kolmogorov, 1968, Chapter 8). Extensions of the asymptotic properties to cases when the component variables z_i are not equidistributed are also given by Cramér (1928).

We finally observe that the convergence properties of the above expansions, regarded as infinite series, have been discussed by various authors. The results obtained are certainly interesting from a purely mathematical point of view. However, when applying one of the expansions to a probability distribution we cannot discuss the question of convergence or divergence without supposing that all moments have known finite values. Thus it is generally not the convergence theory of the expansions which is statistically relevant, but the question of the asymptotic properties of the sum of a limited number of terms.

REFERENCES

CHARLIER, C. V. L. (1905). Über das Fehlergesetz. *Ark. Mat. Astr. Fys.* **2**, No. 8.

CHARLIER, C. V. L. (1914). Contributions to the mathematical theory of statistics 5. *Ark. Mat. Astr. Fys.* **9**, No. 25.

CRAMÉR, H. (1925). On some classes of series used in mathematical statistics. *Trans. 6th Congr. Scand. Math.* 399–425.

CRAMÉR, H. (1928). On the composition of elementary errors. *Skand. Aktuarietidskrift* **11**, 13–74, 141–80.

CRAMÉR, H. (1970). *Random Variables and Probability Distributions*, 3rd edition. Cambridge University Press.

EDGEWORTH, F. Y. (1905). The law of error. *Trans. Camb. Phil. Soc.* **20**, 36–65, 113–41.

EDGEWORTH, F. Y. (1907). On the representation of statistical frequency by a series. *J. R. Statist. Soc.* **70**, 102–6.

GNEDENKO, B. V. & KOLMOGOROV, A. N. (1968). *Limit Distributions for Sums of Independent Random Variables*, 2nd edition. Cambridge, Mass.: Addison-Wesley.

SÄRNDAL, C.-E. (1971). The hypothesis of elementary errors and the Scandinavian School in statistical theory. *Biometrika* **58**, 375–91.

[*Received October* 1971]

On the multiplicity of a stochastic vector process

Perspectives in Probability and Statistics
(Papers in honour of M. S. Bartlett on the occasion of his sixty-fifth birthday)
Ed. by J. Gani. Applied Probability Trust. University of Sheffield, 187–194 (1975)

Abstract

This note deals with a q-dimensional stochastic vector process $x(t) = \{x_1(t), \cdots, x_q(t)\}$, satisfying certain stated general conditions. For such a process, there is a representation (1) in terms of stochastic innovations acting throughout the past of the process. The number N of terms in this representation is called the multiplicity of the $x(t)$ process, and is uniquely determined by the process. For a one-dimensional process ($q = 1$) it is known that under certain conditions we have $N = 1$. For an arbitrary value of q, this note gives conditions under which we have $N \leqq q$.

1. Introduction

Let $x(t) = \{x_1(t), \cdots, x_q(t)\}$ be a q-dimensional stochastic vector process, where the $x_i(t)$ are real-valued random variables on some fixed probability space, defined in the interval $A \leqq t \leqq B$. Here A may be finite or $-\infty$, while B may be finite or $+\infty$. All $x_i(t)$ will be supposed to have zero mean values and finite variances. The Hilbert spaces $H(x)$ and $H(x, t)$ are spanned in the customary way by the random variables $x_i(u)$ according to the relations

$$H(x) = S\{x_i(u),\ i = 1, \cdots, q,\ A \leqq u \leqq B\},$$

$$H(x, t) = S\{x_i(u),\ i = 1, \cdots, q,\ A \leqq u \leqq t\}.$$

We shall suppose that the mean square (m.s.) limits $x_i(t \pm 0)$ exist for all i and t, and that there is always m.s. continuity to the left, so that $x_i(t - 0) = x_i(t)$. The Hilbert space $H(x)$ will then be separable. We shall further suppose that in the case of a finite A we have $x_i(A)=0$, while if $A=-\infty$ we have $H(x,-\infty)=0$, so that the $x(t)$ process is purely non-deterministic. For the literature on multiplicity theory, we refer to the recent publication, [1].

If z is any element of $H(x)$, we denote by $z(t)$ the projection of z on the subspace $H(x,t)$. Then $z(t)$ determines a stochastic process with zero mean values and orthogonal increments. By $H(z)$ we denote the cyclic subspace of $H(x)$ spanned by all the $z(t)$. We then have $Ez^2(t)=F(t)$, where $F(t)$ is a non-decreasing function of t in (A,B), which is always continuous to the left, and such that $F(A)=0$, $F(B)=Ez^2$.

It is then known that there is a number N uniquely defined by the $x(t)$ process and called the *multiplicity* of the process. N may be a finite positive integer or $+\infty$, and has the following properties.

There exists a sequence of random variables $z_1,\cdots,z_N$ in $H(x)$ such that

$$x_i(t)=\sum_{n=1}^{N}\int_A^t g_{in}(t,u)dz_n(u) \tag{1}$$

for $i=1,\cdots,q$ and all t in (A,B). The g_{in} are real-valued non-random functions, and the integrals are m.s. integrals. We have

$$Ex_i^2(t)=\sum_{n=1}^{N}\int_A^t |g_{in}(t,u)|^2dF_n(u),$$

where $F_n(u)=Ez_n^2(u)$. The cyclic subspaces $H(z_n)$ are mutually orthogonal, and the space $H(x)$ is identical with the vector sum of all the $H(z_n)$, so that we have in the usual Hilbert space notation

$$H(x)=H(z_1)\oplus\cdots\oplus H(z_N). \tag{2}$$

Finally, every F_n is absolutely continuous with respect to the preceding F_{n-1}.

The representation (1) shows how the random variables $x_i(t)$ are additively built up by the N-dimensional *innovation elements* $\{dz_1(u),\cdots,dz_N(u)\}$ acting throughout the past of the process, $A\leqq u\leqq t$. This is a canonical representation of the $x(t)$ process in the sense that any two representations satisfying the above conditions must have the same number N of terms. Moreover, if $F_n^{(1)}$ and $F_n^{(2)}$ are the corresponding functions occurring in the two representations, then $F_n^{(1)}$ and $F_n^{(2)}$ are mutually absolutely continuous for every n.

2. The multiplicity problem

It would be very desirable to be able to determine the multiplicity N of a given $\boldsymbol{x}(t)$ process. This problem is largely unsolved, and the object of the present note is to make a modest contribution to the study of the problem.

For a one-dimensional process ($q = 1$) it is known (cf. [1], p. 319) that, if it is assumed that the functions $g_n(t, u)$ and $F_n(u)$ occurring in a canonical representation (1) of the process satisfy certain simple regularity conditions, the multiplicity of the process will have the simplest possible value $N = 1$.

For the case of a number $q > 1$ of dimensions, it seems obvious that it is only under rather special conditions that we may expect to have a multiplicity $N < q$. It will, however, be proved below that under somewhat more complicated regularity conditions we shall have $N \leqq q$. The conclusion seems to be that, in a case where it is only known that our general regularity conditions are satisfied, we should expect to have $N = q$.

For the proof we shall use the following remark (cf. [1], p. 318). Suppose that (1) is a canonical representation of the $\boldsymbol{x}(t)$ process. The Hilbert spaces occurring in the two members of the relation (2) will then be identical. It follows, in particular, that the vector sum of the mutually orthogonal $H(z_n)$ spaces does not contain any element orthogonal to the space $H(\boldsymbol{x})$. Now any non-zero element of the vector sum of the $H(z_n)$ is a random variable y of the form

$$y = \sum_{n=1}^{N} \int_A^B h_n(u) dz_n(u), \tag{3}$$

where

$$0 < Ey^2 = \sum_{n=1}^{N} \int_A^B h_n^2(u) dF_n(u) < \infty. \tag{4}$$

If y is orthogonal to $H(\boldsymbol{x})$, it will be orthogonal to all $x_i(t)$ for $A \leqq t \leqq B$, so that

$$Eyx_i(t) = \sum_{n=1}^{N} \int_A^t h_n(u) g_{in}(t, u) dF_n(u) = 0 \tag{5}$$

for $i = 1, \cdots, q$ and $A \leqq t \leqq B$. It thus follows that, if we are able to find functions $h_1(u), \cdots, h_N(u)$ satisfying the conditions (3)–(5), the representation (1) cannot be canonical, and we have a contradiction.

3. A lemma on Volterra integral equations

Let $R = \{\alpha_{in}(t)\}$, where $i = 1, \cdots, q$ and $n = 1, \cdots, q+1$, be a matrix the elements of which are functions of a real variable t defined in the interval (A, B) and having at most a finite number of zeros in (A, B). We shall say that the matrix R satisfies the *condition* Q if the following properties hold:

(1) Any second order minor of the form

$$D_{in} = \alpha_{in}\alpha_{q,q+1} - \alpha_{i,q+1}\alpha_{qn}$$

with $i < q$, $n < q+1$, has at most a finite number of zeros in (A, B).

(2) The matrix $R^{(1)}$ with elements D_{in}, where $i = 1, \cdots, q-1$, $n = 1, \cdots, q$, has second order minors of the form

$$D_{in}^{(1)} = D_{in}D_{q-1,q} - D_{iq}D_{q-1,n},$$

with $i < q-1$, $n < q$, all of which have at most a finite number of zeros in (A, B).

(3) The corresponding properties hold for the matrix $R^{(2)}$ with elements $D_{in}^{(1)}$, where $i = 1, \cdots, q-2$, $n = 1, \cdots, q-1$, and so on, until we finally arrive at a matrix with a single row of two elements, each of which has at most a finite number of zeros in (A, B).

It will be seen that the condition Q implies that the elements $\alpha_{in}(t)$ of the matrix R are not related by certain simple algebraic identities.

Consider now a system of q integral equations with $q+1$ unknown functions

$$\sum_{n=1}^{q+1}\left[\alpha_{in}(t)\phi_n(t) + \int_A^t \beta_{in}(t,u)\phi_n(u)du\right] = 0 \tag{6}$$

where $i = 1, \cdots, q$ and $A \leqq t \leqq B$. The $\phi_n(t)$ are unknown functions, while the $\alpha_{in}(t)$ and $\beta_{in}(t,u)$ are known. We shall suppose that the α_{in} and β_{in} are bounded and continuous in (A, B), that the α_{in} have at most a finite number of zeros in (A, B), and that the matrix $R = \{\alpha_{in}(t)\}$ with q rows and $q+1$ columns satisfies the condition Q.

Under these conditions it is possible to find functions $\phi_1(t), \cdots, \phi_{q+1}(t)$ which are bounded in (A, B), have at most a finite number of discontinuities and are not all identically zero, so that the system (6) *is satisfied.*

This lemma will be proved by induction. For $q = 1$ the system (6) reduces to the single equation

(7) $$\sum_{n=1}^{2}\left[\alpha_{1n}(t)\phi_n(t)+\int_A^t \beta_{1n}(t,u)\phi_n(u)du\right]=0.$$

From the assumptions made with respect to the α_{1n} and β_{1n} it follows that it is possible to find in the interior of (A,B) an interval, say (a,b), such that α_{11} and α_{12} are both bounded away from zero in (a,b). We now take $\phi_1(t)=\phi_2(t)=0$ outside (a,b), while for $a \leqq t \leqq b$ we determine ϕ_1 and ϕ_2 from the Volterra integral equations

$$\alpha_{1n}(t)\phi_n(t)+\int_a^t \beta_{1n}(t,u)\phi_n(u)du=(-1)^n$$

for $n=1$ and 2. According to the classical theory, these equations have solutions which are bounded and continuous in (a,b) and are not identically zero. The functions ϕ_1 and ϕ_2 so determined in (A,B) will satisfy the equation (7), so that the lemma has been proved for $q=1$.

We now suppose that the lemma has been proved up to the case of a system of $q-1$ equations with q unknown functions, and consider the system (6) of q equations with $q+1$ unknown functions. By our assumptions there will be an interval (a,b) inside (A,B) such that all the $\alpha_{in}(t)$, as well as all the minors of the $\{\alpha_{in}(t)\}$ matrix occurring in the conditions of the lemma, will be bounded away from zero in (a,b). The system (6) will be satisfied if we take $\phi_1(t)=\cdots=\phi_{q+1}(t)=0$ outside (a,b), while for $a \leqq t \leqq b$ we determine the ϕ_n so as to satisfy the system (6) after replacing the lower limit A of the integral by a. In order to show that this is possible, we write

(8) $$\alpha_{in}(t)\phi_n(t)+\int_a^t \beta_{in}(t,u)\phi_n(u)du=\alpha_{in}(t)\psi_{in}(t),$$

so that the system (6) becomes

(9) $$\sum_{n=1}^{q+1}\alpha_{in}(t)\psi_{in}(t)=0$$

for $i=1,\cdots,q$. We then have to show that the $\psi_{in}(t)$ can be determined for $a \leqq t \leqq b$ so that the equations (9) will be satisfied.

Regarding (8) as a Volterra integral equation with $\phi_n(t)$ as the unknown function, we have the solution

$$\phi_n(t)=\psi_{in}(t)+\int_a^t K_{in}(t,u)\psi_{in}(u)du,$$

where $K_{in}(t, u)$ is the resolvent corresponding to the kernel

$$J_{in}(t,u) = -\beta_{in}(t,u)/\alpha_{in}(t).$$

J_{in} and K_{in} are both bounded and continuous for $a \leqq u \leqq t \leqq b$. We thus obtain

$$\begin{aligned} \psi_{in}(t) + \int_a^t K_{in}(t, u)\psi_{in}(u)du &= \phi_n(t) \\ (10) \qquad &= \psi_{qn}(t) + \int_a^t K_{qn}(t, u)\psi_{qn}(u)du \end{aligned}$$

for $i = 1, \cdots, q-1$ and $n = 1, \cdots, q+1$. Now regarding this as an integral equation with the unknown function $\psi_{in}(t)$, we obtain after some calculation the following solution expressing ψ_{in} in terms of ψ_{qn}

$$\psi_{in}(t) = \psi_{qn}(t) + \int_a^t L_{in}(t, u)\psi_{qn}(u)du$$

with

$$L_{in}(t, u) = K_{qn}(t, u) - J_{in}(t, u) - \int_u^t J_{in}(t, s)K_{qn}(s, u)ds.$$

L_{in} is bounded and continuous for $a \leqq u \leqq t \leqq b$.

The system (9) is then equivalent to

$$\text{(9a)} \qquad \sum_{n=1}^{q+1} \alpha_{in}(t)\left[\psi_{qn}(t) + \int_a^t L_{in}(t, u)\psi_{qn}(u)du\right] = 0$$

for $i = 1, \cdots, q-1$, and

$$\text{(9b)} \qquad \sum_{n=1}^{q+1} \alpha_{qn}(t)\psi_{qn}(t) = 0.$$

Solving (9b) with respect to $\psi_{q,q+1}$ and inserting the solution in (9a), we have

$$\text{(11)} \qquad \sum_{n=1}^{q}\left[D_{in}(t)\psi_{qn}(t) + \int_a^t M_{in}(t, u)\psi_{qn}(u)du\right] = 0$$

for $i = 1, \cdots, q-1$, with

$$D_{in}(t) = \alpha_{in}(t)\alpha_{q,q+1}(t) - \alpha_{i,q+1}(t)\alpha_{q,n}(t),$$

and a $M_{in}(t, u)$ which is bounded and continuous for $a \leqq u \leqq t \leqq b$, while by hypothesis, $D_{in}(t)$ is continuous and bounded away from zero in (a, b).

Thus (11) constitutes a system of $q - 1$ equations for the q unknown functions $\psi_{qn}(t)$ with $n = 1, \cdots, q$. The system (10) is of the same form as (6), and by hypothesis has a solution with the desired properties. As has been shown, the ψ_{qn} will determine all the ψ_{in} so as to satisfy (9). By (8) and (10), the corresponding ϕ_n will form a solution of the system (6). This completes the proof of the lemma.

4. A theorem on multiplicity

Let $\mathbf{x}(t) = (x_1(t), \cdots, x_q(t))$ be a q-dimensional stochastic vector process of the kind specified in the Introduction. Suppose that $\mathbf{x}(t)$ has a canonical representation of the form (1), such that the following regularity conditions are satisfied:

(R_1) The $g_{in}(t, u)$ and $\partial g_{in}(t, u)/\partial t$ are bounded and continuous for $A \leqq u \leqq t \leqq B$.

(R_2) If $N > q$, the matrix $\{g_{in}(t, t)\}$ of order $q \cdot N$ contains at least one submatrix of order $q \cdot (q + 1)$ satisfying the condition Q in (A, B).

(R_3) The functions $F_n(u)$ defined in the Introduction are absolutely continuous, with derivatives $f_n(u) = F_n'(u)$ having at most a finite number of discontinuities in (A, B).

Then the multiplicity N of the $\mathbf{x}(t)$ process is at most equal to q.

Suppose, in fact, that we have a canonical representation of the form (1) with $N > q$. In order to prove that this is not possible, we have to show that functions $h_n(u)$ satisfying (3)–(5) can be found. By hypothesis, (5) can be differentiated and yields a system of equations

$$(12) \qquad \sum_{n=1}^{N} \left[g_{in}(t, t) h_n(t) f_n(t) + \int_A^t \frac{\partial g_{in}}{\partial t} h_n(u) f_n(u) du \right] = 0.$$

Without restricting the generality we may evidently assume that the matrix of the 'instantaneous response functions' $g_{in}(t, t)$ occurring in condition (R_2) consists of the $g_{in}(t, t)$ with $i = 1, \cdots, q$ and $n = 1, \cdots, q + 1$. Then there must be an interval, say (a, b), in the interior of (A, B) such that the $f_n(t)$ and all the minors of the $g_{in}(t, t)$ matrix occurring in condition Q are bounded away from zero in (a, b). If $N > q + 1$, we then take $h_n(t) = 0$ for all t and $n > q + 1$, while for $n \leqq q + 1$, we take $h_n(t) = 0$ outside (a, b). The system (12) then reduces to a system of equations of the form (6), satisfying conditions corresponding to those stated in Section 3. It thus follows that $h_n(u)$ can be determined so as to satisfy the relations (3)–(5). Consequently our assumption $N > q$ must be false, and we have $N \leqq q$, as was to be proved.

Reference

[1] EPHREMIDES, A. AND THOMAS, J. B. (Editors) (1973) *Random Processes, Multiplicity Theory and Canonical Decompositions.* Dawden, Hutchinson and Ross. Distributed by Wiley, London.

70.

Half a century with probability theory. Some personal recollections

Ann. Probab. 4 (4), 509–546 (1976)
(Russian translation 1979)

CONTENTS

Received November 14, 1975.

We deeply appreciate Professor Cramér's gracious interest in presenting these personal recollections of the development of probability during the half-century, 1920–70. Professor Cramér has had a tremendous influence upon the development of both probability and statistics. His 1946 text alone, *Mathematical Methods of Statistics* [24], has so greatly influenced the studies and research of many during the past 30 years. During 1947–49, he served as Associate Editor of *Ann. Math. Statist.*, the forerunner of this journal. He also served as a Council member of the IMS in 1953–56 and the Rietz Lecturer of the IMS in 1953.—RP

AMS 1970 *subject classification*. Primary 60-03.
Key words and phrases. History, personal recollections.

1. Introduction. The previous Editor of this journal, Professor Ronald Pyke, has kindly suggested that I should write down some personal recollections from the period, approximately consisting of the half-century from 1920 to 1970, when I was actively engaged in work on mathematical probability theory. In attempting to follow this suggestion, I am anxious to make it quite clear that I shall not be concerned with writing a history of probability theory. I am only trying to give my own personal impressions of the work that was being done, and of some of the people who did it. I am going to deal mainly with the types of problems that attracted my personal interest, while many other lines of investigation, possibly even more interesting from a general point of view, will hardly be mentioned.

On account of this strongly subjective character of the paper, it may perhaps be convenient to begin with some brief notes of my personal background, and the reasons for my interest in probability theory.

I was born in Stockholm on September 25, 1893, as the second son of my parents, who were cousins. Our ancestors had for several centuries lived in the old town of Wisby, on the island of Gotland, but my parents lived in Stockholm, where my father was a banker.

When I started my academic studies at the University of Stockholm in 1912, my interest was fairly equally divided between chemistry and mathematics. My first University employment was as a research assistant in biochemistry, during the year immediately before World War I, and my first scientific publications belonged to that field. But I soon came to the conclusion that mathematics was the right subject for me. As my main mathematical teacher and friend, I was happy to have Marcel Riesz, a young Hungarian who had come to Sweden to work in the Mittag-Leffler Institute, and who stayed on in Sweden, and later became Professor at the University of Lund. Through him I was educated according to current standards of mathematical rigor, and was introduced to the modern subject of Lebesgue measure and integration theory. My personal research work was concerned with analytic number theory, where I became familiar with the technique of using Fourier integrals of a type closely similar to those I was to encounter later on when studying the relations between a probability distribution and its characteristic function. I obtained my Ph. D. in 1917, having written a thesis on Dirichlet series, one of the chief analytic tools used in prime number theory.

For a young Swedish mathematician of my generation, who wanted to find a job that would enable him to support a family, it was quite natural to turn to insurance. It was a tradition for Swedish insurance companies to employ highly qualified mathematicians as actuaries, and several of my University friends had jobs of this kind. After starting insurance work in 1918, I advanced in 1920 to a position as actuary of a life insurance company.

It was my actuarial work that brought me into contact with probability problems and gave a new turn to my mathematical interests. From 1918 on, I tried to get acquainted with the available literature on probability theory, and to do some work on certain problems connected with the mathematical aspect of insurance risk. These problems, although of a special kind, were in fact closely related to parts of probability theory that would come to occupy a central position in the future development, and I will say some introductory words about them already here.

The net result of the risk business of an insurance company during a period of, say, one year, could be regarded as the sum of the net results of all the individual insurances. If these were supposed to be mutually independent, the connection with the classical "central limit theorem" in probability theory was evident. But there was also the possibility of regarding the total risk business as an economic system developing in time, and in every instant subject to random fluctuations. Systems of this kind had been considered in some pioneering works which today appear as forerunners of the modern theory of stochastic processes.

Both these problems caught my interest already at this early stage. In the following chapter I shall try to give a brief account of the general situation in probability theory, and also of the history of those particular problems, up to 1920.

2. Probability theory before 1920.

2.1. *General remarks; foundations.* It was rather a confused picture that met the eyes of a young man educated in pure mathematics, according to the standards of rigor current since the early part of the present century. There was the great classical treatise of probability theory written by Laplace and first published in 1812. It offered an interesting and stimulating reading, but was entirely nonrigorous from a modern mathematician's point of view. And it was surprisingly uncritical both with regard to the foundations and the applications of the theory. Incidentally, it may be interesting to mention that the Emperor Napoleon, who used Laplace as Minister and Senator, later criticized him, saying that Laplace had introduced "l'esprit des infiniments petits" into the practical administration.

And the French followers of Laplace, even such high class mathematicians as Poincaré and Borel, did not seem to have produced a connected and well organized theory, built on satisfactory foundations. With very few exceptions, books and papers on probability problems were only too obviously lacking in mathematical rigor.

The work of the Russian school, with Tchebychev, Markov and Liapounov, which was perfectly rigorous and of high class, was at this stage still not very well known outside their own country.

The situation may be characterized by some quotations. The British economist Keynes, in his book [61], says about the classical probability theory that "there is about it for scientists a smack of astrology, of alchemy." And the German mathematician von Mises asserts in his paper [98] of 1919 that "today, probability theory is not a mathematical discipline," while the French probabilist Paul Lévy, talking in his autobiography [85] of his first acquaintance with probability theory as a young man, says that "in a certain sense this theory did not exist; it had to be created."

By Laplace and his followers, the classical definition of the probability of an event in terms of the famous "equally possible cases" was regarded as generally applicable, even when it seemed impossible to give a clear explanation of the nature of these "cases." There had been some attempts to overcome this difficulty, but they were not convincing.

Some authors had tried to work out a definition of probability based on the properties of statistical frequency ratios. In his paper [98] of 1919 and his book [99], von Mises gave a definition of this kind, founded on an axiomatic basis. He considered a series of independent trials, made under similar conditions, and postulated the existence of limiting values for the frequencies of various observed events, as well as the invariance of these limits for any appropriately selected subsequence of trials. He had enthusiastic followers and also severe critics. Paul Lévy later expressed the opinion [85, page 79] that it is "just as impossible as squaring the circle" to obtain a satisfactory definition in this way. Personally I was interested in von Mises' work, but took a critical attitude, looking forward to something more satisfactory. It was to come, but the time was not yet ripe for it about 1920.

2.2. *The central limit theorem.* The term "central limit theorem" was introduced by G. Pólya in his paper [103] of 1920, to which I shall return in the following chapter. In present day terminology the theorem asserts that, under appropriate conditions, the probability distribution of the sum of a large number of independent random variables will be approximately normal (or Gaussian). In a very special case this had been proved already in 1733 by de Moivre [100]. The general theorem was given by Laplace, but his proof was incomplete. If the terms in the sum are considered to be small "elementary errors," the theorem was regarded as giving an explanation of the occurrence of the normal distribution in connection with errors of observation.

After a number of unsuccessful attempts by various authors to find a correct proof of the general theorem stated by Laplace, a promising approach was made by Tchebychev [115, 116], who used the method of moments. The first complete proof was given in 1901 by Liapounov [86], who worked with the analytic tool known today under the name of characteristic functions. The work of Liapounov was very little known outside Russia, but I had the good luck to be allowed to see some notes on his work made by the German mathematician Hausdorff, and these had a great influence on my subsequent work in the field.

2.3. *Pioneering work on stochastic processes.* During the first ten years of the present century several works were published which, to a modern reader, appear as forerunners of the theory of stochastic processes created during the 1930's. All these works deal with the temporal development of some variable system exposed to random influences.

Bachelier, in his paper [2] of 1900, analysed the fluctuations of stock market prices, and was led to an important particular case of a stochastic process. In 1905, the same process was encountered by Einstein in his famous study [37] of the Brownian movement. This process is designed to give a mathematical description of the continuous, although extremely irregular path of a particle suspended in a fluid and subject to random molecular shocks.

Between these works of Bachelier and Einstein, the Uppsala Thesis [92] of Filip Lundberg appeared in 1903, written in Swedish. Lundberg studies the essentially discontinuous variations of the accumulated amount of claims incurred by an insurance company. He is thus led to a stochastic process of a type entirely different from the continuous Brownian movement process. The "Poisson process," well known today for its numerous applications, comes out as a very special case of Lundberg process, arising when all sums due under the claims are equal. But for his general case, developed also in some later works, he uses a functional equation which is a particular case of the famous "forward equation," introduced in 1931 by Kolmogorov [73] for a general class of continuous processes, and in 1940 by Feller [43] for the corresponding class of discontinuous processes, which includes the Lundberg process of 1903 as special case.

In 1909 the Poisson process was introduced by Erlang [39] in the study of telephone traffic problems, and by Rutherford and Geiger [113] in the analysis of radioactive disintegration.

All these pioneers used mathematical methods more or less lacking in rigor. But they developed a wonderful ability of dealing intuitively with concepts and methods that would have to wait until the 1930's before being placed on rigorous foundations.

3. A decade of preparation: 1920 to 1929.

3.1. *Stochastic processes and limit theorems.* In the summer of 1920 I spent some time in Cambridge, England, working with number theory under the guidance of the great mathematician G. H. Hardy. There I met an American of my own age, Norbert Wiener, who says in his autobiography *I Am a Mathematician* [122, page 64] that it was quite a coincidence that he should have met both Paul Lévy in France and me on this European trip, since our work has always had close relations to his own. Already at this early stage, Norbert made the impression of being "quite a character." At this first meeting we had no opportunity to talk probability, but it was only three years afterwards in 1923 that he published his famous paper "Differential space" [118] where, several years before the basic works of Kolmogorov, he introduced a probability measure in a function space, thus giving a rigorous theory of the Brownian movement stochastic process, also known today as the Wiener process. Among other results he was able to show that almost all the trajectories of the process considered are continuous functions without a derivative. Unfortunately the paper was quite difficult to read, and many people interested in the field—including myself—did not see its true significance until the great works of the 1930's had paved the way.

Incidentally I may mention that my own first probabilistic paper was a brief note [12] of 1919 on the Poisson process, written in Swedish, where I deduced the expression for the relevant probability distribution from a simple set of necessary and sufficient conditions, well known today. In this connection I studied the Lundberg risk process, but my publications on that subject belong to a later period.

With respect to the central limit theorem, I was greatly interested in G. Pólya's paper [103] of 1920 and J. W. Lindeberg's [87] of 1922. Pólya, who was a Hungarian friend of Marcel Riesz and often visited Sweden, introduced in his paper the name "central limit theorem," as I have already mentioned. He referred to the work of Liapounov and indicated a proof based on the use of characteristic functions, pointing out the analogy with the methods currently used in prime number theory. He also discussed the method of moments used by Tchebychev and, in a more general setting, by Stieltjes.

Lindeberg, in his 1922 paper, gave a complete proof of the central limit theorem under more general conditions than Liapounov. He introduced the famous Lindeberg condition of which I shall have more to say below. I was happy to make his personal acquaintance at a mathematical congress in Helsingfors in the summer of 1922. He was Professor at the University of Helsingfors, and owned a beautiful farm in the eastern part of the country. When he was reproached for not being sufficiently active in his scientific work, he said, "Well, I am really a farmer." And if somebody happened to say that his farm was not properly cultivated, his answer was "Of course, my real job is to be a professor." I was very fond of him, and saw him often during the following years.

However, for the risk problems in which I was interested, it was not enough to know that a certain probability distribution was approximately normal; it was necessary to have an idea of the magnitude of the error involved in replacing the distribution under consideration by the normal one. Liapounov had given an upper limit for this error. I studied his method and gave in a paper [13] of 1923, a simplified version of his proof, together with a numerical estimation of the error, implying a slight improvement of Liapounov's result. For the particularly important case of identically distributed variables, Liapounov's theorem in my 1923 version runs as follows:

Let $x_1, x_2, \cdots, x_n$ be independent and identically distributed random variables, such that every x_i has zero mean, standard deviation σ, and a third absolute moment β_3.

Let G_n be the distribution function of the normed sum

$$z_n = \frac{x_1 + x_2 + \cdots + x_n}{\sigma n^{\frac{1}{2}}}, \tag{3.1.1}$$

while Φ is the normal distribution function

$$\Phi(x) = \frac{1}{(2\pi)^{\frac{1}{2}}} \int_{-\infty}^{x} e^{-t^2/2}\, dt .$$

Then for all $n > (\beta_3/\sigma^3)^2$

$$|G_n(x) - \Phi(x)| < \frac{3\beta_3}{\sigma^3} \cdot \frac{\log n}{n^{\frac{1}{2}}} . \tag{3.1.2}$$

For the case when the x_i are not required to be identically distributed, there is a corresponding, slightly more complicated proposition.

However, it soon became clear that the estimation of the remainder given by (3.1.2) was far from sufficient for the numerical applications to risk problems that I had in view, and a great part of my work during the 1920's was concerned with attempts to find an improved estimation, preferably in the form of some asymptotic expansion of the remainder for large values of n.

In the literature on mathematical statistics available at the time, two kinds of expansion in series of a difference like $G_n(x) - \Phi(x)$ had been considered. As pointed out by Gnedenko and Kolmogorov in their book [48], they had both been introduced by Tchebychev, although they are often referred to respectively as the Edgeworth and the Charlier expansions. I shall here only give the explicit expression of the former, which was discussed by Edgeworth in [36], in a purely formal way. The Charlier series is a rearrangement of the terms of the Edgeworth one, which in some respects is simpler to deal with, but does not provide the best possible estimation. Charlier in his paper [11] of 1905 had claimed for his series certain asymptotic properties, but his proof was entirely wrong, being based on an incorrect use of what we now call the inversion formula for characteristic functions. Moreover, the assertion which Charlier claimed to have proved is true only in a modified form and under more restrictive conditions than those given by him (cf. Cramér, [33]). In fact, no rigorous discussion of the asymptotic properties of those series had been made, so that the problem was still open. In a preliminary paper [14] of 1925, and a more definitive one [16] of 1928 I attacked it, and gave a solution valid under certain conditions (cf. also Cramér [19] and Gnedenko-Kolmogorov [48] Chapter 8). For the particular case of identically distributed variables, my main theorem of 1928 runs as follows. (In the order of the error term, the O of my original statement is here replaced by o, as given by Esseen [40] in 1944.)

Let the x_i of (3.1.1) *be independent and identically distributed, each with zero mean, standard deviation σ and finite moments $\alpha_1 = 0$, $\alpha_2 = \sigma^2$, $\alpha_3, \cdots, \alpha_k$, where $k \geqq 3$. Let each x_i have the distribution function $F(x)$ and the characteristic function*

$$f(t) = \int_{-\infty}^{\infty} e^{itx}\, dF(x) . \tag{3.1.3}$$

Suppose that

$$\text{(3.1.4)} \qquad \limsup_{t\to\infty} |f(t)| < 1 .$$

Then we have the expansion

$$\text{(3.1.5)} \qquad G_n(x) = \Phi(x) + (2\pi)^{-\frac{1}{2}} e^{-x^2/2} \sum_{j=1}^{k-2} \frac{p_j(x)}{n^{j/2}} + o(n^{-(k-2)/2})$$

uniformly in x, where $p_j(x)$ *is a polynomial in x of degree* $3j - 1$, *the coefficients of which depend only on the moments* $\alpha_3, \cdots, \alpha_{j+2}$. *In particular we have*

$$p_1(x) = \frac{\alpha_3}{3!}(1 - x^2) .$$

Since $k \geqq 3$, it will be seen that the condition (3.1.4) has enabled us to remove the factor $\log n$ in the upper limit for the error term in Liapounov's theorem. An important particular case when the condition (3.1.4) is satisfied occurs when the given distribution function F contains an absolutely continuous component not identically zero.

In my 1928 paper [15] I also treated the case of nonidentically distributed variables, as well as the corresponding expansion for the probability density $G_n'(x)$. I further showed that, in a case when the condition (3.1.4) is not satisfied, a simple integrated average of the series in (3.1.5) over a small interval containing the point x still has the required asymptotic properties.

In this connection, I may mention that many years afterwards, I studied in a paper [29] of 1963 the asymptotic expansion of $G_n(x)$ in the case when the limiting distribution is nonnormal and stable. Under appropriate conditions there is still an expansion corresponding to (3.1.5), but of a more complicated structure.

3.2. *Lévy's book of* 1925; *my own plans for a book*. While I was deeply engaged in my work on the asymptotic expansions, the book *Calcul des probabilitiés* by Paul Lévy [80] appeared in 1925. Although I could not quite agree with him in respect of the foundations, I at once realized that this was a major event in the development of mathematical probability theory. It seemed clear that here was a first attempt to present the theory as a connected whole, using mathematically rigorous methods. It contained the first systematic exposition of the theory of random variables, their probability distributions and their characteristic functions. Although I had used these concepts in my own work for several years, Lévy's account of the theory brought much that was new to me. He also gave a discussion of the central limit theorem and of the stable probability distributions, as well as a very interesting chapter on kinetic gas theory.

During a visit to England in 1927, I saw my old teacher and friend G. H. Hardy. When I told him that I had become interested in probability theory, he said that there was no mathematically satisfactory book in English on this subject, and encouraged me to write one. I was more than willing to follow up his suggestion, but it seemed evident that it would take a fairly long time. In fact, it was not until ten years afterwards, in 1937, that my Cambridge Tract [19] was ready to appear.

3.3. *Foundations.* With respect to the foundations of probability theory, the attitudes of von Mises and Lévy were fundamentally different, and I did not feel able to agree with either of them. In the 1919 paper [98] of von Mises there was, however, one general statement with which I wholeheartedly agreed, although it seemed to me that he had not followed up the consequences of it when building his own collective theory.

On page 53 of that paper, von Mises expressed the opinion that probability theory is "a natural science of the same kind as geometry or theoretical mechanics." It is the object of this theory to describe certain observable phenomena, "not exactly, but with some abstraction and idealization."—In other words, probability theory is to be regarded as a *mathematical model* of a certain class of observable phenomena.

In a Swedish paper [15] of 1926 I referred to this statement of von Mises. I expressed my agreement and made some further comments, from which I should like to quote here:

> The probability concept should be introduced by a purely mathematical definition, from which its fundamental properties and the classical theorems are deduced by purely mathematical operations. ...Against such a mathematical theory, no objection can be valid except on mathematical grounds. On the other hand, it should be emphasized that the mathematical theory does not *prove* anything about the real events that will occur. Probability formulas are just as unable to dictate the behavior of real events as are the formulas of classical mechanics to prescribe that the stars must attract one another according to the Newton law. It is only experience that can guide us here and show if our mathematical model yields an acceptable approximation of our observations.

I still feel that these remarks are fairly reasonable, and am glad to have put them in print seven years before the definite formalization of probability theory by Kolmogorov.

3.4. *The new Russian school.* All through the latter part of the 1920's, it was clear that a strong new activity in mathematical probability was taking place in the Soviet Union. In a remarkable paper [7] of 1927, S. N. Bernstein discussed the extension of the central limit theorem to sums of random variables which are not required to be independent. He introduced an important method of dealing with such cases, to which I shall return later.

But the main Russian works of this period were written by two quite young mathematicians, A. Ya. Khintchine and A. N. Kolmogorov, who were to be leaders in the coming development of the field. In a joint paper [70] of 1925 they proved the famous "three series theorem," giving necessary and sufficient conditions for the convergence of a series, the terms of which are independent random variables. The probability that such a series is convergent can only be equal to zero or one, which is a particular case of the so called "zero-one law," discovered at the same time.

In his paper [71] of 1928, Kolmogorov proved his celebrated inequality for sums of independent random variables, which is a highly refined generalization of the well-known elementary inequality due to Tchebychev. Let $x_1, \cdots, x_n$ be independent random variables with zero means and finite (not necessarily equal) standard deviations. Then Kolmogorov's inequality asserts that

$$P(\max_{j=1,\cdots,n} \textstyle\sum_1^j x_i \geqq k) \leqq \dfrac{E(\sum_1^n x_i)^2}{k^2}. \tag{3.4.1}$$

The proofs of this and other similar inequalities are based on an expert use of conditional probabilities and expectations, foreshadowing the general theory of this part of the subject that Kolmogorov was soon to give. The inequality (3.4.1) is an invaluable tool in all investigations concerning sums of independent random variables.

In 1929 Kolmogorov gave in [72] a proof of the so called law of the iterated logarithm, previously found by Khintchine [62] in a particular case. For the case of a normed sum z_n of n independent and identically distributed variables as defined by (3.1.1), it follows from the central limit theorem that, for any function $h(n)$ tending to infinity with n, the probability of the relation $z_n > h(n)$ will tend to zero as n tends to infinity. Still, if we consider the infinite sequence $z_1, z_2, \cdots$, we may expect sometimes to observe very large values. The law of the iterated logarithm gives a precise expression of this vague statement by asserting that the relation

$$\limsup_{n\to\infty} \frac{z_n}{(2 \log \log n)^{\frac{1}{2}}} = 1 \tag{3.4.2}$$

will hold with a probability equal to one. It is interesting to observe that the particular case of this proposition proved by Khintchine in his paper [62] of 1924 was concerned with the frequency of digits in dyadic fractions. This was considered as a purely measure—theoretic problem, which had already attracted my interest before I had taken up probability theory. The general statement proved by Kolmogorov made a great impression, and prepared the way for the identification of probability with measure that he was soon to give.

4. The great changes: 1930 to 1939.

4.1. *The Stockholm group.* In the present chapter it will be even more impossible than in the preceding one to follow a strictly chronological order. Among the great number of new ideas that were coming forward during the 1930's there are certain main groups, the development of which will have to be considered separately, always from my personal point of view. I will begin with a few words on my own activities during the beginning of this "heroic period" of mathematical probability theory.

On the initiative of the Swedish insurance companies, a professorship for "Actuarial Mathematics and Mathematical Statistics" had been founded at Stockholm University, and in the fall of 1929 I was nominated to be its first holder. From the beginning and all through the years, I was fortunate to work with a group of ambitious and well-qualified students. We followed with keen interest the important new works that were forthcoming abroad, and tried to give some contributions of our own to the development of the field. During the first years we were mainly following up the work of the 1920's with respect to limit theorems and risk processes, where important new results seemed to be within reach.

In the fall of 1934 our group had the good fortune to receive a new member from abroad. It was during the bad days of the Nazi regime in Germany, when so many outstanding scientists were leaving that country. Will Feller, who had been turned out from the University of Kiel, came to join our group, and stayed on in Stockholm for five years. He made a great number of Swedish friends, collaborating with economists and biologists as well as with the members of our probabilistic group. He had studied in Göttingen and was well initiated in the great traditions of this mathematical center. We tried hard to get a permanent position for him in Sweden, but in those years before the war this was next to impossible, and it was with great regret that we saw him leave for the United States, where an outstanding career was awaiting him. In the following sections I shall have more to say about his work during the years he spent with us.

4.2. *Foundations*. Looking back towards the beginning of a new era in mathematical probability theory, it seems evident that the real breakthrough came with the publication in 1933 of Kolmogorov's book [75] *Grundbegriffe der Wahrscheinlichkeitsrechnung*. In this book he laid the foundations of an abstract theory, designed to be used as a mathematical model for certain classes of observable events. The fundamental concept of the theory is the now familiar, classical concept of a probability space (Ω, A, P), where Ω is a space of points ω which are denoted as elementary events, while A is a σ-algebra of sets in Ω, and P is a probability measure defined for all A-measurable events, i.e. for all sets S belonging to A.

The concepts of random variable $x = x(\omega)$ and stochastic process $x(t) = x(t, \omega)$, where t belongs to some parameter space T, are introduced in the way well known today, which in 1933 represented a remarkable innovation. It was made clear that, viewed in this way, a stochastic process defines a probability distribution in the space X of all functions $x(t)$ of the variable t. For any finite set of points $t_1, \cdots, t_n$, the n-dimensional joint distribution of the random variables

$x(t_1), \cdots, x(t_n)$ is called a finite-dimensional distribution of the $x(t)$ process. The family of all these distributions satisfies certain evident conditions of consistency. One of the main propositions given in Kolmogorov's book states that, if a family of finite-dimensional distributions is given and satisfies the consistency conditions, there exists a stochastic process corresponding to the given distributions. Further, the probability that the function $x(t)$ belongs to a set S in the function space X is uniquely determined by the finite-dimensional distributions for all Borel sets in X, i.e. for all sets S belonging to the smallest σ-algebra of sets in X containing all sets of functions satisfying a finite set of inequalities of the form $a_i < x(t_i) < b_i$. (This applies to the case of a real-valued $x(t)$; the extension to the complex-valued case being obvious.) In this way, rigorous foundations were established for the investigations of stochastic processes, which were rapidly increasing in number and importance.

However, it soon appeared that in many cases one encounters sets of functions interesting from the point of view of the applications, which are not Borel sets. Thus for the Wiener process and for the Lundberg risk process, both mentioned above in 2.3 and 3.1, it would be desirable to find the probability of the set of all functions $x(t)$ such that $x(t) < a$ for all t with $0 < t < b$, but these sets are not Borel sets, and their probabilities are not uniquely defined by the finite-dimensional distributions. For such cases it is necessary to modify the general definitions. In the book [120] by Paley and Wiener, a possible modification is shown for the Wiener process, and a similar method can be used for the risk process. For the general case Doob has analyzed the question in a series of penetrating works, summed up in his great book [35] of 1953. But all these works rest ultimately on the foundations laid down by Kolmogorov.

Kolmogorov's book of 1933 also contains a chapter on the subject of conditional probabilities and expectations, where these concepts are introduced and treated by a radically new method.

Kolmogorov's book still ranks as the basic document of modern probability theory. If in 1920 it might be said (cf. above, 2.1) that this theory was not a mathematical subject, it was impossible to express such an opinion after the publication of this book in 1933.

4.3. *Markov processes.* In his paper [73] published in 1931, two years before the "Grundlagen" book, Kolmogorov investigated a general class of stochastic processes, which later received the name of Markov processes, owing to the fact that they form a natural generalization of the classical concept of Markov chains. Consider a process $x(t)$, where t is a real parameter representing time. If, for any $t_0 < t_1$, the conditional distribution of $x(t_1)$, relative to the hypothesis $x(t_0) = a$, is independent of any additional information about the values assumed by $x(t)$ for $t < t_0$, then $x(t)$ is called a Markov process. Kolmogorov showed that the probability distributions associated with a Markov process satisfy certain functional equations, which under appropriate continuity conditions reduce to partial differential equations of parabolic type, and that these equations uniquely determine the corresponding distributions.

In our Stockholm group, it was particularly Feller who followed up this general theory of Kolmogorov with two papers ([42] and [43], the latter written during his first year in the United States) dealing with the partial differential equations of a continuous process and the integro-differential equations encountered in the discontinuous case. It is well known that the subject of Markov processes has since this time developed into a very extensive field of research.

Personally, while admiring the 1931 paper by Kolmogorov and its continuation by Feller, I did not take up any research on the general Markov process theory, perhaps owing to the fact that I have never learned to feel quite at home with partial differential equations. But there was a particular class of Markov processes that, already in the early 1930's, presented a direct interest to those of us who were working with the stochastic processes of risk theory. I am referring to the processes with independent increments, which will be discussed in the following section.

4.4. *Processes with independent increments.* In 1932, Kolmogorov [74] gave an expression for the characteristic function of a random variable $x(t)$ associated with a stochastic process satisfying the following conditions:

(1) $x(t)$ has zero mean and a finite second order moment; $x(0) = 0$.

(2) For $0 \leqq t_0 < t_1 < \cdots < t_n$ the differences $x(t_i) - x(t_{i-1})$ are independent random variables.

(3) The probability distribution of $x(t_i) - x(t_{i-1})$ depends only on the difference $t_i - t_{i-1}$.

Such a process is known as a process with stationary and independent increments. Under more general conditions, without assuming the existence of any finite moments, these processes were investigated in an admirable paper by Paul Lévy [81], published in 1934. He gave the classical general expression for the characteristic function of $x(t)$. An alternative expression which is sometimes more easily manageable was given in 1937 by Khintchine [67].

For our Stockholm group these works appeared as revelations and were eagerly studied. Already from the 1932 formula of Kolmogorov it followed that the Wiener process and the Lundberg risk process come out as opposite particular cases of the general expression. In fact, under the above conditions (1)—(3), the characteristic function

$$f(z, t) = E(e^{izx(t)})$$

is given by the Kolmogorov formula

$$\log f(z, t) = t\left(-\tfrac{1}{2}\sigma^2 z^2 + \int_{-\infty}^{\infty} \frac{e^{izu} - 1 - izu}{u^2}\, dK(u)\right),$$

where $\sigma^2 \geqq 0$ is a constant, while $K(u)$ is a bounded and never decreasing function which is continuous at $u = 0$.

It is clear that if $K(u)$ is identically zero we obtain the Wiener process. Lévy made in [81] the interesting remark that if $x(t)$ is a function of t that is almost certainly continuous, we necessarily have this case.

On the other hand, if $\sigma^2 = 0$ and $K(u) = \lambda \int_{-\infty}^{u} v^2 \, dG(v)$, where λ is a positive constant and $G(u)$ is a distribution function, we have a risk process where the claims occur according to a Poisson process with the parameter λ, while the amounts of the claims are independent random variables, each with the distribution function G. The variation of $x(t)$ is here essentially discontinuous. It was this case that we had particularly studied in Stockholm. $x(t)$ is here the accumulated amount of claims up to the time t, and the "ruin problem" which is particularly important for the insurance applications, is concerned with the probability of having $x(t) < a + bt$ during the whole interval $0 < t < T$. In a thesis [114] of 1939, C. O. Segerdahl, a member of our group, studied this problem and proved certain important inequalities for the probability of ruin.

4.5. *Infinitely divisible distributions; arithmetic of distributions.* If a random variable x can be represented in the form $x = x_1 + x_2$, where x_1 and x_2 are independent, Lévy proposed in 1934 to say that the distribution of x is the product of the distributions of x_1 and x_2, and contains each of these as a factor. If there is no nontrivial representation of this form, the distribution of x is said to be indecomposable.

If $x(t)$ is the random variable associated with a process with stationary and independent increments, it was shown in the 1934 paper [81] of Lévy that $x(t)$ can be represented as the sum of an arbitrary number of independent variables, all having the same distribution. The distribution of $x(t)$ is then said to be infinitely divisible, and the Lévy formula gives the general expression for a distribution of this kind. The normal and Poisson distributions, as well as the stable distributions, all belong to this class.

In his 1934 paper Lévy had expressed the conjecture that any factor of a normal distribution must itself be normal. He had repeated this conjecture in some subsequent papers, saying that he regarded this as very probably true, but had not been able to prove it. In the beginning of 1936 I had the good luck of finding a proof of this conjecture, which I published in [18]. In his autobiography of 1970, Lévy says [85, page 111] that he regretted not having found the proof himself, since it was a direct application of the theory of characteristic functions which he had systematically used in his own work. Soon afterwards Raikov [105] proved the corresponding statement for the Poisson distribution.

In 1937 Khintchine [68] proved the general theorem that any distribution can be represented as the product of an infinitely divisible distribution and an at most enumerable number of indecomposable distributions. In general this factorization is not unique.

4.6. *Limit theorems.* The works of Liapounov and Lindeberg mentioned above in 3.1 give sufficient conditions for the validity of the central limit theorem for sums of independent random variables. In an important book of 1933, Khintchine [64] gave an account of the main results so far known in this field. However, the problem of finding conditions which are both necessary and sufficient still remained open, as well as the case when the existence of finite moments of the variables in the sum is not assumed. The general problem attacked during the 1930's by several authors can be thus expressed:

If $x_1, x_2, \cdots$ are independent random variables, it is required to find conditions under which there exist constants a_n and b_n such that the probability distribution of the normed sum

$$(x_1 + \cdots + x_n - a_n)/b_n$$

tends to the normal distribution as n tends to infinity.

Important contributions to the investigation of this problem were given by Lévy and Khintchine, but it was Feller who first published the complete solution, in a paper [41] consisting of two parts, written in 1935 and 1937 while he was working as a member of our Stockholm probabilistic group.

Feller showed that the required necessary and sufficient conditions can be expressed by appropriate modifications of the sufficient condition given in 1922 by Lindeberg. In the particular case when the x_i have distribution functions F_i with zero means and finite standard deviations σ_i, such that $\sigma_i/s_n \to 0$ as $n \to \infty$, uniformly for $i = 1, 2, \cdots, n$, where as usual $s_n^2 = \sigma_1^2 + \cdots + \sigma_n^2$, we can take $a_n = 0$ and $b_n = s_n$, and Feller proved that the original Lindeberg condition

$$\lim_{n\to\infty} \frac{1}{s_n^2} \sum_1^n \int_{|x|>\varepsilon s_n} x^2\, dF_i(x) = 0$$

for any given $\varepsilon > 0$ is both necessary and sufficient for the convergence to the normal distribution. He also gave more elaborate conditions for the case when the x_i have no finite moments.

The outstanding Italian mathematician F. P. Cantelli was engaged in actuarial work on similar lines to those of several members of our Stockholm group. In a paper [10] of 1933, he had given an improved form of the iterated logarithm law mentioned above in 3.4. In the course of a correspondence with him, I considered a modification of the problem, where a "most possibly precise" result could be obtained. This was published in my paper [17] in 1934. For the classical law of the iterated logarithm as expressed by (3.4.2), a statement of this character was to be given by Feller [44] in 1943. I will quote here for comparison first the Feller theorem of 1943, and then my own of 1934, in both cases without giving detailed statements of the conditions under which our results are shown to hold.

Using the above notation, Feller proved that the probability

$$P(|x_1 + \cdots + x_n| < \mu_n s_n \quad \text{for all sufficiently large } n)$$

is equal to 1 or 0 according as the series

$$\sum_1^\infty \frac{\sigma_n^2}{s_n^2} \mu_n e^{-\frac{1}{2}\mu_n^2}$$

is convergent or divergent.

On the other hand, in my 1934 paper I had somewhat modified the problem. Instead of a single sequence of independent random variables $x_1, x_2, \cdots$, I considered the double sequence of variables

$$\begin{array}{c} x_{11}, \\ x_{21}, x_{22}, \\ \cdots \\ x_{n1}, x_{n2}, \cdots, x_{nn}, \end{array}$$

all supposed to be independent, with zero means and finite standard deviations σ_{ij}. Writing $s_n^2 = \sigma_{n1}^2 + \cdots + \sigma_{nn}^2$, I showed that the probability

$$P(|x_{n1} + \cdots + x_{nn}| < \lambda_n s_n \text{ for all sufficiently large } n)$$

is equal to 1 or 0 according as the series

$$\sum_1^\infty \frac{1}{\lambda_n} e^{-\frac{1}{2}\lambda_n^2}$$

is convergent or divergent.

A similar, but much more far-reaching generalization of the whole problem of limiting distributions for sums of independent random variables was given by Khintchine [67] in 1937, followed by works due to Gnedenko [47] and other Russian authors. I shall here only indicate the main lines of their work, referring for more details to the important book [48] by Gnedenko and Kolmogorov. Consider a double sequence of random variables

$$\begin{array}{c} x_{11}, x_{12}, \cdots, x_{1k_1}, \\ \cdots \\ x_{n1}, x_{n2}, \cdots, x_{nk_n}, \\ \cdots \end{array}$$

the variables in each line being supposed to be mutually independent, and let z_n denote the sum of the variables in the nth line. Suppose that we have for any $\varepsilon > 0$

$$\lim_{n\to\infty} P(|x_{nk}| > \varepsilon) = 0$$

uniformly for $k = 1, \cdots, k_n$. Then Khintchine in 1937 proved the following fundamental result: In order that F should be the limiting distribution function for the variables $z_n - b_n$ associated with such a double sequence, where the b_n are appropriately chosen constants, it is necessary and sufficient that F be infinitely divisible. Thus the class of all possible limit laws associated with sums of the type here considered coincides with the class of all infinitely divisible laws. It is hardly necessary to say that in our Stockholm group we followed these new ideas of our Russian colleagues with the greatest interest.

4.7. *Characteristic functions.* The use of an analytic tool more or less equivalent to that known to us under the name of characteristic functions goes back to Lagrange, Laplace and Cauchy. As mentioned above, it was used by Liapounov for his proof of the central limit theorem. But the first systematic account of the theory was given in the 1925 book of Lévy [80] referred to in 3.2. I had used this method for my work on asymptotic expansions, and throughout the 1930's we tried in Stockholm to develop the subject further. My proof of Lévy's conjecture for the normal distribution mentioned in 4.5 was an outcome of this work. H. Wold and I published in 1936 a joint paper [34] on multidimensional distributions, where generalizations of characteristic functions and the central limit theorem were discussed.

In the same year 1936 I had concluded the work on a book on mathematical probability, which G. H. Hardy in 1927 had encouraged me to write. It took the form of a Cambridge Tract [19] entitled "Random Variables and Probability Distributions," published in the early part of 1937. It was based on Kolmogorov's foundations, and its main contents were a full account of the theory of probability distributions in finite-dimensional spaces and their characteristic functions, with applications to subjects such as the central limit theorem, the allied asymptotic expansions, and stochastic processes with independent increments.

With respect to the general theory of characteristic functions I was able to give some complements to Lévy's work. But in this connection I had made a regrettable error which had to be corrected in later editions of the book, of 1963 and 1970. As the fact is of some interest, I shall give a brief account of it here.

The well-known "continuity theorem" for characteristic functions asserts that a sequence of distribution functions F_n converges to a distribution function F in every continuity point of the latter, if and only if the corresponding characteristic functions $f_n(t)$ converge for every t to a limit which is continuous for $t = 0$. In the statement of this theorem given in the first edition of my Tract, I required only convergence of the $f_n(t)$ in some finite interval $|t| < c$. However, a letter from Khintchine informed me of my error. It is sufficient to point out that it is possible to find two characteristic functions $f_1(t)$ and $f_2(t)$ which are equal for $|t| < 1$, but not identically equal. Take, in fact,

$$f_1(t) = f_2(t) = 1 - |t|$$

for $|t| \leqq 1$, and let $f_1(t)$ be periodic with period 2, while $f_2(t) = 0$ for $|t| > 1$. It is easily seen that both of these are characteristic functions. A sequence the members of which are alternatively equal to f_1 and f_2 thus converges for $|t| < 1$, but not for all t, which proves the falsehood of the statement in my first edition.

In this connection I may add that in a paper [21] of 1939 I gave some theorems on the representation of functions by Fourier integrals which, among other results, include a simplification of an earlier theorem by Bochner, giving a necessary and sufficient condition for a function to be the characteristic function of some probability distribution.

4.8. *Stationary processes.* I must now go some years back in time, and take up an important new thread of the development during the 1930's. In 1934 Khintchine published a basic paper [65] where he introduced the class of stationary stochastic processes. He pointed out that a Markov process cannot be used in a case when the past history of the system under consideration has an essential influence on any prediction with respect to its future development, as e.g. in statistical mechanics. As a convenient tool for the study of such systems he introduced the class of stationary processes.

Khintchine gave definitions both for the class of strictly stationary processes, and for the class which I shall here simply call stationary. A process $x(t)$ with a continuous time parameter t is called strictly stationary if the associated finite-dimensional distributions are all invariant under a translation in time. It is called stationary, if the same invariance holds for its moments of the first and second order. When reviewing the results of Khintchine's basic paper [65] and of the subsequent developments, I shall in the sequel in general consider a complex-valued process $x(t)$ which is continuous in mean square, and such that

$$Ex(t) = 0\,, \qquad Ex(t)\overline{x(u)} = r(t-u)\,.$$

The covariance function $r(t)$ is then continuous for all real t, and Khintchine showed that it admits the spectral representation

$$r(t) = \int_{-\infty}^{\infty} e^{itu}\, dF(u)\,, \tag{4.8.1}$$

where the spectral function $F(u)$ is real, never decreasing and bounded. From this representation he deduced various properties of $r(t)$. He also proved that the time average

$$\frac{1}{T}\int_0^T x(t)\, dt$$

converges in quadratic mean as $T \to \infty$, and pointed out that this is equivalent to the mean ergodic theorem of von Neumann.

In a somewhat earlier paper [63], Khintchine had considered the case of a strictly stationary process with discrete time, or a sequence of random variables $\cdots, x_{-1}, x_0, x_1, \cdots$, satisfying the condition of strict stationarity. For this case he proved that the time average

$$\frac{1}{N}\sum_1^N x_n$$

converges even with probability one, which for this process is the equivalent of Birkhoff's ergodic theorem. The corresponding result for continuous time was proved somewhat later by Kolmogorov [76].

There are interesting relations between Khintchine's theory of stationary processes and the earlier work by Norbert Wiener on generalized harmonic analysis.

Wiener published in 1930 an extensive paper [119] on this subject, where he dealt with complex-valued functions f such that the limit

$$\lim_{T\to\infty} \frac{1}{2T} \int_{-T}^{T} f(u+t)\overline{f(u)}\, du = r(t) \tag{4.8.2}$$

exists for all real t. The limiting function r has properties quite similar to those of the covariance function of a stationary process. In particular it is continuous for all t if it is continuous for $t = 0$, and in this case it admits a spectral representation of the form (4.8.1). It seems reasonable to expect that the Wiener relation (4.8.2) should, in general, be satisfied by the sample functions of a stationary process.

Wiener's paper, quite like his 1923 paper on differential space, is not easy to read, but Masani has given in [95] a very readable account of it from a modern point of view. He shows, in particular, that Wiener gives an example of what was later to be called a normal stationary process such that, with probability one, any sample function of the process satisfies the Wiener relation (4.8.2). However, I should like to point out that it is also easy to find an example where this property does not hold. Let, in fact, a real-valued and normal stationary process have the covariance function $r(t) = 1 - |t|$ for $|t| \leqq 1$, and let $r(t)$ be periodic with period 2. Then it can be shown that, for any given t, there is a set of sample functions of this process having positive measure, and such that the first member of the Wiener relation (4.8.2) does not converge to any finite limit as $T \to \infty$.

Khintchine's paper [65] on stationary processes had a very significant impact. It seemed clear that this new type of stochastic process would provide a convenient mathematical model not only for statistical mechanics, but also in other fields, such as meteorology and economics. In particular it opened up new possibilities for the investigation of phenomena showing a tendency to periodic behavior, and for applications to information theory.

In our Stockholm group the subject was taken up by Herman Wold, who in his thesis [123] of 1938 dealt with stationary processes with discrete time, i.e. stationary sequences of random variables x_n. He established their covariance and spectral properties, and gave extensive applications to various statistical problems. His most remarkable result was the proof, for the class of processes considered, of an important decomposition theorem which has later been shown to hold also, with due modifications, for more general types of processes. He proved that, for a stationary sequence x_n, there is a unique decomposition

$$x_n = u_n + v_n\,,$$

where u_n and v_n are mutually uncorrelated stationary sequences such that, in present-day terminology, u_n is purely nondeterministic, while v_n is deterministic. Further, u_n can be represented in the form

$$u_n = \sum_{-\infty}^{n} c_{n-i} z_i\,,$$

where the z_i are mutually uncorrelated random variables, while the c_i are non-random constants.

In a paper [22] of 1940, I investigated the covariance properties of a stationary vector process $\mathbf{x}(t) = (x_1(t), \cdots, x_q(t))$ with continuous time t. I proved that there is a spectral representation of the mutual covariance functions

$$r_{mn}(t) = Ex_m(t+u)\overline{x_n(u)} = \int_{-\infty}^{\infty} e^{itu}\, dF_{mn}(u)\,, \tag{4.8.3}$$

where the F_{mn} are complex-valued functions of bounded variation, such that the increments ΔF_{mn} over any interval form a nonnegative Hermitian matrix. This implies that the matrix function $\mathbf{F}$ with the entries F_{mn} is never decreasing, and I proved that, like a never decreasing function of one variable, it is the sum of three components: one absolutely continuous, one purely discontinuous, and one singular. I also proved the corresponding properties for a vector process with discrete time.

4.9. *Paris, London and Geneva*, 1937–1939. In the spring of 1937 I was invited to Paris, is order to give some lectures at the Sorbonne. Of the French probabilists I had met Fréchet before, but this was my first acquaintance with Paul Lévy, and with a number of the younger generation, such as Doeblin, Dugué, Fortet and Loève. In earlier years Fréchet had been an outstanding mathematician, doing pathbreaking work in functional analysis. He had taken up probabilistic work at a fairly advanced age, and I am bound to say that his work in this field did not seem very impressive to me. On the other hand, already in 1937 it seemed clear that Lévy was to be one of the leaders in the development of probability theory, particularly after the publication of his book [83] *L'Addition des Variables Aléatoires*. Incidentally, he mentions in the preface to this book that it was on receiving my letter in the beginning of 1936 containing my proof of his conjecture about the normal distribution that he made the decision to write the book. Among the younger men, Doeblin had already done outstanding work, and it was a great loss to science when he was killed during the first months of the war.

Later in the same year 1937 there was a conference for probability theory in Geneva. Feller and I attended from Stockholm, and it was quite exciting to see such a large group of eminent probabilists assembled. Among new acquaintances, there were Steinhaus from Poland, Hopf from Germany, and Jerzy Neyman, at that time still working in England. Several colleagues from the Soviet Union had accepted the invitation and announced lectures, but to our great disappointment none of them turned up. Neyman gave a lecture on his theory of confidence intervals, which was then something quite new. At this early stage his ideas had not yet received their final expression, and both Fréchet and Lévy took a very critical attitude. I had already read his paper [101] on the subject, and had come to the conclusion that his basic ideas were quite sound, as, indeed, the later development showed them to be. Feller lectured about axiomatics, and I gave a talk [20] on the "large deviations" connected with the central limit theorem. If, using the notation of 3.1, we allow x to tend to infinity with n, both $G_n(x)$ and $\Phi(x)$ tend to unity, so that the assertion of Liapounov's theorem becomes trivial. In my paper, I considered the ratios

$$\frac{1 - G_n(x)}{1 - \Phi(x)} \quad \text{and} \quad \frac{G_n(x)}{\Phi(x)},$$

where x and n both tend to $+\infty$ in the first case, and to $-\infty$ in the second. I showed that, subject to appropriate conditions, it is possible to find an asymptotic expansion for each case. A first generalization of my results was given by Feller in 1943 who used it for his improved form of the iterated logarithm law [44] mentioned above in 4.6. Very important further generalizations were given by Yu. V. Linnik and his Leningrad group. An account of their work is given in the excellent 1966 book [55] by Ibragimov and Linnik.

In October 1938 I made a visit to England. It was shortly after the Munich conference, and the question of peace or war was already on everybody's mind. In Cambridge I saw my old teacher and friend G. H. Hardy, who lived in Newton's rooms in Trinity College. He expressed his satisfaction with my tract, which was written on his initiative. In London I was received by R. A. Fisher, William Elderton and Egon Pearson, and saw something of the work in mathematical statistics that was being done there. Neyman was already in California.

In the summer of 1939 there was again a conference in Geneva, this time for mathematical statistics. I was happy to make the acquaintance of Sam Wilks and Maurice Bartlett, who both gave interesting lectures. Neyman did not appear, but had contributed a basic paper [102] on the theory of testing statistical hypotheses, recently founded by him in collaboration with Egon Pearson. From these days I remember a conversation with R. A. Fisher. I had expressed my admiration for his geometrical intuition in dealing with probability distributions in multidimensional spaces, and received the somewhat acid reply: "I am sometimes accused of intuition as a crime!"

When leaving Geneva in July 1939, it seemed fairly clear that the war was rapidly approaching.

5. The war years: 1940 to 1945.

5.1. *Isolation in Sweden*. During the war our international contacts were reduced to a minimum. Sweden remained neutral, but surrounded by war: Denmark and Norway were occupied by the Nazis, and Finland was at war with Russia. There were very few opportunities to exchange literature or correspondence with colleagues in England, France and the United States, and none at all with those in the Soviet Union. But we tried to keep our research work going as far as possible.

In his 1940 thesis [93] Ove Lundberg, a member of our Stockholm group, investigated a new class of stochastic processes which he applied to certain statistical problems of non-life insurance. On the foundations laid by him, this subject has had a strong international development among actuaries.

In February 1941 I organized a conference for mathematical probability in Stockholm. We were happy to have some guests from Denmark and Finland; in Norway the German occupation was harder, and none of our colleagues had been allowed to come. Harald Bohr and Børge Jessen of Denmark talked about the foundation of probability and its relations to general mathematical analysis, and Gustav Elfving of Finland about Markov processes. Several members of our Stockholm group gave accounts of their theses which I have mentioned above, and I presented on this occasion a theorem about the spectral representation of a stationary stochastic process which I had just found. If $x(t)$ is the stationary process considered above in 4.8, the covariance function of which has the spectral representation (4.8.1), there is a process with orthogonal increments $z(u)$ such that, with an appropriate definition of the stochastic integral,

$$x(t) = \int_{-\infty}^{\infty} e^{itu}\, dz(u)\,, \tag{5.1.1}$$

where in the usual symbolism

$$E\, dz(u) = 0\,, \qquad E|dz(u)|^2 = dF(u)\,.$$

This shows how $x(t)$ is additively built up by elementary harmonic oscillations $e^{itu}\, dz(u)$, each of which has an angular frequency u, while the amplitude and the phase are random variables determined by $dz(u)$.

In 1942 I published a paper [23] on the spectral representation (5.1.1), explicitly pointing out the relations to Wiener's generalized harmonic analysis. But it was not yet clear to me that what I had done was really to give a probabilistic version of Stone's theorem on the spectral representation of a unitary group in Hilbert space. The method I had used for my proof employed stochastic Fourier integrals, but no Hilbert space theory. The fundamental importance of this theory for the study of stochastic processes did not become known to us until after the war.

A young Danish physicist, N. Arley, had written a thesis on the application of stochastic processes to the theory of cosmic radiation [1], and to my great surprise I was allowed to come to Copenhagen in the spring of 1943 as a member of the examination committee. The thesis was a good piece of work, and I was happy to see my Danish colleagues again, but it was painful to see German troops marching through the streets of Copenhagen.

In 1944 the Uppsala thesis [40] on Fourier analysis of distribution functions by C. G. Esseen appeared. This was a very important work, based on a penetrating investigation of the properties of characteristic functions. In particular Esseen showed that the factor $\log n$ in the above evaluation (3.1.2) of the remainder in Liapounov's theorem can always be omitted, thus generalizing my result mentioned in connection with (3.1.5), which holds only subject to the condition (3.1.4). He also gave estimations of the remainder depending both on n and x, as well as an improvement of my asymptotic expansion of 1928.

While it still seemed clear that the end of the war was far away, I decided to use the years of undesired isolation to write a book. This book [24] was to be ready in 1946, and was entitled *Mathematical Methods of Statistics.* From the preface I quote the following lines:

> During the last 25 years, statistical science has made great progress, thanks to the brilliant schools of British and American statisticians, among whom the name of Professor R. A. Fisher should be mentioned in the foremost place. During the same time, largely owing to the work of French and Russian mathematicians, the classical calculus of probability has developed into a purely mathematical theory satisfying modern standards with respect to rigor. The purpose of the present work is to join these two lines of development in an exposition of the mathematical theory of modern statistical methods, in so far as these are based on the concept of probability.

Thus the book is not to be regarded as a contribution to mathematical probability theory, but rather as a treatise of its applications to modern statistical methods. The book is dedicated to my wife, who had supported and encouraged me all through my work, as she had always done. While I was writing, I sometimes said to her that I hoped this book would be my entrance card to the new world after the war. Perhaps there was something in it: today there are editions of the book in English, Russian, Spanish, Polish and Japanese.

5.2. *International development during the war years.* On the few occasions when there was mail from the United States, we received letters and reprints from Feller, and so were able to follow his work on Markov processes and on the perfection of the iterated logarithm law mentioned above in 4.3 and 4.6.

Mathematicians in war-making countries became often engaged in work with antiaircraft fire control and noise filtration in radar. It appeared that stationary stochastic processes provided an efficient tool for this kind of work. In particular the possibility of making predictions for the future course of such a process, based on observations during the past, was of vital importance. Independently of one another, Kolmogorov in the Soviet Union and Wiener in the United States made important contributions to this subject. They do not seem to have been aware of one another's work until after the war.

In two papers [77, 78] of 1944, Kolmogorov investigated a stationary stochastic process with discrete time. He pointed out that the class of all random variables with finite second order moments constitutes a Hilbert space, if the inner product of two points is defined as the covariance moment of the corresponding random variables. A stochastic process with discrete time can thus be regarded as a sequence of points in Hilbert space, so that the theory of this space becomes available for the study of the process.

For stationary sequences of random variables, Kolmogorov showed that the application of Hilbert space theory makes it possible to deduce in a simple way all results previously known, such as the Wold decomposition and the covariance properties of a vector process with discrete time given in my paper [22] mentioned above in 4.8. Moreover, he used powerful methods of complex function theory to give for the first time a necessary and sufficient condition for a stationary sequence to be purely nondeterministic (regular in Kolmogorov's terminology), and deduced a complete solution of the linear least squares prediction problem. His work was followed up by Zasuchin [127], who considered a stationary vector process with discrete time, and gave a number of important results for such a process.

The fundamental importance of this work by Kolmogorov lies in the fact that he showed how the abstract theory of Hilbert space (as well, of course, as of other types of spaces) could be applied to the theory of random variables and stochastic processes. This had a powerful influence on the whole subsequent development of the theory.

Wiener's war work was related to the problems of linear prediction and filtering for stationary processes, both with discrete and continuous time. For the prediction of the value of a stationary process $x(t)$ with continuous time at $t = h > 0$, based on the observation of the past of the process up to $t = 0$, he introduced a predictor of the form

$$x^*(h) = \int_{-\infty}^{0} x(t)\, dK(t) ,$$

where K is of bounded variation, and showed how K could be determined so as to minimize the mean square error of prediction

$$E(x^*(h) - x(h))^2 .$$

This is not a complete solution of the mathematical problem of linear prediction, but it has important applications to various engineering problems. Wiener's work on this and allied subjects was completed in 1942, and during the following years circulated in mimeographic copies bearing the nickname "the yellow peril," owing to the difficult mathematics involved. It did not become publicly available until 1949, in the book [121].

6. After the war: 1946 to 1970.

6.1. *Introductory remarks.* After the great turmoil of the war years, it gradually became possible again to take up scientific research work on an international basis, and to renew contacts with colleagues in other countries. It was now clear to everybody concerned that mathematical probability theory had passed through a radical change during the twenty-five years between 1920 and 1945. While in 1920 it had hardly deserved the name of a mathematical theory, in 1945 it entered into the postwar world as a well-organized part of pure mathematics with problems and methods of its own, and with an ever growing field of applications to other sciences, as well as to many practical activities. There were intimate mutual relations between the applications and the purely mathematical theory, and already it was hardly possible for an individual research worker to survey the whole field.

The years after the war brought a further powerful development in many different directions. In a review of personal recollections like the present one, it is evidently even more impossible than for the time up to 1945 to give a full account of the development. I shall have to confine myself strictly to those parts of the field in which I was able to be more or less actively engaged, and to those people with whom I had personal contacts.

I will begin by giving a brief account of my contacts with scientific colleagues during the years immediately following the war, and then proceed to a survey of the subsequent development, always from my personal point of view.

6.2. *Paris*, *Princeton*, *Yale*, *Berkeley*, 1946–1947. Soon after the end of the war, I received invitations to the universities named above. In Paris I was to give a series of five lectures, two on statistical estimation and three on stationary processes, in the spring of 1946. Then I was appointed as Visiting Professor in Princeton for the fall term of 1946, at Yale for the spring term of 1947, and in Berkeley for the following summer term.

In Paris I was happy to see again my old friends from 1937, with the exception of the young Doeblin, who had been killed in the beginning of the war. Lévy had had his apartment sacked by the Nazis, who had destroyed his books and papers, but he was already starting new work on stochastic processes, which were soon to give rise to a new important book [84] of 1948.

What I had to say in Paris about statistical estimation was taken from my book on mathematical methods, which was now in the course of being printed. I planned to say something about the controversial subjects of Fisher's fiducial probability and Neyman's confidence intervals, where I definitely sided with Neyman. It was a not altogether pleasant surprise that R. A. Fisher was in Paris and attended my lecture on these topics. Afterwards we had a private discussion which ended better than I had feared, perhaps partly owing to the fact that I happened to know a good eating place, of which there were not so many in Paris less than a year after the end of the war.

In my Paris lectures on stationary processes I gave, among other things, an account of Khintchine's spectral representation (4.8.1) of the covariance function and my own representation (5.1.1) of the process variable itself. It appeared that Loève had independently found the latter representation. In an appendix [90] to Lévy's book of 1948, he gave an account of his important work on this and other related topics, which somewhat later formed a chapter in his own great book [91] of 1955.

In September 1946 I made my first journey to the United States, where I was received in Princeton by Sam Wilks. I gave a course on stochastic processes, and among my audience were K. L. Chung, Ted Harris, G. A. Hunt and Sam

Karlin. It was a great pleasure to work with these intelligent young men. It was the year of the bicentennial of Princeton University, and among the scientific conferences forming part of the programme was one entitled "Problems of Mathematics." Among the great number of outstanding mathematicians attending were my old friends Einar Hille, Will Feller, Jerzy Neyman and Norbert Wiener, whom I was happy to see again. There were also many new acquaintances, including outstanding probabilists and statisticians, such as J. L. Doob, Harold Hotelling and Mark Kac. There was a section on mathematical probability, where all these mathematicians made interesting contributions. I was particularly glad to meet Doob, whose work on stochastic processes I had read and admired. In a paper [24] on "Problems in probability theory," I gave a survey of old and new problems in the field. Following an invitation from Gertrude Cox and Harold Hotelling, I made my first visit to Chapel Hill, where this time I only spent a few days, and met P. L. Hsu and Herbert Robbins.

During the spring term at Yale and the summer term in Berkeley I continued my work and my lectures on stochastic processes. In Berkeley, Neyman was making preparations for the celebrated series of Berkeley Symposia on Probability and Mathematical Statistics. Among the group engaged in this work I was happy to meet people like J. L. Hodges, Erich Lehmann, Henry Scheffé and Betty Scott. They are now all well known for their outstanding scientific work.

6.3. *Work in the Stockholm group.* During my absence from Sweden in 1946 and 1947, Gustav Elfving of Helsingfors acted as my substitute at the Stockholm University. Through him and his fellow countryman Kari Karhunen, our group had valuable contacts with probabilistic research in Finland. In his Helsingfors thesis [59] of 1947, Karhunen gave a systematic treatment of the application of Hilbert space theory to stochastic processes, thus following up the work of Kolmogorov mentioned above in 5.2. After completing his thesis, which contains remarkable new results on stochastic integrals and stationary processes, Karhunen for some time worked in Stockholm as a member of our group. In a paper [60] of 1949 he treated a stationary process with continuous time, and obtained for this case results corresponding to those given by Kolmogorov for the case of discrete time, including the Wold decomposition as well as a necessary and sufficient condition for such a process to be purely nondeterministic. Some similar results were given in a paper [53] by O. Hanner, another member of our group.

In a paper [26] written for the Berkeley Symposium in 1950, I considered some more general classes of stochastic processes, using Hilbert space theory. I gave a derivation of a general type of spectral representation which, in the particular case of a stationary process, seems to be the simplest so far known (cf. Doob [35], page 483).

The 1950 thesis [49] of Ulf Grenander, a member of our Stockholm group, was entitled "Stochastic processes and statistical inference," and turned out to be a pathbreaking work in this difficult and important field. It is well known that Grenander later followed up this work, e.g. in the excellent joint book [50] *Statistical Analysis of Stationary Time Series* of 1957, written in collaboration with Murray Rosenblatt. He has also extended his outstanding work to other related fields, in the joint book [51] *Toeplitz Forms and Their Applications* of 1958, written together with Gabor Szegö, and in numerous other works. It was Ulf Grenander who took the initiative of publishing, in 1958, a "Harald Cramér Volume" [52], containing articles in probability and mathematical statistics written by a great number of my friends. I was very happy to receive this token of friendship. Ulf Grenander became my immediate successor as Professor in the University of Stockholm. He has since then left Sweden, and now belongs to Brown University, in Providence, R. I.

From 1950 on, I became engaged in administrative University work, which absorbed a great part of my time until I was able to leave it in 1961. Still, I managed to write a monograph on risk theory [27], published in 1955, the most important part of which is a study of the ruin problem for the Lundberg risk process (cf. above, 4.4). In the most general case, this problem leads to an integral equation of the Wiener–Hopf type, and the discussion of this equation makes it possible to obtain a number of asymptotic results valuable in the practical applications. By means of the new methods later introduced by Feller, some of these results can be obtained without the use of the Wiener–Hopf equation.

6.4. *Moscow*, 1955. In May 1955 the University of Moscow celebrated its bicentennial, and I was invited to attend, representing the Stockholm University. It was a great event, and it gave me an opportunity to make the personal acquaintance of the Soviet mathematicians, whose work had meant so much for the advancement of probability theory. Unfortunately, Khintchine was ill—he died shortly afterwards—but I met Kolmogorov, who gave the impression of being a great scientific personality, and I had some interesting conversations with him. I was also happy to meet other members of their probabilistic group. There was Dynkin, who was beginning his great work on Markov processes, Gnedenko who in collaboration with Kolmogorov had written the book on limit problems referred to above, Linnik who was beginning his work on large deviations so closely connected with my own work of 1937, Yaglom and Rozanov who were to do outstanding work on stationary processes, and many others. They formed a group of wonderful scientific activity, and were preparing to start their new *Journal of Probability Theory and its Applications*, which soon acquired an internationally leading position in the field.

In connection with this Bicentennial there was also a mathematical conference, and I gave a lecture on my recent work on the ruin problem. Later in the same year, Kolmogorov spent some time in Stockholm as the guest of our university, and gave a series of lectures on limit theorems in probability to our group.

6.5. *Books on probability*. Before the war there were only a small number of books on mathematical probability theory built on modern foundations. Some of these have been mentioned in previous chapters.

After the war the situation radically changed, and there has been a stream of general treatises as well as of monographs on special parts of the field. I am going to give a very brief survey, strictly limited to those books which have had direct influence on the development of my own research work. This means, of course, that many important contributions to the field will not be mentioned. In particular, the extensive literature on Markov processes and on martingales will be passed over in silence, even though I am well aware of the great importance of these subjects. In all cases I give only the year of first publication of a book in its original language.

The first general postwar treatise of probability theory was Feller's [45] which appeared in 1950, and was followed by a second volume in 1966. For the young generation of the 1950's this book was an excellent introduction into an important new field of scientific research. It contained the basic theory as well as a great number of applications, all written in the fascinating personal style of its author. The second volume gives many new results, and simplified proofs of some already known.

Loève's treatise [91] of 1955 entered deeply into the background of mathematical analysis, with chapters on measure and integration. The classical limit theorems and their modern extensions to interdependent variables were fully treated, as well as important classes of stochastic processes.

The Russian treatise [104] of 1967 by Prochorov and Rozanov, is a well-written and important work, taking account of the most recent developments, and showing the high class of the work in our field performed in the Soviet Union.

The early postwar monographs by Gnedenko and Kolmogorov on limit problems, and by Wiener on prediction and filtering for time series, as well as the books by Grenander–Rosenblatt and Grenander–Szegö have already been referred to above.

In 1948 and 1953 there appeared three monographs on the general theory of stochastic processes, of which those written by Lévy and Doob have been briefly mentioned above. The third is a joint work by Blanc–Lapierre and Fortet. They are all three classics in the field. Lévy's book contains in particular a detailed study of the Brownian movement on the line and in the plane. Doob gives a complete account of the theory, including difficult basic measure-theoretic investigations as well as extensive chapters on the most important classes of processes. The book by Blanc–Lapierre and Fortet offers a particular emphasis in its treatment of the applications of stochastic processes to many important physical problems.

Yaglom gave in 1952 a useful introduction to the theory of stationary processes [124]. A complete monograph on this important class of processes was published by Rozanov [110] in 1963. A joint book of 1965 by Linnik and Ibragimov [55] gives in its first part a full account of the work of Linnik and his group on large deviations for sums of independent variables, while the second part deals with stationary processes. Both the Rozanov and the Linnik–Ibragimov book contain important new results, e.g. about the generalization of the central limit theorem to stationary processes. Another monograph by Rozanov [111] of 1968 on infinite-dimensional Gaussian distributions contains an excellent chapter on the problem of equivalence or perpendicularity of normal distributions in function space. In the 4th volume of the great work by Gelfand and Viljenkin on generalized functions there is an interesting chapter on generalized stochastic processes.

Finally, the autobiographies of two leading probabilists should be mentioned here. They are written by Norbert Wiener [122] of 1956, and by Paul Lévy [85] of 1970. Both contain a wealth of scientific material of the highest interest.

6.6. *Stationary and related stochastic processes.* My own research work in probability after 1960 followed mainly two different lines, and I now propose to give a brief account of the work of others and myself on these two lines, without paying too much regard to chronological order. I will begin by talking of the work done during the 1950's and 1960's on the subject of stationary and related stochastic processes.

The development that started with Kolmogorov's and Zasuchin's work during the war, and was continued by Karhunen, Hanner and others during the late 1940's soon led to remarkable further results. By means of Hilbert space theory the properties of the univariate stationary process were generalized to vector processes with discrete and continuous time, and to homogeneous random fields, i.e. processes with several independent parameters, such as the two coordinates of a plane, or the four coordinates of space-time, which satisfy some condition analogous to stationarity in the one-parameter case.

The spectral properties of a stationary vector process have been investigated by the Russian authors whose books have been mentioned in the preceding section, and in a number of papers by Wiener and Masani [96]. For a non-deterministic univariate process with the spectral function F it had been shown by Kolmogorov and Karhunen that the spectral functions of the purely non-deterministic and the deterministic components are respectively identical with the absolutely continuous and the jump-singular parts of F. In the investigation of the corresponding properties for a vector process one encounters difficulties connected with the rank of the never decreasing spectral matrix $\mathbf{F}$ mentioned above in 4.8. If $\mathbf{F}$ has (appropriately defined) maximum rank, the decomposition property for the univariate case can be directly generalized, as shown by Masani [94], but otherwise the situation is more complicated.

In our Stockholm group, B. Matérn wrote a thesis [97] in 1960 on "spatial variation," dealing with homogeneous random fields in the two-dimensional plane, with important applications to estimation problems in forestry.

An important line of investigations started from the attempt to generalize the central limit theorem to stationary processes. In 3.4 above I mentioned the 1927 paper of Bernstein. He showed that the central limit theorem can be extended to sums of weakly dependent variables, and introduced a technique for dealing with such cases. In a paper [109] of 1956, Rosenblatt gave a remarkable definition of weak dependence, leading to the important concept of "strong mixing" for stochastic processes. Consider a process x_n with discrete time, and let $M_{a,b}$ be the smallest complete σ-algebra of events (ω sets) relative to which all x_n with $a \leqq n \leqq b$ are measurable. As a measure of dependence between $M_{-\infty,m}$ and $M_{m+n,\infty}$ Rosenblatt introduces the upper bound of $|P(A \cap B) - P(A)P(B)|$ for all events $A \in M_{-\infty,m}$ and $B \in M_{m+n,\infty}$. If the upper bound of this quantity for all m tends to zero as n tends to infinity, this implies that the degree of dependence between any two sections of the x_n sequence is always small when the sections lie far apart, and the process is then said to be strongly mixing. This is a stronger form of the ordinary mixing condition used in ergodic theory. The generalization to the case of continuous time is immediate. If this condition is satisfied, and some further conditions relative to the behavior of moments of the form $E(\sum_m^{m+n} x_i)^2$ and $E|\sum_m^{m+n} x_i|^3$ for large n are imposed, it can be shown that the sum $\sum_m^{m+n} x_i$ is, for any fixed m, asymptotically normal as $n \to \infty$. For the particular case when the x_n sequence is strictly stationary, it has proved convenient to introduce a "uniformly strong mixing" condition due to Ibragimov, which simplifies the proof of the central limit theorem and makes it possible to generalize the law of the iterated logarithm. These questions are discussed in the book [55] by Linnik and Ibragimov and in papers by Volkonski and Rozanov [117] and by Reznik [106]. In lectures at the Universities of Aarhus in 1967 and Copenhagen in 1969, I gave surveys of these works, and was able to complete them in some places.

All these results are obtained by using the estimation technique originally introduced by Bernstein in his paper [7] of 1927. For the particular case of normal processes it is possible to deduce more precise results, as shown e.g. in an important paper [79] by Kolmogorov and Rozanov.

When taking up work at the Research Triangle Institute of North Carolina in 1962, I became engaged in research on certain problems connected with the extreme values of stationary processes, which had applications in spacecraft navigation. As my assistant I had Ross Leadbetter, a young New Zealander, with whom I found it very stimulating to work. Our collaboration led to some promising results, and it was suggested that we should develop it and write a joint monograph on the problems involved. In planning the book we soon came to the conclusion that it would be desirable to include a fairly broad account of the general theory of stationary processes, with special emphasis on the properties of their sample functions. In particular we wanted to discuss the analytic properties of the sample functions, such as continuity and differentiability, and study the random variables expressing the number of crossings in a given time interval between a sample function and a fixed level or curve. We were able to build our work on previous research by a large number of authors, among which I may quote here Beljaev [3–6], Bulinskaja [9], Kac and Slepian [56], Rice [107, 108], Volkonski and Rozanov [117] and Ylvisaker [125, 126]. In particular the work of Rice had a basic importance for the whole field of problems concerned. For a normal stationary process which satisfies the strong mixing condition, Volkonski and Rozanov had proved the important theorem that the crossings between a sample function and a very high level approximately form a Poisson process. I had been able to show that this property holds already under simpler conditions, and that it leads to interesting conclusions concerning the magnitude and distribution of the extreme values of the sample functions. The corresponding propositions were included in our book, which appeared in 1967 under the title *Stationary and Related Stochastic Processes*. Many of our results have since been improved by other authors, e.g. by Beljaev, who wrote an introduction to the Russian edition of the book, by S. M. Berman in a series of significant papers on extreme values, and by G. Lindgren, who in a 1972 thesis [88] at the University of Lund gave a number of results concerning the configuration and distribution of maxima and minima of sample functions.

6.7. *Structure problems for a general class of stochastic processes.* From 1958 on, I had tried to generalize some of the results obtained for stationary vector processes by Zasuchin, Wiener–Masani and others, as referred to above. It seemed clear that the possibilities of a generalization would be very restricted in the case of all results connected with spectral representations. On the other hand, that part of the field which Wiener and Masani have called the "time domain" analysis of the processes seemed to be open for a fairly wide generalization to nonstationary cases.

In a paper [28] presented to the Berkeley Symposium in 1960, I discussed these questions, both for the time domain and the spectral analysis. In the latter case, I could only give some tentative results for special classes of processes. With regard to the time domain analysis, however, it was possible to proceed much further. It turns out that the situation is quite different for processes with discrete and continuous time.

Consider an arbitrary vector process with discrete time, say $\mathbf{x}_n = (x_{n1}, \cdots, x_{nq})$, where n runs through all negative and positive integers. Suppose that the components have zero mean values and finite second order moments, and define the concepts of deterministic and purely nondeterministic processes in the well-known way. Then, without imposing any further conditions, the generalization of the Wold decomposition goes through, and we have

$$\mathbf{x}_n = \mathbf{u}_n + \mathbf{v}_n ,$$

where $\mathbf{u}_n$ is purely nondeterministic and $\mathbf{v}_n$ is deterministic, while $\mathbf{u}_n$ and $\mathbf{v}_n$ are mutually orthogonal. Further, the nondeterministic component $\mathbf{u}_n$ can be linearly represented in terms of innovations, in the form

$$\mathbf{u}_n = \sum_{i=-\infty}^{n} \mathbf{c}_{ni} \mathbf{z}_i ,$$

where $\mathbf{z}_i = (z_{i1}, \cdots, z_{ir_i})$ is a column vector of order $r_i \leqq q$, while $\mathbf{c}_{ni}$ is a matrix of order $q \times r_i$. The innovation components z_{ij} are mutually orthogonal for all i and j.

In the case of a vector process $\mathbf{x}(t) = (x_1(t), \cdots, x_q(t))$ with continuous t, there is a similar decomposition

$$\mathbf{x}(t) = \mathbf{u}(t) + \mathbf{v}(t)$$

with a purely nondeterministic $\mathbf{u}(t)$ and a deterministic $\mathbf{v}(t)$, which are mutually orthogonal. However, in this case the representation of $\mathbf{u}(t)$ in terms of innovations turns out to be more complicated. If we assume that the mean square limits $\mathbf{u}(t+0)$ and $\mathbf{u}(t-0)$ exist for all t, it can be shown that the Hilbert space spanned by all the components of $\mathbf{u}(t)$ is separable. With the aid of Hilbert space geometry it is then possibe to show that we have for all t

$$\mathbf{u}(t) = \int_{-\infty}^{t} \mathrm{G}(t, v)\, d\mathbf{z}(v) ,$$

where $\mathbf{z}(v) = (z_1(v), \cdots, z_N(v))$ is a column vector of order N, the components of which are mutually orthogonal stochastic processes with orthogonal increments, while $\mathrm{G}(t, v)$ is a nonrandom matrix of order $q \times N$. The number N is uniquely determined by the given $\mathbf{x}(t)$ process, and may be any finite positive integer, or equal to $+\infty$. Thus, while in the discrete time case, the innovation vectors $\mathbf{z}_i$ are at most of the order q of the given vector process, the innovation process $\mathbf{z}(v)$ that occurs in the continuous case may have any order, even an infinite one.

Both in the discrete and the continuous time case, the representation of the nondeterministic component $\mathbf{u}_n$ or $\mathbf{u}(t)$ immediately gives an explicit solution of the linear least squares prediction problem. I have discussed the properties of this representation in some further papers [29, 31, 32]. It would be important to be able to determine the multiplicity N corresponding to a given $\mathbf{x}(t)$ process, and I have given some contributions to the study of this problem, but no complete solution seems to be known so far.

For the case of a normal process the representation of $u(t)$ was given by Hida [54] at the same time as my paper [28] was read before the 1960 Berkeley Symposium. Further interesting contributions have been given by Kallianpur and Mandrekar [57, 58], and quite recently by Rozanov [112]. A number of papers on this and allied subjects have been collected in a volume [39] edited by Ephremides and Thomas.

6.8. *Travels and work*, 1961–1970. In the early summer of 1961 I retired from my work in the university administration and became a free scientist. At the Paris meeting of the International Statistical Institute I had a conversation with Gertrude Cox, who invited me to come and work at the Research Triangle Institute in North Carolina, where I was to spend several months in each of the years 1962, 1963 and 1965. I also acted as Visiting Professor, in 1963 at Columbia, 1965 at Yale, and 1966 in Berkeley. In the two preceding sections I have already said something about the work on stochastic processes in which I was engaged during these years. In all these places I was happy to work with old and new friends.

In the summer of 1962 we had the International Mathematical Congress in Stockholm. My wife and I arranged a probabilistic lunch in our home for a number of distinguished friends, among whom were Doob, Hunt, Ito, Kappos, Kolmogorov, Linnik, Masani, Rényi, Rosenblatt, Takacs and Urbanik.

Having received an invitation to an All-Soviet Conference for Probability and Mathematical Statistics, to be held in October 1963, I came fairly directly from the United States to the Soviet Union. The meeting took place in Tbilisi (Tiflis), and had an excellent programme, including lectures by those Russian probabilists whose work I have referred to above in this paper, and also contributions by young members, who made a strong impression of scientific interest and vitality. Kolmogorov lectured about the application of probability methods to the analysis of poetical works, Yaglom, Ibragimov and Rozanov on different aspects of stochastic processes, and Linnik on statistical tests. I talked about stochastic processes as curves in Hilbert space, giving some contributions to the theory of multiplicity mentioned in the preceding section. After the meeting I also had the occasion to lecture in Moscow and Leningrad.

In 1967 I attended a conference for the applications of statistical extreme values at Faro in Portugal, with good representation both from theoretical and various practical fields. Afterwards I acted as Visiting Professor at the University of Aarhus in Denmark, and in 1969 at the University of Copenhagen. For 1970 I received an invitation from John Tukey to give the first S. S. Wilks Lecture [32], at the inauguration of the new Fine Hall building in Princeton. In this connection, my wife and I visited Chapel Hill, Princeton, Providence and Storrs, and were happy to be received by friends in all these places.

And here, at 1970, I will put an end to these scattered recollections from the development of mathematical probability theory during half a century.

REFERENCES

[1] Arley, N. (1943). Stochastic processes and cosmic radiation. Ph. D. Thesis, Copenhagen.

[2] Bachelier, L. (1900). Théorie de la spéculation. *Ann. École Norm. Sup.* **17** 21–86.

[3] Beljaev, Yu. K. (1959). Analytic random processes. *Teor. Verojatnost. i Primenen* **4** 437–444.

[4] Beljaev, Yu. K. (1960). Local properties of the sample functions of stationary Gaussian processes. *Teor. Verojatnost. i Primenen* **5** 128–131.

[5] Beljaev, Yu. K. (1960). Continuity and Hölder's conditions for sample functions of stationary Gaussian processes. *Proc. Fourth Berkeley Symp. Math. Statist. Prob.* 23-33, Univ. of California Press.

[6] Beljaev, Yu. K. (1966). On the number of intersections of a level by a Gaussian stochastic process. *Teor. Verojatnost. i Primenen* **11** 120-128.

[7] Bernstein, S. N. (1927). Sur l'extension du théorème limite du calcul des probabilités aux sommes de quantités dépendantes. *Math. Ann.* **97** 1-59.

[8] Brockmeyer, E., Halstrøm, H. L. and Jensen, Arne. (1948). The life and works of A. K. Erlang. *Trans. Danish Acad. Tech. Sci.* no. 2, 277 pages.

[9] Bulinskaja, E. V. (1961). On the mean number of crossings of a level by a stationary Gaussian process. *Teor. Verojatnost. i Primenen* **6** 474-477.

[10] Cantelli, F. P. (1933). Considerazione sulla legge uniforme dei grandi numeri. *Giorn. Ist. Ital. Attuari* **4** no. 3.

[11] Charlier, C. V. L. (1905). Ueber das Fehlergesetz. *Ark. Mat. Astr. Fys.* **2** no. 8.

[12] Cramér, H. (1919). Bidrag till utjämningsförsäkringens teori. Insur. Inspect. Sthlm.

[13] Cramér, H. (1923). Das Gesetz von Gauss und die Theorie des Risikos. *Skand. Aktuarietidskr.* 209-237.

[14] Cramér, H. (1925). On some classes of series used in mathematical statistics. *Sixth Scand. Math. Congr. Copenhagen* 399-425.

[15] Cramér, H. (1926). Sannolikhetskalkylen i den vetenskapliga litteraturen. *Nordisk. Stat. Tidskr.* **5** 1-32.

[16] Cramér, H. (1928). On the composition of elementary errors. *Skand. Aktuarietidskr.* 13-74, 141-180.

[17] Cramér, H. (1934). Su un teorema relativo alla legge uniforme dei grandi numeri. *Giorn. Ist. Ital. Attuari* **5** no. 1.

[18] Cramér, H. (1936). Ueber eine Eigenschaft der normalen Verteilungsfunktion. *Math. Z.* **41** 405-414.

[19] Cramér, H. (1937). Random Variables and Probability Distributions. *Cambridge Tracts in Math.* **36**.

[20] Cramér, H. (1938). Sur un nouveau théorème—limite de la théorie des probabilités. *Actualités Sci. Indust.* **736** 5-23.

[21] Cramér, H. (1939). On the representation of a function by certain Fourier integrals. *Trans. Amer. Math. Soc.* **46** 191-201.

[22] Cramér, H. (1940). On the theory of stationary random processes. *Ann. of Math.* **41** 215-230.

[23] Cramér, H. (1942). On harmonic analysis in certain functional spaces. *Ark. Mat. Astr. Fys.* **28B** no. 12.

[24] Cramér, H. (1945). *Mathematical Methods of Statistics.* Almquist and Wiksells, Uppsala. ((1946), Princeton Univ. Press.)

[25] Cramér, H. (1947). Problems in probability theory. *Ann. Math. Statist.* **18** 165-193.

[26] Cramér, H. (1950). A contribution to the theory of stochastic processes. *Proc. Second Berkeley Symp. Math. Statist. Prob.* 329-339, Univ. of California Press.

[27] Cramér, H. (1955). Collective risk theory. Skandia Insurance Co., Stockholm.

[28] Cramér, H. (1960). On some classes of non-stationary stochastic processes. *Proc. Fourth Berkeley Symp. Math. Statist.* **2** 57-77, Univ. of California Press.

[29] Cramér, H. (1961). On the structure of purely nondeterministic stochastic processes. *Ark. Mat.* **4** 249-266.

[30] Cramér, H. (1963). On asymptotic expansions for sums of independent random variables with a limiting stable distribution. *Sankhyā Ser. A* **25** 13-24.

[31] Cramér, H. (1964). Stochastic processes as curves in Hilbert space. *Teor. Verojatnost. i Primenen* **9** 169-179.

[32] Cramér, H. (1971). Structural and statistical problems for a class of stochastic processes. S. S. Wilks lecture, Princeton Univ. Press.

[33] CRAMÉR, H. (1972). On the history of certain expansions used in mathematical statistics. *Biometrika* **59** 205-207.

[34] CRAMÉR, H. and LEADBETTER, M. R. (1967). *Stationary and Related Stochastic Processes.* Wiley, New York.

[35] CRAMÉR, H. and WOLD, H. (1936). Some theorems on distribution functions. *J. London Math. Soc.* **11** 290-294.

[36] DOOB, J. L. (1953). *Stochastic Processes.* Wiley, New York.

[37] EDGEWORTH, F. Y. (1905). The law of error. *Proc. Cambridge Philos. Soc.* **20** 36-141.

[38] EINSTEIN, A. (1906). Zur Theorie der Brownschen Bewegung. *Ann. Physics* IV **19** 371-381.

[39] EPHREMIDES, A. and THOMAS, J. B. (eds.) (1973). *Random Processes.* Dowden, Hutchinson, Ross, Stroudsburg.

[40] ESSEEN, C. G. (1945). Fourier analysis of distribution functions. *Acta Math.* **77** 1-125.

[41] FELLER, W. (1935, 1937). Ueber den zentralen Grenzwertsatz der Wahrscheinlichkeitsrechnung. *Math. Z.* **40** 521-559; **42** 301-312.

[42] FELLER, W. (1936). Zur Theorie der stochastischen Prozesse. *Math. Ann.* **113** 113-160.

[43] FELLER, W. (1940). On the integro-differential equations of purely discontinuous Markoff processes. *Trans. Amer. Math. Soc.* **48** 488-575.

[44] FELLER, W. (1943). The general form of the so-called law of the iterated logarithm. *Trans. Amer. Math. Soc.* **54** 373-402.

[45] FELLER, W. (1950, 1966). *An Introduction to Probability Theory and Its Applications* **1** and **2**. Wiley, New York.

[46] GELFAND, I. M. and VILJENKIN, N. JA. (1961). *Generalized Functions* **4**. Moskva.

[47] GNEDENKO, B. V. (1939). On the theory of limit theorems for sums of independent random variables. *Izv. Akad. Nauk SSSR Ser. Mat.* 181-232, 643-647.

[48] GNEDENKO, B. V. and KOLMOGOROV, A. N. (1949). Limit theorems for sums of independent random variables. *Gosudarstv. Izdat. Tehn.-Teor. Lit.*, Moscow-Leningrad.

[49] GRENANDER, U. (1950). Stochastic processes and statistical inference. *Ark. Mat.* **1** 195-277.

[50] GRENANDER, U. and ROSENBLATT, M. (1956). *Statistical Analysis of Stationary Time Series.* Almquist and Wiksells, Stockholm.

[51] GRENANDER, U. and SZEGÖ, G. (1958). *Toeplitz Forms and Their Applications.* Univ. of California Press.

[52] GRENANDER, U. (ed.) (1959). *Probability and Statistics, the Harald Cramér Volume.* Almquist and Wiksells, Stockholm and Wiley, New York.

[53] HANNER, O. (1949). Deterministic and nondeterministic stationary random processes. *Ark. Mat.* **1** 161-177.

[54] HIDA, T. (1960). Canonical representations of Gaussian processes and their applications. *Mem. Coll. Sci. Univ. Kyoto, A* **33** 109-155.

[55] IBRAGIMOV, I. A. and LINNIK, YU. V. (1965). Independent and stationarily connected variables. *Izad. "Nauka,"* Moscow.

[56] KAC, M. and SLEPIAN, D. (1959). Large excursions of Gaussian processes. *Ann. Math. Statist.* **30** 1215-1228.

[57] KALLIANPUR, G. and MANDREKAR, V. (1970). On the connections between multiplicity theory and O. Hanner's time domain analysis of weakly stationary stochastic processes. *Essays in Probability and Statistics* 385-396. Univ. of North Carolina Press.

[58] KALLIANPUR, G. and MANDREKAR, V. (1965). Multiplicity and representation theory of purely nondeterministic stochastic processes. *Teor. Verojatnost. i Primenen* **10** 553-581.

[59] KARHUNEN, K. (1947). Ueber lineare Methoden in der Wahrscheinlichkeitsrechnung. *Ann. Acad. Sci. Fenn. Ser. A* **37** 1-79.

[60] KARHUNEN, K. (1949). Ueber die Struktur stationärer zufälliger Funktionen. *Ark. Mat.* **1** 141-160.

[61] KEYNES, J. M. (1921). *A Treatise on Probability.* Macmillan, London.

[62] KHINTCHINE, A. YA. (1923). Ueber dyadische Brüche. *Math. Z.* **18** 109-116.

[63] KHINTCHINE, A. YA. (1933). Zur mathematischen Begründung der statistischen Mechanik. *Z. Angew. Math. Mech.* **13** 101–103.

[64] KHINTCHINE, A. YA. (1933). *Asymptotische Gesetze der Wahrscheinlichkeitsrechnung.* Ergebnisse der Mathematik und ihrer Grenzgebiete. Springer, Berlin.

[65] KHINTCHINE, A. YA. (1934). Korrelationstheorie der stationären stochastischen Prozesse. *Math. Ann.* **109** 604–615.

[66] KHINTCHINE, A. YA. (1936). Sul dominio di attrazione della legge di Gauss. *Giorn. Ist. Ital. Attuari* **7** 3–18.

[67] KHINTCHINE, A. YA. (1937). A new derivation of a formula of P. Lévy. *Bull. Moskva Univ.* **1** 1–5.

[68] KHINTCHINE, A. YA. (1937). On the arithmetic of distribution laws. *Bull. Moskva Univ.* **1** 6–17.

[69] KHINTCHINE, A. YA. (1937). Zur Theorie der unbeschränkt teilbaren Verteilungsgesetze. *Mat. Sb.* **44** 79–119.

[70] KHINTCHINE, A. YA. and KOLMOGOROV, A. N. (1925). Ueber Konvergenz von Reihen, deren Glieder durch den Zufall bestimmt werden. *Mat. Sb.* **32** 668–677.

[71] KOLMOGOROV, A. N. (1928). Ueber die Summen durch den Zufall bestimmter unabhängiger Grössen. *Math. Ann.* **99** 309–319.

[72] KOLMOGOROV, A. N. (1929). Ueber das Gesetz des iterierten Logarithmus. *Math. Ann.* **101** 126–135.

[73] KOLMOGOROV, A. N. (1931). Ueber die analytischen Methoden in der Wahrscheinlichkeitsrechnung. *Math. Ann.* **104** 415–458.

[74] KOLMOGOROV, A. N. (1932). Sulla forma generale di un processo stocastico omogeneo (un problema di B. de Finetti). *R. Accad. Lincei* **15** 805–808, 866–869.

[75] KOLMOGOROV, A. N. (1933). *Grundbegriffe der Wahrscheinlichkeitsrechnung.* Ergebnisse der Mathematik und ihrer Grenzgebiete. Springer, Berlin.

[76] KOLMOGOROV, A. N. (1938). A simplified proof of the Birkhoff-Khintchine ergodic theorem. *Uspehi. Mat. Nauk* **5** 52–56.

[77] KOLMOGOROV, A. N. (1941). Stationary sequences in Hilbert space. *Bull. Moskva Univ.* **2** no. 6.

[78] KOLMOGOROV, A. N. (1941). Interpolation and extrapolation of stationary random sequences. *Izv. Akad. Nauk SSSR* **5** 3–14.

[79] KOLMOGOROV, A. N. and ROZANOV, YU. A. (1960). On strong mixing conditions for stationary Gaussian processes. *Teor. Verojatnost. i Primenen* **5** 204–208.

[80] LÉVY, P. (1925). *Calcul des Probabilités.* Gauthier-Villars, Paris.

[81] LÉVY, P. (1934). Sur les intégrales dont les éléments sont des variables aléatoires indépendantes. *Ann. Ecole. Norm. Sup. Pisa* (2) **3** 337–366.

[82] LÉVY, P. (1935). Propriétés asymptotiques des sommes de variables aléatoires indépendantes ou enchainées. *J. Math. Pures Appl.* (9) **14** 347–402.

[83] LÉVY, P. (1937). *Théorie de L'Addition des Variables Aléatoires.* Gauthier-Villars, Paris.

[84] LÉVY, P. (1948). *Processus Stochastiques et Mouvement Brownien.* Gauthier-Villars, Paris.

[85] LÉVY, P. (1970). *Queleques aspects de la pensée d'un mathématicien.* Blanchard, Paris.

[86] LIAPOUNOV, A. M. (1901). Nouvelle forme du théorème sur la limite des probabilités. *Mem. Acad. Sci. St. Petersbourg.* **12** (no. 5) 1–24.

[87] LINDEBERG, J. W. (1922). Eine neue Herleitung des Exponentialgesetzes in der Wahrscheinlichkeitsrechnung. *Math. Z.* **15** 211–225.

[88] LINDGREN, G. (1972). On wave-forms in normal random processes. Thesis, Lund.

[89] LINNIK, YU. V. See Ibragimov and Linnik.

[90] LOÈVE, M. (1948). Fonctions aléatoires du second ordre. Note à Lévy, *Processus Stochastiques et Mouvement Brownien* [84] 299–352. Paris.

[91] LOÈVE, M. (1955). *Probability Theory.* Van Nostrand, Princeton. (3rd ed., 1963.)

[92] LUNDBERG, F. (1903). Approximerad framställning av sannolikhetsfunktionen. Återförsäkring av kollektivrisker. Thesis, Uppsala.

[93] LUNDBERG, O. (1940). On random processes and their application to sickness and accident statistics. Thesis, Stockholm.

[94] MASANI, P. (1959). Cramér's theorem on monotone matrix-valued functions and the Wold decomposition. *Probability and Statistics: The Harald Cramér Volume* (Ulf Grenander, ed.) 175-189. Almquist and Wiksell, Stockholm; Wiley, New York.

[95] MASANI, P. (1966). Wiener's contributions to generalized harmonic analysis, prediction theory and filter theory, Norbert Wiener 1894-1964. *Bull. Amer. Math. Soc.* **72** 73-125.

[96] MASANI, P. and WIENER, N. (1957). The prediction theory of multivariate stochastic processes. *Acta Math.* **98** 111-150, and (1958), **99** 93-137.

[97] MATÉRN, B. (1960). Spatial variation. *Swed. Forestry Res. Inst.* **49** 1-144.

[98] VON MISES, R. (1919). Grundlagen der Wahrscheinlichkeitsrechnung. *Math. Z.* **5** 52-99.

[99] VON MISES, R. (1928). *Wahrscheinlichkeit, Statistik und Wahrheit.* Springer, Wien.

[100] DE MOIVRE, A. (1733). Miscellanea analytica. Second supplement, London.

[101] NEYMAN, J. (1937). Outline of a theory of statistical estimation based on the classical theory of probability. *Philos. Trans. Roy. Soc. London* **236** 333-380.

[102] NEYMAN, J. (1942). Basic ideas and some recent results of the theory of testing statistical hypotheses. *J. Roy. Statist. Soc.* (N.S.) **105** 292-327.

[103] PÓLYA, G. (1920). Ueber den zentralen Grenzwertsatz der Wahrscheinlichkeitsrechnung und das Momentenproblem. *Math. Z.* **8** 171-181.

[104] PROCHOROV, YU. V. and ROZANOV, YU. A. (1967). *Probability Theory: Fundamental Concepts, Limit Theorems, Random Processes.* Izdat. "Nauka," Moscow.

[105] RAIKOV, D. (1938). On the decomposition of Gauss and Poisson laws. *Izv. Akad. Nauk. SSSR, Ser. Mat.* 91-124.

[106] REZNIK, M. K. (1968). The law of the iterated logarithm for some classes of stationary processes. *Teor. Verojatnost. i Primenen.* **13** 606-621.

[107] RICE, S. O. (1945). Mathematical analysis of random noise. *Bell System Tech. J.* **24** 46-156.

[108] RICE, S. O. (1958). Distribution of the duration of fades in radio transmission. *Bell System Tech. J.* **37** 581-635.

[109] ROSENBLATT, M. (1956). A central limit theorem and a strong mixing condition. *Proc. Nat. Acad. Sci.* **42** 43-47.

[110] ROZANOV, YU. A. (1963). Stationary random processes. *Gosudarstv. Izdat. Fiz.-Mat. Lit.*, Moscow.

[111] ROZANOV, YU. A. (1968). Infinite-dimensional Gaussian distributions. *Proc. Steklov Inst. Math.* **108** 1-136.

[112] ROZANOV, YU. A. (1974). *Theory of innovation processes* 1-128. Izdat. "Nauka," Moskva.

[113] RUTHERFORD, E. and GEIGER, H. (1908). An electrical method of counting the number of particles from radioactive substances. *Proc. Roy. Soc.* **A81** 141-161.

[114] SEGERDAHL, C. O. (1939). On homogeneous random processes and collective risk theory. Thesis, Stockholm.

[115] TCHEBYCHEV, P. L. (1867). On mean values. *J. Math. Pures Appl.* **12** 177-184.

[116] TCHEBYCHEV, P. L. (1890). Sur deux théorèmes relatifs aux Probabilités. *Acta Math.* **14** 305-315.

[117] VOLKONSKI, V. A. and ROZANOV, YU. A. (1959, 1961). Some limit theorems for random functions. *Teor. Verojatnost.* **4** 186-207 and **6** 202-215.

[118] WIENER, N. (1923). Differential space. *J. Math. and Phys.* **2** 131-174.

[119] WIENER, N. (1930). Generalized harmonic analysis. *Acta Math.* **55** 117-258.

[120] WIENER, N. (1934). Fourier transforms in the complex domain (with R. Paley). *Amer. Math. Soc. Coll. Publ.* **19**.

[121] WIENER, N. (1949). *Extrapolation, Interpolation and Smoothing of Stationary Time Series.* With engineering applications. Wiley, New York.

[122] WIENER, N. (1956). *I Am a Mathematician.* The Later Life of a Prodigy. Doubleday, New York.

[123] WOLD, H. (1938). A study in the analysis of stationary time series. Thesis, Stockholm.
[124] YAGLOM, A. M. (1952). An introduction to the theory of stationary random functions. *Uspehi Matem. Nauk* (N.S.) **7**, no. 5(51), 3–168.
[125] YLVISAKER, N. D. (1965). The expected number of zeros of a stationary Gaussian process. *Ann. Math. Statist.* **36** 1043–1046.
[126] YLVISAKER, N. D. (1966). On a theorem of Cramér and Leadbetter. *Ann. Math. Statist.* **37** 682–685.
[127] ZASUCHIN, V. (1941). On the theory of multi-dimensional stationary random processes. *Dokl. Akad. Nauk USSR* **33** 435–437.

SJOTULLSBACKEN 15
BLOCKHUSUDDEN
115 25 STOCKHOLM
SWEDEN

71.

On the multiplicity of a stochastic vector process

Ark. Mat. **16** (1), 89–94 (1978)

1. Introduction

The object of this note is to prove a theorem on the multiplicity of a stochastic vector process. Before stating the theorem, it will be necessary to give some introductory remarks.

We shall be concerned with complex-valued random variables defined on a fixed probability space. Any random variable x considered will be assumed to have zero mean and a finite variance:

$$Ex = 0, \quad E|x|^2 < \infty.$$

It is well known that all random variables satisfying these conditions can be regarded as elements in a Hilbert space, if the inner product of two elements x and y is defined by the relation

$$(x, y) = Ex\bar{y}.$$

A stochastic vector process of finite dimensionality q will be denoted

$$\mathbf{x}(t) = \{x_1(t), \ldots, x_q(t)\}$$

where the components $x_n(t)$ are random variables depending on a real parameter t, which may be regarded as representing time. By $H(\mathbf{x})$ we denote the Hilbert space spanned by the random variables $x_n(u)$ for all n and all real u:

$$H(\mathbf{x}) = \mathscr{S}\{x_n(u),\ n = 1, \ldots, q,\ -\infty < u < +\infty\}$$

while $H(\mathbf{x}, t)$ is the subspace of $H(\mathbf{x})$ spanned by the same $x_n(u)$ for all $u \leqq t$. The projection operator in $H(\mathbf{x})$ with range $H(\mathbf{x}, t)$ will be denoted by P_t.

For any element z of $H(\mathbf{x})$ we write

$$z(t) = P_t z.$$

$z(t)$ will then be the random variable of a stochastic process with orthogonal increments such that

$$Ez(t) = 0, \quad E|z(t)|^2 = F(t),$$

where the variance function $F(t)$ is never decreasing, everywhere continuous to the left provided (A) and (B) below are satisfied, and such that $F(-\infty)=0$, $F(+\infty)=E|z|^2$. The Hilbert space spanned by the $z(u)$ for all real u is a cyclic subspace of $H(\mathbf{x})$ which we denote by $C(z)$, while $C(z, t)$ is the subspace spanned by the $z(u)$ for all $u \leqq t$. It is known that every element of $C(z)$ will be of the form

$$\int_{-\infty}^{\infty} c(t)\, dz(t),$$

where $c(t)$ is a nonrandom function belonging to $L_2 F(t)$, and the stochastic integral is defined as an integral in quadratic mean.

In the set of all z, and in the set of the corresponding variance functions $F(t)$, we introduce a partial ordering by writing $z_1 \succ z_2$ and $F_1 \succ F_2$ whenever F_2 is absolutely continuous with respect to F_1. Every variance function defines in the well-known way a measure on the real axis, and F_2 is absolutely continuous with respect to F_1 if and only if every set of F_1 measure zero is also of F_2 measure zero. If simultaneously $F_1 \succ F_2$ and $F_2 \succ F_1$ we say that F_1 and F_2 are equivalent, and also the corresponding z_1 and z_2. All variance functions F equivalent to a given F_1 will be said to form the equivalence class of F_1.

We shall consider a stochastic vector process $\mathbf{x}(t)$ satisfying the following two conditions (A) and (B):

(A) The limits $x_n(t \pm 0)$ exist for all n and t as limits in the norm of $H(\mathbf{x})$, and $x_n(t-0)=x_n(t)$.

(B) The limiting space $H(\mathbf{x}, -\infty)$ contains only the zero element of $H(\mathbf{x})$.

It follows from (A) that the Hilbert space $H(\mathbf{x})$ is separable, and consequently the subspace $H(\mathbf{x}, t)$, which is never decreasing when t increases, has at most an enumerable number of discontinuities. When (B) is satisfied, the $\mathbf{x}(t)$ process is called purely nondeterministic (or linearly regular).

When (A) and (B) are satisfied, it is known that it is possible to find a finite or infinite sequence of elements of $H(\mathbf{x})$

(1) $$z_1 \succ z_2 \succ \ldots \succ z_N$$

such that the corresponding cyclic subspaces $C(z_n)$ are mutually orthogonal, and such that $H(\mathbf{x})$ is the vector sum of all these orthogonal cyclic subspaces:

(2) $$H(\mathbf{x}) = C(z_1) \oplus \ldots \oplus C(z_N).$$

The number N and the equivalence classes of the corresponding variance functions $F_1, \ldots, F_N$ are uniquely defined by the given $\mathbf{x}(t)$ process, and N is called the multiplicity of the process. Applying the projection P_t we obtain the relation

$$H(\mathbf{x}, t) = C(z_1, t) \oplus \ldots \oplus C(z_N, t)$$

for every t. As $x_n(t)$ is an element of $H(\mathbf{x}, t)$, this gives the representation

$$x_n(t) = \sum_{r=1}^{N} \int_{-\infty}^{t} c_{n,r}(t, u)\, dz_r(u).$$

This shows how the components of the $\mathbf{x}(t)$ process can be regarded as additively built up by N-dimensional innovation elements $\{dz_1(u), \ldots, dz_N(u)\}$ associated with every instant u in the "past and present" from the point of view of the instant t.

The basic facts of multiplicity theory given above are stated here in the form in which they occur in my papers [1—3].

2. A theorem on multiplicity

Theorem. *Let* $\mathbf{x}(t)$ *be a given vector process satisfying* (A) *and* (B), *and suppose that the component* $x_n(t)$, *regarded as a one-dimensional process, has finite multip licity* N_n. *Then the multiplicity of the* $\mathbf{x}(t)$ *process is* $N \leqq \sum_{n=1}^{q} N_n$.

I have stated this theorem without proof in a paper of 1961 [1, p. 265]. Although the theorem has been referred to in the literature [e.g. 4, p. 228], no proof has been published, as far as I know. It may be of some interest to give a complete proof. As stated in my paper just quoted, the proof is slightly more involved than may possibly be expected. The proof will depend on two Lemmas.

Lemma 1. *Let* $z_1, \ldots, z_n$ *and* w *be elements of* $H(\mathbf{x})$ *such that the cyclic subspaces* $C(z_1), \ldots, C(z_n)$ *are mutually orthogonal. Then it is possible to find an element* z_{n+1} *of* $H(\mathbf{x})$ *such that* $C(z_{n+1})$ *is orthogonal to* $C(z_r)$ *for* $r=1, \ldots, n$, *and such that the vector sum*

$$C(z_1) \oplus \ldots \oplus C(z_{n+1}) \tag{3}$$

includes the cyclic subspace $C(w)$.

Let w_r be the projection of w on $C(z_r)$. Then define z_{n+1} by

$$w = w_1 + \ldots + w_n + z_{n+1},$$

$$P_t w = P_t w_1 + \ldots + P_t w_n + P_t z_{n+1}.$$

Here w_r and $P_t w_r$ belong to $C(z_r)$, while z_{n+1} is orthogonal to $C(z_r)$ for all $r=1, \ldots, n$. It follows [1, p. 259] that $C(z_{n+1})$ is orthogonal to all these $C(z_r)$.

Now every element of $C(w)$ is the limit of a convergent sequence of finite linear combinations of the $P_t w$. According to the above relation every such linear combination is a sum of finite linear combinations of elements of $C(z_1), \ldots, C(z_{n+1})$, and since these subspaces are mutually orthogonal, the limit of any convergent sequence of this kind will be a sum of elements of the same subspaces. Thus the vector sum (3) includes every element of $C(w)$, and the Lemma is proved. Note that, in the case when w belongs to the vector sum of $C(z_1), \ldots, C(z_n)$, the z_{n+1} defined in the above proof is equal to zero.

It will be seen without difficulty that, by repeated application of this Lemma to the vector process $\mathbf{x}(t)$ satisfying the conditions of the Theorem, we can find $N=\sum_{n=1}^{q} N_n$ elements $z_1, \ldots, z_N$ of $H(\mathbf{x})$, some of which may be equal to zero, such that the corresponding cyclic subspaces are all mutually orthogonal, and the condition (2) will be satisfied. But the condition (1) may not be satisfied, and in order to complete the proof of the Theorem we shall require the following Lemma.

Lemma 2. *If $z_1, \ldots, z_N$ are a finite number of elements of $H(\mathbf{x})$ such that the cyclic subspaces $C(z_n)$ are all mutually orthogonal, and the relation* (2) *is satisfied, we can find N elements $w_1, \ldots, w_N$ of $H(\mathbf{x})$, some of which may be equal to zero, such that the cyclic subspaces $C(w_n)$ are all mutually orthogonal, and we have*

$$w_1 \succ w_2 \succ \ldots \succ w_N \tag{4}$$

$$H(\mathbf{x}) = C(w_1) \oplus \ldots \oplus C(w_N). \tag{5}$$

This Lemma being proved, it will be seen that the truth of the Theorem follows from the remark made before stating the Lemma.

Consider the variance functions F_n corresponding to the z_n given in the Lemma. The never decreasing function $\mathbf{F}(t)=\sum_{n=1}^{N} F_n(t)$ defines a measure on the real axis, which we shall use throughout the proof of the Lemma. Expressions like 'measurable', 'almost everywhere (a.e.)', etc., will always refer to the $\mathbf{F}$ measure.

We have $\mathbf{F} \succ F_n$ for every n, so that the Radon—Nikodym derivative $F_n' = dF_n/d\mathbf{F}$ will be defined a.e. and measurable. Note that, if t is a discontinuity point of some of the F_n, we shall have $F_n'(t)>0$ for every F_n which is discontinuous at t, and $F_n'(t)=0$ for the other F_n. For $n, r=1, \ldots, N$ we now define a function $g_{n,r}(t)$ by writing for $n \geqq r$

$$g_{n,r}(t) = \begin{cases} 1 & \text{if } F_n'(t) > 0 \text{ and there are exactly } r-1 \text{ positive} \\ & \text{among } F_1'(t), \ldots, F_{n-1}'(t), \\ 0 & \text{otherwise,} \end{cases}$$

and $g_{n,r}(t)=0$ for all t when $n<r$. Then $g_{n,r}(t)$ will be defined a.e. and measurable. It will be seen that $g_{n,r}(t)$ has the following three properties a.e.

(6) For given values of n and t, we have $g_{n,r}(t)=1$ for at most one r.

(7) For given values of r and t, we have $g_{n,r}(t)=1$ for at most one n.

(8) If $g_{n,r}(t)=1$ and $n \geqq r > 1$, there is exactly one $m<n$ such that $g_{m,r-1}(t)=1$.

The sum of stochastic integrals

$$w_r = \sum_{n=1}^{N} \int_{-\infty}^{\infty} g_{n,r}(u)\, dz_n(u)$$

is a well-defined element of $H(\mathbf{x})$, and we have

$$w_r(t) = P_t w_r = \sum_{n=1}^N \int_{-\infty}^t g_{n,r}(u)\, dz_n(u),$$

$$G_r(t) = E|w_r(t)|^2 = \sum_{n=1}^N \int_{-\infty}^t g_{n,r}(u)\, dF_n(u).$$

We shall now show that the sequence $w_1, \ldots, w_N$ satisfies the conditions of the Lemma.

In order to show that the cyclic subspaces $C(w_1), \ldots, C(w_N)$ are orthogonal, it will be sufficient to show that $w_r(t) \perp w_s(u)$ for $r \neq s$ and all t, u. Clearly we may suppose $t \leqq u$, and then have by (6)

$$Ew_r(t)\overline{w_s(u)} = \sum_{n=1}^N \int_{-\infty}^t g_{n,r}(v) g_{n,s}(v)\, dF_n(v) = 0.$$

Further we evidently have, since all the w_r are elements of $H(\mathbf{x})$,

$$H(\mathbf{x}) \supset C(w_1) \oplus \ldots \oplus C(w_N).$$

In order to show that the two members of this relation are identical, we have to show that every element of $H(\mathbf{x})$ belongs to the vector sum of the $C(w_r)$. Now by hypothesis every element of $H(\mathbf{x})$ is of the form

$$y = \sum_{n=1}^N \int_{-\infty}^{\infty} c_n(t)\, dz_n(t)$$

where $c_n(t)$ belongs to $L_2 F_n$. By (6) we have for every n and t

$$\sum_{r=1}^N g_{n,r}(t) = 1 \quad \text{or} \quad 0.$$

In the set of values of t where this sum is equal to zero, it follows from the definition of $g_{n,r}(t)$ that $F_n'(t) = 0$, so that the integral of $c_n(t)\,dz_n(t)$ over this set is zero. Thus we have

$$y = \sum_{r=1}^N \sum_{n=1}^N \int_{-\infty}^{\infty} c_n(t) g_{n,r}(t)\, dz_n(t).$$

Now define a function $b_r(t)$ by writing

$$b_r(t) = \begin{cases} c_n(t) & \text{when} \quad g_{n,r}(t) = 1, \\ 0 & \text{when} \quad g_{n,r}(t) = 0. \end{cases}$$

By (7) this $b_r(t)$ is uniquely defined a.e., and we have

$$y = \sum_{r=1}^N \int_{-\infty}^{\infty} b_r(t) \sum_{n=1}^N g_{n,r}(t)\, dz_n(t) = \sum_{r=1}^N \int_{-\infty}^{\infty} b_r(t)\, dw_r(t).$$

As it is easily seen that $b_r(t)$ belongs to $L_2 G_r$, this shows that y is an element of the vector sum of the $C(w_r)$, so that (5) is satisfied..

It remains to show that (4) is also satisfied. We consider the variance functions $G_r(t)$ of the w_r and have to show that $G_r \succ G_{r+1}$ for $r=1, \ldots, N-1$. Let S be a set of positive measure such that

$$\int_S dG_r(t) = \sum_{n=1}^{N} \int_S g_{n,r}(t)\, dF_n(t) = 0. \tag{9}$$

It will then be proved that

$$\int_S dG_{r+1}(t) = \sum_{n=1}^{N} \int_S g_{n,r+1}(t)\, dF_n(t) = 0. \tag{10}$$

In every term of the sum in the last member, the integral extended over that part of S where $g_{n,r+1}(t)=0$ is obviously zero. If $g_{n,r+1}(t)=1$, there is by (8) exactly one $m<n$ such that $g_{m,r}(t)=1$. But it follows from (9) that, for every m, the integral

$$\int dF_m = \int F'_m\, d\mathbf{F}$$

extended over that part of S where $g_{m,r}(t)=1$, and thus $F'_m(t)>0$, is equal to zero. Consequently this set is a null set for the **F** measure, and the integral

$$\int dF_n = \int F'_n\, d\mathbf{F}$$

extended over this set, is also equal to zero. As this holds for all n and for all $m<n$, the relation (10) follows, and the proof of the Lemma is completed.

As already remarked above, the truth of the Theorem immediately follows from the two Lemmas.

References

1. Cramér, H., On the structure of purely nondeterministic stochastic processes. *Arkiv för Matematik* **4** (1961), 249—266.
2. Cramér, H., Stochastic processes as curves in Hilbert space. *Teorija Verojatnostei i eio Primenenija*, **9** (1964), 193—204.
3. Cramér, H., *Structural and statistical problems for a class of stochastic processes.* S. S. Wlksi Lecture, Princeton University Press 1971.
4. Kallianpur, G. and Mandrekar, V., Multiplicity and representation theory of purely nondeterministic stochastic processes. *Teorija Verojatnostei i eio Primenenija*, **10** (1965), 208—236.

Harald Cramér
Sjötullsbacken 15
115 25 Stockholm
Sweden

72.

On some points of the theory of stochastic processes

Sankhyā Ser. **A 40**, Part 2, 91–115 (1978)

I want to begin by expressing my heartfelt thanks to the Indian Statistical Institute and its Director, Professor G. Kallianpur, for the invitation to come and give these lectures, and for the kind and generous hospitality offered to me during my visit to India.

The first lecture gives a historical introduction to the theory of stochastic processes, while the second contains a comparison between Wiener's generalized harmonic analysis and the theory of stationary stochastic processes due to Khintchine and his followers. In the third lecture the theory of Hilbert space is applied to deduce the fundamental proposition concerning the multiplicity of a stochastic vector process of finite order. Finally, the fourth lecture develops some results in multiplicity theory which may possibly be new.

I

1.1. During the first ten years of the present century there appeared some remarkable pioneering works dealing with particular cases of what has later been called stochastic processes with a continuous time. The authors of these papers did not use rigorous mathematical methods, but they had an intuition which enabled them to reach results that would have to wait about thirty years before being placed on logically satisfactory foundations.

In present-day terminology these works are concerned with a random variable $x(t)$ defined on a fixed probability space and depending on a continuous time parameter t. Let P denote the probability measure, and introduce the ordinary functions

$$F(y, t) = P\{x(t) \leqslant y\},$$

$$f(y, t) = \frac{\partial F(y, t)}{\partial y},$$

which characterize the probability distribution of $x(t)$.

*Lectures delivered at the Indian Statistical Institute, Calcutta, 21–24 February 1977.

In a paper Bachelier (1900) analyzed the fluctuations of stock market prices. For the price $x(t)$ of a certain object at time t he was led to the equation

$$\frac{\partial f}{\partial t} = a(t)\,\frac{\partial f}{\partial y} + b(t)\,\frac{\partial^2 f}{\partial y^2} \qquad \ldots \quad (1)$$

which is satisfied by a normal (Gaussian) distribution, $a(t)$ and $b(t)$ being simply related to the mean and the variance of $x(t)$. In the particular case when $a(t)$ and $b(t)$ are constant, $a(t) = 0$ and $b(t) = \frac{1}{2}\sigma^2$, the equation reduces to

$$\frac{\partial f}{\partial t} = \tfrac{1}{2}\,\sigma^2\,\frac{\partial^2 f}{\partial y^2},$$

and $x(t)$ will be normally distributed with zero mean and variance $\sigma^2 t$.

In a famous paper on the Brownian movement Einstein (1906) studied the continuous but extremely irregular path of a small particle immersed in a fluid and subject to random molecular shocks. Denoting by $x(t)$ the abscissa of the particle at time t, he deduced by physical arguments the above partial differential equation for the probability density $f(y, t)$ and gave the corresponding normal distribution of $x(t)$. He also stated that the increments of $x(t)$ over disjoint time intervals are independent, and gave an expression for the variance $\sigma^2 t$ in terms of the physical constants relevant to the process under consideration.

Between the works of Bachelier and Einstein there appeared the Uppsala thesis of F. Lundberg (1903), written in Swedish. While Bachelier and Einstein were concerned with a random variable $x(t)$ varying continuously with time, Lundberg considered a case of essentially discontinuous variation. Letting $x(t)$ denote the accumulated amount of claims incurred by an insurance company up to time t, he gave the integrodifferential equation

$$\frac{\partial F}{\partial t} = \int_{-\infty}^{\infty} [F(y-u, t) - F(y, t)]\, dG(u), \qquad \ldots \quad (2)$$

where $G(u)$ is the distribution function for the amount of a single claim, in the case when this can be assumed to be independent of time. In the particular case when all claims are equal to one, he showed that the probability of n claims occurring in a time interval of length t will be

$$\frac{t^n}{n!}\; e^{-t},$$

so that we have what is today known as a Poisson process. For the general case he used his equation for the study of various problems related to insurance practice.

To a modern reader it is interesting to note that both the Bachelier equation (1) and the Lundberg equation (2) appear as particular cases of the general functional equations for Markov processes, given by Kolmogorov (1931) and further Studied by Feller (1936, 1940).

The Poisson process was applied in 1909 by Erlang to telephone traffic problems and by Rutherford and Geiger to the analysis of radioactive disintegration.

1.2. The methods of classical probability theory were obviously insufficient to deal rigorously with problems of the kind considered in the pioneering works related above. The first one to point the way to a satisfactory mathematical treatment of such problems was Norbert Wiener. In his famous paper "Differential Space" Wiener (1923) introduced for the first time a probability distribution in a functional space, giving a rigorous theory of the Brownian movement stochastic process, also known today as the Wiener process. The 1923 paper was quite difficult to read, but somewhat later, in their joint book (Wiener and Paley, 1934), Wiener gave a simpler deduction of which I shall say a few words here.

Let $y_0, y_1, y_2, \ldots$ and $z_1, z_2, \ldots$ be independent random variables, all normal (0, 1). Set for $0 < t < 1$

$$x_N(t) = y_0 t + \frac{1}{\pi\sqrt{2}} \sum_{n=1}^{2^N} \frac{y_n \sin 2n\pi t + z_n(1-\cos 2n\pi t)}{n}.$$

Then $x_N(t)$ has a well-defined normal distribution for every N. Allowing N to tend to infinity, it can be shown without difficulty that, with a probability equal to one, $x_N(t)$ converges uniformly in t to a random variable $x(t)$ such that the increments of $x(t)$ over disjoint t intervals are always normal and independent, with zero means and a variance proportional to the length of the corresponding interval. This is the stochastic process encountered by Bachelier and Einstein.

Regarding $x(t)$ as a function of the real and non-negative variable t, Wiener was able to prove that, with probability one, $x(t)$ is a continuous function of t without a derivative. Even if this process can only be taken as a first approximation to a realistic mathematical model of the Brownian movement, it gives a striking picture of the irregularity of the phenomenon.

1.3. Wiener's differential space work was mathematically rigorous, but was only concerned with one very special case of a stochastic process. The decisive turning point for probability theory as a whole, including the theory of stochastic processes, came with the two great works of Kolmogorov (1931 and 1933). The first of these contains the theory of what we now call Markov processes, while the second gives general foundations of mathematical probability theory.

Kolmogorov introduced the now familiar concept of a probability space $(\Omega, \mathscr{A}, P)$, where Ω is a space of points ω, denoted as elementary events, while $\mathscr{A}$ is a σ-algebra of sets in Ω, and P is a probability measure defined for all P-measurable events, i.e., for all sets S in $\mathscr{A}$. The concepts of a random variable $x = x(\omega)$ and a stochastic process $x(t) = x(t, \omega)$, where t is a parameter belonging to some parameter space T, were introduced in the now familiar way, which in 1933 represented a radical innovation.

In this way it became possible to regard a stochastic process as defined by a probability distribution in the space of functions $x(t)$ of the variable t, varying over the space T.

If the parameter space T is the real axis, and if for every $s < t$ the conditional distribution of $x(t)$, relative to the hypothesis $x(s) = a$, is independent of any information about the course of the process before time s, the process is called a Markov process. Kolmogorov proved that the probability distributions connected with a Markov process satisfy certain functional equations, which under appropriate continuity conditions reduce to partial differential equations of parabolic type. He showed that these differential equations uniquely determine the relevant probability distributions. The relations encountered by Bachelier, Einstein and Lundberg in their pioneering works are particular cases of the general equations given by Kolmogorov. His work was completed by Feller (1936 and 1940), both in respect of the solvability of the partial differential equations and of the integrodifferential equations obtained in the discontinuous case.

Processes with independent increments form a particular class of Markov processes. Both the continuous Bachelier-Einstein-Wiener and the discontinuous Lundberg process belong to this class. General expressions for the probability distributions connected with a process of this class have been given by Kolmogorov (1932), and under more general conditions by Lévy (1934).

The general theory of Markov processes has had a great development during recent years, largely owing to the work of Dynkin and his Moscow group.

II

2.1. The classical definition of a Markov process may be somewhat unprecisely expressed by saying that the past history of the process is without influence on its future course. In many important fields of application we are, however, concerned with systems for the temporal development of which the whole past history is relevant. As examples we may think of statistical mechanics, meteorology and economics. Two closely related attempts to deal with cases of this character were made by Wiener (1930) and by Khintchine (1934).

I now propose to give a brief review of each of these attempts, and then add some remarks on their mutual relations. It will be convenient from now on to deal with *complex-valued* functions $x(t)$ of the real variable t, which will always be regarded as representing time.

It is interesting to note that Wiener (1930) in his paper on "Generalized Harmonic Analysis", later developed, e.g., in Wiener (1934), entirely leaves the stochastic point of view applied in his previous work on the Brownian movement process. He is now concerned with individual complex-valued functions $x(t)$, not with random variables, and he considers only functions $x(t)$ such that the limit

$$r(h) = \lim_{T \to \infty} \frac{1}{2T} \int_{-T}^{T} x(t+h)\, \overline{x(t)}\, dt \qquad \ldots \quad (3)$$

exists for all real h.

It then follows that, for all real $h_1, \ldots, h_n$ and any complex $z_1, \ldots, z_n$

$$\sum_{j,k=1}^{n} r(h_j - h_k)\, z_j\, \bar{z}_k$$

is real and $\geqslant 0$, so that $r(h)$ is non-negative definite. Further, if $r(h)$ is continuous at $h = 0$, it is everywhere continuous, and we have

$$r(h) = \int_{-\infty}^{\infty} e^{ihu}\, dF(u), \qquad \ldots \quad (4)$$

where $F(u)$ is real, never decreasing and bounded.

Wiener regards $x(t)$ as representing a signal, of which $r(h)$ gives the spectrum and $F(u)$ the power distribution. In the simple particular case when $x(t)$ expresses an elementary harmonic oscillation

$$x(t) = \sum_{n=1}^{N} c_n e^{\lambda_n i t} \qquad \ldots \quad (5)$$

we have

$$r(h) = \sum_{1}^{N} |c_n|^2 e^{ih\lambda_n} = \int_{-\infty}^{\infty} e^{ihu}\, dF(u)$$

with

$$dF(u) = \begin{cases} |c_n|^2 & \text{at } u = \lambda_n \\ 0 & \text{otherwise.} \end{cases}$$

For any $x(t)$ satisfying Wiener's fundamental relation (3), a function $z(u)$ can be defined but for an additive constant by the integral

$$z(u) - z(v) = \frac{1}{2\pi} \int_{-\infty}^{\infty} \frac{e^{-iut} - e^{-ivt}}{-it}\, x(t)dt,$$

the integrand of which is in $L_2(-\infty, \infty)$. Formally we should then have the reciprocal relation

$$x(t) = \int_{-\infty}^{\infty} e^{itu}\, dz(u), \qquad \dots \quad (6)$$

expressing $x(t)$ as the Fourier-Stieltjes transform of $z(u)$, and this is easily seen to hold in the simple case when $x(t)$ is given by (5). In this case we also have

$$dF(u) = |dz(u)|^2. \qquad \dots \quad (7)$$

But in the general case $z(u)$ will not be of bounded variation, and the two last relations may not hold. Using very complicated analysis, Wiener was able to deduce some similar relations, such as

$$x(t) = \lim_{\varepsilon \to 0} \int_{-\infty}^{\infty} e^{itu} \frac{z(u+\varepsilon) - z(u-\varepsilon)}{2\varepsilon}\, du$$

and

$$F(b) - F(a) = \lim_{\varepsilon \to 0} \int_{a}^{b} \frac{|z(u+\varepsilon) - z(u-\varepsilon)|^2}{2\varepsilon}\, du,$$

which may be regarded as surrogates for the simple relations (6) and (7).

2.2. I now pass to the work on stationary stochastic processes by Khintchine (1934). He considers a complex-valued stochastic process $x(t)$ such that for all real t and h

$$Ex(t) = 0, \qquad E|x(t)|^2 < \infty$$

and

$$Ex(t+h)\,\overline{x(t)} = r(h),$$

where the covariance function $r(h)$ is independent of t. Such a process is sometimes called "stationary in the wide sense". Khintchine also defined a class of processes called "strictly stationary", which I am not going to deal with here.

It is well known that the properties of the covariance function $r(h)$ are very similar to those of Wiener's $r(h)$ as defined by (3). Thus the covariance function is non-negative definite, and if it is continuous at $h = 0$ it will be everywhere continuous, and given by the integral representation (4), where the spectral distribution function $F(u)$ is real, never decreasing and bounded. Khintchine showed that there are close relations between the theory of this class of stochastic processes and the ergodic theory of statistical mechanics.

Somewhat later it was shown by Cramér (1942) and Loève (1948) that it is possible to define a stochastic process $z(u)$ with orthogonal increments such that the relations (6) and (7) will hold, with an appropriate definition of the stochastic elements. These relations, which in the Wiener approach based on individual functions can only be shown to hold in simple particular cases, will thus in the stochastic approach be generally valid. The spectral representation (6) shows how the stochastic variable $x(t)$ is additively built up by elementary harmonic oscillations

$$e^{itu}\, dz(u),$$

each of which has an angular frequency u, while the amplitude and the phase are random variables determined by the increment $dz(u)$.

2.3. Comparing the approaches of Wiener and Khintchine, it seems natural to presume that, under fairly general conditions, the individual sample functions of a stationary stochastic process will satisfy Wiener's fundamental relation (3). Masani (1966) in his valuable account of Wiener's work on generalized harmonic analysis, points out that in Wiener's 1930 paper there is an example of what was later to be called a stationary process related to the Brownian movement process and such that, with a probability equal to one, its sample functions satisfy (3).

A somewhat more general example can be given in the following way. Let $x(t)$ be a real-valued and normal stationary process such that $Ex(t) = 0$, $Ex^2(t) = 1$, and suppose that for $h > 0$ we have

$$r(h) = O(h^{-\alpha}) \quad \text{with} \quad \alpha > 0 \quad \text{for} \quad h \to \infty,$$

$$r(h) = 1 - O\left(\log \frac{1}{h}\right)^{-\beta} \text{ with } \beta > 1 \text{ for } h \to 0.$$

It is then known that $x(t)$ can be taken such that, with probability one, its sample functions are everywhere continuous, and it can be shown without difficulty that the Wiener fundamental relation (3) is satisfied by almost all sample functions.

But it can also be shown that this property does not hold generally. We shall, in fact, give an example of a stationary process $x(t)$ such that for no real h it will be true that almost all its sample functions satisfy the Wiener relation (3).

Let $x(t)$ be real-valued, normal and stationary as in the previous example, with $Ex(t) = 0$, and suppose that the covariance function $r(h)$ is periodic with period 2, and that for $|h| \leqslant 1$ we have

$$r(h) = 1 - |h|$$

$$= \frac{1}{2} + \frac{4}{\pi^2} \sum_1^\infty \frac{\cos(2n-1)\pi h}{(2n-1)^2}.$$

(The example can be generalized; the main thing is that $r(h) \geqslant 0$ always, and that $r(h)$ does not tend to zero as $h \to \infty$.) As in the previous example $x(t)$ can be taken such that, with probability one, its sample functions are everywhere continuous. Write now

$$y(T) = \frac{1}{2T} \int_{-T}^{T} [x(t+h)x(t) - r(h)]dt.$$

We have to show that, for any given h, it will not be true that the random variable $y(T)$ will tend to zero with probability one as $T \to \infty$. By some calculation we find, using the normality of the $x(t)$ process

$$Ey^2(T) = \frac{1}{4T^2} \int_{-T}^{T} \int_{-T}^{T} [r^2(t-u) + r(t-u+h)\, r(t-u-h)]dt\, du$$

$$= \frac{1}{2T} \int_{-2T}^{2T} \left(1 - \frac{|v|}{2T}\right) [r^2(v) + r(v+h) r(v-h)]dv$$

$$> \frac{1}{4T} \int_{-T}^{T} [r^2(v) + r(v+h) r(v-h)]dv.$$

The integrand in the last member is bounded and periodic with period 2, and it follows that, for any given $\eta > 0$, there is a T_η independent of h such that for all $T > T_\eta$ we have

$$Ey^2(T) > \frac{1}{4} \int_{-1}^{1} [r^2(v) + r(v+h) r(v-h)]dv - \eta.$$

Introducing here the Fourier expansion of $r(h)$ we obtain by some calculation

$$Ey^2(T) > \frac{1}{4}+\frac{8}{\pi^4}\sum_1^\infty \frac{\cos^2(2n-1)\pi h}{(2n-1)^4}-\eta \geqslant \frac{1}{4}-\eta.$$

Further, by the normality of $x(t)$, it will be seen without difficulty that there is a constant K independent of h such that for all T

$$Ey^4(T) < K.$$

(A detailed calculation shows that we can take $K = 156$.)

Suppose now that $y(T) \to 0$ with probability one as $T \to \infty$. Then for any given positive δ and ε, and for all sufficiently large T, we have

$$P(|y(T)| \geqslant \delta) < \varepsilon,$$

and thus

$$\begin{aligned}\frac{1}{4}-\eta < Ey^2(T) &= \int_{|y(T)|<\delta} y^2(T)dP + \int_{|y(T)|\geqslant\delta} y^2(T)dP \\ &< \delta^2 + \left[\int_{|y(T)|\geqslant\delta} y^4(T)dP \cdot \int_{|y(T)|\geqslant\delta} dP\right]^{\frac{1}{2}} \\ &< \delta^2 + [Ey^4(T)\cdot\varepsilon]^{\frac{1}{2}} \\ &< \delta^2 + (K\varepsilon)^{\frac{1}{2}}.\end{aligned}$$

But this is evidently impossible if δ, ε and η are sufficiently small. Thus it follows that the Wiener relation (3) will not be satisfied for almost all sample functions of this $x(t)$ process. This shows that the analogy between Wiener's generalized harmonic analysis and Khintchine's stationary stochastic processes is not quite so close as might possibly have been expected.

III

3.1. In his two papers Kolmogorov (1941a; 1941b) applied Hilbert space theory to stationary stochastic processes with discrete time, showing that this approach yields important new results and simple proofs of those already known. His work has been followed up by many authors, and extended to very general classes of stochastic processes. In the sequel I propose to give some examples of the application of this method.

I shall begin by recalling some parts of general Hilbert space theory, and then proceed to apply them to stochastic processes. The definition and simplest properties of Hilbert space will be assumed to be known. We shall consider a Hilbert space H which is *separable*, and thus contains a complete orthogonal sequence of elements.

A2–2

Suppose that, for every real t, there is a subspace $H(t)$ of H such that the family of all $H(t)$ is never decreasing, and that the limiting space $H(-\infty)$ only contains the zero element of H, while $H(+\infty)$ is identical with the whole space H:

$$H(t) \subset H(u) \text{ for } t < u, \quad H(-\infty) = 0, \quad H(+\infty) = H,$$

and let P_t denote the projection operator in H wth range $H(t)$.

For any element z in H we shall consider the subspaces $H(z)$ and $H(z, t)$ spanned by the elements $P_u z$ according to the following definitions

$$H(z) = \mathfrak{S}(P_u z \text{ for all real } u),$$

$$H(z, t) = \mathfrak{S}(P_u z \text{ for all } u \leqslant t).$$

$H(z)$ is called the cyclic subspace of H determined by the element z. Then $F(t) = \|P_t z\|^2$ will be a real and never decreasing function of the real variable t, such that $F(-\infty) = 0$ and $F(+\infty) = \|z\|^2$. We shall say that $F(t)$ determines the spectral type of the element z.

We now introduce a partial ordering between the elements z and the corresponding $F(t)$, saying that z_1 dominates z_2, and F_1 dominates F_2, and writing

$$z_1 \succ z_2 \quad \text{and} \quad F_1 \succ F_2,$$

whenever the function $F_2(t)$ is absolutely continuous with respect to $F_1(t)$. If F_1 dominates F_2 and at the same time F_2 dominates F_1, we say that F_1 and F_2 are equivalent, and similarly for z_1 and z_2. The set of all F equivalent to a given F_1 is called the equivalence class of F_1.

Using only the simplest properties of projections in H, it can then be proved that there exists a finite or infinite sequence $z_1, z_2, \ldots, z_N$ of elements in H such that the corresponding cyclic subspaces $H(z_1), \ldots, H(z_N)$ are mutually orthogonal, and that we have

$$z_1 \succ z_2 \succ \ldots \succ z_N$$

and

$$H = H(z_1) \oplus H(z_2) \oplus \ldots \oplus H(z_N),$$

where the second member denotes the vector sum of the mutually orthogonal cyclic subspaces $H(z_n)$. Applying the projection operator P_t we obtain

$$H(t) = H(z_1, t) \oplus H(z_2, t) \oplus \ldots \oplus H(z_N, t)$$

for every real t.

This representation of H is *canonical* in the sense that, for any sequence $w_1, w_2, \ldots, w_M$ of elements in H such that the cyclic subspaces $H(w_n)$ are mutually orthogonal, and that we have

$$w_1 \succ w_2 \succ \ldots \succ w_M,$$
$$H = H(w_1) \oplus H(w_2) \oplus \ldots \oplus H(w_M),$$

we must have $M = N$ and w_n equivalent to z_n for all n. The number N, which may be finite or infinite, is called the *multiplicity* of the family of subspaces $H(t)$.

3.2. I now proceed to the application of the properties of Hilbert space recalled above to stochastic processes, defined on a fixed probability space. We shall consider a finite-dimensional vector process

$$\boldsymbol{x}(t) = (x_1(t), \ldots, x_q(t))$$

where the components $x_n(t)$ are complex-valued random variables with zero means and finite second-order moments. The parameter t is a real variable belonging to the interval $A \leqslant t \leqslant B$, where A is finite or $-\infty$, while B is finite or $+\infty$.

More general vector processes, of infinite order, have been considered by Kallianpur and Mandrekar (1965), but I shall here confine myself to the finite-dimensional case.

All random variables with zero means and finite second-order moments, defined on the given probability space, can be regarded as elements in a Hilbert space, if the inner product of any two such elements y and z is defined by the covariance moment

$$(y, z) = Ey\bar{z}.$$

The squared norm of the element y will then be equal to $E|y|^2$. If two random variables y and z are almost everywhere equal, we have $E|y-z|^2 = 0$, and we shall then regard y and z as identical, and represented by the same element of the Hilbert space.

The Hilbert space spanned by all the components $x_1(u), \ldots, x_q(u)$ of the given $\boldsymbol{x}(u)$ process for all u in $[A, B]$ will be denoted by $H(\boldsymbol{x})$ and called the Hilbert space of the process. By $H(\boldsymbol{x}, t)$ we denote the subspace of $H(\boldsymbol{x})$ spanned by $x_1(u), \ldots, x_q(u)$ for $A \leqslant u \leqslant t$.

In general the space $H(\boldsymbol{x})$ is not separable, and in order to have separability we must impose some mild regularity conditions on the $\boldsymbol{x}(t)$ process. We shall assume that the quadratic mean limits $x_n(t \pm 0)$ exist for all the components $x_n(t)$ and for all t. It is then known that the space $H(\boldsymbol{x})$ is separable, and that the variances $E|x_n(t)|^2$ are bounded in every finite subinterval of $[A, B]$.

For the sake of formal simplification we shall also assume that $x_n(t-0) = x_n(t)$ for all n and t, so that there is always quadratic mean continuity to the left. This assumption is not essential and does not imply any real restriction of generality.

Finally we shall assume that the limiting space $H(\boldsymbol{x}, -\infty)$ only contains the random variable which is almost everywhere equal to zero, which we denote by writing $H(\boldsymbol{x}, -\infty) = 0$. This means that we are concerned with a *purely nondeterministic* $\boldsymbol{x}(t)$ process. (In the case of a finite lower limit A we may complete the definition by taking all $x_n(t) = 0$ for $t \leqslant A$).

Each component $x_n(t)$ of the $\boldsymbol{x}(t)$ process can be said to generate a "curve" in the Hilbert space $H(\boldsymbol{x})$, and the subspace $H(\boldsymbol{x}, t)$ is the smallest subspace containing all the "arcs" $x_n(u)$ with $u \leqslant t$. As before we denote by P_t the projection operator in $H(\boldsymbol{x})$ with range $H(\boldsymbol{x}, t)$.

For any element z in $H(\boldsymbol{x})$ we define a stochastic process by writing

$$z(t) = P_t z.$$

Then $z(t)$ is a process with orthogonal increments such that

$$Ez(t) = 0, \qquad E|z(t)|^2 = F(t),$$

where $F(t)$ is real and never decreasing, and

$$F(A) = 0, \qquad F(B) = E|z|^2.$$

The cyclic subspace $H(z)$ and its subspaces $H(z, t)$ are defined as before. It is known that any element of $H(z, t)$ can be expressed in the form

$$\int_A^t c(u)\, dz(u),$$

where the integral is defined as a quadratic mean integral, and $c(u)$ is a non-random function such that the integral

$$\int_A^t |c(u)|^2 dF(u)$$

exists and is finite.

The general Hilbert space theorem mentioned in 3.1 above can now be applied and shows that the space $H(\boldsymbol{x})$ of the given $\boldsymbol{x}(t)$ process contains a finite or infinite sequence of elements $z_1, \ldots, z_N$ such that the corresponding cyclic subspaces $H(z_n)$ are mutually orthogonal, and that we have

$$z_1 \succ z_2 \succ \cdots \succ z_N,$$

$$H(\boldsymbol{x}, t) = H(z_1, t) \oplus \ldots \oplus H(z_N, t)$$

for all t. Now every component random variable $x_p(t)$ of the $\boldsymbol{x}(t)$ process is an element of the subspace $H(\boldsymbol{x}, t)$, which is the vector sum of the $H(z_n, t)$. Thus $x_p(t)$ can be represented as a sum of one element from each of the $H(z_n, t)$, and we obtain the fundamental representation formula

$$x_p(t) = \sum_{n=1}^{N} \int_A^t c_{pn}(t, u) dz_n(u). \qquad \dots \quad (8)$$

Here the $z_n(u)$ are mutually orthogonal processes of orthogonal increments, which are independent of p and t, and such that

$$E|z_n(u)|^2 = F_n(u)$$

and

$$F_1 \succ F_2 \succ \dots \succ F_N.$$

The $c_{pn}(t, u)$ are nonrandom functions defined for $A \leqslant u \leqslant t \leqslant B$ and such that the integrals occurring in the expression of the variance

$$E|x_p(t)|^2 = \sum_{n=1}^{N} \int_A^t |c_{pn}(t, u)|^2 dF_n(u)$$

exist and are finite, and have a finite sum for all p and t.

The number N, which is called the *multiplicity* of the given $\boldsymbol{x}(t)$ process, is uniquely determined by the process, and may be finite or infinite.

The relation (8) gives a *canonical* representation of the $\boldsymbol{x}(t)$ process in the sense that, for any two representations of the form (8) satisfying the above conditions, N will have the same value, and the functions F_n will be pairwise equivalent.

The representation (8) shows that the value assumed by any component $x_p(t)$ at time t, and thus also the whole stochastic vector $\boldsymbol{x}(t)$, can be regarded as the cumulative effect of N-dimensional innovation elements

$$dz(u) = (dz_1(u), \dots, dz_N(u)),$$

entering into the process during the time element $(u, u+du)$ in the course of the whole past $A \leqslant u < t$.

A representation of the form (8), based on Hilbert space theory, has been given by Hida (1960) for the case of a normal (Gaussian) process, and by Cramér (1960, 1961, 1964 and 1971) for the general case. This representation gives a simple and explicit solution of the prediction problem for the $\boldsymbol{x}(t)$ process (cf. e.g., Cramér, 1961).

3.3. An important problem that immediately presents itself in connection with this mode of representation of a stochastic process is the *multiplicity problem*, namely the problem to determine the multiplicity N associated with a given $x(t)$ process.

It is known (cf. e.g., Cramér, 1964) that there exist cases of any finite or infinite multiplicity, even if we impose further regularity conditions on the $x(t)$ process, such as continuity in quadratic mean for all t.

However, the examples that have been given of q-dimensional vector processes with multiplicity $N > q$ are somewhat pathological. For some important classes, such as stationary and Markov processes, we have $N \leqslant q$. And for a process with discrete time, $t = nh$ with integral n, we also have $N \leqslant q$. For a one-dimensional process, we thus have in all these cases $N = 1$.

It thus seems plausible that we should be able to find some appropriate general regularity conditions implying $N \leqslant q$, and thus in the one-dimensional case $N = 1$. Clearly it would be important for the applications, say to statistical estimation problems, to find some simple conditions of this type.

In some previous work, (Cramér, 1971) I have proved that we have in fact $N \leqslant q$ if we assume that the $c_{pn}(t, u)$ and the $F_n(u)$ associated with the canonical representation (8) are sufficiently regular. But obviously this is by no means a satisfactory solution of the problem, as it works with conditions expressed in terms of the functions appearing in the canonical representation. We must require a set of conditions bearing directly on the $x(t)$ process itself, without any reference to the canonical representation.

In the sequel I shall try to give a modest contribution to the study of this problem.

IV

4.1. We shall first consider the simple case of a one-dimensional, complex-valued process $x(t)$, defined for all t in a finite interval $[A, B]$, with $x(A) = 0$. Then $x(t)$ is purely nondeterministic, and we shall always assume as in 3.2 that the quadratic mean limits $x(t \pm 0)$ exist, so that $H(x)$ is separable and $E|x(t)|^2$ is bounded in $[A, B]$, and that $x(t-0) = x(t)$.

Let u and t be given such that

$$A < u \leqslant t < B,$$

and let h be a small positive quantity which will be allowed to tend to zero. The random variable

$$P_u x(t) - P_{u-h} x(t) = (P_u - P_{u-h}) x(t)$$

can then be regarded as that part of the variable $x(t)$ that was received as an innovation during the time interval $(u-h, u]$, and the variable

$$(P_{u+h}-P_u)x(t)$$

is related in the same way to the innovation received in $(u, u+h]$.

Further, consider the points

$$t_\nu = A+\nu h$$

for $\nu = 0, 1, 2, \ldots$ as long as $t_\nu \leqslant B$. Let $H^*(x, t_\nu)$ denote the ν-dimensional subspace of $H(x, t_\nu)$ spanned by the ν random variables

$$x(t_1),\ x(t_2), \ldots, x(t_\nu).$$

The projection operator $P^*_{t_\nu}$ with range $H^*(x, t_\nu)$ is then defined throughout the whole space $H(x)$.

We shall now state three conditions which will be imposed on the $x(t)$ process, and will imply that this process has multiplicity $N = 1$. The two first are regularity conditions to be satisfied by the innovations mentioned above, while the third is concerned with the degree of approximation between the $x(t)$ process and the process with discrete time $x(t_\nu)$. In all three conditions it is understood that h tends to zero through positive values, and the constants implied by the O signs are supposed to be independent of u, t and t_ν.

Condition I:

$$E\,|(P_u-P_{u-h})\,x(t)\,|^2 = hf(t, u)+O(h^{3/2}),$$
$$E\,|(P_{u+h}-P_u)\,x(t)\,|^2 = hf(t, u)+O(h^{3/2}),$$

where $f(t, u) > k > 0$ is bounded and has bounded first order partial derivatives for $A < u \leqslant t < B$.

Condition II:

$$E(P_u-P_{u-h})\,x(t)\cdot\overline{(P_u-P_{u-h})x(u)} = hg(t, u)+O(h^{3/2}),$$

where $g(t, u)$ is bounded and has bounded first order partial derivatives for $A < u \leqslant t < B$.

Note that in Condition II we are only concerned with innovations in an interval $(u-h, u]$. Note also that $g(t, u)$ is not assumed to be real and positive; however we have $g(t, t) = f(t, t) > k > 0$.

Condition III: For $t_\nu \leqslant t$ we have

$$E\,|P_{t_\nu}\,x(t)-P^*_{t_\nu}\,x(t)\,|^2 = O(h^2).$$

We can now state the following theorem

Theorem 1 : *If the conditions* I, II *and* III *are satisfied the $x(t)$ process has multiplicity* $N = 1$.

4.2. We shall first show that the conditions are satisfied in some important examples. For the two classes of wide-sense stationary and wide-sense Markov processes it will be shown that the conditions are satisfied under appropriate regularity conditions.

A purely nondeterministic stationary process can be put in the form

$$y(t) = \int_{-\infty}^{t} c(t-w)dz(w),$$

where $c(t)$ is in $L_2(0, \infty)$, while $z(w)$ is a process with orthogonal increments such that

$$Edz(w) = 0, \qquad E|dz(w)|^2 = dw.$$

In order to have a related process defined for t in a finite interval $[A, B]$, we consider the projection of $y(t)$ on the subspace spanned by the $dz(w)$ for $A < w < B$, and so obtain

$$x(t) = \int_{A}^{t} c(t-w)dz(w),$$

$$P_u x(t) = \int_{A}^{u} c(t-w)dz(w),$$

$$E|(P_u - P_{u-h})x(t)|^2 = \int_{u-h}^{u} |c(t-w)|^2 dw,$$

$$E|(P_{u+h} - P_u)x(t)|^2 = \int_{u}^{u+h} |c(t-w)|^2 dw,$$

$$E((P_u - P_{u-h})x(t) \cdot \overline{(P_u - P_{u-h})x(u)}) = \int_{u-h}^{u} c(t-w)\overline{c(u-w)}dw.$$

It will now be seen without difficulty that Conditions I and II are satisfied by $x(t)$ if $c(t)$ is bounded and continuous, has a bounded derivative and is everywhere different from zero. We shall now see that under these conditions also III is satisfied.

For $t_\nu \leqslant t$ we have

$$P_{t_\nu} x(t) = \int_{A}^{t_\nu} c(t-w)dz(w),$$

while any element in $H^*(x, t_\nu)$ is a linear combination of the form

$$L = \sum_{\mu=1}^{\nu} k_\mu x(t_\mu)$$

$$= \int_A^{t_\nu} \sum_{\mu=1}^{\nu} k_\mu c(t_\mu - w) dz(w),$$

where it should be observed that $c(t) = 0$ for $t < 0$. Under the above conditions with respect to $c(t)$ we can now successively determine the coefficients $k_\nu, k_{\nu-1}, \ldots, k_1$ so that L will satisfy the inequality

$$E|P_{t_\nu} x(t) - L|^2 < Kh^2,$$

where K is a constant independent of t_ν and t. Thus a fortiori we must have, since $P^*_{t_\nu} x(t)$ is identical with the projection of $P_{t_\nu} x(t)$ on $H^*(x, t_\nu)$,

$$E|P_{t_\nu} x(t) - P^*_{t_\nu} x(t)|^2 < Kh^2,$$

so that Condition III is satisfied.

As our second example we take a real-valued wide-sense Markov process as defined by Doob (1953, p. 233). Let $x(t)$ be such a process with $Ex^2(t) > k > 0$ for $A < t < B$, and let

$$r(t, u) = Ex(t)x(u),$$

$$\rho(t, u) = \frac{r(t, u)}{[r(t, t) r(u, u)]^{\frac{1}{2}}}$$

be the corresponding covariance function and correlation coefficient. The characteristic properties of a wide-sense Markov process then give

$$P_u x(t) = \frac{r(t, u)}{r(u, u)} x(u)$$

and

$$\rho(t, u-h) = \rho(t, u)\rho(u, u-h).$$

The first of these relations shows immediately, taking $u = t_\nu$, that Condition III is satisfied, since the projections $P_{t_\nu} x(t)$ and $P^*_{t_\nu} x(t)$ are identical. Allowing h to tend to zero, the second relation gives, if appropriate regularity conditions are imposed on $r(t, u)$, a differential equation for $\rho(t, u)$ from which we obtain for $u < t$

$$\rho(t, u) = e^{-\int_u^t b(s) ds},$$

A2-3

where $b(s)$ is non-negative, bounded and continuous. By some calculation we then obtain

$$E\,|(P_u-P_{u-h})x(t)|^2 = \rho^2(t,u)\,r(t,t)\left[1-e^{-2\int_{u-h}^{u} b(s)ds}\right],$$

$$E(P_u-P_{u-h})x(t)\cdot(P_u-P_{u-h})x(u) = r(t,u)\left[1-e^{-2\int_{u-h}^{u} b(s)ds}\right],$$

and an analogous expression for $E\,|(P_{u+h}-P_u)x(t)|^2$. It follows that Conditions I and II are also satisfied, if $b(s)$ is always positive and has a bounded derivative.

4.3. I now proceed to the proof of Theorem 1. Without restricting the generality we may take $A = 0$, so that the $x(t)$ process is given for $0 \leqslant t \leqslant B$, with $x(0) = 0$. We shall assume that $x(t)$ satisfies the Conditions I–III. In the present Section some auxiliary results will be given, while the following Section will complete the proof. We shall repeatedly use some general inequalities for random variables, which are given (without proof) in the Appendix.

For a fixed t, and $0 < u < t$, the random variable

$$z(u) = P_u x(t)$$

defines a stochastic process of orthogonal increments in u. Writing

$$F(t,u) = E\,|z(u)|^2 = E\,|P_u x(t)|^2$$

we have by Condition I

$$\begin{aligned} F(t,u)-F(t,u-h) &= E\,|P_u\,x(t)-P_{u-h}\,x(t)|^2 \\ &= hf(t,u)+O(h^{3/2}), \end{aligned}$$

and similarly

$$F(t,u+h)-F(t,u) = hf(t,u)+O(h^{3/2}).$$

Thus the partial derivative of $F(t,u)$ with respect to u exists and is equal to $f(t,u)$. Writing

$$y(t,u) = \int_0^u \frac{dP_s\,x(t)}{\sqrt{f(t,s)}} \qquad \dots \quad (9)$$

where t is still fixed, while s is the integration variable, it then follows that $y(t,u)$ is a process of orthogonal increments in u such that

$$E\,|dy(t,u)|^2 = du$$

and, since $P_0\,x(t) = 0$,

$$\int_0^u \sqrt{f(t,s)}\,dy(t,s) = \int_0^u dP_s x(t) = P_u x(t). \qquad \dots \quad (10)$$

The integral in the first member is here as usual defined as a quadratic mean integral, in the sense that the ordinary sums approximating the integral converge in q.m. to the random variable $P_u x(t)$ when the maximum length of the intervals in the partition tends to zero.

Further, using the inequalities (21) and (22) from the Appendix, we deduce from Condition III by some calculation that, in the first members of the relations given in Conditions I and II, we may replace the projections P by P^*, and the variable u by t_ν, without changing the second members. Thus we have

$$E\,|(P^*_{t_\nu}-P^*_{t_\nu-h})x(t)\,|^2 = hf(t, t_\nu)+O(h^{3/2}),$$

$$E((P^*_{t_\nu}-P^*_{t_\nu-h})x(t)\cdot\overline{(P^*_{t_\nu}-P^*_{t_\nu-h})x(t_\nu)}) = hg(t, t_\nu)+O(h^{3/2}), \qquad \dots \quad (11)$$

where f and g are the same functions as in Conditions I and II.

4.4. We now regard t and u as fixed so that $0 < u < t < B$, and h as a small positive quantity tending to zero. As before we consider the points $t_\nu = \nu h$, and determine t_μ by the condition $t_\mu \leqslant t < t_{\mu+1}$. For $\nu = 1, 2, \dots,$ we define the random variables

$$\xi(t_\nu, h) = x(t_\nu)-P^*_{t_\nu-h}x(t_\nu)$$
$$= (P^*_{t_\nu}-P^*_{t_\nu-h})x(t_\nu), \qquad \dots \quad (12)$$

$$\eta(t_\nu, h) = \xi(t_\nu, h)\left(\frac{h}{E\,|\,\xi(t_\nu, h)\,|^2}\right)^{\frac{1}{2}}$$

so that

$$E\,|\,\eta(t_\nu, h)\,|^2 = h.$$

By the properties of projections it now follows that the $\eta(t_\nu, h)$ are mutually orthogonal for $\nu = 1, 2, \dots, \mu$. For any $\lambda \leqslant \mu$ the η corresponding to $\nu = 1, 2, \dots, \lambda$ belong to the λ-dimensional subspace $H^*(x, t_\lambda)$, and thus form a complete orthogonal sequence in this space. Thus we have a Fourier expansion

$$x(t_\mu) = \sum_{\nu=1}^{\mu} b(t_\mu, t_\nu, h)\eta(t_\nu, h)$$

with

$$b(t_\mu, t_\nu, h) = \frac{1}{h}\,E\,x(t_\mu)\overline{\eta(t_\nu, h)},$$

and the projection $P^*_{t_\lambda} x(t_\mu)$ is the sum of the λ first terms of this series

$$P^*_{t_\lambda} x(t_\mu) = \sum_{\nu=1}^{\lambda} b(t_\mu, t_\nu, h)\eta(t_\nu, h). \qquad \ldots \quad (13)$$

On the other hand (10) gives

$$P_{t_\lambda} x(t_\mu) = \int_0^{t_\lambda} \sqrt{f(t_\mu, s)}\, dy(t_\mu, s).$$

For any $\lambda_1 < \lambda_2 \leqslant \mu$ we thus have by Condition III

$$E\left| \sum_{\lambda_1 < \nu \leqslant \lambda_2} b(t_\mu, t_\nu, h)\eta(t_\nu, h) - \int_{t_{\lambda_1}}^{t_{\lambda_2}} \sqrt{f(t_\mu, s)} dy(t_\mu, s) \right|^2 = O(h^2). \qquad \ldots \quad (14)$$

Now by (13)

$$(P^*_{t_\nu} - P^*_{t_\nu - h}) x(t_\mu) = b(t_\mu, t_\nu, h)\eta(t_\nu, h),$$

$$E(P^*_{t_\nu} - P^*_{t_\nu - h}) x(t_\mu) \overline{\eta(t_\nu, h)} = h\, b(t_\mu, t_\nu, h),$$

$$b(t_\mu, t_\nu, h) = \frac{E(P^*_{t_\nu} - P^*_{t_\nu - h}) x(t_\mu) \cdot \overline{(P^*_{t_\nu} - P^*_{t_\nu - h}) x(t_\nu)}}{[h\, E\,|(P^*_{t_\nu} - P^*_{t_\nu - h}) x(t_\nu)\,|^2]^{\frac{1}{2}}}$$

By means of Conditions I and II, as transformed by (11), we then obtain

$$b(t_\mu, t_\nu, h) = \frac{g(t_\mu, t_\nu)}{[f(t_\nu, t_\nu)]^{\frac{1}{2}}} + O(h^{\frac{1}{2}}).$$

Writing

$$c(t, u) = \frac{g(t, u)}{[f(u, u)]^{\frac{1}{2}}}, \qquad \ldots \quad (15)$$

$c(t, u)$ will be bounded and have bounded first order partial derivatives for $0 < u < t < B$, and we have

$$b(t_\mu, t_\nu, h) = c(t_\mu, t_\nu) + O(h^{\frac{1}{2}}),$$

the constant implied by the O sign being independent of t_μ and t_ν.

Keeping still t and u fixed, we now let h tend to zero, and at the same time λ_1, λ_2 and μ tend to infinity, in such a way that

$$\mu h \leqslant t < (\mu+1)h,$$
$$\lambda_1 h \leqslant u < (\lambda_1+1)h,$$
$$\lambda_2 h \leqslant u + h^{\frac{1}{2}} < (\lambda_2+1)h.$$

It follows that we have

$$\lambda_2-\lambda_1-1 < h^{-\frac{1}{2}} < \lambda_2-\lambda_1+1,$$

so that the difference $\lambda_2-\lambda_1$ tends to infinity as $h^{-\frac{1}{2}}$.

In the sum and the integral occurring in (14) we then have

$$b(t_\mu, t_\nu, h)-c(t_\mu, u) = O(h^{\frac{1}{2}}),$$

$$\sqrt{f(t_\mu, s)}-\sqrt{f(t_\mu, u)} = O(h^{\frac{1}{2}}).$$

Further the $\eta(t_\nu, h)$ are mutually orthogonal with $E\,|\,\eta(t_\nu, h)\,|^2 = h$, and $y(t_\mu, s)$ is a process of orthogonal increments in s with $E\,|\,dy(t_\mu, s)\,|^2 = ds$. By the inequality (21) of the Appendix we now obtain from (14)

$$E\left|\, c(t_\mu, u) \sum_{\lambda_1+1}^{\lambda_2} \eta(t_\nu, h)-\sqrt{f(t_\mu, u)} \int_{\lambda_1 h}^{\lambda_2 h} dy(t_\mu, s)\right|^2 = O(h^{3/2}). \qquad \ldots \quad (16)$$

Writing

$$X = c(t_\mu, u) \sum_{\lambda_1+1}^{\lambda_2} \eta(t_\nu, h),$$

$$Y = \sqrt{f(t_\mu, u)} \int_{\lambda_1 h}^{\lambda_2 h} dy(t_\mu, s),$$

and applying the inequality (21) of the Appendix, we find

$$E\,|\,X\,|^2-E\,|\,Y\,|^2 = O(h),$$

$$|\,c(t_\mu, u)\,|^2-f(t_\mu, u) = O(h^{\frac{1}{2}}).$$

By Condition I we then have for sufficiently small h

$$|\,c(t_\mu, u)\,|^2 = f(t_\mu, u)+O(h^{\frac{1}{2}}) > \frac{1}{2}\,k > 0,$$

and thus from (16)

$$E\left|\, \sum_{\lambda_1+1}^{\lambda_2} \eta(t_\nu, h)-\frac{\sqrt{f(t_\mu, u)}}{c(t_\mu, u)} \int_{\lambda_1 h}^{\lambda_2 h} dy(t_\mu, s)\right|^2 = O(h^{3/2})$$

and further by the same argument as above

$$E\left|\, \sum_{\lambda_1+1}^{\lambda_2} \eta(t_\nu, h)-\int_{\lambda_1 h}^{\lambda_2 h} \frac{\sqrt{f(t_\mu, s)}}{c(t_\mu, s)}\, dy(t_\mu, s)\right|^2 = O(h^{3/2}).$$

Between 0 and $t_\mu = \mu h$ there are

$$\frac{t_\mu}{(\lambda_2-\lambda_1)h} = \frac{t_\mu}{h^{\frac{1}{2}}+O(h)} = O(h^{-\frac{1}{2}})$$

disjoint intervals of the same length as $(\lambda_1 h, \lambda_2 h)$, each of which gives an upper bound of the above form. Summing over all these intervals and applying the inequality (20) of the Appendix and the definition (9) of $y(t, s)$, we obtain

$$E\left|\sum_{1}^{\mu} \eta(t_\nu, h) - \int_{0}^{t_\mu} \frac{dP_s x(t_\mu)}{c(t_\mu, s)}\right|^2 = O(h^{\frac{1}{2}}).$$

Now when $\mu \to \infty$, the upper limit $t_\mu = \mu h$ of the integral tends to t from below, and thus by hypothesis $x(t_\mu)$ converges in quadratic mean to $x(t)$. Owing to the continuity properties of $c(t_\mu, s)$ it then follows that

$$\sum_{1}^{\mu} \eta(t_\nu, h) \xrightarrow{\text{q.m.}} \int_{0}^{t} \frac{dP_s x(t)}{c(t, s)} = z(t), \qquad \dots \quad (17)$$

where $z(t)$ is a random variable belonging to $H(x, t)$. Since the $\eta(t_\nu, h)$ are all mutually orthogonal, $z(t)$ defines a process of orthogonal increments, and we have

$$\int_{0}^{t} c(t, u) dz(u) = \int_{0}^{t} dP_u x(t) = x(t). \qquad \dots \quad (18)$$

This representation of $x(t)$ has thus all the properties of a canonical representation as defined in 3.2, and it follows that the $x(t)$ process has multiplicity $N = 1$, so that Theorem 1 is proved.

The modifications to be introduced in the above expressions for the case of an arbitrary finite interval $[A, B]$ are evident.

4.5. We shall now consider the case of an $x(t)$ process defined for all real t, and prove the following theorem.

Theorem 2 : *Let $x(t)$ be a purely nondeterministic stochastic process defined for all real t and such that the q.m. limits $x(t \pm 0)$ exist, and $x(t-0) = x(t)$ for all t. Suppose that, in any finite interval $[A, B]$, $x(t)$ satisfies Conditions* I *and* II, *and $y(t, A) = x(t) - P_A x(t)$ satisfies Condition* III. *Then the $x(t)$ process has multiplicity $N = 1$.*

(Note that it has been shown in 4.2 that these conditions are satisfied for a sufficiently regular stationary process.)

The functions $f(t, u)$ and $g(t, u)$ of Conditions I and II and $c(t, u)$ as defined by (15) will then be defined for all real t and $u \leqslant t$, and will be bounded and have bounded first order partial derivatives in any finite interval.

For $A < u < t < B$ and sufficiently small $h > 0$ we then have

$$P_u y(t, A) - P_{u \pm h}\, y(t, A) = P_u x(t) - P_{u \pm h} x(t)$$

so that $y(t, A)$ satisfies all Conditions I to III in $[A, B]$, with the same functions $f(t, u)$, $g(t, u)$ and $c(t, u)$.

By Theorem 1 and the expressions (17) and (18) we then have for $A < t < B$

$$y(t, A) = x(t) - P_A\, x(t) = \int_A^t c(t, u)\, dz(u, A), \qquad \dots \quad (19)$$

where $z(u, A)$ is a process of orthogonal increments in u such that

$$z(u, A) = \underset{h \to 0}{\mathrm{l.i.m.}} \sum_{A < t_\nu \leqslant u} \eta(t_\nu, h),$$

denoting by l.i.m. a limit in quadratic mean. Taking $A < 0$ it follows that we have

$$\begin{aligned} z(u, A) &= z(0, A) + \underset{h \to 0}{\mathrm{l.i.m.}} \sum_{0 < t_\nu \leqslant u} \eta(t_\nu, h) \\ &= z(0, A) + z(u, 0) \end{aligned}$$

for $u > 0$, and

$$z(u, A) = z(0, A) - \underset{h \to 0}{\mathrm{l.i.m.}} \sum_{u < t_\nu \leqslant 0} \eta(t_\nu, h)$$

for $A < u \leqslant 0$.

It follows that $z(u, A)$ is the sum of a random variable $z(0, A)$ independent of u and a process of orthogonal increments independent of A, which we may denote by $z(u)$. Thus for $A < u < t < B$ we have $dz(u, A) = dz(u)$, and consequently from (19)

$$x(t) = P_A\, x(t) + \int_A^t c(t, u)\, dz(u).$$

Now this holds for any $A < 0$ and any $B > A$. Further $P_A\, x(t)$ tends to zero in q.m. as $A \to -\infty$, since the $x(t)$ process is purely nondeterministic. It follows that also the integral in the second member tends to a limit in q.m., and we obtain for all t the expression

$$x(t) = \int_{-\infty}^t c(t, u)\, dz(u)$$

possessing all the properties of a canonical representation, so that Theorem 2 is proved.

4.6. We finally proceed to the case of a finite-dimensional stochastic vector process

$$\boldsymbol{x}(t) = (x_1(t), \dots, x_q(t))$$

defined for all real t. From a general proposition on the multiplicity of a stochastic vector process due to Cramér (1978) we then immediately obtain the following theorem.

Theorem 3 : *If every component $x_p(t)$ of the vector process $\boldsymbol{x}(t)$, regarded as a one-dimensional stochastic process, satisfies the conditions of Theorem 2, the multiplicity of the vector process $\boldsymbol{x}(t)$ is $N \leqslant q$.*

Appendix

The following inequalities for random variables have been repeatedly used in the above deductions. They are easily proved by means of the simple standard inequalities found in current textbooks. The x and y occurring here are all complex-valued random variables with zero means and finite second order moments, defined on a fixed probability space.

$$E|x_1+x_2+\ldots+x_n|^2 \leqslant n(E|x_1|^2+E|x_2|^2+\ldots+E|x_n|^2) \qquad \ldots (20)$$

$$\begin{aligned}(E|x|^2-E|y|^2)^2 &\leqslant 2E|x-y|^2(E|x|^2+E|y|^2)\\ &\leqslant 6E|x-y|^2(E|x|^2+E|x-y|^2) \qquad \ldots (21)\end{aligned}$$

$$\begin{aligned}|Ex_1x_2-Ey_1y_2|^2 \leqslant 3[&E|x_1|^2E|x_2-y_2|^2\\ &+E|x_2|^2E|x_1-y_1|^2+E|x_1-y_1|^2E|x_2-y_2|^2]. \qquad \ldots (22)\end{aligned}$$

References

BACHELIER, L. (1900) : Théorie de la spéculation, *Ann. Ecole Norm. Sup.*, **17**, 21.

CRAMÉR, H. (1942) : On harmonic analysis in certain functional spaces, *Arkiv f. Matematik*, **28B**, No. 12.

———— (1960) : On some classes of non-stationary stochastic processes, *Fourth Berkeley Symp.*, **2**, 57.

———— (1961) : On the structure of purely nondeterministic stochastic processes, *Arkiv f. Matematik*, **4**, 249.

———— (1964) : Stochastic processes as curves in Hilbert space, *Teorija Verojatn. Primen*, **9**, 169.

———— (1971) : Structural and statistical problems for a class of stochastic processes, S. S. Wilks Lecture, Princeton Univ. Press.

———— (1978) : On the multiplicity of a stochastic vector process, *Arkiv f. Matematik*, **16**, 89.

DOOB, J. L. (1953) : Stochastic Processes, Wiley, New York.

EINSTEIN, A. (1906) : Zur theorie der Brownschen bewegung, *Ann. Physics*, **19**, 371.

FELLER, W. (1936) : Zur theorie der stochastischen prozesse, *Math. Ann.*, **113**, 113.

———— (1940) : On the integro-differential equations of purely discontinuous Markoff processes. *Trans. Amer. Math. Soc.*, **48**, 488.

HIDA, T. (1960) : Canonical representations of Gaussian processes and their application, *Mem. Univ. Kyoto, A*, **33**, 109.

KALLIANPUR, G. and MANDREKAR, V. (1965) : Multiplicity and representation theory of purely non-deterministic stochastic processes, *Teorija Verojatn. Primen.* **10**, 553.

KHINTCHINE, A. YA. (1934): Korrelations theorie der stationären stochastischen prozesse, *Math. Ann.* **109**, 604.

KOLMOGOROV, A. N. (1931): Ueber die analytischen methoden der Wahrscheinlichkeitsrechnung, *Math. Ann.*, **104**, 415.

——— (1932): Sulla forma generale di un processo stocastico omogeneo, *R. Accad. Lincei*, **15**, 805 and 866.

——— (1933): Grundbegriffe der Wahrscheinlichkeitsrechnung, *Ergebn. d. Mathematik, Springer, Berlin.*

——— (1941a): Stationary sequences in Hilbert space, *Bull. Moskva Univ.*, **2**, No. 6.

——— (1941b): Interpolation and extrapolation of stationary random sequences, *Izv. Akad. Nauk*, **5**, 3.

LÉVY, P. (1934): Sur les intégrales dont les éléments sont des variables aléatoires indépendantes, *Ann. Ecole Norm. Pisa*, **3**, 337.

LOÉVE, M. (1948): Fonctions aléatoires du second ordre, Note à Lévy, Processus stochastiques et mouvement Brownien, Paris, 299.

LUNDBERG, F. (1903): Approximerad framställning av sannolikhetsfunktionen, Thesis, Uppsala University.

MASANI, P. (1966): Wiener's contribution to generalized harmonic analysis, prediction theory and filter theory, *Bull. Amer. Math. Soc.*, **72**, 73.

WIENER, N. (1923): Differential space, *J. Math. Phys.*, **2**, 131.

——— (1930): Generalized harmonic analysis, *Acta Math.*, **55**, 117.

WIENER and PALEY, R. (1934): Fourier transforms in the complex domain, *Amer. Math. Soc. Colloque. Publ., New York.*

——— (1934): The Fourier Integral, Cambridge University Press.

73.

Mathematical probability and statistical inference

(Pfizer Colloquium Lecture 1980)
Internat. Statist. Rev. **49**, 309–317 (1981)
(French translation 1983)

Some personal recollections from an important phase of scientific development[1]

Summary

The paper gives some personal recollections of the development of mathematical probability theory and its applications to statistical inference during the twenty years between the two world wars, preceded and followed by some brief notes on the development before 1920 and after 1940.

Key words: Central limit theorem; Confidence sets; Fiducial probability; Foundations of probability; Statistical estimation; Stochastic processes.

1 Introduction

It is a great honour for me to have been invited to give this third Pfizer Colloquium, and I am very grateful for the extremely kind words of welcome that I have been allowed to listen to. I am happy to have had this opportunity to come back to the University of Connecticut, from which I have such pleasant memories.

I now propose to give some personal recollections from the development of two important scientific fields: the purely mathematical theory of probability and the methodology of statistical inference. Although intimately and necessarily related, these fields have sometimes developed more or less independently, sometimes in close connection. Examples of both these types of work can be found in course of the intense development that took place during the 20 years between the two world wars, which led to a complete transformation of both these scientific fields.

The main part of my talk will consist of some recollections from this period of 20 years, to which I shall then add some scattered remarks on the subsequent history. I shall be talking from a strictly personal point of view, giving my impressions of those events that appealed to my interest, as I remember them today, many years afterwards.

But I will begin with a brief bird's eye view of the development of these fields up to the end of the first World War.

---

[1] Pfizer Colloquium Lecture, May 12, 1980. Professor Cramér was the Third Annual Pfizer Colloquium speaker of the Department of Statistics, University of Connecticut. This Colloquium was videotaped for archival purposes under the joint sponsorship of Pfizer Research and the American Statistical Association.

2 Probability theory before 1920

The first attempt to use what we now call mathematical probability to predict the values of statistical frequency ratios was, as everyone knows, due to the gamblers. They sometimes found at their loss that practical experience did not agree with their computation of chances. It is well known that, about 1650, one of these unfortunate gamblers consulted Blaise Pascal in Paris, and that this led to the beginning of a scientific study of probability, which soon attracted a great interest.

James Bernoulli, of the famous Swiss mathematical family, gave an account of this new theory in his posthumous work of 1713: 'Ars Conjectandi' or 'The Art of Conjecture'. It may be interesting to recall that the words he actually used were: 'Ars Conjectandi sive Stochastica', and I believe this is the first time that one encounters in this connection the term 'stochastic', which has since then been so generally used. Bernoulli gave in his great book the mathematical theory, as far as it was then developed, and pointed out the possibility of its applications to various questions, not only in gambling, but also in social science, in economics and in meteorology. He proved his well-known theorem, leading from a known probability to a predicted approximate value of a statistical frequency ratio, and also argued in the opposite direction, from an observed frequency ratio to an estimate of the corresponding probability. I believe it is fair to say that he was the first to see the possibility of using probability theory as a tool for drawing valid inferences from statistical data.

Almost exactly a hundred years after the Ars Conjectandi, in 1812, appeared the great work of Laplace: 'Théorie Analytique des Probabilités'. It covered a broad field of both mathematical theory and statistical applications, and still offers a stimulating reading, but it was often regrettably nonrigorous from a mathematical point of view, and it was surprisingly uncritical in respect of the foundations and the applications of the mathematical theory. For Laplace the probability definition based on the famous 'equally possible cases' occurring in the simple games of chance was evidently applicable everywhere. And his main theoretical contribution, the highly important proposition known to us as 'The Central Limit Theorem' of probability theory, was stated without a complete proof.

During a part of his career, Laplace served as a minister of the Emperor Napoleon, who afterwards criticized him, saying that he introduced 'l'esprit des infiniments petits' (the spirit of the infinitely small) into the practical administration.

Nevertheless, the work of Laplace served as a point of departure for most research work in probability and statistics during the whole century following the publication of the Théorie Analytique. I shall here only say a few words about two main lines of investigation during this time, which became of particular importance for the subsequent development. I am referring first to the work on the central limit theorem stated by Laplace, and then to the discussion about the foundations of probability theory.

It was clearly seen by the nineteenth-century mathematicians that the central limit theorem stated by Laplace lacked a rigorous proof, and that it would be of the utmost importance to find one. A great number of unsuccessful attempts were made until finally, in 1901, the Russian mathematician Liapounov succeeded in giving a complete proof, valid under certain conditions. He worked with the important analytic tool known today as characteristic functions of the probability distributions, but his work remained fairly unknown for some time, and it was not until considerably later, in the 1920's, that further research on this line was done.

The foundations of probability theory have been broadly discussed by mathematicians, statisticians and philosophers, and have always been a strongly controversial topic. Some philosophers tried to give a clear meaning to the famous equally possible cases on which Laplace had based his general definition, but without reaching any convincing results. Finding this way impossible, other authors tried to find a radically different probability definition in terms of the empirically observed long run stability of statistical frequency ratios. The most advanced attempt on these lines is due to von Mises, a German mathematician later working in the USA. He based his theory on two axioms concerning the limiting behaviour of statistical frequency ratios in an unlimited sequence of trials. He had enthusiastic followers, and also severe critics. I shall quote one of the latter, the great French probabilist Paul Lévy who said, looking back in his autobiography of 1970, that it is 'just as impossible as squaring the circle' to reach a satisfactory probability definition in this way.

3 A decade of preparation

The work of von Mises appeared in 1919, on the threshold of a new era in probability theory. I was then a young man studying this theory and its applications, and of course his work attracted my great interest.

Although taking a strongly critical attitude to his axiomatics, I found in his paper a statement of principle with which I whole-heartedly agreed—and still agree. He said that probability theory should be 'a natural science of the same kind as geometry or theoretical mechanics', designed to describe certain observable phenomena connected with random experiments, 'not exactly, but with some abstraction and idealization'. To me it seemed natural to express this view by saying that probability theory should be regarded as a *mathematical model* of this class of phenomena. However, I did not think that von Mises had followed up the consequences of his statement strictly when building his own system. This was to be done later by Kolmogorov, whose work brought the definite formalization of probability theory, to which I shall return later.

During the 1920's I followed with a keen interest the powerful new international development of mathematical probability theory, and I now proceed to give some recollections from this period. It was not until somewhat later, during the 1930's, that I became actively engaged in the applications to statistical inference.

In a paper of 1920 George Pólya, then working in Switzerland, later in California, introduced the name 'central limit theorem of probability theory' that has since then been universally accepted. He recalled the work of Liapounov and discussed its relations to other parts of mathematical analysis. Soon afterwards, in 1922, the Finnish mathematician J.W. Lindeberg applied a new method to the central limit theorem, and proved its validity under more general conditions than those given by Liapounov. The Lindeberg condition was later to play an important part in the attempts to find conditions which are both necessary and sufficient.

As is generally known, the central limit theorem asserts that, under certain conditions, the probability distribution of a properly normalized sum of independent random variables tends to the normal, or Gauss–Laplace distribution as the number of terms tends to infinity. Liapounov had given an upper limit for the error committed in replacing, for a finite number of terms, the actual distribution by the limiting one. This limit was quite insufficient for the applications to insurance risk theory, in which I was then interested, and I asked myself if it would not be possible to reach a closer approximation by regarding the normal distribution as the first term of some asymptotic expansion. In two papers of 1925 and 1928 I proved that the expansion known as Edgeworth's series, which had been studied by Edgeworth in a purely formal way, really did have the asymptotic properties that I required. These results were found by the same method of characteristic functions that had been used by Liapounov.

In a remarkable paper of 1927, the Russian mathematician S.N. Bernstein discussed the extension of the central limit theorem to sums of random variables which are not required to be independent. The method introduced by him for dealing with this problem has later led to further important results.

About the same time the French mathematician Paul Lévy published the first of his many important works on probability theory. It was his 'Calcul des Probabilités' of 1925, which contained the first systematic treatise of random variables, their probability distributions and their characteristic functions. With this work, probability theory was beginning its transition from a collection of more or less picturesque examples to a connected and important mathematical theory, a transition which was soon to be definitely completed.

All through the latter part of the 1920's it was evident that a strong new development in probability theory was taking place in the Soviet Union. I have already mentioned the work of Bernstein. The two great mathematicians A.Y. Khintchine and A.N. Kolmogorov, who were quite young men in the 1920's, already now started their work, although their main pathbreaking contributions belong to the 1930's. In a joint paper of 1925 they proved the famous 'three series theorem', giving necessary and sufficient conditions for the convergence of a series whose terms are independent random variables. The probability that such a series is convergent can only be equal to zero or one, which is a particular case of the so called 'zero or one law' discovered at the same time. In 1929 Kolmogorov gave a proof of the now well-known 'law of the iterated logarithm', previously found by Khintchine in the particular case when the random variables concerned are simply the digits of a dyadic or decimal fraction. This had been regarded as a purely measure-theoretic problem, and the generalization due to Kolmogorov prepared the way for the identification of probability with measure that he was soon to give.

4 The breakthrough of the 1930's

Looking back towards the beginning of the new era in probability theory, it now seems evident that the real breakthrough came with the publication in 1933 of Kolmogorov's book 'Grundbegriffe der Wahrscheinlichkeitsrechnung'. In this book he laid the axiomatic foundations of an abstract theory, designed to be used as a mathematical model of certain observable events connected with repeated random experiments. The fundamental concept of this new theory is the now familiar concept of a probability space (Ω, A, P), where Ω is a space of points ω denoted as elementary events. Further, A is a σ-algebra of sets S in Ω, which are regarded as the conceptual counterparts of observable events, while P is a probability measure defined for all sets belonging to A. The concept of a random variable $x = x(\omega)$ is introduced in the now familiar way, which in 1933 represented a radical innovation, as an A-measurable function of the elementary event ω. The whole theory of random variables and probability distributions is developed on this foundation. The concept of conditional probability is treated in an entirely new way. This book of 1933 by Kolmogorov still ranks as one of the basic documents of modern probability theory.

When all these new ideas appeared, I had recently been nominated as the first holder of a chair for Actuarial Mathematics and Mathematical Statistics in the University of Stockholm. I was working with a small group of ambitious and intelligent students, and we followed the development abroad with a keen interest. Will Feller, who had been turned out from a German university by the Nazis, came to join our group in 1934, and stayed on in Stockholm for five years, giving highly valuable contributions to our work, before he left us in 1939 to take up his great career in the USA.

During the first part of the 1930's, we were in our group primarily interested in the probabilistic work of the French and Russian mathematicians. The theory of characteristic functions seemed to open a great field of new possibilities. The addition of independent random variables corresponds to a multiplication of their characteristic functions, and it was natural to regard the probability distribution of the sum as a symbolic product of the factors consisting of the component distributions. A probability distribution that can be represented as the symbolic product of any number of mutually identical factors is said to be infinitely divisible. The normal, the Poisson and the stable distributions all belong to this class, which had been specially studied by Lévy and Khintchine. In an admirable paper of 1934 Lévy gave a general expression for the characteristic function of an infinitely divisible distribution. He expressed in this paper the conjecture that any factor of a normal distribution must itself be normal, saying that he regarded this as very probably true, but had not been able to prove it. I had the good luck of finding a proof of this conjecture, based on the properties of characteristic functions, and somewhat afterwards it was shown by Raikov in the Soviet Union that the Poisson distribution has the same property.

The conditions for the validity of the central limit theorem due to Liapounov and Lindeberg were generalized by Lévy, Khintchine and Feller, working in close contact. It was Feller who was the first to find conditions that are both necessary and sufficient, and his work on this problem was done and published during his stay in Stockholm.

In my Cambridge Tract 'Random Variables and Probability Distributions' of 1937 I summed up our work in the Stockholm group, basing it on the Kolmogorov axioms. The main parts of this little book were concerned with the theory of probability distributions and characteristic functions, applying them to the central limit theorem and its allied asymptotic expansions.

A further considerable part of our work in the Stockholm group during the 1930's was related to the theory and applications of stochastic processes, at that time a quite new subject. In a famous paper of 1923, Norbert Wiener had given a rigorous deduction of the probabilistic model of the Brownian molecular motion introduced by Einstein. But his paper was quite difficult to read, and we did not see its significance until the great work of Kolmogorov had paved the way. Kolmogorov treated in 1931 the class now known as Markov processes, of which the Brownian motion process is a particular case. He introduced the differential equations which, under appropriate continuity conditions, determine their probability distributions. Somewhat later Feller completed this work under more general conditions, when the basic relations are of the integro-differential type. Other members of our group were interested in the special kind of Markov processes known as processes of independent increments, which are closely related to the infinitely divisible distributions and have important applications to insurance risk theory.

A Markov process arises when a random variable develops in time in such a way that, at any given moment, its future probabilistic properties are completely determined by its actual present value, whereas its previous history is irrelevant. In a basic paper of 1934, Khintchine pointed out that this assumption is not valid in many important applications, such as those occurring in meteorology, economics and sociology, where the whole prehistory of the process must be taken into account. As a convenient probabilistic model for some of these cases, he introduced the class now known as stationary processes, where the relevant probability distributions are more or less invariant under a translation in time. He gave a representation of the autocovariance function of such a process as a Fourier–Stieltjes integral, and somewhat later I gave a representation of the random variable itself associated with such a process as a stochastic integral of the same type. Herman Wold, who at that time was one of my students, treated in his thesis of 1938 the class of stationary time series, and proved for these a decomposition theorem which has since been extended to a wide class of stochastic processes.

It was during these years, about the middle of the 1930's, that I became actively interested in the work of British and American statisticians. Of course I had already long ago made acquaintance with the literature on frequency curves, correlation and regression associated with the names of Karl Pearson, G.U. Yule and others. But I had a definite—and perhaps somewhat exaggerated—impression that this was fairly superficial and not very interesting. The early work of R.A. Fisher during the 1920's was entirely different. It was clearly designed to make it possible to draw valid inferences from statistical data, and it was always stimulating and interesting, even in cases where I could not agree with him. His work on multidimensional probability distributions, on statistical estimation and the use of the maximum likelihood method made a great impression. But he used a probability theory which still lacked a rigorous foundation, and he admitted himself that some of the statements of his admirable 1925 paper on statistical estimation were not yet completely proved. Of course to me this seemed to be a challenge to go further and supply such proofs whenever possible.

But in the beginning of the 1930's it became evident that the situation in British statistics was highly controversial, to say the least. Fisher strongly criticized the use of the famous Bayes theorem for the estimation of an unknown parameter in a probability distribution, and advocated his own maximum likelihood method. While emphasizing the difference between the concepts of probability and likelihood he still, somewhat surprisingly in my view, regarded them both as 'alternative measures of rational belief'. And he said that from a known statistical sample we can 'express our incomplete knowledge of the population in terms of likelihood', thus obtaining 'a definite probability statement about the unknown parameter'. This is his famous 'fiducial distribution' of the parameter concerned.

I found it entirely impossible to follow Fisher on these new lines, which seemed to me based on a definite mathematical slip. In some applications it is legitimate to regard an unknown parameter as determined itself by a random experiment, and the Bayes method is then clearly applicable. But in other cases—and I believe they are the majority—the parameter is simply a constant having a fixed but unknown value, and Fisher's deduction of a fiducial distribution must then be in clear contradiction with modern probability theory.

At this time, in the middle of the 1930's, Jerzy Neyman was still working in England. He had been educated in the atmosphere of Polish mathematical traditions, and he developed a method of statistical estimation by confidence intervals, or more generally confidence sets, based on what he called 'modernized classical probability theory'. His work on this method was hotly criticized by Fisher. I followed the discussion in the publications of the Royal Statistical Society and elsewhere, and it soon seemed clear to me that Fisher's fiducial argument was wrong and that, duly corrected, it would lead to something very like the Neyman theory. Fisher stood also in strong opposition to the joint work of Neyman and Egon Pearson on the testing of statistical hypotheses that was being carried out during this period. It was not until somewhat later, when Neyman's comprehensive account of 1939 had come into my hands, that I became properly acquainted with this highly important theory, which has since developed in such a remarkable way, as described in the well-known excellent book by Erich Lehmann.

My Cambridge Tract and my proof of Lévy's conjecture rendered me some invitations to give lectures in various places, and I was glad to have this opportunity of meeting some of the authors whose work I had followed with such great interest. In the spring of 1937 I was invited to Paris, to give some lectures at the Sorbonne. Of the French probabilists, I knew Fréchet ever since my first international mathematical congress long ago, in 1920. This time I met Paul Lévy and several members of the younger generation, such as Dugué, Fortet and Loève. It seemed clear that Lévy was to be one of the leaders of the development of probability theory, and his important book 'Théorie de l'addition des variables aléatoires' was just being published.

Later in the same year 1937 there was a conference for probability and mathematical statistics in Geneva, and it was quite exciting to meet so many famous scientists. Unfortunately the Russian colleagues, who had accepted the invitation, had not been able to come—something which has often occurred since then. Feller and I attended from Stockholm, and for me it was particularly interesting to meet Neyman, who gave an account of his new method of estimation by confidence sets. In the middle of his talk he was interrupted by Fréchet and Lévy, who wanted to criticize. I happened to be chairman of that meeting, and had to use my poor ability of talking French to quiet them down and let him finish his talk. Having previously read his main paper in the *Proceedings of the Royal Society*, I was convinced that his ideas were sound, and I believe that his French opponents afterwards came to the same conclusion.

In the fall of 1938 I was invited to London, where I was received by some old actuarial friends, and met Fisher and Egon Pearson for the first time. Neyman was already in California. I gave some talks on my work on probability, avoiding the controversial statistical inference topics. It was shortly after the Munich agreement, and the question of peace or war was on everybody's mind. In the summer of 1939, hardly a month before the outbreak of the war, there was again a conference in Geneva, this time for applied probability. Fisher was there, and I remember saying some complimentary words to him about his geometrical intuition in dealing with multi-dimensional distributions, and receiving the somewhat acid reply: 'I am sometimes accused of intuition as a crime'. Among new acquaintances there were Sam Wilks and Maurice Bartlett. There was just enough time to get back to Sweden before the Nazi attack on Poland, which started the second World War.

5 The war years

During the war years we were quite isolated in Sweden. We were surrounded by war: Denmark and Norway were occupied by the Nazis, and Finland was at war with Russia. In our small probabilistic group in Stockholm we tried to keep the work going on, but it was almost impossible to get access to foreign scientific publications. In this state of isolation and insecurity I decided to take up an old plan of writing a new book.

In the years before the war it had seemed to me that the continental mathematicians and the Anglo-Saxon statisticians were working without sufficient mutual contact, and that it might be useful to try to join both these lines of research. Already in 1937 I had fairly advanced plans for a book on modern statistical methods, based on mathematical probability theory. There was a German scientific editor who was interested in the project, and I still have a proposed table of contents for such a book written in German. But my strong anti-Nazi feelings made me reluctant to publish in Germany, I discarded the plan, and nothing more than the table of contents was written.

In 1942, when the end of the war still seemed to be far away, I took up the old plan again, and began writing a book in English. In the summer of 1945 I had a manuscript ready to be printed, under the title 'Mathematical Methods of Statistics'. The book was published jointly by the Swedish editors Almqvist and Wiksell, and by the Princeton University Press. It contained three parts: a purely mathematical introduction, a theory of random variables and probability distributions, and their applications to statistical inference. It was dedicated to my wife Marta who, as always, had helped and encouraged me all through my work. While I was writing it, I sometimes said to her that I hoped this book would be my entrance card into the new world after the war. And certainly it did provide us with dear friends in a number of countries.

I have often regretted that in the probabilistic part of this book I did not give the general Kolmogorov theory, but restricted it to distributions in finite-dimensional Euclidean spaces. This made it perhaps easier to read, but at the same time it made it impossible to include a satisfactory discussion of the convergence of sequences of random variables, and to enter upon the subject of stochastic processes. In the statistical inference part I tried to give a systematic treatment of Fisher's work on sampling distributions and statistical estimation, supplying rigorous proofs of his results and completing them on some points. I included a chapter on Neyman's confidence regions, trying to make it clear to the reader, without being too explicit about it, that I took his part in the dispute over Fisher's fiducial probability. Also I gave an account of the Neyman–Pearson theory, afterwards finding that I should have entered more fully into this important subject.

My books, and my work on stochastic processes, represented the main outcome of what I had learned during the 20 years between the world wars. This period had involved a complete change of structure for mathematical probability theory and its applications to statistical inference. I have tried to give you some personal recollections from this important work, and I will now finish up with some very brief comments on subsequent events.

6 Notes on further development

During the war years, the importance of a highly advanced statistical methodology for industrial and military applications became ever more clear. Mathematicians and statisticians in war-making countries were engaged in the development and use of such methods.

When, after the war, some of the results obtained became known, it appeared that problems of antiaircraft fire control and radar had given rise to some very important research. The theory of stationary stochastic processes, which I have mentioned before, provided an efficient tool for this kind of work. Independently of one another, Kolmogorov in Moscow and Norbert Wiener at the Massachusetts Institute of Technology had made important work on these lines.

In two small but extremely important notes published during the war, Kolmogorov had pointed out that the mathematical theory of Hilbert space could be successfully applied to the study of random variables and stationary stochastic processes. Shortly after the war his work was further developed by some of his students and by the Finnish mathematician Kari Karhunen, who for some time worked as a member of our Stockholm probabilistic group.

This work had a powerful influence on the development of the theory and applications of stochastic processes, for stationary as well as other classes. A path-breaking work along these lines was the 1950 thesis on 'Stochastic processes and statistical inference' by Ulf Grenander, at that time a member of our Stockholm group, later my first successor as Professor at Stockholm University and now, as everybody knows, at Brown University where he has developed his research work in an admirable way.

During a visit to Paris soon after the war, in the spring of 1946, I was happy to see some of my old friends again. Lévy had had his apartment sacked by the Nazis, who had destroyed his books and papers, but he was already taking up new work on stochastic processes, which was to give important results. In one of my Paris lectures I talked about statistical estimation, and I had planned to say something about the controversial questions of confidence intervals and fiducial probability. It was a somewhat unpleasant surprise that Fisher was in Paris and attended my lecture. I expected some unfavourable comments, but at the end of the lecture he only said that he did not know enough French to have understood what I was saying, and wanted to have a private talk. This took place in the same evening, and ended quite well even though I did not hide my opinion.

In the fall of 1946 I went for the first time to the USA where I saw some old friends, such as Feller, Neyman, Wilks and others, and met a great number of new ones on the occasion of the Princeton Bicentennial. I was particularly interested in meeting Doob, whose work on stochastic processes I had always admired. On the invitation of Gertrude Cox and Harold Hotelling I made a brief visit to Chapel Hill, a place to which I have often come back. In Berkeley, where I was invited for a summer term, Neyman was making preparations for his celebrated series of Berkeley Symposia, and it was stimulating to follow his work and to meet the strong group of his young collaborators at the Berkeley Statistical Laboratory. They are now all well known for their outstanding scientific work.

In 1955 I attended the Bicentennial of the Moscow University, representing the Stockholm University. It was a great event, where in the first evening we went to the Bolshoi Opera and saw the whole Soviet Government of that time assembled on the podium. For me it was a great opportunity to make the personal acquaintance of those Soviet mathematicians whose work had meant so much for the advancement of probability theory. Unfortunately Khintchine was ill, and died shortly afterwards. But I met Kolmogorov, who made the impression of being a great scientific personality, and whom I am happy to have seen several times afterwards. I also met Dynkin, who was beginning his work on Markov processes, which he is now carrying on in the USA. I met Linnik whose work has had so close relations to my own; I met Gnedenko who with Kolmogorov had written the important book on probability limit problems, and many others. They were just making preparations for starting the new *Journal for Probability Theory and its Applications*, which has since then become such an internationally important publication.

During the 1950's I was heavily engaged in University administration work, which took a great part of my time. Since I was able to leave this kind of work in 1961, I have done some research on stationary processes, and have also tried to generalize some known important parts of their theory to more general classes of stochastic processes. I have spent quite a substantial part of my time on travels, visiting the United States, the Soviet Union, India and various European countries. Everywhere I found an intense research work going on in the fields of which I have been talking today. New lines of research were taken up, and new results were reached in the work with old problems. But it was no longer possible for one single man to follow and really understand more than a very small fraction of all the great work being carried on.

I will end my talk here, expressing my thanks for your attention, and my sincere hopes for a powerful future development of research and international collaboration in the fields of probability theory and statistical methodology.

Résumé

Dans cet article je présente quelques souvenirs personnels du développement de la théorie de probabilité mathématique et ses applications à l'inférence statisque pendant les vingt années entre les deux guerres mondiales, précédés et suivis de quelques notes sommaires sur le développement avant 1920 et après 1940.

[*Received May* 1981]

Bibliography

Most of the articles have been reproduced from offprints. In a few cases the page numbering is different from that in the original publications as given in the Bibliography. This is due to the fact that at that time, many journals changed the page make-up to produce the offprints.

The works of Harald Cramér in chronological order

1. Sur une classe de séries de Dirichlet
   Thesis, Stockholm University. Almquist & Wiksell, Uppsala, 51 pp. (1917)
2. Etudes sur la sommation des séries de Fourier
   Ark. Mat. Astr. Fys. **13** (20), 1–21 (1918)
3. Un théorème sur les séries de Dirichlet et son application
   Ark. Mat. Astr. Fys. **13** (22), 1–14 (1918)
4. Über die Herleitung der Riemannschen Primzahlformel
   Ark. Mat. Astr. Fys. **13** (24), 1–7 (1918)
5. Über die Nullstellen der Zetafunktion
   Math. Z. **2** (3/4), 237–241 (1918)
6. Studien über die Nullstellen der Riemannschen Zetafunktion
   Math. Z. **4** (1/2), 104–130 (1919)
7. Nombres premiers et équations indéterminées
   Ark. Mat. Astr. Fys. **14** (13), 1–11 (1919)
8. Bemerkung zu der vorstehenden Arbeit des Herrn E. Landau
   Math. Z. **6** (1/2), 155–157 (1920)
9. Some theorems concerning prime numbers
   Ark. Mat. Astr. Fys. **15** (5), 1–33 (1920)
10. On the distribution of primes
    Proc. Cambridge Phil. Soc. **20**, 272–280 (1921)
11. Über die Zetafunktion auf der Mittellinie des kritischen Streifens
    (with E. Landau)
    Ark. Mat. Astr. Fys. **15** (28), 1–4 (1921)
12. Über das Teilerproblem von Piltz
    Ark. Mat. Astr. Fys. **16** (21), 1–40 (1922)
13. Sur un problème de M. Phragmén
    Ark. Mat. Astr. Fys. **16** (27), 1–5 (1922)
14. Ein Mittelwertsatz in der Primzahltheorie
    Math. Z. **12**, 147–153 (1922)

15. Über zwei Sätze des Herrn G. H. Hardy
Math. Z. **15**, 201–210 (1922)
16. Contributions to the analytic theory of numbers
Proc. 5th Scand. Math. Congress, Helsingfors 1922, pp. 266–272
17. Ein Satz über Dirichletsche Reihen
Ark. Mat. Astr. Fys. **18** (2), 1–7 (1923)
18. Das Gesetz von Gauss und die Theorie des Risikos
Skand. Aktuarietidskr. **6**, 209–237 (1923)
19. Die neuere Entwicklung der analytischen Zahlentheorie (with H. Bohr)
Enzykl. d. Math. Wissensch. II, C 8, 722–849 (1923)
20. Remarks on correlation
Skand. Aktuarietidskr. **7**, 220–240 (1924)
21. On some classes of series used in mathematical statistics
Proc. 6th Scand. Math. Congr. Copenhagen 1925, 399–425
22. Sur quelques points du calcul des probabilités
Proc. London Math. Soc. **23**, LVIII–LXIII (1925)
23. Some notes on recent mortality investigations
Skand. Aktuarietidskr. **9**, 73–99 (1926)
24. On an asymptotic expansion occurring in the theory of probability
J. Lond. Math. Soc. **2**, 262–265 (1927)
25. On the composition of elementary errors
Skand. Aktuarietidskr. **11**, 13–74, 141–180 (1928)
26. On the mathematical theory of risk
Försäkringsaktiebolaget Skandia 1855–1930, Parts I and II, Stockholm 1930, 7–84
27. The risk problem. Exordial review
Proc. Trans. 9th Internat. Congr. Actuaries, IV, 163–172 (1931)
28. The theory of risk in its application to life insurance problems
Proc. Trans. 9th Internat. Congr. Actuaries, II, 380–390 (1931)
29. Su un teorema relativo alla legge uniforme dei grandi numeri
Giorn. Ist. Ital. Attuari **5**, 3–15 (1934)
30. On the development of the mortality of the adult Swedish population since 1800 (with H. Wold)
Nordic Statistical Journal **5**, 3–22 (1934)
31. Sugli sviluppi asintotici di funzioni di ripartizione in serie di polinomi di Hermite
Giorn. Ist. Ital. Attuari **6**, 141–157 (1935)
32. Mortality variations in Sweden: A study in graduation and forecasting (with H. Wold)
Skand. Aktuarietidskr. **18**, 161–241 (1935)
33. Prime numbers and probability
Proc. 8th Scand. Math. Congr. Stockholm 1934, 1–9
33. Prime numbers and probability
Proc. 8th Scand. Math. Congr. Stockholm, 1–9 (1934)

34. Über die Vorausberechnung der Bevölkerungsentwicklung in Schweden
Skand. Aktuarietidskr. **18**, 35–54 (1935)
35. Über eine Eigenschaft der normalen Verteilungsfunktion
Math. Z. **41**, 405–414 (1936)
36. Some theorems on distribution functions (with H. Wold)
J. London Math. Soc. **11**, 290–294 (1936)
37. On the order of magnitude of the difference
between consecutive prime numbers
Acta Arith. **2** (1), 23–46 (1936)
38. Sur un nouveau théorème-limite de la théorie des probabilités
Actual. Sci. Indust. (736), 5–23 (1938)
39. On the representation of a function by certain Fourier integrals
Trans. Amer. Math. Soc. **46**, 191–201 (1939)
40. On the theory of stationary random processes
Ann. Math. **41** (1), 215–230 (1940)
41. On harmonic analysis in certain functional spaces
Ark. Mat. Astr. Fys. **28 B** (12), 1–7 (1942)
42. A contribution to the theory of statistical estimation
Skand. Aktuarietidskr. **29**, 85–94 (1946)
43. Problems in probability theory
Ann. Math. Statist. **18** (2), 28–39 (1947)
44. On the factorization of certain probability distributions
Ark. Mat. **1** (7), 61–65 (1949)
45. A contribution to the theory of stochastic processes
Proc. Second Berkeley Symp. Math. Statist. Prob.
University of California Press
Berkeley and Los Angeles, 329–339 (1951)
46. On some questions connected with mathematical risk
Univ. California Publ. Stat. **2** (5), 99–123 (1954)
47. Collective risk theory: A survey of the theory from the point of view
of the theory of stochastic process
7th Jubilee Volume of Skandia Insurance Company Stockholm, 51–92 (1955)
48. Ein Satz über geordnete Mengen von Wahrscheinlichkeitsverteilungen
Teor. Verojatnost. i Primenen. **I**, 19–24 (1956)
49. On the linear prediction problem for certain stochastic processes
Ark. Mat. **4** (6), 45–53 (1959)
50. On some classes of nonstationary stochastic processes
Proc. Fourth Berkeley Symp. Math. Statist. Prob., Vol. II, 57–78
University of California Press, Berkeley and Los Angeles 1961
(Reprinted in Random Processes)
51. On the structure of purely non-deterministic stochastic processes
Ark. Mat. **4** (19), 249–266 (1961)
(Reprinted in Random Processes)

52. Model building with the aid of stochastic processes
Bull. Inst. Internat. Statist. **39**, 2, 3–30 (1961)
(Reprinted 1964 in Technometrics **6**, 133–159)
53. On the maximum of a normal stationary stochastic process
Bull. Amer. Math. Soc. **68** (5), 512–516 (1962)
54. Décompositions orthogonales de certains processus stochastiques
Ann. Fac. Sci. Clermont **8**, 15–21 (1962)
55. On the approximation to a stable probability distribution
Studies in Mathematical Analysis and Related Topics
(Essays in Honor of Georg Polya). Ed. by G. Szegö et al.,
Stanford University Press, Stanford, California, 70–76 (1962)
56. On asymptotic expansions for sums of independent random variables
with a limiting stable distribution
Sankhyā Ser. **A 25**, 13–24 (1963)
57. Stochastic processes as curves in Hilbert space
Teor. Verojatnost i Primenen. **9** (2), 193–204 (1964)
(Reprinted in Random Processes)
58. The moments of the number of crossings of a level by a stationary normal
process (with M. R. Leadbetter)
Ann. Math. Statist. **36**, 1656–1663 (1965)
59. A limit theorem for the maximum values of certain stochastic processes
Teor. Verojatnost. i Primenen. **10**, 137–139 (1965)
60. On the intersections between the trajectories of a normal stationary stochastic
process and a high level
Ark. Mat. **6** (20), 337–349 (1966)
61. On stochastic processes whose trajectories have no discontinuities
of the second kind
Ann. Mat. Pura Appl. **71**, 85–92 (1966)
62. On extreme values of certain stochastic processes
Research Papers in Statistics
Festschrift for J. Neyman, ed. by F. N. David
Wiley, London New York Sydney, 73–78 (1966)
63. A contribution to the multiplicity theory of stochastic processes
Proc. Fifth Berkeley Symp. Math. Stat. Prob., Vol. II, Part 1
University of California Press, Berkeley and Los Angeles, 215–221 (1967)
(Reprinted in Random Processes)
64. Historical review of Filip Lundberg's works on risk theory
Skand. Aktuarietidskr. **52**, Supplement (3–4), 6–12 (1969)
65. On streams of random events
Skand. Aktuarietidskr. **52**, Supplement (3–4), 13–23 (1969)
66. Structural and statistical problems for a class of stochastic processes
(The first Samuel Stanley Wilks Lecture)
Princeton University Press, Princeton, 3–30 (1974)
(Reprinted in Random Processes)

67. On distribution function–moment relationships in a stationary point process
(with M. R. Leadbetter and R. J. Serfling)
Z. Wahrscheinlichkeitstheorie und verw. Gebiete **18**, 1–8 (1971)
68. On the history of certain expansions used in mathematical statistics
Biometrika **59**, 205–207 (1972)
(Reprinted in Studies in the History of Statistics and Probability, Vol. II, ed. by M. G. Kendall and R. L. Plackett, Griffin, London 1977)
69. On the multiplicity of a stochastic vector process
Perspectives in Probability and Statistics
(Papers in honour of M. S. Bartlett on the occasion of his sixty-fifth birthday)
Ed. by J. Gani. Applied Probability Trust. University of Sheffield, 187–194 (1975)
70. Half a century with probability theory.
Some personal recollections
Ann. Probab. **4** (4), 509–546 (1976)
(Russian translation 1979)
71. On the multiplicity of a stochastic vector process
Ark. Mat. **16** (1), 89–94 (1978)
72. On some points of the theory of stochastic processes
Sankhyā Ser. **A 40**, Part 2, 91–115 (1978)
73. Mathematical probability and statistical inference
(Pfizer Colloquium Lecture 1980)
Internat. Statist. Rev. **49**, 309–317 (1981)
(French translation 1983)

Acknowledgements

The editor would like to thank the publishers and copyright-holders of Harald Cramér's papers for granting permission to reprint them here.

The numbers following each source correspond to the numbering of the articles in the Table of contents.

Acta Arithmetica: 37
American Mathematical Society: 39, 53
Annales Scientifiques de l'Université de Clermont: 54
Annals of Mathematics: 40
Applied Probability Trust, Sheffield: 69
Biometrika: 68
Cambridge University Press: 10
Hermann, Éditeurs des Sciences et des Arts: 38
Institut Mittag-Leffler: 2, 3, 4, 7, 9, 11, 12, 13, 17, 41, 44, 49, 51, 60, 71
Institute of Mathematical Statistics: 43, 58, 70
International Statistical Institute: 52, 73
Istituto Italianico degli Attuari: 29, 31
The London Mathematical Society: 22, 24, 36
Princeton University Press: 66
Sankhya: 56, 72
Scandinavian University Press: 18, 20, 23, 25, 32, 34, 42, 64, 65
Springer-Verlag: 5, 6, 8, 14, 15, 35, 67
Stanford University Press: 55
Teorija Verojatnostjej i ee Primenenija: 48, 57, 59
Teubner Verlag: 19
University of California Press: 45, 46, 50, 63
John Wiley & Sons, Ltd: 62